The LIFE *and*
SCIENTIFIC LEGACY *of*
GEORGE PORTER

Published by

Imperial College Press
57 Shelton Street
Covent Garden
London WC2H 9HE

Distributed by

World Scientific Publishing Co. Pte. Ltd.
5 Toh Tuck Link, Singapore 596224
USA office: 27 Warren Street, Suite 401-402, Hackensack, NJ 07601
UK office: 57 Shelton Street, Covent Garden, London WC2H 9HE

British Library Cataloguing-in-Publication Data
A catalogue record for this book is available from the British Library.

Cover photo © The Godfrey Argent Studio

While every effort has been made to contact the publishers of reprinted papers prior to publication, we have not been successful in some cases. Where we could not contact the publishers, we have acknowledged the source of the material. Proper credit will be accorded to these publications in future editions of this work after permission is granted.

ISBN 1-86094-660-7
ISBN 1-86094-695-X (pbk)

Printed in Singapore by Mainland Press

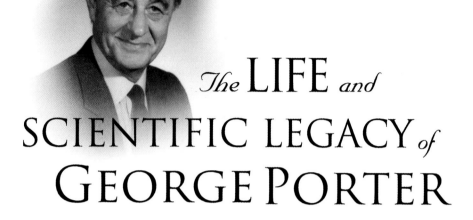

The LIFE and SCIENTIFIC LEGACY of GEORGE PORTER

Editors

David Phillips
James Barber

Imperial College London, UK

Imperial College Press

Courtesy of Peter Branch

ACKNOWLEDGMENTS

We would like to thank all of the colleagues and friends of George Porter who have contributed this volume, and others who have made useful suggestions. In particular however, we wish to thank Lady Porter for her helpful advice on some aspects of George Porter's early life and career, and for providing many photographs.

PREFACE

This volume, dedicated to the life and memory of George Porter, is a collection of his prominent original papers, each with a commentary by invited scientists, chosen from amongst the many who were influenced by his work. Each commentator has also contributed a paper of their own to demonstrate how their own science has progressed. Some of these are original papers, some reprinted from journals, with permission.

George Porter made many significant contributions, both in science, in the politics of science, and in its promotion. The volume is arranged thematically, beginning with a potted history of George Porter, tracing his modest origins, through his undergraduate career, wartime service in the Royal Navy, and to Cambridge as a research student. It was here that he developed, with Norrish, the technique of flash photolysis for which he is justly famous, and which earned them both a share in the 1967 Nobel Prize. These days are dwelt upon by many of the contributors to the volume, since flash photolysis opened up the study of many research areas, including free radical spectroscopy and kinetics; the study of excited electronic states, firstly triplet states, then singlet states, and the field of photosynthesis, both natural and artificial.

The individual contributions commenting upon George Porter's papers vary from being very personal reminiscences about the way work was conducted, through to knowledgeable scientific assessments of the difficulties of some of the early work on flash photolysis and laser flash photolysis. We also include an appraisal of some of George Porter's contributions in the field of science communication, and his influence on politics and the appreciation of science, principally in the UK, but with an international dimension also.

This volume is intended as a tribute to his qualities as a man, as a scientist and as a friend to many of us who knew him well. We hope it does him justice.

David Phillips, OBE
James Barber, FRS
Editors

CONTENTS

The Biography of

GEORGE PORTER

(6th December 1920–31st August 2002)

by David Phillips

This brief biography is based upon the several obituaries written by the author for Lord Porter, including that jointly authored with Professor Graham Fleming for the Royal Society Biographical Memoirs.

Early Days

Born in Ash House, Stainforth, near Doncaster, on 6th December 1920, George was the only child of John Smith Porter and Alice Ann (née Roebuck) Porter.

A very young George Porter.

Porter before the age of five.

An early photograph of him shows his characteristic physiognomy, with prominent eyebrows retained throughout his life and revealed again in a posed photograph taken in Ryhl when he was not yet five years old.

*All photographs courtesy of Lady Porter, unless indicated otherwise.

1

John Porter's father was a miner. He himself was a builder, educated only up to the age of thirteen, as was common then, but he seems to have been an enlightened individual, serving on the local council, and on the governing body of Thorne Grammar School (to which his son would go) as Chairman for several years.

The Porter family on vacation.

The family is depicted on holiday at about the time John Porter provided his young son with a chemistry set and an old bus in which to experiment. This sowed the seeds of George Porter's eventual love of chemistry as a subject, and the practical aspects in particular.

Porter is said to have enjoyed Chemistry and English most at school; certainly his command of the spoken and written language, and his prowess at his scientific pursuits would be consistent with this view. He was greatly influenced at Thorne Grammar School by his science teacher (Moore), and English teacher (Todd); the mathematics master although "brilliant" did not stay long, due to his penchant for alcohol. It is not clear whether this was an abiding influence upon Porter! He clearly also was developing his talents for theatricality, the evidence being his performance in school plays.

Porter appearing in the school play, Thorne Grammar School.

Porter left Thorne Grammar School with high school certificates in Chemistry, Physics and Mathematics, plus an Ackroyd Scholarship to the University of Leeds. The scholarship was worth forty pounds per annum, and he was thus reliant upon his father's support for payment of his rent. His chosen subject at Leeds was Chemistry, and M. G. Evans, who set him on the path of physical chemistry, particularly influenced him. He graduated with honours in 1941

The new graduate, Leeds 1941.

Wartime Service

Before completion of his degree, Porter had been recruited into the Hankey Scheme, which was a secret project to provide the UK with radiophysicists. He attended a crash course at Aberdeen University, Department of Natural Philosophy and commended in particular the lectures given by Professor Carroll. In September 1941, Porter began service with the Royal Naval Volunteer Reserve, and spent from September to December of that year at the signals school in Portsmouth. From January 1942 to September 1943 he was Radar Group Officer.

Porter served in both the Western Approaches and Mediterranean theatres. He described his duties as those of a kind of radio mechanic, with all sets breaking down every day, and few spares available. Keeping the radar equipment operational on board destroyers thus called for great improvisational skills; for example, Porter discovered that the large yellow and black resistors known as "tigers" could be repaired with boot polish (a fact reported by Brian Thrush, Cambridge). There seemed little doubt that Porter's experimental skills were put to good use during his wartime service and developed to great advantage in his coming experimental research at Cambridge. His work called for physical as well as intellectual skills, since as he said, he had to "hop from one ship to another in mid-Atlantic". He saw some action, witnessing the sinking of three or four U-boats. He was also at the invasion of North Africa, Operation Torch.

Garrison & Deakin   Doncaster

Naval officer.

With the RN, Belfast. Porter is in the front row, fifth from left.

Later in this period, his duties reverted to those of a more traditional executive officer. In 1943–1944, he was shore-based as an Instruction Officer at the Radar Training School, Belfast. In 1945 Porter was detailed to join Force X, for the invasion of Japan, but this never materialised because of the surrender of Japan, and he then endured a period of inaction and few duties while waiting for demobilisation.

Studies at Cambridge

Upon demobilisation in 1945, Porter began his PhD work with R. G. W. Norrish in Cambridge supported by a grant from the Anglo-Iranian Oil Company, and residing in Emmanuel College.

Porter's original problem was concerned with the methylene radical using a technique based on the time to destroy a metal mirror at a variable distance from the source of gaseous free radicals (the Paneth mirror technique). At the time, it was felt that direct observation of such short-lived (typically 10^{-3} s) intermediates was impossible, but as will be described in the bulk of this volume, Porter's development of flash photolysis changed this view overnight.

In papers in the Royal Society archives, Porter described how he conceived the idea for the flash photolysis method, and this will be found in the contribution in the present volume by Graham Fleming.

The goal of the technique was the spectroscopic detection and study of free radicals in the gas phase. The first experiment on the photolysis of acetone with a 4000 J flash resulted in total destruction of the acetone and the deposition of filaments of carbon throughout the 1 m long sample cell! Perhaps unwittingly, this Porter experiment created the very form of carbon, C_{60} which was to lead to a

Porter's contemporaries at Cambridge.

Nobel Prize for one of Porter's young colleagues in Sheffield, Harry Kroto. In the first single author paper on flash photolysis, in 1949, several radicals and transient species, including new spectra for ClO and CH_3CO radicals were described, followed by studies ranging from the explosive reaction of hydrogen and oxygen and the kinetics of HO radicals, fragmentation of organic molecules, the explosive combustion of hydrocarbons to the study of such phenomena as knock, carbon formation and atom recombination reactions. The students involved in this work included Brian Thrush, Frank Wright, Margaret Christie, Kerro Knox and M. A. Khan.

Porter was awarded his PhD in 1949 for a thesis entitled "The Study of Free Radicals Produced by Photochemical Means". He married Stella Jean Brooke, a diminutive and beautiful ballet teacher in the same year. Porter met Stella at a dance at the London College of Dance which she and Mary Norrish attended in Holyport in 1946. Mary had asked her father to enlist some suitable research students as guests, and Porter, being very sociable and one of Norrish's PhD students, was happy to oblige.

In 1949, Porter was also appointed to a University Demonstratorship in Physical Chemistry. Porter's love of life was already well to the fore at that time. Contemporaries report his enthusiastic participation in life at Emmanuel College, citing his membership of the then active Emmanuel Singers and notable performances of Gilbert and Sullivan, including a role as "Patience" and "Buttercup"! Stella shared this exuberance, and together they made an exceptional couple.

Porter was himself an accomplished dancer and his nickname later on of "flash Porter", which he disliked strongly, referred mainly to his science, but was not an inappropriate reference to his prowess on the dance floor. Porter was also a member of the Cambridge University Cruising Club and his love of sailing remained with him throughout his life.

The elegant young couple, Stella and George, celebrating at the Dorchester Hotel in London.

He stayed on in Cambridge as a Demonstrator (1949–1952) and Assistant Director of Research (1954–1956) in the Department of Physical Chemistry. He was a fellow of Emmanuel College from 1952–1954, and was made an honorary fellow in 1967.

While in Cambridge, Porter and his student Maurice Windsor successfully recorded the triplet spectra of aromatic molecules in solution. The triplet state had been convincingly identified as the origin of phosphorescence in 1944 by G. N. Lewis and Michael Kasha but it was not clear whether triplet states were formed at all following excitation of aromatic singlet states or whether the triplet state simply decayed very rapidly, since phosphorescence in solution had not been observed. Porter and Windsor were lucky; the concentration of oxygen (which along with other paramagnetic species quenches triplet states) could be reduced by degassing to the point that triplet lifetimes were in the hundreds of microsecond range, detectable with the flash photolysis apparatus available, which had a time resolution of twenty-five microseconds.

The first experiments were made on anthracene solutions in chloroform, with the idea that the heavy atoms in the solvent might enhance the triplet yield. No triplets were found and the solvent was switched to hexane. Following degassing, the experiment was immediately successful and the spectrum of the triplet state of anthracene was obtained. Soon after, Franklin Wright and Porter were able to record triplet-triplet spectra in the gas phase leading to the observation of the benzyl radical while Porter and Windsor detected the benzosemiquinone radical, ushering in the study of solution phase photochemical reactions. Later, in collaboration with Robert Livingston (a sabbatical visitor from the University of Minnesota) Porter and Windsor recorded the transient spectra of chlorophylls a and b, pheophytin and coproporphrin dimethyl ester, beginning the study of the photochemistry and photophysics of photosynthetic systems that would dominate the latter part of Porter's career. By the time he gave his Chemical Society Tilden Lectures in 1958–1959 (the first of many prize lectures to come), Porter was able to give an authoritative overview of the many roles of the triplet state in chemistry.

Manchester and Sheffield

Porter left Cambridge in 1954 to spend a year as Assistant Director of the British Rayon Research Association in Wythenshawe, Manchester, where he studied the phototendering of dyed cellulose fabrics in sunlight, a problem he traced to a triplet-initiated hydrogen abstraction reaction. Porter had been attracted to the British Research Association by John Wilson, whose enthusiasm and powers of persuasion convinced Porter that he could teach industry how to do research. At this stage, relations were also a little strained with Norrish. At British Rayon however, Porter found the 9-to-5 regimen of industrial life totally alien and was eager to return to academia.

In 1955 he moved to the University of Sheffield as their first Professor of Physical Chemistry. During his eleven years in Sheffield, Porter steadily improved the department, starting with Physical Chemistry, then, on the retirement of R. D. Haworth in 1966, he organised the appointment of a new Chair of Organic Chemistry and of the just-created Chair of Inorganic Chemistry. By the time he left Sheffield, Porter had achieved a high standing for the department over the whole sweep of chemical science. In 1963 (until 1966) he became Firth Professor of Chemistry (he joked that had there been adequate funding, there would have been a "Thecond" Chair) and head of the Chemistry Department.

George and Stella always greatly enjoyed inviting guests, both great and small, to their home. When established as Professor of Physical Chemistry in Sheffield, they together transformed the local perception of senior academics, the so-called "Herr Professor" world, by their unstinting hospitality and un-stuffiness. At their frequent "at-homes", Porter delighted in showing off his ex-wardroom trick of balancing a half-pint tankard of beer on his forehead while standing, and drinking it without touching it with his hand or arms by various contortions involving lying down and using both feet.

The research work on flash photolysis, progress from millisecond to microsecond time resolution was very rapid, but there, despite much effort, things stalled. With microsecond resolution many fundamental processes and electronic states were out of reach to absorption spectroscopy, in particular all excited singlet states had lifetimes three orders of magnitude or more too short for direct observation. Fluorescence lifetimes could be studied with low energy spark excitation, but the fate of excited states, their chemistry and pathways of deactivation could not be followed. The problem was in the physics of electrical discharges used to produce pulses of light in the lamps used in flash photolysis, which prevented the simultaneous production of high intensity and short duration. Microseconds seemed to be about the limit. However, everything changed with Maiman's demonstration of the ruby laser in 1960. Porter immediately recognised the potential of this new tool, but it took a number of years before the laser was successfully applied to flash photolysis because of the need to develop the associated fast detection and recording methods.

The first application of the Q-switched ruby laser, with pulse duration of about 20 ns, did not have nanosecond resolution, but was very likely the first observation and correct interpretation of a multiphoton dissociation process. Jeff Steinfeld, an American postdoc who came to Sheffield with the intention of studying iodine dissociation by conventional flash photolysis, excited phthalocyanine vapour with both the Q-switched (20 ns) pulse, and the un-Q-switched laser output. Only the Q-switched pulse produced dissociation and "this strongly suggests that a process involving two photons is responsible for the dissociation". Similar experiments on chlorophyll vapour were far less successful and resulted in laborious cleaning of the decomposed products from the sample cell.

The Royal Institution

During his happy and extremely productive years in Sheffield, Porter had begun his association with the Royal Institution and his high academic standing, charismatic lectures, and remarkable energy, clearly recommended him as successor to Sir Lawrence Bragg, to provide much needed new leadership to an institution still recovering from the bitter fights surrounding the dismissal of Bragg's predecessor, E. N. da C. Andrade.

Porter's association with the Royal Institution of Great Britain in London began in November 1960, when he gave his first Friday Evening Discourse (then still a white tie and tails event), under the title "Very Fast Chemical Reactions". In 1963, he was invited by the Committee of Managers to take up the part-time post of Professor of Chemistry at the RI. In the next couple of years, while still in his full-time post at Sheffield, he gave two more Discourses ("The Laws of Disorder", January 1964, and "The Chemical Bond since Frankland", May 1965). The latter was in honour of Frankland, who in the previous century had also been RI Professor of Chemistry (until 1868). One of the duties of the Professors at the RI was to participate in the Schools Lectures Programme which had been initiated by Sir Lawrence Bragg in 1953, following Michael Faraday's Lectures for Juveniles. Porter responded to this challenge with enthusiasm delivering schools lectures from 1963 to 1965 ("Chaos and Chemical Equilibria", 1963 and 1965; "Electrons in Molecules", 1964; and "Patterns of Chemical Change", 1965). It should be remembered that at this time, the devotion of time and energy to "popularisation" of science was not widely thought to be a worthy pursuit for a serious scientist; Porter characteristically ignored the prevailing view, and in 1965–1966 went further by appearing on BBC Television in a series of ten lectures ("The Laws of Disorder", subtitled "An introduction to chemical change and thermodynamics"). These lectures, each shown twice a week were produced by Alan Sleath, who remained a good friend of Porter and the Institution for many years. The great success of his popular lectures, as well as his eminence as a scientist, placed Porter in prime position to replace Bragg as Director of the Royal Institution, which he duly did on Bragg's retirement in 1966.

Porter was Director of the Royal Institution for just under twenty years, and he often said they were the happiest and most rewarding days of his life. At the outset, he was faced with the problems which had best all Resident Professors/ Directors from the time of its founding in 1799, those of chronic under-funding, high cost of maintenance, and above all, a cumbersome and almost unworkable system of management. In 1812, management of the Royal Institution had been reformed such that overall responsibility was in the hands of a Committee of Managers to whom the Director reported; there was however a second committee, the Committee of Visitors whose function was to vet the activities of the Committee of Managers. There was thus built-in conflict, and this had come to a head in the late 1940s when the then Director A. C. Andrade had

Porter in "The Laws of Disorder" TV series, 1965.

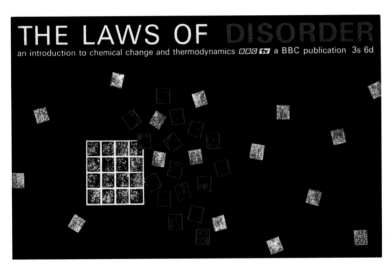

THE LAWS OF DISORDER

an introduction to chemical change and thermodynamics ⬛⬛⬛ 📺 a BBC publication   3s 6d

This pamphlet accompanies
a series of lectures on
BBC 1 given by

Professor George Porter

Produced by
Alan Sleath

Sundays 1.15 pm
starting Sunday 17 January
repeated in the late evening
of the following Monday

The drawings are by
John Griffiths

The

laws

of

disorder

*An introduction
to chemical change
and thermodynamics*

Contents

		page
	Introduction	*page 2*
1	The static and dynamic worlds	4
2	Molecules in motion	6
3	Entropy	8
4	The second law	10
5	Equilibrium – the limit of disorder	12
6	Molecules at work	14
7	Times of change	16
8	Patterns of change	18
9	Chemistry and light	20
10	Chaos and evolution	22

Front cover and contents page of the BBC Booklet for "The Laws of Disorder".

tried without success to initiate reform. Andrade's brief period as Director and his enforced removal left a bitter division in the RI between Managers and Visitors, which took all of the skills of Lawrence Bragg to calm down. Notwithstanding Bragg's immense presence and reputation, it did not prove possible during his Directorship to again attempt change of the governance of the Institution. Porter was however determined to achieve this, and with diplomacy and the assistance of the then Secretary of the Royal Institution, Professor Val Tyrrell, an old colleague from the Sheffield Department of Chemistry, he was able by the early 1980s to merge the Committees of Managers and Visitors into a single Council, and obtain for the Director the executive powers necessary in a modern organisation.

Porter was no less successful in obtaining funding for the modernisation of the RI. There were two major appeals for funding, the Faraday Centenary Appeal of 1967, and that in 1976 under the Chairmanship of Sir Monty Finniston. The former allowed major improvements to the fabric of the RI, with the provision of a new theatre (The Bernard Sunley Theatre), refurbishment of some laboratories, the restoration of Michael Faraday's Magnetic Laboratory, the building of secure archives and a strong room to house some of the treasures of the RI, as well as essential maintenance such as replacement of electrical wiring etc. The second successful appeal was in support of the Schools Lectures Programme, which was to be expanded, endowments for a Dewar Fellowship and the Fullerian Professorship, as well as new positions, such as the Wolfson Chair in Natural Philosophy, first occupied in 1980 (by the author). Further work on the fabric of the building was also supported, such as provision of a new lift to replace the historic hydraulic lift,

An RI school's lecture of the early Porter period. The lecturer is Professor Ronald King, Professor of Metal Physics, who had worked with Neville Mott before coming to the RI; the assistant is the legendary Bill Coates, who had joined the RI in 1948. The experiment is a demonstration of pyrophoric lead, which would certainly be in violation of present day health and safety regulations! (Courtesy of the Royal Institution.)

necessary because of the demise of the London Hydraulic Power Company and consequent cessation of the high-pressure water supply required for its operation. Despite the undoubted success of these appeals, the financial position of the Institution continued to be (and is today), precarious, since as a private institution it was not in receipt of state funding available for example to Universities. The one exception to this was the invitation to the members of the Davy Faraday Research Laboratory of the RI to compete for research council funding on a competitive basis, and during Porter's reign as Director, research funding was obtained at a level high enough to prosecute world-class research, and contribute also to the running of the institution as a whole.

Foreword

Sir Montague Finniston   *Sir George Porter*

The historic role and international reputation of the Royal Institution have stemmed equally from its skilful exposition, to a very wide audience, of the latest advances of science and technology and from the succession of brilliant scientists who have worked there and whose discoveries have transformed our lives.

Today the Royal Institution continues to make a unique contribution to the scientific life of the country, both through its own programme of research and by its exposition of science to many varied audiences. In a highly industrialised society dependent for its prosperity on an informed approach to science, this explanatory role is of the greatest importance. As the 'British repertory theatre of science' the Institution conveys the excitement and importance of science, stimulating interest and correcting misapprehension, with skills derived from long experience.

Young people especially must be encouraged to turn their energies and talents towards science and technology. The celebrated lectures for schoolchildren attract large and eager audiences increased on occasions to national proportions by presentation on television.

Apart from the young, scientists from different fields and interested laymen, among them businessmen wanting to enlarge their knowledge of science and its applications to industrial and managerial problems, are brought together at the Evening Discourses and by informal lectures and discussion groups.

The task of maintaining and expanding educational and research programmes, as well as fulfilling important responsibilities for its historic buildings, imposes on the Institution heavier financial burdens than it can shoulder alone. The Managers have therefore decided to launch an Appeal for £750 000 to finance the developments proposed in this booklet. The outstanding success of the Faraday Centenary Appeal in 1967 was reassuring evidence that the Institution has many generous friends, who appreciate the significance of its work and the need to uphold its independence. As the scale of private patronage has inevitably diminished, the Institution has increasingly to look to industry and trusts for support. We are now seeking to ensure, with your help, that the Royal Institution has the means to serve present and future generations with the same high distinction that has characterised its service in the past.

Monty Finniston.
Chairman of
Appeal Committee

George Porter
Director of
the Royal Institution

Foreword to 1976 appeal document by Sir Montague Finniston and Sir George Porter (Courtesy of the Royal Institution).

It was during his Directorship of the Royal Institution that Porter's contribution to the study of fast reactions and photochemistry and photobiology really flowered. In 1967, he shared the Nobel Prize for Chemistry with Ronald Norrish and Manfred Eigen. Their share of the Prize was given for Norrish and Porter's development of flash photolysis in Cambridge. Porter was knighted in 1972, following his two-year term of office as President of the Chemical Society, during which time he steered the amalgamation of the "old" Chemical Society, the Royal Institute of Chemistry, the Faraday Society and the Society for Analytical Chemistry, to form the Royal Society of Chemistry.

While in Sheffield, Porter had already investigated the use of the laser to study nanosecond phenomena in phthalocyanines, but it was upon his arrival in the Royal Institution that laser-based research really took off. The first RI system was a Q-switched ruby laser, built by Mike Topp, which boasted a 20 ns pulse duration, and a pump-probe configuration achieved by splitting the laser beam and routing the probe pulse along an optical delay line to provide the variable time-lag between excitation and interrogation of the sample. Wavelength variation in the probe pulse was achieved by using the laser pulse to excite fluorescence in a dye emitter. With this system, the spectra and decay kinetics of excited singlet states of molecules became accessible, and pioneering studies on aromatic hydrocarbons were carried out.

Porter receiving the Nobel Prize, Stockholm 1967.

PREPRINT

LES PRIX NOBEL

EN 1967

FLASH PHOTOLYSIS AND SOME OF ITS APPLICATIONS

NOBEL LECTURE

BY

GEORGE PORTER

COPYRIGHT © THE NOBEL FOUNDATION 1968

Cover of Porter's Nobel Address, 1967 (courtesy of the Nobel Foundation).

Porter's "race against time" continued at the RI into the picosecond time domain with the development of mode-locked lasers which provided a stable train of pulses of light each of a few picoseconds duration; femtosecond lasers were to be developed later, such that by 1985, the first colliding pulse laser system was built in the RI. In the forty years from 1947 the timescale of experimental studies on fast reaction had progressed by twelve orders of magnitude. Femtosecond lasers are now routine instruments rather than instruments requiring immense skills in the operators. Attosecond lasers are now under development. Although he did not live to see an attosecond laser, Porter remarked towards the end of his life that he hoped to, since if one took the second (the time of a human heart-beat) as the unit of time, he would then have witnessed in his lifetime the shortening of the scale of times by as many orders of magnitude as is the age of the Universe in comparison with one second.

This shortening of timescales was not a goal in itself, the research being driven by the need to study primary chemical events, such as electron transfer, molecular rearrangements, dissociations, and external processes such as energy transfer, molecular motions, diffusion, etc. which occur on these fast timescales. It was at this time that photobiology began to feature highly in the science at the Royal Institution. In the late seventies, the Porter group contained some exceptionally talented individuals who led particular interests. These included several contributors to the present volume, Graham Fleming (now at UC Berkeley) with fundamental interests in photophysics, and who shared an interest in photobiology with Godfrey Beddard (now at University of Leeds), and Anthony Harriman, (now at University of Newcastle), who was developing the work in "artificial" photosynthesis. Some of this seminal work was reviewed in the Bakerian Lecture delivered by Porter in November 1977, and published in the *Proceedings of the Royal Society* in 1978 as a model of clarity and thoughtfulness. The "Z" scheme for plant photosynthesis is outlined, and the individual steps leading to charge separation and water splitting in photosystem II detailed. The structural features of the plant which prevent back electron transfer reactions which bedevil *in vitro* attempts to emulate photosynthesis were further discussed with regard to light harvesting mechanisms in chloroplasts and algae, in which pigments of decreasing energy are arranged spatially such that the photon energy is effectively funnelled down to the reaction centre without significant self-quenching, and consequent inefficiency. Water oxidation using manganese complexes was described as a demonstration of a model system in which photosynthesis might be reproduced artificially in the laboratory. This was always a goal of Porter's, to harness the "free" energy of sunlight as a replacement for fossil fuels. He was very optimistic about success here, and would in the early days often introduce seminars on the subject by saying that,

"Nature is miserably inefficient in converting light to useable energy (about 1%) and surely science properly applied could do better."

The gradual realisation of the scientific and technical difficulties of achieving the goal had by the mid-eighties caused a change in this introduction, which now became

"Nature is very inefficient in converting light into useable energy, but how presumptuous it would be for Mankind to expect to improve on the Almighty."

In 1980, David Phillips, who moved his research from Southampton, took up the Wolfson Chair in Natural Philosophy. Although there was a common use of mode-locked lasers, the groups retained their separate identities and the research in the Davy Faraday Research Laboratory covered a broader spectrum of photochemistry, particularly with the addition of interest in "supersonic jet" spectroscopy. Porter was particularly generous in making the newcomers feel welcome, and providing basic facilities.

The Royal Institution then and now is concerned with the public appreciation of science, and the tradition was reinforced during the Porter era. He was an enthusiastic lecturer himself, and gave dozens of schools lectures and many Friday Evening Discourses. At his best, he was without equal as a communicator; as all professionals, he rehearsed his performances to perfection. He took a great interest in the Schools Programme, and would often attend rehearsals or first lectures of a series, and then join in the informal criticism of the lecture over drinks with the lecturer. Many speakers may have found this uncomfortable, but the advice was always meant kindly, and always led to an uplift in the next performance. Porter was a passionate believer in the powers of simple demonstrations to enhance a presentation, and cement a point in the mind of the audience. He insisted that all Friday Evening Discourse speakers use demonstrations, but none did so better than he. Perhaps his most poignant performance was at a farewell Discourse in 1985, in the presence of HRH The Duke of Kent, the President of the Royal Institution, and at which a bust of Sir George, as he then was, was presented to him.

In the present days of ubiquitous and too often trivial Powerpoint slide shows, his lesson still has meaning. In addition to expanding the Schools Lectures Programme, particularly the introduction of Primary Schools Lectures, during his Directorship the extremely successful Mathematics Master Classes programme was launched, and Master Classes in Technology were also initiated in collaboration with the University of Sussex. Many series of Science Seminars in Schools were introduced; the Royal Institution began to offer Schools Lectures outside London, particularly in the northwest and southwest of England.

Arguably his greatest personal contribution to the presentation of science to the young, and lay public, was his introduction of television to the Institution, and the broadcasting by the BBC throughout his Directorship of the Christmas Lectures. Porter gave two of these himself, in 1969–1970 "Time Machines" and in 1976–1977, "The Natural History of a Sunbeam", the latter based upon Michael Faraday's "Chemical History of a Candle". Other television

Porter's last Discourse at the RI. Demonstration of the Faraday cage. The assistant is Melanie Thody, then assistant to Bill Coates. (Courtesy of Peter Branch.)

Doing what he loved best, expounding science (courtesy of Peter Branch).

Bill Coates assisting. Note the "historic" projector. (Courtesy of Peter Branch.)

Presentation of the bust. Front, from left to right: David Phillips, (the Wolfson Professor at the RI), Alan Boulstridge (Bursar), Irene McCabe (RI librarian and archivist), Lady Porter, The Duke of Kent (President of the RI) and Porter. (Courtesy of Peter Branch.)

A well-earned refreshment after the Discourse (courtesy of Peter Branch).

Eric Ash, then Secretary of the Royal Institution, who was also Rector of Imperial College and instrumental in moving Porter to Imperial College in 1986 (courtesy of Peter Branch).

Stephen Bragg, son of Sir Lawrence Bragg, Porter's predecessor as Director of the RI (courtesy of Peter Branch).

Cake cutting. Behind Porter are Henriette Allul, a post-graduate student, and Patience and David Thompson. David Thompson is the grandson of J. J. Thompson; Patience (née Bragg), the granddaughter of William Bragg and daughter of the Thompson progeny thus have four Nobel Laureates amongst their ancestors. Lady Porter is seen at the rear. (Courtesy of Peter Branch.)

Margaret Judith Wright, always known as Judith, Porter's talented and devoted assistant. She moved with Porter from Sheffield and continued to be his PA throughout his time at the RI, and also acted in this capacity for his successor, John Meurig Thomas. (Courtesy of Peter Branch.)

programmes from the Royal Institution included the "Controversy" series of debates (1971–1975), which he chaired. Between 1966 and 1981, he was a judge on the BBC Young Scientist of the Year competition. He enjoyed being on the media, particularly television, and always ensured he was well groomed before an appearance, although on one occasion this was somewhat redundant, since his secretary had neglected to tell him the BBC feature was a radio rather than television broadcast.

The Royal Society

Porter became President of the Royal Society in 1985 succeeding Sir Andrew Huxley. This was a time of serious under-funding of UK science and Porter used his position to argue with great energy and eloquence for increased funding by the government. A typically vigorous comment was this from 1987, "From various Ministers I have been told in turn that there is too much science, that this country can leave it to others, and that the importance of Nobel Prizes went out with Harold Wilson. To answer their views by declaring our belief in the intrinsic value of natural knowledge is met with blank incomprehension." He warned that Britain was "well prepared to join the third world of science," and commented that, "The country is run by people who have no scientific education whatever.

CHRISTMAS LECTURES

*The one hundred and forty-seventh course of
six lectures for young people*

SIR GEORGE PORTER, F.R.S.

on

THE NATURAL HISTORY
OF A SUNBEAM

SATURDAY Dec. 18; TUESDAY Dec. 21; THURSDAY Dec. 23;
WEDNESDAY Dec. 29; FRIDAY Dec. 31, 1976; TUESDAY Jan 4, 1977
at 3 p.m.

Course tickets: Members £3.50, non-Members £6.00, Members' children (aged
10-17) £1.25, non-Members' children (aged 10-17) £2.50.

Single tickets (*available on day of lecture only*): Members 75p, non-Members £1.25,
Members' children (aged 10-17) 25p, non-Members' children (aged 10-17) 50p.

THE ROYAL INSTITUTION
21 ALBEMARLE STREET, LONDON, W1X 4BS

Cover of booklet for 1976 BBC TV Christmas lectures (courtesy of Lady Porter and the Royal Institution).

The Christmas Lectures for young people at the Royal Institution were started in 1826 at a time when there was little or no organised education in science at schools or universities. They were thus a new venture in the teaching of science. Michael Faraday gave the series on nineteen occasions between 1827 and 1860 and established the tradition of illustrating the lectures by a large number of experiments. Perhaps his most famous series was on The Chemical History of a Candle.

Many other famous scientists have given the lectures which have been held annually since 1826 with only one short break during the Second World War. Their long tradition and the technique which has been developed in the management and presentation of the demonstrations have helped to make the lectures so well known. This is now the one hundred and forty-seventh course 'adapted to a juvenile auditory'; the lectures will be televised and broadcast in colour on B.B.C.2.

Professor Sir George Porter has been Director and Fullerian Professor of Chemistry of the Royal Institution and Director of the Davy Faraday Research Laboratory since 1966. He was educated at Thorne Grammar School, Leeds University and Emmanuel College, Cambridge. Between 1941 and 1945 he served as Radar Officer in the Royal Navy where his work gave him an interest in electronic pulse techniques. On completion of his doctorate degree he spent a further five years in the Physical Chemistry Department of the University of Cambridge. He was appointed Professor of Physical Chemistry in the University of Sheffield in 1955 and became Firth Professor and Head of Department in 1963. He is an Honorary Fellow of Emmanuel College, Cambridge, Honorary Professor of Physical Chemistry in the University of Kent and Visiting Professor in the Department of Chemistry, University College London. In 1960 he was elected a Fellow of the Royal Society and shared the Nobel Prize for Chemistry in 1967. He was knighted in 1972 and has received many honours and awards in recognition of his work.

His research interests are in the field of fast reactions, photochemistry, photobiology and solar energy. He is also interested in scientific education and the presentation of science to non-specialists and was elected the first President of the National Association for Gifted Children in 1975. He gave a series of lectures on BBC television in 1965 on *The Laws of Disorder* and made four films on this subject in the ICI Films for Schools series. He has been associated with the BBC *Young Scientists of the Year* programmes since their introduction and gave the 1969/70 series of Royal Institution Christmas Lectures on *Time Machines*. He has also lectured widely in other countries.

Details of TV Christmas lectures (courtesy of Lady Porter and the Royal Institution).

Synopsis of "The Natural History of a Sunbeam" (courtesy of Lady Porter and the Royal Institution).

THE NATURAL HISTORY OF A SUNBEAM

The world about us is forever undergoing wondrous chemical changes. Each spring, parts of the earth which were apparently lifeless become green and vegetation appears as if from nowhere. Tiny seeds grow into vast forests and small eggs grow into animals of the most intricate design. The chemical changes which have taken place over the course of evolution are even more remarkable. A primitive planet of earth, air, fire and water, somehow spontaneously brought forth the complex and delicate chemical substances of which life is made, followed by life itself, in ever increasing complexity. The evolutionary process continues beyond the individual man, and an outside observer would have noticed how, over the last few centuries, the earth grew large cities, roads and railways whilst machines, bigger than any bird, flew over its surface at great speed.

How does it all happen? Order is being created out of chaos and this seems to happen spontaneously. Yet we know that in our ordinary lives this is not the way things happen and that, if we leave things to themselves, they all too easily run down, lose their order and become chaotic. It is the same with chemical reactions, they go spontaneously to a state of greater chaos. Things burn but never unburn, wood disappears as smoke into the air and we would be astounded if the opposite happened and a log of wood suddenly fell out of the sky. But is this any more remarkable than what happens when a tree is created 'out of thin air'?

Changes such as this cannot go of their own accord; they must be driven by some outside power and that outside power is the light of the sun. The elements of the earth can be made into almost any conceivable chemical substance, no matter how large or complicated it is, provided we have enough energy to force the changes to come about. Sunlight provides that energy in abundance to keep us warm, to grow our food and to provide our fossil fuels – gas, oil and coal. Life, from its earliest beginnings to the most advanced technological civilisations of today, is made possible not merely by chemical reactions but by *photochemical* reactions, chemical changes which are driven by light. The book of Genesis records that, on the first day, God said 'Let there be light'. Without that light on the first day the rest could not have happened.

In these lectures, we shall look first at the photochemistry of the creation, then at the processes by which the sun maintains our lives today and finally we shall look to the future when fossil fuels are gone and when, once again, we may have to live on our income of energy from the sun.

There are exceptions, and of course Mrs. Thatcher is one." He considered her genuinely interested in science, but that some of her ministers were definitely "anti-science". Porter had courted Mrs. Thatcher while still at the Royal Institution.

He argued that "we must stop agonising over whether basic science is exploitable," but admitted fault on the part of scientists as being too narrow. Indeed, he used his Presidency to push for reform in the school curriculum to produce more broadly-educated citizens, a situation which has come to pass in the UK.

Porter also used the platform of the Presidency to push for human rights

Porter entertaining an Oxford chemist, Prime Minister Margaret Thatcher at the Royal Institution.

of dissidents in the Soviet Union, China, and Burma. In 1986, he led a delegation to the USSR Academy and brought up specifically the health and "exile" in Gorki of Nobel Laureate Physicist Sakharov, offering to travel to Gorki to meet him. The offer was declined, but Sakharov was realeased shortly thereafter. Porter was heavily involved in the release of the Chinese astrophysicist, Fang Lizhi, and broadcast on the imprisonment of Aung San Suu Kyi in Burma. Despite fundamental differences with totalitarian regimes, Porter kept scientific contacts open, signing an "Agreement on Scientific Cooperation and the Exchange of Scientists between the Academy of Science and the Royal Society of London" with A. P. Aleksandrov on his 1986 visit, and kept a Fellowship program running with China after the Tiananmen Square massacre.

Although much of his time as PRS was taken up with impassioned pleas for public support of basic science, Porter found the time to make incisive and prescient comments on the societal impact of science, the responsibilities of scientists, restraints (by government) on the pursuit of knowledge and on the responsibility of scientists to science in which he covered the area of academic fraud.

His term as President ended on November 30, 1990 with many areas of his greatest concern showing significant improvement, for example the treatment of dissident scientists in the Soviet Union and the funding of UK science. There is little doubt that Porter's efforts had influence in both areas.

Imperial College

In 1985, When Porter became President of the Royal Society, he had decided this role was incompatible with the full-time position as Director of the Royal

Institution, and he thus resigned as Director of the RI in late 1985. However, he wished very strongly to carry on research, and given the small size of the laboratory facilities at the RI, recognised that in order to provide research space for the incoming Director, John Meurig Thomas, he would have to re-establish his research elsewhere. Eric Ash, later Secretary of the RI, and to become Treasurer of the Royal Society, was then Rector of Imperial College, and was responsible for Porter's move to Imperial. Porter was at first in the Department of Biology, in the laboratories of Jim Barber, with whom he had a long and successful collaboration in the field of photosynthesis. Porter took most of his research group with him to Imperial College, including David Klug, who was to become a driving force in the development of femtosecond techniques at Imperial. All of the picosecond flash photolysis apparatus, and much of the basic photochemical equipment moved from the Davy Faraday Laboratories at the RI in early 1986. In 1989, David Phillips became Professor of Physical Chemistry at Imperial, and Jim Barber moved from Biology to become Head of Biochemistry at Imperial College. Since Biochemistry and Chemistry occupied contiguous space, this seemed like an ideal opportunity to bring photochemical research in the two Departments together, and concentrate research efforts in ultrafast techniques. The Porter/Klug Group moved from Biology to good, new laboratories in Chemistry, and a Centre for Photo-molecular Science was set up, with Porter as its Chairman.

Scientists from Chemistry, Biochemistry, Biology, and Physics formed the core of this very successful virtual grouping, which held over the subsequent decade twice-annual one-day meetings on a variety of photo-molecular topics. Porter greatly enjoyed his involvement in the research and played an active role in the meetings until a couple of years before his death. Porter also played a figurehead role in the very successful International Conference on Photochemistry organized at Imperial College in August 1995, welcoming the 450 or so participants, which included several Nobel Laureates (Jean Marie Lehn, Rudi Marcus, Ahmed Zewail, Sherry Rowland), and delivering the opening plenary lecture. The intense science was matched by the extremely hot weather of that week.

Porter was already of normal retirement age when he moved to Imperial College, but he took a very active interest in the science of his group. This period saw the development of femtosecond lasers, and there were some key experiments done during this late period. In particular, the femtosecond laser built by James Durrant, David Klug and the group was used in some ultrafast studies on the primary steps in photosynthesis on samples provided by Jim Barber's group. Porter maintained a strong interest in this aspect of the work on samples provided by Jim Barber's group, but inevitably, his capacity to take on new subjects declined with age. He remained active until two years before his death. His colleagues from around the UK were very pleased to be able to honour him on the occasion of his 80th birthday with a lunch at Imperial College.

Porter received the Order of Merit in 1989 and was made a life peer in 1990, taking the title Baron Porter of Luddenham, after the tiny village in Kent where

Porter at his first "home" in Beit Quad, in Imperial College London.

Opening address, International Conference on Photochemistry, Imperial College London, August 1995 (courtesy of Christopher Phillips).

The Porters and members of the Department of Chemistry, Imperial College, celebrating an honour to the Head of Department, David Phillips (courtesy of the Department of Chemistry, Imperial College).

the family had a weekend cottage. He used his maiden speech in the Lords of May 1991 to criticise again the falling level of support for science.

Political Influence

Porter will never be forgotten for his impact on science, but equally, he made major lasting contributions in raising public awareness of science, and through the media, bringing science issues to the fore. His Nobel Prize ensured he was listened to by all, particularly the politicians with whom he sparred in his later years. Porter was a great champion of fundamental research, or as he put it, "research which has not yet been applied." He argued that without a sound scientific base from which applications would emerge, a nation's prosperity would inevitably decline. In his memorable Dimbleby lectures of 1988, he beguiled the audience by the following quotation,

"(Since the war) in the rivalry of skills, England alone has hesitated to take part. Elevated by her wartime triumphs she seems to have looked with contempt on the less dazzling achievements of her philosopher. Her artisans have quitted her service, her machinery has been exported to distant markets, the inventions of her philosophers, slighted at home, have been eagerly introduced abroad, her scientific institutions have been discouraged and even abolished, the articles which she supplied to other states have been eagerly introduced abroad, her scientific institutions have been discouraged and even abolished, the articles which she supplied to other states have been gradually manufactured by and transferred themselves to other nations. Enough we trust, has been said to satisfy every lover of his country that the sciences and the arts of England are in a wretched state of

depression, and their decline is mainly owing to the ignorance and supine-ness of the Government."

He then revealed that these words were spoken by Sir David Brewster just after the Napoleonic War, and used this as a theme to outline his views on present day ills in science in the UK. There followed a passionate plea that "curiosity-driven" research should be funded more generously, and particularly, that young scientists in their most creative phases, should be enabled with funding to develop their ideas without the necessity of demonstrating an end product or application. This was a continuing theme of his, and it really came to the fore during the five years of his Presidency of the Royal Society, when he spoke out often on public policy issues concerning science (the Dimbleby lectures were given during this period). In his Presidential Address to the Royal Society in 1987, he warned against the "over-management" of research, citing as a prime example of a successful scientific centre, the Laboratory of Molecular Biology in Cambridge under Max Perutz, who was on record as saying

"My laboratory was often held up as a model of a centre of excellence, but this is not because I ever managed it. I tried to attract talented people by giving them independence... had I tried to direct other people's work, the mediocrities would have stayed, and the talented ones would have left. The laboratory was never mission orientated."

This encapsulates the ongoing tensions in UK science, the methods by which the relatively small amount of research funding can be used to best effect without stifling talent through over-management. Porter clearly felt there was a grave danger of this, quoting again in his 1987 Presidential Address the then President of the Mexican Academy of Scientific Research,

"The most tangible evidence of Third World Science is the early preparation and export of outstanding scientists and the production of an avalanche of experts and documents in the politics of science rather than the production of scientific works."

Porter warned that this could be the fate of the UK, a thought widely reported in the UK press. In his farewell address to the Royal Society in 1990, Porter bemoaned the fact that the research councils held an effective monopoly over basic research funding, and advocated a new "Science Foundation" operating at arm's length from the Royal Society, but drawing on the judgment of its Fellows to distribute grants. The dream was never realised.

Porter not only championed "response mode" research, and particularly help for young researchers, but was strongly opposed to central facilities, such as the Rutherford Appleton Laboratories, and Interdisciplinary Research Centres which he felt, soaked up too much of the available funding. In this, he was perhaps a little extreme, in that some research, for example the physics of plasmas using ultra high power lasers, could feasibly only be carried out in such large central groupings, and the High Power Laser Facility at RAL has been responsible for much world-class research. Porter's dislike of RAL had a personal overtone because an application he had made to SERC for a copper vapour laser was peer reviewed at a rating not high enough to secure funding. A referee had made the suggestion that the work could be carried out at RAL, which irritated Porter, and was widely reported in the press in support of his case for the need for more funding, and a rebalancing of the way funds were utilised by the research councils. Porter made another plea, this time in support of chemistry, at the 150th Annual Chemical Congress in 1991, pointing out that SERC funding for chemistry research had dropped to half its level of just one year previously, and yet the UK was known to be extremely strong in chemistry.

In his maiden speech in the House of Lords in May 1991 and subsequently, Porter returned to the theme of funding of "blue skies" research by the research councils, and the relationship between the councils and university scientists. He was active in the House of Lords until shortly before his death.

All of his life Porter had championed the public awareness of science, particularly among the young, and during his unique tenure in 1985 as President of the Royal Society, Director of the Royal Institution, and President of the British Association, he was able to bring together these three institutions to attempt to co-ordinate their efforts, and to act as a focus for the many other efforts in this regard already existing in the UK. Thus COPUS (Committee for the Public Understanding of Science) was born, and chaired by Porter throughout his period as PRS. This move was catalysed by the Royal Society Bodmer report. Although the nature of the committee has now changed somewhat, the publicity and new project brought by COPUS will act as a lasting legacy to Porter's dedication to this vital area.

Despite his great achievement in science and in public life, he never lost his sense of fun, and his devotion to his families, both personal and scientific. When George and Stella moved to the Royal Institution in 1967, the flat they occupied quickly became their home, though they used to spend weekends in their own beautiful house in Luddenham, Kent, from where they also enjoyed sailing. This had been a pursuit of Porter's from Cambridge days and was greatly enhanced by the purchase of a new boat, appropriately named "Annobelle" with some of the 1967 Nobel Prize money.

In 1967, the Royal Institution was as ever under-funded, and somewhat austere. George and Stella set about livening it up, enlisting the aid of all staff in helping them in cleaning and painting the interior of the building. This initial informality with staff later became less prominent, but the feeling was always created of a "family" working together at the RI. Stella would often invite students and staff to lunch in the flat, and was always there on social occasions, both informal and formal. In time, the formal Friday Evening Discourses were to become glittering events, in which Lady Porter played an enthusiastic role as hostess at the Dinners which at that time preceded the 9.00 pm lecture, and at

Aboard "Annobelle" on the Thames, with the Houses of Parliament where George sat as Lord Porter.

Sailing in the Channel with one of his last research students, Ben Crystal.

the drinks party which followed in the flat, referred to by Stella as "The George and Dragon". Many a Discourse speaker, stressed by the ordeal of performing to a strict timetable in formal evening wear late on a Friday evening, found welcome relaxation in their company afterwards in the elegant surroundings of the flat which had been occupied previously by Davy, Faraday, and all successive Directors of the Royal Institution. Both George and Stella delighted in regaling guests with historical anecdotes about their predecessors.

George Porter could occasionally be somewhat serious on social occasions, but Stella did not hesitate to chide him if he were overdoing this. She was however fiercely loyal to her husband and would not tolerate any criticism of him from others.

Porter travelled extensively throughout the world, and where possible, Lady Porter accompanied him. They were an elegant and admired couple who were exemplary ambassadors for Britain, for UK Science, and indeed for science in general.

Porter died in Canterbury on 31st August 2002. His wife, and two sons, John and Andrew Porter, survived him.

Acknowledgments

We would like to thank all the colleagues and friends of George Porter who have contributed to this volume, and others who have made useful suggestions. In particular however, we wish to thank Lady Porter for her helpful advice on some aspects of Porter's early life and career, and for providing many photographs.

George and Stella Porter.

THE PORTER MEDAL

At the valedictory Royal Institution scientific meeting in 1986, friends and colleagues of George Porter collected a substantial fund, which permitted the establishment of a "Porter Medal". This was to be awarded to the scientist who had made the most substantial contribution to the field of photochemistry in the views of the officers of the European Photochemistry Association, The Inter-American Photochemistry Society, and the Japan Photochemistry Association. The then handsome silver medal was designed by Porter himself, and is presented bi-annually at the IUPAC Symposium on Photochemistry.

Porter himself was the first recipient. Where possible he travelled to the IUPAC Symposiums to present the medal to the winners. In his last years though, he was unable to carry this out, and the duty was taken on by David Phillips.

The most recent winner, in 2004 and the first to be awarded since the death of Lord Porter in August 2002, was most appropriately Professor Graham Fleming, a PhD student of Porter's in the 1970s.

The Porter Medal being presented to Professor Nobaru Mataga, Osaka University, Japan, at the IUPAC Symposium on Photochemistry in Helsinki.

The special medal being presented in 1995 to Professor Tito Scaiano, University of Ottawa, Canada.

AWARDS AND APPOINTMENTS

Honours and Awards

1955	Corday–Morgan Medal of the Chemical Society
1960	Fellow of the Royal Society
1967	Nobel Prize for Chemistry
1968	Silvanus Thompson Medal of The British Institute of Radiology
1971	Davy Medal of the Royal Society
1972	Knight Bachelor
1974	Foreign Associate of the National Academy of Sciences, Washington
1977	1976 Kalinga Prize (UNESCO) for the Popularisation of Science
	Bose Medal of Bose Institute, Calcutta
1978	Robertson Memorial Lecture and Prize of the National Academy of Sciences
	Communications Award of the European Physical Society
	Rumford Medal of The Royal Society
1980	Faraday Medal of The Chemical Society
1981	Longstaff Medal of The Royal Society of Chemistry
	Freeman and Liveryman, Honoris Causa, of the Salters' Company City of London
1987	Melchett Medal of the Institute of Energy
1988	Porter Medal for Photochemistry (first recipient)
1989	Member of the Order of Merit
1990	Created Baron Porter of Luddenham in Birthday Honours
1991	Society Medal of the Society of Chemical Industry
	Michael Faraday Medal of Royal Society
	Ellison–Cliffe Medal of the Royal Society of Medicine
1992	Copley Medal of Royal Society
1993/1994	Master, Salters' Company

Honorary Degrees

1968	Honorary DSc University of Sheffield, UK
	Honorary DSc University of Utah, Salt Lake City, USA
1970	Honorary DSc University of East Anglia, UK
	Honorary DSc University of Surrey, UK
	Honorary DSc University of Durham, UK
1971	Honorary DSc University of Leicester, UK
	Honorary DSc University of Leeds, UK
	Honorary DSc Heriot-Watt University, UK
	Honorary DSc City University, UK

1972	Honorary DSc University of Manchester, UK
	Honorary DSc University of St. Andrews, UK
	Honorary DSc University of London, UK
1973	Honorary DSc University of Kent at Canterbury, UK
1974	Honorary DSc University of Oxford, UK
1980	Honorary DSc University of Hull, UK
1981	Honorary DSc of the Instituto Quimico de Sarria, Barcelona, Spain
1983	Honorary DSc University of Pennsylvania, USA
	Honorary DSc University of Coimbra, Portugal
	Honorary D. Univ. The Open University, UK
1984	Honorary DSc University of Lille, France
1985	Honorary DSc University of the Philippines
1986	Honorary DSc University of Notre Dame, USA
	Honorary DSc University of Bristol, UK
	Honorary DSc University of Reading, UK
1987	Honorary DSc Loughborough University, UK
1988	Honorary DSc Brunel University, UK
	Laurea in Chimica Honoris Casusa, University of Bologna, Italy
1990	Honorary DSc University of Cordoba, Argentina
	Honorary DSc University of Liverpool, UK
1993	Honorary LLD Cantab, UK
1995	Honorary DSc University of Bangalore, India
	Honorary DSc University of Buckingham, UK
	Honorary DSc University Bath, UK

Fellowships and Honorary Positions

1966	Honorary Professor of Physical Chemistry, University of Kent
1967	Honorary Fellow of Emmanuel College Cambridge
1968	Honorary Member of the New York Academy of Sciences
1970	Honorary Fellow of the Institute of Patentees and Inventors
	Honorary Member of the Leopoldina Academy
1974	Member of the Pontifical Academy of Sciences
	Corresponding Member of the Gottingen Academy of Sciences
1975	Honorary Fellow of the Royal Scottish Society of Arts
1976	Frew Fellow, Australian Academy of Science
1978	Honorary Member, Chemical Society, Spain
	Foreign Corresponding Member of La Real Academia de Ciencias, Madrid
1979	Honorary Member of the Societé de Chimie Physique (now Societé Francaise de Chimie)

	Foreign Honorary Member of the American Academy of Arts and Sciences
	Member of the Academie International de Lutece
1980	Member of the European Academy of Arts, Sciences & Humanities
	Honorary Research Professor, Dalian Institute of Chemical Physics of the Academy of Sciences of China
1982	Honorary Member of the Chemical Society of Japan
1983	Honorary Fellow of the Royal Society of Edinborough
	Foreign Member, Academy of Sciences, Lisbon
1985	Foreign Member, Institute of Fundamental Studies, Sri Lanka
	Honorary Member, Association for Science Education
1986	Honorary Fellow of Queen Mary College, London
	Honorary Member, American Philosophical Society
	Foreign Fellow, Indian National Academy of Sciences (INSA, Delhi)
	Honorary Professor, Institute of Technology, Beijing, China
	Honorary Member of the Society for Free Radical Research
1987	Honorary Fellow, The Royal Medical Society, Edinburgh
	Honorary Fellow, Imperial College, London
	Honorary Member, Fondation de la Maison de la Chimie, Paris
	Foreign Member, European Academy of Arts, Sciences and Humanities, Paris
	Foreign Member of the Academia Nazionale Dei Lincei, Italy
	Honorary Member, Indian Academy, Maharashtra
1988	Honorary Member of the Royal Institution
	Professor Honoris Causa, Royal Institution
	Honorary Foreign Member, Hungarian Academy of Sciences
	Honorary Member, Japan Academy
1989	Honorary Member, Indian Science Academy, Bangalore
	Honorary Member, Royal Society of Medicine, Edinboro
	Member, Academia Europa
	Honorary Member, Soviet Academy of Sciences
1990	Honorary Member, Indian Academy of Sciences
	Honorary Member, Scientific Society of Argentina
1991	Honorary Fellow, Royal Society of Chemistry, London
	Honorary Fellow, Science Museum of London
	Distinguished Visiting Professor, University of Exeter (1991–1992)
	Honorary Fellow, Society of Radiation Physics
1992	Honorary Fellow, University of N. Lancashire
1995	Honorary Fellow, King's College, London

	Honorary Member of KAST (Korean Academy of Science and Technology)
	Honorary Fellow, Tata Institute, Bombay
1996	Honorary Member of KIAS (Korean Institute for Advanced Studies)

Presidencies and Appointments

1968–1972	President, Comite International de Photobiologie
1970–1972	President, The Chemical Society
1973–1974	President, Faraday Division of the Chemical Society
1972–1974	Trustee, British Museum
1975–1980	First President, The National Association for Gifted Children
1977–1982	President, The Research and Development Society
1985	President, Association for Science Education
1985–1986	President, British Association for the Advancement of Science
1985–1990	President of The Royal Society of London
1986–	Trustee, The Exploratory, Bristol
1986	Trustee, Glynn Research Institute
1986–1995	Chancellor, University of Leicester
1987–1989	President, London International Youth Science Fortnight
1987–1989	President, Scitech
1987–	Vice-President of the British Video History Trust
1991–	President, British Mathematics Olympiad
1987–1991	Member of the Cabinet office Advisory Council on Science and Technology (ACOST)
1989–1991	Trustee, National Energy Foundation
1991	President, National Energy Foundation
1990–1994	Member of the House of Lords Select Committee on Science and Technology
1991–1992	Second Warden, Upper Warden (1992–1993) and Master (1993–1994) Salters' Company
1994–	Almoner of Christ's Hospital

LECTURES AND PROGRAMMES

Principal Named Lectures

1958	Tilden Lecturer of the Chemical Society
1960	P. O'Reilly Lecture, University of Notre Dame
1968	The Silvanus Thompson Memorial Lecture
1969	Liversidge Lecturer of the Chemical Society
	Farkas Memorial Lecture, Hebrew University, Jerusalem
	Elsie O. and Philip D. Sang Exchange Lectureship at Illinois Institute, Chicago
1970	John Dalton Lecture, Manchester
	Liversidge Lecture, University of Essex and University College of South Wales and Monmouthshire
1971	Irvine Memorial Lecture, The University of St. Andrews
1972	Seaver Lecture, University of Southern California, USA
	Leverhulme Lecture of S.C.I., Liverpool
	Bruce–Peller Lecture, Royal Society of Edinburgh
	A. M. Tyndall Memorial Lecture, University of Bristol
1973	The Selby Memorial Lecture, University College, Cardiff
	Brunel Lecture, Brunel University
	Reginald Mitchell Memorial Lecture, Association of Engineers
	Joseph Larmor Lecture, University of Belfast
1974	Phi Lamda Upsilon Priestly Lecturer, Pennsylvania State University, USA
	Burton Memorial Lecture, Queen Elizabeth College Student Chemistry Society
	Henry Tizard Lecture, Westminster School
	Haden Memorial Lecture, Institution of Heating and Ventilation Engineers
	Purves Lectures, McGill University, Canada
1975	Walker Memorial Lecture, Edinburgh University
	Farrington Daniels Memorial Lecture, ISES
1976	Robbins Lectures, Pomona College, California, USA
	Godfrey Frew Lecture, The University of Adelaide, Australia
1977	Bakerian Lecture of the Royal Society
	Pahlavi Lectures, Pahlavi University, Iran
	The Cecil H. and Ida Green Visiting Professorships, The Vancouver Institute, Canada
1978	Romanes Lecture, University of Oxford
	Charles M. and Martha Hitchcock Lectures at the University of California, Berkeley, USA

	Robertson Memorial Lecture, National Academy of Sciences, Washington
1979	Ahron Katzir–Katchalasky Lectures, Weizmann Institute, Israel
	Goodman Lecture, Aitchison Memorial Trust, London
1980	The Nuffield Lecture, The Nuffield Foundation, Chelsea College
1981	Swift Lecture, California Institute of Technology, USA
1983	Edgar Fahs Smith Lecture, University of Pennsylvania, USA
	George M. Batemen Lecture, Arizona State University, USA
	F. J. Toole Lecturer, University of New Brunswick, Canada
	Quain Lecture, University College, London
1984	Dorab Tata Memorial Lectures, Bombay
	Reddy Lectures. Osmania University, Hyderabad
1985	The Julian H. Gibbs Lectures, Amherst College, Massachusetts, USA
	The Hampton Robinson Lecture, Texas A & M University, USA
	The Humphrey Davy Lecture, The Académie des Sciences, Paris
	Vollmer Fries Lecture, Rensselaer Polytechnic Institute, Troy, USA
	Royal Institution of Great Britain Lecture, University of Arizona, USA
1986	9th J. T. Baker Nobel Laureate Lecture, Stanford University
	Neil Graham Lecture, Toronto University
1987	Lee Kuan Yew Distinguished Visitor, National University of Singapore
	Melchett Lecture, Institute of Energy
	Redfearn Lecture, Leicester University
	Special Gresham Lecture, Bishopsgate Institute
1988	Dimbleby Lecture "Knowledge Itself is Power" BBC Television
	McGovern Lecture (Sigma Sci, Orlando, USA)
	Fawley Lecture. Southampton
	Koimura Shimbun Lecture, Nagoya, Japan
1989	9th Einstein Memorial Lecture, Israel
	Cecil and Ida Green Lectures, Galveston, Texas
1990	Ramon Alceras Memorial Lecture, Madrid
	Darwin Lecture, Darwin College, Cambridge
1991	Linus Pauling 90th Birthday Lecture, Caltech, Pasadena
	Maiden Speech, House of Lords, 23 May
	Gerald Walters Memorial Lecture, University of Bath
	R.S.C. 150th Anniversary Address
	Faraday Bicentennial Addresses at RI, Leeds, Newcastle etc.
	SCI Medal Address, Liverpool
	Science Centre Young People's Lecture (for IC series), Singapore
1992	Michael Faraday Award Lecture of The Royal Society

5th Ellison–Cliffe Lecture and Gold Medal of RSM
3rd Rayleigh Lecture, Harrow School
Centenary Lecture at the University of Chicago
Bernal Lecture, Birkbeck College, London

1993   Birle Memorial Lecture, Science Centre, Hyderabad
1994   Campaign for Resource Lecture to University Court, Bristol
1995   Lee Kuan Yew Video Lecture, National University of Singapore

Rajiv Gandhi Memorial Lecture, Puna
Sidney Chapman Memorial Lecture, Fairbanks, Alaska

Television and Radio Programmes

1960	Eye on Research, "Quick as a Flash"
1973	"A Candle to Nature", pilot for "Horizon" series
1965–1966	"Laws of Disorder", ten programmes (r. 4 times) BBC
1969–1970	"Time Machines", six Christmas Lectures from RI, BBC (r. twice)
1971–1975	"Controversy" series of debates, Chairman, BBC
1976–1977	"The Natural History of a Sunbeam", six Christmas Lectures from RI
1966–1981	Judge, Young Scientists of the Year, BBC
1982	Contributor, Open University Course on "Photochemistry: Light, Chemical Change and Life"
1984	"Man of Action", Radio
1987	"Conversation Piece", Radio
1988	Dimbleby Lecture
1990	Desert Island Discs (with Sue Lawley)

Films and Videotapes

The Laws of Disorder – a series of four I.C.I. (Milbank) films for Schools.
"Why Chemistry" – videotape for Salters' Company and I.C.I
"George Porter" – three autobiographical tapes for RS 1996

ARTICLES BY PORTER

Reprinted from *Proc. Roy. Soc. A* **200**, 284–300 (1950) with permission from The Royal Society

Flash photolysis and spectroscopy
A new method for the study of free radical reactions

By G. PORTER, *Department of Physical Chemistry, University of Cambridge*

(*Communicated by R. G. W. Norrish, F.R.S.—Received* 9 *August* 1949)

(PLATES 6 TO 8)

Photochemistry provides us with one of the most generally useful methods of studying the reactions of free radicals and atoms, but the concentration of these intermediates in the usual photochemical systems is too low to allow the use of direct physical methods of investigation such as absorption spectroscopy.

To overcome this difficulty a new technique of flash photolysis and spectroscopy has been developed, using gas-filled flash discharge tubes of very high power. The properties of these lamps as spectroscopic and photochemical sources have been studied and details are given of their construction, spectra, duration of flash, and luminous efficiency in the photochemically useful region. An apparatus is described which produces a very great photochemical change, in some cases over 80 %, in one-thousandth of a second and in a gas at several cm. pressure contained in an absorption tube 1 m. long, and which photographs the absorption spectrum at high resolution in one twenty-thousandth of a second at short intervals afterwards.

Examples of the rapidly changing spectra of substances undergoing reaction, including the spectra of some of the intermediate radicals involved, are shown. These include the recombination of chlorine atoms, the absorption spectra of S_2 and CS obtained during the photochemical decomposition of carbon disulphide and new spectra attributed to the ClO and CH_3CO radicals.

INTRODUCTION

Investigations into the mechanism of chemical reactions have revealed that in very many cases intermediate substances are involved which, although they exist only for a very short time, determine the course and rate of the changes which take place. Much information has been obtained about these intermediates, which are usually free radicals or atoms, by induction from the kinetics of the overall change, but the indirectness of this method has so far rendered it incapable of giving the kinetic details of some of even the simplest radical reactions. We have much to learn, for example, about the combination of two methyl radicals. The mirror technique, developed by Paneth and his colleagues, is another powerful means of study, but suffers from similar limitations in so far as it depends on inference from the final products with the metal mirror, and it is also accompanied by some rather severe experimental difficulties (Norrish & Porter 1947).

The only direct methods of investigation which have been applied to the problem are spectroscopy and mass spectroscopy, and, of the two, the former is potentially the more powerful because it enables the reaction to be studied in a static system as well as giving information about the structure of the radical, and more certain identification. Furthermore, spectroscopy is ideally suited to free radical studies, as the majority of free radicals have a transition involving the ground state in the easily accessible visible or near ultra-violet regions and, as pointed out by Wieland (1947), this often forms an extremely sensitive method of detection.

[284]

Flash photolysis and spectroscopy

Although the spectra of a large number of diatomic radicals are known from their emission bands in flames and discharge tubes, very little information about their chemical reactions can be obtained in this way because of the difficulty of estimating concentrations from emission spectra. Absorption spectroscopy, on the other hand, makes it possible to estimate concentrations and follow the reactions of the absorbing molecule without interfering with the reacting system, but unfortunately to observe a radical in absorption the concentration must be higher than is usually obtained by any method other than the electrical discharge. High equilibrium concentrations can be obtained by thermal decomposition, but it is not possible to change the temperature of the gas rapidly enough to use this method for kinetic studies. The electrical discharge method has been used very successfully by Oldenberg (1935) in a detailed study of the OH radical and also by White (1940). For the study of the chemistry of intermediate compounds, however, the method is limited to radicals which survive the violent conditions in a discharge tube, that is, virtually, to diatomic radicals, and to the pressures under which the discharge will take place. A further disadvantage is that the complexity of the reactions occurring in the discharge tube makes interpretation of the results impossible in all but the very simplest systems. Very few polyatomic radical spectra are known, and even in these cases the identity of the radical is doubtful.

Photochemical decomposition provides the best general method of preparation of free radicals as the overall reaction is relatively quite simple, the initial act and the radicals produced are better known than in any other type of reaction and nearly all radicals, both simple and complex, can be produced by photochemical means. No free radical has ever been detected in absorption in a photochemical reaction however, because the method has one great disadvantage; the concentration of radicals produced by even the most intense light sources is very low indeed. If this difficulty could be overcome, and free radicals could be followed spectroscopically in photochemical systems, an ideal combination would result. This communication describes how such a method has been developed, making use of the fact that illumination of the system for a longer time than the half-life of the radicals is not necessary so that a flash technique can be used. It is shown that a modification of a type of discharge lamp now in use as a photographic source is capable of producing a partial pressure of free radicals higher than has ever been obtained by photochemical or any other methods.

There is one other difficulty associated with kinetic absorption spectroscopy of all kinds, that of producing an image on the photographic plate in a short time, from a source of continuum. Other workers have resorted to the integration of a large number of short exposures, but this is not feasible in the apparatus described here, first, because with such a high percentage decomposition the absorption tube has to be removed and cleaned after each flash with many of the substances used, and secondly, because the high energy makes it necessary to cool the lamp between flashes, so that if several thousand flashes had to be used each spectrum would require an experiment of several days.

An attempt was made to overcome this difficulty by making use of the great sensitivity and rapid response time of photomultiplier tubes, and a rapid scanning

G. Porter   286

system was built into a 10·5 ft. grating spectrograph for this purpose. If this method is to be suitable, a high resolution coupled with good signal/noise ratio is necessary. The noise, caused by statistical fluctuations in the photocurrent, can only be decreased by decreasing the band width and hence the resolution, and even with the brightest light sources available no satisfactory compromise was reached. Fortunately, investigations into the properties of flash discharge lamps as photochemical sources showed that they could be made to produce a very intense continuum over the whole spectral region, and a lamp was eventually designed which would record photographically down to 2000 Å in a large Littrow spectrograph in less than 10^{-4} sec.

The procedure adopted was to produce a high percentage decomposition of the reactant by a high intensity flash in a lamp alongside the 1 m. absorption tube and to record the absorption spectrum of the products as the reaction progressed by means of flashes from another lamp at the opposite end of the tube from the spectrograph. The requirements for these two lamps are slightly different. Both must be of very high intensity and must be capable of accurate synchronization, but whereas a maximum energy output in the region producing chemical change is the main consideration with the photolysis lamp, a continuous spectrum covering the whole ultra-violet region is required from the spectrographic source. The duration of the photolysis flash should be not greater than the half-life of the radicals which are to be studied, and the spectroflash must be shorter still if several snapshots of the changing radical spectra are to be obtained.

Of the several ways of obtaining a brief flash of light, high pressure gas-filled discharge tubes showed the best promise for this type of experiment. Very high energies can be dissipated in one flash, and by arranging the pressure to be high enough to prevent breakdown at the operating voltage until a trigger pulse is applied, accurate synchronization can be obtained without any power loss at a switch or spark gap. No previous investigation of the possibilities of this type of lamp for photochemical and spectroscopic purposes seems to have been carried out, and there follows a brief account of the properties of high-energy flash discharge lamps as emitters in the photographic ultra-violet.

FLASH DISCHARGE TUBES AS PHOTOCHEMICAL AND SPECTROSCOPIC SOURCES

The lamps used were all fundamentally very simple. They consisted of a gas-filled tube with an electrode at each end, across which the discharge condenser was connected, and some means of triggering the flash was provided near the centre of the tube, either in the form of an external coil or a third internal electrode. The main difference in design from the ordinary continuous source was in the pressure of the gas filling which must be high enough to prevent the discharge taking place until the triggering pulse is applied. Unless the lamp is to be flashed very frequently cooling is not necessary. It may be constructed in the shape and length most suitable for the particular application, and the other factors which can be varied are gas filling, gas pressure, capacity, voltage, and the material of the tube and electrodes.

Construction

Almost any ordinary continuous discharge lamp may be modified for flash work if the power per flash is not too high. Above about 100 J, however, it is necessary to use quartz tubes and large electrodes of tungsten or similar material with large current capacity seals of the type used in high-power mercury discharge lamps. Murphy & Edgerton (1941) have described the construction of lamps which were capable of dissipating 500 J per flash, and lamps of this kind are now produced commercially for photographic purposes. The maximum energy per flash used in the experiments described here was 10,000 J, and at these high values two difficulties appear. First, the discharge decomposes the material of the tube and electrodes as well as liberating any occluded gas which was not removed before filling. Although a small amount of impurity such as oxygen has very little effect on the output of the lamp, it has a very pronounced effect on the firing characteristics. The pressure of the original gas filling has to be high enough to prevent firing until the trigger pulse is applied, and after a few high-power flashes sufficient impurity may have been liberated to prevent the discharge being initiated by the application of this pulse, and evacuation of the tube and refilling is necessary. These troubles were partly overcome by intense heating of the tube during evacuation and by discharging the maximum capacity across the lamp and evacuating and refilling several times. At the highest powers used it was still necessary to refill the lamps frequently.

The other difficulties were caused by mechanical weaknesses in the seals and tubes. The high pressure developed is sometimes sufficient to explode the lamp, and the rapid heating sets up strains in the seals which may finally crack. By careful construction and the use of the latest type of tungsten-molybdenum quartz seals* these difficulties were eliminated up to powers where refilling after nearly every flash was necessary. The crazing of the tubes does not appear to weaken them unduly, and poisoning of the gas filling remains the most important energy limitation.

Duration of the flash

Edgerton found that when high capacities were used the current passed by the discharge tube tended to a maximum corresponding to a resistance of $0.2\,\Omega$ for a tube measuring 30×1.4 cm. That a corresponding maximum in light output is reached is shown by the oscillograph recordings in figure 1. These were taken by using a 1 P 28 photomultiplier cell which is sensitive over the whole region between 2000 and 8000 Å. The lamp was 1 m. × 1 cm. in size, was krypton filled at a pressure of 5 cm. Hg and the voltage was 4000 V. It is seen that the maximum output is reached very rapidly, and that this is followed by fairly constant emission whose duration is proportional to the capacity, and then an exponential decay. The secondary pulse is probably due to the circuit inductance.

The duration of the flash may be decreased by using shorter or wider tubes which decreases the resistance, but this reduces the power which can be dissipated per flash. For a given capacity the duration of the light pulse is proportional to the length

---

* These seals were supplied by Messrs Siemens and I am indebted to Dr J. N. Aldington and Mr A. J. Meadowcroft for information about the construction of photographic flash tubes.

of the tube. In figure 9 the short pulse from a 15 cm. long krypton-filled tube produced by the discharge of a $50\,\mu$F condenser charged to 4000 V is shown superposed
on a $480\,\mu$F flash from the longer tube described above.

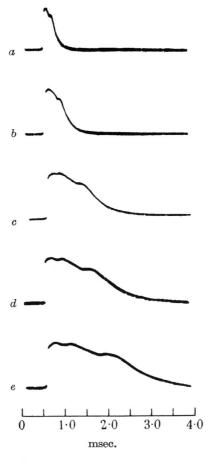

FIGURE 1. The effect of capacity on the shape and duration of the light pulse.
Capacities in a to e were 48, 96, 240, 336 and $480\,\mu$F respectively.

Light output

In order to get some idea of the amount of photochemical change which could be
obtained in a single flash and the efficiency of flash discharge lamps as emitters in
the photochemically useful region, some experiments were carried out using a uranyl
oxalate actinometer. The arrangement was exactly the same as would eventually
be used in the complete apparatus. A metre-long lamp of the type described above
was placed alongside a 1 m. quartz reaction vessel 2·5 cm. in diameter, both being
surrounded by a cylindrical reflector coated on the inside with magnesium oxide
which has a very high reflectivity in the ultra-violet. The reaction vessel was filled
with 350 ml. of the oxalate solution, and the photochemical change produced by one
flash was estimated by permanganate titrations using the methods recommended
by Leighton & Forbes (1930). The maximum temperature rise recorded on a thermo-

couple placed in the solution was 0·6° C, and as the temperature coefficient is only 1·03 per 10° C this was a negligible effect.

Using a solution of 0·001 M-uranyl nitrate and 0·005 M-oxalic acid 11·6 % of the oxalate was decomposed by a single flash of 480 μF capacity which corresponds to the reaction of 2·6 × 10^{-4} g.moles in less than 2 msec. Repetition of this determination showed that the output per flash was constant to within 2 %. If we assume a quantum yield of 0·6 and that the average wave-length responsible for photochemical change is 3500 Å, the total energy necessary to produce this change is 145 J. The energy stored in the condenser is $\frac{1}{2}CV^2$ or 3840 J, so that the luminous efficiency of the whole system in the region between about 4800 and 2000 Å is 3·8 %. As the concentration of the oxalate is increased this figure also increases, and in figure 2 the efficiency, obtained as above, is plotted against the concentration of oxalate, the uranyl nitrate-oxalic acid ratio being always the same. By plotting reciprocals an extrapolation to infinite concentration corresponding to maximum absorption may be made which leads to a figure of 16·5 %. This gives a lower limit for the luminous efficiency of the flash source in the region of absorption of the uranyl ion and is a remarkably high figure, including, as it does, losses of all kinds, such as those due to incomplete discharge of the condenser and the impedance of the connexions. For an average wave-length of 3500 Å it corresponds to an output of 10^{21} quanta/flash or 5 × 10^{23} quanta/sec.

It is also necessary to know how the output of the lamp varies with the capacity per flash, and in figure 3 the photochemical change expressed in ml. of 0·005 N-permanganate per 25 ml. of oxalate is plotted for different capacities per flash, the total energy input into the lamp being the same in each case.

The exposures were as follows:

capacity	number of flashes	change/25 ml. (ml. 0·005 N-KMnO$_4$)
4	120	1·41
24	20	2·25
48	10	2·40
96	5	2·74
240	2	2·94
480	1	2·94

It is clear that only above a capacity of about 200 μF does the lamp reach maximum efficiency in this region, and that at higher capacities the efficiency is fairly constant.

Flash spectra

Whilst the main requirement for the spectrum of the photochemical source is that a higher percentage of the emission should be in the region to be used for the photolysis, the spectroflash must give a continuous spectrum over the whole region to be photographed with as few interfering lines as possible. Under ordinary conditions most gases give mainly a line spectrum in the discharge tube, but the continuum emitted by hydrogen is well known, and attempts were first made to produce this continuum at high intensity in a flash discharge. It was found, however, that a purer and more intense continuum could be obtained under flash conditions from

G. Porter   290

the rare gases than from hydrogen. Except where otherwise stated the spectra shown
were produced in a 12 mm. diameter quartz tube with tungsten electrodes in side
arms 15 cm. apart, and were taken via the 1 m. absorption tube by means of a large
Littrow (Hilger E 1) spectrograph using a slit width of 0·02 mm.

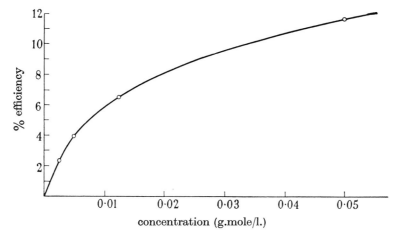

FIGURE 2. Energy efficiency of photolysis of uranyl oxalate against concentration.

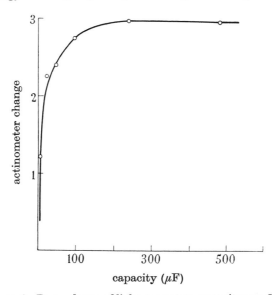

FIGURE 3. Dependence of light output on capacity per flash.

Figure 4, plate 6, shows the effect of increasing the energy per flash on hydrogen-
and krypton-filled tubes. In the case of hydrogen the spectrum at low capacities is
a weak continuum overlaid only by very intense Balmer lines, the secondary molecular spectrum of hydrogen being absent. The Balmer lines are broadened and
partially reversed, this being especially noticeable in the Hγ line. As the capacity
per flash is increased the intensity of the continuum increases, but a relatively sharp
line spectrum appears covering the whole region. Wave-length measurements show
that this is almost entirely the spectrum of silicon and oxygen, presumably due to

Flash photolysis and spectroscopy

the decomposition of the tube. It appears, with slight differences, in both quartz and Pyrex tubes and with tungsten or aluminium electrodes; in the latter case a few additional lines appear due to the decomposition of the electrodes. As the capacity per flash is increased the relative intensity of this line spectrum to the continuum is further increased. The spectrum in figure 4 a, which is a single flash, is of just sufficient intensity to record down as far as 2900 Å, but any further increase of capacity renders the line spectrum too intense to be tolerated. If the lamp is run continuously at normal current densities after the flash, but without refilling, no trace of the line spectrum appears.

The krypton spectra in the same figure show that as the capacity is increased the relative intensity of the line spectrum to the continuum is decreased. In figure 4 f the line spectrum is entirely that of krypton, and even at high capacities there are relatively very few other lines present. The krypton continuum is about six times as intense as that of hydrogen under the same conditions, and at high capacities there is less interference from the line spectrum.

Figure 5 d, e and f, plate 6, which are all the spectra of single flashes, show more clearly how the relative intensity of continuum to line spectrum of krypton increases with capacity, the 73 μF flash being a very good continuum of sufficient intensity to record clearly down to 2000 Å. A few lines due to silicon appear above this capacity, and if the tube is flashed continuously the group of six lines between 2506·9 and 2528·5 Å are reversed, these being the only discontinuities over the whole visible and ultra-violet region. The other spectra in figure 5 show the effect of voltage and pressure. Increasing the voltage has a similar effect to increasing the capacity, but the gas pressure has very little effect on the spectrum. These parameters cannot be varied independently over a much greater range than that shown, as the pressure is determined within these limits for a given voltage by the firing potential.

The spectra of other gas fillings are shown in figure 6, plate 7. It is seen that the only difference between the spectra of argon, krypton and xenon flashed under the same conditions is in the respective rare gas line spectrum, the intensity and distribution of the continuum being almost identical in the three cases. Figure 6 a is the spectrum of a 1 m. lamp krypton-filled with the addition of a little liquid mercury. Although the tube was flashed cold, so that at the beginning of the flash the mercury pressure was very low, the krypton line spectrum has entirely disappeared and has been replaced by the much more intense spectrum of mercury. This has the effect of shifting the energy distribution further into the ultra-violet, and as an example of this the photochemical decomposition of acetone produced by this lamp was three times that obtained with a lamp filled with pure krypton and flashed under the same conditions.

The origin of the continuum in high-current density gaseous discharges is not known with certainty, but it seems probable that it is due to retardation and recombination radiation of the electrons in the positive column. These results show that above a certain capacity, which depends on the dimensions of the lamp, the peak light output tends to a limiting value, and this is accompanied by a rapid increase in the intensity of the continuum relative to the line spectrum of the gas and a maximum luminous efficiency in the region studied.

G. Porter   292

Triggering and synchronization

An internal electrode, sealed through the tube half-way between the main electrodes, was found to be more reliable than an external coil as a means of applying the trigger pulse. The delay between the making of the contacts in the primary of the induction coil and the firing of the lamp was constant to within a few μsec. in the apparatus described in the next section.

The above investigations are sufficient to show that this type of flash discharge lamp amply fulfils the requirements stated in the introduction for photochemical and spectroscopic purposes. In most of the experiments to be described here the photochemical source was a 1 m. quartz tube 1 cm. in diameter with 4·84 mm. ($\frac{3}{16}$ in.) tungsten electrodes through molybdenum seals and a small central trigger electrode. The filling was pure krypton at 5 cm. pressure, this being chosen because the mercury/krypton type had a rather variable output depending to some extent on the position of the mercury in the tube and its temperature. When discharged with 500 μF at 4000 V this lamp had an effective flash time of 1·5 msec.

The source of continuum chosen was a lamp of the same tube and electrode type but with the electrodes in side arms 15 cm. apart and with a flat quartz plate sealed firmly into the end. The usual filling was 10 cm. pressure of krypton which gives the purest continuum over the whole region, though argon was occasionally used for the region below 2700 Å where it is slightly better. When flashed with 70 μF at 4500 V this lamp gave a continuum of sufficient intensity to record in 50 μsec. over the whole spectrum down as far as 2200 Å. Both these lamps had a life of many hundred flashes, though refilling was necessary from time to time.

DESCRIPTION OF THE EXPERIMENTAL ARRANGEMENT

The complete apparatus is shown schematically in figure 7. The 1 m. photolysis flash lamp lies alongside the quartz reaction vessel in the reflector already described, which is constructed in two semi-cylinders to facilitate inspection and removal. The spectroflash lamp is at one end of the reaction vessel and the spectrograph slit at the other, and a small detachable mirror placed near the slit enables the iron arc or other comparison spectrum to be taken without disturbing the alinement. A current of air is blown into the centre of the reflecting cylinder and escapes at either end, and the high-voltage trigger pulse is led on to the centre electrode of the lamp via a glass insulator through this case.

The vacuum apparatus consists of the usual pumping arrangements and pressure gauges, storage bulb, purification train and gas-analysis apparatus. Provision is made for filling the lamps *in situ*, although frequent refilling is necessary unless they are heated vigorously during evacuation.

Between the reaction vessel and spectrograph is the wheel responsible for synchronization of the shutter, photolysis flash, spectroflash and oscilloscope time base. The scattered light from the photolysis flash which recorded on the plate was about one-quarter as intense as that from the spectroflash, so that for exploratory work it is possible to work without a shutter, but for intensity measurements it is necessary

Flash photolysis and spectroscopy

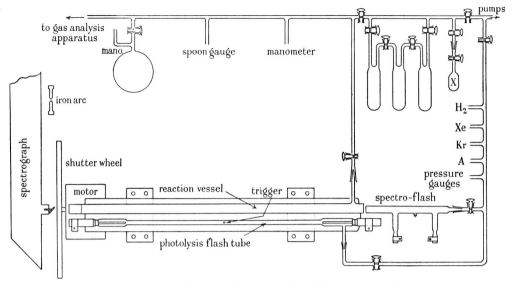

FIGURE 7. Diagram of the experimental arrangement.

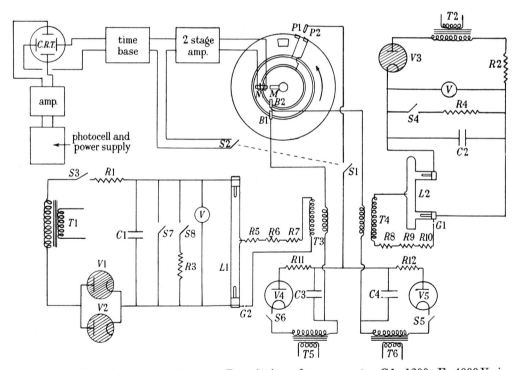

FIGURE 8. Electrical circuit diagram. Description of components: $C1$, $1200\,\mu\text{F}$, $4000\,\text{V}$, in units of 1 to $4\,\mu\text{F}$; $C2$, $125\,\mu\text{F}$, $4500\,\text{V}$, in units of 1 to $4\,\mu\text{F}$; $C3$ and $C4$, $1\,\mu\text{F}$, $1000\,\text{V}$; $V1$, $V2$ and $V3$, mercury rectifiers, $8000\,\text{V}$, R.M.S. type CV 128; $V4$ and $V5$, rectifiers, $2000\,\text{V}$ R.M.S.; V, electrostatic voltmeter, $5000\,\text{V}$; $T1$, 200/3000, 4000, $5000\,\text{V}$, $10\,\text{A}$; $T2$, 200/$5000\,\text{V}$, $4\,\text{A}$; $T3$ and $T4$, high-ratio induction coils; $T5$ and $T6$, 200/$1000\,\text{V}$, $2\,\text{A}$; $R1$, $2000\,\Omega$, $400\,\text{W}$; $R2$, $40{,}000\,\Omega$; $200\,\text{W}$; $R3$, $30{,}000\,\Omega$, $400\,\text{W}$; $R4$, $4000\,\Omega$, $400\,\text{W}$; $R5$, to $R10$, $60{,}000\,\Omega$, $5\,\text{W}$; $R11$ and $R12$, $30{,}000\,\Omega$, $3\,\text{W}$.

G. Porter   294

to eliminate it completely. A synchro-motor rotates the wheel via gears giving
8, 60, 200 and 600 r.p.m., and fine adjustment of the time interval between flashes
is obtained by altering the distance apart of the two contacts on the wheel which are
responsible for the trigger pulses.

The electrical arrangement is shown in figure 8. Apart from brushes, direct contacts to the wheel have been avoided so that the speed of rotation is uniform. The
contacts at $P1$ and $P2$, which supply trigger pulses to the photolysis flash and
spectro-flash respectively, are 0·1 mm. apart and platinum tipped, and the 1000 V
primary pulse is sufficient to cross the air gap between them. $P1$ is fixed but $P2$

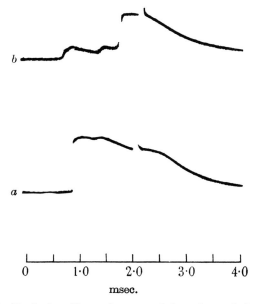

FIGURE 9. Typical oscillograph traces of the pulse and shutter timing.

moves radially over the circumference of the wheel and has a scale attached enabling
time intervals between 2×10^{-4} and 0·5 sec. to be selected, longer intervals being
obtained by manual switching. The synchronization pulse for the cathode-ray tube
time base is induced by a small permanent magnet M on the wheel shaft in a small
telephone coil, N, amplified, differentiated and rectified and passed to the hard-valve
time base.

A long slot, the height of the spectrograph slit, is cut in the wheel and a thin
aluminium slide moves on a scale over this enabling the whole or part of the photolysis flash to be eliminated. Calibration of the time scales and regular checking was
necessary as there is a considerable interval introduced between contact and the
initiation of the flash. This was carried out by means of a photocell inside the spectrograph connected via an amplifier and switch $S2$ to the cathode-ray tube. The
oscilloscope trace is photographed, and, so that only one scan shall appear, the switch
$S2$ is ganged to the main trigger switch $S1$. The timing of two typical experiments
is illustrated in figure 9. In $9a$ the timing between the beginning of the two flashes
is 1·1 msec. and in $9b$ it is 1·5 msec. The latter also shows how the photolysis flash
can be eliminated quite sharply by introducing the shutter.

The trigger pulses for the two flash tubes are obtained by discharging a 1 μF condenser charged to 1000 V through the primary of an induction coil. The condenser $C\,3$ produces the trigger pulse for the photolysis flash tube when it is discharged via the primary of $T\,3$, the brush $B\,1$, the gap $P\,1$ and the switch $S\,1$. The pulse from the secondary of transformer $T\,3$ discharges through the lamp $L\,1$ via the gap $G\,2$ and resistances $R\,6$, 5 and 7. The gap avoids the necessity of a direct connexion of the lamp electrodes to the rest of the circuit, and the resistance chain is necessary to prevent the discharge short-circuiting to earth via the trigger electrode and $T\,3$. The spectroflash lamp has identical trigger arrangements employing $C\,4$, $T\,4$, $B\,2$, $P\,2$, $S\,1$, $G\,1$ and $R\,8$, 9 and 10.

$C\,1$ is the main condenser for the photoflash which is a variable capacity having a maximum value of 1200 μF. It is made up of units of 1 to 4 μF and is housed in a separate room with remote controls. The connecting leads are of 1·27 cm. ($\frac{1}{2}$ in.) copper strip, the main leads from the whole bank are of 2·54 cm. (1 in.) diameter copper pipe in order to keep all resistances down to a minimum, and precautions were taken during construction to reduce the inductance as far as possible. $C\,2$ is a similar condenser of 125 μF maximum capacity which supplies the power for the spectroflash. The master switch which fires both lamps and triggers the time base is the ganged switch $S\,1$ and $S\,2$.

The procedure used to photograph the spectrum a fraction of a second after photolysis is as follows. The reaction vessel is filled with the substance to be studied and the spectrograph shutter is opened. If a time check is required, the oscilloscope camera shutter is opened and the photocell, oscilloscope and trigger circuits are switched on. The shutter slide and spectroflash trigger contacts are set at the positions required to produce the desired time interval, the filament circuits of all valves are made and switches 4, 7 and 8 are opened. $S\,7$ is a safety-door switch, and $S\,4$ and $S\,8$ are remote discharge switches. The synchromotor is started, and switches 5 and 6 made for a few seconds to charge condensers 3 and 4. Switches 3 and 4 are now made and the condenser banks charged to exactly the operating voltage. All is now ready, and switch $S\,1/2$ is pressed when the following train of events occurs. The oscilloscope scan is tripped by the pulse induced in N and the wheel turns until $P\,1$ is opposite the standing contact when $C\,3$ discharges through $T\,3$ and fires $L\,1$. As the wheel turns farther the shutter clears the slit of the spectrograph, and when $P\,2$ comes opposite the standing contact $C\,4$ discharges through $T\,4$ and fires $L\,2$. A series of spectra at increasing time intervals is obtained by repeating this procedure for different positions of $P\,2$.

Several different spectrographs have been used, but the most generally suitable is the large Littrow type as it combines high dispersion with short exposures. One flash of the type described is sufficient to record over the whole photographic region with a slit width of 0·02 mm.

RESULTS

The amount of photochemical decomposition obtained with a few typical compounds is given below. The mercury/krypton filling was used with acetone and

G. Porter   296

ketene and pure krypton with the other substances; in each case the volume of gas was 500 ml. and the illumination a single flash of 4000 J:

gas	pressure (cm.)	% decomposition
NO_2	4	nearly 100
CH_3COCH_3	3	15
CH_3COBr	3	15
$CH_3COCOCH_3$	2·5	50
CH_2CO	10	40

If the lifetime of the intermediates involved in these reactions is comparable with the duration of the flash, the amount of decomposition obtained in this way makes possible the direct study of radical reactions by pressure change and the other physical methods which, hitherto, it has only been possible to apply to the overall change. The effect of high intensities on the nature of photochemical change has been mentioned in a previous note (Norrish & Porter 1949) and will be the subject of another communication.

The spectrographic technique has been applied to several photochemical reactions, and in most cases the amount of change was sufficient to be easily measurable by absorption changes. Three such systems will be mentioned here as illustrations of the potentialities of the method.

Atomic chlorine and its reactions

With a flash of 4000 J, 1 cm. of chlorine showed a Budde effect of 2·2 cm. Control experiments with inert gas showed that there was no detectable direct heating effect from the lamp. In 0·5 cm. of chlorine the absorption spectrum of the chlorine molecules almost completely disappeared, a direct demonstration of dissociation into atoms. The pressure of atoms must have been nearly 1 cm. and over 80 % of the total, a much higher concentration of atoms than has ever been obtained by photochemical or any other methods. The recombination of chlorine atoms may be studied in this way and figure 10, plate 7, shows the spectrum at increasing times after a flash of 2000 J, with a pressure of 1 cm. chlorine. The original decomposition is about 50 % in this case, and the half-life of the chlorine atom immediately after the flash, judged by visual comparison with the spectrum of the chlorine molecule at different pressures, is about 30×10^{-3} sec. Accurate estimations will have to take into account the temperature dependence of the absorption coefficient (Gibson & Baylis 1933).

When carbon monoxide was added to the system there was no permanent pressure change even after several flashes, and no phosgene or intermediate radicals were detected spectroscopically. If the mixture was illuminated in the ordinary way by a small mercury lamp, phosgene was rapidly formed, and the reaction could be taken almost to completion. On flash illumination of the mixture containing phosgene the pressure increased again and the phosgene was once more decomposed. The pressure change/flash was small at first, increased as more chlorine was formed and finally decreased when most of the phosgene had disappeared. The reaction was quite reversible and could be taken almost to completion either way by simply changing the intensity of the illumination. Two factors probably play a part in changing the

mechanism of the reaction at high intensities. First, owing to the much higher concentration of atoms and radicals, interradical reactions become very frequent, so
that the relative probability of the reaction

$$COCl + Cl = CO + Cl_2 \qquad (1)$$

to the reaction

$$COCl + Cl_2 = COCl_2 + Cl \qquad (2)$$

will be greatly increased. Secondly, owing to the higher temperature, the reverse of
reaction (2) will occur more readily.

No evidence of $COCl$ or Cl_3 radicals was obtained in the chlorine or chlorine-carbon
monoxide systems, but when phosgene was present an intense continuous absorption
over the whole region studied appeared. It has not yet been decided whether this
is due to an intermediate or to absorption by 'hot' phosgene molecules.

When oxygen was present a new banded spectrum appeared in the region of
2800 Å. This spectrum was also obtained with chlorine and oxygen alone, but not
when nitrogen or inert gases were substituted for the oxygen. It is shown in figure 11 a,
plate 7, and consists of a regular system of bands degraded to the red with a few
weaker bands probably belonging to the $v'' = 1$ progression. The bands appear to be
single headed and have a very simple rotational structure, and it seems most probable
that a diatomic molecule is responsible. The most reasonable choice under these
conditions is the ClO radical which has been frequently discussed as an intermediate
in chlorine-sensitized oxidations. The half-life of the radical was found to be about
4×10^{-3} sec. in the presence of 1 cm. chlorine and 10 cm. oxygen. A Birge-Sponer
extrapolation gives a dissociation energy to the products in the upper state of
108 kcal./g.mole. If the products of dissociation are normal chlorine atoms and ^1D
oxygen atoms this would lead to a dissociation energy of the radical to atoms in the
ground state of 63 kcal./g.mole.

Diacetyl

Figure 12 a, plate 8, shows the absorption spectrum of diacetyl at 2·5 cm. pressure
before photolysis, there being two distinct regions of absorption, one between 3800
and 4500 Å and the other below 3000 Å. The amount of decomposition can be judged
from figure 12 e, which is taken several minutes after a flash of 4000 J. Figure 12 b is
taken during the flash when photochemical decomposition is not complete. The
reappearance of the absorption in the long wave-length region may be due to
recombination of radicals, but is more probably due to a temperature effect. The
increased absorption at lower wave-lengths is of too short a duration to be a temperature effect, however, the half-life being less than 1 msec., and it is difficult to
find any other interpretation than that it is due to some intermediate substance of
short life formed in the decomposition. The acetyl radical, which is probably formed
in good yield in the reaction, might be expected to give an absorption in this region
without any obvious fine structure. An analysis of the products not condensed in
solid CO_2/ether gave: ethane 27 %, carbon monoxide 62 %, and methane 11 %.

Acetyl bromide on photolysis showed a similar increase of absorption in the same
region, but in this case, owing to the smaller percentage decomposition, it was not
possible to discriminate between this and the long-duration temperature effect on
the acetyl bromide spectrum sufficiently readily to obtain a lifetime measurement.

G. Porter   298

Carbon disulphide

This substance is an example of those whose decomposition mechanism cannot easily be elucidated by analysis of the products alone. The final products of photolysis are solid sulphur and a polymer of composition $(CS)_n$, little being known with certainty about the existence of gaseous, unpolymerized CS.

The spectrum of carbon disulphide at 2 cm. pressure after illumination with a flash of 4000 J is shown in figure 13, plate 8. The decrease in the intensity of the continuum in c is due to clouding of the window by the solid products. The bands of the S_2 molecule are seen in b and continued at lower wave-lengths in i. The latter spectrum also shows three clear new bands with prominent heads at 2575·5, 2509·2 and 2444·5 Å, which agree closely with the wave-lengths of the 0, 0, 1, 0 and 2, 0 emission bands of carbon monosulphide measured by Jevons (1928). The 0, 0 and 0, 1 bands in absorption are shown enlarged in figure 14 a and b, plate 8, respectively. Weak bands around 2588·6 and 2523·2 Å correspond to the wave-lengths of the first two bands of the $v'' = 1$ progression, and the band head at 2504·8 Å is probably the 0, 0 band of the $^1\Sigma-^1\Sigma$ system observed by Crawford & Shurcliffe (1934), though its occurrence in absorption is not in agreement with their contention that a different lower state is involved in this system; as both are observed here the transition must be from the ground state in each case.

The spectra so far discussed were all taken a few msec. after the flash, and when the time interval was extended in order to measure the lifetime of the CS molecule no decrease in absorption was observable until several seconds had elapsed. The spectra in figure 13d to 13h show the decreasing absorption with time in 1 cm. of carbon disulphide, and the 0, 0 band is still faintly visible after 5 min. So persistent was this spectrum that it was observed as an impurity due to a little carbon disulphide which had dissolved in the tap grease, and it is surprising that it has not previously been observed in absorption.

The spectrum of S_2 appeared at maximum intensity immediately after the flash and disappeared much more rapidly than CS. These observations suggest the following as the most likely reaction scheme, the times given being those measured when the pressure of carbon disulphide originally present was 0·5 cm.:

	approximate time of half-reaction
$CS_2 + h\nu = CS + S$	—
$S + CS_2 = S_2 + CS$	less than $1·5 \times 10^{-3}$ sec.
$nS_2 = S_{2n}$	10^{-1} sec.
$nCS = (CS)_n$	60 sec.

As the ground states of S and CS are triplet and singlet respectively one must be produced in the excited state from singlet carbon disulphide, and the above lifetimes indicate that it is the sulphur atom which is so liberated.

REMARKS

Only one important difficulty in the method has appeared in the course of a wide range of investigations, the fact that the adiabatic nature of the reaction produces a change in the spectra of the parent molecules themselves. It is usually possible

299   *Flash photolysis and spectroscopy*

to discriminate between the two effects by lifetime measurements as was shown in the case of diacetyl, which is a particularly difficult example, but for quantitative absorption measurements it will often be necessary to measure the temperature and make allowance for the changing absorption coefficient. On the other hand, the phenomenon suggests the possibility of thermal and kinetic measurements on radical reactions by direct pressure observations.

The results which have been described are preliminary only, and the conclusions must be verified by more detailed investigations. The main purpose here has been to illustrate the power of this method for the investigation of fast reactions, and for this reason a wide range of substances has been studied. It is believed that the results are sufficient to show that the methods of flash photochemistry and spectroscopy provide a valuable weapon for the study of the more elusive of chemical compounds.

I am extremely grateful to Professor R. G. W. Norrish, F.R.S., for his support in this work from the beginning and for much encouragement and valued advice throughout. Thanks are also due to the Anglo Iranian Oil Company for a grant for research, part of which was applied to this work.

REFERENCES

Crawford, F. H. & Shurcliffe, W. A. 1934 *Phys. Rev.* **45**, 860.
Gibson, G. E. & Baylis, N. S. 1933 *Phys. Rev.* **44**, 188.
Jevons, W. 1928 *Proc. Roy. Soc.* A, **117**, 351.
Leighton, W. G. & Forbes, G. S. 1930 *J. Amer. Chem. Soc.* **52**, 3139, 5309.
Murphy, P. M. & Edgerton, H. E. 1941 *J. Appl. Phys.* **12**, 848.
Norrish, R. G. W. & Porter, G. 1947 *Faraday Soc. Discussion. The Labile Molecule*, **2**, 142.
Norrish, R. G. W. & Porter, G. 1949 *Nature*, **164**, 658.
Oldenberg, O. 1935 *J. Chem. Phys.* **3**, 266.
White, J. U. 1940 *J. Chem. Phys.* **8**, 79.
Wieland, K. 1947 *Faraday Soc. Discussion. The Labile Molecule*, **2**, 172.

DESCRIPTION OF PLATES 6 TO 8

Plate 6

FIGURE 4. Effect of capacity on flash spectra of krypton and hydrogen. *a*, Hydrogen: 1 flash at 73 μF; *b*, 6 flashes at 16 μF; *c*, 50 flashes at 2 μF; Krypton: *d*, 1 flash at 125 μF; *e*, 2 flashes at 16 μF; *f*, 50 flashes at 2 μF.

FIGURE 5. Effect of voltage, capacity and pressure on flash spectra of krypton. All spectra are of single flashes.

	capacity (μF)	pressure (cm.)	voltage (V)		capacity (μF)	pressure (mm.)	voltage (V)
a	16	10	2000	*g*	33	3	4000
b	16	10	3000	*h*	33	6	4000
c	16	10	4000	*i*	33	9	4000
d	73	10	4000	*j*	33	15	4000
e	33	10	4000				
f	16	10	4000				

Porter   *Proc. Roy. Soc. A, volume* 200, *plate* 6

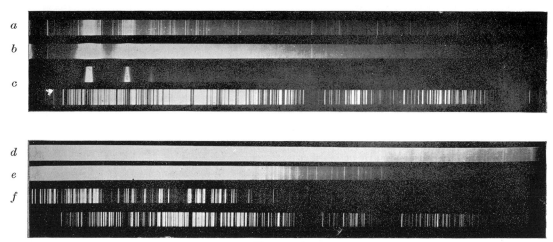

FIGURE 4

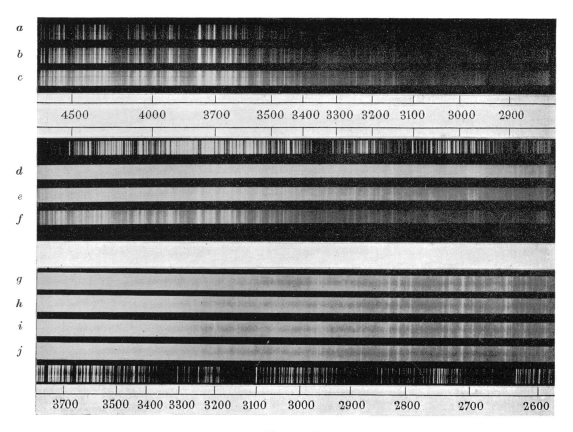

FIGURE 5

Porter   *Proc. Roy. Soc. A, volume* 200, *plate* 7

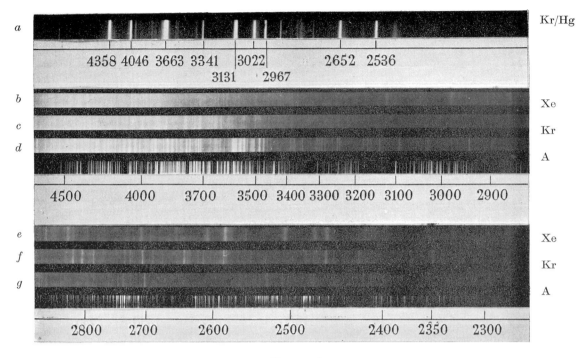

FIGURE 6

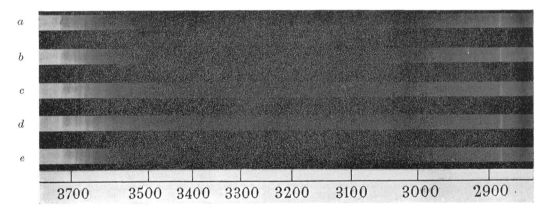

FIGURE 10

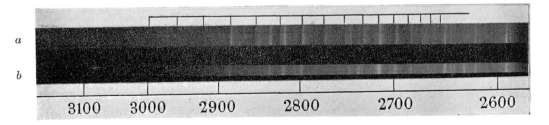

FIGURE 11

Porter   *Proc. Roy. Soc. A, volume* **200**, *plate* 8

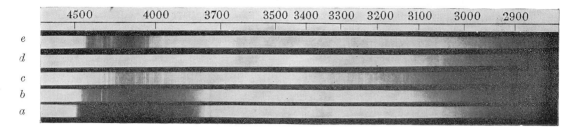

FIGURE 12

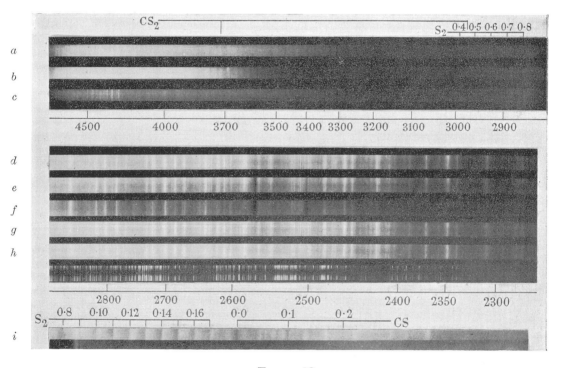

FIGURE 13

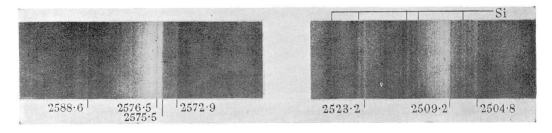

FIGURE 14

Plate 7

FIGURE 6. Effect of gas filling on flash spectra. *a*, 1 m. lamp, voltage 4000 V, capacity 1000 μF; krypton pressure 5 cm. *b* to *g*, 15 cm. lamp, voltage 4000 V, capacity 33 μF, pressure 10 cm.

FIGURE 10. Recombination of chlorine atoms. *e* was taken through 1 cm. chlorine before the flash and *d*, *c*, *b* and *a* were taken at intervals afterwards of 12, 17, 60 msec. and 60 sec. respectively.

FIGURE 11. Spectrum attributed to ClO. *b* is a comparison spectrum taken before the flash.

Plate 8

FIGURE 12. Photolysis of diacetyl. *a* was taken before and *b*, *c* and *d* 1·2, 2·0 and 3·3 msec. respectively after the beginning of the flash. *e* was taken 2 min. later.

FIGURE 13. Photolysis of carbon disulphide. *a*, *b* and *c* were taken before, during and after the flash. *h* is a comparison through a vacuum. *g*, *f*, *e* and *d* were taken before and 2 msec., 1 and 5 min. afterwards.

FIGURE 14. The 0·0 and 0·1 absorption bands of CS.

PRINTED IN GREAT BRITAIN AT THE UNIVERSITY PRESS, CAMBRIDGE

(BROOKE CRUTCHLEY, UNIVERSITY PRINTER)

Reprinted from *Proc. Roy. Soc. A* **216**, 165–183 (1953) with permission from The Royal Society

Studies of the explosive combustion of hydrocarbons by kinetic spectroscopy

I. Free radical absorption spectra in acetylene combustion

By R. G. W. Norrish, F.R.S., G. Porter and B. A. Thrush

(*Received* 16 *August* 1952)

[Plate 1]

The explosive oxidation of acetylene, initiated homogeneously by the flash photolysis of a small quantity of nitrogen dioxide, has been investigated by flash spectroscopy. The absorption spectra of OH, CH, C_2 (singlet and triplet), C_3, CN and NH, a number of which have not previously been observed, are described, and the relative concentrations, at all times throughout the explosion, are given. Four stages have been distinguished in the explosive reaction:

1. An initial period during which only OH appears.
2. A rapid chain branching involving all the diatomic radicals.
3. Further reaction, occurring only when oxygen is present in excess of equimolecular proportions, during which the OH concentration rises exponentially and the other radicals are totally consumed.
4. A relatively slow exponential decay of the excess radical concentration remaining after completion of stages 2 and 3.

The duration of stage 1 is 0 to 3 ms. In an equimolecular mixture at 20 mm total pressure, containing 1·5 mm NO_2, the durations of both stage 2 and stage 3 are approximately 10^{-4} s and the half-life of OH in stage 4 is 0·28 ms. A preliminary interpretation of these changes and of the radical reactions is given.

The light emission accompanying the high-temperature combustion of hydrocarbons consists mainly of the characteristic band spectra of the free radicals OH, CH, C_2 and, in the presence of nitrogen, CN and NH as well as the flame bands attributed to CHO. The numerous investigations of the occurrence of these spectra in flames and explosions, and the contribution which such observations have made to theories of combustion, have been summarized in the useful monograph by Gaydon (1948). While they have led to a better understanding of the types of combustion which occur under different conditions and the existence or otherwise of temperature equilibrium, the information which such studies have provided about the chemical kinetics of combustion has been rather disappointing. It is usually found that all these spectra appear together in the reaction zone of flames,

166   R. G. W. Norrish, G. Porter and B. A. Thrush

and the differences observed between one hydrocarbon and another are remarkably slight. Because of the narrowness of the reaction zone it is virtually impossible to differentiate between the spectra at the beginning and at the end of the reaction, although Gaydon & Wolfhard (1947) have had some success in this respect by using low-pressure flames. Our lack of knowledge of the concentration and reactions of the free radicals responsible for these spectra is reflected in the fact that, with the exception of OH, they are absent from most of the mechanisms of combustion which have been proposed so frequently during the last two decades.

Most investigators in the field of combustion spectroscopy would probably agree that the difficulties in the interpretation of the part played by these free radicals are to a large extent due to the fact that the spectra are observed almost exclusively in emission. As a result, the intensity of a band system is determined by the temperature and conditions of chemiluminescent excitation as much as by the concentration of the radical itself and is furthermore a summation over different stages of the reaction. On the other hand, the obvious advantages of absorption spectroscopy are subject to the well-known experimental limitations of this technique and, as far as we are aware, of the radicals mentioned above, only OH and, more recently, NH have been observed in absorption during combustion. This paper describes the observation of all the possible diatomic radicals in absorption as well as spectra of more complex molecules, and the recording of these spectra throughout the course of the combustion.

The experimental method of flash photolysis and spectroscopy by which these spectra were obtained has been fully described in previous communications (Porter 1950 a, b). Kinetic studies of the OH radical in hydrogen/oxygen explosions have also been reported (Norrish & Porter 1952), as well as a preliminary account of the absorption spectra of the CN and NH radicals in explosions of ethylene with oxygen-nitrogen dioxide mixtures (Porter 1952). Our main purpose in this work was to investigate the occurrence of these three radicals in hydrocarbon combustion and to attempt the observation of other radicals, such as C_2 and CH, about which little is known. For this purpose acetylene was chosen as the most suitable hydrocarbon, because it shows C_2 and CH very strongly in emission, and the photochemical initiation was effected, as in the hydrogen-oxygen work, by means of a small addition of nitrogen dioxide to the acetylene-oxygen mixture.

EXPERIMENTAL

The principle of the method is to decompose the nitrogen dioxide photo-chemically into oxygen atoms and nitric oxide by means of a flash alongside the reaction vessel. This also results in the liberation of heat, by degradation of the absorbed radiant energy, and explosive reaction occurs homogeneously throughout the entire volume. A second flash, giving a continuous spectrum, is used to record the absorption spectra at any required time interval after initiation. Reference may be made to the original papers for experimental details, and all that will be necessary here is an account of the modifications introduced for the present problem.

Norrish et al.   *Proc. Roy. Soc. A, volume* **216**, *plate* 1

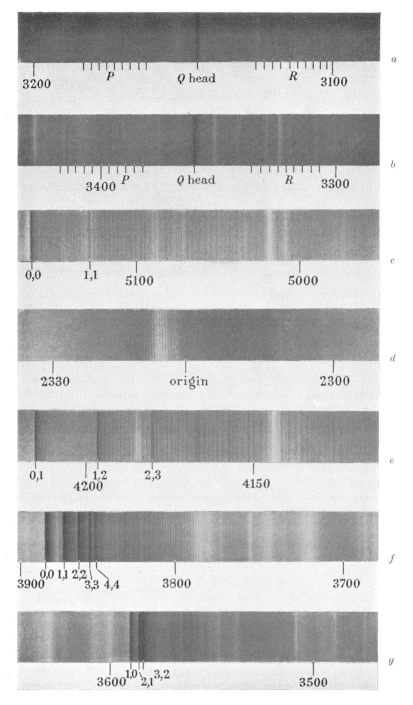

FIGURE 9. (*a*) CH $^2\Sigma^+$–$^2\Pi$ system. (*b*) NH $^3\Pi$–$^3\Sigma$ system. (*c*) C$_2$ 'Swan' bands $^3\Pi$–$^3\Pi$ (0,0) progression. (*d*) C$_2$ 'Mulliken' bands $^1\Sigma$–$^1\Sigma$. (*e*) CN violet system $^2\Sigma$–$^2\Sigma$ (0,1) progression. (*f*) CN violet system $^2\Sigma$–$^2\Sigma$ (0,0) progression. (*g*) CN violet system $^2\Sigma$–$^2\Sigma$ (1,0) progression.

Studies of the explosive combustion of hydrocarbons. I   **167**

The earlier work on flash-initiated explosions had shown that the total duration of the explosion, including the induction period, was frequently less than 1 ms, and the necessity therefore arose of adapting the apparatus for shorter times. The modifications introduced were as follows:

Flash-lamps

The duration of the photolysis flash was reduced in three ways. First, the length was reduced from 100 to 50 cm; secondly, the voltage was increased to 6000 V, which made it possible to reduce the capacity, and hence the duration, without reducing the energy; thirdly, a further reduction in capacity was possible, owing to the fact that a lower energy flash was sufficient to initiate explosion in acetylene-oxygen-nitrogen dioxide mixtures. Throughout these investigations the capacity used was 35 μF, giving an energy of 630 J/flash and an effective duration of 5×10^{-5} s. The spectroscopic flash-lamp eventually used was identical with those previously described, attempts to improve upon this design by operation at 20 000 V having met with only limited success, owing to frequent explosion and unreliable firing characteristics.

Trigger and timing circuit

Operation of a mechanical trigger wheel at short times involves inconveniently high speeds of revolution. Furthermore, it has been found that with careful alinement it is possible to eliminate virtually all the scattered light from the photolysis flash without the necessity of a shutter, which was the main purpose of this device. It was therefore more convenient to use electronic methods, the photolysis flash being made to fire the spectroflash via a photocell, delay unit and thyratron. A comparison of the two methods shows that there is little difference in reproducibility, but the electronic method is more suitable for the shorter times for the reasons given. The circuit diagram of the timing unit is shown in figure 1.

The various time intervals were obtained by adjustment of the pre-set controls in the delay unit, but the actual measurement of the interval was in every case

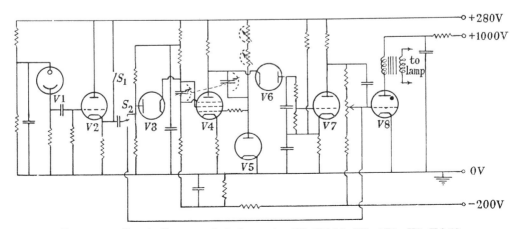

Figure 1. Circuit diagram of timing unit. $V1$, VA16; $V2$, 6J5; $V3$, EA50; $V4$, 6F32; $V5$, EB34; $V6$, EA50; $V7$, EC52; $V8$, MT57.

168   R. G. W. Norrish, G. Porter and B. A. Thrush

made on a cathode-ray oscillograph using a photocell placed near to both lamps. This was found to be essential at the shorter times, otherwise a considerable scatter in the delay time resulted, owing to a variable delay between the application of the trigger current and the main discharge. As before, the experiments were carried out in an arbitrary order, so that any changes observed are a function of the time interval only.

Acetylene was taken from a cylinder and redistilled *in vacuo*. Oxygen and nitrogen dioxide were prepared and purified as previously described. The mixtures were made up in a 2 l. spherical bulb and a minimum of 30 min was allowed for complete mixing. The spectrograph was a Littrow type instrument with inter-changeable glass and quartz optics (Hilger E 478). Ilford H.P. 3, Selochrome, and Q. 3 plates were used in the visible, ultra-violet and far ultra-violet regions respectively, and a slit width of 0·02 mm was used throughout.

Intensity measurements

For extinction measurements of the continuous spectra, plate densities were determined by a microphotometer and compared with measurements on known densities from plate calibrations in the usual way. This method is unsuitable for complex line spectra because, owing to imperfect resolution, such measurements would not be a true indication of relative concentrations and also because the apparent intensity of a single line is very dependent on temperature. It was found, however, that the intensity distribution in a band was the same within the accuracy of measurement, in any two spectra, taken in different mixtures which had the same overall intensity. This is caused by the fact that over nearly the whole range of conditions under which the radicals appear the temperatures are sufficiently high so that small temperature variations from one mixture to another have little effect on the rotational level population. By the use of an optical comparator, it was possible to match equal intensities quite accurately, using the whole of the $v'' = 0$ sequence for this purpose.

In order that the intensities should be a true measure of concentration it was necessary to obtain spectra which bore a known concentration relationship. This problem was solved by a method which should be universally applicable to partly resolved band spectra. Series of spectra are taken under identical con-ditions using absorption paths of different lengths. Now for spectra of equal intensity at a given wave-length

$$\log I_0/I = \epsilon_1 c_1 l_1 = \epsilon_2 c_2 l_2,$$

where ϵ, c and l are the extinction coefficients, concentrations and path lengths for the two exposures. It is to be noted that, if $\epsilon_1 = \epsilon_2$, this applies even when, owing to incompletely resolved line contours, Lambert's and Beer's laws are not sepa-rately applicable, for, as long as the true line contour is the same in both cases, the apparent contour is also the same, being determined only by the number of mole-cules in the absorbing path. In this case, for equal intensities at all wave-lengths $c_1/c_2 = l_2/l_1$, and a true scale of concentrations is simply constructed.

In our experiments the method is not rigidly applicable, owing to the fact that the changes in concentration are accompanied by temperature and composition variations which affect the line contour and therefore the value of ϵ apparent. Nevertheless, owing to the fact already mentioned, that over a small range of conditions the relative temperature changes are insignificant, the line contour, and therefore the values of ϵ, are almost constant, and the method may be used with reasonable accuracy. Two vessels 50 and 25 cm in length were used, and comparisons of intensities were made during the radical decay where temperature changes are relatively small. By successive comparisons an intensity scale was obtained which was directly proportional to the radical concentration, and all other spectra were referred to this by means of a comparator.

RESULTS

Total pressures of 1 or 2 cm of mercury were used, the explosion with higher pressures being violent enough to break the quartz end plates of the reaction vessel. Initial periods were found during which only the NO bands and a weak continuum were present, and which increased as the relative amount of nitrogen dioxide was reduced. The maximum initial periods observed were of the order of 3 ms; if the nitrogen dioxide pressure was then further reduced, no explosion occurred, no absorption spectra appeared and there was very little pressure change. Only a trace of nitrogen dioxide remained after the flash, even when no explosion took place, and its spectrum was entirely absent during the explosion.

The time interval over which free radical spectra were observed was 1 or 2 ms, but most of this was occupied by the decay, the initial rise between first detection and the time at which the intensity maximum was reached being usually less than 10^{-4} s. It will be seen that in some cases it has been possible to obtain spectra at intervals of 2×10^{-5} s during this rise which, although taken in different explosions, show little scatter and appear in a regular order. When an excess of acetylene was present and carbon appeared as a product, the scatter became more noticeable, but it was still possible, by taking a larger number of spectra, to measure the form of the intensity-time curve with reasonable accuracy. It was necessary to clean the reaction vessel between each explosion in the region of carbon deposition, even when the carbon was hardly visible, as otherwise the reduced transparency of the reaction vessel resulted in increased induction periods and a complete irreproducibility in the time-intensity measurements.

Preliminary investigations showed that the appearance of the band spectra was very dependent on the relative proportions of acetylene and oxygen, and that a complete change of the whole spectrum occurred in the region of equimolecularity, if nitrogen dioxide was calculated in terms of an equivalent amount of oxygen. When oxygen was in excess the spectrum of OH appeared at high intensity and some CN was also present. On the acetylene-rich side OH was barely detectable but C_2 and CN appeared very strongly and CH was also present. The NH radical was observed only in mixtures very near to equimolecular. The change from one type of spectrum to the other was very marked when the mixture ratio was changed by 1 or 2 % on either side of equimolecular proportions. In addition to the five diatomic radicals mentioned a weak band at 4051 Å was observed, as well as absorption from upper vibrational levels of NO, and a number of continuous

170   R. G. W. Norrish, G. Porter and B. A. Thrush

spectra. As most of the absorption spectra are new, reproductions are given in figure 9, plate 1. In general, owing to the high temperatures, the appearance of the band systems of the diatomic molecules is very similar to that of their well-known emission spectra. Wave-length measurements of the band heads reported agreed within the accuracy of measurement with those given in the literature for the emission bands (Pearse & Gaydon 1950) and will not, therefore, be tabulated.

Description of spectra

OH: $^2\Sigma$–$^2\Pi$ system

The 0,0, 1,0, 2,0 and 3,0 sequences were observed, but not the 0,1 sequence. The intensity distribution between rotational and vibrational levels was very similar, at a given overall intensity, to that observed in our related experiments on hydrogen-oxygen explosions (Norrish & Porter 1952). Furthermore, the total intensity in oxygen-rich mixtures was the same as that in hydrogen-oxygen explosions at approximately the same pressure, whereas, in emission from the inner cone of low-pressure flames, the intensity of OH is 500 times greater in acetylene than in hydrogen flames (Gaydon 1948). This is strong evidence that nearly all the OH radiation from such flames is of non-thermal origin.

C_2 Swan bands

This familiar system has not hitherto been detected in absorption in combustion processes. It was reported by Klemenc, Wechsberg & Wagner (1934) in absorption during the thermal decomposition of carbon suboxide, the 1,0 band at 4737 Å being observed, but in spite of many attempts these authors were unable to repeat the observation (private communication).

The following band heads were identified in our spectra: the 2,0, 3,1, 4,2; the 1,0, 2,1, 3,2, 4,3, 5,4; the 0,0, 1,1, 2,2; and the 0,1, 1,2, 2,3. A careful search was made for the high-pressure bands of C_2 which have been interpreted as arising from an inverse predissociation from the $v' = 6$ level of the Swan band system (Herzberg 1946). If this is the case, one might expect an increase in intensity in absorption, owing to the effect of increased line width on incompletely resolved spectra, followed by an abrupt disappearance of higher bands, a phenomenon observed under similar conditions, for example, in the spectrum of SH (Porter 1950b). No such effect was observed, and the intensities of the band heads decreased quite regularly up to the last observed head, which was that of the 5,4 band, but owing to extended rotational stucture the higher band heads were rather difficult to locate, and this cannot be taken as conclusive evidence against such a predissociation.

C_2 Mulliken bands

The absorption spectrum of this transition, which arises from the $^1\Sigma$ state, has not previously been observed, and there are very few examples known of absorption from upper electronic states. The spectrum appeared quite strongly, the lines of the P and R branches being considerably more intense than in the spectrum of CH, but it was difficult to photograph, owing to the very strong continuous absorption near 2300 Å where this spectrum appears.

Studies of the explosive combustion of hydrocarbons. I   **171**

The appearance of both singlet (Mulliken) and triplet (Swan) systems in absorption at similar intensity shows that the C_2 molecule and diradical have very similar energies and raises the question as to which is the ground state. Herzberg & Sutton (1940) have predicted that the $^1\Sigma$ state lies 5600 cm^{-1} above the $^3\Pi_u$ state; the temperature in our system is of the order of 3000° K, and, allowing for the lower wave-length of the singlet system, a separation of this order is in accordance with the observed intensities, if the f value of the singlet transition is near unity. It should be possible to obtain confirmation of this point by taking the spectra at different temperatures and using the intensity distribution in the higher vibrational levels of the Swan bands as a comparison.

C_3 *band at* 4051 Å

A single diffuse line at 4051 Å was observed very weakly in rich mixtures, along with the intense continuum in this region. This is probably identical with the 4050 Å group discovered in emission from comets by Swings, Elvey & Babcock (1941), and later in the laboratory by Herzberg (1942), and attributed by the latter to CH_2. On the basis of convincing evidence obtained by using the ^{13}C isotope, it has recently been attributed to the C_3 molecule (Douglas 1951), and the conditions of its occurrence in our experiments support the assignment to a polyatomic carbon molecule.

Doublet systems of CH

Three systems of this radical are known, all of which occur in emission in flames and none of which has been observed in absorption in the laboratory, though the 4315 and 3900 Å systems are well known in stellar sources. The only band to appear strongly in our experiments was the 3143 Å band, the other two being only just detectable. This is the opposite of flame-emission spectra, where the 4315 Å system is much the strongest and the 3143 Å band is only observed in the hottest flames. The close packing of lines in the Q head of the 3143 Å band increases the sensitivity of the absorption method, but the individual lines of the P and R branches of this band were also appreciably stronger than those of the other two systems.

All three systems have the same lower $^2\Pi$ level, and it has been suggested, on the grounds of an expected high concentration of CH in flames and the failure to detect it in absorption, that the $^2\Pi$ state is not the ground level but lies above an unobserved $^4\Sigma$ state (Gaydon 1948, 1951). Now that all these systems have been observed in absorption the original reason for this suggestion disappears, and it can be said that the $^2\Pi$ state cannot lie more than a few thousand wave numbers above the ground state, the f values of these transitions being low (Herzberg 1951). Other considerations indicate that the $^2\Pi$ is, in fact, the lowest level (Porter 1951).

CN *and* NH

The violet $^2\Sigma$–$^2\Sigma$ system of CN was the strongest and most persistent feature of all our spectra. The bands identified were the 0,1, 1,2, 2,3, 3,4, 4,5; the 0,0, 1,1, 2,2, 3,3, 4,4; and the 1,0, 2,1 and 3,2. The 0,0 band of this system has been observed in absorption by Kistiakowsky & Gershinowitz (1933) and by White (1940) in thermally dissociated cyanogen.

172   R. G. W. Norrish, G. Porter and B. A. Thrush

The line-like Q branches of NH at 3360 and 3370 Å and the extended triplets of the P and R branches appeared in mixtures near to equimolecular, the intensity being similar to that of the 3143 Å band of CH.

Continuous spectra

The only other discrete spectra observed were the bands of the hot NO molecule which were always present during the induction period. In particular, the spectra of formaldehyde and of aromatic molecules such as benzene were absent at all times. In acetylene-rich mixtures a strong continuous absorption was present, which was associated, in part at least, with the deposition of solid carbon. The spectrum is complex, the wave-length-intensity distribution being a function of time, and consists of at least two parts with the following characteristics:

(*a*) A continuum, with a maximum at 3800 Å, whose intensity varies with time in the same way as the carbon radicals.

(*b*) A continuum whose intensity increases from 3000 Å to shorter wave-lengths and shows an approximate proportionality to λ^{-4}, and which reaches maximum intensity after 3 ms and decays for a period of several minutes.

Detailed measurements of these spectra, and their interpretation, will be given in later parts.

Dependence of radical concentration on mixture ratio

A series of mixtures were investigated in which the acetylene pressure was varied, the partial pressures of nitrogen dioxide and oxygen being constant at 1·5 and 10 mm respectively. The intensities of the various absorption spectra were measured in the manner described throughout the explosion and the maximum

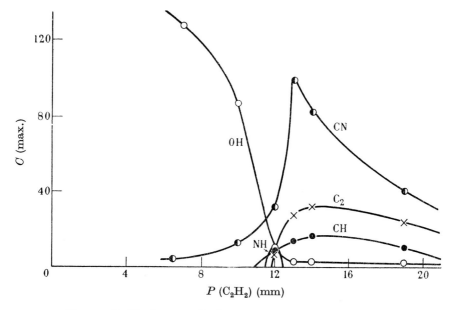

FIGURE 2. Maximum radical concentration against $P(C_2H_2)$;
$P(O_2) = 10$ mm, $P(NO_2) = 1·5$ mm.

Studies of the explosive combustion of hydrocarbons. I   **173**

intensity of each radical obtained in this way for each mixture. The results are plotted in figure 2, the acetylene partial pressures investigated being 7, 10, 12, 13, 14 and 19 mm Hg.

No significance should be attached to the relative concentrations between one radical and another, except as an approximate indication of intensity. The complete change in radical concentration which occurs about equimolecular proportions of acetylene and oxygen is very striking. The decrease in the concentration of CN, C_2 and CH as the acetylene pressure is increased beyond 14 mm Hg was accompanied by a greatly increased continuous absorption and the deposition of visible quantities of solid carbon.

Dependence of radical concentration on time

The OH and CN radicals were observed at some point during the explosion of every mixture investigated, and the concentrations of these two radicals throughout the explosions are plotted on a logarithmic scale in figures 3 and 4 respectively. When excess oxygen was present, and no carbon was deposited, the reproducibility was good, and it has been possible to measure the OH concentration at intervals of 2×10^{-5} s throughout the development of the explosion, the induction period in

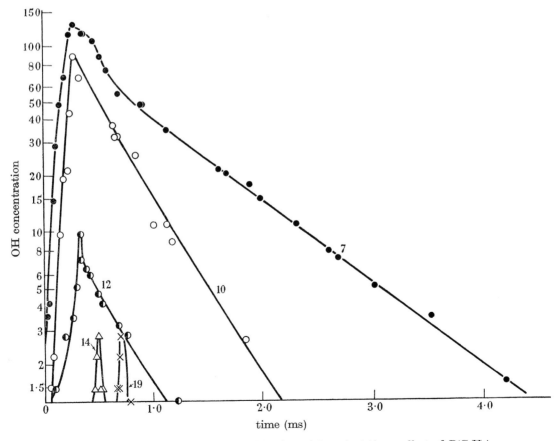

FIGURE 3. OH concentration (logarithmic scale) against time: effect of $P(C_2H_2)$; $P(O_2) = 10$ mm, $P(NO_2) = 1 \cdot 5$ mm. Numbers on curves show pressure in mm.

R. G. W. Norrish, G. Porter and B. A. Thrush

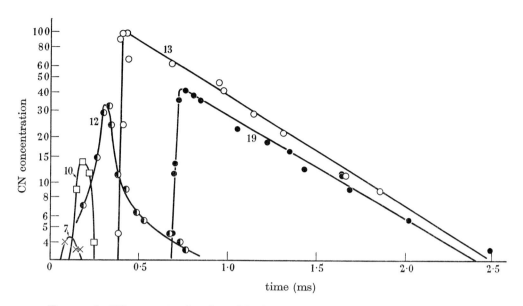

FIGURE 4. CN concentration (logarithmic scale) against time: effect of $P(C_2H_2)$; $P(O_2) = 10$ mm, $P(NO_2) = 1·5$ mm. Numbers on curves show pressure in mm.

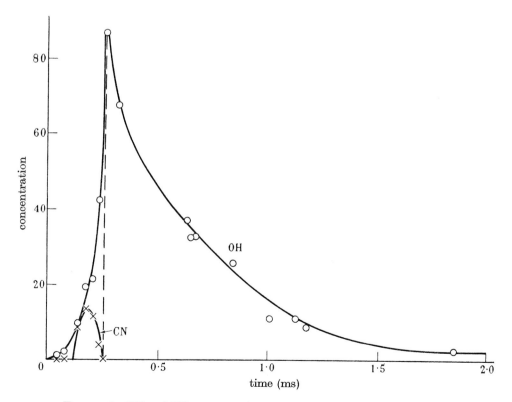

FIGURE 5. OH and CN concentrations against time; $P(C_2H_2) = 10$ mm, $P(O_2) = 10$ mm, $P(NO_2) = 1·5$ mm.

Studies of the explosive combustion of hydrocarbons. I   **175**

the mixture containing 7 mm acetylene being less than $30\,\mu$s. The rate of decay of the two radicals at a given concentration towards the end of the reaction depends markedly and in opposite senses on the mixture ratio, the decay rate of OH being increased by increasing acetylene/oxygen ratios, and the CN decay rate being simultaneously decreased. The final decay is exponential in both cases. A comparison of the curves for the two radicals reveals two other important facts, which are shown more clearly in figure 5, where the intensities of OH and CN during the explosion of the mixture containing 10 mm C_2H_2 are plotted. First, the OH radical appears before the CN radical is detected. This point is fully confirmed in the later results where the OH concentration does not rise to such high values, and the phenomenon was encountered in every case studied. Secondly, the maximum CN concentration is reached 10^{-4} s before the OH maximum, and the CN decay is complete before the OH decay begins, the same being true in the 7 mm acetylene mixture (figures 3 and 4).

The C_2 and CH radicals were only detectable in the richer mixtures, and in all cases their intensity-time curves were very similar. The occurrence of these radicals in the mixtures containing 13 and 19 mm acetylene is shown in figures 6 and 7 respectively. Owing to the rapid rise and the carbon deposition, there is some scatter of the points, but as each set of intensities for a given time is recorded on the same spectrum the points are more significant than is indicated by the curves themselves, and, by comparison of corresponding points, it is possible to say quite definitely that the peak concentrations of CN, C_2 and CH, and also of C_3 when present, are reached at the same time and that the radicals are formed concurrently. For the same reason, one can say that the OH radical appears earlier and the intensity falls to zero before the decay of the carbon radicals begins. In all other mixtures the same conclusions apply, and no distinction whatever could

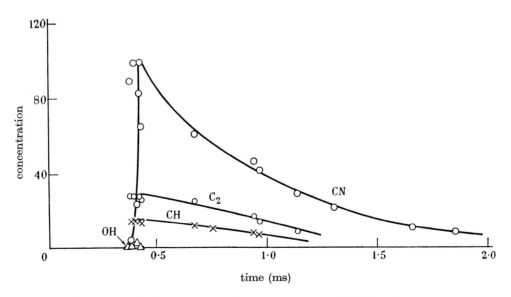

FIGURE 6. Radical concentrations against time; $P(C_2H_2) = 13$ mm, $P(O_2) = 10$ mm, $P(NO_2) = 1.5$ mm.

176   R. G. W. Norrish, G. Porter and B. A. Thrush

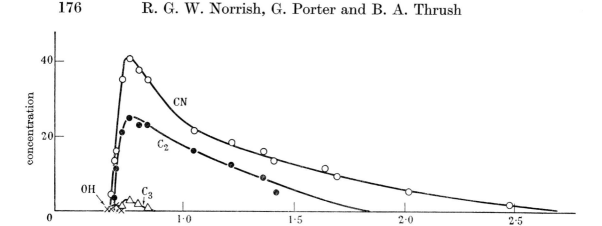

FIGURE 7. Radical concentrations against time; $P(C_2H_2) = 19$ mm,
$P(O_2) = 10$ mm, $P(NO_2) = 1.5$ mm.

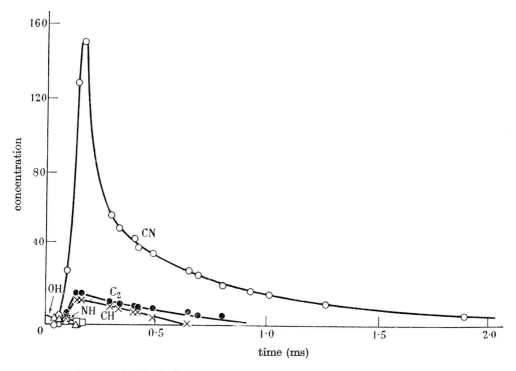

FIGURE 8. Radical concentrations against time; $P(C_2H_2) = 14.5$ mm,
$P(O_2) = 10$ mm, $P(NO_2) = 3$ mm.

be made between the times of appearance and rates of growth of the carbon radicals.

Effect of nitrogen dioxide pressure

Although the relationship of CN to C_2 and CH in time was very close there was a strong dependence of the intensities of these radicals on nitrogen dioxide pressure, as is shown in figure 8. Conditions here are identical with figure 6 except that

1·5 mm of the O_2 has been replaced by an equivalent amount of NO_2. The first effect of this change is greatly to reduce the initial period. Secondly, although the carbon radicals all appear together, the rate of growth and the peak intensities reached are increased for CN and considerably reduced for C_2 and CH. The effect of doubling the NO present is to increase the peak CN concentration by 50 % and to reduce peak C_2 and CH concentrations by 40 and 13 % respectively. The OH radical appears at the same intensity, but in the mixture containing 3 mm NO_2 its early appearance is particularly noticeable, and it appears at maximum intensity before any other radical is observed and rapidly falls to zero before the carbon radical maximum is reached. The NH radical, which was absent in figure 6, now appears and reaches maximum intensity after the OH but before CN. Here, as in all other cases, the NH radical only appears when both CN and OH are present.

Effect of nitrogen, water and carbon dioxide

The addition of nitrogen to the explosive mixture had the inert-gas effects, observed in hydrogen/oxygen explosions, of increasing the induction period and reducing the maximum concentration of radicals observed. For example, in the mixture containing acetylene, oxygen and nitrogen dioxide at pressures 14, 10 and 3 mm, the addition of 17 mm of nitrogen increased the induction period from 0·15 to 1·6 ms and decreased the maximum radical concentration to about one-half.

The influence of water on hydrogen/oxygen mixtures was the same as that of an inert gas, but in acetylene explosions small additions of water had a very specific and striking effect on the appearance of the radicals. This is illustrated by the data in table 1, which refer to the peak radical concentrations observed in the 13 mm acetylene, 10 mm oxygen, 1·5 mm nitrogen dioxide mixture, to which varying amounts of D_2O have been added. The intensities refer to the lighter isotopic species only.

TABLE 1. EFFECT OF ADDED D_2O ON MAXIMUM RADICAL INTENSITY

D_2O (mm)	OH	CN	C_2	CH	NH	time to max. rate (sec)
0	3	98	28	15	0	$4·8 \times 10^{-4}$
2·0	3·5	62	17	13	2	3×10^{-4}
3·5	37	24	5	9	12	2×10^{-4}
5·0	100	11	0	0	0	5×10^{-5}

Comparison with figure 2 shows that the addition of water has very nearly the same effect as the addition of an equivalent amount of oxygen in displacing the position at which the carbon radicals appear. More exact comparison shows that, in small amounts, it is equivalent to rather less than $\frac{1}{2}O_2$. Water has the additional effect of reducing the initial period, the times given in table 1 being those between the beginning of the flash and the time of maximum rate. Water therefore plays an important chemical role in the initial reactions, for its physical effect would be to increase, slightly, the thermal capacity of the system and therefore the initial period.

178   R. G. W. Norrish, G. Porter and B. A. Thrush

The addition of carbon dioxide to the above mixture had the same effect as the addition of water in replacing the carbon radicals by OH, and quantitatively it, also, was equivalent to slightly less than $\frac{1}{2}O_2$. On the other hand, it resulted in a greatly increased initial period, in the same way as nitrogen.

DISCUSSION

These investigations, incomplete as they must be at this stage, form a beginning to a complete kinetic analysis of hydrocarbon explosions. Our immediate interest will be to discover the relationship of the diatomic radicals observed to the combustion process.

The concentration of any radical R at a time t during the explosion will be given by an expression of the form

$$[R]_t = \int_0^t \sum_1 k_1(A)^a(B)^b \dots \mathrm{d}t - \int_0^t \sum_2 k_2(R)^r(C)^c(D)^d \dots \mathrm{d}t = \int_0^t K_1 \, \mathrm{d}t - \int_0^t K_2 \, \mathrm{d}t,$$

where K_1 and K_2 are the sums of the rate terms of all reactions by which R is formed and destroyed, respectively, (A), (B), etc., being the concentrations of the other molecules taking part. These concentrations are all functions of time and the rate constants, k, being functions of temperature, are also functions of time. The evaluation of this expression is clearly not possible at present, and we must begin by seeking qualitative relationships which may lead to some simplification.

The general form of the curves is explained as follows. The flash liberates oxygen atoms and a certain amount of heat and immediately afterwards the system is at a temperature T_i. As long as T_i is above a certain critical value, further exothermic reactions will occur resulting in heat liberation at a rate greater than heat dissipation to the wall, and consequently in a further temperature rise. For a given radical, which is to be observed, terms K_1 must exceed terms K_2, and the radical concentration then grows rapidly. In general, the radical growth will be accompanied by chain branching and further heat liberation resulting in explosive reaction. Eventually the reactants A, B, etc., will be consumed and terms K_1 will become smaller in spite of the high temperature. When $K_1 = K_2$ the peak radical concentration is reached. K_2 now becomes greater than K_1 and the radical concentration falls eventually to zero.

In most cases the decay of the radical is many times slower than the initial rise in concentration, and this must be interpreted as meaning that the observed decay rate is in fact that of the terms K_2, terms K_1 being negligible owing to complete removal of one reactant. The latter parts of the decay curves of CN and OH give very satisfactory unimolecular plots, the half-lives being 0·38 ms for CN in the 13 mm acetylene mixture and 0·63 ms for OH in the 7 mm mixture. It is probable that this implies an approximate constancy of composition and of temperature during this period and a zero value of terms K_1, owing to complete consumption of one reactant.

The interpretation of the maximum radical concentrations in figure 2 is now possible. It was shown by Bone (1932) that the explosion of acetylene and oxygen in equimolecular proportions occurred according to the overall equation

$$C_2H_2 + O_2 = 2CO + H_2,$$

and that excess oxygen resulted in water formation whilst excess acetylene gave carbon deposition. A naïve interpretation of the results of figure 2 would be the statement that the OH radical is formed only during the combustion of H_2 in excess oxygen and the radicals C_2, CH and CN result only from the cracking of excess acetylene and carbon formation, after consumption of the oxygen according to the above equation. This statement is, however, in discord with many of the facts recorded in the concentration—time curves and ignores the reactions by which the radicals are removed. These curves show that in every case a low maximum radical concentration is accompanied by a proportionately high rate of removal, indicating that the low concentration is to a great extent, if not entirely, due to increased K_2 terms rather than decreased K_1 terms. Anticipating part of the later discussion, the observed peak concentrations are to be interpreted as follows. In mixtures of all proportions investigated, both OH and the carbon radicals are formed to some extent during the chain-branching reactions which occur as the explosion develops. They react rapidly together and with the original reactants until one reactant is consumed and the other remains in excess. The concentration of carbon radicals or OH may then increase further by the cracking of any remaining acetylene or by the subsequent reaction between hydrogen and oxygen, though the former is clearly not of great importance if acetylene is increased much above equimolecular proportions, for the concentration of carbon radicals then falls again, owing to the lower temperature. When water or carbon dioxide are present in the original mixture, they are rapidly reduced by the carbon radicals or by the material from which these radicals are formed in the same way as oxygen or OH.

Time relationship of the radical concentrations

We may now inquire whether the chain-branching processes in which the radicals take part are so coupled that all the radical concentrations increase proportionately. It has already been noted that this is the case for the radicals C_2, CH and CN, and the reactions by which they are formed must therefore proceed from a common source or, alternatively, the formation of one results in the rapid reaction to form the others. So close is this correspondence that we shall frequently refer to these radicals together as the carbon radicals, and if, as is often the case, only CN is observed, owing to its high extinction and also to reasons of reactivity, which will be understood later, we shall infer that the other carbon radicals are probably being formed at a proportionate rate. The justifications for this assumption, which has important consequences, are as follows:

(*a*) The concentrations of C_2, CH and CN in figure 2 fall in proportion as the acetylene is decreased from 13 to 12 mm, and the absence of C_2 and CH in weaker mixtures is therefore expected, for a further decrease, in proportion to CN, results in a concentration of C_2 and CH below the limit of detectability.

(*b*) In all cases when CH and C_2 are observed, their rate of formation is proportional to that of CN.

(*c*) The rate of decay of C_2 and CH increases as the proportion of oxygen is increased in the same way as CN, indicating that the decreased concentration is, in part at least, caused by increased rate of removal.

180   R. G. W. Norrish, G. Porter and B. A. Thrush

The fact that C_2 and CH are formed concurrently rather than consecutively is important in view of the reports of a number of investigators that one appears before the other in emission in the extended reaction zones of flames. Gaydon (1948) has pointed out that C_2 usually appears lower in the flame than CH and has inferred that this is the order of formation, possibly owing to carbon breakdown to C_2 followed by reaction with OH. Although diffusion processes may alter the concentration distribution in a flame, it seems to us that a more probable explanation is to be found in conditions of excitation in different parts of the reaction zone, for there can be little doubt that the difference in times of formation during the explosion reaction is insufficient to explain the spatial resolution of these two radicals in flames.

The appearance of the OH radical is quite distinct from the carbon radicals. It invariably occurs before all other radicals at the end of the induction period, and there is no question of this always being caused by different limits of detectability, for in some cases (e.g. figure 8) its intensity has fallen considerably before any other radical spectrum appears. We must conclude that in all mixture ratios the early part of the reaction proceeds via a chain-branching process, which involves OH formation, but does not produce the carbon radicals in detectable amounts. The reaction therefore has at least two stages, and further examination of figure 5 shows that a third stage is involved when excess oxygen is present. The complete decay of CN, and, by inference, of the other carbon radicals also, is followed by further chain branching involving the formation of OH. Now the early disappearance of the carbon radicals implies, by our previous discussion, that the reactant from which they are formed has been consumed, and this must be acetylene, carbon or some other carbon compound. As there is an excess of oxygen, some of it must react with hydrogen, which will now be present if carbon has been consumed, and the third phase of the reaction, in which OH alone is observed and reaches a very high concentration, is therefore to be identified with the final hydrogen-oxygen reaction.

The explosive reaction between acetylene and oxygen, in the presence of a small amount of NO, can therefore be divided into the following four stages, the figures given in brackets being the approximate duration of each stage in the mixture containing 10 mm C_2H_2, 10 mm O_2 and 1·5 mm NO_2:

Stage 1. An initial period towards the end of which the OH radical concentration increases rapidly (10^{-4} s).

Stage 2. A rapid reaction involving C_2, CH, OH and, in the presence of NO, also CN and NH, the semi-stationary concentration of these radicals remaining low until the consumption of one reactant is nearly complete (10^{-4} s).

Stage 3. (Which only occurs when oxygen is present in excess of equimolecular proportions.) A further reaction during which the OH concentration rises exponentially and the carbon radicals are totally consumed (10^{-4} s).

Stage 4. A relatively slow exponential decay of the excess radical concentration remaining after the completion of stages 2 and 3 (half-life of OH = 0·28 ms).

The early part of stage 1 may be similar in mechanism to the slow isothermal reaction, but the rotational temperature of OH is very high, even when it first appears, and the reactions in the latter part probably involve a different mechanism. Formaldehyde and glyoxal, the products of slow oxidation, were never observed. The marked influence of water vapour on the induction period may be due to chain initiation by the dissociation of water at temperatures above about 1000° C. In the slow oxidation of acetylene the influence of water on the induction period does not seem to have been studied, but a heterogeneous reaction between acetylene and water to form acetaldehyde occurs at 300° C (Bone & Andrew 1905). Water has a negligible effect on the induction period during ethylene oxidation (Bone, Haffner & Rance 1933).

Stage 3 is undoubtedly the reaction of the hydrogen produced in stages 1 and 2 with the remaining oxygen, and there is a close correspondence between the intensity-time curves of this stage with those obtained with equivalent quantities of hydrogen. The maximum intensity of OH in the mixture containing 7 mm acetylene is approximately double that which would be obtained from 7 mm of hydrogen and a slight excess of oxygen under the same conditions. The decay of OH in stage 4 is to be attributed to its reaction with hydrogen and carbon monoxide, and the details of the OH reactions in stages 3 and 4 are best arrived at by a study of these reactions individually (Norrish & Porter 1952). The decay of the carbon radicals in stage 4 will be discussed in connexion with the continuous spectra and carbon formation.

The essentially new part of our mechanism is stage 2, and we shall, therefore, examine the reactions occurring during this process in a little more detail.

The C_2 and CH radicals

The proportionality in rates of formation of these two radicals shows either that one is formed from the other very rapidly or that they arise from a common source. There can be little doubt that the connecting reaction is

$$C_2 + OH = CH + CO, \tag{1}$$

for the results show that the radicals C_2 and OH react together so rapidly that they are never present in significant amounts at the same time. The rapid disappearance of both C_2 and CH in mixtures containing oxygen is then explained if CH reacts with O_2 or OH.

The carbon radicals could be formed either:

(1) by decomposition of acetylene, or

(2) by breakdown of small carbon particles formed from acetylene.

This important question is closely connected with carbon formation and will be discussed in this context.

The CN and NH radicals

At the very earliest part of the initial period which can be observed the decomposition of NO_2 into NO and oxygen is complete, and the radicals CN and NH must therefore be formed by the subsequent reactions of NO. The close correspondence in time between the formation and disappearance of CN and C_2 has been noted, but there are several significant differences in the effect of concentration on their relative intensities, the most useful being given by a comparison of

182   R. G. W. Norrish, G. Porter and B. A. Thrush

figures 6 and 8. As expected, more CN is formed in the mixture containing more NO_2, but the amount of C_2 formed is reduced by 40 %. This large reduction cannot be caused by removal of acetylene, for even if all the NO reacts with acetylene to form CN, this only corresponds to the removal of 10 % of the acetylene. We conclude that either:

(1) C_2 is very rapidly removed by the reaction

$$C_2 + NO = CN + CO, \tag{2}$$

until the NO is consumed; or

(2) the C_2 and CN are formed from a common source X, and the reaction of X with NO occurs more readily than its reaction to form C_2.

It is clear that the formation of C_2 from CN is not the important mechanism.

The NH radical is always associated with the presence of both OH and CN. This is shown as a function of mixture ratio in figure 2, of time in figure 8, and of added water in table 1. This is very strong evidence for the reaction

$$CN + OH = CO + NH. \tag{3}$$

As in our hydrogen-oxygen work, the formation of NH by direct reaction of NO with H_2, H or OH does not seem to occur. Unlike the other radicals, the disappearance of NH is rapid in mixtures of all proportions, showing that it reacts, not only with O_2 or OH, but also with the carbon radicals, presumably by

$$NH + C_2 = CN + CH \tag{4}$$

and

$$NH + CH = CN + H_2. \tag{5}$$

For simplicity we have written the above reactions in terms of the observed molecules, but it must be remembered that a high concentration of atoms is also probably present, and the reactions as written indicate the overall changes, which may proceed via atoms. It is now a simple matter to write a complete reaction scheme for stage 2 by the addition to the above of several equations of equal thermodynamical probability, but the discussion has been confined to reactions for which there is direct experimental evidence. The outstanding question which remains is what proportion of the oxidation mechanism proceeds via stage 2. The high intensities of the absorption spectra suggest that this may be by no means insignificant, but the quantitative answer cannot be given until absolute concentration determinations of the radicals concerned become available.

We are indebted to the Government Grants Committee of the Royal Society for the loan of the Littrow spectrograph used in this work. We are also indebted to the Anglo-Iranian Oil Company for financial support.

REFERENCES

Bone, W. A. 1932 *Proc. Roy. Soc.* A, **137**, 243.
Bone, W. A. & Andrew, G. W. 1905 *J. Chem. Soc.* **87**, 1232.
Bone, W. A., Haffner, A. E. & Rance, H. F. 1933 *Proc. Roy. Soc.* A, **143**, 16.
Douglas, A. E. 1951 *Astrophys. J.* **114**, 466.
Gaydon, A. G. 1948 *Spectroscopy and combustion theory*. London: Chapman and Hall.

Gaydon, A. G. 1951 *Disc. Faraday Soc.*, Hydrocarbons, **10**, 108.

Gaydon, A. G. & Wolfhard, H. G. 1947 *Disc. Faraday Soc.*, The labile molecule, **2**, 161.

Herzberg, G. 1942 *Astrophys. J.* **96**, 314.

Herzberg, G. 1946 *Phys. Rev.* **70**, 762.

Herzberg, G. 1951 *Molecular spectra and molecular structure*, **1**, 386. New York: D. Van Nostrand.

Herzberg, G. & Sutton, R. B. 1940 *Canad. J. Res.* A, **18**, 74.

Kistiakowsky, G. B. & Gershinowitz, H. 1933 *J. Chem. Phys.* **1**, 432.

Klemenc, A., Wechsberg, R. & Wagner, G. 1934 *Z. phys. Chem.* A, **170**, 97.

Norrish, R. G. W. & Porter, G. 1952 *Proc. Roy. Soc.* A, **210**, 439.

Pearse, R. W. B. & Gaydon, A. G. 1950 *The identification of molecular spectra*. London: Chapman and Hall.

Porter, G. 1950 *a* *Proc. Roy. Soc.* A, **200**, 284.

Porter, G. 1950 *b* *Disc. Faraday Soc.*, Spectroscopy and molecular structure, **9**, 60.

Porter, G. 1951 *Disc. Faraday Soc.*, Hydrocarbons, **10**, 108.

Porter, G. 1952 *Boll. Sci. Chim. Industr. Bologna*, **1**.

Swings, P., Elvey, C. T. & Babcock, H. W. 1941 *Astrophys. J.* **94**, 320.

White, J. U. 1940 *J. Chem. Phys.* **8**, 79, 459.

PRINTED IN GREAT BRITAIN AT THE UNIVERSITY PRESS, CAMBRIDGE

(BROOKE CRUTCHLEY, UNIVERSITY PRINTER)

Reprinted from *Proc. Roy. Soc. A* **315**, 163–184 (1970) with permission from The Royal Society

Proc. Roy. Soc. Lond. A. **315**, 163–184 (1970)
Printed in Great Britain

Nanosecond flash photolysis

By G. Porter, F.R.S. and M. R. Topp

*Davy Faraday Research Laboratory of The Royal Institution,
21 Albemarle Street, London W1X 4BS*

(*Received 23 July* 1969)

[Plates 1–3]

A flash photolysis system, using a pulsed laser as source, has been designed and used to study events having a duration of a few nanoseconds; an improvement over conventional flash techniques by a factor of a thousand.

The apparatus incorporates both spectrographic and photoelectric monitoring techniques which are easily interchangeable and, apart from the laser itself, it is readily constructed from standard components.

Its applications to the observation of the absorption spectra of excited singlet states, short-lived excited triplet states and chemical events in the nanosecond time region are described.

Introduction

As flash photolysis techniques have been increasingly applied to a variety of problems, it has become clear than at extension of the method to shorter times would have very wide applications.

The time resolution of a flash photolysis system is limited first by the duration of the initiation flash and secondly by the response time of the diagnostics. In the double flash spectrographic recording system (Porter 1950) both limitations are primarily determined by the flash duration, although another factor—which becomes increasingly important at shorter times—is the reproducibility of the delay between the initiating and monitoring flashes. Many attempts have been made in recent years to reduce the duration without a consequent decrease in the energy of the conventional electronic flash discharge tube; they have met with only marginal success (Boag 1968). The advent of the pulsed laser has provided a new source with great potentialities for flash photolysis studies.

The principal components of a flash photolysis apparatus, apart from a spectrograph or monochromator, are an initiating photolysis flash source, a monitoring flash source, a unit which introduces a delay between these two flashes and, for subsequent kinetic work at selected wavelengths, a monitoring source with photo-electric detection system. The substantial changes which have been made in each of these components in the present work will now be described.

G. Porter and M. R. Topp

EXPERIMENTAL

The laser flash

In order that useful information may be derived from a flash photolysis experiment, it is usually necessary that the photolytic and monitoring pulses be of shorter duration than the processes to be studied. For nanosecond events, the giant-pulsed laser provides a means of generation of a burst of ultraviolet radiation sufficiently energetic for the purposes of flash photolysis, of duration less than or equal to 20 ns.

The ruby laser employed in this work was equipped with a $6\frac{1}{2}$ in $\times \frac{1}{2}$ in (165 mm $\times$ 13 mm) parallel-ended ruby rod, and was originally obtained from G. and E. Bradley Limited. With vanadyl phthalocyanine solution in nitrobenzene used as a Q-switch (Sorokin, Luzzi, Lankard & Pettit 1964) and with a totally internally reflecting quartz prism as the 100 % reflector, the laser generated a pulse containing about 1.5 J of red light of wavelength 694.3 nm (10^{18} photons), whose half-peak duration was less than 20 ns, and whose pulse shape was roughly Gaussian (Müller & Pflüger 1968). The exponential 'tail' of the gas discharge flash was absent so that the real improvement in the time resolution is considerably greater than is indicated by a comparison of half-peak durations.

The red laser pulse was passed through a crystal of ammonium dihydrogen phosphate, from which emerged, in addition to residual red light, a pulse containing about 80 mJ of ultraviolet radiation at 347.1 nm, the second harmonic of the ruby frequency. It was of approximately the same duration as the red laser pulse.

We have found this pulse to be quite adequate for the creation of transient concentrations of up to 5×10^{-4} mol l^{-1} in a 1 cm path which is sufficient for spectroscopic purposes.

The monitoring source

Initially, we modified a commercially available nanosecond spark source for our spectroscopic flash (The Fischer Nanolite (Fischer 1961)). This source gave a pulse of 13 ns duration at half-peak intensity, in an atmosphere of pressurized oxygen, sufficiently bright to expose completely an Ilford HPS or Kodak Royal-X Pan film with a single shot, using a small Hilger spectrograph (Porter & Topp 1967).

This source was triggered by focusing the red laser beam onto the spark gap, but the irreproducibility of the firing was quite large unless the electrodes were freshly cleaned (Pendleton & Guenther 1965). Subsequently it was found that if the Nanolite were replaced by a spark-gap powered by a more conventional low-inductance capacitor, with pin-heads as the electrodes, the discharge could be triggered photoelectrically using the ultraviolet laser pulse, probably due to the presence of zinc in the electrode material. (Work function of zinc = 3.32 eV or wavelengths shorter than 372 nm (Dillon 1931).)

With this arrangement, the pulse duration in pressurized oxygen was 25 to 30 ns with a jitter of only a few tens of nanoseconds. This was reproducible up to about 100 flashes, when burning of the electrodes became appreciable.

Nanosecond flash photolysis   165

The larger delays in this spark apparatus were furnished by a passive delay unit—a small spark gap (1 kV) was triggered by the laser pulse, from which the resulting electrical pulse was fed into a system of B.I.C.C. delay cables (55 ns m^{-1}). These were capable of producing delays in steps of 50 ns from 50 to 1500 ns. The output end of the delay system was connected to a third electrode in the primary spark gap of a Marx–Bank cascade capacitor unit (Marx 1924; Lewis, Jung, Chapman, Van Loon & Romanovski 1966) whereupon the pulse was amplified to about 12 kV, and could be used to trigger the main monitoring spark. The over-all minimum delay was 20 ns and the jitter of the order of 20 to 30 ns.

Using this system, we were able to observe the excited singlet state absorption spectrum of coronene (half-life 280 ns). Although a useful apparatus in some ways, the spark apparatus lacked the nanosecond sensitivity and reproducibility that we were seeking.

Apart from the problem of a rather large jitter, on a nanosecond scale at least, the spark apparatus suffered from another major drawback. There were present, in the emission spectrum of the monitoring spark, many gaseous emission lines together with several from the electrode material. Owing to the necessity for a short duration and intense light output, we were unable to use the inert gases, and had to use oxygen, which is notable for its line emission.

It was not possible, unless fairly substantial absorption was present in our sample, to distinguish between real absorption and regions of low-intensity emission between the lines. The narrower the absorption, the worse the problem, and microdensitometry of the photographic plates obtained on this apparatus was difficult to interpret. It should be borne in mind that this is, to a greater or lesser degree a disadvantage of most gaseous spectroscopic flashes, although oxygen is a particularly extreme example.

The necessity to eliminate this line structure in the monitoring background spectrum led us to consider the possibility of using a fluorescent substance as a monitoring source. Fluorescence emission can be extremely broad in its spectral distribution and is, of course, free from sharp emission lines. The use of the fluorescent monitoring technique has solved both this problem, and the problem of jitter, without the addition of any electronic component to the laser unit, and has enabled us to dispense with the rather troublesome spark system. In this technique, a part of the laser beam is split off from the photolysis beam, and is used to excite fluorescence in the monitoring cell which, in its turn, is used as the background continuum for the spectrographic technique.

The monitoring source is self-synchronizing with the exciting laser pulse, the inherent delay being easily calculable, and usually about equal to the fluorescence lifetime of the monitor substance (Berlman 1965). The duration of this monitoring flash depends upon the fluorescence lifetime of the monitor and the laser pulse lifetime. Since there is no need for external synchronization, the relatively trouble-free and inexpensive passive Q-switch may be used. A table of some common scintillators and their relevant properties is given in table 1. It can be seen that,

G. Porter and M. R. Topp

TABLE 1. PROPERTIES OF SCINTILLATORS

scintillator	$\Delta\lambda$	ϕ_F	τ_F	ϵ_{347}	ϵ_{265}
PPF (Cx)	420–340	1.0	1.2	15 000	3 500
terphenyl (Cx)	400–315	0.93	0.95	0	28 000
PPO (Cx)	440–333	1.0	1.4	0	8 000
PBD (Cx)	430–310	0.89	1.35	0	20 000
POPOP (MeOH)	510–380	0.9	1.65	52 000	6 000
TPB (Cx)	590–390	0.5	1.76	36 000	(3 000)
BBOT (Cx)	530–395	0.74	1.1	40 000	(2 000)
DPS (benzene)	480–370	0.8	1.2	50 000	(3 000)

$\Delta\lambda$, fluorescence bandwidth at 10 % of maximum intensity (nm); ϕ_F, fluorescence quantum yield; τ_F, fluorescence decay time (ns); ϵ_{347}, ϵ_{265}, molar decadic extinction coefficients at 347 and 265 nm. Cx, cyclohexane; MeOH, methanol.

with a suitable choice of scintillators, alone or mixed, it should be possible to cover the whole of the spectral range of interest in the photographic work. The figures given in the extreme right-hand column show that the technique is also suitable for use with a far-u.v. laser such as the fourth harmonic from a Nd (III) glass laser (265 nm). An allowance has to be made for the re-absorption of the emitted fluoresence by the ground-state scintillator, a reservation which places some limit on the scintillators which may and may not usefully be used together. Tetraphenyl-butadiene (TPB) in cyclohexane solution is very stable to large amounts of laser radiation at 347.1 nm and, in addition, has a short fluorescence decay time, a broad emission spectrum, and a fairly high quantum yield of fluorescence.

The delay unit

In the normal flash spectrographic technique, two flashes are used, usually of several microseconds duration, the delay between which is usually accomplished by an active electronic delay unit. For delays of less than about 1 μs, however, the reproducibility in the delays deteriorates owing to limitations in circuit response time, and, for delays of such an order, a passive delay unit is to be preferred. For nanosecond processes, one requires a delay system which can impose reproducible delays separated by only a few nanoseconds.

Since light travels at the finite rate of 30 cm ns⁻¹, it is possible to delay one half of the laser beam behind the other, by interposing a suitable path difference behind them, using the first half for photolysis and the second half to trigger a monitoring flash. In this way, we have been able to achieve our nanosecond delays, from zero to 150 ns; for times longer than this the flash photoelectric technique was used. The accuracy of the optical delay itself is determined by the positioning of the mirrors, which can be performed simply by using a metre rule for all but picosecond work. Figure 1 shows an oscillograph record of the two flashes separated by an optical delay of 60 ns.

Nanosecond flash photolysis   167

The optical delay would be less valuable if used with a spectroscopic source which had an appreciable uncertainty in triggering. Because of the extreme reproducibility of the delays, and of the laser pulse shape and energy, it has been possible to observe transient absorption changes in solution over times considerably shorter than the pulse duration.

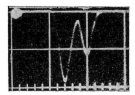

FIGURE 1. Oscillograph of an optical delay of 60 ns. Horizontal scale 100 ns/division.

The flash spectrographic apparatus

The complete apparatus is shown in figure 2. The mixed laser beams are passed through a filter of copper sulphate solution (F_2) which transmits only the ultra-violet pulse. This is then passed through the quartz beam splitter (S) which is set for 50/50 reflexion transmission, the optimum for the operation of the apparatus.

The beam splitter consists of two right-angle quartz prisms pressed together in dry contact so that they form a cube. A beam passing normally through a side of this cube will be totally internally reflected by the hypotenuse face, as it impinges at an angle greater than the critical angle. However, at just below the critical angle of incidence, the beam emerges from the hypotenuse face at a glancing angle, after considerable internal reflexion in the prism, and impinges on the corresponding face of the second prism. At such angles of incidence, the faces are very highly reflective, and several reflexions between them contribute to both the reflected and transmitted beams through the beam splitter, in addition to the initial partial internal reflexion in the first prism.

The reflected beam is passed through a weak (25 cm) lens into the reaction vessel (V), a quartz irradiation cell (usually 1.25 cm) situated on the 'waist' of the light beam focused by the lens. The focusing is by no means sharp, the angle of focusing being 0.03 radians. For photolysis with the red beam, the copper sulphate filter can be diluted slightly to allow 10 % transmission.

The transmitted laser beam is passed from the beam splitter through a telescope system of quartz lenses and is fed into the optical delay line. For the very shortest delays, a mirror was placed before the telescope. The returning laser beam from the optical delay line is collected by the same telescope and, after being turned by the beam splitter, is focused strongly into the cell containing the scintillator solution (C). The fluorescence, reinforced by reflexion off a plane mirror fused to the rear wall of the 1 mm cell, is collected by the same strong lens and, by symmetry, it is channelled through the irradiation cell in the same manner as the original

168   G. Porter and M. R. Topp

photolysis beam. The spectrograph (G) placed behind the cell receives light transmitted through the irradiated volume of the cell. Surplus laser radiation is eliminated
by the use of a filter of concentrated biphenylene solution (F_4) before the spectrograph. This compound neither fluoresces nor has any detectable transient absorption
to interfere with the spectroscopic measurements; furthermore it is transparent to
wavelengths greater than 360 nm.

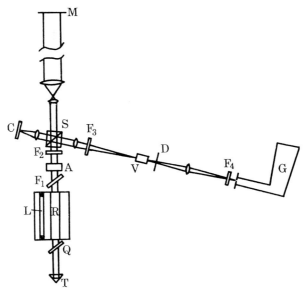

FIGURE 2. Nanosecond spectrographic apparatus. A, ADPh crystal; C, scintillator cell;
    D, aperture stop; F, filters; F_1, Wratten 29, transmits $\lambda \geqslant 630$ nm; F_2, $CuSO_4$, attenuates
    694 nm; F_3, biphenylene, control of intensity of 347 nm, F_4, biphenylene, u.v. laser
    cut-off; G, spectrograph; L, xenon-filled flash lamp; M, mirror; Q, passive Q-switch;
    S, beam splitter; T, quartz t.i.r. prism; V, reaction vessel; R, ruby.

It is possible to record, with a single shot, a complete absorption spectrum within
the limits set by the scintillators. By this means, we have been able, by simple
variation of the optical path difference, to time-resolve the absorption spectra of
the excited singlet states of several aromatic molecules (Porter & Topp 1968).
    One great advantage of the scintillator-monitor is that one can create an effective
point source of light without having to comply with any of the stringent mechanical
requirements which would be necessary for an active spectral source such as a spark
or a flash lamp. The point source allows optimum recollimation by the collector
lens, thus allowing more light to be focused onto the spectrograph slit.

The flash photoelectric apparatus

Owing to the limitation of the length of the optical delay system, and its unsuit-
ability for long delays, as well as the need for kinetic measurements at a single
wavelength once the nanosecond transients have been assigned, the apparatus was
developed further to allow photoelectric monitoring. For ease of conversion it was

desirable to use, as far as possible, the existing components of the photographic apparatus.

Since the bandwidth of response of the photomultiplier system must be about 1000 times greater than that used for conventional microsecond flash photolysis, it follows that, in order to maintain the same signal/noise ratio in the detection system, the light intensity falling onto the photomultiplier cathode must be of the order of 1000 times greater than that of a conventional monitoring source. This intensity is conveniently provided by the laser flash-pumping lamp which has a duration of $1200\,\mu s$. The laser pulse occurs at the peak intensity of the lamp, or a little after it.

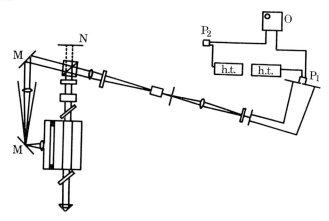

FIGURE 3. Nanosecond photoelectric apparatus. h.t., power supplies; M, mirrors; N, safety screen; O, oscilloscope; P_1 and P_2, photomultipliers.

Figure 3 shows the arrangement of our apparatus for photoelectric recording. Light from the laser flashlamp is leaked out of a hole drilled in the reflector, through the beam splitter and into the spectrograph. The scintillator cell and the lens of the spectrographic system are removed, and the photographic plate holder on the rear of the spectrograph is replaced by a travelling photomultiplier (P_1) on a calibrated slide. The photomultiplier has its own slit arrangement, which gives a resolution of 1.5 nm on the small Hilger spectrograph. The oscilloscope (O) is triggered by a second photomultiplier (P_2) which is activated by the laser pulse.

The application of these techniques to the two principal types of excited molecule —the lowest excited singlet and triplet states—will now be described.

ABSORPTION SPECTRA AND KINETICS OF EXCITED SINGLET STATES

The lowest excited singlet (S_1) states of polyatomic molecules have lifetimes too short for detection by conventional flash methods though they can, of course, often be observed in emission as fluorescence. The positions of higher singlet states to which strong transitions could occur in absorption are also generally unknown since, because of the parity selection rule, these would generally be those states to

170   G. Porter and M. R. Topp

which transitions from the ground state are forbidden. We have therefore sought, throughout the spectra, new absorptions having decay rates identical with those of the corresponding fluorescence and with the growth rate of the triplet state.

Using the spectrographic apparatus, we have investigated solutions of polynuclear aromatic hydrocarbons. In most cases, a new absorption was observed which decayed within the 70 ns time-spread of the delay sequences; the simultaneous growth in several cases of the well-known triplet absorption was observed. Examples of the records obtained are presented in figures 4 and 5, plates 1 and 2.

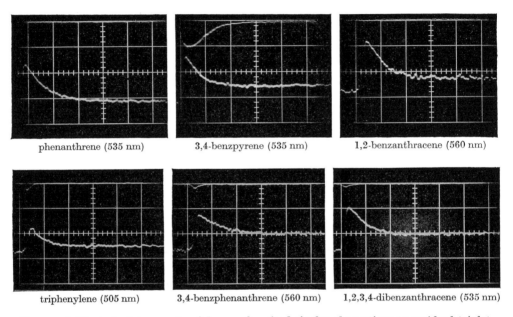

phenanthrene (535 nm)   3,4-benzpyrene (535 nm)   1,2-benzanthracene (560 nm)

triphenylene (505 nm)   3,4-benzphenanthrene (560 nm)   1,2,3,4-dibenzanthracene (535 nm)

FIGURE 6. Photoelectric records of decay of excited singlet absorptions to residual triplet. Horizontal scale, 100 ns/division; vertical scale, transmitted light intensity.

Using the kinetic apparatus, the decays of these same absorptions were monitored at wavelengths corresponding to the singlet absorption maxima; examples of these decays are presented in figure 6. In every case, the decays to the new base corresponding to residual triplet absorption were found to be accurately first order, with lifetimes corresponding, within experimental error, to the fluorescence lifetimes observed from the same samples. In each case, the lower trace is the absorption and the upper trace shows the total amount of fluorescence and scattered laser light received by the spectrograph with the monitoring light shuttered off. The low level of this extraneous light demonstrates the usefulness of the monochromatic exitation beam. Only in the case of 3,4-benzpyrene is the fluorescence of comparable intensity to the monitoring light. Subtraction of the scattered light from the total light signal enables accurate calculation of the decay kinetics to be made even in this case.

Porter & Topp   *Proc. Roy. Soc. Lond. A, volume* **315**, *plate* 1

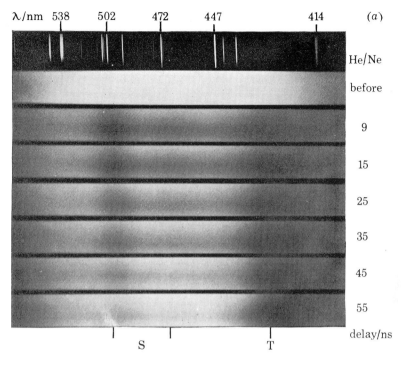

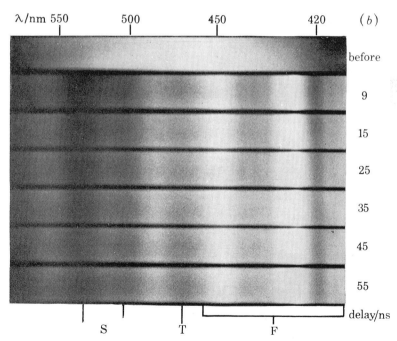

FIGURE 4. Time-resolved excited singlet and triplet absorption spectra:
(*a*) triphenylene in benzene; (*b*) 3,4-benzpyrene in cyclohexane.

Porter & Topp   *Proc. Roy. Soc. Lond. A, volume* **315**, *plate* **2**

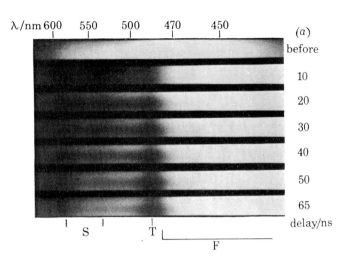

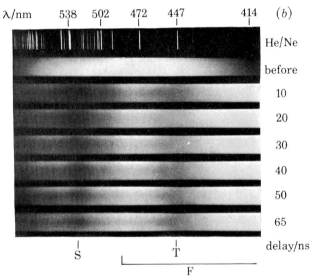

FIGURE 5. Time-resolved excited singlet and triplet absorption spectra:
(a) 1,2-benzanthracene; (b) 1,2,3,4-dibenzanthracene.

TABLE 2. HIGHER SINGLET LEVELS AND CORRELATION OF LIFETIMES WITH FLUORESCENCE DATA

molecule	solvent	singlet S_1 absorption/nm	level of higher S state/cm^{-1}	triplet T_1† absorption/nm	singlet absorption decay time/ns	fluorescence decay time/ns
phenanthrene (d_{10})	Cx	(545), 515	48200	520, 510 481, 454	61.1	63.5
phenanthrene (h_{10})	PMMA	(545), 515	48200	520, 510, 481 454	65	67.2
triphenylene	Benzene	500, 465, 433	48800	428	44.2	43.0
	PMMA	500, 465, 433	48800	428	45.0	44.0
1,2-benzanthracene	Cx	550, 495	44200	540, 490, 461	52.7	49.4
	PMMA	550, 495	44200	540, 490, 461	51.7	52.5
1,2,3,4-dibenzanthracene	Cx	540, 500	45700	445‖	51.2	53.5
	PMMA	540, 500	45700	445‖	50.3	52.5
pyrene	Cx	515, (480), 470	48000	520, 483, 416	296‡	261‡
	PMMA	515, (480), 470	48000	520, 483, 416	326	319
coronene	Dx	(600, 570), 530, 495, (465)	42500	525, 480	319	307
	PMMA	(600, 570), 530, 495, (465)	42500	525, 480	390	380
3,4-benzpyrene	Cx	(590), 535, 510	43700	480§	49.1	57.5
	PMMA	(590), 535, 510	43700	480§	39.4	40.5
3,4-benzphenanthrene	Cx	595, 525	43200	517	68.5	70
	PMMA	595, 525	43200	517	76	81
3,4,9,10-di-benzpyrene	Cx	575, 552	40400	495‖	140	143
1,2,5,6-di-benzanthracene	PMMA	(570)	(42900)	532, 480	(40.0)¶	37.5

† Except where stated, data correlate with those from Porter & Windsor (1958).
‡ 3×10^{-4} mol l^{-1}.
§ (Craig & Ross 1954).
‖ Experimentally determined.
¶ Growth of triplet absorption.
Dx, dioxan; PMMA, polymethylmethacrylate.

172   G. Porter and M. R. Topp

Table 2 summarizes the results of these investigations in both solid and fluid solution. The value for the singlet lifetime of phenanthrene is larger than that reported previously (Birks & Munro 1967; Porter & Topp 1968). In general, the spectra in polymethylmethacrylate solution were indistinguishable from those in liquid solution although, as will be seen, it has not been possible to assign accurately the absolute spectra. The levels of the higher states have been calculated by direct addition of the S_1–S_n transition energies to the known values of the energies of S_1.

Fluorescence lifetime determinations

Owing to the presence of strong transient absorptions in solution, the wavelength of monitoring of the fluorescence emission profile had to be carefully selected.

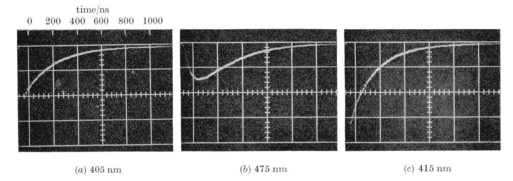

time/ns
0   200 400 600 800 1000

(a) 405 nm   (b) 475 nm   (c) 415 nm

FIGURE 7. Effects of reabsorption on the fluorescence profile of pyrene.

Figure 7 shows the fluorescence profile of pyrene observed at three different wavelengths in cyclohexane solution. The pyrene concentration was sufficiently low ($3 \times 10^{-4}\,\mathrm{mol\,l^{-1}}$) to exclude appreciable excimer formation. At 405 nm, which is close to the peak of the fluorescence emission spectrum, the observed fluorescence lifetime is close to the value obtained with low irradiation intensities. However, at 475 nm the presence of excited singlet state absorption distorts the profile, and the observed lifetime is longer. At the triplet absorption peak of 415 nm the decay time is shorter, owing to the ingrowth of reabsorption with time.

Effect of oxygen

When our hydrocarbon solutions were air-saturated, lifetimes of both singlet and triplet were reduced. Fluorescence decay times were shortened to several tens of nanoseconds, except in viscous solvents, but the triplet decay times were greater than this by a factor varying between three and ten. An example is shown in figure 8. The upper trace (a) depicts the decay of absorption at 535 nm due to the first excited singlet state of perdeutero-phenanthrene in aerated cyclohexane. The singlet absorption is seen to follow approximately the laser pulse profile, while the triplet (b) decays with a pseudo-first order lifetime of 262 ns. Details of oxygen

Porter & Topp   Proc. Roy. Soc. Lond. A, volume 315, plate 3

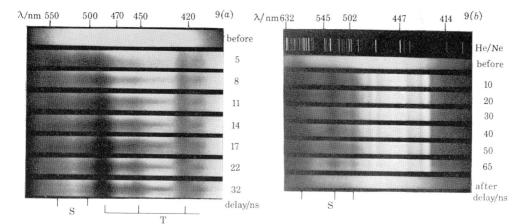

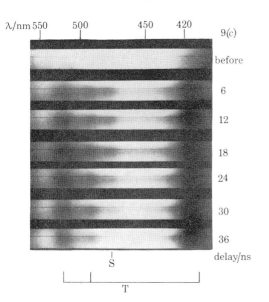

FIGURE 9. Time resolved excited singlet and triplet absorption sprectra of oxygen-quenched excited states: (*a*) phenanthrene–cyclohexane; (*b*) coronene–dioxan; (*c*) pyrene–cyclohexane.

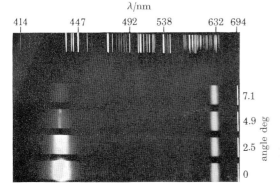

FIGURE 14. Laser emission from scintillator BBOT (ethanol). Variation of laser output with angle of cell windows to direction of excitation beam.

quenching rates of singlet and triplet states and their dependence on energy levels will be given elsewhere.

Figure 9 (*a*), plate 3, shows the spectrographic sequence of these changes in phenanthrene. The triplet absorption increases in intensity whilst the singlet absorption at longer wavelengths decays within the laser pulse duration of about 20 ns. In this way, the singlet–singlet absorption decays have been time-resolved for those compounds whose natural fluorescence decay times are too long for the spectrographic apparatus. Figures 9 (*b*) and (*c*), plate 3, show the decays of coronene and pyrene singlet absorptions under these conditions.

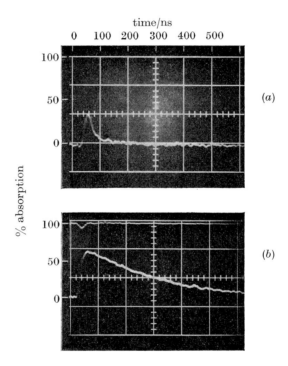

FIGURE 8. Demonstration of oxygen-quenching of (*a*) excited singlet (535 nm), (*b*) triplet (480 nm) absorptions in phenanthrene-d_{10} in cyclohexane. Horizontal scale 100 ns/ division.

Extinction coefficients

The residual absorption seen in figure 6 is due to the triplet state which has been formed by decay of the excited singlet state. In several cases, the extinction co-efficient of the triplet–triplet absorption is known at these wavelengths and, using literature values of triplet yields formed from the singlet, values of the extinction coefficients of the singlet–singlet absorptions may be calculated.

We have assumed that, to a first approximation, the total amount of excitation has been applied instantaneously at a point in time ('zero time') corresponding to the peak of the actual laser pulse. The logarithm of the optical density when plotted

174   G. Porter and M. R. Topp

against time yields a straight line from whose slope has been calculated the absorption decay time. The intercept of this line with 'zero time' chosen as above yields a value of the initial singlet state optical density, corresponding to an initial singlet concentration (S_0), with an error not greater than 20 %. At very long times (about $1\,\mu s$) after the photolysis pulse, all the absorption is assigned to the triplet, concentration (T_∞). From these values of optical density, the following relationship may be calculated:

initial optical density $= D_0 = (S_0)\,\epsilon_S$ ($\epsilon_S =$ singlet extinction coefficient),

residual optical density $= D_\infty = (T_\infty)\,\epsilon_T$ ($\epsilon_T =$ triplet extinction coefficient),

and $(T_\infty) = \phi_T(S_0)$ where ϕ_T is the quantum yield of triplet formation. Thus

$$\epsilon_S = \frac{\phi_T \epsilon_T D_0}{D_\infty}.$$

The experimental results are averaged over several determinations. The ratio $(D_0)/(D_\infty)$ determined for several solutions is listed in table 3. Where possible, calculation of the singlet extinction coefficient has been carried out.

TABLE 3. RELATIVE EXTINCTION COEFFICIENTS OF TRANSIENT
SINGLET AND TRIPLET ABSORPTIONS

molecule	λ_{max}	D_0/D_∞	ϕ_T[†]	ϵ_T[‡] (λ_{max})	ϵ_S
phenanthrene	535	8.9 ± 0.4	0.85 ± 0.02	2000	$1.5 \pm 0.2 \times 10^4$
1,2-benzanthracene	560	9.6 ± 0.3	0.77 ± 0.02	3450	$2.5 \pm 0.2 \times 10^4$
1,2,3,4-dibenz-anthracene	535	4.4 ± 0.4	—	—	—
pyrene	475	7.0 ± 0.7	0.38 ± 0.02	—	—
coronene	525	3.3 ± 0.4	0.56 ± 0.01	320	590
3,4-benzphenanthrene	560	7.2 ± 0.7	—	300	—
triphenylene	505	6.9 ± 0.6	(0.93)	~ 0	—

† Horrocks & Wilkinson (1968)
‡ Land (1968); Porter & Windsor (1958).

It can be seen from the ratio of initial and final optical densities that there is considerable overlap between the singlet and triplet spectra and this has prevented the observation of triplet growth versus singlet decay in most of these solutions. In 1,2,5,6-dibenzanthracene, no absorption decay was discernible on either the spectrographic or the kinetic apparatus but the triplet absorption was seen to increase with time at precisely the same rate as the fluorescence decay. Recent work by Thomas (1969) has established, in the case of 1,2-benzanthracene, that the singlet and triplet absorption spectra are very similar.

Microdensitometry

The spectrographic records obtained from the laser apparatus were analysed on a microdensitometer. From these records were obtained the total absorption spectra of the transient species in solution.

Where the singlet lifetime is far greater than the laser pulse duration, then the absorptions recorded immediately after it had passed were taken as true absorption spectra for the excited singlet states. In such cases, the triplet contribution to the total amount of absorption is considered negligible. The absorption spectra of pyrene and coronene excited singlet states obtained in this way are shown in figure 10.

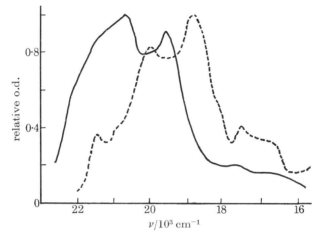

FIGURE 10. Absorption spectra from microdensitometer traces:
————, pyrene; - - - -, coronene.

In the majority of the molecules studied, this assumption was not valid. Because of shorter singlet lifetime, the absorption spectra could not be corrected for triplet absorption. Figure 11 shows uncorrected spectra of the singlet absorption of 1,2-benzanthracene, 3,4-benzpyrene, 3,4-benzphenanthrene, triphenylene and 1,2,3,4-dibenzanthracene.

Microdensitometry of sequences of absorption spectra taken at different times after photolysis show clearly the redistribution of absorption which takes place as the singlet absorption is replaced by the triplet. Figure 12 shows three absorption spectra taken at different times during the lifetime of the oxygen-quenched singlet state of perdeuterophenanthrene. The data were taken directly from figure 9. The presence of an isosbestic point at about $20\,300\,\mathrm{cm}^{-1}$ shows that the decaying transient gives rise directly to the growing one. Moreover, the presence of the isosbestic point makes possible an estimation of the singlet absorption extinction coefficient at this point. The value obtained is $\epsilon_{560} = 9000\,\mathrm{l\,mol^{-1}\,cm^{-1}}$.

Difference spectra

In order to obtain an estimate of the shape of the singlet–singlet absorption peaks in the vicinity of large triplet absorptions, difference spectra were employed. By subtracting the absorption spectrum recorded immediately after the laser pulse

176 G. Porter and M. R. Topp

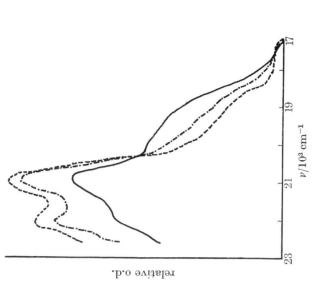

FIGURE 12. Time-resolved absorption spectra from micro-densitometer traces. Phenanthrene-aerated cyclohexane. Delays: ———, 5 ns; ········, 22 ns; - - - - -, 32 ns.

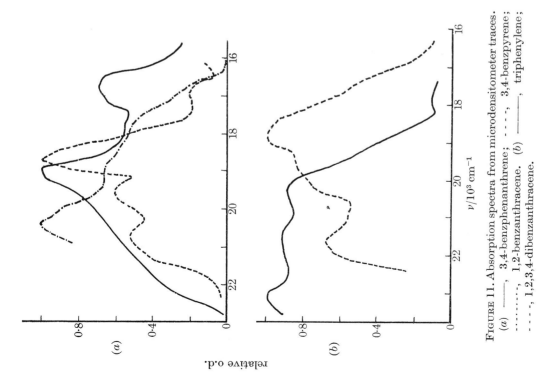

FIGURE 11. Absorption spectra from microdensitometer traces. (a) ———, 3,4-benzphenanthrene; - - -, 3,4-benzpyrene; ········, 1,2-benzanthracene. (b) ———, triphenylene; - - - -, 1,2,3,4-dibenzanthracene.

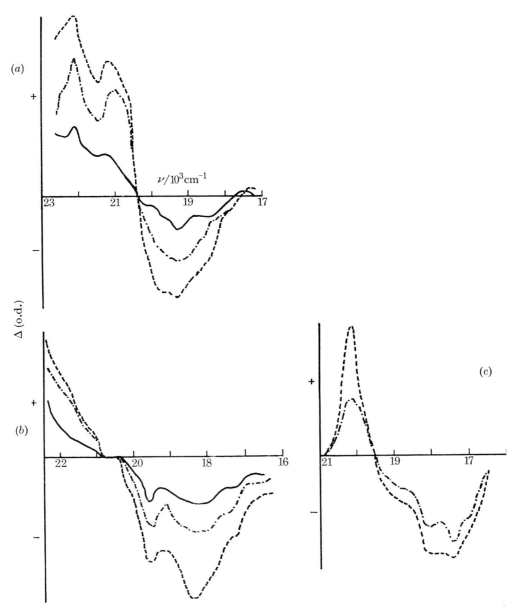

FIGURE 13. Absorption spectra from microdensitometer traces; sequences of difference spectra.
(*a*) Phenanthrene-aerated cyclohexane: ———, o.d. (11 ns)–o.d. (5 ns); · · · · ·, o.d. (22 ns)–
o.d. (5 ns); – – – – –, o.d. (32 ns)–o.d. (5 ns); (*b*) 1,2,3,4-dibenzanthracene–cyclohexane:
———, o.d. (30 ns)–o.d. (10 ns); · · · · ·, o.d. (40 ns)–o.d. (10 ns); – – – – –, o.d. (65 ns)–o.d.
(10 ns). (*c*) 3,4,9,10-dibenzpyrene-aerated cyclohexane: · · · · · · ·, o.d. (40 ns)–o.d. (10 ns);
– – – – –, o.d. (65 ns)–o.d. (10 ns).

178   G. Porter and M. R. Topp

from a spectrum taken at a later time, one may plot a difference spectrum, whose ordinate is a measure of the difference between the triplet and singlet extinctions. A sequence of such difference spectra for (a) the same phenanthrene example, (b) 1,2,3,4-dibenzanthracene, and (c) 3,4,9,10-dibenzpyrene (oxygen quenched), is shown in figure 13.

Intersystem crossing rate constants

These may be measured with some accuracy from traces obtained on a kinetic apparatus, although the present laser pulse duration does not permit accurate extrapolation. However, this limitation is only temporary, as will be discussed at the end of the paper.

The ratios of the initial and residual optical densities of the singlet states have already been discussed, and several results presented in table 3. Consider the relations

$$D_0/D_\infty = \epsilon_S(S_0)/\epsilon_T(T_\infty) \quad \text{and} \quad D_0/D_\infty = \epsilon_S/\epsilon_T \phi_T = R.$$

By altering ϕ_T, for example by adding oxygen, it is possible to obtain two values of R.

Hence $$R/R' = \phi'_T/\phi_T.$$

But $$\phi'_T = (k_{\text{i.s.c.}} + k')/(k_D + k') \quad \text{and} \quad \phi_T = k_{\text{i.s.c.}}/k_D;$$

where k' is the resultant increase in intersystem crossing rate constant as a result of the perturbation applied, $k_{\text{i.s.c.}}$ is the normal intersystem crossing rate constant and k_D is the normal decay rate constant of singlet.

Whence

$$k_{\text{i.s.c.}} = \frac{k'R'}{R(k_D+k')/(k_D) - R'}. \tag{1}$$

Such a perturbation occurs when air is admitted to the solution. Coronene, under anaerobic conditions, has a fluorescence decay rate constant of $2.5 \times 10^6 \, \text{s}^{-1}$. The ratio R is found to be 3.3 (see table 3). Owing to the finite triplet decay on the time scale used, in aerated solution, it is necessary to take the triplet optical density, D_∞, as that at which the decay ceases to have any contribution due to singlet. In aerated solution, the ratio R' is found to be 2.1, whereas the fluorescence decay rate constant is $2.5 \times 10^7 \, \text{s}^{-1}$. From equation (1), we find that $k_{\text{i.s.c.}} = 1.55 \times 10^6 \, \text{s}^{-1}$. Division of this rate constant by the observed natural decay rate constant of the singlet state gives the triplet quantum yield as $1.55/2.5 = 0.6 \pm 0.04$. This agrees with published data (Horrocks & Wilkinson 1968). Another example, even though less accurate, as a result of the short fluorescence lifetime in aerated solution, is that of 1,2-benz-anthracene. In this case, $k_D = 2.0 \times 10^7 \, \text{s}^{-1}$; $k' = 3.9 \times 10^7 \, \text{s}^{-1}$; $R = 9.6$ and $R' = 8.7$. Equation (1) gives, for this set of results, $k_{\text{i.s.c.}} = 1.7 \times 10^7 \, \text{s}^{-1}$. The quantum yield of triplet formation is found to be 0.85 with an error of $\pm 10 \%$ (literature value 0.77 ± 0.22).

Materials

All aromatic compounds and solvents could be obtained sufficiently pure for use in our experiment with one exception. The phenanthrene to be used in the fluid solvents resisted attempts at removal of anthracene as impurity. We are grateful to Dr D. Lavalette for the pure sample of perdeuterophenanthrene which we used for our fluid solution work. All solutions in polymethylmethacrylate were provided by Dr M. West.

TRIPLET STATE OF KETONES AND QUINONES

Although the lifetimes of the triplet state of many organic molecules are long enough, in both gases and liquids, for easy detection by conventional flash methods, there are other triplet states, particularly the reactive triplets of some carbonyl compounds, which have submicrosecond lifetimes and have not, therefore, hitherto been observed. The wavelength of the frequency doubled laser flash at 347 nm is very suitable for photolysis of most aromatic ketones and quinones and we have investigated two important cases, benzophenone and derivatives of benzoquinone.

Benzophenone

Benzophenone photochemistry has been studied extensively by various workers, and is fairly well understood. The lowest triplet state, which is of $n - \pi^*$ character, readily abstracts hydrogen from solvent molecules, forming the long-lived ketyl radical with unit quantum yield:

$$\phi_2\mathrm{CO} \xrightarrow{h\nu} {}^1\phi_2\mathrm{CO}^* \rightarrow {}^3\phi_2\mathrm{CO}^*,$$
$$^3\phi_2\mathrm{CO}^* + \mathrm{RH} \rightarrow \phi_2\dot{\mathrm{C}}\mathrm{OH} + \dot{\mathrm{R}}.$$

It has been calculated that the lifetime of the triplet state in an alcoholic solvent is of the order of several tens of nanoseconds (Beckett & Porter 1963), and the direct observation of the triplet state under these conditions should, therefore, be possible. There is a difficulty inherent in benzophenone transient spectra in that both the triplet and the transient reaction products have similar spectra (Bell & Linschitz 1963). Fortunately, however, the extinction coefficient of the triplet state is the larger.

Using the flash-photoelectric apparatus for studies of benzophenone in various solvents we observed an absorption at 535 nm, the peak of the triplet–triplet absorption. We have determined the lifetimes of the benzophenone triplet state in various solvents which are listed in table 4. Beckett & Porter (1963) assigned a lifetime for the benzophenone triplet state in isopropanol of 60 ns using results based on naphthalene quenching and assuming diffusion controlled rates. Our direct measurement agrees moderately well with their estimate.

From earlier argument, the ratio of the initial and final optical densities is equal to
$$D_0/D_\infty = \epsilon_\mathrm{A}/\epsilon_\mathrm{B}\phi_\mathrm{B} = R,$$

180   G. Porter and M. R. Topp

where A is the initial transient species and B is the residue. It can be seen from the table that there are two apparent modes of reaction of the benzophenone triplet. The top four solvents have comparable values of R, and decay rates which closely correlate with predicted relative rates of the hydrogen abstraction reactions. The benzenoid solvents, however, exhibit quite different behaviour. In the case of benzene-h_6 and benzene-d_6, there is the possibility that the hydrogen abstraction may still occur, although it is energetically unfavourable, but then the absorption should appear, as in the other cases, as ketyl radical. It seems that chemical reaction other than hydrogen abstraction is taking place. This reaction has been interpreted as electrophilic attack upon the aromatic ring by the benzophenone triplet to form an addition product (Schuster & Topp 1969).

TABLE 4. BENZOPHENONE TRIPLET IN VARIOUS SOLVENTS

	triplet state		
solvent	lifetime/ns	decay constant, k/s^{-1}	D_0/D_∞
isopropanol	46 ± 6	2.2×10^7	1.7 ± 0.2
ethanol	104 ± 15	9.7×10^6	1.5 ± 0.2
dioxan	200 ± 20	5.0×10^6	1.5 ± 0.2
cyclohexane	300 ± 20	3.3×10^6	2.0 ± 0.3
benzene-h_6	2500 ± 300	4.0×10^5	$9.0 \pm 0.5\dagger$
benzene-d_6	3450 ± 300	2.9×10^5	$12.5 \pm 1.0\dagger$
benzene-f_6	435 ± 40	2.3×10^6	$15.0 \pm 2.0\dagger$

† Results obtained in collaboration with Professor D. I. Schuster.

The similarity of the R values for the first four solvents, in spite of widely different triplet lifetimes, can only be rationalized if the quantum yield of reaction is approximately unity in all four cases. The quantum yields of final product vary by as much as a factor of 4 (Beckett & Porter 1963) but this must be due to subsequent reactions. R is, therefore, a measure of the ratio of the extinction coefficients of the triplet and ketyl radical and the value of R in cyclohexane is 2.0 ± 0.3. Estimates of the extinction coefficient of the radical at 535 nm of 5000 (Beckett & Porter 1963) and 3220 (Land 1968) lead to triplet extinction coefficients at the same wavelength of 10000 and 6440.

Energy transfer

The relatively high energy of the benzophenone triplet ($24\,400\ cm^{-1}$) compared with triplets of many aromatic hydrocarbons has made possible its use as a donor in the study of intermolecular energy transfer reactions. Porter & Wilkinson (1961) investigated the transfer of energy from benzophenone to naphthalene (triplet energy $21\,300\ cm^{-1}$) in benzene. The transfer reaction competes with the chemical reaction in benzene for the triplet energy. A study of the dependence of the relative rates of these processes on naphthalene concentration enabled Porter & Wilkinson to estimate the second-order rate constant for the energy transfer process.

This type of system is ideal for study with the ruby laser, as the donor molecule (benzophenone) can be selectively excited by monochromatic light at 347 nm. Moreover, the process, competing with the chemical reaction of the benzophenone triplet in benzene, has a submicrosecond duration.

We studied the energy transfer process with both the spectrographic and the photoelectric apparatus, and found a strong absorption due to the benzophenone triplet (535 nm) present immediately after the photolysis pulse. This decayed rapidly and was replaced by an absorption due to the naphthalene triplet (425 nm). We used a naphthalene concentration substantially higher than that of the benzophenone triplet produced on photolysis and this resulted in the observation of pseudo-first-order kinetics. For example, on photolysis an initial benzophenone triplet optical density of 0.28 (3×10^{-5} mol l^{-1}), in the presence of 8.15×10^{-4} mol l^{-1} naphthalene, decayed with a first-order rate constant of 4.3×10^6 s^{-1}. The total decay constant of the benzophenone triplet may be expressed as $k = k_d + k_q(N)$ where k_d is the decay constant observed in the absence of naphthalene (4.0×10^5 s^{-6}), k_q is the second-order constant for energy transfer and (N) is the naphthalene concentration, assumed constant. We found $k_q = 4.7 \pm 0.4 \times 10^9$ l mol^{-1} s^{-1}, a value higher than that of Porter & Wilkinson who indicated that their indirect estimation might be too low.

Derivatives of benzoquinone

Benzoquinone and some of its derivatives were studied by Bridge & Porter (1958) who found the semiquinone radical at 410 nm in all cases. With duroquinone they also observed a short lived transient with band maxima at 490 nm which they assigned to the triplet state. Some subsequent workers have questioned this assignment and given good reasons for attributing the 490 nm band to a reversible tautomeric rearrangement (Herman & Schenck 1968; Wilkinson, Seddon & Tickle 1967).

TABLE 5. DUROQUINONE AND CHLORANIL TRIPLET DECAY CONSTANTS

quinone	solvent	λ_{max}/nm triplet	k/s^{-1}
duroquinone	benzene	490	3.2×10^5
chloranil	ethanol	500	8.4×10^5
chloranil	cyclohexane	500	5.0×10^5

Apart from this, there is little information about the triplet state of benzoquinones and since phosphorescence is very weak or absent, but quantum yields of reaction high, it appears that triplet state lifetimes are likely to be small and probably in the nanosecond region.

In collaboration with D. R. Kemp we have investigated the transient absorption spectra of benzoquinone derivatives in various solvents. While the possibility remains that with duroquinone in some solvents the 490 nm transient is that of a tautomer, we have established with reasonable certainty that the triplet state is the species responsible for this band from duroquinone in benzene. Furthermore, we have observed an almost identical absorption from chloranil in ethanol and cyclohexane where tautomerisation is impossible and have shown in the ethanol

182   G. Porter and M. R. Topp

solution, by energy transfer experiments, that this is a triplet state and that it
reacts to from the radical. The triplet state maxima and first-order decay constants
are given in table 5 and full details of this work will appear elsewhere (Kemp &
Porter 1969).

DISCUSSION

In the spectrographic and in the photoelectric apparatus, the limiting factor of
time resolution is the duration of the laser pulse, which is, at present, about 20 ns.
Should the duration of the laser pulse be reduced, the limiting factor in the spectrographic technique would be the fluorescence lifetime of the scintillator solution.
Similarly, the photoelectric technique would be time-limited by the response time
of the photomultiplier (1.5 ns) and the rise time of the oscilloscope circuits (2 ns).

It is now possible, through the techniques of regeneration switching, or pulse
transmission mode operation (p.t.m.) of a laser to generate pulses sufficiently short,
and of sufficient energy (1 J) for the above parameters to become the limiting factor.
Such techniques are also applicable for use with neodymium lasers, and deep u.v.
(265 nm) irradiation with ultra-short pulses obtained by quadrupling the fundamental frequency is available.

In order to produce the same amount of transient absorption with shorter flashes
and without changing the sensitivity of the monitoring techniques, a corresponding
increase in the pulse intensity is necessary.

While many of the consequences of increased laser intensity are quite straightforward, there is one which merits further discussion.

Stimulated emission in the scintillator cell

It is apparent that stimulated emission plays some part in the output of light from
the scintillator cell, even with the 20 ns laser pulse. For example, when POPOP is
irradiated in the scintillator cell position shown in figure 2, the emitted pulse duration is about 25 % shorter than the irradiating pulse duration. In the cases of the
scintillators POPOP, BBOT and DPS, it has been found that the emission spectrum
is strongly dependent on the laser intensity.

When high laser intensities are used, there is a marked dependence of the
fluorescent output on the angle of the scintillator cell windows to the direction of
the exciting laser beam. In orientations where the cell faces are far from normal to
the laser beam direction, then the fluorescent output received at the spectrograph
is of normal efficiency as determined by the collection ratio of the lens system.
Moreover, the emission spectrum under such conditions agrees closely with that
published by Berlman (1965), obtained at low light intensities. However, as the
cell face is brought into a position normal to the laser beam direction, then the emission spectrum develops a very intense, and very sharp line at the wavelength of
the 1–0 vibronic fluorescence transition band, in the case of BBOT in ethanol
situated at about 435 nm as is shown in figure 14, plate 3. This phenomenon is
attributed to 'lasing' in the cell, such as has already been obtained in other cases
by Sorokin et al. (1967). These scintillators do not emit the 'laser' beam in the 0–0

Nanosecond flash photolysis   183

band, presumably owing to reduction of gain by reabsorption of emitted light by the ground state. The effect is illustrated by our observations for these scintillators given in table 6.

These emissions were obtained without the use of a special laser cavity around the scintillator cell. Even when the scintillator cell reflector was removed, the windows possessed sufficient reflectivity to obtain laser threshold level, owing to the high fluorescence efficiency of these molecules.

It might have been hoped that the limitation imposed on time resolution by the fluorescence lifetime of the scintillator would be overcome by use of stimulated emission, especially with a picosecond laser pulse. However, if the spectrum always shows the extreme narrowing which has been observed in these three cases, then sacrifice of the polychromaticity may prove not worthwhile. Stimulated emission under such circumstances can easily be prevented by mixing two scintillators with similar characteristics. In this way the total amount of light emitted from the scintillator remains unchanged.

TABLE 6. LASER WAVELENGTHS OF SCINTILLATORS

scintillator	solvent	1–0 band/nm	laser wavelength/nm	ϵ_0 at 0–0
POPOP	EtOH	418	419	500–1000
BBOT	EtOH	433	435	500–1000
DPS	Cx	408	409	1000

ϵ_0 is the ground state extinction coefficient.

During the last two years, several other laboratories have developed laser techniques for nanosecond flash photolysis. Novak & Windsor (1967, 1968) used a laser-induced spark in a gaseous medium as a spectroscopic background continuum. Time-resolution was achieved by an image-converter scanning technique. In this way the apparatus combined both spectroscopic and photoelectric techniques. Apparatus using only the point by point photoelectric method but capable of more quantitative measurements was used by Bonneau, Faure & Joussot-Dubien (1968) and Thomas (1969). The former obtained kinetics and a point by point plot of the naphthalene transient singlet–singlet absorption spectrum. Thomas used a laser flash apparatus to verify observations made using a nanosecond pulse radiolysis technique.

Each of these modifications of the applications of lasers to flash photolysis has its particular advantages. The method described here is remarkably simple and reproducible. Apart from the laser itself, the equipment is available in most laboratories and is even less complex than that used in a conventional flash photolysis apparatus. In the spectrographic double-flash method, electronic components, with their intrinsic limitations on time resolution, are completely eliminated. There are of course, other limitations at the present time, particularly of wavelengths available for pulsed lasers, but in view of the rapid development of laser technology, these limitations may reasonably be regarded as only temporary.

184   G. Porter and M. R. Topp

We thank the Science Research Council and the U.S. Air Force Office of Scientific Research, for support of this work.

REFERENCES

Beckett, A. & Porter, G. 1963 *Trans. Faraday Soc.* **59**, 2038.
Bell, J. A. & Linschitz, H. 1963 *J. Am. Chem. Soc.* **85**, 528.
Berlman, I. B. 1965 *Handbook of fluorescence spectra of aromatic molecules*. New York: Academic Press.
Birks, J. B. & Munro, H. 1967 *Progr. Reaction Kinetics*, **4**, 239.
Boag, J. W. 1968 *Photochem. Photobiol.* **8**, 565.
Bonneau, R., Faure, J. & Joussot-Dubien, J. 1968 *Chem. Phys. Lett.* **2**, 65.
Bridge, N. K. & Porter, G. 1958 *Proc. Roy. Soc. Lond.* A **244**, 259, 276.
Craig, D. P. & Ross, I. 1954 *J. chem. Soc.* p. 1589.
Dillon, J. H. 1931 *Phys. Rev.* **38**, 408.
Fischer, H. 1961 *J. opt. Soc. Am.* **51**, 543.
Herman H. & Schenck, G. O. 1968 *Photochem. Photobiol.* **7**, 255.
Horrocks, A. R. & Wilkinson, F. 1968 *Proc. Roy. Soc. Lond.* A **306**, 257.
Kemp, D. R. & Porter, G. 1969 To be published.
Land, E. J. 1968 *Proc. Roy. Soc. Lond.* A **305**, 457.
Lewis, R. N., Jung, E. A., Chapman, G. L., Van Loon, C. S. & Romanovski, T. A. 1966 *I.E.E.E. Trans. Nuc. Sci.* April, p. 84.
Marx, E. 1924 *Elektrotech. Z.* **45**, 652.
Müller, A. & Pflüger, E. 1968 *Chem. Phys. Lett.* **2**, 155.
Novak, J. R. & Windsor, M. W. 1967 *J. chem. Phys.* **47**, 3075.
Novak, J. R. & Windsor, M. W. 1968 *Proc. Roy. Soc. Lond.* A **308**, 95.
Pendleton, W. K. & Guenther, A. H. 1965 *Rev. scient. Instrum.* **36**, 1546.
Porter, G. 1950 *Proc. Roy. Soc. Lond.* A **200**, 284.
Porter, G. & Topp, M. R. 1967 *Nobel Symposium 5—Fast Reactions and Primary Processes in Reaction Kinetics*, p. 158. London and New York: Interscience.
Porter, G. & Topp, M. R. 1968 *Nature, Lond.* **220**, 1228.
Porter, G. & Wilkinson, F. 1961 *Proc. Roy. Soc. Lond.* A **264**, 1.
Porter, G. & Windsor, M. W. 1958 *Proc. Roy. Soc. Lond.* A **245**, 238.
Schuster, D. I. & Topp, M. R. To be published.
Sorokin, P. P., Luzzi, J. J., Lankard, J. R. & Pettit, G. D. 1964 *IBM Jl Res. Dev.* **11**, 130.
Thomas, J. K. 1969 *J. chem. Phys.* **51**, 770; also private communication.
Wilkinson, F., Seddon, G. M. & Tickle, K. 1968 *Ber. Bunsen-Ges. phys. Chem.* **72**, 315.

Reprinted from *Chem. Phys. Lett.* **29**, 469–472 (1974) with permission from Elsevier

Volume 29, number 3   CHEMICAL PHYSICS LETTERS   1 December 1974

TIME RESOLVED FLUORESCENCE IN THE PICOSECOND REGION

G. PORTER, E.S. REID and C.J. TREDWELL

Davy Faraday Research Laboratory of The Royal Institution,
London W1X 4BS, UK

Received 12 August 1974

Time resolved fluorescence studies of four organic molecules in solution show that, contrary to an earlier report, the relaxed fluorescent S_1 states are formed within 10 ps of excitation even when excitation is to higher states. Fluorescence lifetimes for fluorescein and its halogen substituted derivatives vary over a factor of 40 owing to enhanced intersystem crossing. Measured radiative lifetimes agree with those calculated from integrated absorption intensities and are relatively constant.

1. Introduction

Alfano and Shapiro [1] have recently reported a sub-nanosecond 'grow in' time for the spontaneous fluorescence from a dye (erythrosin) dissolved in water or methanol. They attribute this to vibrational relaxation in the excited singlet state, prior to fluorescence, and quote a lifetime of 30–40 ps for the process. Their data do not, however, eliminate the possibility of other processes such as solvent orientation relaxation [2] which could cause time dependent spectral shifts. Also, their quoted vibrational relaxation rates are smaller than those measured for other dye molecules in solution [3] or calculated from gas-phase data [4,5]. We report further picosecond studies of

the spontaneous fluorescence of erythrosin and other molecules in solution in none of which do we observe induction periods of fluorescence longer than the time resolution of the apparatus (10 ps).

2. Experimental

The experimental arrangement is shown in fig. 1. A mode-locked Nd^{3+} glass laser-oscillator produces, at each firing, a train of 50–80 pulses of 1060 nm radiation, each of 5–10 ps duration, as measured by two-photon fluorescence. These pulses are frequency-doubled with 10–15 % efficiency in a temperature-tuned caesium dihydrogen arsenate crystal (CDA).

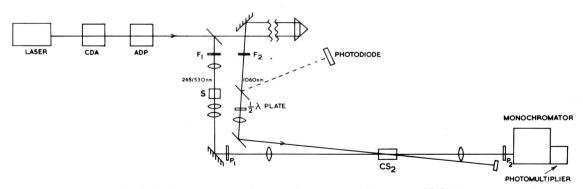

Fig. 1. The laser arrangement for measuring picosecond fluorescence lifetimes.

Volume 29, number 3   CHEMICAL PHYSICS LETTERS   1 December 1974

The pulses produced at 530 nm are again frequency-doubled with about 3–4 % efficiency in a temperature-tuned crystal of ammonium dihydrogen phosphate (ADP). A beam splitter (90 % reflection at 265 nm, 4 % at 530 nm) separates the laser beam into two portions. One portion is filtered (F_1) to excite the sample (S) with either 265 nm or 530 nm radiation.

The fluorescence from the sample is collected and focussed through a shutter, comprising crossed polarisers (P_1, P_2) and a 1 cm cell of CS_2, as originally described by Duguay and Hansen [6]. Any fluorescence passing through this shutter is incident upon an f-4 monochromator, set at 10 nm bandpass, and is detected by a 1P28 photomultiplier. The shutter is opened by the second portion of the beam which is sent through a variable optical delay, filtered (F_2) and focussed slightly into the CS_2 cell, thus allowing the temporal variation of the fluorescence to be monitored over a series of firings of the laser. The operation of this shutter has been described most fully by Duguay and Mattick [7].

The energy of the switching pulses is monitored via a photodiode as a check on the performance of the laser. Extinction ratios of up to 50 000 : 1 can be obtained using Polaroid HN22 polarisers which give signal/noise ratios of 5:1 for the fluorescence signals measured here. The extinction ratios for these polarisers are wavelength sensitive and the figures quoted are the optimum obtainable. The zero time is determined by switching the excitation pulse itself with solvent alone in the sample cell. The plot thus obtained represents a convolution of the excitation and switching pulses and has a half width of 10 ps, indicating that both the 530 nm and 265 nm pulses have pulse widths similar to that of the 1060 nm pulses. Optimum signal/noise ratios at peak transmission in these experiments are 50:1.

The delay position corresponding to the peak transmission point is taken as the zero time and subsequent times are calculated from this zero point. It should be noted that the zero time for the 530 nm pulse occurs 25 ps before that of the 265 nm pulse which indicates

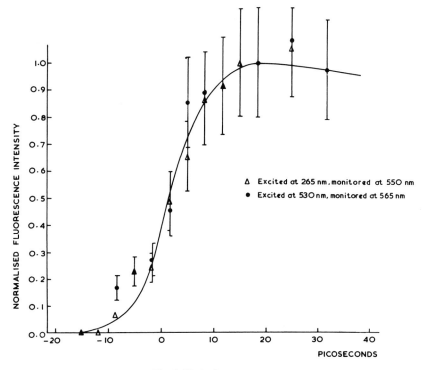

Fig. 2. Eosin fluorescence.

Volume 29, number 3   CHEMICAL PHYSICS LETTERS   1 December 1974

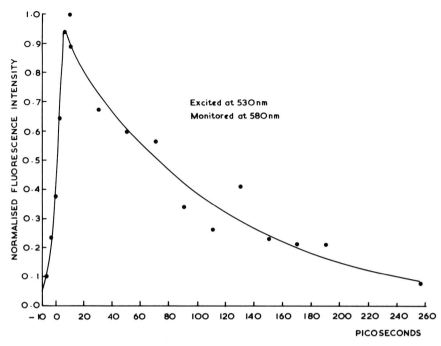

Fig. 3. Erythrosin fluorescence.

a delay between the two pulses. This delay has yet to be explained but must be taken into account when comparing data at the two wavelengths. Fluorescence lifetimes from 0.01–1 ns can be measured using this technique.

Eosin and erythrosin were purified chromatographically and dissolved in distilled water at pH7. Perylene was obtained from Sigma Ltd. and dissolved in spectroscopic grade cyclohexane without further purification.

3. Results

We have measured the picosecond time-resolved spontaneous fluorescence intensities of eosin, erythrosin, and fluorescein in aqueous solution, and perylene in cyclohexane. Fig. 2 shows the results obtained from eosin up to 50 ps after excitation. Each data point represents the mean of three separate firings of the laser and the error bars represent the total observed scatter from that mean. Results obtained from excitation at 530 nm and 265 nm are shown together with a computed curve of the convolution of the instrument response function and a fluorescence lifetime of 0.9 ns.

Most of the scatter can be attributed to variations in the average peak power in the switching beam which alters the transmission of the shutter. Similar results were obtained across the fluorescence spectrum from 520 nm to 610 nm, and by monitoring with the first polariser either parallel or at right angles to the polarisation of the exciting beam.

Fig. 3 shows the results obtained from erythrosin where, again, instrument-limited rise-times were obtained. A fluorescence lifetime of 110±20 ps was obtained by a least squares fit of the experimental data to an exponential decay; this fit is shown as the smooth curve. Similar instrument-limited rise-times were obtained from all molecules studied and the measured lifetimes are summarised in table 1. Care was taken to minimize the influence of rotational depolarisation [8] effects which are observable in certain circumstances in these experiments.

4. Discussion

Our results indicate that the normal S_1 fluorescence is observed within 10 ps of excitation into either the

Volume 29, number 3   CHEMICAL PHYSICS LETTERS   1 December 1974

Table 1
Fluorescence lifetimes of fluorescein derivatives

Compound	τ_{exp} (ns)	Φ_{exp} [b)	τ_{rad} (ns)	τ_{rad} (calc) [a) (ns)
fluorescein (Fl)	3.6 [a)	0.92	3.9	4.11
eosin (Fl Br$_4$)	0.90	0.19	4.7	4.26
erythrosin (Fl I$_4$)	0.11	0.02	5.5	4.36

[a) See ref. [11];   [b) see refs. [10,12].

S_1 state or a higher electronic state. We can therefore conclude that internal conversion rates are $> 10^{11}$ s^{-1}, in agreement with earlier predictions [9]. Vibrationally excited S_1 states are formed following internal conversion from higher electronic states; fluorescence from these levels is possible although we would expect some broadening and a wavelength shift compared with fluorescence from thermally equilibrated levels. We observed no change in the fluorescence spectrum of either eosin in aqueous solution or perylene in cyclohexane, as a function of time after excitation into high electronic levels (S_2 in the case of perylene). The nature or magnitude of the expected spectral shifts are not known but the absence of any spectral changes indicates that vibrational relaxation is also complete in $< 10^{-11}$ s. Similar results were obtained from all the molecules studied, including erythrosin. We therefore suggest that the 'grow in' time observed by Alfano and Shapiro was a fortuitous combination of experimental fluctuations.

The measured fluorescence lifetimes of erythrosin and eosin, along with published quantum yield data [10], allow an estimate of the radiative lifetime of these two fluorescein derivatives. The experimental values are in reasonable agreement with values calculated from the integrated absorption intensities [11] (table 1). The variation in lifetime with halogen substitution is consistent with heavy atom enhanced intersystem crossing and with published triplet quantum yields [12].

Acknowledgement

We wish to thank the Science Research Council for the award of Studentships to E.S.R. and C.J.T. We should also like to thank Dr. Colin Lewis for many helpful discussions, and Mr. D. Madill for technical assistance.

References

[1] R.R. Alfano and S.L. Shapiro, Opt. Commun. 6 (1972) 98.
[2] W.R. Ware, S.K. Lee, G.J. Brant and P.P. Chow, J. Chem. Phys. 54 (1971) 4729.
[3] M.R. Topp, P.M. Rentzepis and R.P. Jones, Chem. Phys. Letters 9 (1971) 1.
[4] S.J. Formosinho, G. Porter and M.A. West, Proc. Roy. Soc. A333 (1973) 289.
[5] G.S. Beddard, O.L.J. Gijzeman, G.R. Fleming and G. Porter, Proc. Roy. Soc. (1974), to be published.
[6] M.A. Duguay and J.W. Hansen, Appl. Phys. Letters 15 (1969) 192.
[7] M.A. Duguay and A.T. Mattick, Appl. Opt. 10 (1971) 2162.
[8] A.C. Albrecht, Progr. Reaction Kinetics 5 (1970) 301.
[9] M. Kasha, Discussions Faraday Soc. 9 (1950) 14.
[10] G. Weber and F. Teale, Trans. Faraday Soc. 53 (1957) 646.
[11] P.G. Seybold, M. Gouterman and J. Collis, Photochem. Photobiol. 9 (1969) 229.
[12] P.G. Bowers and G. Porter, Proc. Roy. Soc. A299 (1967) 348.

Reprinted from *J. de Chimie Physique*, 1517–1522 (1964)

THE HIGH RESOLUTION ABSORPTION SPECTROSCOPY
OF AROMATIC FREE RADICALS,

by G. PORTER and B. WARD.

[*Department of Chemistry, The University, Sheffield* 10.]

SUMMARY

The flash photolysis of aromatic molecules at high resolution, using long path lengths, has enabled us to observe a wealth of new spectra. In particular the weak, long wavelength transition of benzyl and substituted benzyl radicals has been observed. Their assignment to benzyl, rather than, for example, to an isomeric tropyl radical is confirmed and the structure of the spectrum is discussed. Brief reference is made to other transient spectra which we have recently recorded, including the spectrum of the phenyl radical.

Introduction.

The first recorded spectrum of an aromatic free radical was that of triphenyl methyl by GOMBERG [1] in 1900. In this classical communication GOMBERG stated « I wish to reserve the field for myself ». After the respectable period of more than half a century, we may perhaps be permitted to look further into the problem.

Apart from the work of GOMBERG and a few similar spectra of rather stable free radicals in solution, polyatomic free radical spectroscopy is a very recent affair made possible mainly by two experimental techniques: flash photolysis [2] and matrix stabilization [3]. Matrix stabilization was introduced by LEWIS and LIPKIN in the early forties but was applied to radicals which were fairly stable even in ordinary solutions. It was shown to be generally applicable to most radicals, with a suitable choice of conditions, by NORMAN and PORTER [4] in 1954. The method of flash photolysis was introduced in 1949 and has since been used for the study of all kinds of transient species in all types of medium.

Both of these methods have been useful in the study of aromatic free radicals and about 140 of these have now been characterised. The radicals are of interest for several reasons; they have fairly sharp spectra in easily accessible regions of the spectrum; they are reasonably stable and above all the possibility of studying a large range of related compounds, which is unique to aromatic molecules, makes identification and interpretation relatively easy.

The first class of aromatic free radicals to be extensively studied by spectroscopy comprised those radicals of which benzyl is the prototype. This class includes most of the stable free aromatic radicals such as triphenyl methyl and Würster's salts and those radicals isoelectronic with benzyl, namely phenoxyl and anilino. PORTER and WRIGHT [5] showed that flash photolysis of toluene, phenol and aniline vapours and many of their derivatives produced transients having characteristic absorption spectra around 3 000 Å, which were assigned to the benzyl, phenoxyl and anilino radicals respectively. These radicals are formed by the loss of an atom or group of atoms β to the ring following predissociation of the parent molecule in the excited state. Further studies of these radicals in rigid matrices [6] and in solution have been carried out. CHILTON and PORTER [7] showed that ionizing radiations were a powerful alternative method of preparation. Flash photolysis has recently been employed to study the spectra of both acidic and basic forms of phenoxyl and anilino radicals in solution [8], and to determine acidity constants and rates of reaction. Similar studies on such related radicals as the semiquinones [9] and ketyls [10] have also been carried out.

High Resolution Studies.

Experimental.

Photochemical studies of aromatic molecules in the gas phase have been hindered by the low vapour pressures of many compounds, by the low quantum yields of the photolytic processes and by the short lifetimes and weak absorptions of many of the resulting

G. PORTER AND B. WARD

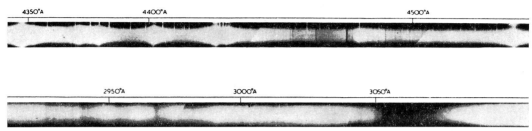

Fig. 1. — The visible and UV absorption spectra of the benzyl radical.

transients. These difficulties have been overcome to a certain extent by using a more sensitive flash photolysis apparatus whose main points are :

1) a reaction vessel containing a multiple reflection mirror system ([11]) giving path lengths of up to 10 metres;

2) fairly high energies (4 000 J. in 50 μsec.) and short delays, and

3) the spectra are photographed on a 21 ft. grating spectrograph so that weak banded absorptions are well resolved.

Many new spectra have been observed at high resolution and much new information obtained about the primary photochemical processes in aromatic molecules. Here we describe some studies of benzyl and related radicals, and mention briefly some new spectra which have been obtained recently.

The Benzyl Radical ΦCH₂.

Benzyl has been the subject of numerous experimental and theoretical investigations and the 3 000 Å transition is well characterized. A photograph at high resolution (fig. 1) shows that the spectrum is predissociated yet contains some potentially analysable structure. In overall appearance it resembles that of a monosubstituted benzene.

Some halogenated toluenes were photolysed and the short wavelength transitions of the resulting benzyl radicals observed. These spectra resembled those of the unsubstituted radical but were shifted to the red (see table I).

Intensity measurements on the spectra showed that the half lives of the radicals were about 100 μsec. It may be noted that the three different species obtained with o, m and p substituents confirm the benzyl structure and eliminate the possibility that isomerisation to tropyl has occurred. Ortho and

meta bromo toluenes behaved in an anomalous manner since we failed to detect any benzyl radicals on photolysis of these compounds. Benzyl bromide is also known to give a negative result ([5]) and it is possible that the increased spin-orbit interaction arising from the presence of the heavy bromine atom results in rapid deactivation of the excited molecules by intersystem crossing.

TABLE I

Molecule	Radical	λ_{max} (Å)
$C_6H_5CH_3$	$C_6H_5CH_2$	3 053
$C_6H_5CHCl_2$	C_6H_5CHCl	3 098
$C_6H_5CCl_3$	$C_6H_5CCl_2$	3 106
$o\text{-}ClC_6H_4CH_3$	$o\text{-}ClC_6H_4CH_2$	3 153
$m\text{-}ClC_6H_4CH_3$	$m\text{-}ClC_6H_4CH_2$	3 130
$p\text{-}ClC_6H_4CH_3$	$p\text{-}ClC_6H_4CH_2$	3 073,5

An emission system from toluene and many of its derivatives in electric discharges was first reported by SCHULER ([12]), who tentatively assigned it to the benzyl radical, although it occurred at 4 500 Å. This apparent discrepancy was resolved when it was realized, as a result of calculations by LONGUET-HIGGINS and POPLE ([13]), that the expected spectrum of benzyl should consist of an allowed transition at short wavelengths and a forbidden one at longer wavelengths. Since emission occurs from the lowest electronically excited state, the high energy transition at 3 000 Å will not appear in emission but the emission system at 4 500 Å should appear in absorption. Eventually PORTER and STRACHEN ([14]), using the matrix stabilization technique with long pathlengths, observed the long wavelength transition in absorption.

Confirmation of the assignment of the emission and absorption systems around 4 500 Å to benzyl

THE HIGH RESOLUTION ABSORPTION SPECTROSCOPY OF AROMATIC FREE RADICALS   1519

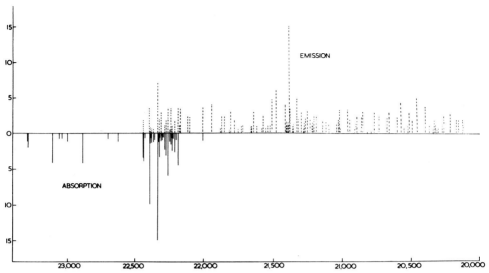

Fig. 2. — Comparison of the visible absorption and emission spectra of the benzyl radical.

has now been obtained following a study of toluene and some of its derivatives. The flash photolysis with an 8 m. pathlength of 14 mm. toluene vapour resulted in the appearance of a very sharp banded spectrum, with a maximum at 4 477 Å, which is shown in figure 1. Identical but weaker spectra were given by benzyl chloride, ethyl benzene and benzyl methyl ether. The wavelengths and wavenumbers of the bands are given in table III, which includes visual estimates of the band intensities. Part of the spectrum was found to coincide with bands in the emission system previously attributed to benzyl (see fig. 2) and therefore it is evident that the absorption and emission systems are due to the same species. Definite evidence that this species is benzyl was given by a study of some halogenated toluenes. Photolysis of these compounds gave spectra similar to that from toluene and the results are summarised in table II.

TABLE II

Parent	Radical	λ_{max} (Å)
$C_6H_5CH_3$	$C_6H_5CH_2$	4 477,2
$C_6H_5CH_2Cl$	$C_6H_5CH_2$	4 477,2
$C_6H_5CHCl_2$	C_6H_5CHCl	4 493,4
$C_6H_5CCl_3$	$C_6H_5CCl_2$	4 517,1
$p\text{-}ClC_6H_4CH_3$	$p\text{-}ClC_6H_4CH_3$	4 600,0

The only photochemical process which is consistent with these results is a β bond fission of the parent molecule :

$$C_6H_5CH_2X \quad \rightarrow \quad C_6H_5CH_2 + X$$

to give the benzyl radicals shown in the table. Since benzal chloride and p-chloro toluene give different radical spectra, the possibility of an isomerization to tropyl is again ruled out - in agreement with the conclusions drawn from our observations on the short wavelength transition.

Careful examination of the plates showed that in molecules containing more than one kind of β bond, the fission was confined solely to one particular bond. For example, the spectrum from benzyl chloride contained no bands due to α-chloro benzyl.

No other transient spectra were given by toluene between 2 800 and 6 800 Å, although several attempts were made to detect a forbidden doublet-quartet transition of benzyl, which is of some theoretical interest.

Attempted Analysis of the Spectrum.

The benzyl radical has been successfully treated as a 7 π electron system having C_{2v} symmetry by several workers. Following Mulliken's recommendations, molecular orbital theory predicts that the spectrum under discussion is due to a $^2A_2 - {}^2B_1$ transition, whereas valence bond theory suggests that it is $^2B_1 - {}^2B_1$. Since the vibronic interactions will be different in the two cases, a correct analysis of the spectrum would, in principle, allow the correct symmetry to be assigned to the lowest excited state.

The transition is symmetry allowed on group theoretical grounds but it is predicted to have a very small transition moment. One might expect, then, that a weak 0-0 band should appear in a region which is common to both emission and absorption. Analyses of the emission spectrum have been attempted by several workers who chose the following 0-0 bands:

SCHULER ([12]) . . . .   21,805 cm^{-1}
LEACH ([15]) . . . .   22,002 cm^{-1}
WALKER ([16, 17]) . . . .   22,324 cm^{-1}

If a mirror image symmetry is postulated for the absorption and emission spectra, then the 0-0 band should lie in the region 22,200-21,700 cm^{-1}.

TABLE III

Absorption Spectrum of the Benzyl Radical.

Wavelength (Å) in Air	Vacuum Wavenumber (cm^{-1})	Intensity	Difference (cm^{-1})
4 543,83	22 001,7	1	0
4 506,90	22 182,0	4	180
4 503,90	22 196,7	0	195
4 502,16	22 205,3	3	204
4 499,54	22 218,2	0	217
4 498,15	22 225,1	2	223
4 497,28	22 229,4	1	228
4 496,34	22 234,1	2	232
4 494,94	22 241,0	1	239
4 492,33	22 253,9	3	252
4 492,01	22 255,5	6	254
4 489,36	22 268,6	3	267
4 487,29	22 278,9	2	277
4 485,45	22 288,0	0	286
4 483,80	22 296,3	1	295
4 482,51	22 302,7	0	301
4 482,31	22 303,7	2	302
4 480,62	22 315,1	0	313
4 479,35	22 318,4	3	317
4 478,52	22 322,5	1	321
4 477,25	22 328,9	10	327
4 472,67	22 351,8	0	350
4 470,09	22 364,6	1	363
4 468,45	22 372,8	1	371
4 467,22	22 379,0	0	377
4 466,30	22 383,6	1	382
4 465,06	22 389,9	6	388
4 457,98	22 425,4	0	424
4 455,96	22 435,6	4	434
4 455,04	22 440,2	4	439
4 419,56	22 620,4	1	619
4 405,13	22 694,4	0	693
4 368,73	22 883,5	4	882
4 346,90	22 998,5	1	997
4 339,78	23 036,2	0	1 035
4 335,95	23 056,5	0	1 055
4 326,47	23 107,0	4	1 105
4 292,99	23 287,2	1	1 286
4 291,97	23 292,8	2	1 291

Intensities estimated visually on a 1-10 scale.

(see fig. 2). This would seem to rule out 22,324 cm^{-1}, which is the strongest band in absorption. LEACH and his co-workers have been able to analyse almost completely the emission spectrum with 22,002 as their 0-0 band, whereas SCHULER was able to make only a partial analysis with 21,805 cm^{-1}. The fact that 22,002 cm^{-1} is the band of longest wavelength in the absorption spectrum supports the assignment of LEACH. Frequency differences are given in table III.

The spectrum is composed of very sharp line like bands and broader diffuse ones, but the physical significance of this is not clear. In the region of the strongest band, the structure is quite complicated and here some of the features may be rotational heads and not separate vibrational structure. The extent of the spectrum is rather small, which probably means that there is little change of shape on excitation. This was confirmed by a simple M.O. calculation based on bond orders which predicted only a slight lengthening of the ring along the C_2 axis.

One might expect the spectrum to analyse in terms of frequencies similar to those active in toluene, but this is not the case. There is very little regularity in the spectrum and no obvious progressions are apparent. A frequency of 180 cm^{-1} is the most common feature but it is not known whether this is a fundamental or not, since experiments with and without a diluent gas proved inconclusive. One of the most puzzling aspects is the absence of the very strong 327 cm^{-1} frequency in combination. The strong band at 882 cm^{-1} is probably the totally symmetric ring breathing frequency. In common with many other substituted benzenes, benzyl exhibits a difference frequency of 61 cm^{-1} which is usually attributed to a 1-1 transition of low frequency.

A consistent and complete analysis of the spectrum has not proved possible. One explanation of the complexity might be that the symmetries of the radical are different in the two states, in which case many more vibrations will become allowed. The spectrum may also be complicated by the presence of olefinic type vibrations of the CH_2 group which is partially conjugated to the ring.

Phenoxyl and Anilino Radicals.

The transitions of these radicals resemble those of benzyl since they have the same number of π electrons. Perturbation theory has some success in predicting the relative intensities of and energy differences between the long and short wavelength transitions of the members of the isoelectronic series.

The Phenoxyl Radical $\Phi 0$.

The long wavelength transition of phenoxyl, previously observed by LAND and PORTER [18] in the gas phase, was photographed at high resolution in the hope that some fine structure might be visible, similar to that in benzyl. This was not realized and the spectrum consisted of two broad diffuse bands having maxima at 3 800 and 3 950 Å. It seems that this type of spectrum is characteristic of phenoxyl type radicals. Similar spectra were given by methyl substituted phenols, showing that phenoxyl and not benzyl type radicals were formed. Band maxima of some phenoxyl radicals in the gas phase are given in the following table IV.

TABLE IV

Parent	Radical	λ_{max} (Å)
C_6H_5OH	C_6H_5O	3 800, 3 950, 6 000
$C_6H_5OCH_3$	C_6H_5O	3 800, 3 950, 6 000
$C_6H_5OC_2H_5$	C_6H_5O	3 800, 3 950
o-MeC_6H_4OH	o-MeC_6H_4O	3 900
p-MeC_6H_4OH	p-MeC_6H_4O	4 000
2,6 $diMeC_6H_3OH$	2,6 $diMeC_6H_3O$	3 890

A transient diffuse spectrum in the region 5 300-6 100 Å was detected after the flash photolysis of phenol and anisole. It is not due to the phenyl or cyclopentadienyl radicals (see later) and it is best assigned to an n-π^* transition of phenoxyl. A similar transition of 2,4,6 tritertiary butyl phenoxyl is known in solution [18].

The Anilino Radical ΦNH.

Experiments with aniline showed that the long wavelength transition of anilino in the region of 4 000 Å was even more diffuse than that of phenoxyl and devoid of any spectroscopic interest. The lack of structure in the spectra of phenoxyl and anilino compared to benzyl is rather surprising. Anilines do, however, give sharp spectra which cannot be assigned to anilino and these will be discussed briefly later.

The Phenyl Radical Φ.

Many unsuccessful attempts have been made to detect the phenyl radical spectroscopically, although it is well established as an intermediate in organic chemistry. We have recently obtained a common spectrum from benzene and halogenated benzenes in the region 4 300-5 300 Å, which we attribute to the phenyl radical. The spectrum has been almost completely analysed in terms of frequencies which are very similar to the main frequencies in the benzene spectrum. The transition occurs from a $\pi^6 n$ ground state configuration to a $\pi^5 n^2$ exited state configuration.

The Cyclopentadienyl Radical C_5H_5.

A common spectrum between 2 900 and 3 500 Å is given on photolysis by phenol, phenolic ethers, aniline, nitrobenzene and cyclopentadiene. Similar spectra are obtained from halogen substituted derivatives of phenol, aniline and nitrobenzene. On the basis of the substituent effects and particularly that the same spectrum is obtained irrespective of the position of the halogen substituted, these spectra are assigned to the cyclopentadienyl radical and its halogen substituted derivatives. The spectra are very extensive and well resolved, and the two strongest bands are identical with those reported by THRUSH [19] upon photolysis of cyclopentadiene and ferrocene.

Details of the spectra of phenyl, cyclopentadienyl and its derivatives, and of other transient species derived from aromatic molecules, will be published shortly.

BIBLIOGRAPHY

(1) M. GOMBERG. — J.A.C.S., 1900, 12, 757.
(2) G. PORTER. — Proc. Roy. Soc., 1950, A 200, 284.
(3) G. N. LEWIS and D. LIPKIN. — J.A.C.S., 1942, 64, 2801.
(4) I. NORMAN and G. PORTER. — Proc. Roy. Soc., 1955, A 230, 399.
(5) G. PORTER and F. J. WRIGHT. — Trans. Faraday Soc., 1955, 51, 395.
(6) G. PORTER and E. E. STRACHEN. — Trans. Faraday Soc., 1958, 54, 431.
(7) H. T. J. CHILTON and G. PORTER. — J. phys. Chem., 1959, 63, 904.
(8) E. J. LAND and G. PORTER. — Trans. Faraday Soc., 1963, 59, 2016, 2027.
(9) N. K. BRIDGE and G. PORTER. — Proc. Roy. Soc., 1958, A 244, 259, 276.
(10) A. BECKETT and G. PORTER. — Trans. Faraday Soc., 1963, 59, 2 038, 2051.
(11) J. U. WHITE. — J. Opt. Soc. Amer., 1942, 32, 285.
(12) H. SCHULER. — Z. Naturf., 1952, 79, 421.
    H. SCHULER and J. KUSJAKOW. — Spectrochim. Acta, 1961, 17, 356.
(13) H. C. LONGUET-HIGGINS and J. A. POPLE. — Proc. Phys. Soc., 1955, 68, 591.
(14) G. PORTER and E. E. STRACHEN. — Spectrochim. Acta, 1958, 12, 299.
(15) S. LEACH, L. GRAJCAR and J. ROBERT. — To be published.
(16) T. F. BINDLEY and S. WALKER. — Trans. Faraday Soc., 1962, 58, 217.
(17) A. T. WATTS and S. WALKER. — J. chem. Soc., 1962, p. 4323.
(18) E. J. LAND, G. PORTER and E. E. STRACHEN. — Trans. Faraday Soc., 1963, 57, 1885.
(19) B. A. THRUSH. — Nature, 1955, 178, 155.

1522   G. PORTER AND B. WARD

DISCUSSION

G. Giacometti. — Pr. PORTER's evidence of the existence of benzyne in his flash photolized systems is very beautiful indeed and his tentative suggestion of a para-structure as

gains support the organic work by VAN TAMELEN (JACS, 1963) who prepared Dewar's benzene (bi-cyclo-butene). A not planar structure for para-benzyne seems to be at least much reasonable.

G. Porter. — The « Dewar » form of benzene is, of course, a quite different substance from para-phenylene or para-benzyne which contains only four hydrogen atoms. The stability of Dewar benzene is not surprising since the molecule is surely non-planar and is not a resonance structure of benzene, indeed the name « Dewar benzene » is not correct for the stable molecule which has recently been prepared. Para-benzyne, if it exists, is likely to be planar or very nearly so.

C. A. McDowell. — Since VAN TAMELEN and his colleagues have shown by N.M.R. and other physical evidence that they have undoubtly synthesised the Dewar form of benzene this surely removed any conceptional difficulty about accepting a structure such as is proposed for p-phenylene, i.e. :

In fact on theoretical grounds one could argue that there may be better reasons for the existence of p-phenylene than the Dewar form of the benzene molecule.

S. Leach. — Phenyl must have C_{2v} symmetry, formally at least, and so I would expect a 0,0 band to appear, perhaps very weakly, this band being strictly forbidden for the 2 600 Å benzene transition. Furthermore, the $e_g{}^+$ vibrations in benzene should have their degenerescence lifted in C_{2v} symmetry and should each give rise to two vibrations of a_1 and b_1 symmetries for C_{2v}. The doublet structure which you mentioned as observing in phenyl might be due to this splitting of the degeneracy. However one should expect this behaviour only for bands derived from $e_g{}^+$ type vibrations but not totally symmetrical progressions.

G. Porter. — I think it likely that the doublet structure which is present in the bands of the phenyl spectrum is rotational fine structure since it is similar in all the bands including that at the origin.

S. Leach. — I would like to ask Pr. PORTER whether he observed any phenyl transitions to shorter wavelengths, in particular, in the benzene region itself?

In photolysis experiments on benzene in an argon matrix at 20 °K, carried out at Berkeley in 1957, I observed the formation of two unstable species. One gives a series of broad bands in the 3 300-2 700 Å region, which I believe may be due to the hexatrienyl diradical. The other gives a series of well resolved bands in the 2 600-2 200 Å region, distinct from the benzene bands in the same region, which might possibly be due to the phenyl radical. Both spectra disappear on warming to about 70 °K.

G. Porter. — We observed no higher energy transitions of phenyl but the 2 600-2 200 Å region is difficult to study. We have never observed the 5 000 Å system of phenyl in our low temperature stabilisation work.

Faraday Soc. Discussion **14**, 23–34 (1953) – Reproduced by permission of the Royal Society of Chemistry

STUDIES OF FREE RADICAL REACTIVITY BY THE METHODS OF FLASH PHOTOLYSIS

THE PHOTOCHEMICAL REACTION BETWEEN CHLORINE AND OXYGEN

BY GEORGE PORTER AND FRANKLIN J. WRIGHT
Department of Physical Chemistry, University of Cambridge

Received 15th April, 1952

The reaction of chlorine atoms with oxygen has been studied by the flash photolysis and flash spectroscopy method of Porter.[1] The chloric oxide radical ClO is readily formed at 293° K and its concentration has been followed throughout the reaction by quantitative measurements of its absorption.

The initial reactions are $Cl + O_2 = ClOO$ and $Cl + ClOO = 2ClO$. The rate constant of removal of chlorine atoms by oxygen, to form both Cl_2 and ClO, is 46 times the rate of removal in nitrogen to form Cl_2.

The decomposition of ClO to Cl_2 and O_2 occurs relatively slowly, is bimolecular with respect to ClO, and the rate is independent of Cl_2, O_2 and total gas pressure. The rate constant of this reaction is

$$7 \cdot 2 \times 10^4 \; \epsilon_s \exp (0 \pm 650/RT) \text{ l. mole}^{-1} \text{ sec}^{-1},$$

where ϵ_s is the molar extinction coefficient of ClO at 2577 Å. The value of ϵ_s is greater than 310 and probably less than 3000. The mechanism of this reaction is discussed in terms of the intermediate Cl_2O_2.

The extensive literature on the reactivity of chlorine atoms, produced by photochemical dissociation of the molecule, gives little evidence of a direct reaction with oxygen. The well-known inhibiting effect of oxygen on photochemical chlorinations is usually attributed to the removal by oxygen of hydrogen atoms, COCl radicals, hydrocarbon radicals, etc., rather than of chlorine atoms, but the rate expressions do not allow an unequivocal choice of mechanism. In a mixture of chlorine and oxygen alone there is no apparent photochemical change and no transient reactions have been detected.

It was therefore somewhat unexpected when a new absorption spectrum, which was clearly that of a diatomic molecule, was discovered at high intensity during the investigation of chlorine + oxygen mixtures by the flash technique.[1] The spectrum was attributed to the ClO radical, and a vibrational analysis, as well as a determination of the dissociation energy of the molecule have already been given.[2] The methods of flash photolysis and flash spectroscopy were originally developed to make possible the production of labile molecules in a concentration high enough for absorption spectroscopy to be applied to the study of their kinetics, and we here describe such an investigation of the formation and reactions of the ClO radical. Direct information is also obtained in this way about the role of the oxygen in the photochemical reactions of chlorine.

EXPERIMENTAL

The absorption spectra of 20 free radicals, about half of them new, or previously unknown in absorption, have now been obtained in this laboratory. A detailed kinetic investigation by the flash technique involves more difficulties, however, than simply recording the spectrum of a free radical. Many hundred spectra must be taken under different conditions, flash intensities and time intervals must be accurately reproduced and the temperature rise, which is a result of the adiabatic nature of the reaction, must be eliminated.

The number of intensity measurements necessary for a complete kinetic investigation by flash photolysis is best reduced by the use of photocell recording. The relative merits of this method, which is being used for other problems, have already been discussed,[2] but in the present case, owing to the fact that the mechanism of the reaction was quite unknown, and also that a rather complex band system was being investigated, it was thought advisable to examine the whole spectral region throughout the investigation so as to be able to detect the presence of other chlorine oxides, changes in the intensity distribution of the ClO spectrum and intensity changes in the spectrum of the chlorine molecule.   The use of photographic recording introduces more scatter into intensity measurements but the relationships eventually obtained are particularly significant as they are the result of many experiments carried out in a quite arbitrary order.

A full description of the method, and of the apparatus used for this investigation has been given elsewhere.[1]   The only modification necessary was to enclose the reaction vessel and the photolysis flash tube in a furnace, so that the temperature dependence of the reaction rates could be determined.   The furnace, which was of the same dimensions as the original reflector, completely enclosed the reaction vessel, and was coated internally with magnesium oxide.   To facilitate removal it consisted of two semi-cylindrical portions wound separately, and the temperature was measured by three thermocouples at the centre and at either end.   An investigation of the properties of the photolysis flash lamp at temperatures up to 350° C showed that the firing characteristics are a function of the concentration of the inert gas filling rather than of the pressure and that the output is not noticeably affected by the rise in temperature.   This, fortunately, makes it possible to use the same gas filling throughout and therefore to avoid intensity variations which occur from one filling to another.[3]

PREPARATION OF GASES.—Chlorine was taken from a cylinder and redistilled several times *in vacuo*.   Oxygen was prepared electrolytically, dried over $CaCl_2$, freed from traces of hydrogen by passing over 30 cm of platinized asbestos at 350° C and finally dried over P_2O_5.   Nitrogen in which not more than 0·05 % oxygen could be tolerated, was prepared by heating sodium azide which gives a pure product.[4]

INTENSITY AND CONCENTRATION MEASUREMENTS.—The spectrograph was a Littrow (Hilger E.1) instrument ; 25 spectra were recorded on each plate, and the intensities were measured on a non-recording microphotometer.   The output of the spectroflash was constant within the accuracy of the microphotometric measurements.   Each plate was calibrated separately and methyl ethyl ketone, which has a continuous spectrum in the same region as the ClO radical, was used for this purpose.   The ketone was made up to about 1 atm. pressure with carbon dioxide and a range of ketone partial pressures chosen to give the same densities as the particular ClO concentrations being studied.   The intensity of the ClO spectrum could then be expressed in terms of the pressure of ketone having the same extinction at a given wavelength, the plate sensitivity, path length and incident intensities being constant.   The extinction curve of methyl ethyl ketone vapour was measured on a Unicam spectrophotometer.

Chlorine itself has a significant absorption in the region of the ClO spectrum, and this must be allowed for in the intensity determination.   It will be shown later that the amount of ClO formed is so small that the chlorine concentration does not change significantly during the experiment, and therefore this correction is a constant one.   If Beer's law is obeyed we have, for a given wavelength,

$$\ln(I_0/I_1) = \epsilon_1 c_1 l + \epsilon_2 c_2 l,$$

where I_0 is the incident intensity, I_1 is the intensity transmitted by the mixture of Cl_2 and ClO, l is the path length, ϵ_1 and c_1 are the extinction coefficient and concentration of ClO, and ϵ_2 and c_2 the same quantities for chlorine.   Now if ketone, at concentration c_3 also transmits the intensity I_1, the incident intensity being unchanged, and the extinction coefficient being ϵ_3,

$$\epsilon_1 c_1 + \epsilon_2 c_2 = \epsilon_3 c_3.$$

If I_2 is the intensity through chlorine alone, and also through ketone at concentration c_4

$$\ln (I_0/I_2) = \epsilon_2 c_2 l = \epsilon_3 c_4 l$$

and
$$\epsilon_3(c_3 - c_4) = \epsilon_1 c_1. \qquad (1)$$

The values of $c_3 - c_4$, which are determined experimentally from the calibration spectra, are therefore proportional to the ClO concentration if Beer's law applies.

ϵ_3 being known, $\epsilon_1 c_1$ can be determined but the values of ϵ_1 and c_1 cannot at present be obtained separately.   It is therefore necessary to express the concentrations of ClO as some function of its extinction coefficient and all concentration measurements are expressed as $\epsilon_s c$, where c is the true pressure of ClO in mm/Hg and ϵ_s is the molar decadic extinction coefficient at 2577 Å. This wavelength is in the continuous region and should therefore be unchanged if a different resolving power is used.

VALIDITY OF BEER'S LAW.—Beers' law has been assumed to apply both to the ketone and the ClO in the derivation of eqn. (1). No deviations were expected for methyl ethyl ketone, which was chosen for the lack of fine structure in its spectrum,[5] and the law was verified by measuring the extinction spectrophotometrically over the range of total pressures used in the calibrations. The following precautions were taken to assure the applicability of the law to the ClO spectrum. Firstly, possible deviations due to pressure effects were eliminated by keeping the total pressure many hundred times greater than the pressure of ClO. Except in one case, which is discussed separately, no comparisons are made at different temperatures. Finally, errors which might be caused by incomplete resolution of fine structure were first reduced by taking all measurements in the continuum or near to the heads of the predissociated bands where most of the absorption is due to lines of a width greater than the resolving power of the spectrograph. Each position was then checked by plotting $c_3 - c_4$ for a number of ClO spectra taken at different concentrations against the same quantity for a different wavelength. In one case only, the 13, 0 band at 2751 Å, there was a small deviation at high concentrations. In all other cases a linear plot was obtained and as this included measurements in the continuum, where the spectrum is completely continuous, the validity of Beer's law is confirmed for the measurements to follow, in which the 13, 0 band is not used.

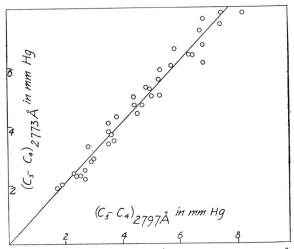

FIG. 1.—Plot of $c_3 - c_4$ at 2773 Å against $c_3 - c_4$ at 2797 Å.

From the gradient of these plots, one of which is shown in fig. 1, the relative extinction coefficients at the different wavelengths were obtained so that all measurements could be expressed in terms of ϵ_s. For the three bands which were used in addition to the continuum they are as follows :

band	12,0	11,0	10,0	continuum
wavelength λ (Å)	2773	2797	2824	2577
$\epsilon_\lambda/\epsilon_s$	1·50	1·44	1·44	1·00

REDUCTION OF THE ADIABATIC TEMPERATURE EFFECT.—Unless precautions are taken, the heat liberated by the reactions of the atoms and radicals formed may produce a temperature rise of over 1000°, the rate of thermal diffusion to the walls being less than that of the chemical reactions. Preliminary experiments showed that the ClO radical was still present in pressures of oxygen or nitrogen as high as 1 atm, and in most of the experiments to follow the total pressure used was 600 mm the pressure of chlorine being about 5 mm. Under these conditions the temperature rise, estimated by calculation and by use of the concentration effect, is 1 or 2° C and it will be shown later that a temperature rise of 100° C has no effect on the measured constants.

The photolysis flash was operated from 168 μF at 4000 V unless otherwise stated and the absorption tube was 1 m in length and 2 cm diam.

26   FLASH PHOTOLYSIS

RESULTS

A typical series of spectra, taken at increasing times after the flash, is shown in plate 1. It was first nessesary to ascertain that the whole of the spectrum being measured was that of ClO and that there was no overlying spectrum of another molecule. It was found that the rates determined in each part of the spectrum were the same, as would be expected in view of the relationships between the extinction coefficients mentioned previously. A search was then made for spectra in other regions, in particular those of the ClO_2 and Cl_2O molecules which have high extinction coefficients in the region of 2800 Å. None was found. Finally there was no difference between the results of experiments on newly mixed gases and those which had been flashed up to 50 times.

Under all conditions, the rate of reaction of ClO was very much less than the rate of its formation during the flash. Further, at the shortest times, immediately after the flash, the observed rate of reaction of ClO at a given concentration was the same as at longer times at the same concentration, showing that the reactions by which ClO is formed occur in a time which is short compared with the duration of the flash. It is therefore possible to divide the investigation into two parts: the dark reactions of ClO occurring after the flash and the photochemical reactions by which ClO is formed during illumination. The times involved in a typical experiment are illustrated by fig. 2 which gives the

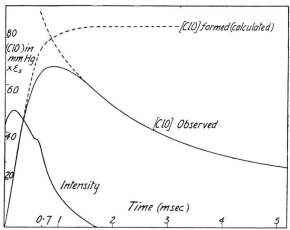

FIG. 2.—Light intensity and ClO concentration against time.

intensity against time and the ClO concentration against time curves. The latter, at times greater than 1·4 msec gives the rate law for the dark reaction and will be studied first, the reactions during the first msec being discussed later.

THE KINETICS OF ClO DISAPPEARANCE.—The concentration of ClO as a function of time was determined for a given mixture by recording a number of spectra, in an arbitrary order, at different times after the flash. Measurements were made at several wavelengths and concentrations determined, via the ketone calibrations, in the manner described.

DEPENDENCE ON ClO CONCENTRATION.—Fig. 3 shows the relationship between the reciprocal of the ClO concentration and the time, using a mixture of 10 mm Cl_2 and 600 mm O_2. A linear plot is obtained and, in the course of this work, over 60 such graphs were drawn from measurements on a wide variety of mixtures, all of which showed a direct proportionality within the experimental error. We conclude that, under all conditions described below,

$$- \, d(ClO)/dt = k(ClO)^2.$$

If k' is defined by the equation

$$- \, d((ClO)\epsilon_s)dt = k'((ClO)\epsilon_s)^2,$$

the value of k' obtained from the gradient of fig. 3 is 4·9 mm^{-1} sec^{-1} ϵ_s units^{-1}, and the absolute rate constant is given by

$$k = k'\epsilon_s \text{ mm}^{-1} \text{ sec}^{-1} = 1.70 \times 10^4 \, k'\epsilon_s \text{l. mole}^{-1} \text{ sec}^{-1}.$$

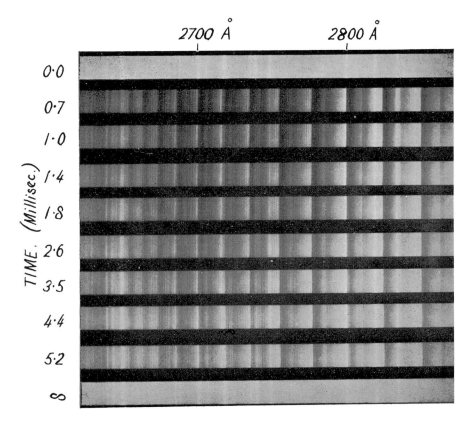

Bimolecular Disappearance of Chloric Oxide (ClO)

PLATE 1.

[To face page 26.

DEPENDENCE ON OXYGEN PRESSURE.—The total pressure of oxygen plus nitrogen was kept constant and the relative pressure of oxygen varied. Owing to the similar specific heats and collision diameters of the two gases, oxygen pressure is the only significant variable. The rate constants for different mixtures (pressures in mm Hg), determined as before, are given in table 1. About 10 spectra were used in each determination of these and all subsequent rate constants. The values of k' are constant to within 10 % whilst the oxygen pressure is changed by a factor of 60.

TABLE 1.—RATE CONSTANT k' AT DIFFERENT OXYGEN PRESSURES (P IN MM Hg)

$P(Cl_2)$	$P(O_2)$	$P(N_2)$	$P(O_2 + N_2)$	k' (mm^{-1} sec^{-1} ε_s units^{-1})				mean k'
10	600	0	600	4·0	3·7	3·8	4·9	4·1
10	100	500	600	4·3	4·1	4·3	—	4·2
10	10	590	600	3·8	4·7	3·5	4·1	4·0

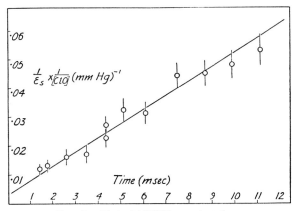

FIG. 3.—Plot of $1/(ClO)\varepsilon_s$ against time.

DEPENDENCE ON CHLORINE PRESSURE.—The results given in table 2 show that the rate constants are invariant over a tenfold range of chlorine pressure.

TABLE 2.—RATE CONSTANT k' AT DIFFERENT CHLORINE PRESSURES (P IN MM Hg)

$P(Cl_2)$	$P(O_2)$	k' (mm^{-1} sec^{-1} ε_s units^{-1})				mean k'
20	600	4·3	3·7	3·8	—	3·9
10	600	4·0	3·7	3·8	4·9	4·1
5	600	4·5	4·4	4·0	4·0	4·2
2	600	3·9	3·7	4·3	—	4·0

DEPENDENCE ON TOTAL PRESSURE.—At low total pressures the reaction is no longer even approximately isothermal but reference to the next section shows that the effect of temperature rise on the rate can be ignored over a wide range and it was therefore possible to vary the total pressure by a factor of 10 without introducing a temperature change large enough to affect the rate. The results given in table 3 show that over this range the value of k' is independent of total pressure.

TABLE 3.—RATE CONSTANT k' AT DIFFERENT TOTAL PRESSURES (P IN MM Hg)

$P(Cl_2)$	$P(O_2)$	$P(N_2)$	P(total)	k' (mm^{-1} sec^{-1} ε_s units^{-1})			mean k'
10	100	500	610	3·8	4·3	5·2	4·4
10	100	300	410	4·9	4·9	6·1	5·3
5	50	100	155	4·0	4·8	4·4	4·4
10	100	0	100	3·1	3·2	3·4	3·2
5	50	0	55	4·8	4·5	5·2	4·8

122   The Life and Scientific Legacy of George Porter

DEPENDENCE ON TEMPERATURE.—The rate constants measured at higher temperatures are given in table 4, the pressures being referred to 293° K.

TABLE 4.—RATE CONSTANT k' AT DIFFERENT TEMPERATURES (P IN MM Hg)

$T°$ K	$P(Cl_2)$	$P(O_2)$	k' (mm^{-1} sec^{-1} ε_s units^{-1})			mean k'
293		mean of values in tables 1, 2 and 3				4·2
433	5	300	4·0	4·2	4·0	4·1
473	10	400	6·2	6·3	7·3	6·6

The values of k' are constant in the temperature range 293° K to 433° K but increase slightly at 473° K. Measurements at higher temperatures showed a further increase but they are of doubtful significance owing to the changed intensity distribution (see results on temperature dependence of ClO formation). If there were a true rate increase at higher temperatures it might imply a change in mechanism at about 450° K but it is also possible that the ratio of ϵ_s to $\int_0^\infty \epsilon_\lambda d\lambda$ is no longer constant. In view of this uncertainty we cannot say anything about the higher temperature mechanism at present.

In the range between 293° K and 433° K there is no significant change in the intensity distribution and the measured rates are constant. Assuming a possible variation in k' of $\pm$ 40 %, in order to allow for any slight changes in the intensity distribution, we conclude that if the temperature coefficient of ClO removal is expressed as exp ($- E/RT$) then the activation energy in the range of temperature between 293° K and 433° K is 0 $\pm$ 650 cal. The introduction of a $T^{\frac{1}{2}}$ dependence of the non-exponential term would reduce the value of E by 350 cal.

The results of this section are summarized as follows:
the rate law of ClO removal in the dark reaction is

$$- d(ClO)/dt = k(ClO)^2(O_2)^0(Cl_2)^0(N_2)^0 ;$$

the mean value of $k' = k/\epsilon_s$ at 293° K is 4·2 mm^{-1} sec^{-1} ϵ_s units^{-1};
the rate constant k, in the temperature range 293° K to 433° K, is

$$7·2 \times 10^4 \; \epsilon_s \exp (0 \pm 650/RT)\text{l. mole}^{-1} \text{ sec}^{-1}.$$

ABSOLUTE VALUE OF ϵ_s AND k.—In view of the very intense absorption by the ClO radical it seemed probable that a proportional decrease in the intensity of the chlorine molecule spectrum would be observed and that this would give the absolute concentrations of ClO and values for ϵ_s and k. Experiments designed for this purpose have shown no such decrease and therefore, at present, it is only possible to give a lower limiting value for ϵ_s.

In order to increase the sensitivity of the method 10 similar spectra were taken of chlorine in the presence of a high concentration of ClO, each accompanied by a blank of the chlorine alone, the voltage being 6000 V in this case. In mixtures of 5 mm Cl_2 with 700 mm O_2 a slight decrease in Cl_2 absorption was observed which control experiments, using nitrogen in place of oxygen, showed to be due to the temperature-concentration effect alone. A mixture of 1 mm Cl_2 with 700 mm O_2 showed no decrease whilst calibration spectra with known chlorine pressures showed that 0·05 mm pressure decrease would have been detected. The average value of $(ClO)\epsilon_s$ was 31 mm which gives the minimum value for ϵ_s of 310.

It is helpful to have some idea of the maximum possible value of ϵ_s for the purpose of discussion and we have two reasons for supposing that it is not greatly different from the above minimum. Firstly an examination of the known extinction coefficients of similar molecules and radicals having partly continuous spectra make it improbable that the value of ϵ_s, which is by no means the maximum extinction coefficient, would exceed 3000. Secondly, if chlorine atoms recombine at a rate equal to or less than the three-body collision rate in the presence of nitrogen the later investigations on the relative rate constants of this reaction and ClO formation, coupled with the fact that the formation of ClO is not observable after 1·5 msec lead to a lower limit for the ClO concentration which is again within a factor of 10 of the above minimum. We hope to be able to determine ϵ_s experimentally by other methods; it seems very probable, however, that the true values of the constants are not more than a factor of 10 greater than the following experimental minima:

$$\epsilon_s > 310, \quad k_{293} > 2·2 \times 10^7 \text{ l. mole}^{-1} \text{ sec}^{-1}.$$

GEORGE PORTER AND FRANKLIN J. WRIGHT   29

THE INITIAL PHOTOCHEMICAL REACTION

The times involved are too short for the investigation of the ClO concentration changes during the flash, though a simple modification of the apparatus would make this possible. A different approach was used here, however, the reactions during the flash being studied by measurements of the concentration $((ClO)\epsilon_s)$ formed during a given time, the time chosen being 0·7 msec. During this time a fraction of the ClO formed will have reacted by the mechanism already studied and, using the known rate constant, it is possible to correct for this and to calculate, for any measured concentration at time 0·7 msec, the total ClO formed during this period. The calculation was performed as follows:

(i) The intensity against time curve was taken from an oscillograph of the flash and a graph of total light output against time was constructed, from the area beneath, in arbitrary units.

(ii) As the concentration changes are small the light absorbed, and therefore, owing to the high rate of ClO formation, the ClO present at a given time will follow the same curve if the reactions by which the ClO is removed are ignored. Taking these reactions into account it is possible to calculate the true ClO concentration at any time by using the known rate constant k'. A step method was found most convenient and the calculation was simplified by the almost linear nature of the intensity curve up to 0·7 msec. A corrected curve is shown in fig. 2.

(iii) For the series of observed concentrations at time 0·7 msec the total ClO formed is determined in this way.

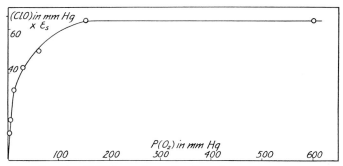

Fig. 4.—(ClO) formed as a function of oxygen pressure.

It was found that even at the highest concentrations the correction was only about 15 % and no significant errors are likely to be involved in these corrections.

RELATION BETWEEN ClO FORMED AND OXYGEN PRESSURE.—The concentration of ClO formed during the first 0·7 msec was determined for a range of oxygen pressures, the total pressure being kept constant at 600 mm by the addition of nitrogen. The chlorine pressure was 5 mm and the flash intensity and other conditions were kept as constant as possible. The concentrations, after correction, are plotted as a function of oxygen pressure in fig. 4. It will be seen that only when the oxygen concentration is reduced to about 1/50 of the nitrogen concentration is the ClO formed reduced by $\frac{1}{2}$. This suggests that the rate constants of the reactions in oxygen and nitrogen respectively are also in this ratio and a more detailed derivation of this relation will be given later.

EFFECT OF TEMPERATURE ON ClO FORMATION.—The intensity of ClO at 0·7 msec was investigated as a function of temperature between 293° K and 593° K using a mixture of 5 mm of chlorine and 300 mm of oxygen measured at 293° K. At the highest temperatures the rotational structure became very extended and the bands were less distinct, but intensity measurements in the continuum and even in the diffuse bands at our standard wavelengths were independent of temperature to within $\pm$ 20 %. The only true measure of relative concentrations under these conditions is $\int_0^\infty \epsilon_\lambda d\lambda$ which could not be evaluated exactly, but intensity comparisons at a number of wavelengths showed that the value of the integral was almost constant and could hardly have varied by a factor of more than 2 over the whole temperature range. If the temperature coefficient of ClO formation is expressed as $\exp(-E/RT)$ the value of E is therefore 0 ± 0.8 kcal.

DISCUSSION

The reactions studied occur exclusively in the homogeneous gas phase. This is shown by the nature of the observations, which are made on the gas near to the centre of the vessel, the independence of the rates on the total pressure and by simple calculation, which gives times of diffusion to the wall greatly in excess of the time intervals observed. The problem is further simplified by our accurate knowledge of the bond energies of ClO (63 kcal)[2] Cl_2 (57 kcal) and O_2 (117 kcal)[6] and the heats of formation of the stable chlorine oxides.[7] Consider, for example, the reactions

$$Cl + O_2 = ClO + O \qquad -54 \text{ kcal} \qquad (1)$$

$$Cl + ClO = Cl_2 + O \qquad -6 \text{ kcal} \qquad (2)$$

$$2ClO = ClO_2 + Cl \qquad -7 \text{ kcal} \qquad (3)$$

$$2ClO = Cl_2O + O \qquad -32 \text{ kcal} \qquad (4)$$

Reaction (1) is very endothermic and cannot be a significant mechanism of ClO formation which has nearly zero temperature dependence. Reaction (3) and (4) are similarly eliminated as mechanisms of ClO removal for which the activation energy is zero within a few 100 cal. Reaction (2) must be considered a little more carefully for, if the oxygen atom formed reacts with ClO, a chain mechanism is possible and with suitable activation energies for the termination reaction, this might lead to a temperature independent rate constant for ClO removal. Apart from the fact that no such termination step can be found which gives the observed rate law, and also that the rate constant is too high to involve reaction (2) as a propagation step, the use of chlorine atoms in this mechanism is entirely incompatible with the fact that ClO is formed very rapidly by their reaction with oxygen but the rate of removal of ClO is completely independent of oxygen pressure. The probable fate of any oxygen atoms formed by this or any other reaction would be a reaction with chlorine by the reverse of (2).

We are now in a position to consider the few remaining possibilities which, in conjunction with the observed rate expressions, give the mechanism in some detail.

THE MECHANISM OF ClO FORMATION.—Reaction (1) having been excluded, the only possible reaction by which ClO can be formed is

$$2Cl + O_2 = 2ClO, \qquad (5)$$

although this may proceed in two stages as follows:

$$Cl + O_2 = ClOO \qquad (6)$$

$$Cl + ClOO = 2ClO. \qquad (7)$$

The radical ClOO is not to be confused with the stable radical O—Cl—O.

Using this mechanism it should now be possible to interpret the results of the experiments on $O_2 + N_2$ mixtures given in fig. 4. As nitrogen can play no chemical role its effect must be ascribed to the reaction

$$2Cl + N_2 = Cl_2 + N_2^* \qquad (8)$$

and we must also consider the reaction

$$2Cl + O_2 = Cl_2 + O_2^* \qquad (9)$$

which again may proceed via the intermediate ClOO.

The reactions by which ClO is removed are automatically eliminated by the method of calculating the total ClO formed. Ignoring the intermediate ClOO for the moment the only reactions by which Cl atoms are removed will now be (5), (8) and (9). Then

$$d(2ClO)/dt = k_5(Cl)^2(O_2) \text{ and } d(Cl_2)/dt = k_9(Cl)^2(O_2) + k_8(Cl)^2(N_2).$$

When the removal of Cl atoms is complete we have

$$(2ClO) = \int_{t=0}^{t=\infty} k_5 (Cl)^2 (O_2) dt$$

and

$$(Cl_2)_f = \int_{t=0}^{t=\infty} k_9 (Cl)^2 (O_2) dt + \int_{t=0}^{t=\infty} k_8 (Cl)^2 (N_2) dt,$$

where $(Cl_2)_f$ is the total (Cl_2) formed from Cl atoms. For any given mixture (O_2) and (N_2) are constants and $\int_{t=0}^{t=\infty} (Cl)^2 dt$ has the same value in both expressions. Therefore

$$\frac{(2ClO)}{(Cl_2)_f} = \frac{k_5(O_2)}{k_9(O_2) + k_8(N_2)}. \tag{i}$$

If $(2ClO)$ and $(Cl_2)_f$ refer to the concentrations at a given time, say 0·7 msec, then the same expression is obtained by eliminating the integral

$$\int_{t=0}^{t=0\cdot7} (Cl)^2 dt.$$

For all $(O_2)/(N_2)$ ratios, twice the number of Cl atoms formed during the flash is a constant and is equal to

$$(2ClO) + (Cl_2)_f = K \text{ (say)}. \tag{ii}$$

This again applies to any time interval if the reactions removing Cl atoms are fast.

When $(N_2) = 0$, and putting $(2ClO) = (2ClO)_{max}$,

then

$$\frac{(2ClO)_{max}}{K - (2ClO)_{max}} = \frac{k_5}{k_9}. \tag{iii}$$

Eliminating K and $(Cl_2)_f$ from (i), (ii) and (iii) we obtain

$$\frac{(ClO)_{max}}{(ClO)} = 1 + \frac{k_8(N_2)}{k_5 + k_9(O_2)}.$$

A plot of $(ClO)_{max}/(ClO)$ against $(N_2)/(O_2)$, where $(ClO)_{max}$ is the ClO formed in the absence of nitrogen, should therefore give a straight line of slope $k_8/(k_5 + k_9)$. The results already shown in fig. 4 are plotted in this way in fig. 5 and a linear plot is obtained confirming the mechanism suggested. Further, the gradient of this line gives the above ratio of rate constants and is found to be 1/46. The sum of k_5 and k_9 is the rate constant of removal of Cl atoms by oxygen and k_8 by nitrogen, i.e.

$$- d(Cl)/dt = 2(k_5 + k_9)(Cl)^2(O_2) \text{ in oxygen,}$$

and

$$- d(Cl)/dt = 2(k_8)(Cl)^2(N_2) \text{ in nitrogen.}$$

Therefore the reaction of chlorine atoms with oxygen to form Cl_2 and ClO occurs at a rate 46 times that of the reaction in nitrogen to form Cl_2.

This result has several interesting consequences. Firstly, there is every reason to suppose that the recombination of chlorine atoms in a gas such as nitrogen occurs at every termolecular collision. The collision diameters of oxygen and nitrogen being very similar, for example in the recombination of other halogen atoms,[8] it follows that reactions (5) and (9) as written are insufficient to account for the rate and these reactions must involve the formation of a relatively stable complex, which can only be ClOO. The inclusion of this intermediate leads to the same rate expressions if reaction (6) is reversible and the equilibrium maintained. It is not possible to say whether this implies an activated complex ClOO* with a lifetime 46 times greater than the ClN_2^* complex or equilibrium involving stabilization by a third body and bimolecular dissociation. In either case, if the collision efficiency of Cl atom recombination in nitrogen at 1 atm. pressure is taken as 1/900, the equilibrium constant $(Cl)(O_2)/(ClOO)$ is less than 20 atm even if the reaction of Cl with ClOO occurs at every collision.

The very small temperature coefficient of ClO formation is fully in accordance with our mechanism and confirms that reactions (1), (2), (3) and (4) are unimportant in the photochemical part of the reaction. If reaction (9) occurs to a significant extent it follows that $E_9 = E_5 \pm 0.8$ kcal.

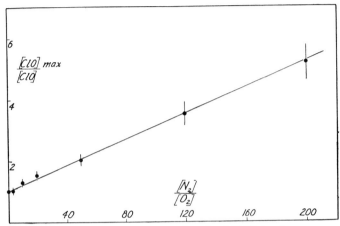

FIG. 5.—$(ClO)_{max}/(ClO)$ against the $(N_2)/(O_2)$ ratio.

THE MECHANISM OF ClO REMOVAL.—The rate law shows conclusively that the subsequent reactions by which ClO is removed are unaffected by the pressure of any gas other than ClO. Reasons have already been given for excluding chlorine atoms, and reactions (3) and (4), and the only remaining possibility is

$$2ClO = Cl_2 + O_2. \tag{10}$$

There is a strong objection to this reaction in its simple form in that all known double decompositions of this type are associated with high energies of activation, a useful empirical rule being that the activation energy is about 1/4 of the sum of the energies of the bonds broken.[9] There is one important difference in the radical reaction, however, and that is the possibility of dimerization. We must therefore consider the reactions

$$2ClO = Cl_2O_2 \tag{11}$$

$$Cl_2O_2 = 2ClO \tag{12}$$

$$Cl_2O_2 = Cl_2 + O_2 \tag{13}$$

which lead to the rate expression

$$-\,\mathrm{d}(ClO)/\mathrm{d}t = \frac{k_{11}k_{13}}{k_{12} + k_{13}}\,(ClO)^2,$$

if we assume a stationary concentration of Cl_2O_2. Two limiting cases may be considered.

(i) $k_{13} \gg k_{12}$.

Reaction (11) now becomes rate determining and a zero activation energy is quite probable. We then have the difficulty of explaining the very low rate constant, for it has been shown that the rate is independent of total pressure and a third-body collision cannot therefore be necessary. We should have to explain the low rate entirely by means of a steric factor which could hardly be greater than 10^{-3}, whereas such rate constants as are known for simple radical recombinations have steric factors very near to unity.[10] This explanation therefore seems improbable.

In this connection it should be mentioned that the possibility of formation of a " stable " Cl_2O_2 molecule is not entirely eliminated by the experimental evidence, in which case the observed rate would be that of formation of Cl_2O_2 which might then decompose more slowly to Cl_2 and O_2. This would only be true if Cl_2O_2 had a very low extinction over the whole region investigated and had a lifetime of the order of seconds. In addition exactly the same objections apply as were given in the last paragraph.

(ii) $k_{12} \gg k_{13}$.

The effective rate constant is now $k_{11}k_{13}/k_{12}$ and the observed activation energy will be $E_{11} - E_{12} + E_{13}$. The difference $E_{12} - E_{11}$ is equal to the heat of formation of Cl_2O_2 from two ClO radicals and must be very nearly equal to E_{13}, the activation energy of dissociation to Cl_2 and O_2 in order to explain the observed temperature independence. This is not unlikely if both energies are small and the low rate is then explicable.

We therefore think that the reaction is best explained in terms of an equilibrium between ClO and Cl_2O_2, the latter decomposing to Cl_2 and O_2. It is then possible to see why ClO might behave differently from both NO and OH which are in many ways similar radicals. The former has a very unstable dimer which dissociates to 2NO much more readily than to N_2 and O_2 whilst the latter forms a stable dimer which dissociates only very slowly. ClO is an intermediate, semistable radical because its dimer is also of intermediate stability. We are investigating this reaction further by the transition state method but it seems doubtful whether, in the absence of data about the Cl_2O_2 molecule, anything much more quantitative can be said.

THE CHLORINE + OXYGEN REACTION IN PHOTOSENSITIZED OXIDATIONS.—Our results show unequivocally that Cl atoms react rapidly with oxygen, and these reactions must occur to some extent in the presence of other gases. This is in accordance with the complete rate expressions of Thon [11] and of Bodenstein and Schenk [12] for the $H_2 + O_2 + Cl_2$ system. We have carried out some preliminary experiments on ClO formation in the presence of hydrogen and have found that in excess oxygen (10 mm Cl_2, 10 mm H_2 and 300 mm O_2) the ClO formed and the rate of its reactions are not greatly changed and that little HCl is formed. In excess hydrogen (10 mm Cl_2, 25 mm O_2 and 350 mm H_2) no ClO was observed and most of the Cl_2 reacted permanently.

Excess carbon monoxide, on the other hand, resulted in a very high ClO concentration and ClO is almost certainly the long-lived intermediate which resulted in the slow approach to the steady state observed by Bodenstein, Brenschede and Schumacher.[13] In this, and similar oxidations photosensitized by chlorine, the ClOO radical may play an important part, in addition to ClO. For example, the following mechanism for the sensitized oxidation of CO_2 would now appear probable :

$$Cl + O_2 \rightleftharpoons ClOO$$

$$ClOO + CO \rightarrow CO_2 + ClO$$

$$ClO + CO \rightarrow CO_2 + Cl.$$

A more detailed discussion of these reactions will be given elsewhere along with an investigation of the reactions of ClO with other gases.

One of us (F. J. W.) is indebted to the Anglo-Iranian Co. for financial support during the tenure of which this work was carried out.

34   FLASH PHOTOLYSIS

[1] Porter, *Proc. Roy. Soc. A*, 1950, **200**, 284.
[2] Porter, *Faraday Soc. Discussions*, 1950, **9**, 60.
[3] Christie and Porter, *Proc. Roy. Soc. A*, 1952, **212**, 398.
[4] Justi, *Ann. Physik*, 1931, **10**, 983.
[5] Duncan, Ells and Noyes, *J. Amer. Chem. Soc.*, 1936, **58**, 1454.
[6] Gaydon, *Dissociation Energies* (Chapman and Hall, 1947).
[7] Goodeve and Marsh, *J. Chem. Soc.*, 1939, 1332.
[8] Rabinowitch and Wood, *Trans. Faraday Soc.*, 1936, **32**, 907.
[9] Glasstone, Laidler and Eyring, *The Theory of Rate Processes* (1941).
[10] see, for example, Dodd, *Trans. Faraday Soc.*, 1951, **47**, 56.
[11] Thon, *Z. physik. Chem.*, 1926, **124**, 327.
[12] Bodenstein and Schenk, *Z. physik. Chem. B*, 1933, **20**, 420.
[13] Bodenstein, Brenschede and Schumacher, *Z. physik. Chem. B*, 1937, **35**, 382.

PRINTED IN GREAT BRITAIN AT
THE UNIVERSITY PRESS
ABERDEEN

Reprinted from *Proc. Roy. Soc. A* **264**, 1–18 (1961) with permission from The Royal Society

Energy transfer from the triplet state

By G. Porter, F.R.S. and F. Wilkinson

Department of Chemistry, University of Sheffield

(*Received* 10 *April* 1961)

[Plates 1 and 2]

Energy transfer from the triplet level of a donor molecule resulting in quenching of the donor and elevation of the acceptor molecule from its singlet ground state to a triplet state has been observed between a number of donor-acceptor pairs in fluid solvents. In most cases the mechanism of the transfer has been unequivocally established by observation of the triplet state absorption spectra of both species. Energy transfer from excited singlet donors to the triplet state of the acceptor is not observed.

When the energy of the acceptor triplet is considerably lower than that of the donor, transfer is diffusion-controlled but there is no evidence for long-range resonance transfer of the kind found in the analogous singlet energy transfer processes. As the triplet energies become comparable the transfer probability is reduced and no quenching is observed by molecules with triplet levels higher than that of the donor. Transfer of triplet energy between pairs of aromatic hydrocarbons has been illustrated and it has been established that complex formation between donor and acceptor cannot be important under the conditions of these experiments.

The longer a molecule persists in an excited state the greater its chance of meeting another which can accept its excitation energy. In most molecules the ground state is a singlet and the lowest excited state is a triplet which has a lifetime several orders of magnitude greater than that of excited singlet states. It is therefore hardly surprising that the triplet state is frequently postulated as an intermediate in energy transfer processes, although the role played by the triplet state has been established for only a very few cases.

Quenching of the triplet state occurs both by chemical reaction and by physical processes. Chemical reactions include electron transfer, hydrogen atom abstraction from the solvent, dissociation and addition of oxygen. In these reactions spin conservation is always possible. Paramagnetic molecules, e.g. O_2, NO, aromatic triplets, transitional and inner transitional metal ions have been characterized as physical quenchers. The quenching efficiency is not proportional to magnetic susceptibility and occurs by the process

$$A^*(\text{triplet}) + Q(\text{multiplet}) \rightarrow A(\text{singlet}) + Q(\text{multiplet})$$

with overall spin conservation and without any change in the multiplicity of Q (Porter & Wright 1959).

When Q is a singlet state, spin conservation is only possible if there is a change in the multiplicity of Q. The energy transfer process represented as

$$A^*(\text{triplet}) + Q(\text{singlet}) \rightarrow A(\text{singlet}) + Q^*(\text{triplet})$$

is spin-allowed. Such a process has been postulated by Terenin & Ermolaev (1956) to explain the sensitized phosphorescence of naphthalene and its derivatives by light absorbed by benzophenone and similar compounds in rigid media at $-195\,°C$.

[1]

2   G. Porter and F. Wilkinson

The concentrations required to bring about transfer in rigid media were very high and the work has been questioned because of the possibility of complex formation (McGlynn, Boggus & Elder 1960).

This paper deals with flash photolysis studies of the same process in solution at room temperature and provides unequivocal proof of the occurrence of this process in a variety of systems. During the course of this work Bäckström & Sandros (1958, 1960) by quite different methods have also provided clear evidence of this process.

Experimental

The flash photographic apparatus was the same as that used by Porter & Jackson (1961). Spectra were recorded on Ilford Selochrome and H.P. 3 plates in a medium quartz Hilger spectrograph. Each plate was calibrated internally by means of a 7 step neutral filter of rhodium on quartz. Band positions of transient absorption peaks were subject to errors of ± 5 Å at 2500 Å which increased to ± 20 Å at 6000 Å.

The flash photoelectric apparatus was originally very similar to that described by Porter & Wright (1958). During the course of this work several modifications were made in order to allow measurements at shorter times after irradiation. Kinetic analysis of the oscillographic traces, which gave a direct measure of any transient during its formation and decay, was only made when the scattered light pulse from the photolysis lamps had decayed so as to give no measurable deflexion under working conditions, i.e. when the intensity of the scattered light at the particular wavelength being investigated was less than 1 % of the monitoring light intensity. The modifications, which included the use of a zirconium arc lamp as the monitoring light source, addition of nitrogen to the lamp fillings and improved circuit design to reduce the flash duration (Gover & Porter 1961) made accurate kinetic measurements possible 50 μs after the photolysis flash.

The reaction vessels were designed to fit both types of flash photolysis apparatus. They were constructed from quartz, Pyrex or soda-glass and were essentially similar to those described by Porter & Windsor (1954). Some of them had outer jackets so that 0·6 cm of filter solution could be placed around the reactants. The reaction vessels were thoroughly cleaned before use and grease traps were introduced below the tap to ensure that no grease could contaminate the solution.

Solutes

Napththalene (B.D.H. micro-analytical reagent) was used without further purification. 1-Bromonaphthalene was fractionated. 1-Iodonaphthalene was treated with sodium sulphite to remove iodine, washed, dried and fractionated. The fraction collected was kept over calcium in the dark. The anthracene was the same as that used by Porter & Windsor (1954). Phenanthrene was freed from anthracene by treatment with maleic anhydride (Kooyman & Farenhorst 1953) in purified benzene and then recrystallized from spectroscopic alcohol. Triphenylene, 1:2-benzanthracene and benzophenone were recrystallized from spectroscopic ethyl alcohol. Diacetyl was fractionally distilled in a nitrogen atmosphere under reduced pressure (100 mm), and stored in the dark. Iodine was re-sublimed and ethyl iodide was passed through a column of activated alumina and fractionated.

Energy transfer from the triplet state   3

Solvents

Hexane was spectroscopic grade and was not further purified. Analar benzene was extracted with concentrated sulphuric acid, washed, dried and fractionated. Hopkins and Williams G.P.R. ethylene glycol was filtered and used without further purification at first. Later supplies of this solvent were less pure, judged by ultra-violet transmission and were fractionally distilled. The fraction which was used still did not have as good transmission properties as the original samples and gave an increased decay rate of triplet states. This did not affect the accuracy of measurement of quenching rate constants since care was taken to ensure that solutions used to investigate any particular pair of compounds were made up from the same stock of solvent.

Thorough outgassing of all solutions was necessary because of the high efficiency of quenching of the triplet state by oxygen. A freeze, pump, melt, shake, freeze procedure was repeated until the pressure of air above the solution after freezing in liquid nitrogen was less than 0·001 mm. When ethylene glycol was used as solvent the bulk of the solution was not frozen but a small reservoir was cooled to liquid-nitrogen temperature to prevent loss of solute or solvent.

The absorption spectra of all solutions investigated were measured before and after flashing. In order to check that no change in concentration had occurred as a result of the outgassing procedure, spectra were recorded before and after outgassing.

Filter solutions

When mixtures of donor and acceptor compounds were flashed an attempt was made to filter the light from the photolysis flash so that light was only absorbed by the donor compound. Soda-glass or Pyrex reaction vessels were used in some cases. Frequently a very highly concentrated solution of the acceptor compound was used in the outer jacket of the reaction vessel. This filter interfered very little with the higher wavelength absorption of the donor compound. Checks were made to ensure that such filters were working efficiently by flashing the acceptor compound alone in a filtered reaction vessel when the amount of light absorbed was shown to be negligible.

RESULTS

Unless otherwise stated all the results refer to thoroughly outgassed solutions. No attempt was made to control thermostatically the reaction vessels but the temperature was always between 20 and 25 °C. Rate constants given are the average of at least two separate runs, four traces from each run being analyzed in most cases.

Flash photolysis of single compounds in solution

Each solute was studied first in the absence of other solutes. The triplet-triplet absorption spectra observed are given in table 1 and the first-order decay constants in table 2.

The band positions and relative intensities are in good agreement with those measured by Porter & Windsor (1958). No transient absorption was detected when

4   G. Porter and F. Wilkinson

solutions 10^{-5} to 10^{-3} M in 1-iodonaphthalene were flashed in either hexane or ethylene glycol. In the case of phenanthrene Porter & Windsor reported two additional bands at 5200 Å and 5085 Å. The present authors also observed these bands when they used the same sample of phenanthrene but they were not obtained from a sample of phenanthrene which had been extensively purified.

TABLE 1. BAND MAXIMA OF TRIPLET-TRIPLET ABSORPTION SPECTRA

compound	solvent	$\lambda_{max.}$(Å)	$\nu_{max.}$(cm^{-1})	relative intensity
naphthalene	hexane	4100	24390	1·07
		3890	25706	0·55
		3690	27100	0·15
	benzene	4150	24100	1·0
		3920	25510	0·60
		3720	26880	0·15
	ethylene glycol	4150	24100	1·0
		3920	25510	0·55
		3780	26880	0·10
1-bromonaphthalene	hexane	4190	23870	1·0
		3980	25130	0·55
		3750	26670	0·20
	ethylene glycol	4200	23800	1·00
		3980	25130	0·60
		3780	26460	0·25
1-iodonaphthalene	hexane ethylene glycol	no detectable absorption		
anthracene	hexane	4650	21510	v. weak
		4230	23640	1·00
		4000	25000	0·30
	ethylene glycol	4690	21320	v. weak
		4250	23530	1·00
		4010	24940	0·30
phenanthrene	hexane	4800	20830	1·00
		4500	22220	0·60
		4280	23360	0·25
		3990	25060	0·10
	ethylene glycol	4850	20620	1·00
		4530	21980	0·60
		4280	23420	0·25
		4000	25000	0·10
triphenylene	hexane	4280	23360	1·00
		4050	24570	0·90
	benzene	4280	23360	1·00
		4070	24570	0·90
	ethylene glycol	4280	23360	1·00
		4070	24570	0·90
1:2-benzanthracene	benzene	5380	18590	0·10
		4880	20490	1·00
		4580	21830	0·65
		4350	22990	0·50
		4020	24880	0·30
pentacene	hexane	4920	20330	0·30
		4600	21740	0·20
		3850	25970	0·10
		3000	33330	1·00

Energy transfer from the triplet state 5

No decomposition was detected of naphthalene, phenanthrene, triphenylene, or 1:2-benzanthracene. Slight decomposition occurred with 1-bromonaphthalene, 1-iodonaphthalene, anthracene and pentacene.

The values of k_1, the first-order rate constant for triplet state decay, were found to be independent of the initial concentration of the solute except for triphenylene. In this case, above about 10^{-3} M, the transient absorption spectrum changed from fairly sharp peaks to a broad general absorption upon which the normal triplet-triplet spectrum seemed to be superimposed. The singlet-singlet absorption spectrum was examined at concentrations above 10^{-3} M but no new bands nor any deviation from Beer's law were observed. Above 10^{-3} M the decay of the transient absorption measured at the triplet maxima was complex. The values given for triphenylene in tables 1 and 2 refer to solutions of concentration less than 10^{-3} M.

TABLE 2. THE FIRST-ORDER RATE CONSTANTS OF TRIPLET STATE DECAY

	$10^{-3}k_1(\mathrm{s^{-1}})$	
	hexane	ethylene glycol
naphthalene	11 ± 1	$1 \cdot 0 \pm 0 \cdot 1$
1-bromonaphthalene	12 ± 2	$1 \cdot 2 \pm 0 \cdot 2$
anthracene	$1 \cdot 1 \pm 0 \cdot 1$	$0 \cdot 26 \pm 0 \cdot 01$
phenanthrene	$10 \cdot 7 \pm 0 \cdot 8$	$1 \cdot 1 \pm 0 \cdot 1$
triphenylene	18 ± 3	$1 \cdot 0 \pm 0 \cdot 1$
1-benzanthracene	$6 \cdot 3 \pm 0 \cdot 5$ (benzene)	—
pentacene	9 ± 1	—

Further outgassing, by the procedure outlined in the experimental section, had no effect on the measured rate constants. The values given in table 2 for naphthalene and anthracene are in good agreement with those obtained by Porter & Wright (1958). However, towards the end of this research a few runs were made using different outgassing procedures. In one run, for example, after the normal outgassing, the whole solution of naphthalene in hexane was distilled from one side arm to another and back a dozen times. The reaction vessel was pumped out after every distillation. A value of $2 \cdot 8 \times 10^3\,\mathrm{s^{-1}}$ was obtained for k_1 compared with $1 \cdot 1 \times 10^4\,\mathrm{s^{-1}}$ given in table 2 for naphthalene in hexane. The value of k_1 in this case increased with the number of times the solution was flashed.

Porter & Windsor (1958) assigned the transient they observed upon flashing benzophenone in liquid paraffin to the triplet state because of the similarity between its absorption spectrum and that observed by McClure & Hanst (1955) as a transient in a rigid solvent. A detailed examination by the present authors (Porter & Wilkinson 1961) indicates that the majority of the transient absorption in fluid media with bands at ~ 3300 and 5440 Å is due to the ketyl radical $\phi_2\mathrm{\overset{\cdot}{C}{-}OH}$ which is formed from the triplet state by hydrogen atom abstraction from the solvent.

Diacetyl phosphoresces in solution and also gives a transient absorption below 3300 Å when flashed. An estimate of the transient lifetime, from the photographic apparatus and a 10^{-2} M solution of diacetyl in benzene gave $\tau = 300 \pm 100$ μs. The phosphorescence lifetime was also measured for 10^{-2} M diacetyl in benzene on a flash photoelectric apparatus available in our laboratories. This gave a value for the

6   G. Porter and F. Wilkinson

mean life of the phosphorescence decay of $600 \pm 50 \, \mu s$. The significant difference between these two values indicates that the absorption and the emission do not arise from the same species.

Scattered light from the photolysis flash was negligible and good fluorescence spectra were recorded photographically by a single flash from many of the solutions studied.

TABLE 3. QUENCHING RATE CONSTANTS

	donor	acceptor	solvent	k_Q(l. mole^{-1} s^{-1})
1	phenanthrene	naphthalene	hexane	$2 \cdot 9 \pm 0 \cdot 7 \times 10^6$
			ethylene glycol	$2 \cdot 3 \pm 0 \cdot 8 \times 10^6$
2	triphenylene	naphthalene	hexane	$1 \cdot 3 \pm 0 \cdot 8 \times 10^9$
3	phenanthrene	1-bromonaphthalene	hexane	$1 \cdot 5 \pm 0 \cdot 8 \times 10^8$
			ethylene glycol	$1 \cdot 5 \pm 0 \cdot 8 \times 10^7$
4	phenanthrene	1-iodonaphthalene	hexane	$7 \pm 2 \times 10^9$
			ethylene glycol	$2 \cdot 1 \pm 0 \cdot 2 \times 10^8$
5	naphthalene	1-iodonaphthalene	ethylene glycol	$2 \cdot 8 \pm 0 \cdot 3 \times 10^8$
6	1-bromonaphthalene	1-iodonaphthalene	ethylene glycol	$8 \pm 4 \times 10^7$
7	benzophenone	naphthalene	benzene	$1 \cdot 2 \times 10^9$
8	diacetyl	1:2-benzanthracene	benzene	$3 \pm 2 \times 10^9$
9	phenanthrene	iodine	hexane	$1 \cdot 4 \pm 0 \cdot 6 \times 10^{10}$
10	anthracene	iodine	hexane	$2 \cdot 4 \pm 0 \cdot 2 \times 10^9$

Systems showing no quenching, for which upper limits were obtained

	donor	acceptor	solvent	
1	naphthalene	phenanthrene	hexane	$\leqslant 2 \times 10^4$
			ethylene glycol	$\leqslant 1 \times 10^5$
2	naphthalene	triphenylene	hexane	$\leqslant 5 \times 10^4$
3	1-bromonaphthalene	phenanthrene	ethylene glycol	$\leqslant 5 \times 10^4$
4	naphthalene	benzophenone	benzene	$\leqslant 1 \times 10^4$
5	1:2-benzanthracene	diacetyl	benzene	$\leqslant 5 \times 10^4$
6	anthracene	phenanthrene	ethylene glycol	$\leqslant 5 \times 10^3$
7	anthracene	naphthalene	hexane	$\leqslant 4 \times 10^4$
8	anthracene	1-iodonaphthalene	ethylene glycol	$\leqslant 2 \times 10^4$
9	anthracene	ethyl iodide	hexane	$\leqslant 1 \cdot 6 \times 10^4$
10	phenanthrene	ethyl iodide	hexane	$\leqslant 3 \times 10^5$

Flash photolysis of mixed solutes

Previous workers who have studied triplet energy transfer have dealt almost exclusively with carbonyl compounds as energy donors. In order to establish the generality of this type of process it seemed important to investigate whether it occurred when both energy donor and acceptor were unsubstituted aromatic hydrocarbons. Having established that transfer of this type did occur the next step was to investigate the factors upon which the efficiency of the transfer process depends.

The results obtained for every pair of compounds studied are summarized in table 3. In the account which follows the donor is written first.

(i) *Phenanthrene and naphthalene*

This pair of compounds was studied in hexane, with a soda-glass reaction vessel as filter, and in ethylene glycol, with a concentrated solution of naphthalene in hexane in the outer jacket of a Pyrex reaction vessel as filter. At the highest

Energy transfer from the triplet state   7

concentrations of naphthalene used only a trace of triplet naphthalene was formed by direct excitation through the soda-glass filter. The concentrated solution of naphthalene in hexane was a very efficient filter and no triplet naphthalene was detected when solutions of naphthalene were flashed through this filter. Figure 5, plate 1 shows the effect of adding 6.9×10^{-3} M of naphthalene to a 2.3×10^{-3} M solution of phenanthrene in ethylene glycol. The triplet state absorption of phenanthrene has been completely suppressed and has been replaced by the absorption spectrum of triplet naphthalene.

Although the triplet-triplet absorption maxima of phenanthrene and naphthalene are quite separate, there is overlap of the weak tails of these bands which interferes with the kinetic analysis when both are present. The quenching rate constants can be estimated, though less accurately, from the series of spectra obtained from the flash photographic apparatus.

The rate of decay of the triplet state can be written as

$$-\mathrm{d}[T]/\mathrm{d}t = k_1[T] + k_2[T]^2 + k_Q[Q][T], \tag{1}$$

where k_1 and k_2 are the first- and second-order rate constants of decay of the triplet donor T, k_Q is the bimolecular quenching rate constant and $[Q]$ is the concentration of the acceptor or quencher.

At low triplet state concentrations the $k_2[T]^2$ term can be neglected and, if the concentration of the quencher which increases the rate of decay of the donor by a factor of 2 is called the half-quenching concentration then

$$k_Q[Q_{\frac{1}{2}}][T] = k_1[T] \quad \text{or} \quad k_Q = k_1/[Q_{\frac{1}{2}}]. \tag{2}$$

The half-quenching concentrations for the quenching of triplet phenanthrene by naphthalene were $4 \pm 1 \times 10^{-3}$ M and $1.0 \pm 0.25 \times 10^{-3}$ M in hexane and ethylene glycol, respectively, k_1 values for phenanthrene were $10.7 \pm 0.8 \times 10^3$ and $2.1 \pm 0.2 \times 10^3 \mathrm{s}^{-1}$, giving quenching rate constants of triplet phenanthrene by singlet naphthalene of $2.9 \pm 0.9 \times 10^6$ and $2.3 \pm 0.8 \times 10^6 \mathrm{l. mole}^{-1}\mathrm{s}^{-1}$ in hexane and ethylene glycol, respectively.

The first-order rate constant for the decay of triplet naphthalene formed by energy transfer was identical with that formed by direct excitation in the presence or absence of phenanthrene.

In these experiments the concentration of naphthalene was comparable with that of phenanthrene so a check was made on the absorption spectra of separate components and that of the mixture to see if there was any indication of complex formation. No new bands were observed and, for all the mixtures studied, the absorption spectrum of the mixture was simply the sum of the separate absorption of the two components.

Fluorescence measurements were made under identical conditions, the same reaction vessel being used with a concentrated solution of naphthalene as filter. Special care was taken to develop the separate plates under similar conditions. The microdensitometer traces of phenanthrene fluorescence were compared for four runs each containing a fixed concentration of phenanthrene in glycol and (a) zero, (b) 1.25×10^{-3} M, (c) 2.5×10^{-3} M, and (d) 5.0×10^{-3} M of naphthalene,

8   G. Porter and F. Wilkinson

The intensity of fluorescence in these four solutions was the same to $\pm 5\%$ and strong fluorescence was observed from phenanthrene even when no triplet phenanthrene absorption was detectable (see, for example, figure 5).

(ii) *Triphenylene and naphthalene*

Although difficulties were experienced in the interpretation of the flash photolysis of triphenylene alone at concentrations above 10^{-3} mole/l., $1 \cdot 01 \times 10^{-3}$ M solutions of triphenylene in hexane containing varying amounts of naphthalene were investigated in a soda-glass reaction vessel. The production of the triplet-triplet absorption of naphthalene under these conditions showed that energy transfer was taking place. The effect of adding $1 \cdot 15 \times 10^{-4}$ M of naphthalene is illustrated in figure 1.

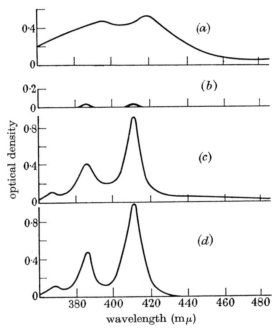

FIGURE 1. Transient absorption spectra following photolysis, in hexane solution of (a) $1 \cdot 01 \times 10^{-3}$ M triphenylene, (b) $1 \cdot 15 \times 10^{-4}$ M naphthalene, (c) $1 \cdot 01 \times 10^{-3}$ M triphenylene, and $1 \cdot 15 \times 10^{-4}$ M naphthalene. In (d) the solution was the same as in (b) but the filter was removed.

Because of the overlap of the triplet absorption spectra of triphenylene and naphthalene, photoelectric measurements were not useful and the quenching constant was estimated from the photographic records with the following results:

$$[Q_{\frac{1}{2}}] = 2 \pm 1 \times 10^{-5}\,\text{M}, \quad k_1 = 1 \cdot 8 \pm 0 \cdot 3 \times 10^4\,\text{s}^{-1}$$

and therefore $k_Q = 1 \cdot 3 \pm 0 \cdot 8 \times 10^9$ l. mole^{-1} s^{-1}.

(iii) *Phenanthrene and 1-bromonaphthalene*

This pair of compounds was studied in hexane and in ethylene glycol with a concentrated solution of naphthalene used as filter. Energy transfer was confirmed

Porter & Wilkinson   *Proc. Roy. Soc. A, volume* 264, *plate* 1

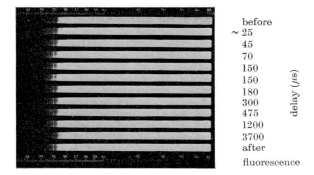

(a) 6.9×10^{-3}M naphthalene

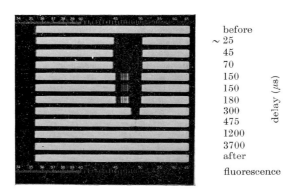

(b) 2.3×10^{-3}M phenanthrene

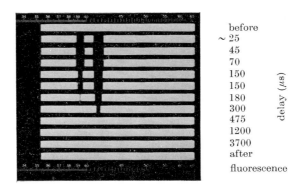

(c) 6.9×10^{-3}M naphthalene $+ 2.3 \times 10^{-3}$M phenanthrene.

FIGURE 5. Absorption spectra illustrating energy transfer from the triplet state of phenanthrene to naphthalene (ethanol solution).

Porter & Wilkinson   *Proc. Roy. Soc. A, volume* 264, *plate* 2

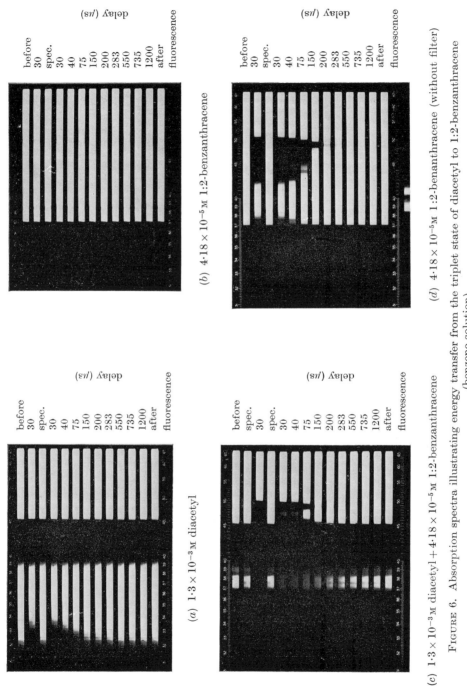

(a) $1·3 \times 10^{-3}$ M diacetyl

(b) $4·18 \times 10^{-5}$ M 1:2-benzanthracene

(c) $1·3 \times 10^{-3}$ M diacetyl $+ 4·18 \times 10^{-5}$ M 1:2-benzanthracene

(d) $4·18 \times 10^{-5}$ M 1:2-benanthracene (without filter)

FIGURE 6.   Absorption spectra illustrating energy transfer from the triplet state of diacetyl to 1:2-benzanthracene (benzene solution).

Energy transfer from the triplet state 9

by the appearance of the triplet-triplet absorption of 1-bromonaphthalene. By the same method as before the following values were derived:

Hexane

$$[Q_{\frac{1}{2}}] = 1 \cdot 0 \pm 0 \cdot 5 \times 10^{-4}\,\mathrm{M}, \quad k_1 = 10 \cdot 7 \pm 0 \cdot 8 \times 10^3\,\mathrm{s}^{-1},$$

$$k_Q = 1 \cdot 5 \pm 0 \cdot 8 \times 10^8\,\mathrm{l.\,mole^{-1}\,s^{-1}}.$$

Ethylene glycol

$$[Q_{\frac{1}{2}}] = 2 \pm 1 \times 10^{-4}\,\mathrm{M}, \quad k_1 = 2 \cdot 1 \pm 0 \cdot 2 \times 10^3\,\mathrm{s}^{-1}$$

$$k_Q = 1 \cdot 5 \pm 0 \cdot 8 \times 10^7\,\mathrm{l.\,mole^{-1}\,s^{-1}}.$$

(iv) *Phenanthrene and 1-iodonaphthalene*

In the case of phenanthrene with naphthalene and 1-bromonaphthalene the fact that energy transfer had taken place was illustrated by the appearance of the triplet-triplet absorption spectra of the acceptors. However, 1-iodonaphthalene does not show a triplet-triplet absorption and this type of illustration was not possible. On the other hand, the effect of 1-iodonaphthalene on the decay of triplet phenanthrene can be studied accurately on the flash photoelectric apparatus since there is no interference from the acceptor triplet. This was done in the solvents hexane and ethylene glycol with a concentrated solution of naphthalene in hexane as filter.

Various concentrations of 1-iodonaphthalene were added to phenanthrene solutions and the rate of decay of triplet phenanthrene was measured from first-order plots which were accurately linear at low triplet concentrations. The first-order rate constant obtained from these plots (k_A) decreased each time the solution in hexane was flashed, indicating that the 1-iodonaphthalene was being destroyed and therefore only the values from the first flashes of each run were used to determine the bimolecular rate constant k_Q for energy transfer. This effect was not observed for phenanthrene and 1-iodonaphthalene in ethylene glycol where the concentrations of 1-iodonaphthalene were approximately 100 times greater.

The rate of decay of the triplet state is given by

$$-\mathrm{d}[T]/\mathrm{d}t = \{k_1 + k_Q[Q]\}\,[T] = k_A[T];$$

therefore,
$$k_A = k_1 + k_Q[Q]. \tag{3}$$

Plots of k_A against $[Q]$ were linear and from the slopes the values $k_Q = 7 \pm 2 \times 10^9$ and $2 \cdot 1 \pm 0 \cdot 2 \times 10^8\,\mathrm{l.\,mole^{-1}\,s^{-1}}$ were obtained for the quenching rate constants in hexane and ethylene glycol, respectively. The plot of k_A against $[Q]$ for triplet phenanthrene with 1-iodonaphthalene as quencher in ethylene glycol is shown in figure 2.

Again the fluorescence yield of phenanthrene was unaffected by the presence of the 1-iodonaphthalene and the absorption spectrum of the mixture was the sum of the absorption spectra of the separate solutes.

(v) *Naphthalene and 1-iodonaphthalene*

The absorption spectra of naphthalene and 1-iodonaphthalene are so similar that it is not possible to filter the system selectively. Measurements were made with a quartz reaction vessel and ethylene glycol as solvent. The concentration of naphthalene was 10^{-3} M and the highest concentration of 1-iodonaphthalene used was 4×10^{-5} M so that most of the light was absorbed by the naphthalene. Figure 3 shows the plot of k_A against the concentration of 1-iodonaphthalene in glycol which leads to the value $k_Q = 2 \cdot 8 \pm 0 \cdot 3 \times 10^8$ l. mole^{-1}s^{-1}.

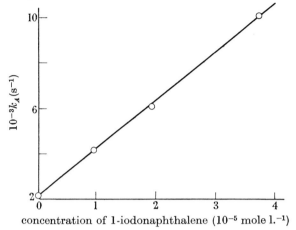

FIGURE 2. Plot of k_A, the decay constant of triplet phenanthrene, against $[Q]$, the concentration of 1-iodonaphthalene.

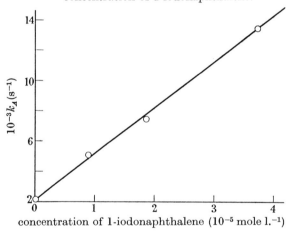

FIGURE 3. Plot of k_A, the decay constant of triplet naphthalene, against $[Q]$, the concentration of 1-iodonaphthalene.

(vi) 1-Bromonaphthalene and 1-iodonaphthalene

Here again selective filtration was not possible and a quartz reaction vessel was used. Two concentrations of 1-iodonaphthalene were studied and the concentration of 1-bromonaphthalene was always at least ten times greater than that of 1-iodonaphthalene. The quenching rate constant derived was $8 \pm 4 \times 10^7$ l. mole^{-1}s^{-1}.

(vii) Benzophenone and naphthalene

Studies on this system are described in detail elsewhere (Porter & Wilkinson 1961). The change in the transient absorption observed upon flashing 10^{-2} M benzophenone in benzene was studied as the amount of added naphthalene was varied in steps from

Energy transfer from the triplet state   **11**

10^{-4} to 10^{-2} M. As naphthalene was added the amount of absorption by the ketyl radical decreased and absorption by triplet naphthalene appeared. To ensure that the naphthalene itself was not irradiated a soda-glass reaction vessel was used.

As the triplet state of the donor is not observed, direct measurements of quenching are not possible. The relevant reactions are as follows:

$$\phi_2 CO(T) \rightarrow \phi_2 CO(S), \tag{1}$$

$$\phi_2 CO(T) + RH \rightarrow \phi_2 C\!-\!OH + R^{\cdot}, \tag{2}$$

$$\phi_2 CO(T) + Q(S) \rightarrow \phi_2 CO(S) + Q(T), \tag{3}$$

where RH = solvent, Q = acceptor naphthalene, and T and S refer to triplet and ground states.

At the concentration of naphthalene which reduces the amount of ketyl formed by one half

$$k_3[Q_{\frac{1}{2}}] = k_1 + k_2[RH].$$

The half-quenching concentration was found to be 4×10^{-4} M. Using the value of Bäckström & Sandros for the lifetime τ_B of triplet benzophenone in benzene we obtain

$$\tau_B = \frac{1}{k_1 + k_2[RH]} = 1\cdot9 \times 10^{-6} \quad \text{and} \quad k_3 = k_Q = 1\cdot2 \times 10^9 \, \text{l. mole}^{-1}\text{s}^{-1}.$$

Measurements of ketyl concentrations were made $70 \, \mu s$ after the flash, and this may make the above estimate of k_Q somewhat low.

(viii) *Diacetyl and 1:2-benzanthracene*

Solutions of diacetyl, 1:2-benzanthracene and mixtures of the two in benzene solution were flashed with a concentrated solution of 1:2-benzanthracene used as filter. Figure 6, plate 2 shows

(i) energy transfer to form the triplet state of 1:2-benzanthracene;

(ii) the absence of fluorescence when the triplet state of 1:2-benzanthracene is formed by energy transfer, proving that its formation bypasses the first excited singlet state.

The value of the quenching constant derived from these photographic results was $k_Q = 3 \pm 2 \times 10^9$ l. mole^{-1}s^{-1} in good agreement with the value $3\cdot8 \times 10^9$ l. mole^{-1}s^{-1} measured by a quite different method by Bäckström & Sandros (1958).

(ix) *Quenching by iodine*

The effect of three different concentrations of iodine, the highest being 10^{-5} M, on the rate of decay of triplet phenanthrene in hexane was studied on the flash photoelectric apparatus. The concentration of phenanthrene was $5\cdot0 \times 10^{-3}$ M throughout these experiments and a concentrated solution of iodine in hexane was used as a filter. The k_1 values for solutions which contained iodine decreased with the number of times the solution was flashed. This indicated that the iodine was

G. Porter and F. Wilkinson

being used up and therefore only the rate values obtained from the first flashes were used. The value of k_Q obtained in this way was $1 \cdot 4 \pm 0 \cdot 4 \times 10^{10} \, \mathrm{l. \, mole^{-1} \, s^{-1}}$.

Similar measurements were carried out using anthracene as the donor, the concentration of anthracene being $1 \cdot 39 \times 10^{-4} \, \mathrm{M}$. Three different concentrations of iodine were used, the highest being $5 \times 10^{-6} \, \mathrm{M}$. Again iodine was consumed on flashing and only the first flash was used for rate measurements. A value of $k_Q = 2 \cdot 4 \pm 0 \cdot 2 \times 10^{9} \, \mathrm{l. \, mole^{-1} \, s^{-1}}$ was obtained for the quenching rate constant of anthracene by iodine in hexane.

(x) *Systems which showed no quenching of the triplet state*

For a number of the pairs of compounds described so far, the lifetime of the triplet state of the acceptor compound, formed by energy transfer, generally in the presence of a high concentration of the donor compound, was found to be identical with that of the triplet state of the acceptor alone formed by direct excitation. Thus k_Q for triplet naphthalene with phenanthrene as quencher in ethylene glycol was $\leqslant 2 \times 10^{4} \, \mathrm{l. \, mole^{-1} \, s^{-1}}$. Upper limits of other k_Q values obtained in this way are included in table 4.

No quenching was observed in the following systems:

(*a*) Anthracene and phenanthrene

The decay of triplet anthracene in ethylene glycol was unaffected by the presence of $1 \cdot 9 \times 10^{-3} \, \mathrm{M}$ of phenanthrene.

Thus $k_Q \leqslant 5 \times 10^{3} \, \mathrm{l. \, mole^{-1} \, s^{-1}}$.

(*b*) Anthracene and naphthalene

The decay of triplet anthracene in hexane was unaffected by $1 \cdot 2 \times 10^{-3} \, \mathrm{M}$ of naphthalene.

Thus $k_Q \leqslant 4 \times 10^{4} \, \mathrm{l. \, mole^{-1} \, s^{-1}}$.

(*c*) Anthracene and 1-iodonaphthalene

The decay of triplet anthracene in ethylene glycol was unaffected by $9 \cdot 3 \times 10^{-4} \, \mathrm{M}$ of 1-iodonaphthalene.

Thus $k_Q \leqslant 2 \times 10^{4} \, \mathrm{l. \, mole^{-1} \, s^{-1}}$.

(*d*) Anthracene and ethyl iodide

The decay of triplet anthracene in hexane showed only slight quenching in the presence of $1 \cdot 24 \times 10^{-2} \, \mathrm{M}$ of ethyl iodide. A concentrated solution of naphthalene in hexane was used as filter for these experiments. The filter was not perfect and the ethyl iodide decomposed to give iodine which then quenched the triplet anthracene. This was clear since both the absorption of iodine and the amount of quenching increased as the solution was flashed.

The upper limit derived for the quenching rate constant was $1 \cdot 6 \times 10^{4} \, \mathrm{l. \, mole^{-1} \, s^{-1}}$.

(*e*) Phenanthrene and ethyl iodide

Again a concentrated solution of naphthalene was used as a filter and the ethyl iodide slightly decomposed each time the solution was flashed. The rate of decay of triplet phenanthrene in hexane was only very slightly quenched in the presence of $1 \cdot 24 \times 10^{-2} \, \mathrm{M}$ of ethyl iodide which gave the value derived for

$$k_Q \leqslant 3 \times 10^{5} \, \mathrm{l. \, mole^{-1} \, s^{-1}}.$$

Energy transfer from the triplet state   13

DISCUSSION

This work has not been directly concerned with the mechanism of triplet state decay in pure fluid solvents but two points may be referred to in this connexion. First, the excellent agreement between our earlier values of k_1 and those of Porter & Wright (1958) is probably an illustration of the contention of Livingston, Jackson & Pugh (1960) that a constant level of impurity such as oxygen can be obtained with a high degree of reproducibility. Our later values of k_1 were significantly lower and this indicates that the earlier values related to pseudo first-order processes. The quenching efficiency of aromatic molecules with lower triplet levels found in this work suggests one possible source of quenching impurity.

Secondly, our finding that naphthalene, 1-chloronaphthalene and 1-bromonaph-thalene have comparable triplet lifetimes whilst for 1-iodonaphthalene no triplet absorption could be detected seems very significant. In rigid media the phosphor-escence lifetimes of the halogenated naphthalenes decrease regularly from 2·6 s for naphthalene to 0·0028 s for 1-iodonaphthalene and the rate constant for radia-tionless decay might be expected to show a similar heavy atom effect. A possible explanation of our findings is that the true first-order radiationless process is only predominant in 1-iodonaphthalene and the rate process in the other naphthalenes is mainly pseudo first order and therefore leads to approximately constant decay rates.

Our measurements of quenching constants are not subject to these uncertainties. k_Q values were determined from the slope of the linear plots of k_A against, $[Q]$ and any change in the value of k_1 affects the intercept but not the slope of this plot. The presence of impurities in the solvent will therefore not affect the value of k_Q provided the level of impurity is constant and this is shown to be the case by the constancy of k_1 values.

Evidence for energy transfer from the triplet state

Energy levels and other relevant data for the compounds investigated are given in table 4.

The energy level diagram appropriate to the pairs phenanthrene-naphthalene, triphenylene-naphthalene, phenanthrene-bromonaphthalene, phenanthrene-iodonaphthalene, and benzophenene-naphthalene is shown in figure 4.

Our results provide unequivocal proof of the occurrence, in these pairs of com-pounds, of the process

donor (triplet) + acceptor (singlet) → donor (singlet) + acceptor (triplet)

since both the quenching of the donor triplet and the formation of the acceptor triplet have been observed directly. The only energetically possible alternative

donor (upper singlet) + acceptor (singlet) → donor (singlet) + acceptor (triplet)

is not only improbable on grounds of spin conservation and the short lifetime of the upper singlet but is experimentally excluded by our observation that, in all cases where fluorescence was observed, the fluorescence yield of the donor was unchanged by the presence of acceptor.

14   G. Porter and F. Wilkinson

Only in the cases of iodonaphthalene and iodine as acceptors were the triplet states not observed spectroscopically. Iodonaphthalene quenched the triplet state of molecules whose triplet state lay higher (e.g. naphthalene and phenan-

TABLE 4. ENERGIES OF FIRST EXCITED SINGLET AND TRIPLET STATES, PHOSPHORESCENCE LIFETIMES (τ), LIFETIMES IN SOLUTION $(\tau_{solution})$ AND PHOSPHORESCENCE YIELDS (Φ)

compound	long-wavelength limit of singlet absorption (cm^{-1})	first triplet level (cm^{-1})	τ (s)	$\tau_{solution}$ $(10^{-4}$ s)	Φ
benzophenone	27800	24400[2]	0·006[6]	$1·6 \times 10^{-2}$ (benzene)[8]	0·84[6]
triphenylene	28200	23500[3]	15·9[2]	0·56 (hexane)	0·6[6]
phenanthrene	28900	21600[3]	3·3[3]	0·93	0·23
naphthalene	31200	21300[2]	2·6[3]	0·91	0·09[6]
1-bromonaphthalene	31200	20700[2]	0·018[3]	0·83	0·55[4]
1-iodonaphthalene	31200	20500[4]	0·0025[3]	—	0·70[4]
diacetyl	21750	19700[2]	0·00225[3]	0·60 (benzene)	—
1:2-benzanthracene	20600	16500[3]	0·3[3]	1·59 (hexane)	0·001
anthracene	26000	14700[2]	0·09[7]	9·1 (benzene)	0·0001
iodine	33774[1]	11888[5]	—	—	—

References

(1) Elliot (1940)
(2) Lewis & Kasha (1944)
(3) McClure (1949)
(4) Terenin & Ermolaev (1958)
(5) Brown (1931)
(6) Gilmore, Gibson & McClure (1952)
(7) McGlynn, Padbye & Kasha (1955)
(8) Bäckström & Sandros (1960)

The values of Φ for which references are not given have been calculated from the ϕ_p/ϕ_f ratios in reference (3) by assuming $\phi_p + \phi_f = 1$.

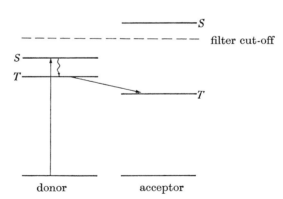

FIGURE 4. Energy levels of donor and acceptor.

threne) but not of molecules with lower triplet states (e.g. anthracene) and it clearly operates by the same mechanism. The mechanism of quenching by iodine is less certain since iodine quenched the triplet states of all molecules investigated. But the lowest triplet level of iodine is of lower energy than those of all other molecules

Energy transfer from the triplet state   15

investigated so that this behaviour is fully in accordance with a triplet energy transfer mechanism. Attempts to establish a negative result between pentacene and iodine were not successful owing to the very low solubility of pentacene and extensive decomposition. The mechanism of quenching by iodine is, however, complicated by the fact that chemical changes occur which result in the consumption of iodine. Whether this is a separate quenching mechanism, a reaction which follows triplet quenching or a process which is quite distinct from the quenching cannot yet be decided.

The possibility that quenching by iodine occurs as a result of heavy atom enhancement of spin-orbit coupling must be considered. Previous work on the effect of heavy ions such as Zn^{2+}, Ga^{3+} and Pb^{2+} (Porter & Wright 1958) and our observation, that ethyl iodide has a negligible effect on triplet lifetime, seem to exclude this possibility. Finally, the well-known tendency of iodine to form change transfer complexes with aromatic molecules suggests a further mechanism by which iodine might operate. However, the concentration of iodine in the mixtures studied was so low that, even if it were all complexed, it would have resulted in a reduction of the concentration of free donor molecules by less than 1 %.

TABLE 5. VISCOSITIES AND DIFFUSION CONTROLLED RATE CONSTANTS AT
25 °C IN THREE SOLVENTS

	$\eta(P)$	k_d(l. mole^{-1} s^{-1})
hexane	0·00326	$2·0 \times 10^{10}$
benzene	0·00647	$1·0 \times 10^{10}$
ethylene glycol	0·199	$3·3 \times 10^{8}$

Similar considerations rule out complex formation as a significant factor in the other systems studied. For example, 4×10^{-5} M iodonaphthalene increased the rate of disappearance of triplet phenanthrene in ethylene glycol by a factor of 5. The ground-state concentration of phenanthrene in this experiment was 2×10^{-3} M and hence, if all the iodonaphthalene were complexed, this would change the phenanthrene concentration by only 2 %. Even if the complex triplet had an absorption spectrum similar to that of triplet phenanthrene and decayed very rapidly it would have negligible effect on the observed decay rate of triplet phenanthrene.

Finally, the 'trivial process' of light emission and reabsorption can be excluded in our experiments since the fraction of triplet molecules which decay by a radiative process in fluid solvents is entirely negligible.

Mechanism of the energy transfer

The rate constants of energy transfer given in table 3 fall into three groups.

(a) When the triplet level of the donor is considerably greater than that of the acceptor the rate is quite close to the encounter rate calculated from the equation of Debye (1942) $$k_d = 8RT/3000\,\eta\,\text{l. mole}^{-1}\text{s}^{-1},$$

where k_d is the diffusion controlled rate constant and η is the solvent viscosity. This is illustrated by comparing the values of k_d in table 5, which have been calculated from this equation, with the rate constants in table 3, for systems where the triplet levels are well separated.

16   G. Porter and F. Wilkinson

(*b*) When the triplet levels of donor and acceptor lie very close, but the energy of the acceptor is below that of the donor, quenching occurs with a rate considerably less than the encounter rate. This is most marked in the phenanthrene + naphthalene system, where the quenching rate is less than the encounter rate, even in the viscous solvent ethylene glycol. The quenching rate by bromonaphthalene, which has a slightly lower triplet level, is higher and quenching by iodonaphthalene is very nearly diffusion controlled.

(*c*) When the triplet level of the donor is below that of the acceptor no quenching is observed.

There are no indications that quenching efficiency decreases as the separation of the acceptor triplet below that of the donor triplet increases.

Energy transfer therefore occurs between molecules during an encounter at normal collisional separation. Under these conditions overlap of the orbitals of the two molecules occurs and an exchange transfer mechanism is possible. In the region of overlap the electrons are indistinguishable and the acceptor may emerge, by a 'collision of the second kind' in an electronically excited state.

According to Dexter (1953) not only has spin momentum to be conserved in such a collision but M_D^* must equal M_A^* and $M_D = M_A$, where M_D^* and M_A^* refer to the multiplicity of the donor and acceptor molecules in their excited states and M_D and M_A to multiplicities in their ground states. This condition is satisfied in the cases being considered here.

Dexter gives the probability of energy transfer by an exchange mechanism between two molecules as

$$P_{DA} = \frac{4\pi^2}{h} Z^2 \int f_D(E)\, F_A(E)\, \mathrm{d}E,$$

where $\int f_D(E)\, F_A(E)\, \mathrm{d}E$ is a type of overlap integral between the emission spectrum of the donor and the absorption spectrum of the acceptor. Z^2, which has the dimensions of energy squared, is a quantity that cannot be directly related to optical experiments. The separation and concentration dependences are hidden in Z^2 which varies approximately as $Y(e^4/K^2R_0^2)\exp(-2R/L)$ where R_0 is the effective average Bohr radius of the unexcited states of D and A, L is the effective average Bohr radius of the excited and unexcited states of D and A, e is the charge on the electron, K is the dielectric constant of the medium, R is the distance between the two molecules, and Y is a dimensionless quantity $\ll 1$ which takes account of cancellation as a result of sign changes in the wave functions.

Transfer times between neighbouring molecules can be very short but they increase rapidly with molecular separation—for example, in a typical case, the transfer time is increased by a factor of the order 10^2 when the molecular separation is increased by one molecular diameter. Exchange transfer would therefore lead to values of quenching constants not exceeding those calculated for diffusion controlled encounters between molecules having normal cross sections.

Energy transfer between molecules in singlet states often occurs over greater distances with correspondingly greater rate constants. For example, the rate constants of energy transfer from 1-chloroanthracene to perylene in various solvents

Energy transfer from the triplet state   **17**

lie in the range $(1 \cdot 4 \text{ to } 2 \cdot 5) \times 10^{11}$ l. mole^{-1} s^{-1} (Bowen & Livingston 1954), and similar transfer occurs in rigid solvents with a transfer distance of 34 Å (Bowen & Brockle-hurst 1955). These high transfer efficiencies are interpreted in terms of inductive resonance or dipole-dipole transfer and the relevance of this mechanism to triplet energy transfer deserves consideration.

Förster (1948) derived a formula from which quantitative predictions can be made provided energy transfer occurs after thermal equilibrium has been established.

$n_{D* \to A*}$, the transfer frequency, is given by

$$n_{D* \to A*} = \frac{9 \times 10^6 (\ln 10)^2 c x^2}{16 \pi^4 n^2 N^2 R_{DA}} \int \epsilon_e^D(\bar{\nu}) \, \epsilon_a^A(\bar{\nu}) \frac{\mathrm{d}\bar{\nu}}{\bar{\nu}^2}, \tag{1}$$

where c is the velocity of light, x is an orientation factor, n is the refractive index, N is the Avogadro number, R_{DA} is the distance between the two molecules D and A, $\epsilon_a^A(\bar{\nu})$ is the molar extinction coefficient of the acceptor A and $\epsilon_e^D(\bar{\nu})$ represents the intensity of emission of D measured in the same units as the extinction coefficient.

If R^0 is defined as the critical transfer distance for which excitation transfer and spontaneous deactivation are of equal probability

$$n_{D* \to A*} = \frac{1}{\tau^D} \left(\frac{R_0}{R_{DA}} \right)^6, \tag{2}$$

where τ^D is the actual mean lifetime of the energy donor D. Substituting, and using numerical values of the constants, we obtain

$$R_0^6 = \frac{1 \cdot 69 \times 10^{-33} \, \tau^D}{n^2} \frac{J(\bar{\nu}),}{\bar{\nu}_0^2} \tag{3}$$

where $J(\bar{\nu}) = \int_0^\infty \epsilon_e^D(\bar{\nu}) \, \epsilon_e^A(\bar{\nu}) \, \mathrm{d}\bar{\nu}$ is called the overlap integral, and $\bar{\nu}^0$ is the wave number of the zero-point emission of the donor.

For typical cases of sensitized fluorescence R_0 values from 30 to 100 Å are calculated.

The values obtained for $J(\bar{\nu})$ will obviously be much smaller when singlet-triplet transitions are substituted in equation (3) because of the very low transition probabilities for singlet-triplet transitions. Calculations of R_0 values were made for the pairs of compounds which have been studied experimentally. The molar extinction coefficients were calculated from the following equation

$$\frac{1}{\tau} = \frac{8000 \pi n^2 (\ln 10) c}{\Phi N} \int \frac{(2 \bar{\nu}_0 - \bar{\nu})^3}{\bar{\nu}} \, \epsilon(\bar{\nu}) \, \mathrm{d}\bar{\nu}, \tag{4}$$

where τ is the phosphorescence lifetime, Φ is the quantum yield of phosphorescence, $\bar{\nu}_0$ is the position of the zero-zero band expressed in wave-numbers and is the median position between the phosphorescence and the singlet-triplet absorption maxima on a wave-number scale. The singlet-triplet spectra were generally drawn as the mirror image of the phosphorescence on a wave-number scale.

The data available to make these calculations are not very extensive, most of the information used being from the sources given in table 4, but the values given in

18   G. Porter and F. Wilkinson

table 6 show the magnitude of transfer distances predicted by this theory for triplet energy transfer.

Clearly these distances, which in most cases are much less than collision diameters, are so small that the theory used for their calculation is no longer valid. The calculations show, however, that long-range dipole-dipole transfer is not likely to play a significant role in triplet energy transfer processes of the type being considered, and this is in accordance with our experimental findings and with the conclusions of Terenin & Ermolaev (1958).

TABLE 6. CALCULATED VALUES OF R^0 (FÖRSTER THEORY)

donor	acceptor	solvent	R° (Å)
phenanthrene	naphthalene	hexane	0·11
triphenylene	naphthalene	hexane	0·11
phenanthrene	1-bromonaphthalene	hexane	0·40
	1-iodonaphthalene	hexane	0·62
naphthalene		hexane	0·52
1-bromonaphthalene		hexane	1·26
benzophenone	naphthalene	benzene	0·18
diacetyl	1:2-benzanthracene	benzene	0·19
phenanthrene	iodine	hexane	3·84

We conclude that triplet energy transfer to singlet molecules occurs only during encounters at normal collision distances and probably occurs by an exchange transfer mechanism. The efficiency of this process approaches unity as the triplet level of the acceptor falls below that of the donor.

REFERENCES

Bäckström, H. L. J. & Sandros, K. 1958 *Acta Chem. Scand.* **12**, 823.
Bäckström, H. L. J. & Sandros, K. 1960 *Acta Chem. Scand.* **14**, 48.
Bowen, E. J. & Livingston, R. 1954 *J. Amer. Chem. Soc.* **76**, 6300.
Bowen, E. J. & Brocklehurst, B. 1955 *Trans. Faraday Soc.* **51**, 774.
Brown, W. G. 1931 *Phys. Rev.* **38**, 1187.
Debye, P. J. W. 1942 *Trans. Electrochem. Soc.* **82**, 205.
Dexter, D. L. 1953 *J. Chem. Phys.* **21**, 836.
Elliot, A. 1940 *Proc. Roy. Soc.* A, **174**, 273.
Förster, Th. 1948 *Ann. Phys., Lpz.,* **2**, 55.
Gilmore, E. H., Gibson, G. E. & McClure, D. S. 1952 *J. Chem. Phys.* **20**, 829.
Gover, T. A. & Porter, G. 1961 In course of publication.
Kooyman, E. C. & Farenhorst, E. 1953 *Trans. Faraday Soc.* **49**, 58.
Lewis, G. W. & Kasha, M. 1944 *J. Amer. Chem. Soc.* **66**, 2100.
Livingston, R., Jackson, G. & Pugh, A. C. 1960 *Trans. Faraday Soc.* **56**, 1635.
McClure, D. S. 1949 *J. Chem. Phys.* **17**, 905.
McClure, D. S. & Hanst, P. L. 1955 *J. Chem. Phys.* **23**, 1772.
McGlynn, S. P., Boggus, J. D. & Elder, E. 1960 *J. Chem. Phys.* **32**, 357.
McGlynn, S. P., Padbye, M. E. & Kasha, M. 1955 *J. Chem. Phys.* **23**, 593.
Porter, G. & Jackson, G. 1961 *Proc. Roy. Soc.* A, **260**, 13.
Porter, G. & Wilkinson, F. 1961 In course of publication.
Porter, G. & Windsor, M. W. 1954 *Disc. Faraday Soc.* **17**, 178.
Porter, G. & Windsor, M. W. 1958 *Proc. Roy. Soc.* A, **245**, 238.
Porter, G. & Wright, M. R. 1958 *J. Chim. phys.* **55**, 705.
Porter, G. & Wright, M. R. 1959 *Disc. Faraday Soc.* **27**, 18.
Terenin, A. W. & Ermolaev, V. L. 1956 *Trans. Faraday Soc.* **52**, 1042.
Terenin, A. W. & Ermolaev, V. L. 1958 *J. Chim. phys.* **55**, 698.

PRINTED IN GREAT BRITAIN AT THE UNIVERSITY PRESS, CAMBRIDGE

(BROOKE CRUTCHLEY, UNIVERSITY PRINTER)

Trans. Faraday Soc. **51**, 1469–1474 (1955) – Reproduced by permission of the Royal Society of Chemistry

PRIMARY PHOTOCHEMICAL PROCESSES IN AROMATIC MOLECULES

† PART 3.   ABSORPTION SPECTRA OF BENZYL, ANILINO, PHENOXY AND RELATED FREE RADICALS

By George Porter * and Franklin J. Wright

Physical Chemistry Dept., University of Cambridge

Received 14*th March*, 1955

Photolysis, in the vapour phase, of toluene, ethyl benzene, benzyl chloride and other benzyl derivatives, results in the formation of a common transient species, with a characteristic narrow-banded electronic absorption spectrum, which must be identified with that of the free benzyl radical. Anilino, phenoxy, *p*-xylyl and similar free radicals have been detected in the same manner. Molecules from which such spectra have been observed are all characterized by the possibility of photolytic fission of a relatively weak bond in the side chain to yield a radical which is stabilized by resonance between benzenoid and quinonoid canonical forms. The lifetimes were always less than 10^{-4} sec which, at the concentrations employed, indicates a high collisional efficiency of recombination.

The resonance stabilization which is to a large extent responsible for the existence, at normal temperatures, of triphenylmethyl, and similar radicals of the type first described by Gomberg,[1] results in a reduction of the bond dissociation energy in many aromatic molecules,[2] although it may not be sufficient to give detectable equilibrium concentrations of free radicals. It is generally found that the probability of photolytic bond fission is greatest at the weakest bond in a molecule, and we might therefore expect that a common photochemical dissociation process in aromatic molecules will be that which results in the formation of two radicals possessing, together, a greater resonance energy than that of the parent molecule. The simplest examples of this behaviour would be provided by toluene,

---

* present address : Chemistry Dept., University of Sheffield.

† Parts 2 and 3 of this work were supported by a grant from the Department of Scientific and Industrial Research.

aniline and phenol which might yield the benzyl, anilino and phenoxy radicals respectively. Each of these radicals has both benzenoid and quinonoid canonical forms, e.g.,

which result in a considerable resonance stabilization. The resonance energy of benzyl has been given as 24·5 kcal/mole which corresponds to the low C—H bond dissociation energy [3] in toluene of 77·5 kcal/mole.

There is considerable evidence for the existence of the benzyl radical, and for its low reactivity with other molecules. Thus it was one of the radicals detected by Paneth and Lautsch,[4] using the mirror technique, and its stability is the basis of the toluene carrier technique of Szwarc.[5] Nevertheless, it does not exist in significant equilibrium concentration at normal temperatures; it has therefore never been detected by physical methods, and its spectrum is unknown. Indeed, very few polyatomic free radical spectra, other than those which can be obtained in equilibrium, have been recorded, and in the gas phase only four, all of them triatomic, can be considered established.[6] Quite recently Schuler and his colleagues [7] have detected a number of emission spectra during the passage of a mild electrical discharge through aromatic vapours, and some of these are very probably attributable to free radicals. In particular, the " V spectrum ", which has also been reported by Walker and Barrow,[8] has been assigned to $C_6H_5C\cdot$, $C_6H_5CH\cdot$, $C_6H_5CH_2$ " or perhaps a form containing less hydrogen and perhaps charged ".[8]

In the course of our study of the triplet states of aromatic molecules in the vapour phase (part 2) [9] we investigated the primary products of photolysis of monocyclic compounds. No triplet state spectra were found, and benzene itself showed no transient species and no dissociation. Toluene and a number of other compounds gave sharp band spectra which, as we hope to show, are to be attributed to benzyl and other free radicals of the type discussed. The experimental conditions were identical with those used during our investigations [9] of the triplet state. The path length was 1 m, and all experiments were conducted in the presence of inert gas (carbon dioxide or nitrogen) at a pressure at least one hundred times greater than that of the aromatic vapour.

RESULTS

Provided a sufficient excess of inert gas was present benzene itself showed no decomposition and no new bands were observed. Most other compounds, whether they gave new band spectra or not, were decomposed photochemically to compounds which gave rise to a continuous absorption lasting several minutes. This absorption was not observable until a few milliseconds after the end of photolysis; it reached a maximum after about 10^{-1} sec and then decayed slowly over a period of several minutes. This behaviour has frequently been observed in other flash photolysis work and is characteristic of the condensation of a supersaturated vapour to small particles which scatter the light whilst they remain in the light path and which eventually condense on the wall. These continuous spectra are insufficiently specific for identification of the products and have not been studied further. Fortunately the time interval which elapses, before their appearance causes nearly complete extinction, is long enough to permit the detection and study of most free radicals and other intermediates of interest.

The phenyl derivatives bromobenzene, chlorobenzene, fluorobenzene, nitrobenzene, and benzonitrile, like benzene itself, gave no new spectra during or immediately after photolysis. Reaction occurred in most cases, evidenced by the later appearance of continuous absorption, and it is possible that the primary decomposition of most of these molecules resulted in the formation of the phenyl radical. If so, the phenyl radical must differ from those radicals whose spectra are described later, either in being more reactive or in having no transition of comparable intensity above 2700 Å. Although the spectral region investigated was 2200 to 5000 Å, spectra below 2700 Å could have been obscured by the absorption of the parent molecule.

THE BENZYL RADICAL

The flash photolysis of toluene vapour at 2 mm pressure, in the presence of 700 mm of carbon dioxide or nitrogen resulted in the appearance of a new banded spectrum which was observed only during irradiation by the photolysis flash. The spectrum consisted of one very sharp band at 3052 Å and several much weaker bands showing no obvious regularity. The 3053 Å band was 3 Å wide at half-maximum extinction, and had a rather sharp head degraded to longer wavelengths. Its appearance was therefore different from that of triplet spectra which all gave broader bands with an intensity distribution nearly symmetrical about the centre.

Investigation of related compounds immediately confirmed that this spectrum could not be identified with the triplet state of toluene. The same spectrum was observed with benzyl chloride, ethyl benzene, o-chlorotoluene, benzylamine, diphenyl methane, and benzyl alcohol. In the three latter molecules the spectrum was very weak and only the 3053 Å band was detectable. The high resolving power of the spectrograph and the sharpness of the band made it possible to compare the wavelengths of the spectra to better than 0·2 Å, using an optical comparator, and there can be no doubt that the same species is responsible for the absorption in each case. Spectra taken before and during photolysis are shown in fig. 1 and positions of the bands are given in table 1. Except with o-chlorotoluene, which will be discussed separately, no other spectra were observed during photolysis of these molecules.

Since no common spectrum can arise from each of these parent molecules the new bands must arise from a dissociation product. The intensity of the spectra did not increase with successive flashes and the absorption cannot therefore be attributed to secondary photolysis of a product. Except for o-chlorotoluene, all the molecules being considered have the generic formula $C_6H_5CH_2X$ and possible primary photochemical processes giving rise to a common species are as follows:

$$C_6H_5CH_2X \rightarrow C_6H_5CH_2 + X \qquad (1)$$

$$C_6H_5CH_2X \rightarrow C_6H_5CH + HX \qquad (2)$$

$$C_6H_5CH_2X \rightarrow C_6H_5 + CH_2X \qquad (3)$$

Ionic species, and processes involving fission of several bonds, are eliminated on energetic grounds.

There are strong arguments against supposing that the spectrum is that of the phenyl radical formed by reaction (3), since it has been shown that no spectrum is obtained from compounds such as chlorobenzene and bromobenzene. Dissociation occurs in these molecules, and the C—Cl and C—Br dissociation energies are less than that of C—CH$_3$ so that the phenyl radical should be observed during photolysis of the phenyl compounds at least as readily as from the benzyl derivatives. Secondly, if this spectrum were attributed to phenyl a second dissociation process would have to be invoked to account for the later results. Reaction (2) might be supported owing to the possibility of formation of a spin-paired molecule containing a divalent carbon atom. No analogous process could occur, however, in compounds such as phenol which, it will shortly be shown, behave in a similar manner. Definite evidence against type 2 decomposition is provided by recent work of Porter and Strachan who have shown that toluene, benzal chloride and benzotrichloride each give a different radical spectrum. We can therefore identify the dissociation process with reaction (1) and the observed spectrum with the benzyl radical.

The case of o-chlorotoluene requires special consideration. Two band systems were observed, one of which was identical with that assigned to benzyl and the second, of comparable intensity, consisted of four bands at longer wavelengths. At first sight the appearance of benzyl during photolysis of o-chlorotoluene is contrary to expectation, but a careful consideration of the primary products of photolysis provides the explanation. There are two bonds which, having low dissociation energy, might be expected to break on excitation of this molecule, firstly the C—Cl bond and secondly a C—H bond of the methyl group. The former will result in a tolyl radical of similar stability to phenyl, having little additional resonance energy. The o-tolyl radical is, however, isomeric with the more stable benzyl, and construction of a molecular model shows that migration of a hydrogen atom from the side chain to the ring may occur very easily. The converse transfer will be so endothermic that when equilibrium between the two forms is reached we shall expect the benzyl form to be greatly predominant. The second spectrum will then be assigned to the product of the second dissociation process, i.e. to *o*-chlorobenzyl. Its position, to long wavelengths of benzyl, is typical of the influence of chlorine substituted in the benzene ring.

TABLE 1.—SPECTRA OBSERVED DURING PHOTOLYSIS AND THEIR PROBABLE ASSIGNMENTS

molecule	$\lambda(\text{Å})$	$\nu(\text{cm}^{-1})$	I relative	radical
$C_6H_5CH_3$	3068	32,590	2	$C_6H_5CH_2$
	3053	32,760	10	
	2966	33,720	3	
$C_6H_5CH_2CH_3$	3068	32,590	2	$C_6H_5CH_2$
	3053	32,760	10	
	2966	33,720	3	
$C_6H_5CH_2Cl$	3100	32,260	1	$C_6H_5CH_2$
	3068	32,590	2	
	3053	32,760	10	
	2966	33,720	3	
	2936	34,060	2	
	2918	34,270	1	
$C_6H_5CH_2NH_2$	3053	32,760	—	$C_6H_5CH_2$
$C_6H_5CH_2OH$	3053	32,760	—	$C_6H_5CH_2$
$C_6H_5CH_2C_6H_5$	3053	32,760	—	$C_6H_5CH_2$
o-Cl . $C_6H_4CH_3$	3153	31,720	2	o-ClC$_6$H$_4$CH$_2$
	3157	31,680	4	
	3161	31,640	5	
	3165	31,600	6	
	3068	32,590	2	$C_6H_5CH_2$
	3053	32,760	10	
	2966	33,720	3	
	2936	34,060	1	
p-CH$_3$C$_6$H$_4$CH$_3$	3100	32,260	—	p-CH$_3$C$_6$H$_4$CH$_2$
$C_6H_5NH_2$	3008	33,250	—	$C_6H_5NH_2$
C_6H_5OH	3920	34,250 (wide)	—	C_6H_5O
$C_6H_5OCH_3$	2920	34,250 (wide)	—	C_6H_5O
C_6H_5SH	3100 to shorter wavelengths	32,260 limit	—	C_6H_5S

ANILINO AND PHENOXY

Photolysis of aniline vapour at 0·2 mm pressure in the presence of 600 mm of carbon dioxide resulted in the appearance of a strong band at 3008 Å which, like benzyl, was only present during irradiation by the photolysis flash (fig. 2). Although no reduction in the absorption of the aniline molecule was detectable the extinction at the maximum at 3008 Å was greater than at the maximum of the aniline spectrum near 2935 Å, showing that the extinction coefficient of the new species was many times greater than that of aniline. There were signs of structure at lower wavelengths which were too weak to measure reliably. The 3008 band was rather more diffuse than the strong band of benzyl and about double the width. Though similar in appearance to many of our triplet spectra, recent experiments in rigid glass by Norman and Porter [10] have shown that a similar spectrum is obtained whose lifetime is much too long for the triplet state. It must, therefore, be a dissociation product and by analogy we shall assign this spectrum to the anilino radical. In this case no confirmation by photolysis of other molecules has yet been obtained. Dimethyl aniline gave negative results and negative results were also obtained

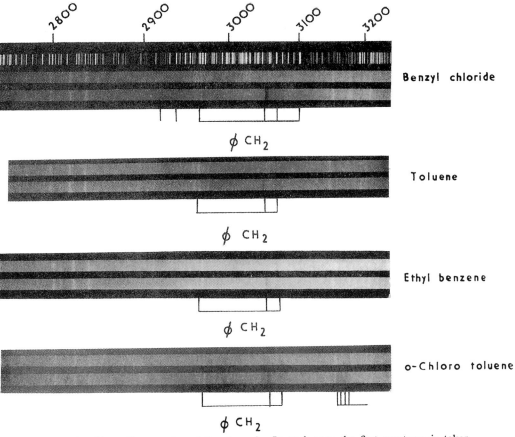

FIG. 1.—Absorption spectra of free benzyl. In each case the first spectrum is taken before, and the second during, photolysis.

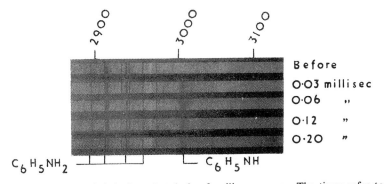

FIG. 2.—Spectra recorded during photolysis of aniline vapour. The times refer to the intervals between the peaks of the two flashes.

with phenylhydrazine (using nitrogen as inert gas) which might have been expected to yield the same product on photolysis. The non-appearance of a certain radical in cases like this cannot, however, give much indication of whether it is formed in the primary act. All the radicals with which we are concerned have lifetimes so short that they are only just detectable by the present techniques and a slight additional reactivity would be sufficient to reduce their lifetime below the experimental limit of detection. It should also be noted that whilst hydrogen abstraction by anilino from aniline produces no change, abstraction from phenylhydrazine produces a different species.

Photolysis of phenol resulted in a transient spectrum consisting of one diffuse band, about 100 Å wide, with a maximum at 2920 Å. In spite of its diffuseness the intensity distribution was characteristic, with a definite maximum, and it was possible to obtain clear evidence of the identity of this spectrum by photolysis of anisole. A similar transient absorption was obtained which density comparisons showed was identical with that from the photolysis of phenol, and which is therefore to be assigned to the phenoxy radical.

OTHER MOLECULES

Other molecules giving negative results, in addition to those already mentioned, were benzaldehyde, acetophenone, benzophenone, o-bromotoluene, o-toluidine, diphenyl, and dibenzyl. α-Methyl naphthalene gave a triplet spectrum (part 2) but no radical was observed. Two other substances gave transient spectra which we believe are to be assigned to radicals. Thiophenol showed a continuous absorption, which was present only during irradiation, beginning at 3400 Å and extending to shorter wavelengths with no definite maximum. p-Xylene gave a sharp band at 3100 Å, similar in appearance to that of benzyl. These are probably the spectra of thiophenoxy and p-xylyl respectively.

The positions of all spectra found in this investigation are recorded in table 1. The exact agreement between measurements from different molecules arises from the fact that it was possible to establish the identity of the spectra to a greater accuracy than the band centre position could be defined.

DISCUSSION

The lifetimes of all radical spectra observed were less than the time resolution of the method, i.e. 10^{-4} sec and therefore no kinetic investigations have been possible. It is worthy of note, however, that the observations indicate a high efficiency of radical recombination. All reactions of benzyl and similar free radicals with the parent molecule, except that which regenerates the same species, would be too endothermic to occur at the observed rate at room temperature and the radicals must disappear either by dimerization or by recombination with a hydrogen atom. Even at the highest concentration of radicals no decrease in absorption by the parent molecule was detectable which sets an upper limit to the radical concentration of about 5 % of that of the molecule, i.e. 0·1 mm for benzyl and 0·01 mm for aniline. This leads to a lower limit for the collision efficiency of recombination of radicals, or radical and atom, of the order of 10^{-2} and 10^{-1} for benzyl and anilino respectively. These radicals therefore show a reactivity with respect to recombination which is typical of free radicals such as methyl, though their reactivity with other molecules must be far less.

The appearance of the spectra is interesting. In each case a single band is predominant and the width of this band increases in the order benzyl, anilino, phenoxy, whilst the separation from the spectrum of the parent molecule decreases. The very sharp line-like structure of benzyl and p-xylyl is also found in triphenyl methyl [11] even in solution, and seems to be characteristic of this type of molecule. It is probably a reflection of the loose coupling between the odd electron and the rest of the molecule which results in an " atomic line " spectrum with little vibrational excitation. The increasing diffuseness in anilino and phenoxy implies a shorter lifetime of the upper state such as could result from an increasing probability of dissociation.

There is little doubt that these investigations could be extended to a number of similar molecules. Unfortunately the radical lifetimes are beyond the time resolution of the technique employed and, therefore, not only are kinetic studies

1474   PRIMARY PHOTOCHEMICAL PROCESSES

impossible but it is probable that in many cases free radicals are formed which escape detection. As a result of this work, however, similar studies have been initiated in solution and in rigid media where these disadvantages are less evident. Studies of this kind, which will be reported later, fully confirm the assignments given here and extend the observations to a wide range of similar radicals.

NOTE ADDED IN PROOF

The ' V spectrum' referred to in the introduction has recently been reinvestigated by Schüler and Michel [12] who have put forward new evidence supporting the assignment to benzyl. The intensity maximum of this system occurs at 4477 Å (22,330 cm^{-1}). Longuet-Higgins and Pople [13] have predicted that the two lowest transitions in benzyl should occur at 27,900 and 33,700 cm^{-1} and that the latter should be much the stronger. Bingel [14] has reached similar conclusions but predicts the lower transition at 21,500 cm^{-1}.

As pointed out by Schüler and Michel, the emission spectra of polyatomic molecules almost invariably consists of the transition from the lowest excited state only. On the other hand, in absorption, we shall observe most readily the system with highest transition probability. The observation of two different systems from benzyl, at 22,300 and 32,600 cm^{-1}, by Schüler and ourselves respectively is therefore readily understood and is in fair quantitative agreement with the predictions of molecular orbital calculations.

[1] Gomberg, *Ber.*, 1900, **33**, 3150 ; *J. Amer. Chem. Soc.*, 1900, **22**, 757.

[2] Wheland, *The Theory of Resonance* (Wiley, 1944).

[3] Szwarc, *J. Chem. Physics*, 1948, **16**, 128.

[4] Paneth and Lautsch, *J. Chem. Soc.*, 1935, 380.

[5] Szwarc, *Chem. Rev.*, 1950, **47**, 75.   [6] Porter, *J. Phys. Radium*, 1954, **15**, 113.

[7] Schüler, Reinebeck and Köberle, *Z. Naturforschung*, 1952, **7a**, 421, 428.

[8] Walker and Barrow, *Trans. Faraday Soc.*, 1954, **50**, 541.

[9] Porter and Wright, *Trans. Faraday Soc.*, 1955, **51**, 1205.

[10] Norman and Porter, *Proc. Roy. Soc. A*, 1955, **230**, 399.

[11] Anderson, *J. Amer. Chem. Soc.*, 1935, **57**, 1673.

[12] Schüler and Michel, *Z. Naturforschung*, 1955, **10a**.

[13] Longuet-Higgins and Pople, *Proc. Physic. Soc.*, 1955, **68**, 591.

[14] Bingel, *Z. Naturforschung*, 1955, **10a**.

Reprinted from *Proc. Roy. Soc. A* **299**, 348–353 (1967) with permission from The Royal Society

Triplet state quantum yields for some aromatic hydrocarbons and xanthene dyes in dilute solution

By P. G. Bowers

Department of Chemistry, University of British Columbia

And G. Porter, F.R.S.

The Royal Institution, 21 Albemarle Street, London, W. 1

(*Received* 15 December 1966)

Quantum yields of triplet state formation and extinction coefficients of the triplet states have been determined by direct depletion methods for solutions of anthracene, phenanthrene, 1,2,5,6-dibenzanthracene, fluorescein, dibromofluorescein, eosin and erythrosin. The values obtained for the hydrocarbons are in reasonable agreement with those obtained by other workers using energy transfer and heavy atom perturbation techniques.

In all cases which we have studied, the sum of the quantum yields of fluorescence and triplet state formation is equal to unity within the limits of experimental error, showing that radiationless transfer from the excited singlet to the ground state is negligible.

Introduction

In a previous paper (Bowers & Porter 1966) we have shown how it is possible to measure quantum yields for triplet state formation (ϕ_T) using direct flash photolysis, by observing the transient absorption changes under conditions which enable the light absorption to be precisely calculated. In this way it was found that the sum of the fluorescence and triplet yields for the chlorophylls in polar solvents was near unity, but that only a small amount of intersystem crossing occurred in non-polar solvents.

This paper reports some further ϕ_T determinations in dilute solution. Polyacene hydrocarbons have been the subject of several other studies (Lamola & Hammond 1965; Medinger & Wilkinson 1965; Parker & Joyce 1966), and it seemed desirable to verify that our procedure gave similar results. The xanthene dyes were thought to be particularly suited to triplet yield determination by the direct method, because of their intense visible absorption bands. In addition, a knowledge of ϕ_T is the key to evaluation of many of the other rate constants for the complex photochemical processes of these substances.

Experimental

The main features of the method were similar to those which have been described previously for determinations with chlorophyll (Bowers & Porter 1966). A parallel, suitably filtered flash (70 to 200 J) was incident normally on the face of a 5 cm quartz cell of 1 cm square cross-section. Triplet absorption or ground state depletion was monitored perpendicular to the direction of excitation using a conventional photoelectric arrangement. The incident flash intensity was measured by ferrioxalate actinometry.

349   *Triplet state quantum yields*

The dyes were studied as their dianions in boric acid + borax pH 9 buffer solutions. Fluorescein and erythrosin had been purified chromatographically. Aromatic hydrocarbons were recrystallized and sublimed before use. 3-*Me* pentane and liquid paraffin was purified by the procedure of Godfrey & Porter (1966). All samples were subjected to thorough degassing before flashing. For liquid paraffin solutions, the degassing was facilitated by maintaining the liquid at 80 to 100 °C in a water bath.

The following filter combinations were used: anthracene, OX7 glass + saturated $CuSO_4$ (2 cm); phenanthrene, OX7 glass + Pyrex plate (2 mm) + 1·4 M $NiSO_4$ and 0·06 M $CuSO_4$ (2 cm); 1,2,5,6-dibenzanthracene (*DBA*), OX7 glass + saturated $Cu(NO_3)_2$ (2 cm) xanthene dyes, Ilford 303 glass + 2 cm water.

Extinction coefficients

The hydrocarbons were studied by excitation into the weak bands in the near ultra violet, and by monitoring the intensity of triplet-triplet absorption in the visible. Maximum triplet extinction coefficients in each case were found by subjecting a dilute sample (*ca.* 10^{-7} M) to an unfiltered flash of 1000 J, in order to achieve complete conversion to the triplet state. Owing to some decomposition, the initial (i.e. zero-time) optical density change decreased by about 10 % with successive flashes, and was extrapolated to 'zero flash' for computation of the extinction coefficient. Several determinations using samples of different concentrations were carried out for each hydrocarbon.

Measurements on the dyes were made by observing depletion of their ground state absorption bands around 500 nm. The triplet extinction coefficients were estimated by taking the transient depletion spectrum (from a weak filtered flash) over a 30 nm range surrounding the maximum, and comparing this with the ground state absorption spectrum. Then, assuming ϵ_3 (triplet) to be approximately constant over the range, for two wavelengths λ_a and λ_b, giving transient optical density changes $\Delta\mu^a$ and $\Delta\mu^b$,

$$\epsilon_3 = \frac{\Delta\mu^b \epsilon_1^a - \Delta\mu^a \epsilon_1^b}{\epsilon_1^b - \epsilon_1^a}. \tag{1}$$

The relevant triplet extinction coefficients, and the ground state values on which they are based, are summarized in table 1.

TABLE 1. EXTINCTION COEFFICIENTS, $\epsilon \times 10^{-4}$ (M^{-1} cm^{-1}) and λ (nm).

compound	solvent	singlet		triplet	
anthracene	liquid paraffin	0·717	378	6·3 ± 0·5	424
phenanthrene	liquid paraffin	0·022	346	2·4 ± 0·2	480
1,2,5,6-dibenzanthracene	liquid paraffin	0·113	394	2·2 ± 0·3	535
fluorescein	aqueous, pH 9	8·8	489	1·5 ± 0·4	489
dibromofluorescein	aqueous, pH 9	9·5	506	1·8 ± 0·6	506
eosin	aqueous, pH 9	9·9	518	2·8 ± 0·4	518
erythrosin	aqueous, pH 9	*10·0	526	2·6 ± 0·7	526

* Assumed.

Sources of data on singlet states are: fluorescein (Lindquist 1960); dibromofluorescein (S. Emmons 1966, personal communication); eosin (Lindquist & Kasche 1965).

P. G. Bowers and G. Porter   350

Calculation of quantum yields

The amount of triplet state formed per flash was found by extrapolating the triplet decay curve (hydrocarbons), or ground state recovery curve (dyes), back to zero time, and combining the initial optical density changes with the appropriate extinction coefficients.

The absorbed intensity I_a was calculated from the incident intensity I_0 by wavenumber integration

$$\frac{I_a}{I_0} = \frac{\int T_f(\nu)\, R(\nu)\, [1 - T_s(\nu)]\, \mathrm{d}\nu}{\int T_f(\nu)\, R(\nu)\, \mathrm{d}\nu};$$

T_f and T_s are the fractional transmissions of filter and sample respectively, and R is the relative output of the flash lamp (Shaw 1967). Typical curves are shown in figure 1. In the case of the filter used in irradiating the dyes, it was necessary to do

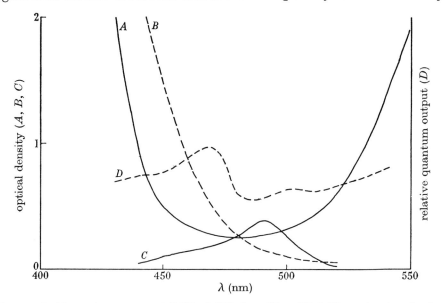

FIGURE 1. Absorption spectra: (A) Ilford 303 glass filter, (B) 0·15 M potassium ferrioxalate, 1 cm path; (C) fluorescein, 8×10^{-7} M, 5 cm path; (D) relative intensity of exciting flash.

a similar integration to calculate I_0 from the fraction of light absorbed by the actinometer solution. None of the compounds studied showed any appreciable permanent decomposition after exposure to about 20 flashes.

The hydrocarbon solutions were 10^{-4} to 10^{-5} M, and the dyes 10^{-6} to 10^{-7} M. These concentrations were chosen to give about 10 % absorption of the incident flash excitation. The value of the apparent triplet quantum yield, $\phi_{\mathrm{app.}}$, measured as the ratio of the amount of triplet formed to the calculated intensity of light absorbed, increased with decreasing intensity in the case of the dyes (figure 2). This was because at the highest intensities the calculated light absorbed corresponded to more than 20 % conversion of the ground state molecules initially present. The reduction in $\phi_{\mathrm{app.}}$ at such significant depletions is due to several factors such as re-circulation, and has been discussed previously (Bowers & Porter 1966). The true

351   *Triplet state quantum yields*

triplet yield was found by extrapolating $\phi_{app.}$ to zero intensity. As figure 2 shows, such a correction was not required for anthracene or the other two hydrocarbons, because the more concentrated solutions were only depleted by about 1 %.

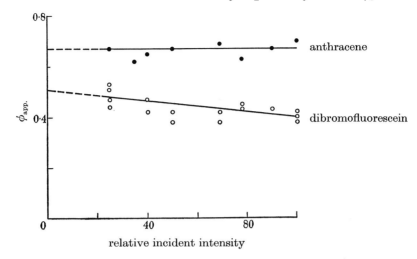

FIGURE 2. Typical plots of the apparent triplet quantum yield against incident intensity.

RESULTS AND DISCUSSION

The triplet extinction coefficients found for the hydrocarbons (table 1) may be compared with several literature values. Porter & Windsor (1958) obtained $7.15 \times 10^4 \mathrm{M}^{-1} \mathrm{cm}^{-1}$ for anthracene in liquid paraffin and more recently Wild & Gunthard (1965) have found $6.7 \times 10^4 \mathrm{M}^{-1} \mathrm{cm}^{-1}$ for glycerol solutions. A value of $2.7 \times 10^4 \mathrm{M}^{-1} \mathrm{cm}^{-1}$ for the maximum triplet extinction coefficient of phenanthrene in a rigid glass at low temperature has been measured by Keller & Hadley (1965). These are all in good accord with our determinations, although Porter & Windsor's value of $6.7 \times 10^4 \mathrm{M}^{-1} \mathrm{cm}^{-1}$ for 1,2,5,6-*DBA* in hexane is about three times as high as the value we obtained.

The method based on equation (1) for determining ϵ_3 for the dyes is not particularly precise, as the error limits show, but since the quantum yields are calculated from $\epsilon_1 - \epsilon_3$, this uncertainty is greatly reduced in the final results.

The triplet yields, generally the mean result of several experiments for each compound, are listed in table 2, together with the fluorescence yields under comparable conditions, where these have been measured. Notice that phenanthrene and 1,2,5,6-*DBA* were studied in 3-*Me* pentane, assuming the same triplet extinction coefficients at maximum absorption as in liquid paraffin.

The figure of 0.58 for anthracene compares with a value of 0.75 found by Medinger & Wilkinson (1965) using heavy atom quenching, and a fluorescence yield of 0.33. Lamola & Hammond (1965), using the method of sensitized *cis-trans* isomerization, obtained 0.76 and 0.89 for phenanthrene and 1,2,5,6-*DBA* respectively. The extent of agreement between these published figures and our values is some assurance that there are no large undetected sources of error in the method.

P. G. Bowers and G. Porter

352

TABLE 2. TRIPLET QUANTUM YIELDS AT 23 °C

compound	solvent	concentration (M)	Φ_T	Φ_F
anthracene	liquid paraffin	2×10^{-5}	0.58 ± 0.10	0.33
phenanthrene	3-Me pentane	5×10^{-4}	0.70 ± 0.12	0.14
1,2,5,6-dibenzanthracene	3-Me pentane	1×10^{-4}	1.03 ± 0.16	—
fluorescein (Fl)	aqueous, pH 9	8×10^{-7}	0.05 ± 0.02	0.92
$FlBr_2$	aqueous, pH 9	8×10^{-7}	0.49 ± 0.07	—
eosin ($FlBr_4$)	aqueous, pH 9	6×10^{-7}	0.71 ± 0.10	0.19
erythrosin (FlI_4)	aqueous, pH 9	9×10^{-7}	1.07 ± 0.13	0.02

There is a growing body of evidence that radiationless $S_1 \to S_0$ conversion plays at most only a minor part in the primary process of the polyacenes. However, until there is a great improvement in techniques it will remain very difficult to account with certainty for the 10 to 15 % of excited molecules covered by experimental error.

The value of ϕ_T for fluorescein is given within rather wide limits, because only a very small amount of depletion could be observed. The effect of halogenation is very marked in increasing the intersystem crossing efficiency (and also in decreasing the triplet lifetime, although detailed studies were not made). In all cases, the fraction of excited singlet molecules which does not undergo fluorescence or intersystem crossing must be quite small.

Previous estimates of ϕ_T for the xanthene dyes have been made from indirect measurements. From a study of their photoreduction by allyl thiourea, Adelman & Oster (1956) quote 0.02, 0.12, 0.09, and 0.05 for the triplet yields of the sodium salts of Fl, $FlBr_2$, $FlBr_4$, and FlI_4 respectively at pH 7. The differences between these figures and our values are much too large to be covered by experimental errors. In more recent work, Koizumi and his co-workers (Ohno, Usui & Koizumi 1965; Ohno, Kato & Koizumi 1966; Momose, Uchida & Koizumi 1965) have found that the ϕ_T values measured in photoreduction experiments appear to vary considerably with the nature of the reductant and with pH, and it seems that the overall photoreduction kinetics are more complicated than was originally thought. In view of this, the low ϕ_T values must be regarded with some suspicion.

In the solvents glycerol and ethanol, Parker & Hatchard (1961) have found ϕ_T for eosin, from delayed fluorescence measurements, to be 0.06 and 0.02 respectively. The corresponding fluorescence yields (0.72 and 0.45) are much higher than in aqueous solution.

REFERENCES

Adelman, A. H. & Oster, G. 1956 *J. Am. Chem. Soc.* **78**, 3977.
Bowers, P. G. & Porter, G. 1966 *Proc. Roy. Soc.* A **296**, 435.
Godfrey, T. S. & Porter, G. 1966 *Trans. Faraday Soc.* **62**, 7.
Keller, R. A. & Hadley, S. G. 1965 *J. Chem. Phys.* **42**, 2382.
Lamola, A. A. & Hammond, G. S. 1965 *J. Chem. Phys.* **43**, 2129.
Lindquist, L. 1960 *Ark. Kemi.* **16**, 79.
Lindquist, L. & Kasche, V. 1965 *Photochem. Photobiol.* **4**, 923.
Medinger, T. & Wilkinson, F. 1965 *Trans. Faraday Soc.* **61**, 620.
Momose, Y., Uchida, K. & Koizumi, M. 1965 *Bull. Chem. Soc. Japan* **38**, 1601.

353   *Triplet state quantum yields*

Ohno, T., Kato, S. & Koizumi, M. 1966 *Bull. Chem. Soc.* Japan **39**, 232.

Ohno, T., Usui, Y. & Koizumi, M. 1965 *Bull. Chem. Soc. Japan* **38**, 1022.

Parker, C. A. & Hatchard, C. G. 1961 *Trans. Faraday Soc.* **57**, 1894.

Parker, C. A. & Joyce, T. A. 1966 *Chem. Commun.* 234.

Porter, G. & Windsor, M. W. 1958 *Proc. Roy. Soc.* A **245**, 238.

Shaw, G. 1967 Ph.D. Thesis, University of Sheffield.

Wild, U. & Gunthard, H. H. 1965 *Helv. Chem. Acta* **48**, 1843.

Reprinted with permission from *Nature* **174**, 508 (1954)

Trapped Atoms and Radicals in a Glass 'Cage'

THE preparation of highly reactive free radicals and atoms in non-equilibrium concentrations high enough for their direct observation by physical methods has hitherto depended on the use of special rapid techniques such as that of flash photolysis. Here we describe a general method for the preparation of these species under conditions which permit their observation at high concentrations for an indefinitely long period of time. The method involves the photolytic dissociation of a substance dissolved in a transparent rigid solvent at very low temperatures. The only related work which we have been able to find is that of Lewis and Lipkin, who showed that certain rather stable radicals, such as triphenyl methyl, could be detected after photolysis of the parent saturated molecule[1].

The essential conditions to be fulfilled if a reactive species, such as an iodine atom, is to be permanently isolated are : (1) diffusion to the vicinity of another atom must be prevented by using a very viscous solvent ; (2) reaction with the solvent must not occur ; such reactions are usually accompanied by appreciable activation energies and may therefore be avoided by working at a sufficiently low temperature. We have found that both conditions are fulfilled, for most radicals so far investigated, by using glasses formed from a mixture of hydrocarbons or from ether, alcohol and *iso*pentane at the temperature of liquid nitrogen.

The method was first tested in a very simple manner by irradiating a 10^{-3} M solution of molecular iodine in a hydrocarbon glass at 86° K., using a conventional high-pressure mercury arc. The absorption spectrum of the iodine molecule occurs at a wave-length considerably less than that corresponding to the energy of dissociation to normal atoms, and it was believed that this energy might be sufficient to allow the atoms to diffuse apart through the rigid solvent by a process of local softening of the glass. Rapid dissipation of this energy would then result in a system of trapped iodine atoms. In complete accordance with this reasoning, we have found that, on irradiation, such an iodine solution becomes colourless throughout the visible and quartz ultra-violet regions of the spectrum and remains so indefinitely at 86° K. On softening the glass, by warming, the colour returns as the atoms recombine.

Further experiments have established that the phenomenon is a very general one. The results of photolysis of a number of other substances in glassy solvents at liquid nitrogen temperatures are summarized below.

(1) *Ethyl iodide.* On irradiation, the spectrum of ethyl iodide decreases in intensity; but no other spectrum appears. On warming, molecular iodine is formed and recooling has no further effect. We conclude that the iodine atoms have been trapped as before, and it seems probable that trapped ethyl radicals were also present.

(2) *Carbon disulphide.* The first four bands of the $v'' = 0$ progression of CS appear at very high intensity after several minutes irradiation and can be observed for many hours. The spectrum is removed instantly on warming and cannot be detected during irradiation of an ordinary solution at room temperatures. In the gas phase, carbon disulphide disappears by polymerization in a few minutes[2].

(3) *Chlorine dioxide.* The spectrum of chlorine dioxide, which, is mainly diffuse in the glassy solvent, is rapidly destroyed and replaced by a second diffuse spectrum the position of which agrees exactly with that of the ClO radical[3]. It has been shown[4] that the flash photolysis of gaseous chlorine dioxide results in the formation of ClO, the life-time of which at comparable concentrations to ours is only a few milliseconds.

(4) *Aromatic compounds.* Porter and Wright have recently shown that the gas-phase photolysis of many single-ringed aromatic molecules results in the appearance of transient banded spectra with life-times less than 10^{-1} sec., which are attributed to free radicals such as benzyl[5]. We have observed similar spectra from the same series of compounds for several hours after photolysis in rigid solvents. These spectra vanished completely on warming the glass, and were therefore readily distinguished from the spectra of permanent products.

In all these examples the spectra were observed, with undiminished intensity, several hours after photolysis; they were removed by softening the glass and did not reappear on cooling.

This method should have many applications. It may be used for trapping the primary products of photochemical or radiation chemical processes for identification. It makes possible the measurement of properties such as infra-red spectra of free radicals, which has not hitherto been possible owing to the need for rapid recording. Bearing in mind the possibility of using still lower temperatures and other solvents, the method should be applicable to practically all free radicals which can be produced by photolysis.

Full details of our techniques and results will be submitted for publication in the *Transactions of the Faraday Society*. One of us (I. N.) is grateful to the Fulbright Commission for a fellowship during the tenure of which this work was carried out.

IRWIN NORMAN
GEORGE PORTER

Physical Chemistry Department,
  University of Cambridge.
    June 22.

[1] Lewis, G. N., and Lipkin, D., *J. Amer. Chem. Soc.*, **64**, 2801 (1942).
[2] Porter, G., *Proc. Roy. Soc.*, A, **200**, 284 (1950).
[3] Porter, G., *Dis. Farad. Soc.*, **9**, 60 (1950).
[4] Lipscomb, F. J., Norrish, R. G. W., and Porter, G., *Nature* (in the press).
[5] Porter, G., and Wright, F. J. (in course of publication).

Printed in Great Britain by Fisher, Knight & Co., Ltd., St. Albans.

Trans. Faraday Soc. 1595–1604 (1958) – Reproduced by permission of the Royal Society of Chemistry

PRIMARY PHOTOCHEMICAL PROCESSES IN AROMATIC MOLECULES

PART 4.—SIDE-CHAIN PHOTOLYSIS IN RIGID MEDIA

BY G. PORTER AND E. STRACHAN

Chemistry Dept., The University of Sheffield
British Rayon Research Association

Received 14th April, 1958

Two processes of side-chain fission have been identified in the photolysis of substituted aromatic molecules in rigid solvents at low temperatures. One involves fission of a β bond to yield two radicals and the other results directly in the formation of two molecules, one of which is styrene. Quantum yields at 2537 Å in one example of each process, were $1 \cdot 1 \times 10^{-2}$ and $3 \cdot 6 \times 10^{-2}$ respectively. The radical products have been identified spectroscopically and conditions necessary for their stabilization have been investigated. Generalizations concerning the relative probability of dissociation of different bonds in equivalent positions are applicable to the forty molecules which have been studied.

The primary photolytic bond dissociation processes in aromatic compounds have received relatively little ,attention owing to the low quantum yields of dissociation and the difficulties of analysis of the complex products which are often formed. Previous studies have been based mainly on analysis of the gaseous products of reaction and the work most relevant to the present discussion is that of Hentz, Sworski and Burton [1, 2] who found evidence for fission in the side chain of toluene with a quantum yield of gaseous products of about 1 %.

The method which we have used here is one which eliminates the complexities of secondary reactions. It is based on the observations of Norman and Porter [3] who found that toluene and related molecules dissociate photochemically in rigid media and that aromatic free radicals remain trapped in the matrix at low temperatures, and can be observed spectroscopically. The spectra were first observed following flash photolysis of aromatic vapours by Porter and Wright (part 3) [4] and were attributed to benzyl and its derivatives. It has recently been shown that they are also formed during photolysis of ordinary solutions at normal temperatures.[5] Whilst further studies in gases and liquids will be necessary for the interpretation of the kinetics of the radical reactions, the matrix isolation technique is in many ways more suitable for the identification of the radical spectra and of primary photochemical processes owing to the elimination of most secondary reactions.

EXPERIMENTAL

The apparatus and procedure used in the main part of this investigation were identical with those described by Norman and Porter,[3] the cell length being 1·5 cm. In later experiments a different arrangement incorporating a cell 20 cm in length was used to detect weaker transitions.[6] Spectra were recorded by means of the Hilger medium and small quartz instruments and scanned by microdensitometer. All solvents were spectroscopically pure and were carefully dried in order to prevent crystallization of the glass. The aromatic compounds available as commercial products were purified by fractional distillation, recrystallization or vacuum sublimation and boiling points or melting points in every case were within one degree of those reported in the literature. Some of the particular methods used were: toluene, ethyl benzene, *iso*propyl benzene, benzyl chloride, benzyl cyanide, anisole, phenetole purified by the method of Vogel,[7] benzyl alcohol according to Mathews,[8] aniline according to Knowles,[9] o-xylene purified by fractional distillation, m- and p-xylenes by fractional crystallization.   Phenyl ethyl chloride was prepared from the corresponding alcohol by the method of Norris and Taylor.[10]

A high-intensity light source was used in most experiments involving the 1·5 cm cell. This was a 1 kw high-pressure mercury arc, type ME/D (combined with a 2-cm water filter), which emits most lines of mercury, except for the 2537 line which is reversed, and also a continuous spectrum throughout the visible and ultra-violet region.   For quantum-yield determinations a low pressure mercury-vapour lamp, combined with a 1-cm filter of 4 N acetic acid was used from which the only significant radiation in the region of aromatic absorption was at 2537 Å.   The 20-cm cell was irradiated from the side by two U-tube low pressure mercury-vapour lamps ; the length of each limb was 20 cm and the whole apparatus was surrounded by a reflector.

The solvents generally used to form glasses were E.P.A. (ether, isopentane and ethanol in proportions 5 : 5 : 2) and M.P. (methyl cyclohexane and isopentane in proportions 2 : 3). Solutions were always outgassed and sealed off before irradiation and all experiments were performed at − 197°C.

RESULTS

Four types of primary dissociation process have been identified in this work.   These are as follows.

(i) RING FISSION

This has previously been suggested to explain the formation of hexatriene during photolysis of benzene in rigid glasses.[11,3.]   Spectra very similar to that of hexatriene are also observed in the permanent products of photolysis of toluene and related molecules in rigid glasses and these are probably substituted hexatrienes formed by ring fission followed by abstraction of two atoms from the solvent.   The quantum yield of this process in benzene is approximately 0·01 at 2537 Å (Anderton and Porter, unpublished work).

(ii) ELECTRON EJECTION

This was first clearly established by Lewis and his collaborators in a number of aromatic molecules.[12]   We have found spectra of radical cations which must result from this process, after photolysis of many amines.   Unlike processes (i), (iii) and (iv), electron ejection (photo-oxidation in the terminology of Lewis) does not readily occur in non-polar glasses (e.g. M.P.), and it is normally encountered only in basic molecules.

(iii) SIDE-CHAIN DISSOCIATION AT THE β BOND

This is exemplified by the formation of the benzyl radical from toluene.   Benzyl and its derivatives have strong banded spectra in the near u.-v. region, which are formed on irradiation of rigid solutions, remain as long as the glass is kept rigid and disappear completely and irreversibly on warming.[3]

(iv) SIDE-CHAIN DISSOCIATION INTO TWO MOLECULES, ONE OF WKICH IS STYRENE

This occurs rather generally in compounds of the type $C_6H_5CHXCH_2Y$.   Styrene is observed in the rigid glass before warming and is therefore formed in the primary act.

A fifth process, dissociation into two ions, was discussed by Lewis but we have found no evidence for this reaction in any of the systems considered here.

Processes (i) and (ii) will be discussed further in later communications and the present work is concerned mainly with processes (iii) and (iv).

ASSIGNMENTS OF RADICAL SPECTRA

In part 3 arguments were given for assigning the common transient spectrum, observed on photolysis of toluene and other benzyl compounds in the gas phase, to benzyl and it was mentioned that more definite evidence was available from our work in rigid solvents. This will now be given.   Common transient spectra, satisfying all energetic requirements, could arise from the primary photochemical dissociation of molecules of formula $C_6H_5CH_2X$ in only three ways :

$$C_6H_5CH_2X \longrightarrow C_6H_5CH_2 + X, \tag{1}$$

$$C_6H_5CH_2X \longrightarrow C_6H_5CH + HX, \tag{2}$$

$$C_6H_5CH_2X \longrightarrow C_6H_5 + CH_2X, \tag{3}$$

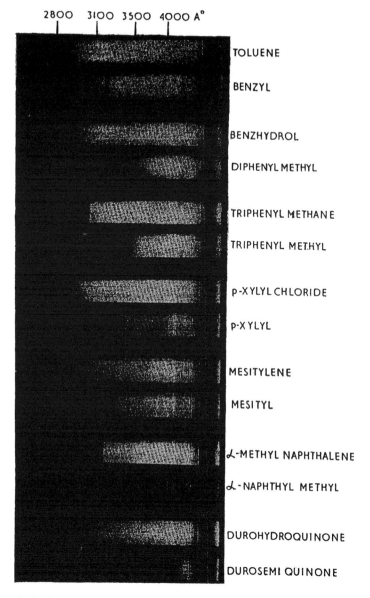

FIG. 1.—Radical spectra resulting from the photolysis of aromatic molecules at − 197°C. In each case the first spectrum is recorded before and the second after irradiation of the rigid solution for 15 min.

[*To face page* 1597

If we consider the series $C_6H_5CH_3$, $C_6H_5CH_2X$, $C_6H_5CHX_2$ and $C_6H_5CX_3$ process (iii) would result in one common spectrum, process (ii) could result in no more than two radical spectra and process (i) in no more than three radical spectra from the four molecules. We have investigated three such series in which X was a chlorine atom, a methyl radical and a phenyl radical respectively. The wavelength maxima of the radical absorption spectra in E.P.A. were as follows.

molecule	max. (Å)	radical
$C_6H_5CH_3$	3187	$C_6H_5CH_2$
$C_6H_5CH_2Cl$	3187	$C_6H_5CH_2$
$C_6H_5CHCl_2$	3231	C_6H_5CHCl
$C_6H_5CCl_3$	3238	$C_6H_5CCl_2$
$C_6H_5CH_2CH_3$	3222	$C_6H_5CHCH_3$
$C_6H_5CH(CH_3)_2$	3242	$C_6H_5C(CH_3)_2$
$C_6H_5C(CH_3)_3$	3242	$C_6H_5C(CH_3)_2$
$C_6H_5CH_2C_6H_5$	3355	$C_6H_5CHC_6H_5$
$C_6H_5CH(C_6H_5)_2$	3415	$C_6H_5C(C_6H_5)_2$

In each series three distinct spectra were observed, providing excellent confirmation that the dissociation process involved is β bond fission. The spectra can be assigned with some confidence to the above radicals with the exception of the 3231 Å band observed from benzal chloride. This spectrum was rather weak and, although definitely different from any other band of the series, its position and appearance were rather sensitive to concentration and other factors and its assignment should be regarded as uncertain pending further examination.

Further confirmation of the assignments is afforded by the fact that one spectrum in the series, that of triphenyl methyl, was previously well known and established being, in fact, the first free radical spectrum to be observed by Gomberg in 1900.[13] The colour of this radical is caused by a weaker absorption band at 5110 Å and we have detected this band in the present work by use of the 20 cm cell. Finally we have considered the relationship of the absorption bands of benzyl and its derivatives to the emission spectra of Schuler and his collaborators [14] and the molecular orbital calculations of Dewar, Longuet-Higgins and Pople [15, 16] and of Bingel.[17] This led us to search for the predicted weak long-wavelength system of benzyl and we have recently recorded this spectrum by use of the 20 cm cell.[6] Comparison of the data from all these sources gives a consistent account of the spectrum of benzyl and final confirmation of our assignments.[6]

Typical free-radical spectra, obtained after 15 min irradiation are shown in fig. 1. The solvent was E.P.A. in each case except α-methyl naphthalene for which the solvent was M.P. The principal absorption maxima of all spectra which we have assigned to neutral free radicals, and the molecules from which they have been observed after photolysis at $-197°C$ are given in table 1. The intensities are visual estimates referred to the strongest band as 10. The wavelength measurements refer to radicals in E.P.A. in all cases except the two naphthyl methyl radicals which were in M.P.

In most cases the bands were sharp and maxima could be estimated to $\pm 5\,\text{Å}$ whilst comparison could be made to about $\pm 1\,\text{Å}$. The basis of the assignments has been described and although it has not always been possible to obtain a radical from more than one molecule, the close resemblances of the radical spectra in position, band width and structure are strong evidence that radicals of the benzyl type are involved and this is usually sufficient to define the carrier. The spectra all fall in a quite narrow spectral region and some chance coincidences may be expected. The only clear case of coincidence of different radical spectra is found in the three xylyl radicals but, even here, there are distinct differences in band structure; the 3230 Å band in m-xylene, for example, is a doublet.

Only spectra attributed to neutral radicals are recorded in table 1 but in all the aromatic amines other transient spectra which we have assigned to the radical cations appeared on photolysis. Our reasons for not attributing these spectra to neutral radicals must be mentioned briefly. The neutral radicals formed from aniline and its N-methyl derivatives were identified in the same manner as benzyl. Thus the three aniline derivatives

1598   PRIMARY PHOTOCHEMICAL PROCESSES

TABLE 1.—FREE RADICAL SPECTRA

radical	parent molecule	$\lambda A°$	$v\,cm^{-1}$	relative intensitie
$C_6H_5CH_2$	$C_6H_5CH_3$, $C_6H_5CH_2Cl$, $C_6H_5CH_2OH$, $C_6H_5CH_2NH_2$, $C_6H_5CH_2CN$, $C_6H_5CH_2COOH$, $C_6H_5CH_2CH_2NH_2$, $C_6H_5CH_2CH_2OH$, $C_6H_5CH_2CH_2CH_2OH$.	3187	31380	10
		3082	32450	3
		3047	32810	3
$C_6H_5CH.CH_3$	$C_6H_5CH_2CH_3$, $C_6H_5CHOHCH_3$,	3222	31040	10
	$C_6H_5CHNH_2CH_3$, $C_6H_5CH_2CH_2OH$.	3167	31580	5
		3129	31960	6
		3083	32440	6
$C_6H_5C(CH_3)_2$	$C_6H_5CH(CH_3)_2$, $C_6H_5C(CH_3)_3$	3242	30840	
$(C_6H_5)_2CH$	$(C_6H_5)_2CH_2$, $(C_6H_5)_2CHOH$	3355	29810	10
		3305	30260	8
		3240	30860	5
		3180	31450	3
		3122	32030	4
$(C_6H_5)_3C$	$(C_6H_5)_3CH$	3415	29290	10
		3358	29780	8
		3303	30270	2
$C_6H_5CH_2CHC_6H_5$	$(C_6H_5CH_2)_2$	3625	27580	
$o—CH_3C_6H_4CH_2$	$o—CH_3C_6H_4CH_3$	3230	30960	10
		3170	31550	1
		3100	32260	1
$m—CH_3C_6H_4CH_2$	$m—CH_3C_6H_4CH_3$	3230	30960	10
		3100	32260	6
$p-CH_3C_6H_4CH_2$	$p-CH_3C_6H_4CH_3$ $p-CH_3C_6H_4CH_2Cl$	3230	30960	10
		3170	31550	1
		3100	32260	3
		3047	32810	2
$1,3(CH_3)_25CH_2$, C_6H_3	$1,3,5(CH_3)_3C_6H_3$	3249	30770	10
		3109	32170	6
		2964	33730	2
C_6H_5CHCl	$C_6H_5CHCl_2$	3231	30950	10
		3190	31350	5
		3100	32260	4
$C_6H_5CCl_2$	$C_6H_5CCl_3$	3238	30880	10
		3108	32180	8
(naphthalene-CH_2 structure)	(naphthalene-CH_3 structure)	3700	27030	7
		3554	28140	5
		3500	28570	7
		3424	29210	10
(naphthalene-CH_2 structure)	(naphthalene-CH_3 structure)	3840	26040	7
		3647	27420	5
		3687	27120	4
		3500	28570	8
		3424	29210	10
C_6H_5NH	$C_6H_5NH_2$	3109	32060	

G. PORTER AND E. STRACHAN   1599

TABLE 1.—(*Cont.*)

radical	parent molecule	λA°	vcm⁻¹	relative intensities
$C_6H_5NCH_3$	$C_6H_5NHCH_3$ $C_6H_5N(CH_3)_2$	3166	31590	
$C_6H_5CH_2NC_6H_5$ or $C_6H_5CHNHC_6H_5$	$C_6H_5CH_2NHC_6H_5$	3745 3635	26700 27510	10 8
C_6H_5O	C_6H_5OH, $C_6H_5OCH_3$ $C_6H_5OCH_2CH_3$	2870 limit	34840	
$p\text{-}HOC_6H_4O$	$p\text{-}HOC_6H_4OH$	4140	24150	10
	$O=\langle \rangle=O(+RH)$	3550	28170	6
(OH, 2,6-dimethyl phenol radical)	(OH, dimethyl structure)	4220 4041 3300	23700 24740 30300	10 4 8
(OH anthracene structure)	(O anthraquinone structure) (+ RH)	5100 3830 3531	19610 26110 28320	10 8 5

can give only two aniline type radicals and the near u.-v. spectra were therefore assigned as follows.

molecule	max. (Å)	radical
$C_6H_5NH_2$	3109	C_6H_5NH
$C_6H_5NHCH_3$	3166	$C_6H_5NCH_3$
$C_6H_5N(CH_3)_2$	3166	$C_6H_5NCH_3$.

These spectra were observed in both the polar E.P.A. glass and the non-polar M.P. glass. Other bands were observed in the visible region which were different for each of the three molecules and could not therefore be spectra of the two anilino radicals. Furthermore the visible bands were not observed in gas-phase flash photolysis and were absent or very weak in M.P. glass. Irradiation of N : N'-dimethyl p-phenylene diamine in E.P.A. yielded the well-known spectrum of the radical cation [18] at very high intensity whilst no such spectra were observed after irradiation in the M.P. glass. Other spectra showing these characteristics are therefore also attributed to radical cations and will be discussed in a later communication.

SIDE-CHAIN DISSOCIATION INTO TWO MOLECULES

A number of substances, after photolysis in E.P.A., gave a common banded spectrum which was still present after warming the glass. The spectrum was similar to that of styrene reported in the literature, and its identity was confirmed by comparison with a solution of pure styrene in our solvents both at room temperature and in the glass at − 197°C. The substances from which styrene was formed on irradiation in the glass were ϕCH_2CH_3, $\phi CH(OH)CH_3$, $\phi CH(NH_2)CH_3$, ϕCH_2CH_2OH, $\phi CH_2CH_2NH_2$, ϕCH_2CH_2Cl and ϕCH_2CH_2Br. Some of these also gave radical products which have been discussed ; others such as ϕCH_2CH_2Cl and ϕCH_2CH_2Br gave no radical products and styrene was the only spectrum observed. Permanent products were also observed

1600   PRIMARY PHOTOCHEMICAL PROCESSES

in the spectra of other compounds, some of which, although they have not been identified, were probably not of the hexatriene type, but no compounds, other than those mentioned above, showed the styrene spectrum.   The only substances investigated which gave neither transient spectra not styrene were benzyl bromide and duroquinone.

DETERMINATION OF QUANTUM YIELDS

Quantum yields of processes (iii) and (iv) were determined in a representative example of each type of decomposition.   For process (iii) we had only one possibility—the photolysis of triphenyl methane to give triphenyl methyl—since extinction coefficients of none of the other radicals are yet known.   For process (iv) phenyl ethyl bromide was chosen because styrene is the only product detected in the photolysis of this molecule and extinctions can therefore be determined without interference from other products.

The source of radiation in both cases was the filtered low-pressure mercury arc which gave essentially monochromatic light at 2537 Å.   The incident intensity was determined by conventional uranyl oxalate actinometry, and the concentrations of products were found by photometry of the photographic absorption spectra.

β BOND DISSOCIATION IN TRIPHENYL METHANE

A solution of triphenyl methane (2·44 g/l.) in E.P.A. was irradiated for 30 min and the concentration of triphenyl methyl determined from its extinction at 3450 Å using the extinction coefficients given by Chu and Weissman.[19]   (These authors used a toluene + triethylamine solvent and the extinction coefficients may therefore differ slightly from those in E.P.A.)   Three determinations of quantum yield gave the values $1·10 \times 10^{-2}$, $1·13 \times 10^{-2}$ and $1·08 \times 10^{-2}$ and a mean value,

$$\phi = 1·11 \times 10^{-2}.$$

STYRENE FORMATION FROM PHENYL ETHYL BROMIDE

A 10^{-3} M solution of phenyl ethyl bromide in E.P.A. was irradiated at the temperature of liquid nitrogen for 2 h and the styrene formed was determined spectrophotometrically after warming the glass to room temperature.   The whole spectrum was measured to confirm its identity and quantitative estimation was based on the first absorption maximum at 2905 Å.   The parent substance does not absorb at this wavelength.   The concentration of styrene was determined by direct comparison with extinction measurements made on solutions of pure styrene in the same solvent.   Three determinations of quantum yield gave the values $3·6 \times 10^{-2}$, $4·2 \times 10^{-2}$ and $3·0 \times 10^{-2}$ and a mean value,

$$\phi = 3·6 \times 10^{-2}.$$

The similarity of absorption intensities of the benzyl type radicals obtained from different molecules at comparable optical densities and times of irradiation and also of the amounts of styrene formed from the molecules listed in the last section suggest that the quantum yields determined probably represent the order of magnitude of quantum yields of processes (iii) and (iv) in most other molecules investigated.

MATRIX REQUIREMENTS FOR STABILIZATION

All experiments on the stabilization of aromatic radicals have been carried out in glasses at $-197°$C.   The low temperature has the twofold effect of ensuring high rigidity of the glass and so lowering the rate of diffusion and also of reducing the rate of chemical reactions which have a finite activation energy.   It is interesting to enquire to what extent a low temperature is necessary for stabilization of benzyl type radicals in matrices which remain rigid at higher temperatures.   Two systems were investigated in order to throw light on this question.

(i) Toluene was irradiated in liquid paraffin (nujol) at $-78°$C at which temperature the glass is quite rigid.   No benzyl radical nor any other transient species were observed.

(ii) α-Methyl naphthalene was irradiated in a matrix of polymethyl methacrylate. Solutions of the α-methyl naphthalene in polymethyl methacrylate were cast into films from chloroform and, after outgassing for several days, the film was cut into discs 1 mm thick and 1 cm in diameter.   Five of these discs were placed together to form a cylinder 0·5 cm long, and this was irradiated in liquid nitrogen in the same manner as our solutions in E.P.A. or M.P.   After irradiation in liquid nitrogen, spectra appeared which were identical with those found in the similar experiment in E.P.A. and described in table 1. On raising the temperature to $-78°$C the spectra disappeared, although the matrix is of course still rigid in a macroscopic sense, at much higher temperatures.   These experiments show that high viscosity of the matrix is not a sufficient condition for stabilization of benzyl type radicals, even at − 78°C and that low temperatures are also necessary.

Photolysis of toluene in polymethyl methacrylate in a similar manner was unsuccessful because of the poor transparency of the polymer in the region of toluene absorption.   No benzyl radical bands were observed but a region of absorption with maximum at 350 mμ, and half-width 35 mμ, appeared after 5 min irradiation and disappeared completely when the polymer was allowed to come to room temperature.   An identical spectrum was obtained from the photolysis, at liquid nitrogen temperature, of polymethyl methacrylate alone.   That the transient spectrum was not a product of photochemical dissociation of chloroform, from which the film had been cast, was established by the fact that the identical spectrum was obtained when the polymethyl methacrylate film was cast from ethyl acetate solution and also the photolysis of chloroform in E.P.A. at − 197°C gave no transient spectrum.   The 350 mμ spectrum is therefore to be attributed to a primary product of photodecomposition of polymethyl methacrylate which is unstable at room temperature.   Its assignment must await further investigations on related molecules.

DISCUSSION

Of the forty molecules investigated all except benzyl bromide undergo side-chain fission by process (iii) or (iv) or both, on irradiation in the near ultra-violet region.   The fission processes which have been identified are summarized in table 2. Of the two primary products of dissociation, only one has been observed in each case and the other is inferred on the grounds that the process written is probably the only one which is energetically possible.

In both types of dissociation, energy is transferred intramolecularly from an excited π electron to the β bond.   Light of wavelength 2537 Å is quite close to the origin of the first singlet-singlet transition in the benzene derivatives and there is therefore little excess energy of vibration in the excited molecule.   Side-chain fission must therefore occur by a predissociation mechanism, i.e. a crossing to a second electronic state which is repulsive in the bond parameter concerned.   For β bond fission the state concerned is probably the triplet formed by combination of the ground doublet states of the two radicals.   The potential energy curves for this process, in particular for toluene, have recently been discussed by one of us.[20]

Comparison of β bond fission processes in the series of related molecules given in table 2 allows some interesting conclusions to be reached concerning the relative probabilities of dissociation of different bonds at equivalent positions.   In order to make such comparisons we shall assume that the extinction coefficients of the benzyl-type radicals are not greatly different so that the optical densities observed are a fair measure of radical concentrations.   The spectra in the cases to be considered are so similar as to be nearly indistinguishable so that this assumption is reasonable.   Now if we study the radicals which are found after dissociation of those molecules which possess more than one kind of β bond we find, in nearly all cases, that only one of these bonds is dissociated.   Furthermore all the data of table 2, including dissociation of the amines as well as the hydrocarbons, are in accordance with the statement that the probability of separation of a radical at the β bond lies in the order:

$$\left[\begin{array}{l} \text{OH, NH}_2\text{, Cl, CN,} \\ \text{COOH, CH}_2\text{NH}_2\text{, CH}_2\text{OH,} \end{array}\right] > \text{H} > \left[\text{CH}_3\text{, C}_2\text{H}_5\text{, C}_6\text{H}_5\text{, C}_6\text{H}_5\text{CH}_2\right].$$

This result is somewhat unexpected.   It would not have been surprising if no regularities had been found since small differences in potential energy curves could greatly affect crossing probabilities.   But, if, as we find, the bonds can be arranged in an order of dissociation probability which is applicable to a large number of different molecules it might be expected that this order would be the order of bond dissociation energies.   This is not the case.   For example, the bond energies of $C_6H_5CH_2$—H, $C_6H_5CH_2$—CH_3 and $C_6H_5CH_2$—$CH_2C_6H_5$ in the gas phase are 77·5, 63 and 47 kcal/mole respectively [21] and there is little doubt

1602   PRIMARY PHOTOCHEMICAL PROCESSES

TABLE 2.—SIDE-CHAIN FISSION PROCESSES

$$C_6H_5CH_3 \rightarrow C_6H_5CH_2 + H$$
$$C_6H_5CH_2Cl \rightarrow C_6H_5CH_2 + Cl$$
$$C_6H_5CH_2OH \rightarrow C_6H_5CH_2 + OH$$
$$C_6H_5CH_2CN \rightarrow C_6H_5CH_2 + CN$$
$$C_6H_5CH_2NH_2 \rightarrow C_6N_5CH_2 + NH_2$$
$$C_6H_5CH_2COOH \rightarrow C_6H_5CH_2 + COOH$$
$$C_6H_5CH_2CH_2NH_2 \rightarrow C_6H_5CH_2 + CH_2NH_2$$
$$\rightarrow C_6H_5CH{=}CH_2 + NH_3$$
$$C_6H_5CH_2CH_3 \rightarrow C_6H_5CHCH_3 + H$$
$$\rightarrow C_6H_5CH{=}CH_2 + H_2$$
$$C_6H_5CHOHCH_3 \rightarrow C_6H_5CHCH_3 + OH$$
$$\rightarrow C_6H_5CH{=}CH_2 + H_2O$$
$$C_6H_5CHNH_2CH_3 \rightarrow C_6H_5CH . CH_3 + NH_2$$
$$\rightarrow C_6H_5CH{=}CH_2 + NH_3$$
$$C_6H_5CH_2CH_2OH \rightarrow C_6H_5CH_2 + CH_2OH$$
$$\rightarrow C_6H_5CH . CH_3 + OH$$
$$\rightarrow C_6H_5CH{=}CH_2 + H_2O$$
$$C_6H_5CH_2CH_2CH_2OH \rightarrow C_6H_5CH_2 + CH_2CH_2OH$$
$$C_6H_5CH_2CH_2Cl \rightarrow C_6H_5CH{=}CH_2 + HCl$$
$$C_6H_5CH_2CH_2Br \rightarrow C_6H_5CH{=}CH_2 + HBr$$
$$C_6H_5CH(CH_3)_2 \rightarrow C_6H_5C(CH_3)_2 + H$$
$$C_6H_5C(CH_3)_3 \rightarrow C_6H_5C(CH_3)_2 + CH_3$$
$$C_6H_5CHCl_2 \rightarrow C_6H_5CHCl + Cl$$
$$C_6H_5CCl_3 \rightarrow C_6H_5CCl_2 + Cl$$
$$C_6H_5CH_2CH_2C_6H_5 \rightarrow C_6H_5CH_2CHC_6H_5 + H$$
$$C_6H_5CH_2C_6H_5 \rightarrow (C_6H_5)_2CH + H$$
$$(C_6H_5)_3CH \rightarrow (C_6H_5)_3C + H$$
$$o\text{-}, m\text{- and } p\text{-}CH_3C_6H_4CH_3 \rightarrow o\text{-}, m\text{- and } p\text{-}CH_3C_6H_4CH_2 + H$$
$$p\text{-}CH_3C_6H_4CH_2Cl \rightarrow p\text{-}CH_3C_6H_4CH_2 + Cl$$
$$1:3:5(CH_3)_3C_6H_3 \rightarrow 1:3:(CH_3)_25, CH_2 . C_6H_3$$
$$C_6H_5NH_2 \rightarrow C_6H_5NH + H$$
$$C_6H_5NHCH_3 \rightarrow C_6H_5NCH_3 + H$$
$$C_6H_5N(CH_3)_2 \rightarrow C_6H_5NCH_3 + CH_3$$
$$C_6H_5CH_2NHC_6H_5 \rightarrow C_6H_5CH_2NC_6H_5 + H$$
or
$$C_6H_5CHNHC_6H_5 + H$$
$$C_6H_5OH \rightarrow C_6H_5O + H$$
$$C_6H_5OCH_3 \rightarrow C_6H_5O + CH_3$$
$$C_6H_5OC_2H_5 \rightarrow C_6H_5O + C_2H_5$$
$$HOC_6H_4OH \rightarrow HOC_6H_4O\cdot + H$$
$$2:3:5:6\text{-}(CH_3)_41:4(OH)_2C_6 \rightarrow 2:3:5:6(CH_3)_41, (OH)4(O), C_6 + H$$
$$\alpha \text{ and } \beta \ CH_3C_{10}H_9 \rightarrow \alpha \text{ and } \beta \ CH_2C_{10}H_9$$

that the dissociation energy of a β C—H bond in the gas phase is greater, in all the molecules of table 2 than that of a β C—C bond. The probability of dissociation found in this work is, however, consistently greater for the β C—H bond.

There is evidence, very limited at present, that the same will be found to hold for the photolysis of these compounds in solution but not in the gas phase. Thus Porter and Windsor [5] found that the flash photolysis of solutions of diphenyl methane in paraffin at normal temperatures gave the diphenyl methyl radical but Porter and Wright [4] found that flash photolysis of vapours of both ethyl benzene and diphenyl methane gave only the benzyl radical. It seems that photolysis in the gas phase may, for equivalent bonds, occur in accordance with bond-energy (gas-phase) considerations but that in solution, and in rigid solvents at low temperatures, other factors are important. There are two obvious factors which should be considered. First, bond-dissociation energies are almost unknown in solution and may be very different from those in vapours owing to differences in solvation energies of both the parent molecules and the radical products. Secondly, the activation energy of dissociation in solution may exceed the bond energy owing to a cage effect and the additional energy required to separate the dissociation products in the presence of the solvent molecules. This would be greater for the larger radicals and might account for the greater probability of separation of the smaller hydrogen atom.

Of the radical dissociation processes listed in table 2 there is one which cannot result directly from β bond fission. This is the formation of the $C_6H_5CHCH_3$ radical from $C_6H_5CH_2CH_2OH$. The most probable explanation of this observation is that, in addition to fission of the β bond to give benzyl and a type (iv) dissociation to give styrene, this molecule undergoes a γ bond fission by ejection of an OH radical and that the $C_6H_5CH_2CH_2$ radical so formed rapidly isomerizes to the more stable radical $C_6H_5CHCH_3$. Perhaps the most exceptional result in all the molecules investigated is the absence of any radicals from the photolysis of benzyl bromide. This confirms earlier results of Porter and Wright in the gas phase and is now well established since we made many attempts to detect radicals from this substance. The C—Br bond is one of the weakest investigated and therefore our failure to detect benzyl radicals from this substance is remarkable. It may be that dissociation of the β bond does occur but that, because the C—Br bond is exceptionally weak, sufficient energy remains in the benzyl radical after dissociation for this " hot " radical to react with the solvent before thermal equilibration can occur. Alternatively it may be that the increased spin-orbit interaction resulting from the presence of the heavy bromine atom results in rapid deactivation by inter-system crossing.

All molecules which give styrene as a primary product in the rigid glass are characterized by the fact that they can form a 4-centre transition state, dissociation of which leads directly to styrene and a second molecule :

$$
\begin{array}{ccc}
C_6H_5CH\!=\!=\!=\!=\!CH_2 & \rightarrow & C_6H_5CH = CH_2 \\
\ \ |\qquad\quad | & & \\
\ \ X\, -\,-\,-\,Y & + & X\!-\!\!-\!\!-\!\!-\!Y
\end{array}
$$

Although intermolecular processes are excluded in the rigid glass the reaction could take place by fission of a single—probably a β—bond followed by reaction of radical X within the solvent cage to give styrene and XY. There is no means of distinguishing between this cage reaction and a true intramolecular mechanism in the present experiments but a distinction would be possible in the vapour phase where only the true intramolecular mechanism could be operative. Two molecules of structure $C_6H_5CHXCH_2Y$ did not show styrene in the products of dissociation. In both cases styrene formation would have necessitated the separation of a somewhat more complex molecule, viz., C_6H_6 and CH_3OH.

Hydroquinones dissociate by a β bond fission in the same way as phenols. The semiquinone radical is also formed from quinones but this occurs by hydrogen abstraction from the solvent and is therefore quite separate from the other processes discussed in this paper. It has recently been the subject of a detailed investigation.[22]

This work forms part of the programme of fundamental research undertaken by the British Rayon Research Association.

1604   PRIMARY PHOTOCHEMICAL PROCESSES

[1] Hentz and Burton, *J. Amer. Chem. Soc.*, 1951, **73**, 532.
[2] Hentz, Sworski and Burton, *J. Amer. Chem. Soc.*, 1951, **73**, 578.
[3] Norman and Porter, *Proc. Roy. Soc. A*, 1955, **230**, 399.
[4] Porter and Wright, *Trans. Faraday Soc.*, 1955, **51**, 1469.
[5] Porter and Windsor, *Nature*, 1957, **180**, 187.
[6] Porter and Strachan, *Spectrochim. Acta*, 1958.
[7] Vogel, *J. Chem. Soc.*, 1948, 607, 616, 674.
[8] Mathews, *J. Amer. Chem. Soc.*, 1926, **48**, 562.
[9] Knowles, *Ind. Eng. Chem.*, 1920, **12**, 881.
[10] Norris and Taylor, *J. Amer. Chem. Soc.*, 1924, **46**, 753.
[11] Gibson, Blake and Kalm, *J. Chem. Physics*, 1953, **21**, 1000.
[12] Lewis and Lipkin, *J. Amer. Chem. Soc.*, 1942, **64**, 2801.   Lewis and Bigeleisen,
     J. Amer. Chem. Soc., 1943, **65**, 2424.
[13] Gomberg, *Ber.*, 1900, **33**, 3150.
[14] Schuler and Michel, *Z. Naturforsch.*, 1955, **10a**, 459.
[15] Dewar and Longuet-Higgins, *Proc. Physic. Soc. A*, 1954, **67**, 795.
[16] Longuet-Higgins and Pople, *Proc. Physic. Soc. A*, 1955, **68**, 591.
[17] Bingel, *Z. Naturforsch.*, 1955, **10a**, 462.
[18] Michaelis, Schubert and Granick, *J. Amer. Chem. Soc.*, 1939, **61**, 1981.
[19] Chu and Weissman, *J. Chem. Physics*, 1954, **22**, 21.
[20] Porter, *Chem. Soc., Special Publ.*, 1958.
[21] Swarc, *Chem. Physics*, 1948, **16**, 128 ; 1949, **17**, 431.
[22] Bridge and Porter, *Proc. Roy. Soc. A*, 1958, **244**, 259.

Reprinted with permission from *Nature* **220**, 1228–1229 (1968)

Nanosecond Flash Photolysis and the Absorption Spectra of Excited Singlet States

THE conventional flash photolysis technique is limited, by the flash duration, to times which are usually greater than 1 μs (ref. 1). There is a need to reach times shorter than this, particularly to observe and study the absorption of excited singlet states which typically have lifetimes in the nanosecond range. The appearance of the Q switched laser has provided a means to this end, and photoelectronic detection methods have already been used in several laboratories to observe transients in the nanosecond region. Because of the high bandwidth required, such methods have serious limitations and we have therefore sought to extend the original double flash technique[2], which has been so useful in microsecond flash photolysis, into the nanosecond region. A brief description of our method was given a year ago[3] and here we describe an improved form of the apparatus and its application to the observation of the absorption spectra of excited singlet states of a number of molecules.

The experimental arrangement which has been found most useful for the present application is shown in Fig. 1. A ruby laser with vanadyl phthalocyanine Q switch delivers a pulse which is frequency doubled by an ADPH crystal, filtered to reduce red light and light scattered from the flash lamp, and eventually delivers to the beam splitter a pulse of 3471 Å light, lasting 18 ns and of 70 mJ energy. The beam is divided into two parts, the first of which passes directly to the reaction vessel and acts as the photolysis flash. The second part passes first to an optical delay system which consists of a light path terminated by a movable plane mirror giving a variable delay of up to 100 ns. After reflexion this pulse returns to the beam splitter and thence to a cell containing a fluorescent solution (1,1′, 4,4′ tetraphenyl buta 1,3-diene in cyclohexane) which emits a pulse lasting 18 ns, having a continuous spectrum in the region from 400 to 600 nm. This pulse passes through the irradiated region of the reaction vessel to the slit of the spectrograph. A single laser pulse is adequate both for photolysis and the recording of an image on the spectograph (HPS) plate. Accurately reproducible delays between excitation and monitoring flash are readily obtained in this way, the time resolution being limited only by the duration of the laser pulse.

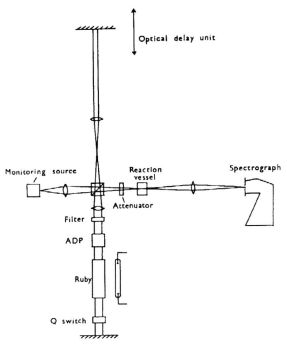

Fig. 1. Nanosecond flash photolysis apparatus.

Transient absorption spectra recorded in this way are shown in Figs. 2 and 3. The solutions were triphenylene and 3,4 benzpyrene in cyclohexane and were outgassed. Both series of spectra show the rise of triplet absorption and the simultaneous decay of new bands. These new transients are assigned to absorption by the lowest excited singlet states, because (a) the lifetimes are equal to the fluorescence lifetimes in the same conditions, and (b) the decay rate is equal to the rate of growth of the triplet states[4] which are simultaneously recorded on the plates. The wavelengths and lifetimes of these spectra and those of five other aromatic hydrocarbons which we have studied in the same way are recorded in Table 1. The solvent was cyclohexane in all cases except coronene for which the solvent was dioxane.

The shortest lifetime transients so far recorded are the singlet decay and triplet growth of phenanthrene, which has a half-life of 19 ns. Although this is comparable with the lifetime of the two flashes, the almost perfect reproducibility of the time delays obtained by using

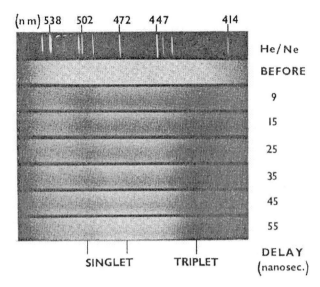

Fig. 2. Sequence of spectra after flash photolysis of triphenylene.

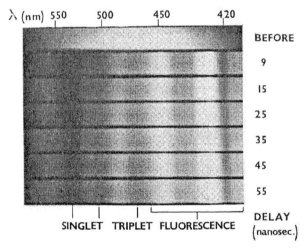

Fig. 3. Sequence of spectra after flash photolysis of 3,4-benzpyrene.

Table 1. WAVELENGTHS AND LIFETIMES OF SPECTRA OF SOME AROMATIC
HYDROCARBONS

Molecule	Singlet S_1 absorption (nm)	Triplet T_1 absorption (nm)	Singlet absorption decay time (nsec)	Fluorescence life-time (nsec)
Phenanthrene	505	485, 460, 425	20 ± 5	19 (ref. 7)
Triphenylene	500, 465	430	35 ± 5	37 (ref. 8)
1,2 benzanthracene	550	480	45 ± 5	44 (ref. 7)
3,4 benzpyrene	535, 510	500, 475	45 ± 5	49 (ref. 7)
1,2,3,4 dibenz-anthracene	535, 495	440	50 ± 5	50 (ref. 5)
Coronene	525	490	>100	300 (ref. 5)
Pyrene	470	510, 490, 415	>100	380 (ref. 6)

a fluorescent source makes it possible to study the concentration changes during the decay of the excitation flash, and the quantitative kinetic study of even shorter transients should therefore be possible by this method even without any reduction in the duration of the flash.

The triplet state absorption spectra are readily observed in aerated solutions with the shorter lifetimes expected as a result of oxygen quenching. The upper states of the singlet–singlet absorptions cannot be immediately assigned on the basis of the normal absorption spectra because parity and other selection rules allow strong transitions from the first excited state to levels which are forbidden from the ground state. Both spectroscopic and kinetic investigation of these transients will therefore provide a rich field for future investigation.

In place of the fluorescent source, we have also used a second spark lasting 20 ns, triggered by ionizing the spark gap with the focused laser beam or, alternatively, a Marx bank circuit in which the first spark gap is ionized by the focused laser beam. Both these systems gave a spectrum of sufficient intensity for single-shot recording on the spectrographic plate but a poorer continuum and less reproducible delays than the fluorescent source.

During the course of this work Novak and Windsor (ref. 5 and personal communication) have described another approach to the problem in which the time resolved spectra are recorded by means of an image converter tube using a laser spark as source. They recorded singlet absorption spectra of coronene, 1,2 benzanthracene and 1,2,3,4 dibenzanthracene which agree with those reported here. Nakata et al.[6] have recently reported a short lived transient after flash photolysis of pyrene which they assigned to an excited singlet state. Our results confirm this assignment.

As in conventional flash photolysis, each of these various modifications of technique will have its advantages for specific problems. The method used here has the general advantage of simplicity and the complete elimination of electronic devices other than the laser itself. The use of optical in place of electronic methods for delay and time measurement will become increasingly advantageous as the work is extended to even shorter times.

We thank the Science Research Council and the European Office of Aerospace Research (USAF) for support of this work.

GEORGE PORTER
MICHAEL R. TOPP

Davy Faraday Research Laboratory,
The Royal Institution,
London.

Received November 15, 1968.

[1] Porter, G., *Z. Electrochem.*, **64**, 59 (1960).
[2] Porter, G., *Proc. Roy. Soc.*, A, **200**, 284 (1950).
[3] Porter, G., and Topp, M. R., in *Nobel Symposium 5 — Fast Reactions and Primary Processes in Reaction Kinetics*, 158 (Interscience, London and New York, 1967).
[4] Porter, G., and Windsor, M. W., *Proc. Roy. Soc.*, A, **245**, 238 (1958).
[5] Novak, J. R., and Windsor, M. W., *J. Chem. Phys.*, **47**, 3075 (1967); *Science*, **161**, 1342 (1968).
[6] Nakato, Y., Yamamoto, N., and Tsubomura, H., *Chem. Phys. Lett.*, **2**, 57 (1968).
[7] Birks, J. B., and Munro, H., *Progress in Reaction Kinetics*, **4**, 239 (Pergamon Press, 1967).
[8] Berlman, I. B., *Handbook of Fluorescence Spectra of Aromatic Molecules* (Academic Press, 1965).

Reprinted from *Proc. Roy. Soc. Lond.* A **315**, 149–161 (1970)

Model systems for photosynthesis
I. Energy transfer and light harvesting mechanisms

By Angela R. Kelly and G. Porter, F.R.S.

*Davy Faraday Research Laboratory of the Royal Institution,
21 Albemarle Street, London W1X 4BS*

(*Received* 7 *July* 1969)

Absorption spectra, triplet state and fluorescence properties of solutions of chlorophylls *b* and *a* and pheophytins *b* and *a* in lecithin have been studied over a range of concentration from 10^{-4} to 5×10^{-2} mol l^{-1}.

Energy transfer has been investigated for the pairs chlorophyll *b* to *a*, pheophytins *b* to *a*, and is adequately described by the Förster inductive resonance mechanism.

We have also examined the self-quenching of fluorescence and triplet state formation for each of the four molecules. With the exception of pheophytin *b*, which forms aggregates at high concentrations, there were no observable quenchers and, in each case, the half quenching concentration is close to that at which energy transfer between like molecules occurs.

Mechanisms for this self-quenching and the possible relevance of these models to photosystem II in photosynthesis are discussed.

Current concepts in photosynthesis (see, for example, a recent review by Boardman (1968)) suggest the participation of two photochemical systems. Many workers believe that the two photosystems represent physically separated pigment arrays each with characteristic absorption spectra. Further evidence for this has been supplied recently by experiments in which chloroplasts are treated with a mild detergent and then fractionated by differential centrifugation. Two particles are separated; the heavier one contains chlorophyll embedded in a lipid matrix and has photochemical properties resembling the function of photosystem II. Properties of interest are its absorption spectrum $\lambda_{max} = 650$ nm (chlorophyll *b*) and 670 nm (chlorophyll *a*) and a fluorescence quantum yield of 0.016 (higher than the 0.003 found for the intact chloroplast). In order to establish a suitable model for photosystem II in photosynthesis it is desirable to reproduce these properties as closely as possible.

Apart from studies of the spectral properties and aggregation effects in various solvents (Brody & Brody 1967; Broyde, Brody & Brody 1968; Amster & Porter 1966) and energy transfer in ether solutions (Watson & Livingston 1950) the principal models for reproducing chloroplast-like conditions have been monomolecular layers. Classical work by Bellamy, Tweet & Gaines (1964) established that energy transfer between like and unlike molecules in diluted monolayers could be described by a Förster mechanism (Förster 1948, 1957). They observed a marked decrease in the fluorescence yield as the separation of molecules within a layer was decreased but suggested no explanation. More recently, Trosper, Park & Sauer

[149]

150   Angela R. Kelly and G. Porter

(1968) studied monolayers of chlorophyll *a* diluted by various chloroplast lipids. They found fluorescence depolarization brought about by a single 'Förster' type energy transfer, and also observed concentration quenching. Absorption spectra of monolayers which have been reported have a red absorption maximum at 685 nm suggesting a possible resemblance to system I particles.

There is an intermediate range of both state and concentration which has not been studied, making it difficult to relate results in monolayers or in the chloroplast to the established properties of dilute solutions. We have, therefore, used another simple model system—a solid solution of the pigments of both chlorophylls and their derivatives, pheophytins—in a lecithin matrix. These studies represent an extension of preliminary work on solid solutions of chlorophyll in cholesterol (Porter & Strauss 1966). The light harvesting process is thought to involve a large number of 'bulk' pigment molecules before the energy is finally trapped. Our experiments have been designed to clarify the conditions for collection of light energy by these traps.

EXPERIMENTAL

Preparation of samples

Lecithin was chosen as the matrix because of its occurrence in the leaf and because it was found to form transparent solid solutions.

Both chlorophyll and lecithin decompose when heated in air, consequently the melting and rapid cooling techniques previously used for preparing glasses were not applicable. Instead, appropriate amounts of the two components were dissolved in chloroform, the volume of this solution was then reduced until it was tacky and a few drops of this mixture were then spread on a microscope slide. The remaining chloroform was evaporated off in a stream of nitrogen and samples allowed to stand in this atmosphere for at least half an hour. This ensured removal of oxygen from the matrix and also allowed equilibration of the solution (important in the cases where aggregation occurred). Measured triplet yields and lifetimes indicated that rediffusion of oxygen into the sample is relatively slow and transients were still readily observed in samples which had been kept in the dark for several days. In order to compare relative yields of triplet formation and of fluorescence over a wide range of concentrations, the path lengths of the solid solutions were varied so that samples having the same optical density were compared in each experiment. Path lengths ranged from 200 μm to less than 5 μm and were measured by focusing methods using a Zeiss microscope with a vernier scale on its fine-focus adjustment.

Flash photolysis

Samples were examined by means of a modified version of the microbeam flash photolysis apparatus previously described (Porter & Strauss 1966). The only significant modification was in the method of excitation; the capillary flash lamp was mounted in a vertical position to one side of the apparatus and the exciting light

Model systems for photosynthesis. I   151

was reflected vertically on to the sample by means of a prism. A lens mounted 10 cm in front of the capillary enabled a sharp one-third size image to be projected onto the plane of the sample. Vertical illumination makes possible the examination of a wider range of samples, including liquid samples in short path length cells, and ensures more uniform and reproducible illumination of the selected area.

Fluorescence

Fluorescence spectra were recorded on an Aminco–Keirs recording spectrofluorimeter using an HTV R 136 photomultiplier tube. The wavelength response of the instrument was calibrated using, as secondary fluorescence standards, quinine sulphate (10^{-4} mol l^{-1} in 0.1 N H$_2$SO$_4$) and a chelate of AlCl$_3$ and pontachrome blue black red (P.B.B.R. reagent, Eastman-Kodak) as described by Augauer & White (1963).

Despite the short path lengths used, sample optical densities were still fairly high and it was found that reproducible measurements of fluorescence spectra of the solid solutions on microscope slides could best be achieved by mounting the sample in the centre of the cavity at an angle of 45° to both exciting and emitted light paths.

Absolute fluorescence yields were determined by plotting the spectra (corrected for instrument response) as relative quanta/wavenumber and determining graphically the area under the curve. Secondary standards used for comparison were quinine sulphate, rhodamine b and a solution of chlorophyll a in ether. Relative fluorescence yields were determined from the peak heights and comparisons of the unknown with a standard were made in every experiment.

Absorption spectra

Absorption spectra were recorded on the Unicam SP 800 spectrophotometer. In the visible region, scatter by lecithin is fairly small and no sample was placed in the reference beam.

Materials

A mixture of pheophytins a and b was kindly supplied by Professor R. B. Woodward. The two components were separated on a Zeokarb resin column by the method of Wilson & Nutting (1963). Some difficulty was experienced with this method; as an alternative, eluting the pheophytin a from icing sugar columns in 5 % benzene/heptane was also tried (J. M. Kelly, private communication). Pheophytin a samples from both sources were used with identical results.

Samples of pure crystalline chlorophylls were obtained from three sources—Koch Light Limited, Sigma Chemical Company and Sandoz Weidemann Limited. Although identical results were obtained with all three, absorption spectral studies showed that the samples from Sigma were most free of pheophytins. Lecithin, obtained from the Nutritional Biochemical Corporation, New York, was 95 % pure. It was further purified, prior to use, by dissolving in a minimum of chloroform and reprecipitating in excess acetone. It was important during this preparation not to

152   Angela R. Kelly and G. Porter

expose the solid to air. The purified lecithin was stored in solution in chloroform in stoppered flasks. A film of lecithin prepared from such a solution is optically clear (transmission of a $50\,\mu$m sample was $> 80\,\%$) down to $350\,$nm.

RESULTS

Absorption spectra

Absorption spectra of the four compounds in dilute solution in lecithin closely resemble the spectra in ether (Seeley & Jensen 1965), but both Soret and red bands are shifted by 6–10 nm to the red. This is in accordance with an observation of Chapman & Fast (1968) on chlorophyll–lecithin–water dispersions.

The spectra of both chlorophylls and of pheophytin a are quite unaffected by increases in concentration over the range 10^{-4} to $5\times10^{-2}\,$mol l^{-1}, indeed both the chlorophylls are extremely soluble in lecithin ($> 10^{-1}\,$mol l^{-1}). The solubility limit for pheophytin a occurred at $\sim 8\times10^{-2}\,$mol l^{-1}. In contrast, marked changes occur in the absorption spectra of pheophytin b at relatively lower concentrations. At concentrations greater than $10^{-3}\,$mol l^{-1} new bands appear in the spectrum at 476 nm which are attributed to the formation of some aggregate. At concentrations up to $4.5\times10^{-3}\,$mol l^{-1} both monomer and aggregate bands increased with increasing concentration but at still higher concentrations only the aggregate bands increased, suggesting a limiting solubility of monomeric pheophytin b in lecithin of 2.3×10^{-3} mol l^{-1}. That no further increase in absorption at 436 nm occurred indicated that there was no absorbance by the aggregate at this wavelength and allowed a difference spectrum to be deduced for this species. This is shown in figure 1.

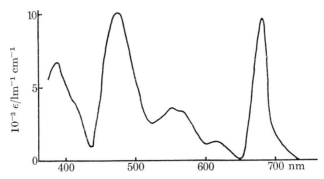

FIGURE 1. Spectrum attributed to aggregate of pheophytin b in lecithin.

Fluorescence spectra

Fluorescence spectra for the four compounds in dilute solution in lecithin show good mirror image relationships with the absorption spectra and are given in figure 2. No changes in fluorescence spectral shapes were detected even at the highest concentrations and the pheophytin b dimer appeared to be non-fluorescent.

Model systems for photosynthesis. I **153**

However, the emissions occur nearly at the limit of detection of our photocell and the possibility of an aggregate emission further to the red cannot be discounted.

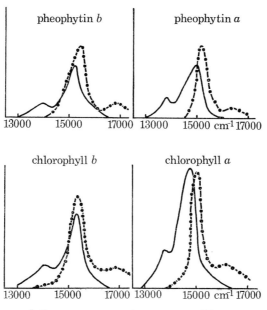

FIGURE 2. Absorption and fluorescence spectra. ———, Fluorescence spectra as relative quantum intensity; —•—, absorption spectra $10^{-4}\,\epsilon/\mathrm{lm}^{-1}\,\mathrm{cm}^{-1}$.

Triplet states

Triplet states of the four compounds were characterized in dilute solution by their difference spectra. These were determined photoelectrically point-by-point over a range of wavelengths 400 to 700 nm. At a number of wavelengths the rate constant for triplet decay was calculated and found to be first order over two to three lifetimes. The same lifetime was found at every wavelength, indicating the presence of only one transient species—the triplet state. The spectra, which are shown in figure 3, resemble closely those reported by Linschitz & Sarkanen (1958) for fluid solutions. Table 1 summarizes the spectral properties of the four pigments in dilute solution in lecithin.

Energy transfer

The fluorescence spectra of chlorophylls *a* and *b* overlap to a considerable extent which makes experimental detection of energy transfer difficult. We have found that this can be overcome conveniently by using flash photolysis to study relative triplet yields of donor and acceptor. No change in the triplet lifetimes of these components is observed under any of our conditions showing that any changes in relative triplet yields are due solely to processes affecting the singlet state.

154   Angela R. Kelly and G. Porter

A filter combination consisting of 1 cm of a solution of p-nitroaniline in propylene glycol and a blue gelatine Wratten 47 B enabled us to excite selectively in the Soret band of the donors in each of the pairs (donor named first) chlorophyll b–chlorophyll a and pheophytin b–pheophytin a. Figure 4 illustrates this for the chlorophylls.

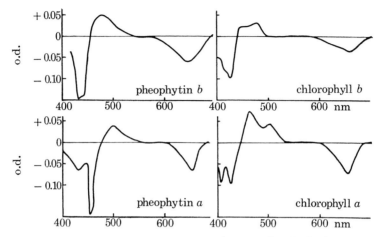

FIGURE 3. Difference spectra observed 30 μs after flash photolysis.

TABLE 1. PHOTOPHYSICAL PROPERTIES OF CHLOROPHYLLS
AND PHEOPHYTINS IN LECITHIN

	absorption spectra		fluorescence		triplet state		1st order decay constant
	λ_{max} nm	$10^{-4}\,\epsilon$ lm^{-1} cm^{-1}	ν_{max} cm^{-1}	quantum yield at ∞ dilution	λ_{max} nm	$10^{-4}\,\epsilon$ lm^{-1} cm^{-1}	s^{-1}
pheophytin b	436 658	19 5.5	15040	0.14 ± 0.02	480	5.15	1980 ± 50
pheophytin a	416 666	10.5 5.5	14750	0.11 ± 0.02	470	1.05	2090 ± 40
chlorophyll b	460 660	14.0 4.5	15380	0.10 ± 0.03	480	7.2	535 ± 30
chlorophyll a	436 676	12.0 8.8	14710	0.26 ± 0.04	470	5.7	1080 ± 40

Furthermore, an examination of the triplet difference spectra reveals that for each pair Δ o.d. (acceptor) is zero just where the singlet depletion of the donor is greatest. This point can then be used as a reference which enables the contribution of the two components to the net optical density change to be calculated at any other wavelength.

A donor concentration of $7 \times 10^{-4}\,\mathrm{mol\,l^{-1}}$ was used throughout and path lengths of the samples were kept approximately the same. This meant that the fraction of exciting light absorbed was similar from one experiment to the next. Relative

Model systems for photosynthesis. I   **155**

triplet yields of donor and acceptor were calculated from the appropriate optical density change. Small variations in the fraction of exciting light absorbed had to be allowed for and this is done by calculating

$$\text{relative triplet yield} = \frac{\Delta\,\text{o.d.}}{\text{fraction of exciting light absorbed by donor}}.$$

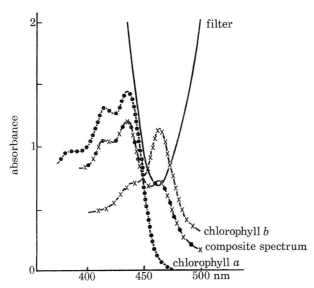

FIGURE 4. Spectra of donor and acceptor and filter used for selective excitation in the energy transfer experiments.

Results for the pheophytins and the chlorophylls are shown in figures 5 and 6 respectively. The yields have been compared with those in the absence of acceptor. For the chlorophylls it was just possible to resolve the components of the composite fluorescence spectra. Relative fluorescence yields fall on the same curve. In table 2 the concentrations of acceptor needed to half-quench the donor are listed along with the corresponding mean separations of the donor–acceptor pair.

These experimental disturbances may be compared with those predicted by the inductive resonance theory of Förster (1948). For weakly interacting chromophores, such as those in the relatively dilute solutions studied here this predicts that

$$R_0^6 = \frac{9000 \ln 10 K^2 \phi_D}{128\pi^5 n^4} \int \frac{f(\nu)\,\epsilon(\nu)\,d\nu}{\nu^4}, \tag{1}$$

where n is the refractive index of the medium $= 1.434$, ϕ_D is the quantum yield of donor fluorescence, and K^2 is a factor which takes account of the orientation of transition dipoles. Like Bennett & Kellog (1964) we have assumed our solutions to be random three dimensional arrays and have taken a value of $K^2 = \frac{2}{3}$. The integral

$$\Omega = \int \frac{f_D(\nu)\,\epsilon_A(\nu)\,d\nu}{\nu^4}$$

156   Angela R. Kelly and G. Porter

represents the overlap of donor emission and acceptor absorption. $f(\nu)$ is the
fluorescence spectral distribution of the donor and must be normalized to unity
on a wave number scale.

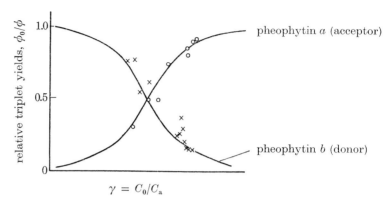

FIGURE 5. Energy transfer from pheophytin b to a. The heavy lines are
theoretical curves (Förster theory).

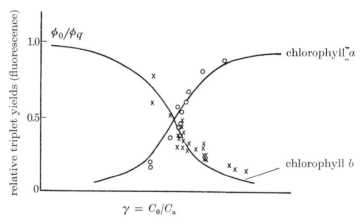

FIGURE 6. Energy transfer from chlorophyll b to a. The heavy lines are
theoretical curves (Förster theory).

TABLE 2

| | C_0 | R_0/nm | |
pair	$\overline{\text{mol l}^{-1}}$	expt.	theory
chlorophyll b (donor)–chlorophyll a	1.53×10^{-3}	6.5	5.45
pheophytin b (donor)–pheophytin a	1.8×10^{-3}	5.5	5.7

This overlap integral was evaluated graphically, calculating the value at suitable
wavenumber intervals and determining the area under the curve by a weighing
procedure. The calculations have since been checked numerically on our Elliott 803
computer by Dr L. K. Patterson. The agreement is excellent.

Model systems for photosynthesis. I   157

Förster theory also predicts the shape of the quenching curves (Förster 1946):

$$\left.\begin{array}{l} \phi_D/\phi' = 1 - \sqrt{\pi}\,\gamma\exp(\gamma^2)(1-\mathrm{erf}\,\gamma), \\ \phi_A/\phi'' = \sqrt{\pi}\,\gamma\exp(\gamma^2)\{1-\mathrm{erf}\,\gamma\}, \end{array}\right\} \tag{2}$$

where ϕ_D is the yield of any process emanating from $^1D^*$ in presence of quencher, ϕ' the yield in absence of quencher, ϕ_A the yield of any process from $^1A^*$ with quenching, ϕ'' the limiting yield after complete transfer and γ is a linear function of concentration $= C_a/C_0$ where C_a is the acceptor concentration and C_0 the critical concentration corresponding to an average of one quencher molecule in a sphere of radius R_0.

By fitting the experimental data to these curves it is possible to make a more accurate estimate of C_0. Good fits were obtained for the pheophytins for both sensitization and quenching (figure 5). Similar plots for the chlorophylls are shown in figure 6. At high concentrations of the acceptor the total yield of both triplet and fluorescence falls as a result of self-quenching which is discussed in the next section. In the low concentration region, where self-quenching is unimportant, the energy transfer between dissimilar molecules is well accounted for by the Förster theory.

Concentration quenching

At concentrations similar to those of the acceptor at which energy transfer occurs from b to a molecules, concentration quenching is observed in solutions of each of the four species alone, resulting in a decrease in both fluorescence and triplet formation. Triplet yields as a function of concentration for chlorophylls a and b and pheophytins a and b are shown in figure 7. The results for both chlorophylls fall on the same curve, that for pheophytin a is displaced to higher concentrations, a result which could be predicted because of the smaller value of the overlap integral for absorption and fluorescence of this molecule.

The Förster expression relating relative yield to concentration which gave a satisfactory account of b to a transfer gives a less satisfactory description of self-quenching as is shown by the broken line graph of figure 8 for chlorophyll b. The experimental yields fall more sharply with concentration than the theoretical curves suggesting a higher power dependence on concentration. The full line in figure 8, which is calculated from the function

$$\phi/\phi_0 = 1/(1+\gamma^2) \tag{3}$$

with $\gamma = C/C_{\frac{1}{2}}$ where $C_{\frac{1}{2}}$ is the half quenching concentration, accounts for the data within the limits of experimental error. Similarly, for chlorophyll a and pheophytin a, the self-quenching data are better described by function (3) than by the Förster expression (equation (2)). The formation of aggregates at higher concentrations of pheophytin b could account for the different shape of quenching curve found in this case, the quenching process probably being a mixture or self-quenching and direct energy transfer to the aggregate.

Simultaneous b to a transfer and concentration quenching

The relationship between energy transfer from chlorophyll *b* to *a* and the self-quenching of chlorophyll *b* is shown in figure 8. The steepest curve (full line), which is best described by the concentration squared dependence, is a pure self-quenching curve for chlorophyll *b*. The broken curve, which is best described by the Förster

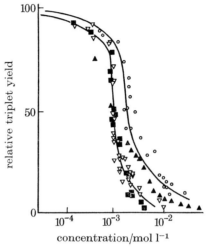

FIGURE 7. Self-quenching of triplet state formation for the four pigments:
$\triangledown$, chlorophyll *a*; ▨, chlorophyll *b*; $\bigcirc$, pheophytin *a*; ▲, pheophytin *b*.

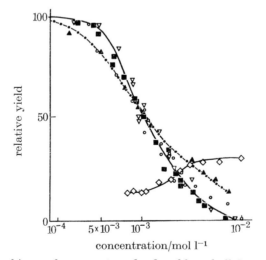

FIGURE 8. Self-quenching and energy transfer for chlorophyll *b*. —●—, Förster function
(equation (2)) with $C_0 = 1.8 \times 10^{-3}$ mol l^{-1}. ——, Empirical function $\phi_0/\phi = 1/(1+\gamma^2)$
with $\gamma = C/C_0$ and $C_0 = 1.3 \times 10^{-3}$ mol l^{-1}. ▲, 6×10^{-4} mol l^{-1} chlorophyll *b*, increasing
amounts of chlorophyll *a*; $\bigcirc$, 6×10^{-4} mol l^{-1} chlorophyll *a*, increasing amounts of
chlorophyll *b*; $\triangledown$, 2×10^{-4} mol l^{-1} chlorophyll *a*; ■, pure self-quenching of chlorophyll *b*;
$\diamondsuit$, relative yield of excitation reaching chlorophyll *a* (6×10^{-4} mol l^{-4}) with increasing
concentrations of chlorophyll *b*.

Model systems for photosynthesis. I   159

expression (2), shows the quenching of a $7 \times 10^{-4}\,\mathrm{mol\,l^{-1}}$ solution of chlorophyll b by varying concentrations of chlorophyll a. The two sets of intermediate points are for cases where both energy transfer to a and self-quenching of b were operative.

TABLE 3. SELF-QUENCHING OF THE SINGLET STATE OF PHEOPHYTINS
AND CHLOROPHYLLS

system	half quenching separations/nm		reference
	experimental	theory	
(a) *solid solutions in lecithin*			
pheophytin b	7.2	5.5	this work
pheophytin a	5.0	4.8	this work
chlorophyll b	7.0	5.55	this work
chlorophyll a	7.3	5.45	this work
(b) *other systems*			
chlorophyll a–cholesterol	6.7	—	Porter & Strauss (1966)
chlorophyll a–ether	6.0–7.0	5.17	Watson & Livingston (1950)†
chlorophyll a monolayers stearicacid diluent	5.0	6.7	Tweet, Gaines & Bellamy (1964)
chlorophyll a monolayer various lipid diluents	6.0–8.0	6.7	Trosper, Park & Sauer (1968)

† Our calculations.

For one of these intermediate cases, that where the chlorophyll a concentration was $6 \times 10^{-4}\,\mathrm{mol\,l^{-1}}$, the yield of triplet chlorophyll a as a function of the concentration of chlorophyll b is also shown. It is seen that the concentration quenching of chlorophyll b is accompanied by a small increase in the energy transferred to chlorophyll a.

DISCUSSION

Although we have worked in a medium in which diffusion is excluded and have mainly studied triplet yields rather than fluorescence, our results are very similar to those described many years ago by Watson & Livingston (1950) for fluid solutions of chlorophylls. These authors found that b to a transfer first became observable at concentrations lower than those required for the onset of self-quenching. Although we have found that the half-quenching concentrations are similar for b to a transfer and for self-quenching, reference to our figure 8 shows that the onset of b to a transfer appears at a lower concentration than that for self-quenching owing to the lower power concentration dependence of the b to a transfer.

Watson & Livingston gave several arguments to support their view that concentration quenching was not a result of ground state dimer formation. We have found no evidence for dimers or other aggregates in our solutions except those of pheophytin b at relatively high concentrations. Furthermore, quenching by collisions of the second kind is excluded in the rigid lecithin matrix of our experiments.

160   Angela R. Kelly and G. Porter

The higher power of concentration dependence of self-quenching over b to a transfer which we have found was also clearly described by Watson and Livingston. These authors ascribed the quenching to some form of energy degradation during an energy transfer process, following earlier suggestions of Vavilov (1943). Our results lend further support to this view; collisional processes are excluded in our systems and the near identity of half quenching concentrations with those for b to a transfer argues strongly for a mechanism of self-quenching in which Förster type long distance transfers play a part.

Concentration quenching of this type occurs in other molecules, and we have observed concentration quenching of β-naphthol in fructose or glucose glasses and in cholesterol using the same method of triplet state yield measurement. The half-quenching concentrations ($10^{-1}\,mol\,l^{-1}$) are much higher than for the chlorophylls, but correspond to critical transfer distances ($1.7\,nm$) which are again close to those calculated by Förster theory.

The dependence of concentration quenching on the second power of the concentration would be explained if an energy transfer process dependent on the first power of the concentration was followed by quenching at a trap whose concentration was also dependent on the first power of concentration. Such a trap could be an impurity or a dimer (although the concentration of dimer is proportional to the second power of monomer concentration, the probability that a molecule to which a transfer has taken place is a dimer is proportional only to the first power of monomer concentration). The dimer, in the sense referred to here, does not necessarily imply a chemically bonded complex but may be merely a statistical association of two molecules close enough to cause quenching when one or other of them is excited. Impurities, which would have to be in the solute in order to explain the concentration dependence, seem to be an unsatisfactory explanation in view of the generality of the phenomenon, the constancy of self-quenching rates when solutes are derived from several sources and the independence of half-quenching concentrations on added impurity.

The simplest explanation of the second power concentration dependence of quenching would be given in terms of a ground state equilibrium of monomers and dimers, the directly excited dimers being rapidly quenched. Although this would not account for the close similarity between concentration quenching and chlorophyll b to a energy transfer, it is impossible to exclude it completely on present evidence. Fluorescence lifetime measurements should give an unequivocal answer on this point. Such measurements will be described in part two of this paper where the mechanism of concentration quenching will be considered further. It is, however, clear from the present work that, whatever the mechanism of radiationless conversion in concentration quenching, it does not result in formation of the triplet state and, therefore, presumably proceeds directly to the ground state.

Finally, we may comment briefly on the relevance of our results to the light harvesting mechanism of photosystem II. Whilst it can be shown that because of the longer lifetime of the triplet state, triplet exciton migration may result in

Model systems for photosynthesis. I   161

more jumps during the excited state lifetime than the corresponding process for singlets, such a process cannot make a contribution to the overall energy harvesting if the triplet state is not significantly populated; this is just what we have found in systems having chlorophyll concentration comparable with those of the chloroplast. Instead, it seems likely that a process of singlet energy transfer among the bulk pigment molecules would lead to collection of light by the traps. Within the trap, the triplet state of chlorophyll may still have an important role.

One of us (A. R. K.) acknowledges a Science Research Council studentship during the tenure of which most of this work was carried out.

REFERENCES

Amster, R. L. & Porter, G. 1966 *Proc. Roy. Soc. Lond.* A **296**, 38.
Augauer, R. J. & White, C. E. 1963 *Anal. Chem.* **35**, 144.
Bennett, R. G. & Kellog, R. E. 1964 *J. chem. Phys.* **41**, 3040.
Bellamy, W. D., Tweet, A. G. & Gaines, G. L. 1964 *J. chem. Phys.* **41**, 2068.
Boardman, N. K. 1968 *Adv. Enzymol.* (ed. Nord), p. 1.
Brody, M. & Brody, S. S. 1967 *Biochim. biophys. Acta* **112**, 54.
Broyde, S. B., Brody, S. S. & Brody, M. 1968 *Biochim. biophys. Acta* **153**, 186.
Chapman, J. & Fast, G. 1968 *Science, N.Y.* **160**, 188.
Förster, T. 1946 *Naturwissenschaften* **33**, 166.
Förster, T. 1948 *Annln. Phys.* **2**, 55.
Förster, T. 1957 *Disc. Faraday Soc.* **17**, 1.
Linschitz, H. & Sarkanen, K. 1958 *J. Am. chem. Soc.* **80**, 4826.
Porter, G. & Strauss, G. 1966 *Proc. Roy. Soc. Lond.* A **295**, 1.
Seeley, G. R. & Jensen, R. G. 1965 *Spectrochim. Acta* **21**, 1835.
Trosper, T., Park, R. B. & Sauer, K. 1968 *Photochem. photobiol.* **7**, 451.
Tweet, A. G., Gaines, G. L. & Bellamy, W. D. 1964 *J. chem. Phys.* **40**, 2596.
Vavilov, S. 1943 *J. Phys. U.S.S.R.* **7**, 141.
Watson, J. & Livingston, R. 1950 *J. chem. Phys.* **18**, 802.
Wilson, G. R. & Nutting, M. D. 1963 *Anal. Chem.* **35**, 144.

Reprinted from *Proc. Roy. Soc. Lond. A* **362**, 281–303 (1978)

The Bakerian Lecture, 1977

In vitro models for photosynthesis

By Sir George Porter, F.R.S.

Davy Faraday Research Laboratory of The Royal Institution,
21 Albemarle Street, London W1X 4BS, U.K.

(*Lecture delivered* 17 *November* 1977 – *Typescript received* 20 *February* 1978)

Attempts to construct, *in vitro*, systems which imitate parts of the photosynthetic process serve two purposes. First, they may confirm, or not confirm, structures and mechanisms proposed on the basis of analyses of the living system. Second, they may lead to a purely photochemical system for the capture and storage of solar energy.

For the latter purpose, the most interesting part of the photosynthetic process is photosystem II, in which water is split by visible light into oxygen and a reduced material. The principal stages of the process are probably (*a*) light harvesting and trapping, (*b*) electron transfer from chlorophyll to a quinone, and (*c*) oxidation of water via an intermediate containing manganese.

Each of these three processes has now been reproduced to some extent *in vitro* but the light harvesting antenna efficiencies are lowered by concentration quenching. Recent progress, including kinetic investigations in the picosecond region and theoretical studies of energy transfer in the antenna are described.

Introduction

There are two approaches to studies of the photosynthetic unit just as there are two approaches to chemistry: the analytical and the synthetic. And, again as in chemistry, there are two purposes in the synthetic approach: first, the confirmation or refutation of hypotheses about the structure and mechanism of the natural system and, second, the practical use of the synthetic product.

The complexity of the whole photosynthetic unit of green plants is comparable with that of a complete cell and is therefore far beyond any possibility of total synthesis at the present time. *In vitro* studies have to be limited to relatively simple models of small parts of the whole but they provide a means of checking separately each part of the proposed mechanism. It is not impossible that, by putting together these separate parts, a much simplified *in vitro* system may be synthesized which is nevertheless capable of carrying out the essential parts of the photosynthetic process and perhaps one which is more suited to some of Man's needs.

Two main categories of the photosynthetic system are found in nature. First, there are the photosynthetic bacteria which reduce carbon dioxide to carbohydrate but are incapable of oxidizing water to oxygen. Instead they use organic and sulphur

282   Sir George Porter

compounds as electron donors and they store relatively little energy. Second, there
are the green plants and algae (green, red and blue-green) which, in addition to
reducing carbon dioxide, oxidize water to oxygen. For practical solar energy
storage in combustible form, the oxidation of water to oxygen is the key; the
nature of the reduced compound, be it hydrogen, carbohydrate or other reduced
material such as hydroquinone, is of less importance provided that it can be made
to regenerate exergonically the original compound upon reaction with oxygen.

The formation of one molecule of oxygen from two molecules of water requires
the transfer of four electrons as does the reduction of one molecule of carbon
dioxide to the level of glucose:

$$2H_2O \longrightarrow O_2 + 4H^+ + 4e,$$
$$4e + 4H^+ + CO_2 \longrightarrow (CH_2O) + H_2O$$
$$\overline{H_2O + CO_2 \longrightarrow (CH_2O) + O_2,} \quad \Delta G = 502\,kJ/mol.$$

In plant and algal photosynthesis, the electron transfers occur in two stages,
each stage of which requires the absorption of a photon so that the overall quantum
requirement is eight photons for each carbon dioxide reduced. The theoretical
quantum yield of 1/8 for O_2 liberation is approached under optimum laboratory
conditions. The overall scheme is far more complex than is suggested by the
equations above and many intermediate stages of oxidation/reduction are involved.
The manner in which two photochemical reactions operate in series was first suggested by Hill & Bendall (1960) and the 'Z' scheme which they proposed, although
it has been and is still being continually refined, is generally accepted today.

The Z scheme (figure 1) is composed of two principal parts, photosystem II
(PS II), photosystem I (PS I) and is followed by the Calvin cycle (or, in C_4 plants,
a modified Calvin cycle). The best understood part is the Calvin cycle by which
two molecules of NADPH and three molecules of ATP, made in the light reactions
of PS I and PS II, reduce one molecule of carbon dioxide to carbohydrate in the
dark. The two systems, PS I and PS II, are thought to be spatially separated in the
chloroplast and can be prepared in partly separated form by mechanical and
chemical processing of chloroplasts. They are linked by a pool of quinone (plastoquinone in the green-plant chloroplast). Each system contains an antenna of
pigment molecules, of which chlorophyll is the principal component, whose function
is to harvest the absorbed light and transfer the energy of this photon to the site
of primary reaction known as the reaction centre or trap. There are typically 300
pigment molecules to each reaction centre.

Structural investigations of the chloroplast by electron microscopy show that it
is composed of lipid membranes in which protein particles are embedded, probably
much as proposed by Singer & Nicholson (1972) for other biological membranes.
There is evidence that PS I and PS II particles are on opposite sides of the membrane
which therefore serves to separate the reduced and oxidized moieties of the electron-
transfer process. The subsequent dark reactions of the Calvin cycle take place in
the stroma, outside the membrane itself.

The Bakerian Lecture, 1977   **283**

After this brief summary of the overall process of photosynthesis, one part of it will now be discussed in more detail. The part which has been selected for detailed experimental study is photosystem II and this was chosen for three reasons: (1) It is logically the first step of green plant photosynthesis, (2) the water oxidation reaction is probably the least understood of all the steps in photosynthesis, and (3) photosystem II alone, by converting water and quinone into oxygen and hydroquinone, provides a cyclic process for the storage and release of solar energy which

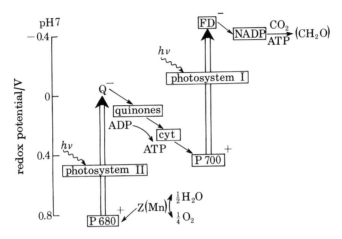

FIGURE 1. The 'Z' scheme for photosynthesis.

is energetically comparable with the complete synthesis of carbohydrate. In principle, the hydrogen gas may be released from a hydroquinone (e.g. electrolytically) with the expenditure of a quantity of energy less than that regained by the subsequent combustion of hydrogen.

The simplest sequence of reactions which accounts for the operation of photosystem II (see figure 2) is as follows:

1. Light harvesting

    (*a*) Several pigments absorb light and transfer energy to chlorophyll *a* (heterogeneous transfer).

    (*b*) Energy is transferred between chlorophyll *a* molecules until it arrives at the reaction centre trap, P 680 (Chl 680) (homogeneous transfer).

2. Quinone reduction

    In the reaction centre an electronically excited molecule P 680* (probably a chlorophyll dimer) reduces a molecule of plastoquinone (probably via an intermediate substance which is given the symbol Q and may also be a quinone)

$$Chl^* + Q = Chl^+ + Q^-.$$

284   Sir George Porter

3. Water oxidation

The chlorophyll radical cation oxidizes water by a process very probably involving manganese and necessarily involving the transfer of four electrons per O_2 liberated. If the donor is designated as Z, the reaction may be written, purely formally, as

$$4Chl^+ + Z = 4Chl + Z^{4+},$$

$$Z^{4+} + 2H_2O = Z + O_2 + 4H^+.$$

The overall scheme of PS II is shown in figure 2 and it forms the basis on which we are attempting to model PS II *in vitro*.

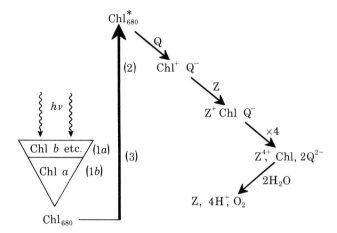

FIGURE 2. Model for photosystem II.

1. THE LIGHT HARVESTING MECHANISM

(a) Kinetic studies in vivo

Picosecond fluorescence and flash photolysis techniques made possible by the mode-locked laser have opened up the possibility of time resolving the primary processes which follow immediately upon light absorption. Most of the work up to the present time has been in two areas (a) transient absorption studies on the isolated reaction centres of photosynthetic bacteria, and (b) fluorescence studies of the light harvesting process in chloroplasts and algae. In the latter, fluorescence arises mainly from photosystem II antennae and is therefore of particular interest here.

(i) Porphyridium cruentum

Light harvesting by ancillary pigments and transmission of the excitation to chlorophyll *a* (process 1 *a*) is illustrated by our recent studies of porphyridium cruentum (Porter, Tredwell, Searle & Barber 1978).

The unicellular red algae, porphyridium cruentum, like other plants, utilises chlorophyll *a* as the last and lowest excitation energy component of the antenna system. In addition, it possesses water-soluble accessory light harvesting pigments, having a non-cyclic tetra-pyrrole structure, that are contained within structures known as phycobilisomes attached to the thylakoid membrane. Phycobilisomes contain three main pigments which, in decreasing order of first singlet excitation

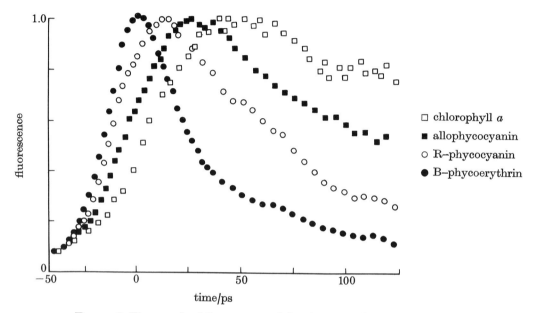

□ chlorophyll *a*
■ allophycocyanin
○ R–phycocyanin
● B–phycoerythrin

FIGURE 3. Time resolved fluorescence of the pigments of intact P. cruentum.

energy, are B-phycoerythrin, R-phycocyanin and allophycocyanin B. The first absorption bands of these three pigments and of chlorophyll *a* are respectively at the wavelengths 561, 622, 651 and 680 nm. Each pigment fluoresces in a wavelength region sufficiently different from the others that the four fluorescences can be studied separately.

Steady state studies indicated that the phycobilisomes preferentially serve photosystem II and the energy transfer path proposed is:

B-phycoerythrin $\longrightarrow$ R-phycocyanin $\longrightarrow$ allophycocyanin $\longrightarrow$ chlorophyll *a*.

We have used our picosecond single-pulse Nd laser to study the kinetics of the fluorescence from these four pigments. The frequency doubled pulse at 530 nm conveniently falls near the maximum absorption of the highest energy pigment, phycoerythrin and there is relatively little direct excitation of the other pigments. The interference filters used to separate the fluorescence of the four pigments had band widths of 9–14 nm and transmission maxima are at 578, 640, 660 and 685 nm respectively.

286   Sir George Porter

The time resolved fluorescence intensities of intact algae at these four wavelengths are shown in figure 3. The rise-time of phycoerythrin fluorescence corresponds to that of the excitation pulse and detector system profile, but the other three pigments show rise times which are significantly longer. The 1/e-times for rise and decay of each of the four fluorescences are given in table 1.

TABLE 1. FLUORESCENCE OF PORPHYRIDIUM CRUENTUM

	wavelength nm	$\tau_{1/e}$ (rise)/ps algae	$\tau_{1/e}$ (decay)/ps algae	phycobilisomes
phycoerythrin	578	0	70 ± 5	70 ± 5
phycocyanin	640	12	90 ± 10	—
allophycocyanin	660	22	118 ± 8	4000
chlorophyll a	685	52	175 ± 10	—

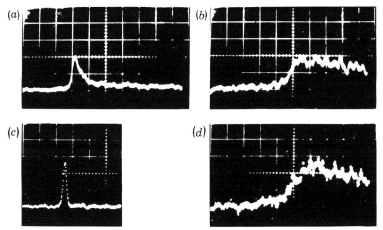

FIGURE 4. Energy transfer from phycoerythrin to allophycocyanin in phycobilisomes. (a) Phycoerythrin fluorescence; (b) allophycocyanin fluorescence; (c) laser pulse profile; (d) rise of allophycocyanin fluorescence on a faster time scale. (a) and (b) time scale 120 ps/cm; (c) and (d) time scale 60 ps/cm. 1 major division ≡ 1 cm.

The decay of phycoerythrin is non-exponential and follows an $\exp(-At^{\frac{1}{2}})$ decay law similar to that found for the decay of chlorophyll a in chlorella. Assuming that the four energy transfer processes, including that from chlorophyll a to the trap, follow this form, kinetic expressions for the four fluorescences have been derived and an excellent fit with the experimental curves of figure 3 is obtained by using, for the four rate constants, the values $A_1 = 0.26\,\mathrm{ps}^{-\frac{1}{2}}$, $A_2 = 0.48\,\mathrm{ps}^{-\frac{1}{2}}$, $A_3 = 0.52\,\mathrm{ps}^{-\frac{1}{2}}$ and $A_4 = 0.40\,\mathrm{ps}^{-\frac{1}{2}}$.

The phycobilisomes were isolated from porphyridium cruentum and their fluorescence studied in a similar manner (Searle et al. 1978). Since chlorophyll is absent from these particles, the lifetime of the allophycocyanin is much longer (4000 ps) and is probably close to that of the pigment in dilute solution. As expected, the lifetime of phycoerythrin is unchanged. The clearly resolved rise time of allophycocyanin in phycobilisomes is shown in figure 4.

This system illustrates the efficiency of interpigment energy transfer in the photosynthetic unit. Quantum yields of fluorescence, calculated from the measured lifetimes are all below 1 % implying an efficiency of energy transfer greater than 99 %. The long lifetime of allophycocyanin in the phycobilisomes compared with that in the intact algae containing chlorophyll is also noteworthy since it shows the virtual absence of concentration quenching in the allophycocyanin antenna.

(ii) Chlorophyll *a* in chlorella and chloroplasts

Most of the fluorescence of whole chloroplasts and chlorella at room temperature is emitted by the light harvesting chlorophyll *a* of photosystem II and it therefore provides a particularly appropriate method for the study of the system of interest here. Immediately picosecond techniques became available, lifetimes of this fluorescence were measured and were found to be much shorter than lifetimes predicted from fluorescence quantum yields as well as being poorly reproducible between laboratories. The problem has now been largely resolved (Campillo, Kollman & Shapiro 1976; Harris, Porter, Synowiec, Tredwell & Barber 1976; Breton & Geacintov 1976) by the recognition that lifetimes and quantum yields are intensity dependent when single pulses of high intensity are used as well as being affected by earlier pulses when a pulse train is employed. The single pulse effect is attributed to singlet–singlet annihilation and the second, multiple pulse effect is probably to be attributed to the formation of triplets in the first process (Porter, Synowiec & Tredwell 1977; Beddard & Porter 1977) followed by singlet–triplet quenching as follows:

$$S_1 + S_1 = T_1 + T_1,$$

$$S_1 + T_1 = T_1 + T_1.$$

When low intensities (below 10^{14} photons/cm^2 per pulse) are used, lifetimes become consistent with those predicted from quantum yields. For example in chlorella, dark-adapted at room temperature, the time for fluorescence to decay to half intensity ($\tau_{\frac{1}{2}}$) is 450 ps and when the traps are closed, with DCMU and pre-irradiation, the lifetime is extended to 1800 ps, corresponding to a quantum yield of fluorescence of 10 %. Similar results have been obtained with isolated particles of photosystem II (Searle *et al.* 1977).

These results and earlier investigations, by steady-state methods, lead to the following conclusions about the light harvesting antennae of photosynthetic units *in vivo*. (1) Heterogeneous energy transfer between donor and acceptor pigments and homogeneous transfer between molecules of the same kind (e.g. allophycocyanin, chlorophyll *a*) often occurs with nearly unit efficiency. (2) In the absence of traps or acceptors, lifetimes approach those of the pigments in low concentration solutions in spite of the fact that the average concentration of chlorophyll *a* in the harvesting unit is as high as 10^{-1} M. (3) The lifetime of the excitation, before transfer to another pigment or trap, is much longer than that of a coherent exciton, and use of a hopping model, involving weak Förster type resonance interaction or the stronger orbital overlap interaction of Dexter is justified.

Support for this last conclusion has recently been obtained by Dr Altmann and Dr Beddard in our laboratory. Using a two-dimensional model of 300 chlorophyll molecules with one trap, with the chlorophyll in a random Poisson distribution, they have used a computer simulation to calculate the average time, for excitation at a randomly chosen site, to reach the trap. Förster transfer by an R^{-6} mechanism was assumed, the transfer parameters being estimated from spectroscopic measurements of the overlap integral. At a chlorophyll concentration of 1 molecule per $4\,\text{nm}^2$, which is close to the average concentration in the chloroplast lamellae, the average half-life of excitation was 500 ps. This result should not be taken as evidence that chlorophyll is randomly distributed in the chloroplast but merely as confirmation that a Förster hopping mechanism in a random array is capable of accounting for the observed rapid excitation transfer to the trap without the need for assumptions involving a special organisation of the pigments. (A film was shown of a computer simulated random walk in a random array of 300 chlorophyll molecules with one trap.)

In vitro models of the light harvesting unit. The simplest system in which to study the individual steps of photosynthesis is a homogeneous fluid solution and, since little is known of the organization within the chloroplast and since the primary steps are so fast that diffusion cannot intervene, this may not be as irrelevant a model as might at first be supposed. Nevertheless, models in which chlorophyll is dissolved in lipid monolayers and multilayers on slides, and in vesicles and liposomes in aqueous suspension are better because chlorophyll, being itself a lipid structure, is miscible with lipids in almost any proportion and therefore it is possible to use concentrations as high as those in the chloroplast. The absence of diffusion over longer times is also important in some studies such as the separation of ionic products after charge separation. Finally, although it is known that at least half of the chlorophyll is attached to protein, the possibility remains that some of it is wholly or partly in a lipid environment.

It might be thought that reproduction of the light harvesting unit *in vitro* would present no difficulties and that a solution of chlorophyll *a* at a concentration comparable with the average concentration in the chloroplast (10^{-1} M) and containing a trap concentration similar to that of the chloroplast (1 acceptor trap to 300 chlorophyll molecules) would form a workable model and that a two-dimensional lipid solution at an equivalent concentration could form an excellent antenna. These expectations have not been realized and it has not so far been possible to construct a system of this kind which transfers energy efficiently to a trap. The reason for this difficulty is the phenomenon of concentration quenching.

Heterogeneous energy transfer between different pigments (chlorophyll *b* to chlorophyll *a* for example) is readily observed *in vitro*, in fluid solvents, in rigid matrices and in lipid multilayers. The relative yield of fluorescence of donor, given by the expression of Förster is

$$\Phi/\Phi_0 = 1 - \sqrt{\pi}\,\gamma \exp \gamma^2 (1 - \mathrm{erf}\,\gamma),$$

where $\gamma = c/c_0$ and c_0 is the critical concentration of acceptor, corresponding to one molecule of acceptor in a sphere of radius R_0, where R_0 is the critical transfer distance at which transfer and fluorescence probabilities are equal (5.8 nm for Chl b to Chl a). Studies of energy transfer in multilayer lipid matrices between chlorophyll b and chlorophyll a agree with this expression both in the form and the absolute magnitude of the quenching curves (Kelly & Porter 1970a).

However, when pure chlorophyll a (or pure b) is investigated, the fluorescence yield falls as the concentration increases. This homogeneous 'concentration quenching' occurs in a similar concentration range to the heterogeneous energy transfer but the form of the quenching curve is different and, in many systems including fluid solvents (Watson & Livingston 1950), monolayers (Costa *et al.* 1972), multilayers, liposomes and vesicles (Beddard, Carlin & Porter 1976), has been found to conform to the empirical expression first found by Watson & Livingston, i.e.

$$\Phi/\Phi_0 = 1/[1 + A\,(\mathrm{Chl})^2].$$

Typical self-quenching data obeying this relation for chlorophyll in lecithin vesicles are shown in figure 5 and half-quenching concentrations in a variety of environments are given in table 2.

The question must be asked 'what is the mechanism of this quenching, and why it is not found in the chloroplast in which the chlorophyll is at even higher concentrations?' For example, the lifetime of chlorophyll a fluorescence in lecithin vesicles at 10^{-1} molar is 200 ps – ten times shorter than the fluorescence lifetime in the chloroplast with the traps closed, and other systems exhibit concentration quenching at even lower concentrations than vesicles.

Mechanisms of concentration quenching. Several mechanisms of concentration quenching are known and are listed in table 3. Over the range of concentrations considered here, chlorophyll in lipid solvents shows no deviation from Beer's law and no change in spectrum, so that the formation of non-fluorescent dimers by mechanism 1 is eliminated (the dimer concentration would be greater than the monomer concentration when the fluorescence yield was reduced by a factor of more than three). Furthermore, the lifetime would be unchanged whereas, in vesicles, $\Phi/\Phi_0 = \tau/\tau_0$. Since the quenching occurs equally readily in rigid solvents, diffusional processes, which could lead to excimer quenching, are also ruled out (mechanism 2). However, excitation diffusion is expected to occur very rapidly and is necessary for light harvesting, and this could lead to rapid quenching if traps were present. Impurities, or chlorophyll dimers, if present, might act as traps but the reproducibility of the quenching rates obtained with different preparations, e.g. the variety of lipids used in the vesicle experiments, makes impurities unlikely

290   Sir George Porter

as a general explanation and the concentration of dimers necessary to give the observed quenching rate would result in changes in the absorption spectrum at high concentrations. It should be emphasized that dimers are known to be formed in poorer solvents such as hydrocarbons and lead to concentration quenching by mechanism 1 at much lower concentrations, but no spectral changes are observed in lipid solutions at the highest concentrations used.

In the general case of chlorophyll in good solvents, where there is no evidence of dimer formation and no diffusion, we have therefore proposed the following mechanism of concentration quenching (Beddard & Porter 1976). Energy migration between chlorophyll molecules proceeds by a Förster hopping mechanism until trapping occurs at a 'statistical pair'. The term 'statistical pair' describes two

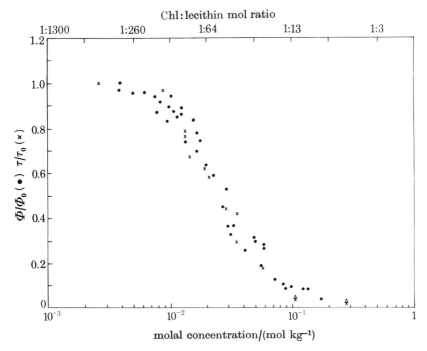

FIGURE 5. Self-quenching of chlorophyll a in lecithin vesicles. Φ/Φ_0 is the yield and τ/τ_0 the lifetime of fluorescence referred to infinite dilution.

TABLE 2. CONCENTRATION QUENCHING IN CHLOROPHYLL a

	$C_{\frac{1}{2}}$ (molal)	Chl a:lipid mol ratio
lecithin multilayers (dry)	1.4×10^{-3}	—
ether solution	1.6×10^{-2}	—
lecithin vesicles	2.8×10^{-2}	1:43
mono-di galactosyl diglyceride 3:1 vesicles	3.9×10^{-2}	1:30
lecithin liposomes	4.6×10^{-2}	1:25
mono-di galactosyl diglyceride 3:1 liposomes	7.0×10^{-2}	1:17

chlorophyll molecules in the random distribution which happen, on purely statistical grounds, to be closer than the average near neighbour distance and are close enough that, when one of them is electronically excited, interaction occurs leading to quenching. This interaction may, in favourable circumstances, lead to collapse to a true equilibrium excimer, though this is not necessary for quenching to occur and no evidence of excimer fluorescence in chlorophyll has yet been found.

TABLE 3. MECHANISMS OF CONCENTRATION QUENCHING

type 1. ground state complex $\quad$ $M + M \rightleftharpoons M_2$

type 2. *molecular* diffusion to $\quad$ $M^* + M \rightleftharpoons MM^* \longrightarrow 2M$
form an excited complex

type 3. *excitation* diffusion to trap $\quad$ $(M^* + M \longrightarrow M + M^*)_n$

$\quad$ (a) trap is extraneous quencher $\quad$ $M^* + Q \longrightarrow M + Q$

$\quad$ (b) trap is dimer or oligomer $\quad$ $M^* + M_2 \longrightarrow M + M_2^* \longrightarrow M + M_2$

$\quad$ (c) trap is statistical pair at R_p $\quad$ $M^* + M \cdot \cdot^{R_p} \cdot \cdot M \longrightarrow M + MM^* \longrightarrow 3M$

The theoretical calculation of the rate of quenching by this mechanism presents considerable mathematical difficulties (though it is relatively easy to calculate random-walk diffusion rates in a *regular* lattice). For the calculation of the rate of trapping in a random distribution of molecules, whether by statistical pair or any other quencher, it is necessary at present to use a Monte Carlo computer simulation. The results of such a calculation for three-dimensional arrays of chlorophyll *a* are shown in figure 6. The only unknown in this calculation is the separation of the statistical pair at which the trapping occurs and we used a Perrin active sphere model here as well as a point molecule approximation. The two curves shown are for statistical pair separations of 10 Å† and 14 Å and are compared with the results of Watson & Livingston for a solution in ether. The experimental points correspond closely to a separation of 10 ± 1 Å.

We are therefore led to the conclusion that the only arrangement which can account for the high-energy transfer probability of the chloroplast is one in which the molecules are closely spaced (and the average concentration in the chloroplast must be 10^{-1} M or greater) and yet no significant proportion of molecules are close enough to their nearest neighbours to permit quenching by excimer or similar interactions. The organization of chlorophyll within the chloroplast is not known, but evidence is accumulating that a high proportion of the chlorophyll is bound in chlorophyll–protein complexes. In one case, admittedly rather remote from the light harvesting unit of photosystem II, the detailed arrangement of chlorophyll has recently been determined. This is a water-soluble complex isolated from photosynthetic bacteria, whose function is not yet known, but whose structure has been determined by X-ray crystallography (Fenna & Matthews 1975). It has been shown that the unit protein contains seven molecules of bacteriochlorophyll, apparently rather randomly arranged in close array but with a nearest neighbour separation

† 1 Å $= 0.1$ nm $= 10^{-10}$ m.

292   Sir George Porter

between centres which is never less than 12 Å. This seems to be the ideal arrangement to effect efficient energy transfer and avoid concentration quenching.

The reason for the failure to construct a model of the light harvesting unit is now
understandable and the question arises as to whether it may be possible, in the
light of this understanding, to construct such a unit. Non-biological methods for

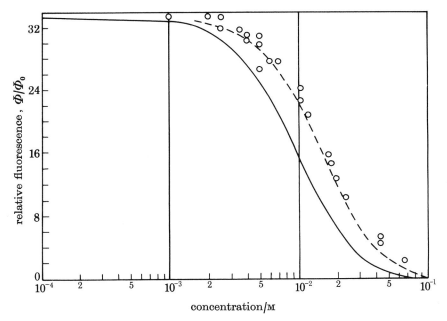

FIGURE 6. Relative fluorescence yield of chlorophyll a as a function of concentration. O,
Experimental data of Watson & Livingston; ——, calculated curve with $R_0 = 70$ Å and
trap $= 14$ Å; - - -, calculated with $R_0 = 70$ Å and trap $= 10$ Å.

the synthesis of protein–chlorophyll complexes are not known. The synthesis of
porphyrins which have substituents causing steric hindrance or the use of micellar
inclusions of chlorophyll are more feasible. There is also the question as to whether,
within the photosynthetic unit *in vivo*, all the chlorophyll is bound in proteins or
whether some of it is dissolved within the lipid membrane. Such an arrangement
would seem to provide an efficient 'lake' of pigment for light harvesting, and
chlorophyll is certainly very soluble in the membrane lipids. The original models
were designed with this possibility in mind and recent work, with chlorophyll in
lipid vesicles, encourages us not to discard this possibility.

Chlorophyll exists generally in a form in which the central magnesium atom is
complexed to at least one other ligand in addition to the four nitrogen atoms of the
porphyrin ring. This ligand may be water, for example, or a carbonyl group of a
second chlorophyll molecule. When dissolved in a lipid membrane, other ligands
of the lipid will have a high probability of complexing with the magnesium of
chlorophyll and, since the lipid molecules are bulky, such complexing will, if

strong enough, ensure that chlorophyll molecules are hindered from a close approach to another chlorophyll. With this in mind, and using vesicles and liposomes in water solution as the nearest practical model to the lipid membranes of the chloroplast, Miss Carlin has studied concentration quenching in systems whose composition is more closely related to those of the chloroplast lipids. In the chloroplast, the composition ratio of galactolipids (mono and di), other lipids (principally phospholipids such as lecithin) and chlorophyll is $3 : 2 : 1$. The galactolipids are particularly well adapted for hydrogen bonding to the carbonyl groups of chlorophyll and coordination with the magnesium atom. This suggestion is supported by the results in table 3 where it is seen that concentration quenching in galactosyl lipids occurs at a concentration significantly greater than in lecithin. The quenching is less in liposomes than in vesicles and, in the best case so far found, liposomes made with a $3 : 1$ mixture ratio of mono to di galactosyl diglyceride, which is close to that of the chloroplast, the half-quenching concentration is not reached until there is one molecule of chlorophyll to 17 molecules of lipid. Whether or not this has relevance to the light-harvesting unit *in vivo*, it suggests that further work on the effect of lipid composition on concentration quenching may be profitable as a route to a synthetic system capable of transferring energy without undue losses by concentration quenching.

Although attempts to synthesize an efficient *in vitro* model of the light harvesting unit have so far failed, the requirements for successful synthesis can now be specified. The antenna should be composed of chlorophyll or other appropriate pigments at an average concentration high enough to ensure rapid transfer by a Förster mechanism, but so arranged that no two of them is close enough to interact into a quenching configuration when one of them is excited. It may be possible to achieve this configuration in several ways, apart from the obvious route of borrowing from the biologically synthesized system. The porphyrins might be synthesized with groups which sterically inhibit close approach of the molecules, a polymeric structure might be synthesized in which the porphyrins are separated at the optimum distance or large molecules, which form coordination complexes with chlorophyll, may serve the same purpose. For the present, the synthetic light harvesting unit will be temporarily abandoned and each absorbing chlorophyll *a* will be its own reaction centre.

2. ELECTRON TRANSFER FROM CHLOROPHYLL TO QUINONE

Electron transfer from electronically excited chlorophyll to quinones has been shown to occur readily and the principal problem is to separate the ions and to compete with the rapid reverse reaction. Both singlet and triplet excited chlorophyll molecules are quenched by quinones and the rate constants of quenching have been determined, for the singlet, by fluorescence lifetime or yield measurements and, for the triplet, by triplet state kinetic absorption spectroscopy following flash photolysis. However, it does not follow, from quenching experiments alone, that separated ions of Chl^+ and Q^- are formed.

294   Sir George Porter

In fluid solvents, the quenching rate constants of chlorophyll a by duroquinone in ethanol are as follows (Kelly & Porter 1970b):

$$Chl^*(S) + Q = (Chl..Q)(S), \quad k_S = 9.2 \times 10^9 \, M^{-1} s^{-1},$$

$$Chl^*(T) + Q = (Chl..Q)(T), \quad k_T = (1.4 \pm 0.4) \times 10^9 \, M^{-1} s^{-1}.$$

The complex (Chl..Q) which, in the first place, will have the same multiplicity as that of the excited chlorophyll as indicated above, may react as follows:

$$(Chl..Q) = Chl + Q, \qquad (a)$$

$$= Chl^+ + Q^-, \quad (b)$$

$$Chl^+ + Q^- = Chl + Q. \qquad (c).$$

There is clear evidence that the complex formed from triplet chlorophyll undergoes reaction (b) in fluid polar solvents; both the chlorophyll cation and the semiquinone radical anion are observed and their subsequent reverse reaction (c) occurs with rate constant $3.5 \times 10^9 \, M^{-1} s^{-1}$ (Kelly & Porter 1970b). On the other hand, there is little evidence for the dissociation of singlet complexes into separate ions. It has been suggested that, in the case of bacteriopheophytin/benzoquinone, this rate of conversion to the ground state by reaction (a) in the singlet complex is much faster than in the triplet complex because of the spin forbidden nature of the triplet process so that it competes effectively with dissociation of the singlet complex into ions (Holten *et al.* 1976; Gouterman & Holten 1977). One must not generalize this result to all donor-acceptor pairs or other solvents since, in other cases, ion-pair formation from singlet species, e.g. dimethylaniline + anthracene, is well established (Weller 1967).

Whether or not ion-pair separation can occur in the reaction of singlet chlorophyll a with quinone in fluid solvents, rigid solvents and lipid solutions are better models of the photosynthetic unit and here the relative importance of singlets and triplets is quite different. Beddard, Porter & Weese (1975) have studied chlorophyll/quinone systems in lecithin, where diffusion is totally inhibited during the lifetimes of excited states. There is no spectral evidence in these experiments of any aggregated species, even at the highest concentrations used ($10^{-1} M$ quinone and $4 \times 10^{-4} M$ Chl$a$).

Singlet states were quenched by quinone and the half quenching concentration for duroquinone was $5 \times 10^{-2} M$. Triplet states were observed by flash photolysis, with lifetimes at 0.9 ms and this lifetime was unchanged when the quinone concentration was varied from 10^{-4} to $10^{-1} M$. The singlet and triplet yields showed the same dependence on quinone concentration. This remarkable difference between singlet and triplet state quenching rates (by a factor of more than 10^5 when the relative lifetimes of the two states are taken into account) compares with a difference of less than 10 in fluid solvents. The explanation of this difference is to be found in the random distribution of molecules, which is frozen in rigid solutions whereas in fluid solvents each molecule has time to sample a variety of near neighbour distances. Consider, in the rigid solution, an excited chlorophyll molecule whose nearest-neighbour quinone is at a distance r. As will be shown below, the dependence of quenching rate on r for an electron transfer between molecules is a very sharp one, approximating to a Perrin active sphere model. If r lies within this sphere, the singlet will be quenched before fluorescence or inter-system crossing occurs, whereas if it is outside this sphere it will fluoresce or form

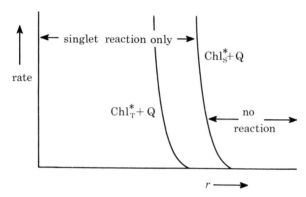

FIGURE 7. The rate of quenching, by quinone, of chlorophyll triplet (Chl_T^*) and chlorophyll singlet (Chl_S^*) as a function of separation r to illustrate why only singlet quenching occurs in rigid media.

triplets but these triplets (now lying outside the quenching sphere and having a lower quenching rate and hence a somewhat larger quenching sphere than the singlets) will be unaffected by the quinone in spite of the longer triplet lifetime. The effect is illustrated qualitatively in figure 7.

This conclusion introduces a problem about the mechanism of charge transfer within the PSU, from the chlorophyll trap to the quinone. Singlets are quenched before triplets can be formed in any significant amount by intersystem crossing and other mechanisms of triplet formation such as singlet–singlet annihilation cannot contribute at the low light intensities of normal photosynthesis. During the lifetime of fluorescence in PS II (500 ps) the triplet yield by intersystem crossing could not exceed 6 %. It seems that only the singlet state remains as the origin of charge separation.

The quenching of singlet chlorophyll by quinones must involve charge separation in some degree. Energy transfer is ruled out, since the excited states of quinones all lie above the singlet excited chlorophyll, and, on the positive side, there is a good correlation between the oxidation potential of the quinone and the rate of singlet quenching as shown in table 4.

An excellent fit to all the experimental data is given (Beddard *et al.* 1975) by a theory based on the following assumptions: (i) Nearest-neighbour interactions between chlorophyll singlet S_1 and quinones predominate over interactions with more distant molecules, (ii) the distribution is random, and (iii) the interaction leading to partial or complete charge-transfer depends on spacial overlap of electron orbitals.

By comparison with the Dexter exchange theory (Dexter 1953), the rate of transfer $k(R)$ at separation R is given by

$$k(R) = A \exp(-2R/L),$$

where A corresponds to the rate of transfer at $R = 0$ and L is determined by the overlap between donor and acceptor orbitals.

TABLE 4. EXPERIMENTAL AND THEORETICAL DATA FOR THE QUENCHING
OF CHLOROPHYLL a FLUORESCENCE BY QUINONES

quinone	oxidation potential/V	$C_{\frac{1}{2}}/10^2 M$	$A/10^{-11} s^{-1}$	L/nm
2,5-dichloro-p-benzoquinone	0.74	2.0	179	0.379
p-benzoquinone	0.71	2.8	64	0.379
2,5-dimethyl-p-benzoquinone	0.60	3.8	39.4	0.342
duroquinone	0.47	4.9	14.3	0.342
plastoquinone-9	0.53	5.8	3.44	0.397
α-tocopherylquinone	0.47	9.4	1.04	0.379

This theory, based on that of Minn & Filipescu (1970) for triplet–triplet transfer, allows values for relative quantum yield and relative lifetime to be calculated. The values of the parameters A and L found by computer fitting to the data are given in table 4. The most probable transfer distances, at half quenching concentration, are 24 Å for benzoquinone, and 20 Å for duroquinone and plastoquinone and the probability of transfer at this concentration falls off very rapidly at smaller or greater distances.

The failure to observe ion products after a few microseconds is not surprising since, in the absence of diffusion or any other acceptors, the ion pair cannot separate and, since the back reaction is exergonic, it will occur rapidly. In rigid media, the distinction between the ion-pair and separated ions is blurred and the important parameter is the lifetime of the singlet pair, of chlorophyll a and plastoquinone, after charge separation, before back charge-transfer. It should be possible to determine this by picosecond studies now in progress. We do know, however, that the time is less than $5 \mu s$ even when the molecules are separated by 20 Å and it seems likely that this problem is overcome *in vivo*, by an even faster transfer from the Q^- to an acceptor or to the Chl^+ from a donor. Since we have chosen to use quinone, which is the final acceptor of PS II, our choice is to interpose an intermediate between Chl and the quinone (the primary acceptor in PS II given the symbol Q is not necessarily a quinone) or to place a donor in permanent close proximity to Chl.

The ultimate donor of PS II is water and the only known intermediate corresponding to Z is some compound containing manganese. The final section describes recent work in which we have attempted to bring about the whole of the PS II reaction using the minimum of components, a chlorophyll or analogue, a quinone, and a manganese complex coupled to water as donor.

3. Models for the water oxidation system incorporating manganese

Apart from the fact that it seems to be involved at the water splitting step in plant photosynthesis, manganese is a promising foundation on which to build a process which we know has to effect the transfer of four electrons from two water molecules and make an oxygen molecule without the production of free radicals or other intermediates which would destroy the system. Manganese forms complexes

TABLE 5. QUENCHING OF PHOTO-EXCITED CHLOROPHYLL a BY A SERIES OF MANGANESE COMPOUNDS

quencher	$\dfrac{10^{-8}k_S}{\text{dm}^3\,\text{mol}^{-1}\text{s}^{-1}}$	$\dfrac{10^{-6}k_T}{\text{dm}^3\,\text{mol}^{-1}\text{s}^{-1}}$	$E_{\frac{1}{2}}/\text{V}$
$\text{Mn(H}_2\text{O)}_6^{\text{II}}$	1.5	1.6	-1.51
$\text{Mn(en)}_3^{\text{II}}$	4.0	4.2	-1.44
$\text{Mn(acac)}_2^{\text{II}}$	3.7	6.5	-1.40
$\text{Mn(bipy)}_3^{\text{II}}$	7.2	16.0	-1.32
$\text{Mn(phen)}_3^{\text{II}}$	9.2	60.0	-1.25
$\text{Mn(acac)}_3^{\text{III}}$	49.0	140.0	$+0.45$
$(\text{Mn(bipy)}_2\text{O)}_2^{\text{III/IV}}$	—	230.0	$+0.90$
$(\text{Mn(phen)}_2\text{O)}_2^{\text{IV/IV}}$	—	400.0	$+1.10$

en $\equiv$ 1,2-diaminoethane, acac $\equiv$ acetylacetone, bipy $\equiv$ 2,2′-bipyridyl, phen $\equiv$ 1,10-phenanthroline

$$(\text{Mn(bipy)}_2\text{O)}_2^{\text{III/IV}} \equiv (\text{bipy}_2\text{Mn}^{\text{III}} \overset{\text{O}}{\underset{\text{O}}{\diamond}} \text{Mn}^{\text{IV}}\text{bipy}_2)^{3+}$$

$$(\text{Mn(phen)}_2\text{O)}_2^{\text{IV/IV}} \equiv (\text{phen}_2\text{Mn}^{\text{IV}} \overset{\text{O}}{\underset{\text{O}}{\diamond}} \text{Mn}^{\text{IV}}\text{phen}_2)^{4+}$$

with many ligands including chlorophyll itself, as well as binuclear complexes containing two manganese atoms bridged by oxygen and these complexes may be prepared in various oxidation states, of which Mn^{II}, Mn^{III} and Mn^{IV} have been described most fully.

We have carried out experiments in which manganese complexes were reacted with excited chlorophyll (Brown, Harriman & Porter 1977). All the manganese complexes quenched both excited singlet and triplet states of chlorophyll a. From the results in table 5 it was shown that:

(i) Singlet state quenching is faster than triplet quenching, as was found also with quinones.

(ii) Quenching of both singlets and triplets increases with the oxidizing power of the manganese complex and shows an excellent linear relation between the log quenching constant and the redox potential in the case of Mn^{II} complexes.

(iii) The most oxidizing complexes (binuclear $\text{Mn}^{\text{III}}/\text{Mn}^{\text{IV}}$ oxygen bridged complexes with bipyridyl and phenanthroline) oxidized ground state chlorophyll and the triplet quenching by these species resulted in the formation of Chl^+.

(iv) The less efficient quenching by Mn^{II} complexes did not result in any detectable transients.

The manganese complexes in the higher oxidation states are therefore acting as electron acceptors, like quinone, as might be expected, but the more reducing manganese (II) shows no evidence that it is acting as a donor. Quenching by Mn^{II} is probably a result of enhanced intersystem crossing by complex formation with this paramagnetic ion (Porter & Wright 1959) and the direction of the dependence of rate constant of quenching on redox potential suggests that, even with Mn^{II} complexes, chlorophyll is the donor and manganese complex the acceptor.

Clearly, direct reaction of excited chlorophyll with manganese complexes, even the most reducing of them, does not appear to be a promising first stage as a model of electron transfer, not only because Chl^- is not observed (once again this could be attributed to a rapid back reaction) but because the dependence of quenching on redox potentials indicates that, in the complex formed with chlorophyll, the charge transfer occurs in the wrong direction. Chl^- formation occurs with donors such as aromatic amines but, even here, the rate constants of triplet quenching are many thousand-fold less than quenching by efficient electron acceptors.

We must, therefore, return to electron transfer from chlorophyll to quinone as the first step as suggested in figure 2 and investigate the possibility of a subsequent reaction of Chl^+ with a manganese complex. But since the back transfer from Q^- to Chl^+ occurs very rapidly it seems necessary to link the manganese complex permanently to the chlorophyll. An attractive possibility is to join them in a single complex, as manganese chlorophyll. This, and the similar complexes of manganese with porphyrins and phthalocyanines, are well known and have previously attracted attention, particularly by Calvin and his school, because of their possible relevance to photosynthesis (Elvidge & Lever 1959; Engelsma, Yamamoto, Markham & Calvin 1962). I should like now to describe some preliminary but very promising work of Dr A. Harriman in our laboratory (Harriman & Porter 1978).

Photochemical reactions of manganese porphyrins and phthalocyanines with quinone

Since it is important to carry out our reactions in water, without the presence of organic solvents, water soluble derivatives of the manganese complexes had to be used. We have used mainly the tetrapyridyl porphyrin in methylated form (formula 1) as well as the free base, and sulphonated phthalocyanines. The coordination number of manganese is six and, in addition to the tetradentate ligand of the porphyrin or phthalocyanine, it is axially ligated to at least one and usually two solvent molecules. In the case where the solvent is water, we immediately have three of the four components of the electron transfer from water to quinone: water, manganese, and porphyrin, bound together in one molecule. Water bound to Mn is acidic and for the monomeric complexes, the acid-base equilibria shown in figure 8 are possible. At pH 9 the Mn^{III} complex is present mainly as the uncharged species.

Manganese complexes of porphyrin and phthalocyanine have absorption spectra through most of the visible region which, at the longer wavelengths, are essentially the $\pi - \pi^*$ bands of the ligands modified by the central manganese atom.

In aerated solution the stable oxidation state of manganese porphyrin is Mn^{III}. This can be reduced to Mn^{II} with dithionite or oxidized to a higher oxidation state, less well characterized but usually assigned to Mn^{IV}, by persulphate or hypochlorite

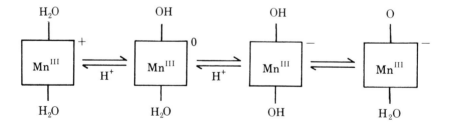

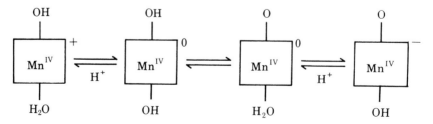

FIGURE 8. Acid/base equilibria in manganese porphyrins.

300   Sir George Porter

in basic solution (pH 9). When the pH of a solution of the Mn^{IV} complex is acidified
to pH 7, it is reduced to Mn^{II}. The Mn^{II} complex slowly changes to Mn^{III} but the
first product is Mn^{II} as is shown clearly by the absorption spectral changes. Preliminary measurements by using a Clark membrane electrode indicate that oxygen

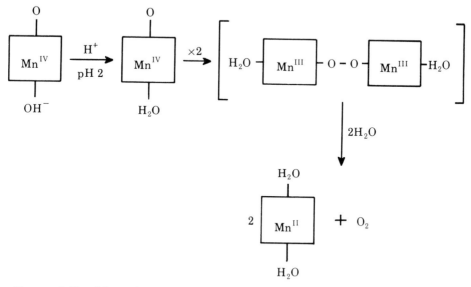

FIGURE 9. Possible mechanism for liberation of oxygen by Mn^{IV} porphyrin complexes.

gas is liberated in this process and that the ratio of reduced manganese to oxygen
is 2.0 ± 0.3. Such a reaction might be expected to occur via an intermediate oxygen-
bridged binuclear complex as shown in figure 9. In fact, most, if not all, complexes
of Mn^{IV} that have been well characterized are known to adopt some kind of
binuclear structure.

Mn^{III}, in outgassed water solution, is photochemically inert to irradiation in the
visible region, while in basic solution, slow photoreduction takes place.

In the presence of quinones such as duroquinone (in ethanol water) and 2, 3-
dicyanoquinone, Mn^{II} is oxidized slowly in the dark to Mn^{III} and this is accelerated
by light. In addition, Mn^{III} undergoes photochemical oxidation to Mn^{IV} and in
both of these oxidations the quinone is reduced to hydroquinone. Flash photolysis
experiments show that the semiquinone radical is formed immediately after the
flash. Since only light of $\lambda < 455$ nm was used in these experiments the photochemical changes observed may be confidently assigned to reactions of the excited
manganese complexes as shown in figure 10. Decomposition of the Mn^{IV} complex
according to the scheme of figure 9 will then complete a cycle in which the only
overall change is a four electron transfer from water to quinone, mediated by two
photoexcited manganese porphyrin complexes, and the liberation of one molecule
of oxygen. This is, of course, the overall reaction of photosystem II *in vivo* but it
has only been brought about by the use of extremes of pH. In order to bring about

Correction: Last para, line 6 should read "... $\lambda > 455$ nm ..."

the whole change at pH 7 it will be necessary to increase the oxidizing potential of the Mn^{IV} complex possibly by a further photochemical step. Energetically, it will be better to use Mn^{III} and Mn^{IV} as the absorbing species in the two photochemical steps since Mn^{II} oxidation to Mn^{III} is exergonic.

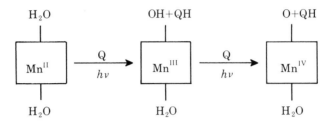

FIGURE 10. Photochemical oxidation of manganese porphyrins by quinones.

Many problems remain to be solved in this reaction scheme. One of the principal difficulties is that oxygen will react back with the reduced products hydroquinone and manganese(II) and this, coupled with the very small amount formed because of solubility limitations on the manganese complexes, probably accounts for our difficulties in measuring any oxygen evolved. Both back reaction and solubility problems may possibly be overcome by incorporating the reactants in micelles or vesicles and experiments on these systems at present in progress are encouraging.

The relevance of this system to the photosynthetic unit *in vivo* is quite unknown at present. There is no evidence that the manganese present in the chloroplast is in the form of a chlorophyll or porphyrin complex, though this possibility cannot be excluded. Energy transfer from the light harvesting unit to a complex of this kind would produce an excited complex identical with that formed by direct absorption of a photon and a reaction sequence similar to those we have begun to investigate here could then follow, resulting in oxygen elimination. Understanding of both the *in vivo* and *in vitro* systems requires a fuller knowledge of the chemistry and photochemistry of manganese complexes, especially in their higher oxidation states.

At the beginning I said that there are two purposes in attempting even the simplest models of parts of the complex apparatus of photosynthesis and I must conclude with a brief reference to the second of these: the possibility of constructing a photochemical system *in vitro*, modelled on plant photosynthesis, for the practical purpose of providing a renewable source of fuel from solar energy. The most useful fuels would be hydrogen to replace natural gas, and a liquid fuel such as methanol. The former would be prepared from water and the latter would have to be based on carbon dioxide as well if the cycle were to be complete. What has so far been achieved, and then only very inefficiently and incompletely, is the conversion of water and quinone into oxygen and hydroquinone. However, if this could be improved upon and made efficient, it is a considerable step forward. By using only four photons, photosystem II stores more than half as much energy in the form

302 Sir George Porter

of oxygen and hydroquinone (assumed to be reacted to water and quinone only) as the whole eight-photon photosynthetic process stores in the form of oxygen and carbohydrate. The reaction of hydroquinone with oxygen occurs spontaneously and rapidly with quinones such as anthraquinone and can be brought about in a practical manner by electrolytic conversion of hydroquinone to quinone and hydrogen, less electrical energy being consumed than is obtained as chemical potential in the hydrogen evolved. Alternatively, we may go further and attempt a further photochemical reaction (PSl) between hydroquinone and carbon dioxide.

Whatever route and whatever chemical storage material seems to be most convenient in the future, the immediate and essential problem is to oxidize water to oxygen by using red light, with the simultaneous formation of a reduced compound which stores a useful fraction of the adsorbed solar energy. The manganese porphyrin (or phthalocyanine) quinone system seems promising and worthy of intensive study as a possible route to the economic storage of solar energy.

I should like to thank all my colleagues whose work I have described in this lecture and, particularly, Dr G. S. Beddard, John Jaffé, Fellow of the Royal Society, who has been associated with me in these studies for a number of years, Dr C. J. Tredwell, who was mainly responsible for developing and operating the picosecond apparatus, and Dr A. Harriman, who has kindly allowed me to describe his unpublished work on the porphyrin manganese–quinone system.

I also wish to thank the Science Research Council, the Directorate-General for Research, Science and Education of the E.E.C., and General Electric Company (Schenectady, U.S.A.) for the financial support which made this work possible.

REFERENCES

Beddard, G. S., Carlin, S. E. & Porter, G. 1976 *Chem. Phys. Lett.* **43**, 27–32.
Beddard, G. S. & Porter, G. 1977 *Biochim. biophys. Acta* **462**, 63–72.
Beddard, G. S. & Porter, G. 1976 *Nature, Lond.* **260**, 366–367.
Beddard, G. S., Porter, G. & Weese, G. M. 1975 *Proc. R. Soc. Lond.* A **342**, 317–325.
Breton, J. & Geacintov, N. E. 1976 *FEBS Lett.* **69**, 86–88.
Brown, R. G., Harriman, A. & Porter, G. 1977 *J. chem. Soc. Faraday Trans. II* **73**, 113–119.
Campillo, A. J., Kollman, V. H. & Shapiro, S. L. 1976 *Science, N.Y.* **193**, 227–229.
Costa, S. M. de B., Froines, J. R., Harris, J. M., Leblanc, R. M., Orger, B. H. & Porter, G. 1972 *Proc. R. Soc. Lond.* A **326**, 503–519.
Dexter, D. L. 1953 *J. chem. Phys.* **21**, 836–850.
Elvidge, J. A. & Lever, A. B. P. 1959 *Proc. chem. Soc.* 195.
Engelsma, G., Yamamoto, A., Markham, E. & Calvin, M. 1962 *J. phys. Chem.* **66**, 2517–2531.
Fenna, R. E. & Matthews, B. W. 1975 *Nature, Lond.* **258**, 573–577.
Gouterman, M. & Holten, D. 1977 *Photochem. Photobiol.* **25**, 85–92.
Harriman, A. & Porter, G. 1978 In the press.
Harris, L., Porter, G., Synowiec, J. A., Tredwell, C. J. & Barber, J. 1976 *Biochim. biophys. Acta* **449**, 329–339.
Hill, R. & Bendall, F. 1960 *Nature, Lond.* **186**, 136–137.
Holten, D., Gouterman, M., Parson, W. W., Windsor, M. W. & Rockley, M. G. 1976 *Photochem. Photobiol.* **23**, 415–423.
Kelly, A. R. & Porter, G. 1970a *Proc. R. Soc. Lond.* A **315**, 149–161.

Kelly, J. M. & Porter, G. 1970*b* *Proc. R. Soc. Lond.* A **319**, 319–329.

Minn, F. L. & Filipescu, N. 1970 *J. chem. Soc.* A, 1016–1020.

Porter, G., Synowiec, J. A. & Tredwell, C. J. 1977 *Biochim. biophys. Acta* **459**, 329–336.

Porter, G., Tredwell, C. J., Searle, G. F. W. & Barber, J. 1978 *Biochim. biophys. Acta* **501**, 232–245.

Porter, G. & Wright, M. R. 1959 *Disc. Faraday Soc.* **27**, 18–27.

Searle, G. F. W., Barber, J., Harris, L., Porter, G. & Tredwell, C. J. 1977 *Biochim. biophys. Acta* **459**, 390–401.

Searle, G. F. W., Barber, J., Porter, G. & Tredwell, C. J. 1978 *Biochim. biophys. Acta* **501**, 246–256.

Singer, S. J. & Nicholson, G. L. 1972 *Science, N.Y.* **175**, 720–731.

Watson, W. F. & Livingston, R. 1950 *J. chem. Phys.* **18**, 802–809.

Weller, A. 1967 *Nobel Symposium 5 – Fast reactions and primary processes in chemical kinetics* (ed. E. Claesson), pp. 413–428. London: Interscience.

Reprinted from *Proc. Roy. Soc.* A **295**, 1–12 (1966) with permission from The Royal Society

Studies of triplet chlorophyll by microbeam flash photolysis

By G. Porter, F.R.S. and G. Strauss*

Department of Chemistry, The University, Sheffield 10

A microbeam flash photolysis apparatus has been developed for use with samples 50 to 250 μm square, and from 5 to several hundred microns thick.

Triplets of chlorophyll a and b were observed in a number of solid solvents, including cholesterol, at room temperature without prior outgassing. In cholesterol the triplet yield decreased with increasing concentration according to the Stern–Volmer law, but the half life of the chlorophyll b triplet was 3 ± 0.2 ms, and independent of concentration. Therefore, the excited singlet state but not the triplet is quenched by a concentration-dependent process. The half-quenching concentration of 2×10^{-3} M, corresponding to a mean intermolecular distance of 95 Å, points to quenching by inductive resonance.

No triplets of chlorophyll appeared on flashing normal or etiolated plant leaves. Leaves treated with cationic detergent gave triplets in a yield of 15%, and exhibited increased fluorescence.

Introduction

The light energy absorbed by photosynthetic structures undergoes many inter-molecular transfers before reaching the site of utilization. Sensitized fluorescence studies (Duysens 1952) have demonstrated that energy is transferred from the accessory pigments to chlorophyll a. Energy transfer between the molecules of 'bulk' chlorophyll a has been inferred from the low polarization of fluorescence resulting from polarized irradiation (Arnold & Meek 1956). The presence of energy sinks, consisting of specially situated, and probably more ordered, molecules of chlorophyll a, has been recognized from their absorption at longer wavelengths; evidence for the transfer of energy to these sinks has come from the emission of polarized fluorescence (of a wavelength characteristic of the sinks, not the bulk chlorophyll) whose plane of polarization depended on the orientation of the chloroplast (Olson, Butler & Jennings 1961).

Parallel studies of chlorophyll *in vitro* have been carried out mainly on relatively dilute homogeneous solutions except for some notable work on monomolecular films (Gaines, Bellamy & Tweet 1964). It is, therefore, relevant to inquire into the behaviour of chlorophyll with regard to optical properties, energy transfer and triplet formation, in conditions of high concentration and in matrices more analogous to those occurring *in vivo*. A comparison with the properties of chlorophyll *in vivo*, and after modification of the structure by various methods, may eventually lead to a realistic model of the chlorophyll organization in the photosynthetic unit.

Our particular interest in the present work has been the importance of triplet formation and energy transfer from the triplet state in solutions of high concentration. In dilute solutions it has recently been shown by flash photolysis studies that the quantum yields of triplet formation in 'wet' solvents are 64 and 88% for

* Present address: School of Chemistry, Rutgers—The State University, New Brunswick, New Jersey 08903, U.S.A.

[1]

2   G. Porter and G. Strauss

chlorophylls *a* and *b* respectively (Bowers & Porter 1966). For the study of solutions of high optical density short absorption paths, often only a few microns thick, are required and for the study of individual parts of a biological system as well as inhomogeneous synthetic specimens the other dimensions of the object may also be measured in microns. We have, therefore, developed a flash photolysis apparatus suitable for the flash irradiation and kinetic spectrophotometric examination of small objects of this kind. Microbeam flash photolysis should have many applications in solid state photochemistry, and particularly in photobiology, of which the present work provides a preliminary example.

EXPERIMENTAL

The apparatus was designed to flash-photolyse samples up to 250 μm square while on the stage of a microscope and to measure transient changes in optical density 50 to 250 μm square and 5 to several hundred microns in thickness. The general arrangement is indicated in figure 1.

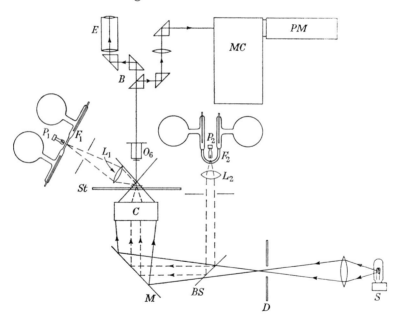

FIGURE 1. Schematic diagram of microflash photolysis apparatus.

Photolysis flash

Capillary flash tubes were used. In the 'normal' arrangement, light from a straight capillary tube F_1 was directed at the sample from above at an angle of 60° to the vertical and a one-third size image of the capillary was produced in the plane of the sample (St) by a quartz lens L_1. Exact positioning of the flash tube was facilitated by a pilot lamp P_1 which projected an image of the tube on to the microscope stage.

A second method of flashing, also indicated in figure 1, was to direct light from a U-shaped capillary flash tube F_2 at the sample from below via the collimating

Triplet chlorophyll and microbeam flash photolysis   3

lens L_2, beam splitter *BS*, and substage condenser *C*. In this method a greatly reduced (1:40) image of the flash tube is formed in the sample plane, so that a selected portion of the sample could be flashed.

Low inductance condensers (Wego, type 1283), coaxial construction of the spark gap, use of a coaxial cable between condenser bank and flash tube and critical damping resulted in a flash duration of 8 μs (half peak) for 50 J input energy.

Photometric system

The microscope was a Leitz instrument with a $12\cdot8\times$ objective, and $6\times$ or $25\times$ eyepieces. The monitoring source *S* was a 12 V, 100 W iodine quartz lamp (Philips), and a diaphragm with movable jaws, *D*, was positioned near the first image of the monitoring light, and formed a sharp image of its edges in the plane of the sample. This allowed a selected and variable rectangular area of the sample to be monitored.

Light reached the sample via the beam splitter, *BS*, a half-silvered glass plate (which was removed except when the flash, F_2, was being used), the plane mirror, *M*, and substage condenser, *C*, thence through the specimen and the objective binocular head to both the eyepiece, *E*, and the monochromator, *MC* (Hilger and Watts). The exit light from the monochromator was measured by the photomultiplier, *PM* (E.M.I., type 6256 B), and the signal from the phototube was displayed on a Tektronix, type 545 A, oscilloscope with type K plug-in calibrated preamplifier. The sensitivity of the arrangement permitted the measurement of transient optical density changes down to $0\cdot5$ %.

Materials

Chlorophyll *a* ('crystalline') and chlorophyll *b* ('purissimum crystallizable') were obtained from Sandoz Limited. Crystalline monoglycerides were obtained by courtesy of Dr A. S. C. Lawrence of this department. Other substances were of reagent quality. A variety of plants were kindly provided by Dr H. W. Woolhouse of the Department of Botany, University of Sheffield, to whom we are also indebted for valuable advice regarding choice of plant species, and for growing etiolated plants.

Preparation of solid solutions

Solid solutions of chlorophyll were prepared by grinding the solid components in an agate mortar. Because of the limited quantities of chlorophyll available, the amounts were estimated rather than weighed out, and concentrations determined later (see below). About 5 mg of the mixture was placed on a microscope slide and melted on a hotplate maintained at a few degrees above the melting point of the solvent. The melt was stirred with a preheated stainless steel spatula, and quickly chilled on a metal block at room temperature. In general, the sample spent ten seconds or less in the melted state. Melting was done in subdued light. With the boric acid solutions, heating was prolonged for another 10 s or so, until most, but not all, of the water of crystallization was driven off. Boric acid solutions were quite transparent when fresh, thus minimizing problems due to scattering of light.

4   G. Porter and G. Strauss

However, they soon became cloudy owing to absorption of moisture, and reproducibility was poor.

The solutions solidified in the form of flattened beads 100 to 300 μm thick, depending on sample size. To prepare thinner samples, the liquid melt was drawn across the slide with a spatula to give a thin smear. Pressing of the melt between glass slides was also effective. Thicknesses down to 5 μm were thus produced.

The thickness of a sample, which usually varied from point to point, was measured at the spot of interest by focusing the microscope first on the top surface (made visible if necessary by a slight scratch) and then on the bare glass surface alongside. The vertical movement was read on the graduated scale of the fine focusing screw of the microscope. Readings were reproducible to ± 3 μm.

Measurement of optical density

The flash apparatus was used as a microspectrophotometer by reading the deflexion on the oscilloscope with and without the sample in the light path. Films of cholesterol and other crystalline materials scattered a considerable fraction of the incident light, thus giving apparent optical densities far too high. These were corrected by subtracting from them the optical densities of the chlorophyll solutions at 470 nm (for chlorophyll a) or 503 nm (for chlorophyll b), where these compounds have practically no absorption. The scattering of pure cholesterol was found to vary by not more than 3 % over the visible spectrum. The validity of the above method of correction was demonstrated by plots of 'net' o.d. against thickness, for different points on the same sample. These gave straight lines passing through the origin. Coloured particles produce additional scattering which changes strongly in the neighbourhood of absorption bands, and is maximal on the long-wavelength side of the band (Jacobs, Holt, Kromhout & Rabinowitch 1957; Latimer & Rabinowitch 1956). The results obtained with solid chlorophyll *solutions* showed that wavelength-dependent scattering was negligible.

The change in optical density, Δ(o.d.), resulting from triplet formation on flashing was given by

$$\Delta(\text{o.d.}) = \log_{10} \frac{I_s}{I_s - T},$$

where I_s is the intensity of light transmitted by the ground state sample, and T is the maximum of the (positive or negative) transient. The concentration of a particular sample was found from the slope of a plot of net optical density versus thickness, divided by the molar extinction coefficient. Values of the latter, known for liquid solutions, were assumed to hold in the solid solutions used here. Checks of the instrument using large specimens were made against a standard Perkin Elmer spectrophotometer.

Determination of triplet yield

Relative yields, or the fraction of chlorophyll converted to the triplet state in a given sample, decreased with increasing thickness of the sample. With samples of high optical density, the light intensity, and hence the local concentration of triplets, varied strongly both vertically and horizontally when the exciting light

Triplet chlorophyll and microbeam flash photolysis   5

entered the sample from above, in a direction 60° to the vertical. The fraction of exciting light absorbed, I_a, was calculated from the optical density measured at the red absorption maximum, corrected for scattering as described above and the relative triplet yield $\Phi_T = \Delta(\text{o.d.})/I_a$ for a given sample, measured at several points of different thicknesses, was then extrapolated to zero thickness, thereby eliminating errors due to scattering and non-uniform distribution.

Encapsulation of green leaves

In order to eliminate possible quenching of triplets by oxygen, plant leaves were encapsulated in various glassy or crystalline materials. A piece of leaf, 1 to 2 mm² in size, was placed on a microscope slide, covered with the powdered substance and cautiously heated on a hotplate. The encapsulating substance was allowed to melt, then was chilled as quickly as possible. Cholestrol, benzhydrol, and glucose were used successfully. Of these, glucose was preferred, since chlorophyll is insoluble in melted glucose, and the possibility of inadvertently bringing any chlorophyll into solution was avoided.

Results

Flash photolysis of solid solutions and plant leaves was carried out using the microbeam flash photolysis apparatus described. In the course of surveying the capabilities of the apparatus many diverse substances were studied in viscous and solid solvents. In addition to the triplet states of the chlorophylls, those of riboflavin and flavin mononucleotide were observed in boric acid, glucose and glycerol and the triplet absorption spectra of fluorescein, napthalene, anthracene, phenanthrene and acridine were observed in boric acid.

Chlorophyll in solid solvents

Solvents found suitable for chlorophyll a and b included: boric acid, cholesterol. benzhydrol, 2,4,6-triphenylphenol, 2,4,6-triphenylaniline. In all of these solvents positive transients having half lives of 0·2 to 3 ms, depending on the solvent, were observed at 500 nm and identified spectroscopically as triplet–triplet absorptions. The transients were observed at room temperature, without any outgassing of the solutions.

Solutions of chlorophyll in benzhydrol solidified in three forms: when chilled rapidly a glass formed which on standing or on scratching became crystalline. Both forms gave chlorophyll triplets in about equal yield; the half life was 0·2 ms in the glass, but 2·5 ms in the crystal. When a liquid melt of the solution was cooled very slowly (over a period of 30 min), no transients were seen. Evidently, a two-phase system had formed during the slow crystallization. A number of substances which dissolved chlorophyll in the melted state, giving deep green liquids, also failed to give transients when solidified. This group included monoglycerides having C_{10}, C_{14}, and C_{18} hydrocarbon chains, the corresponding fatty acids, β-naphthol, and neopentyl alcohol. In all of these it appears that the chlorophyll separated in a second phase.

The last point is illustrated by the following observation: chlorophyll b in C_{14}

6   G. Porter and G. Strauss

monoglyceride gave no transient. A transient was seen however if a small amount
of water was added prior to solidification. Therefore, chlorophyll is soluble in the
monoglyceride–water system, but not in pure monoglyceride. Slow penetration of
water into monoglycerides takes place below their melting point, but above a certain
minimum temperature (Lawrence, Bingham, Capper & Hume 1964). This process
also gives rise to myelin tubes. It was possible to follow the progress of penetration
of water into a solid solution of chlorophyll in monoglyceride by monitoring selected
areas of the sample with the aid of the adjustable rectangular diaphragm. Transients
due to the chlorophyll triplet appeared on flashing of the wet, but not the dry, areas.

Spectra and yields of chlorophyll triplets

Solutions of chlorophyll a and b in cholesterol were flashed and monitored at
intervals from 400 to 700 nm. Transients due to triplet–triplet absorption and
ground state depletion are shown in figures 2 and 3. Also included are the ground
state absorption spectra, measured at the same area of the sample.

The fraction, α, of chlorophyll converted to the triplet was calculated from the
expression

$$\alpha = \frac{\Delta(\text{o.d.})}{(\text{o.d.})_g} \cdot \frac{\epsilon_G}{\epsilon_T - \epsilon_G},$$

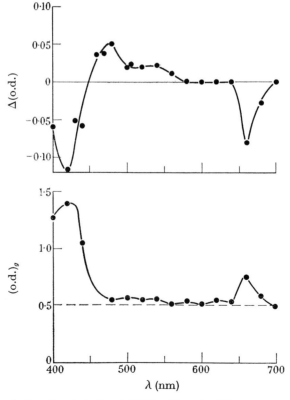

FIGURE 2. Chlorophyll a in cholesterol ($1 \cdot 96 \times 10^{-4}$ M): difference spectrum produced on
flashing, $\Delta(\text{o.d.})$, and ground-state spectrum, measured in flash apparatus $(\text{o.d.})_g$.
– – –, Correction applied for scattering.

Triplet chlorophyll and microbeam flash photolysis   **7**

where Δ(o.d.) is the optical density change observed on flashing, (o.d.)$_g$ is the optical density in the ground state, and ϵ_T and ϵ_G are the molar extinction coefficients of the triplet and ground state respectively.

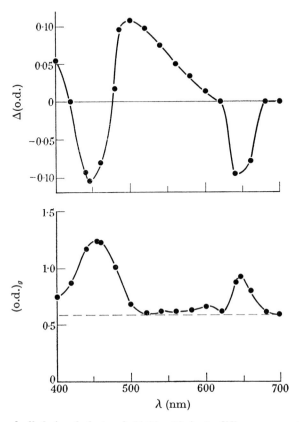

FIGURE 3. Chlorophyll b in cholesterol ($4 \cdot 80 \times 10^{-4}$ M): difference spectrum produced on flashing, Δ(o.d.), and ground-state spectrum, measured in flash apparatus, (o.d.)$_g$. ---, Correction applied for scattering.

Values of α are given in table 1, and are seen to be reasonably constant. The values of ϵ_T were taken from the data of Linschitz & Sarkanen (1958); those of ϵ_G are given by French (1960).

TABLE 1. CALCULATION OF TRIPLET YIELD FROM DIFFERENCE SPECTRUM

(c_G is the concentration of chlorophyll: l is the path length)

sample	c_G (moles/l.)	l (μm)	λ (nm)	Δ(o.d.)	$10^4\epsilon_T$	$10^4\epsilon_G$	α
chl. a in cholesterol	$1 \cdot 96 \times 10^{-4}$	140	400	$-0 \cdot 060$	$2 \cdot 57$	$6 \cdot 18$	$0 \cdot 217$
			462	$+0 \cdot 036$	$3 \cdot 20$	$0 \cdot 13$	$0 \cdot 153$
			505	$+0 \cdot 022$	$1 \cdot 74$	$0 \cdot 18$	$0 \cdot 183$
			530	$+0 \cdot 024$	$1 \cdot 78$	$0 \cdot 38$	$0 \cdot 186$
chl. b in cholesterol	$4 \cdot 80 \times 10^{-4}$	115	400	$+0 \cdot 055$	$1 \cdot 76$	$0 \cdot 13$	$0 \cdot 190$
			485	$+0 \cdot 097$	$2 \cdot 71$	$0 \cdot 76$	$0 \cdot 225$
			550	$+0 \cdot 056$	$1 \cdot 51$	$0 \cdot 64$	$0 \cdot 230$

8   G. Porter and G. Strauss

Concentration quenching

The effect of chlorophyll concentration on the yield and lifetime of the chlorophyll triplets was studied in a series of solutions of chlorophyll b in cholesterol, which ranged from 1.2×10^{-4} to 3.5×10^{-2} M. The initial triplet yield, Φ_T, at constant incident intensity and optical density decreased with increasing concentration, becoming undetectable at 3×10^{-2} M. These results are shown in figure 4

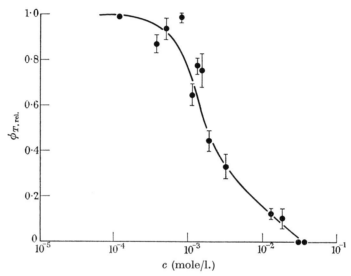

FIGURE 4. Chlorophyll b in cholesterol: relative triplet yield,
$\phi_{T,\text{rel.}}$, as function of concentration.

where the relative yield, $\Phi_{T,\text{rel.}}$, referred to unity at the lowest concentration measured, is plotted against chlorophyll concentration. A plot of $1/\Phi_{T,\text{rel.}}$ against concentration showed that most of the results can be described by a Stern–Volmer relation with $1/\Phi_{T,\text{rel.}} = 1 + 510$ M. The half life of the transients was 3.0 ± 0.2 ms at all concentrations.

Shifts in absorption maxima and red:blue peak ratios

The absorption spectra of chlorophyll a and b depend to some extent on state of orientation and intermolecular distance (Jacobs *et al.* 1954). Going from a dilute solution to a condensed monolayer or crystal has the effect of increasing the wavelength of the blue and red peaks, and of lowering the blue:red peak ratio. Although the peak ratio is the more sensitive parameter, its determination was difficult in the strongly scattering samples under investigation, and impossible in green leaves, which of course contain other pigments besides chlorophyll. The shift of the red absorption maximum is free of these limitations. Values for chlorophylls in various environments, and for leaves (see below) are given in table 2. They show that chlorophyll exists in a more crystalline form in monostearin or slowly cooled benzhydrol than in cholesterol or quickly cooled benzhydrol.

Triplet chlorophyll and microbeam flash photolysis   9

TABLE 2. WAVELENGTH OF THE RED ABSORPTION MAXIMUM OF CHLOROPHYLLS IN VARIOUS MATRICES

sample	transient on flashing	red $\lambda_{max.}$ (nm)
chl. *a* in cholesterol	+	670
in monostearin	−	675
solid (evap. from ether)	−	677
dilute solution in ether	+	622*§
monolayer	.	680†
monolayer (liquid)	.	675‡
monolayer (crystalline)	.	730‡
chl. *b* in cholesterol	+	652
in amorphous benzhydrol	+	648
in crystalline benzhydrol	−	654
in monostearin	−	654
solid (evap. from ether)	−	658
dilute solution in ether	+	644*
green leaf fresh	−	680
boiled in water	−	668
boiled in cationic detergent	+	668

Literature references: *French (1960); †Gaines *et al.* (1964); ‡Jacobs *et al.* (1954); §Bowers & Porter (1966).

Plant leaves

Fresh green leaves were examined for fluorescence and for formation of chlorophyll triplets on flashing. Most of the work was done with leaves of the fern *Leptopteris superba*, in which observation of individual chloroplasts was especially convenient, owing to the presence of only one layer of cells.

Intact leaves gave no observable triplets when flashed in air, water, or glycerol, or when encapsulated in glucose. Treatment of leaves with the cationic detergent dodecyl trimethyl ammonium bromide (30 min at room temperature or 1 min at 100 °C) caused a marked shift in the red absorption peak (table 2), indicating disruption of the chlorophyll structure. Detergent-treated leaves, when dried and encapsulated in glucose, gave large transients, which from the shape of the difference spectrum were identified as being due to the chlorophyll triplet (figure 5). No transients were observed in detergent-treated leaves without encapsulation. From the degree of ground state depletion, the fraction converted to the triplet was estimated at 15 %. Leaves boiled in water or treated with an anionic detergent (sodium dodecyl sulphate), or non-ionic detergent (Triton X-100) gave no transients when subsequently encapsulated. However, boiling with water alone shifted the red absorption peak to the same extent as did treatment with cationic detergent. Boiling with 8 M urea or concentrated LiCl (treatments designed to disrupt chlorophyll–protein complexes by denaturation) also failed to yield triplets after encapsulation.

The fluorescence of fresh and treated leaves was compared visually in a fluorescence microscope. Boiling in water caused no increase, or at most a very slight increase over the fluorescence of the untreated leaf. Treatment with cationic detergent caused an enhancement of fluorescence, estimated as an increase by a factor 2 or 3, which became noticeable even after treatment with cold detergent

10   G. Porter and G. Strauss

for a few minutes. Boiling for progressively longer periods caused no further increase in fluorescence. Only the chloroplasts, not the intervening spaces, were fluorescent.

Etiolated wheat seedlings, grown in the dark and allowed to become pale green by exposure to light for one day, gave no transients either when fresh or after encapsulation. When treated with cationic detergent and then encapsulated they gave transients of about the same order of magnitude as normal leaves.

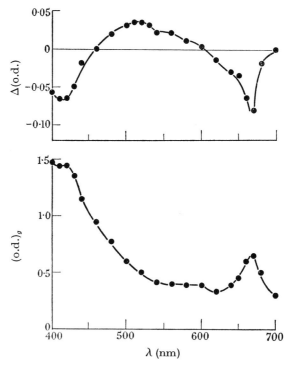

FIGURE 5. Green plant leaf, boiled with cationic detergent, then encapsulated in glucose: difference spectrum produced on flashing, $\triangle$(o.d.), and ground state absorption spectrum $(o.d.)_g$ measured in flash apparatus.

DISCUSSION

Concentration quenching in solid solutions

The constancy of the triplet lifetime of chlorophyll in solid solutions over a wide concentration range shows that no concentration-dependent deactivation of the triplet takes place. From the observed decrease in triplet yield with increasing concentration it follows that the excited singlet state must be the species being quenched. This quenching process, which obeys the Stern–Volmer law, differs from that in liquid solutions of chlorophyll, where the concentration quenching of fluorescence was found to follow the equation $1/\Phi_f = 1 + km^2$ (Watson & Livingston 1950). The latter process was ascribed to collisional quenching.

The present data are comparable, as far as concentration dependence is concerned, to the depolarization of fluorescence of chlorophyll in cyclohexanol solution (Weil 1952; Livingston 1960), which also follows the Stern–Volmer equation.

Triplet chlorophyll and microbeam flash photolysis   11

For both of these processes, occurring in viscous or rigid media, collisional deactivation may be ruled out and singlet exciton transfer cannot occur at concentrations as low as 10^{-4} M. Deactivation of the excited singlet state thus appears to take place by inductive resonance between it and ground state chlorophyll and this view is supported by the value of the observed half-quenching concentration of 2×10^{-3} M, which, in a random solution, corresponds to a mean intermolecular distance of 95 Å, a value close to the limiting distance for inductive resonance energy transfer.

Quenching in plant leaves

The absence of triplet formation on flashing of intact green leaves and of isolated chloroplasts might readily be ascribed to concentration quenching if it were not for the fact that chlorophyll *in vivo* has a fluorescence yield of 2 % (Latimer & Rabinowitch 1956). This unexpectedly high value is to be contrasted with an immeasurably small fluorescence yield of chlorophyll solutions of comparable concentration (0·1 M). Dilute (*ca.* 10^{-6} M) solutions of chlorophyll *a* have a fluorescence yield of 25 % (Forster & Livingston 1952), and a triplet yield of 70 % (Bowers & Porter 1965). If these decay processes of the excited singlet are unimolecular, then a comparable ratio of fluorescence to triplet formation should obtain in chloroplasts. Failure of triplets to appear in chloroplasts in the expected yield of the order of 5 % (a value well above the limit of sensitivity of the apparatus used) therefore must mean that triplets are being quenched more efficiently than singlets, which is just the reverse of the situation in solid chlorophyll solutions.

Energy transfer through the 'bulk' chlorophyll to the energy sinks is considered to take place by exciton migration, rather than by inductive resonance, as shown by the observation that the chlorophyll *b*-sensitized photooxidation of a special form of chlorophyll *a*, located in an energy sink, is efficient even at liquid helium temperatures, where inductive resonance could not operate (Clayton 1963). If so, then the preferential quenching of triplets over singlets becomes understandable, since triplet exciton migration has been shown to be about ten times as efficient as singlet exciton migration in certain organic crystals at low temperatures, owing to the longer lifetime of the triplet (Nieman & Robinson 1962).

The appearance of triplets in leaves treated with cationic detergent can now be interpreted as a blocking of triplet exciton transfer. This may result from a lining-up of detergent and chlorophyll molecules, possibly by attachment of the cationic group to the porphyrin, and of the hydrocarbon part to the phytyl chain. Cationic, but not anionic or non-ionic detergents were found to inhibit photosynthesis in *Chlorella*, and to make the chlorophyll of this alga benzene-extractable (Ke & Clendenning 1956). These results parallel our observations and suggest that a cationic detergent acts by weakening bonds between chlorophyll molecules or between chlorophyll and the structural protein framework.

The increase in fluorescence of chloroplasts resulting from treatment with cationic detergent shows that quenching of the singlet state was also partly abolished, presumably through hindering the migration of singlet excitons. It follows that both singlet and triplet energy migration occur in intact chloroplasts, in

12   G. Porter and G. Strauss

agreement with the conclusions of Franck & Rosenberg (1963), reached on the basis of fluorescence yields with and without photosynthesis.

The magnitude of the triplet yield (15%) in detergent-treated leaves indicates that the triplets originated from the bulk chlorophyll rather than from the energy sinks which constitute only 0·4% of the total chlorophyll content. The triplets observed in yields of *ca.* 0·1% by Witt and coworkers (Müller, Rumberg & Witt 1963) in chloroplasts, after certain treatments, appear to have arisen from the energy sinks, in contrast to our findings. The treatments found effective for the appearance of triplets (heating, chlorophyll depletion, and treatment with digitonin) thus represent methods to prevent triplet quenching in the sinks, or to block energy transfer from the sinks to the acceptors, without however disrupting the flow of energy through the bulk chlorophyll.

One of us (G.S.) thanks the Research Council of Rutgers, The State University (New Jersey) and the United States National Institutes of Health for awards of research fellowships.

REFERENCES

Arnold, W. & Meek, E. S. 1956 *Arch. Biochem.* **60**, 82.
Bowers, P. G. & Porter, G. 1966 *Proc. Roy. Soc.* A (in the press).
Clayton, R. K. 1963 *Ann. Rev. Plant Physiol.* **14**, 159.
Duysens, L. N. M. 1952 Thesis, University of Utrecht.
Forster, L. S. & Livingston, R. 1952 *J. Chem. Phys.* **20**, 1315.
Franck, J. & Rosenberg, J. L. 1963 *Photosynthetic mechanisms of green plants*, NAS–NRC Publication 1145, p. 101. Washington, D.C.
French, C. S. 1960 *Encyclopedia of plant physiology*, vol. 5, part 1, p. 282. Berlin: Springer.
Gaines, G. L., Bellamy, W. D. & Tweet, A. G. 1964 *J. Chem. Phys.* **41**, 538.
Jacobs, E. E., Holt, A. S., Kromhout, R. & Rabinowitch, E. 1957 *Arch. Biochem. Biophys.* **72**, 495.
Jacobs, E. E., Holt, A. S. & Rabinowitch, E. 1954 *J. Chem. Phys.* **22**, 142.
Ke, B. & Clendenning, K. A. 1956 *Biochim. biophys. Acta* **19**, 74.
Latimer, P. & Rabinowitch, E. 1956 *J. Chem. Phys.* **24**, 480.
Lawrence, A. S. C., Bingham, A., Capper, C. B. & Hume, K. 1964 *J. Phys. Chem.* **68**, 3470.
Linschitz, H. & Sarkanen, K. 1958 *J. Am. Chem. Soc.* **80**, 4826.
Livingston, R. 1960 *Encylopedia of plant physiology*, vol. 5, part 1, p. 830. Berlin: Springer.
Müller, A., Rumberg, R. & Witt, H. T. 1963 *Proc. Roy. Soc.* B **157**, 313.
Nieman, G. C. & Robinson, G. W. 1962 *J. Chem. Phys.* **37**, 2150.
Olson, R. A., Butler, W. L. & Jennings, W. H. 1961 *Biochim. biophys. Acta* **54**, 615.
Watson, R. & Livingston, R. 1950 *J. Chem. Phys.* **18**, 802.
Weil, P. 1952 Thesis, University of Minnesota.

PRINTED IN GREAT BRITAIN AT THE UNIVERSITY PRINTING HOUSE, CAMBRIDGE

Reprinted from *Interdisciplinary Science Reviews* **1**, 119–143 (1976) with permission from Maney Publishing

In Vitro Photosynthesis

PROFESSOR SIR GEORGE PORTER, F.R.S. and DR MARY D. ARCHER

The Royal Institution, London, England

A process powered by sunlight, producing fuel through a direct conversion mechanism that does not rely on living matter—this is the definition of an *in vitro* photosynthetic process. Research into the theoretical problems, conversion efficiencies and potential economic importance of such processes is both timely and of long-term economic importance. As fossil fuels become exhausted, new forms of energy will have to be pressed into service. *In vitro* photosynthesis offers the possibility of inexhaustible energy supplies, obtained in a more versatile and valuable form than those from solar/thermal processes. The following article reviews the most important *in vitro* photosynthetic processes and evaluates their respective efficiencies. The authors argue for a vigorous research programme to obtain essential information on which correct future economic decisions may be based.

The process of natural photosynthesis in green plants and some algae (the red and green varieties) is complex but its main result may be written rather simply:

$$CO_2 + H_2O \xrightarrow[\text{Chlorophyll}]{\text{Sunlight}} (CH_2O) + O_2 : \quad \Delta G = +502 \text{ kJ} \qquad (1)$$

Ambient carbon dioxide and water are converted into carbohydrates, represented by (CH_2O), in a thermodynamically extremely unfavourable process. This is achieved photochemically, the necessary driving force being provided by the solar energy absorbed by the photosynthetic pigments, mainly chlorophylls. These act as catalysts in the sense that they are not consumed by the overall process, although they are to some extent chemically involved in the intermediate stages.

Photosynthetic bacteria are more primitive organisms than algae and do not evolve oxygen. They are common in situations where organic decomposition is taking place with the production of reduced compounds. Green and purple sulphur bacteria, for example, make use of hydrogen sulphide:

$$2H_2 + CO_2 \xrightarrow[\text{Bacteriochlorophyll}]{\text{Sunlight}} (CH_2O) + H_2O + 2S \qquad (2)$$

All life on earth is sustained, directly or indirectly, by the Sun, which provides not only heat, but also food and fuel. We derive the chemical energy needed to run our bodies by eating green plants (or by eating animals that have eaten green plants, or possibly even by eating fungi that have grown parasitically on the dead remains of other organisms). Fossil fuels are hydrocarbons, derived from the anaerobic decomposition of vegetable and animal matter. The carbon in them is ultimately derived from the carbon dioxide of the atmosphere. The convenience of these fuels is due to the fact that they are readily stored, and yet have high calorific value, which is rapidly available on

burning the fuel, a process which reverses reaction (1).

The crucial importance of natural photosynthesis in supplying the world's food requires no discussion here. As far as fuel is concerned, not since the late 19th century, when coal usurped the role of wood, has photosynthesis supplied this commodity on a renewable basis. However, with every sign of a serious shortage of oil and gas by the end of this century, interest is awakening in the possible use of solar energy to provide fuel renewably. Thus, we define *in*

Table 1. Direct Conversion Devices for Solar Energy Utilization

Classification	Device	Product
Photobiological	Green plants Photosynthetic algae and bacteria	Chemical free energy
	'Model photosynthetic systems' (membranes, thin films, etc.)	Electrical and/or chemical free energy
Photovoltaic	Photovoltaic cells	Electrical energy
Photoelectro-chemical	Semiconductor–electrolyte systems Photogalvanic cells	Electrical and/or chemical free energy
Photochemical	Any cyclic system in which the energy storing photochemical reaction	
	A → B	
	is followed at a later time by the reverse process	Chemical free energy
	B → A + useful work	

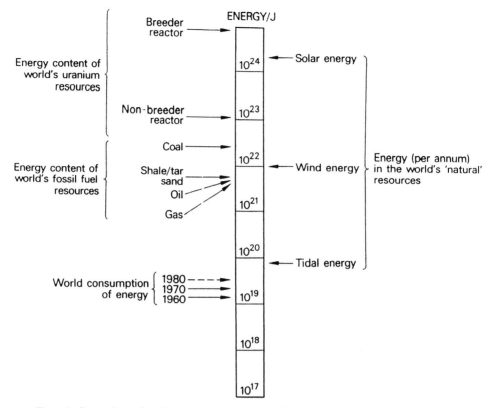

Figure 1. Comparison of world energy consumption and fuel resources with natural energy inputs. (Sources of data: *Energy and Power. Scientific American* **225**, No. 3 (September 1971); *Energy for the Future.* Report of Working Party of the Institute of Fuel (1973).)

vitro photosynthesis as a process driven by sunlight that produces fuel by a non-biological direct conversion process. We shall take a broad view of the term fuel, as shown in Table 1, which summarizes the processes to be discussed. These produce either an energy-rich reaction product or electrical energy; both these forms of energy are more versatile and in many ways more valuable than the rather low temperature heat that is obtained from solar/thermal devices, because heat can neither be stored for any length of time, nor can it be converted into work with 100% efficiency.

The order adopted in Table 1 is the order which we follow in our discussion, since it enables us to show the extent to which increasing understanding of the process of *in vivo* photosynthesis is leading to better prospects for it. First, however, we shall summarize the characteristics of the solar energy received at the Earth's surface, and the factors that limit the efficiency with which it may be converted into work.

SIR GEORGE PORTER, F.R.S., has been Director of the Royal Institution of Great Britain and Fullerian Professor of Physical Chemistry since 1966. Previously he has been Firth Professor of Chemistry at the University of Sheffield. Nobel laureate, 1967, for his work on photochemistry, he is a Past-President of the Chemical Society of London and, among a large number of academic distinctions, is Foreign Associate of the National Academy of Sciences and of the Leopoldina Academy.
Address: The Royal Institution, 21 Albemarle Street, London W1X 4BS, England.

MARY ARCHER read Chemistry at Oxford and did her Ph.D. at Imperial College, London. After post-doctoral work at Oxford on electrochemistry, she joined the Davy Faraday Laboratory of the Royal Institution in 1972, to work on photoelectrochemical phenomena.
Address: The Royal Institution, 21 Albemarle Street, London W1X 4BS, England.

1. INSOLATION AT THE EARTH'S SURFACE

1.1. Global Irradiance

The earth receives a vast amount of energy from the sun.[1,2] Figure 1 compares the annual global supply of solar energy to the Earth's surface, 3×10^{24} J, with global energy consumption and with the energy contained in the world's initial resources of fossil and nuclear fuels. Wind and tidal energy are also shown for comparison; valuable though these may become in favoured locations, it is clear that they do not represent an energy source of magnitude comparable to solar energy. However, the encouragingly large amount of solar energy shown in Fig. 1 falls on a very large area of land and water, and the power density is therefore low.

Figure 2 gives mean insolation data for the world. The amount of solar energy received at any particular location depends mainly on the latitude and the average cloud cover. In cloudy areas, a substantial portion of the total radiation arrives diffusely, rather than directly from the Sun, owing to scattering by clouds. For example, in the United Kingdom, 60% of the total irradiance is diffuse. In such areas, the use of focusing collectors to concentrate direct sunlight is very unlikely to be economic, and flat collectors capable of using diffuse as well as direct radiation must be employed. It is clear that to collect a substantial amount of energy, large areas must be covered with collection devices, and it follows that the cost per unit area and lifetime of the system are critical. At

1975 prices, a 10% efficient solar converter that receives average insolation of 250 W m^{-2} must cost no more than £5 per square metre to be economically competitive with conventional power stations. As insolation varies both diurnally and seasonally, an inbuilt capacity for energy storage is highly desirable; therefore direct conversion devices that produce metastable chemical products, hydrogen for example, have an advantage over those that produce electrical energy, which is expensive to store.

About two-thirds of the solar energy reaching the Earth's surface is 'used' simply to heat it, and about one-third goes to maintain the hydrological cycle, by evaporation of water from oceans and inland waters. Only a very small fraction ($\sim 0.1\%$) is chemically fixed by the process of photosynthesis, and of this annually fixed carbon, only about 0.5% is consumed as food. The scope for improving the efficiency of natural photosynthesis is considerable,[3] but we shall not pursue this important topic here.

1.2. Spectral Irradiance

The mean solar irradiance at normal incidence just outside the Earth's atmosphere is 1353 W m^{-2} and the solar spectrum approximates fairly closely to that of a black body at 6000 K.[4] This spectral distribution is modified in the atmosphere by scattering, and by absorption by ozone, in the ultraviolet (u.v.), and by water vapour and carbon dioxide in the infrared (i.r.), to an extent which depends on the optical path length of the sunlight in the air. Figure 3 gives the spectrum of 'average' British clear sky sunlight, from the data of Henderson and Hodgkiss.[5]

Figure 2. Mean annual intensity of global solar radiation on a horizontal plane at the surface of the Earth, W m^{-1} averaged over 24 h day.

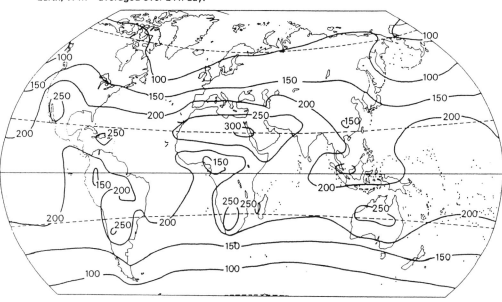

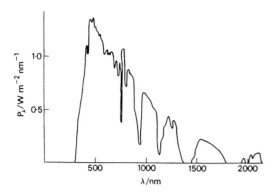

Figure 3. Radiance per unit wavelength of 'average' British sunlight, from data of Ref. 5.

2. LIMITATIONS ON THE EFFICIENCY OF RADIANT ENERGY TRANSDUCERS

2.1 Thermodynamic Limitations

The laws of equilibrium thermodynamics tell us that by carrying out a reversible process, work can be converted entirely into heat, but that heat can only partly be converted into work. The distinction between these two forms of energy is normally perfectly clear: work is completely ordered energy, heat is completely disordered energy. The nature of radiant energy is, however, ambiguous in this respect. We might intuitively conclude that the entropy content of highly monochromatic, highly directional light from a laser was a good deal lower than the entropy content of the broad band, omnidirectional light from an incandescent lamp emitting energy at the same overall rate. That is indeed the case, and by treating radiant energy as a collection of bosons and using the appropriate quantum statistical expression for the entropy of the ensemble, the entropy and free energy content of any arbitrary packet of radiant energy may be defined.[6] The special though important case of black body radiation was first considered by Planck,[7] in which case it may be shown that the temperature of the black body T_{BB} and the entropy and energy flux $\dot{S}_L$ and $\dot{E}_L$ in the emitted radiation are related by the following equation:

$$T_{BB} = \dot{E}_L / \dot{S}_L \qquad (3)$$

More generally, the temperature T_L of an arbitrary pencil of light is defined by

$$T_L = \dot{E}_L / \dot{S}_L \qquad (4)$$

The entropy content of light has a bearing on the maximum efficiency that is achievable by direct energy converters, such as those listed in Table 1. The entropy of the converter will remain constant if the

converter is operated reversibly and increase if it is operated irreversibly. Therefore, if the converter is operated by radiant energy that contains some entropy, output of work cannot occur with 100% efficiency. Both chemical free energy and electrical energy are, thermodynamically speaking, work, and therefore there is a thermodynamic limitation on the efficiency of the devices listed in Table 1.

It may be shown[8-10] that the efficiency of the converter operated at temperature T_S by light of temperature T_L is given by the Carnot expression

$$\eta \leq 1 - T_S/T_L \qquad (5)$$

By similar arguments it may be shown[11-13] that the efficiency of a converter that turns work into light is given by

$$\eta \leq (1 - T_S/T_L)^{-1} \qquad (6)$$

If solar energy provides the energy to drive the converter, then T_L in the direction of the propagating beam is ~6000 K. This beam subtends an angle of 6.8×10^{-5} steradians at the earth. If the sunlight is scattered, either by the atmosphere or by the converter itself, over a solid angle of 4π, then its effective temperature is lowered to ~1350 K.[14] For the common value of 298 K for T_S, (5) implies that solar energy cannot be converted to electric energy or chemical free energy with efficiency greater than ~78%. The remaining 22% is dissipated in the converter and eventually lost in the thermal $T_{BB} = 298$ K radiation from it.

Unfortunately, it is not possible to obtain experimental evidence for the existence of this limit, because achievable efficiencies lie well below it.[9] (The situation is, however, different for light emitting diodes operated at low temperatures.[11]) The most efficient man-made solar direct conversion device is the silicon photovoltaic cell. In full sunlight, present day conversion efficiencies at the maximum power point are of the order of 12–18%. An efficiency of 20–22% is unlikely to be exceeded.[15] The fundamental photoprocesses of photosynthesis operate with red light at an overall efficiency of ~36%,[16] which is still well below the thermodynamic limit, although impressively high by comparison with most man-made devices. The structure and function of photosynthetic units have much to tell us about how this high efficiency is achieved by avoiding wasteful back and side reactions in the natural photosynthetic process.

2.2 The Threshold Wavelength and Spectral Absorbance of the Converter

All quantum converters of light are threshold devices; there is a certain threshold photon energy E_g, which in cases of interest to us is either a semiconductor band gap or the energy of the first excited singlet state of a molecule, below which photons are not absorbed, or if they are, they do not produce the desired effect. Photons of energy $E > E_g$ are not entirely efficient even if they are completely absorbed, since vibrational relaxation almost inevitably occurs in the upper excited state before the charge transfer process can take place; the fraction $(E - E_g)/E$ of the photon's energy is therefore dissipated as heat, and only the fraction E_g/E can be converted to useful work.

The fraction of the incident light that is available for conversion to work η_{av} when a direct converter is irradiated with white light of spectral radiance P_E, power per unit area per unit energy interval, is therefore given by

$$\eta_{av} = E_g \frac{\int_{E_g}^{\infty} A_E P_E \, d \ln E}{\int_0^{\infty} P_E \, dE} \qquad (7)$$

A_E is the fractional optical absorbance of the converter for photons of energy E. η_{av} is therefore a function of P_E, E_g and A_E. The maximum value of A_E is 1.0 for all $E > E_g$, and this leads to the highest possible values of η_{av}. Figure 4, line (a) shows η_{av} as a function of E_g for this case, calculated by using P_E values derived from Fig. 3. η_{av} has a maximum value of 0.51 at $E_g = 1.1$ eV, equivalent to light of wavelength 1127 nm. The band gap in silicon is 1.14 eV at 300 K, which is one reason for the good performance of silicon solar cells. However, most photochemical reactions have thresholds much shorter than the optimal value, and thus cannot make use of a large portion of the solar spectrum. Moreover, the absorption spectrum of a solution of photochemically reactive molecules, or even of a weakly coupled molecular crystal, is generally not such that $A_E = 1$ for all $E > E_g$. Dyes, for example, which by definition absorb visible light strongly, usually have a single, rather broad and symmetrical absorption band in the visible, for which the extinction coefficient is a Gaussian function of E, with a half-width at half-height of the order 0.2–0.25 eV. The optical density D_E and the absorbance A_E for such a system are illustrated in Fig. 5. (The band origin E_g is near the lower edge of the absorption and can be ascertained for any particular system from the overlap of the absorption and fluorescence spectra.)

If the A_E profile of the system tails off at $E \gg E_g$, it is clear from (7) that η_{av} will be reduced. This is illustrated by lines (b) and (c) of Fig. 4, which are calculated for two fairly typical cases. The value of η_{av} becomes disconcertingly low for a photochemical system with a high E_g value and a narrow absorption band. However, it is relatively easy in photochemistry to increase the effective bandwidth of the absorbance by using two or more components. This is how green leaves achieve the very high absorption of visible light illustrated in Fig. 6. The plant pigments (mainly chlorophylls a and b and carotenoids) have overlapping absorption spectra that between them scan the photosynthetically active range 300–700 nm very effectively. In green plant photosynthesis, energy absorbed is degraded to ~700 nm ($E_g = 1.77$ eV), by an energy transfer process discussed in the next

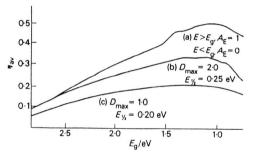

Figure 4. Solar energy available for conversion, η_{av}, as a function of E_g: (a) for an ideal converter; (b), (c) for a photochemical energy converter with a Gaussian absorption spectrum, maximum optical density, D_{max} and half-width at half height, $E_{\frac{1}{2}}$, as shown.

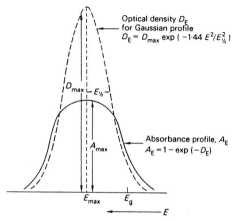

Figure 5. Optical density D_E and absorbance A_E of a Gaussian absorption band, of half-width at half-height $E_{\frac{1}{2}}$, maximum optical density D_{max} ($0 < D_{max} < \infty$) and maximum absorbance A_{max} ($0 < A_{max} < 1$). The band origin is at E_g.

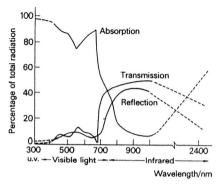

Figure 6. Absorption, reflection and transmission of light by green leaves.

section, and this means that η_{av} in full sunlight for leaves is ~0.33. Above 700 nm, absorbed light is photosynthetically inactive, and exerts only a deleterious heating effect on the leaf. It is noteworthy how steeply the absorption of the leaf falls to a low value in the i.r. The chlorophylls and ancillary pigments in photosynthetic bacteria are somewhat different from those in green plants, and all absorbed energy is degraded to a rather longer wavelength. In the case of purple bacteria, this wavelength is ~870 nm ($E_g = 1.43$ eV) and η_{av} in full sunlight is ~0.47.

It would appear from Fig. 4, line (a) that the threshold energy of 1.8 eV for green plant photosynthesis is too high to be optimal. However, this was calculated for average British sunlight—that is, averaged over clear days only. In cloudy conditions the optimal value of E_g shifts upward from 1.1 eV, because the solar spectrum is more attenuated in the i.r. than in the visible.[17] Recently, Landsberg and Mallinson[18] have performed a vigorous calculation of η_{av} for abrupt junction solar cells as the sky changes from clear to cloudy and have shown that under very cloudy skies, the optimal threshold becomes as high as 2.2 eV. A band gap of about 1.7 eV appears optimal for an average 60% diffuse radiation, as in the United Kingdom. Thus, the threshold energy for green plant photosynthesis appears to be rather closely tailored to average meteorological conditions in temperate latitudes.

The energy threshold is only the first of several operational limitations on the efficiency of direct converters of solar energy. The other factors that control overall efficiency depend on the nature of the converter, and are discussed in the appropriate section below.

3. PHOTOBIOLOGICAL ENERGY CONVERSION

A great deal of research is being carried out into the elucidation of the structure and function of photosynthetic membranes, the better understanding of which might possibly lead to more efficient manmade photochemical systems for harnessing solar energy. Only some of the more recent work will be outlined here, as several excellent reviews are available. A recent book[16] on the bioenergetics of photosynthesis provides an extremely valuable compilation of authoritative contributions. Bacterial photosynthesis[19] and primary processes in bacterial photosynthesis[20] have also been comprehensively reviewed.

3.1. Arrangement of Chlorophyll in Photosynthetic Membranes

Photosynthetic membranes consist mainly of lipids, pigments, chiefly chlorophylls, and proteins, some of which, the intrinsic proteins, are embedded in the

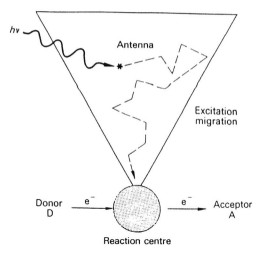

Figure 7. Diagrammatic sketch of a photosynthetic unit. An absorbed photon produces an exciton (*) in the antenna. The exciton migrates to the reaction centre, at which the primary photochemical reaction occurs.

membrane, possibly spanning it, and others which, the extrinsic proteins, are adsorbed on the surface. Most of the chlorophylls and other pigments in the membranes are not involved in any photochemistry, but act as a light-harvesting antenna which transfers absorbed energy to a special part of the photosynthetic unit called the reaction centre, at which an electron transfer then occurs, as sketched in Fig. 7. Each photosynthetic unit in green plants contains about 300 chlorophyll molecules: smaller units are involved in bacterial photosynthesis, each containing about 50 bacteriochlorophyll molecules. The oxidized donor D^+ and the reduced acceptor A^- formed by the primary electron transfer reaction, are converted back to their original form by a series of further electron transfers, not shown in Fig. 7, which culminate in an oxidation reaction, the evolution of oxygen from water in the case of green plant photosynthesis, on the D^+ side and the reduction of carbon dioxide to carbohydrate on the A^- side.

The location of the chlorophyll molecules in photosynthetic membranes is still something of an open question and is of prime importance in view of the role of chlorophyll in capturing and transferring light energy. The absorption and fluorescence spectra of chlorophylls *in vivo* are complex compared with those observed *in vitro*, which is generally agreed to reflect different aggregates in different environments for chlorophyll *in vivo*. The comparison of fluorescence polarization *in vivo* and in stretched chlorophyll films show that the specific orientation of antenna chlorophyll in plants is low.

Anderson[21,22] has postulated that the bulk of the antenna chlorophyll in chloroplasts is in the fixed,

Chlorophyll

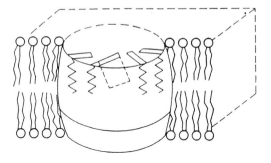

Figure 8. Schematic cross-section of a chloroplast membrane showing an intrinsic protein spanning the membrane, with hydrophilic regions located at the membrane surfaces and a hydrophobic portion (shaded) embedded within the non-polar interior of the lipid bilayer.[21,22]

boundary lipids that form a monomolecular layer attached to the hydrophobic shells of the intrinsic proteins; the suggested arrangement is shown in Fig. 8. The necessary aggregations of the chlorin rings, indicated by spectral data and required for maximum energy migration, would be possible in this model. The orientation of the chlorin rings, almost buried in the folds of the protein, is more securely locked in position than if the chlorophyll were in the rather fluid bulk lipid domain of the membrane.

Very recently, the first direct structure determination of a chlorophyll-containing protein by X-ray crystallography has been reported.[23] This method, applicable in principle to all proteins that can be crystallized, produces very precise information about the location of chlorophyll *in vivo*. The protein concerned, extracted from the green photosynthetic bacterium *Chlorobium Limicola*, consists of three identical subunits, each containing a core of seven bacteriochlorophyll molecules confined within an ellipsoid of axial dimensions $4.5 \times 3.5 \times 1.5$ nm. The orientation of the porphin rings do not conform to any repetitive pattern, although their planes do lie roughly parallel to one another.

3.2. Energy Migration in Photosynthetic Units

One of the intriguing aspects of the primary processes of photosynthesis is the high efficiency with which energy migrates from the light harvesting antenna of pigment molecules to the trap, for this process has not as yet been achieved at anything like the same efficiency *in vitro*. Solutions of chlorophyll *in vitro*,

whether in fluid solvents, rigid matrices, monolayers, multilayers or bilayer vesicles, exhibit the phenomenon of concentration quenching of the excited state at concentrations much lower than those which are present in the chloroplast ($\sim 10^{-1}$ M). To account for this, Porter[24] has proposed that the mechanism of concentration quenching in non-biological chlorophyll systems is Förster energy transfer between chlorophyll molecules followed by capture at a non-fluorescent trap, which is merely a pair of chlorophyll molecules whose separation, although part of the equilibrium statistical distribution, is less than a critical distance R_t. A Monte Carlo method was used by Beddard[24] to calculate fluorescence yields as a function of chlorophyll concentration, and excellent agreement was obtained with experimental data from a wide variety of systems for a trap distance R_t of 10 Å.

These calculations indicate that efficient energy transfer would be impossible in the photosynthetic unit if it resembled a randomly distributed solution of chlorophyll molecules containing, as such a distribution would, a high proportion of chlorophyll *a* molecules with nearest neighbours within 10 Å. It is therefore inferred that the chlorophyll molecules are separated from each other, probably by co-ordination of lipids or proteins to the magnesium atom of the chlorophyll. There is some evidence that galactolipids, which are present in the chloroplast at two or three times the chlorophyll concentration, may fulfil this function. Alternatively, the work of Fenna and Matthews,[23] although it does not refer to a light-harvesting unit, shows how a protein is able to produce an arrangement of chlorophyll molecules with an average near neighbour distance of 12 Å, so allowing efficient energy transfer whilst inhibiting concentration quenching.

Knox[25] has considered excitation energy transfer and migration in photosynthetic systems in great detail and has drawn attention to the large spread, of a factor of over 100, in the estimated rates of pairwise excitation transfer between chlorophyll *a* molecules; considerable further analysis of the experimental data will be necessary, as the pairwise transfer rate is required to construct any model of more extensive energy migration. One valuable experimental probe of energy transfer in photosynthetic units is the *in vivo* fluorescence lifetime and yield of chlorophyll, and its variation with the redox state of the reaction centre. For example, the fluorescent quantum yield increases by up to ten times when the reaction centres are inactivated by chemical reduction, or by saturating light intensities. This corresponds to a quantum yield of excitation energy trapping of ~90% in the functional reaction centres when they are fully operative.

In the past, fluorescence lifetimes of *in vivo* chlorophyll have been determined by indirect means or by nanosecond pulse techniques, which have inadequate time resolution. However, picosecond techniques involving the use of mode-locked lasers and electrooptic shutters, or more recently, streak cameras, now enable direct and reliable measurements to be made. Lifetimes of the order of 50–200 ps have been reported for various algae and chloroplast fragments,[26–28] corresponding to a time of only a few hundred femtoseconds for each pairwise energy transfer.

3.3. Primary Photochemical Events at the Reaction Centre

Another component of the photosynthetic membrane, whose structure is doubtless the key to its successful function, is the reaction centre; electronic excitation migrating from the antenna is trapped here, and an electron transfer from a donor D to an acceptor A ensues. Knowledge of the primary photochemistry of reaction centres, particularly in bacterial photosynthesis, is advancing rapidly, owing to the development of techniques for isolating them from the rest of the membrane of purple bacteria by use of detergents, and to the application of picosecond techniques to the study of their photoprocesses. Each reaction centre in purple bacteria contains three different polypeptides, four bacteriochlorophyll molecules, two bacteriopheophytins, one ubiquinone and one non-heme iron. Two of the BChl molecules are strongly exciton coupled to form a dimer usually known as P870, whose photo-induced one electron oxidation, which shows a maximal absorbance change at ~870 nm, is well established to occur in the primary photochemical process. In chloroplasts two similar dimers of chlorophyll a, P700 and P680, are known to act analogously. The nature of the primary electron acceptor A, both in bacterial and green plant reaction centres, has been less clear, and a variety of candidates have been proposed over the past 15 years.

There are two principal methods by which the nature of A has been studied. Electron spin resonance (e.s.r.) provides valuable information about free radical intermediates. Recent e.s.r. evidence[29–32] indicates that A, in bacterial photosynthesis, is 1:1 complex of ubiquinone and an iron compound (Fe–UQ). However, time resolution in kinetic e.s.r. work is poor compared with the spectral detection of intermediates by laser flash photolysis, and recent picosecond and nanosecond spectroscopic work[32–34] on the reaction centres of the purple bacterium R. *spheroides* has shown that a transient state P^F, which is not simply the excited singlet of P870, forms even at redox potentials at which (Fe–UQ) is chemically reduced, and so cannot act as an electron acceptor. The absorbance changes in the region 300–900 nm accompanying the formation of state P^F have been measured in some detail,[33–35] but no firm conclusions as to the nature of P^F could be reached in the absence of spectral data on some of the other components in the reaction centre which might be involved. However, Fajer and co-workers[36] have recently measured the optical spectrum of the radical anion BPh^- (elec-

trochemically generated in CH_2Cl_2) and have demonstrated an extremely convincing fit between the P^F data and the sum of the spectral data for $BChl_2^+$ and BPh^-. They propose, therefore, that the excited reaction centre $BChl_2^*$ transfers an electron first to BPh, forming BPh^-, which then passes the electron on to Fe–UQ, and have incorporated BPh^- into the known cyclic electron transport processes that occur in R. *spheroides*, the first three steps of which are shown below:

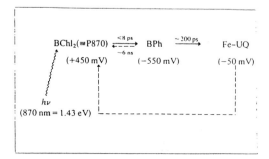

The numbers in parentheses are the reversible half-wave potentials for the couples $BChl_2^+/BChl_2$, BPh/BPh^- and $Fe-UQ/Fe-UQ^-$. The rapidity (<8 ps) with which the state P^F is formed is remarkable. The reactants $BChl_2$ and BPh must be so constrained in the reaction centre that electron transfer from $BChl_2^*$ to BPh requires little or no nuclear motion. The energy efficiency of the primary step ($=1.0/1.43 = 0.70$) is also quite remarkable, although it is followed rapidly by the rather downhill process, energetically speaking, of electron transfer to Fe–UQ, but this has to occur rapidly to beat the wasteful back transfer of the electron from BPh^- to $BChl_2^+$. It seems likely therefore that BPh is the nearest neighbour both of $BChl_2$ and Fe–UQ, but $BChl_2$ and Fe–UQ are farther apart. The electron transferred to Fe–UQ is eventually returned to $BChl_2^+$ in a fairly well established cyclic process which generates the high energy intermediates required in the carbon dioxide fixation cycle.[37]

The structure of the reaction centres remains to be determined in detail, and even their positions in the photosynthetic membrane are not known with certainty. However, the chloroplast thylakoid membrane is known to be highly asymmetric, with the outer surface having more protein bound upon it than the inner one. It is possible that the electron transport carriers (D and A) are also arranged asymmetrically across the membrane, with electron donors D at the inside and electron acceptors A at the outside.[21,38] In this way, the high energy intermediates D^+ and A^- formed by the primary processes might be prevented from reacting with each other, to reform D and A. Direct observations have been made of light-induced electric fields across chloroplast membranes,[39] which support this hypothesis.

4. MODEL SYSTEMS FOR PHOTOSYNTHESIS

Photosynthesis is only one of many vital processes that occur in living cells at membranes. The fabrication and properties of simple artificial membranes of broadly similar structure and composition to natural membranes is helping to elucidate the complex processes that can occur within the latter. In particular, the study of photochemical reactions involving pigment-containing thin films, membranes and micelles has been stimulated by the present world-wide research activity in natural photosynthesis. For example, a detailed study of electron transfer between chlorophyll and quinones in a multilayer lecithin matrix has recently been completed.[40]

4.1. Artificial Membranes

There are at present two principal methods of fabricating thin membranes. Planar membranes may be produced by 'painting' a lipid solution across a vertical orifice. The film thins under the influence of gravity until it is only two molecules thick, as shown in Fig. 9(a), and the membranes are hence known as bimolecular lipid membranes (BLM). Alternatively, spherical vesicular BLM enclosing the aqueous phase, as shown in Figure 9(b), can be formed by mechanical agitation or ultrasonication of a suspension of lipids in water. One advantage of the microvesicles is that they contain a large BLM area per unit volume, and thus are better suited to spectral work. Planar BLM, on the other hand, are better suited to the study of charge transfer across membranes.

Photosensitized charge injection can be observed if pigments are incorporated into a planar BLM, and this has suggested[41] the design of a photoelectrochemical cell of the type shown in Fig. 10, in which there is an asymmetric membrane. Chlorophyll, or

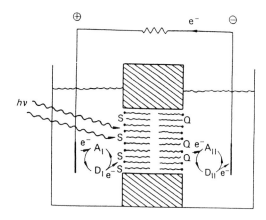

Figure 10. Photoelectrochemical cell containing an asymmetric bilayer membrane. The sensitizer molecule S and the electron acceptor Q must have hydrophilic heads and hydrophobic tails.

another chemically suitable pigment, is incorporated in one side of the membrane, and an electron acceptor, such as ubiquinone, in the other side. Photopotentials of greater than 100 mV can be generated by irradiation of such a BLM. There is evidence that the mechanism of charge transfer through the membrane is electron tunnelling.

With suitable electron donors and acceptors in the aqueous phase, a continuous current passes, owing to the occurrence of Faradaic processes at both interfaces, as shown in Fig. 10. It has been suggested[41,42] that it might be possible to couple membranes of this nature in such a way as to achieve the Faradaic reduction of water to hydrogen at one interface, with concomitant production of oxygen at the other. At the moment, there are many obstacles in the way of this approach. First, asymmetric membranes are

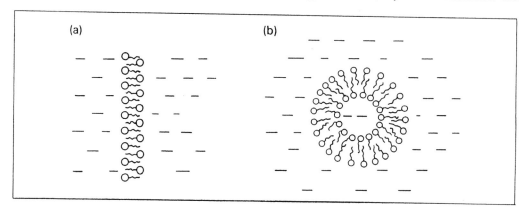

Figure 9. Molecular form of (a) planar BLM, (b) vesicular BLM. Each lipid molecule (O~) has a hydrophilic head (O) and a hydrophobic tail (~).

extremely difficult to fabricate and handle, and they can only be formed in areas of a few mm². Second, the most photosensitive membranes invariably have very high electrical resistivity. Third, although the quantum efficiency of energy transduction can be as high as 0.1, in terms of electrons flowing per photon absorbed, in practical terms the efficiency in terms of electrons flowing per photon incident will be very low. This is because the BLM contains only a sub-monolayer amount of pigment, which absorbs very little of the incident light.

However, none of these problems is necessarily insuperable. It may be possible to develop membranes of much greater mechanical stability than present BLM. For example, it has recently been reported[43] that BLM of the lipid glycerol monooleate can be formed in the interstices of a nitrocellulose polymer film, which is fairly robust. A small group at General Electric, Schenectady are working on *in situ* formation of a supporting substrate for planar BLM.[44]

It has been known for some years that certain algae, which contain the enzyme hydrogenase, will, under suitable conditions, produce molecular hydrogen instead of fixing carbon dioxide to carbohydrate.[45] The hydrogenase enzyme which acts as the electron acceptor A in the algal membrane, is an iron–sulphur protein and a synthetic model compound of similar structure might facilitate the production of hydrogen on one side of an artificial membrane. A manganese complex about which relatively little is known is involved in the oxygen evolution step in green plant photosynthesis, and again a synthetic analogue of such a complex might facilitate the production of oxygen at a photosensitive membrane. It is interesting that a binuclear manganese complex, a dimer of manganese(II) gluconate, has recently been reported to mimic much of the oxidation–reduction chemistry that has been postulated for the manganese complex involved in natural photosynthesis.[46]

4.2. Photochemistry in Micellar Systems

Micelles are spherical or ellipsoidal aggregates of long tailed molecules of type similar to those that form BLM. The head groups of the molecules are charged and this produces quite a sizeable gradient of electric potential across the micelle–water interface.

In a recent series of elegant experiments,[46–50] Henglein, Grätzel and co-workers have investigated photoelectrochemical effects across the aqueous-lipoid interfaces of positively and negatively charged micelles. Electron transfer from photochemically excited electron donors such as phenothiazine, solubilized in a negatively charged micelle, to water, forming an aquated electron, has been observed.[47] This process is shown schematically in Fig. 11(a). The reverse process, transfer of an electron from the aqueous phase to an acceptor, 9-nitroanthracene or pyrene, in the interior of the micelle, which is illustrated in Fig. 11(b), has also been studied.[48] The rate

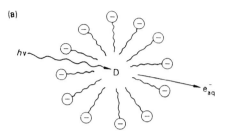

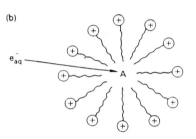

Figure   11(a). The   photoionization   of phenothiazine D in micelles of sodium lauryl sulphate by 347 nm light.[47] The electron tunnels through the double layer into unoccupied electronic redox levels of the system aq/e_{aq}^-. (b) Electron tunnelling from e_{aq}^-, produced by pulse radiolysis, to an acceptor molecule A (9-nitroanthracene or pyrene) in the interior of a   cetyl   trimethyl   ammonium   bromide micelle.[48]

of reaction with acceptors in cationic micelles was considerably faster than with those in anionic micelles.

These reactions are regarded as electron tunnelling processes, and the results are interpreted in terms of the estimated distributions of occupied and unoccupied electronic levels in the redox systems aq/e_{aq}^-, D/D$^+$ and A/A$^-$ involved. If the overlap of the occupied levels of the donor with the unoccupied levels of the acceptor is not good, transfer is slow, irrespective of how thermodynamically favourable it may be. These levels are shifted relative to one another, by the charge on the micellar head group, which determines the direction of the electrical double layer at the interface. Reaction rates are therefore influenced both by the charge on the micelle and the concentration of electrolyte in the aqueous phase.[49,50] The lifetime of normally unstable intermediates may be enormously enhanced by working in a micellar system.[51]

The possibility of effecting the decomposition of water in a heterogeneous micellar system by a

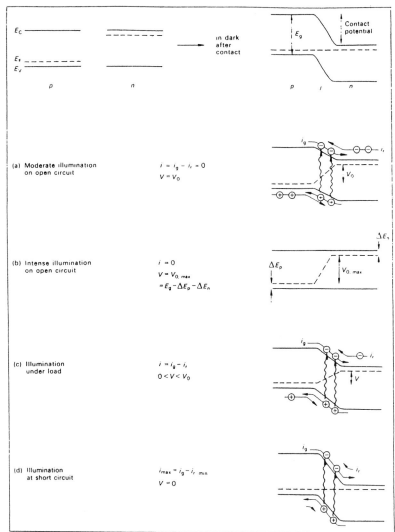

Figure 12. *p–n* junction schematics for a photovoltaic cell under various operational conditions.

suitable series of electron transfers offers certain advantages compared with homogeneous systems.[46] Not only may the rate of geminate recombination of the products of electron transfer be lowered by the double layer, but organic sensitizers which are water insoluble, but lipid soluble, can be used. Moreover, the photon energy required to initiate the tunnelling processes from the donor to the acceptor is much lower than that required for photoionization. It might not be necessary or desirable to keep the sensitizer concentration down to one molecule per micelle, and much higher concentrations are possible.

5. PHOTOVOLTAIC ENERGY CONVERSION

We shall treat the subject of photovoltaic cells before considering photoelectrochemical energy conver-

sion, although in a sense the latter are more like natural photosynthetic systems than the former. We do this because many of the underlying principles of these two topics are the same, and they are more simply introduced by considering photovoltaic cells.

A solar cell is a purely solid state semiconductor device which contains a region of varying chemical composition which is associated with a gradient of electric field. This field gradient separates the charge carriers, electrons and holes, that are produced by irradiating the cell. The gradation in chemical composition may be achieved either by layering two dissimilar materials together or by doping a single semiconductor asymmetrically. The latter method is used to produce silicon solar cells, which are at present by far the most developed of the photovoltaic cells, and this type of cell will be discussed first.

If each dopant atom has one more valence electron than the semiconductor atoms, then these 'spare' electrons can be readily given up to the conduction band, thus greatly reducing the electrical resistance of the material. The effect is pronounced even if only one phosphorus atom is added to silicon for each million silicon atoms. Semiconductors so doped are termed negative-type. If, however, the dopant atoms have one less valence electron than the normal semiconductor atoms, the dopant 'steals' electrons from the valence band, leaving mobile holes, and rendering the semiconductor positive-type. A hole is an electron empty state surrounded by full states, and both electrons and holes are mobile in semiconductors.

5.1. Factors that Limit the Efficiency of Photovoltaic Cells

The mode of action of a silicon solar cell is illustrated in Fig. 12. Contact between àn n-type and p-type semiconductor in the dark produces a contact potential, as shown at the top of Fig. 12. The Fermi level is the same through the p, n and junction regions. On irradiation with photons of energy $E > E_g$, hole-electron pairs are created with a quantum yield of unity. If the radiation is absorbed in a region of varying electric potential, such as the p–n junction, then the electrostatic field tends to separate the mobile hole–electron pairs. On open circuit, no net current flows across the junction, and the generation current i_g is equal to the recombination current i_r, as in Fig. 12(a). A limiting open circuit voltage is obtained under very intense illumination, which destroys the ability of the junction to separate hole-electron pairs. As shown in Fig. 12(b), this limiting voltage $V_{0,\max}$ is given by

$$V_{0,\max} = E_g - \Delta E_p - \Delta E_n \qquad (8)$$

where ΔE_p is the difference in energy between the Fermi level in the p-type material and the top of the valence band, and ΔE_n is the difference in energy between the bottom of the conductor band and the Fermi level in the n-type material. The open circuit voltage efficiency, η_{V_0} is defined

$$\eta_{V_0} = V_0 / E_g \qquad (9)$$

This factor is clearly a function of light intensity, and it has a limiting value of slightly less than unity.

On illumination under load, as illustrated in Fig. 12(c), the cell delivers current $i = i_g - i_r$ at voltage V. At short circuit, shown in Fig. 12(d), the Fermi level is the same throughout the semiconductor, ignoring ohmic effects, so the contact potential again has its maximum value, the recombination current is a minimum, and the short circuit current $i_{\max}$ flows. The rate at which electrons flow will, however, be rather smaller than the rate at which illumination generates hole–electron pairs, because of bulk and surface recombination. This loss is represented by the current efficiency η_i in (10), which is the number of electrons flowing in the circuit per photon absorbed

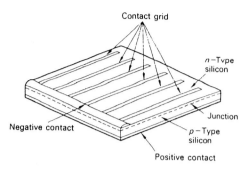

Figure 13. Silicon solar cell.

$$\eta_i = \frac{\text{Flux of electrons at short circuit}}{\text{Flux of photons absorbed by cell}} \qquad (10)$$

Finally the factor η_{iV}, defined by

$$\eta_{iV} = i_{mp} V_{mp} / i_{max} V_0 \qquad (11)$$

where i_{mp} and V_{mp} are the cell current and voltage at the maximum power point, expresses the fact that the current–voltage characteristic of the cell is not rectangular.

Ignoring minor losses due to the reflectivity and ohmic resistance of the cell, the overall conversion efficiency η_i in (10), which is the number of electrons flowing in the circuit per photon absorbed

$$\eta_{pv} = \eta_{av} \eta_{V_0} \eta_i \eta_{iV} \qquad (12)$$

The factor η_{av}, defined by (7), has the same significance as before.

5.2. Silicon Cells

A type of silicon cell in common use for space applications is shown in Fig. 13. Measuring 20 mm square and 0.3 mm thick, it is cut from a single crystal of boron-doped p-type silicon. A shallow junction is formed on one surface by the diffusion of an n-type impurity, usually phosphorus, forming an 'n-on-p' cell, with light incident on the n-side of the junction. The alternative 'p-on-n' cell can be made by diffusing boron into a phosphorus-doped wafer. The front contact is a narrow strip of metal running along one edge and extended into a grid of narrow fingers. The other contact, positive on an 'n-on-p' cell, is a continuous metal film covering the back surface. The front of the cell is usually given an antireflective coating and is protected by a thin glass or quartz cover.

The cells currently being made for terrestrial applications are similar, except that they are made of 50 mm to 75 mm diameter discs, cut from the untrimmed crystal, and are finished to a simpler specification to reduce cost.

The values of the factors in (12) for a 'p-on-n' silicon cell with a resistivity of 1 ohm cm under full sunlight are[15] $\eta_{av} = 0.44$, $\eta_{V_0} = 0.60$, $\eta_i = 0.78$. $\eta_{iV} =$

0.74, giving an energy conversion efficiency at the maximum power point of ~15%. (The theoretical limit is about 22% at 25 °C.[15]) This good working efficiency is due to a combination of factors. First, the silicon band gap (1.1 eV) is reasonably compatible with the spectral radiance of the Sun, as explained in Section 2.2. Second, the charge carriers are quite mobile, and in a carefully manufactured cell there are few traps for either electrons or holes, so that their mean diffusional length is sufficiently long for most of them to reach the collector electrodes; this leads to a high value for η_i. Third, the internal resistance of the cell is low so that η_{iV} is also high. These three conditions must be met in any photovoltaic cell that is to function efficiently. Unfortunately, in the case of silicon, efficient cells can only be made by exercising scrupulous control over all steps in the fabrication, and this causes them to be expensive. Costs may be much reduced, however, by the successful development of the EFG (Edge-defined Film-fed Growth) method for growing simple crystal silicon in a continuous thin ribbon,[53] or of a multiple pass thin film silicon cell.[54-56]

5.3. Cadmium Sulphide Cells

A considerable amount of developmental work has been carried out on the cadmium sulphide cell, which is actually composed of two semiconductors, CdS and Cu_2S, a heterojunction existing between these two materials. These cells are made from thin (~25 μm) vapour-deposited polycrystalline films of CdS, on which a thin Cu_2S layer is chemiplated. In the past, these cells were found to degrade severely when illuminated under open-circuit conditions, but this effect has been much diminished by control of the stoichiometry of the Cu_2S layer. The relatively good efficiency of these cells, open-circuit voltage ~0.45 V, conversion efficiency ~6% in full terrestrial sunlight, is unusual in a polycrystalline device, and may depend on the fact that most of the crystallites extend from the front to the back of the thin film, so that charge carriers do not have to cross many grain boundaries, at which they tend to be trapped.

5.4. Gallium Arsenide Cells

The advantages of gallium arsenide over silicon as a solar cell material are that its energy gap is nearer the optimum, it absorbs most of the sunlight in a much thinner layer (about 2 μm) and it is less temperature-sensitive. Until recently, technical problems prevented the effective exploitation of these advantages, although some gallium arsenide cells have been used by the Russians for high temperature applications in space. But in 1971, IBM claimed to have achieved an efficiency of 18% in cells consisting of a thin layer of p-type gallium aluminium arsenide grown by liquid-phase epitaxy on single crystals of n-type gallium arsenide. So far, only very small cells (about 1 mm^2)

have been made. Because of their good performance at high intensities and temperatures, these cells are suitable for operation in concentrated sunlight, and this may offset their high cost to some extent.

5.5. Other Heterojunction Devices

There have been a few reports of II/VI and III/V heterojunction devices which have approached or even surpassed the performance of silicon cells. p-InP/n-CdS cells with a solar conversion efficiency of 12.5% have been fabricated.[57,58] This good efficiency arises first because of the nature of InP: the band gap is at 1.34 eV, which is optimal for the solar spectrum, and the material is intensely absorbing above this energy. Second, there is an excellent crystal lattice match between InP and CdS, which means that almost fault-free junctions can be grown. p-$CuInSe_2$/n-CdS cells which display current efficiencies, defined as electrons flowing in the short circuit current per photon absorbed, of up to 70% between 550 and 1250 nm, and solar conversion efficiencies of ~5% have been made.[59,60] p-CdTe/n-CdS cells of rather similar performance (current efficiency 85%, solar conversion efficiency 4.0%) have been produced without detailed attention to optimization of cell design, and it has been calculated that p-CdTe/n-$Zn_xCd_{1-x}S$ cells should be capable of a conversion efficiency of 23%.[61]

5.6. Schottky Barrier Solar Cells

Certain types of metal–semiconductor junctions are blocking, that is to say they show rectifying, non-ohmic properties in the dark. On irradiation, such a junction acts to separate hole–electron pairs and can therefore be made the basis of a photovoltaic device. Unlike p–n junction cells, Schottky barrier photovoltaic cells are majority carrier devices and are therefore relatively less susceptible to impurities. Figure 14 shows that a Schottky barrier is generally formed between p-type semiconductors and metals of fairly low work function, and between n-type semiconductors and metals of fairly high work function. Superficial oxide layers can, of course, greatly affect the work function of a metal, so these 'rules' must be cautiously applied. A Schottky barrier solar cell (SBSC) consists of a thin layer of semiconductor sandwiched between two metals, with an ohmic contact on one side, generally deposited as a finger grid so that the semiconductor can be irradiated through it, and a Schottky barrier contact on the other side. The advantage of SBSCs is that they are thin film devices and contain very little material per unit area.

Theoretical calculations[62] showed that the maximum solar energy conversion efficiency of a SBSC is high (~25%) before any cells that approached such a figure had been made. The fabrication of the first efficient 'large area' (~1 cm^2) SBSC has since been reported.[63,64] It consists of a Schottky barrier of

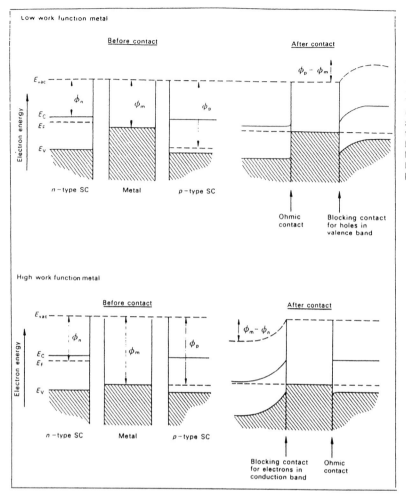

Figure 14. Schematic diagram showing the formation of ohmic and blocking junctions between metals and semiconductors. ϕ_n, ϕ_m, ϕ_p = work functions of n-type semiconductor, metal and p-type semiconductor, respectively. E_{vac} = vacuum level, E_c = conduction band edge, E_v = valence band edge, E_F = Fermi level.

5 nm Cr overlaid with 5–7 nm Cu on 2 ohm cm p-type silicon, with a top (ohmic) contact of Al fingers covered with 69 nm of antireflective silicon oxide coating. The sunlight conversion efficiency of the cell is 9.5%.

6. PHOTOELECTROCHEMICAL ENERGY CONVERSION

Photoelectrochemical cells act similarly to photovoltaic cells in some ways, but they contain electrolyte solution and are rather more complicated in principle, and very much less developed in practice, than purely solid state devices. A photoelectrochemical effect is defined as one in which the irradiation of an electrode/electrolyte system produces a change in the electrode potential (on open circuit) or in the current flowing (on closed circuit). The cause of this may be a photochemical reaction in bulk solution, the products of which are electroactive, in which case the cell is commonly called a photogalvanic cell. Alternatively, it may be that the cell contains a photosensitive membrane or electrode. In all cases, the absorbed light directly or indirectly causes a Faradaic electrode reaction to occur, in which chemical changes take place at the electrodes. Thus, both flow of electrons and chemical change are associated with photoelectrochemical phenomena, and a system may be intended either to produce electric power directly or, more usefully, to produce an energy-rich chemical product which may be stored and converted to electric power at will, with regeneration of the reactants. The latter type of cell, with its inherent storage capacity, accomplishes in one stage what a photovoltaic cell/storage battery combination accomplishes in two. Moreover, it is possible that photoelectrochemical cells could be considerably cheaper than photovoltaic cells. However, no device of practical value has yet been produced, and a great deal of progress will have to be made if these cells are to approach the reliability and efficiency of photovoltaic cells.

6.1. Factors that Limit the Efficiency of Photoelectrochemical Cells

The overall expression for the efficiency of a photo-electrochemical cell appears the same as that for a photovoltaic cell, namely

$$\eta_{pec} = \eta_{av}\eta_{V_0}\eta_i\eta_{iV} \qquad (13)$$

The factor η_{av} has the same significance as before, but the factors η_{V_0} and η_i are here determined by a combination of the effects that operate in photovoltaic cells and purely chemical effects.

η_{V_0} is again defined by (9) as V_0/E_g. The open circuit voltage, however, is not determined simply by the requirement that hole–electron pairs are eliminated as rapidly as they are formed. It is now determined by the requirement that no net Faradaic current flows at the electrodes. However, there may well be anodic and cathodic currents flowing across the electrode–electrolyte interface, implying fluxes of chemical intermediates and products to the electrode and a quite complicated expression for V_0 that cannot be expressed simply in general terms, but will depend on the transport regime in the cell and the electrode kinetics of the various redox couples involved.

η_i in (13) is defined as before by (10). However, as current is carried to the electrodes by ions in a photoelectrochemical device, the factors that determine η_i are again more various than for the photovoltaic case. η_i will have a low value if the energy of the optically excited states is quenched before reaction occurs, or if the electroactive species is produced by photolysis of the bulk solution and is so unstable that only a small fraction diffuses to and reacts at the electrode. Concentration and charge transfer polarization may also severely limit i_{max}. The value of η_i in the photoelectrochemical case is therefore dependent on all the factors that affect photochemical generation rates, electrode kinetics and mass transfer in the cell.

This brief description of loss factors in photoelectrochemical cells is intended only to illustrate the sort of problems that have to be solved when evaluating their performance. A more extended discussion is available in an earlier review[65] and a detailed analysis of the factors that limit V_0 in photogalvanic cells has been completed.[66]

6.2. Semiconductor–Electrolyte Interfaces

The interface between a semiconductor electrode and an electrolyte solution acts like a Schottky barrier, provided the semiconductor is in its depletion mode,[67] that is, provided the space charge layer in the semiconductor is such that majority carriers are depleted at the electrode surface. The electrolyte solution, which in all cases of interest has a far higher density of charge carriers (ions) than does the semiconductor, acts analogously to the metal in a solid state Schottky barrier cell.

If a p-type semiconductor electrode is irradiated with photons of energy $E > E_g$, then some electrons

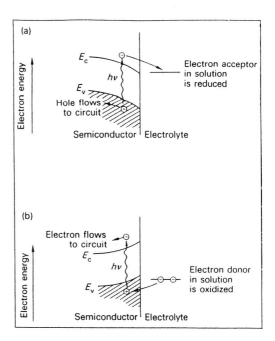

Figure 15. Charge injection on irradiation of a semiconductor/electrolyte interface. (a) Photoreduction of an electron acceptor in solution. (b) Photooxidation of an electron donor in solution.

are promoted to the conduction band, and they can then be transferred to a suitable electron acceptor in solution, as illustrated in Fig. 15(a). An electrochemical reduction has thus been caused to occur. The efficiency of this process varies with the potential of the semiconductor electrode. If the electrode is at a potential negative with respect to its flat band potential (at which no space charge layer exists) then the valence and conduction bands at the electrode surface bend as shown in Fig. 15(a), so that the electrons promoted to the conduction band by the light tend to migrate to the surface, where they can be efficiently captured by the electron acceptors in solution. Figure 15(b) illustrates the opposite process, the oxidation of an electron donor in solution by irradiation of an n-type semiconductor.

The current efficiency η_i of charge injection in the processes shown in Fig. 15 can be reasonably high, and it is possible to design a number of solar photoelectrochemical devices based on them. The most efficient to date is that investigated by Gerischer,[67,68] a thin layer sandwich cell with one semiconductor electrode (cadmium sulphide, cadmium selenide or gallium phosphide), and one transparent tin dioxide electrode. The electrolyte is a mixture of potassium ferrocyanide and ferricyanide:

$$CdS\,(CdSe, GaP)\,|\,Fe(CN)_6^{2-},\, Fe(CN)_6^{4-}\,|\,SnO_2$$

The tin dioxide electrode is so highly doped that it is quasimetallic, and the redox couple behaves reversibly at it. The irradiated semiconductor electrode acts as shown in Fig. 15(b), so that ferrocyanide is oxidized at this electrode. The cathodic process at the tin dioxide electrode produces ferrocyanide at an equal rate by reducing ferricyanide, and so overall no chemicals are consumed by the operation of the cell, which simply produces electric current on illumination with photons of energy greater than E_g (2.4 eV for cadmium sulphide). With a freshly prepared CdS electrode, a conversion efficiency of solar to electric energy of ~9% was achieved.[68] However, the cells show evidence of corrosion. The photocurrent decays continuously with time because of the deposition of sulphur on the surface (14). This competes with the desired process at this electrode (15)

$$CdS + 2h^+ \rightarrow Cd^{2+} \text{ (in solution)} + S \qquad (14)$$

$$Fe(CN)_6^{4-} + h^+ \rightarrow Fe(CN)_6^{3-} \qquad (15)$$

The suppression of such corrosion processes is essential for the future development of these cells, and it is an unfortunate fact that all the semiconductors with band gaps in the optimum range for solar conversion are unstable in an electrochemical environment, particularly when irradiated. Electrodes with a wide band gap tend to be more stable, but cannot make use of much of the solar spectrum.

It is possible to sensitize wide band gap electrodes to light of longer wavelength by coating them with monolayers of certain dyes, but rather thick dye films have to be used if these wavelengths are to be completely absorbed, and this poses several problems. First, charge transfer mobility even within a single dye crystal is very poor because of weak coupling in molecular organic crystals, which leads to narrow

conduction bands compared with inorganic semiconductors such as silicon. Second, if a polycrystalline film is used, then grain boundaries will act as efficient traps for charge carriers. These factors combine to produce a low value for η_i, although η_{V_o} can be quite reasonable. For example, Wang[69] reported light-driven charge transfer across more than 70 molecular layers of pigment molecules in the cell:

$$Al \left| \begin{array}{c} \text{multilayered} \\ \text{zinc tetraphenylporphyrin} \end{array} \right| Fe(CN)_6^{3-}, Fe(CN)_6^{4-} \left| Pt \right.$$

On irradiation with amber light, the coated electrode exhibited a potential of -1.1 to -1.3 V and photocurrents of the order of microamps could be drawn. The porphyrin used in this study is rather similar in structure to a chlorophyll, and Wang drew an analogy between the behaviour of the pigment multilayer and that of the chlorophyll antennae in the chloroplast.

6.3. The Photoassisted Electrolysis of Water at a Titanium Dioxide Electrode

Water is electrochemically a rather unstable liquid, with a thermodynamic breakdown voltage of only 1.23 eV. The electrochemical decomposition of one molecule of water requires two electrons, acting consecutively rather than simultaneously. The end products of electrochemical decomposition are oxygen and hydrogen, which are produced via $H^.$ and $OH^.$ intermediates. One common reaction sequence is shown below for illustration:

At the cathode:
$$\begin{aligned} H^+ + e^- &\rightarrow H^. \\ H^. + H^+ + e^- &\rightarrow H_2 \\ \hline 2H^+ + 2e^- &\rightarrow H_2 \end{aligned}$$

At the anode:
$$\begin{aligned} OH^- &\rightarrow OH^. + e^- \\ OH^. + OH^- &\rightarrow H_2O_2 + e^- \\ H_2O_2 &\rightarrow HOO^. + H^+ + e^- \\ HOO^. &\rightarrow O_2^- + H^+ + e^- \\ O_2^- &\rightarrow O_2 + e^- \\ \hline 2OH^- &\rightarrow O_2 + 2H^+ + 4e^- \end{aligned}$$

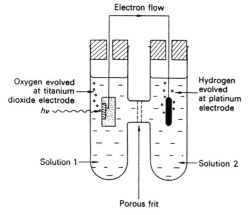

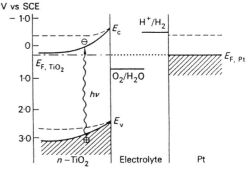

Figure 16. (left) Cell arrangement for the electrochemical photolysis of water as a titanium dioxide electrode.

Figure 17. (right) Relations between the equilibrium redox levels and the energy levels of the electrodes in the cell:[72] single crystal $n\text{-}TiO_2$|0.5 m K_2SO_4, 0.05 m CH_3COOH, 0.5 m CH_3COONa|Pt. Solid lines are for a dark, equilibrium state, and dashed lines are for a steady state under illumination. E_F = Fermi level, E_v and E_c = energy of valence and conduction band edges.

The fact that two electrons are required to decompose one molecule of water suggests that it might be possible to find a photosensitive electrode at which a two quantum photoelectrochemical breakdown of water could occur.

Fujishima and Honda[70,71] first reported a particular case of the reaction shown in Fig. 15(b) for an n-type titanium dioxide (TiO_2) single crystal electrode in water. Their work has attracted much interest, and has been taken up in several laboratories, so that additional data are becoming available.[72-80] Honda's cell is shown schematically in Fig. 16. The TiO_2 electrode is connected to a platinum counter electrode, and both are immersed in aqueous electrolyte solutions. On irradiation of the TiO_2 electrode with light of $E > 3.0$ eV ($\lambda < 415$ nm), oxygen is evolved from this electrode, and hydrogen from the counter electrode, without any electric source of power being connected to the cell.

The reason for this behaviour is illustrated in Fig. 17. Irradiation of TiO_2 (which, in all these experiments is n-type through oxygen deficiency) produces holes in the valence band, and these holes are filled by transfer of an electron from a water molecule, producing $OH^{\cdot}$ adsorbed on the electrode. Molecular oxygen is eventually produced in an overall four quantum step, although it is not yet known whether the mechanism of this reaction is that given above or whether different intermediates are involved.

This anodic process is coupled with the cathodic production of hydrogen at the platinum electrode, provided there are no species more readily reducible than H^+ present in solution 2.

The relative positions of the relevant electronic levels, calculated from various sources[72] are shown in Fig. 17, for a cell containing electrolyte of pH 4.7 throughout. The Fermi level of the TiO_2 electrode in the dark is at 0.17 V (measured against a saturated calomel electrode, SCE), and a depletion layer is formed under the TiO_2 surface in the dark, short-circuited cell. On illumination of the surface of the TiO_2 electrode, electron–hole pairs are formed. Some of the holes migrate to the surface and oxidize water. The space charge layer of the TiO_2 electrode causes the free electrons in the conduction band to migrate to the interior of the semiconductor, before they all recombine with holes. The Fermi levels of both electrodes are thereby shifted upwards in Fig. 17. If E_{F,TiO_2} reaches the height of the equilibrium potential of H^+/H_2 then hydrogen is evolved at the Pt electrode. However, irradiation cannot cause E_{F,TiO_2} to rise above the flat-band potential, which is at -0.5 V vs SCE at pH 4.7,[72] and when this potential is reached, the equilibrium potential of the H^+/H_2 couple is only just reached and there is no space charge layer remaining to separate the electron–hole pairs. Consequently, the current efficiency of the cell is very low ($< 6 \times 10^{-3}$ electrons flowing per incident photon,[72]) although if the TiO_2 electrode is biased anodically, a much higher current efficiency, of the order of 0.1, is observed.

Hydrogen can therefore be produced more efficiently if a small electrochemical bias is applied to the cell, in such a way as to make the TiO_2 electrode positive with respect to the Pt electrode. This enhances the efficiency with which the space charge layer separates electron–hole pairs, or put another way, it moves the potential of the galvanic TiO_2 electrode from near the foot of the anodic photocurrent wave to a point nearer the saturation region. The required electrochemical bias can be applied either by using strong alkali as solution 1 and strong acid as solution 2, or by use of a polarizing circuit. Honda and co-workers[74,75] have used the first method, and Wrighton[76] and Nozik[80] have used both methods simultaneously. If an auxiliary source of electric power is used to polarize the electrodes, then of course its energy requirements must be considered in evaluating the cell performance.

An n-TiO_2 single crystal anode has been combined with a zinc-doped p-GaP cathode to make a photoelectrochemical cell in which both electrodes are light sensitive.[77] The potential of the GaP electrode is shifted anodically by illumination, in the manner typical of p-type semiconductors illustrated in Fig. 15(a), and hydrogen evolution occurs at this electrode at a higher positive potential than in the dark, owing to the reduction of H_2O by photoexcited electrons in the GaP conduction band. However, the cell performance deteriorates rapidly, because GaP ($E_g = 2.25$ eV) dissolves anodically, as do most small band gap conductors.

The use of polycrystalline TiO_2, produced either by chemical vapour deposition[78] or by anodizing sheet titanium[78,79] produces a somewhat smaller photocurrent than does the single crystal material, probably because of the large number of charge carrier traps in the polycrystalline electrode.

The photochemical decomposition of water by means of visible light is a prime target as far as photochemical solar energy utilization is concerned. The Honda cell does achieve this on a laboratory scale. Could it be developed for use on a larger scale?

Materials costs do not appear prohibitive: TiO_2 is readily prepared from a variety of titanium compounds, and a thin film of it is extremely cheap. (TiO_2 coatings are applied to glassware such as bottles to make them scuff-resistant.) Platinum-impregnated electrodes, although reversible to hydrogen, are so costly that a cheaper alternative might have to be sought.

However, the major difficulty is the fact that only light of less than 415 nm wavelength is effective. The laboratory experiments reported have been carried out with high intensity u.v. lamps, which produce u.v. irradiance of $\sim 5 \times 10^{21}$ photons $m^{-2} s^{-1}$. Photopotentials (at the open circuit TiO_2 electrode) of several tenths of a volt are produced by such lamps. However, the photopotential is logarithmically dependent on light intensity, and the mean solar irradiance obtained at sea level in the wavelength region below 415 nm is only $\sim 10^{20}$ photons $m^{-2} s^{-1}$. This would produce a lower photopotential which might be inadequate to drive the cell. It would therefore be necessary to focus sunlight to achieve the necessary light intensities. However, short wavelength light is strongly scattered out of the beam radiation and it is extremely unlikely that such a focused system would prove to be an economical way of producing hydrogen from water.

Therefore, it seems, extremely interesting and encouraging though the laboratory success of this cell may be, it will be necessary either to sensitize the TiO_2 to longer wavelengths or to fabricate other semiconductors of smaller band gap (perhaps a mixed oxide containing some titanium) that show the same behaviour.

6.4. Photogalvanic Cells

Photogalvanic cells contain a solution in which a homogeneous photochemical reaction occurs on irradiation. At least one of the products of this reaction is electroactive, and by arranging the two electrodes appropriately some electric power may be drawn from the illuminated cell with regeneration of the original cell constituents, so that overall the cell operates without consumption of materials. The homogeneous photochemical reaction must be endergonic and the photogalvanic method of harnessing the energy of the reaction is adopted if the back reaction in solution of the products is rapid, so that they cannot be efficiently separated and stored.

The best known of these photogalvanic cells is undoubtedly the iron–thionine cell, first investigated by Rabinowitch.[81] This is based on the reversible photo-bleaching reaction between the thiazine dyes thionine (Th) or methylene blue and ferrous ions, reaction (16)

$$\tfrac{1}{2}Th + Fe^{2+} \underset{Dark}{\overset{Light}{\rightleftharpoons}} \tfrac{1}{2}leu\text{-}Th + Fe^{3+}$$

$$\Delta G^0 = 40 \text{ kJ mol}^{-1} \text{ at pH 2} \qquad (16)$$

The forward reaction is moderately endergonic and the products, leuco-dye and ferric ion, are virtually colourless, whereas the dyes absorb in the region 500–700 nm. Moreover, the photochemical reaction is free of unwanted side reactions, at least in the absence of oxygen. The rapidity of the back reaction is, however, such that the energy-rich products Fe^{3+} and leuco-dye cannot be separated and stored, although attempts have been made to do this by the use of ethereal emulsions, the leuco-dye accumulating in the ether phase.

Both the redox couples involved in reaction (16) are electrochemically moderately reversible, and some electric power can be drawn from the system in a photogalvanic cell without separation of reaction products by photolysing the solution near one of two inert electrodes. The illuminated electrode becomes negative with respect to the dark one, typically by 100–200 mV, and on closing the circuit, current can be drawn continuously. Usually, two platinum elec-

trodes are used.[81-84] Recently, however, it has been shown that if one platinum and one transparent tin dioxide electrode are used, power can be drawn even if the solution is uniformly illuminated.[85]

The explanation of the behaviour of this cell is based on the electrode kinetics of the two reduction-oxidation (redox) couples involved, Th/Leu-Th and Fe^{3+}/Fe^{2+}. In the dark, the thermodynamic equilibrium lies heavily on the left-hand side of (16) and the potential of the unilluminated solution will be given by application of the Nernst equation to either couple. The system is displaced to the right by light, as indicated in (16), but because the back reaction is fast, irradiation produces not a progressive reaction, but a photostationary state in which the two redox couples are no longer in equilibrium. An inert electrode in illuminated solution therefore exhibits a so-called mixed potential, which is not thermodynamically, but kinetically determined, by the requirement that the anodic component of the photocurrent, due to the oxidation of the leucothionine, is equal to the cathodic component, due to the reduction of the ferric ions, so that no net current flows. Since in this type of photogalvanic cell, the redox potential of one couple, thionine/leucothionine, is always shifted negatively with respect to its dark value, while that of the other couple, ferric/ferrous, is always shifted positively, the mixed potential may be shifted negatively, positively or hardly at all from the dark potential, depending on which couple is the more reversible, that is, which reacts the faster, at the electrode.[66] The photopotential in the iron–thionine cell is negative because the thionine/leucothionine couple behaves completely reversibly at a platinum electrode, that is to say, it reacts very rapidly, while the ferric/ferrous couple is largely inhibited by the presence of thionine adsorbed on the electrode surface, and reacts rather slowly. Therefore, at the illuminated electrode, the predominant reaction is the oxidation of photochemically produced leucothionine

$$Leu\text{-}Th \rightarrow Th + 2e^- \qquad (17)$$

At the dark electrode, a cathodic process occurs, which is probably the reverse of reaction (17). We say probably, because there may also be some reduction of ferric ions, depending on their concentration in the dark solution.

Another photogalvanic cell involving the Fe^{3+}/Fe^{2+} couple has recently been reported.[86] This is based on the reversible photooxidation of the coloured complex tris-(bipyridyl)ruthenium(II) by ferric ions:

$$Fe^{3+} + Ru(bipy)_3^{2+} \underset{Dark}{\overset{Light}{\rightleftharpoons}} Fe^{2+} + Ru(bipy)_3^{3+} \qquad (18)$$

In this case, the observed photopotentials are all positive, showing that the Ru couple acts more reversibly at the electrode than the Fe couple, which again may well be partially suppressed by adsorption (of the bipyridyl complexes.)

Table 2. Iron–Thionine Thin Layer Cell Performance Characteristics[85]

Cell composition:

$$\text{Pt sputtered on glass} \left| \begin{array}{c} 10^{-2}\,\text{M Fe}^{2+},\ \sim 10^{-4}\,\text{M thionine} \\ \text{pH 1–3, aqueous solution} \end{array} \right| \text{SnO}_2$$

Cell illuminated through SnO_2 electrode

Voltage efficiency

$$= \frac{\text{Voltage at maximum power}}{\text{Absorbed light energy}} = 0.024$$

Monochromatic quantum efficiency at 578 nm, 185 W m^{-2}

$$= \frac{\text{Electrons flowing at maximum power point}}{\text{Absorbed photon flux}} = 0.62$$

Monochromatic power efficiency at 578 nm

$$= \frac{iV \text{ at maximum power point}}{\text{Absorbed light power}} = 0.015$$

Sunlight engineering efficiency

$$= \frac{\text{Maximum power in sunlight}}{\text{Incident sunlight power}} = 10^{-5}$$

The power that can be drawn from a photogalvanic cell depends not only on the open circuit photopotential, that is on η_{V_0}, but also on η_{iV} and more importantly on n_i. For the latter to have a value of 1, all the product generated by the light must be 'collected' by and react at the illuminated electrode. However, the products are produced in bulk solution, and diffusion to the electrode surface and back reaction in solution are competitive processes. To gain a favourable balance, the ratio of the electrode area to solution volume must be made rather large, which means working with thin layer cells. Data for a thin layer sandwich cell containing iron–thionine solution, one platinum electrode, and one NESA (SnO_2) transparent electrode are given in Table 2. This cell works despite the electrodes being uniformly illuminated, which indicates that the electrode kinetics of one or possibly both redox couples must be different at these two electrodes. The very low value of the sunlight engineering efficiency in Table 2 arises mainly because only about 0.1% of the incident light is absorbed by the cell. It might be possible to use stacked thin layer cells to improve light absorption, but it must be admitted that photogalvanic cells have all in all a long way to go if they are to challenge their obvious competitors.

7. PHOTOCHEMICAL ENERGY CONVERSION

A wide variety of chemical reactions can be initiated by light. Most of these reactions are, however, exergonic, the absorbed light serving merely to overcome the activation energy barrier of a thermodynamically spontaneous process.

Exergonic photochemical reactions are of no interest in connection with solar energy conversion, as the product is poorer in energy than the reactant: in this case the energy of the photons absorbed by reactant molecules is eventually dissipated as heat. However, an endergonic photochemical reaction results in the storage of some of the energy of the absorbed photons in the chemical bonds of the product. Such a reaction may be employed to convert solar energy into chemical energy.

Most of the photochemical processes so far proposed for solar energy utilization may be summed up as in Table 1: a cyclic system is devised in which the endergonic photochemical reaction $A \xrightarrow{h\nu} B$ is followed at a later time by the reverse, exergonic process $B \rightarrow A$. B is thus a material capable of energy storage by means of a photochemical transformation, provided that it is kinetically inert at normal ambient temperatures. B is then a chemical fuel which can be stored. This innate capacity for compact energy storage, for an indefinite time, is the most attractive feature of photochemical solar energy conversion. However, this aspect of 'in vitro photosynthesis' has been relegated to the end of our discussion because it is attended by a fundamental difficulty. As the process $B \rightarrow A$ is spontaneous, it is very often rather fast so B cannot really be stored. On the other hand, if B can be stored, then it generally is not of very high calorific value. There are, however, some possibilities, which were discussed at some length at a Workshop sponsored by the U.S. National Science Foundation in September 1974.[87]

7.1. Factors that Limit the Efficiency of Photochemical Energy Conversion

The conditions that a photochemical reaction $A \xrightarrow{h\nu} B$ must fulfil in order to act as an efficient store of solar energy are well established:[88–90] (1) The reaction should be initiated by a wide bandwidth of 'visible light, that is the reactant should be strongly coloured, and the threshold wavelength should be usually long for a photochemical reaction, many of which are driven only by blue or even shorter wavelength light. (This is because the electronically excited state A^* must be of rather high energy to initiate the photochemical bond-disrupting process.) (2) The photochemical reaction must be endergonic so that the product is richer in energy than the reactant. Compilations of endergonic photochemical reactions have been given.[88,90] A few reactions are listed in Table 3 as examples. (3) The quantum yield of the desired product should be high, and there should be no side reactions leading to unwanted products and fouling the system. (4) The exergonic reaction of the product(s) to reform the reactant(s) should be extremely slow under ordinary conditions but should proceed rapidly under special, controlled conditions, for example at an elevated temperature or in a battery, to liberate thermal or electrical energy. If the reaction is too readily reversible, the product will be

Table 3. Some Endothermic Photochemical Reactions Driven by Visible Light
Adapted from Calvert[88]

Overall chemical reaction	Threshold wavelength/nm	ΔG_{298}/ kJ mol^{-1}	$\Delta G^{\ddagger}_{298}/E_g$ $(= \eta_E)$	ϕ (product) $(= \eta_\phi)$	$\eta_E \eta_\phi$
$NO_2 \rightarrow NO + \frac{1}{2}O_2$	435	37.6	0.137	0.046(NO)	0.06
$NOCl \rightarrow NO + \frac{1}{2}Cl_2$	637	20.4	0.109	2.0(NO)	0.218
$AgCl(s) \rightarrow Ag(s) + \frac{1}{2}Cl_2$	405	109.0	0.37	1.0(Ag)	0.37
$2Fe^{2+} + I_3^- \rightarrow 2Fe^{3+} + 3I^-$	546	39.3	0.179	2.0(Fe^{3+})	0.09
$Fe^{2+} + \frac{1}{2}$thionine $\rightarrow Fe^{3+} + \frac{1}{2}$leucothionine	600	39.8(pH = 2)	0.20	0.5(Fe^{3+})	0.10
$Ce^{3+} + H^+ \rightarrow Ce^{4+} + \frac{1}{2}H_2$	254	158.5	0.337	0.0007(H$_2$)	0.00035
$Fe^{2+} + H^+ \rightarrow Fe^{3+} + \frac{1}{2}H_2$ (in H_2SO_4)	300	74.5	0.187		
Anthracene $\rightarrow$ dianthracene	380	~15	~0.05	~0.25	~0.0125

too unstable to store. Alternatively, if two products are formed, they may be physically separated, stored separately and mixed to reverse the reaction. (5) The materials and any container materials used should be cheap and nontoxic and it would be advantageous if the system were unaffected by the presence of atmospheric oxygen.

There is at present no system known which conforms to all these demanding criteria. Each of the requirements (1)–(4) above can be quantified into a loss factor, yielding the following expression for the overall efficiency of η_{pc} of a photochemical solar energy converter

$$\eta_{pc} = \eta_{av}\eta_E\eta_\phi\eta_{coll} \qquad (19)$$

The first factor has the same significance as before, and the value of E_g used to calculate η_{av} must in this case be the energy of the lowest vibrational level of the excited singlet molecule which initiates the photochemical reaction. There are, however, no circumstances under which all this energy may be stored, because if the product B is to be sufficiently metastable to store, then the thermal back reaction to A must be an activated process. As illustrated in Fig. 18, this implies that the free energy, ΔG, stored by the process $A \rightarrow B$ is appreciably less than the singlet excitation E_g, for the process $A(S_1) \leftarrow A(S_0)$. This is generally the case for endergonic photochemical reactions, as shown by the fourth column of Table 3, which is adapted in abridged form from Calvert's compilation.[88] There are several means by which $A(S_1)$ may lose energy during the course of its reaction to B. First, many reactions occur from the triplet state and the conversion $A(T_1) \leftarrow A(S_1)$ is exergonic. All subsequent reactions of either $A(S_1)$ or $A(T_1)$ are spontaneous and hence also exergonic, often highly so, so that a substantial fraction of the excitation energy E_g is dissipated to the system as heat, particularly if the reaction involves several intermediates. The energy conversion efficiency, η_E in (19), is defined as

$$\eta_E = \Delta G/E_g \qquad (20)$$

The quantum efficiency η_ϕ arises because the quantum yield of the desired product may be less than unity. For example, some fraction of $A(S_1)$ may be quenched to $A(S_0)$ before it reacts. To allow for this,

the quantum efficiency η_ϕ (which is simply the quantum yield ϕ of the desired product) is included in (19).

The collection efficiency term η_{coll} is included to cover the case when B is not sufficiently metastable with respect to reversion to A to be stored indefinitely. Irradiation of A produces, then, not complete conversion to B, but a photostationary state containing some A and some B. Irrespective of how the energy of the back reaction $B \rightarrow A$ is to be harnessed, it might be by using B as fuel cell feed or by separating and storing two products B' and B", the efficiency of the device is lowered by the wasteful back reaction of B to A *in situ*, resulting only in an unwanted heat gain by the system. This produces a collection efficiency term η_{coll} which cannot in general terms be expressed more elegantly than

$$\eta_{coll} = \frac{\text{Rate at which reaction } B \rightarrow A \text{ occurs usefully}}{\text{Rate at which reaction } B \rightarrow A \text{ occurs wastefully}} \qquad (21)$$

The value of η_{coll} obviously depends not only on the chemistry involved, but on the design of the system. It may well, however, limit the overall efficiency η_{pc} to extremely low values. For example, the second process listed in Table 3, the photochemical decomposition of nitrosyl chloride, which seems promising

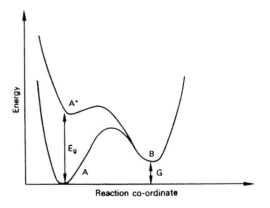

Figure 18. Potential energy profile for an endergonic photochemical reaction of substance A, yielding a metastable substance B.

with its good quantum yield and long threshold wavelength, has long ago been dismissed on this score.[90,91]

7.2. Photochemical Water Decomposition

Cyclic reactions, involving transition metal complexes, which result in the photochemical formation of hydrogen from water have been divided by Balzani[92] into four categories C1–C4, summarized in Table 4, and their minimum energy requirements (per photon of absorbed light) are shown in Fig. 19. C1 requires an X^+/X (for example Ce^{4+}/Ce^{3+}) couple having $E^0 > 1.23$ V, and has a thermodynamic threshold energy requirement of 330 nm. C2 requires a Y^-/Y couple having $E^0 < 0$ V, for example Eu^{2+}/Eu^{3+}, and has a threshold of 370 nm. They have lower threshold energies than the decomposition $H_2O \rightarrow H + OH$ because only one radical is formed in either case.

It has been known for many years that some metal cations, on irradiation in the appropriate region of their absorption spectra, will undergo the photochemical processes of C1 and C2. Among the ions for which these reactions have been demonstrated are the following:

V^{II}, Cr^{II}, Mn^{II}, Fe^{II}, Co^{II}, Ni^{II}, Ce^{III}, Eu^{II}: first step of C1
Fe^{III}, Co^{III}, Cu^{II}, Ce^{IV}, U^{IV}: first step of C2.

Heidt's well-known work[93] on aqueous solutions of Ce^{III}/Ce^{IV} attempted to achieve a balanced system in which the relative concentrations of Ce^{3+} and Ce^{4+} were such that oxidation and reduction proceeded at equal rates on irradiation of the system. However, this type of approach is attended by many difficulties. The relevant regions of the absorption spectra of the

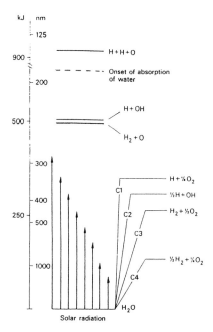

Figure 19. Threshold energies for water photodissociation by cycles C1–C4.[92]

reacting ions are their charge-transfer bands, and these are generally in the blue or u.v., close to or beyond the limit of solar radiation at ground level. Quantum yields for production of hydrogen and oxygen are generally low even at the optimal wavelength: in the case of Heidt's system, for example, the maximum quantum yield of molecular hydrogen was 0.007. These low yields occur because of the ready back reaction of the primary photochemical products of cycles C1 and C2 before they can diffuse apart. Finally, if radicals such as $OH^.$ and $H^.$ are produced in aqueous solution containing polyvalent cations, polymerization of the metal cations tends to occur.

By contrast, no radicals are produced in C3 and C4, and this is energetically advantageous. The first formulation of C3 involves a rather implausible simultaneous transfer of two electrons from the 'catalyst' Z to water, though this might possibly be achieved by use of a binuclear complex in which the metal ions remain bound together on reaction (so that further steps can lead back to the original structure). The second formulation of C3 involves a dihydro transition metal complex, some of which do decompose on irradiation as shown in (i) of Table 4 to yield molecular hydrogen. In this case, L represents an electron-donating ligand such as 1,10-phenanthroline and the reactions are actually slightly exergonic. There are other complexes of the type ML_4^{2+} which are known to react spontaneously with water to yield dihydro complexes, as in equations (ii) and (iii). In this case, L is an electron-withdrawing ligand such as a

Table 4. Classification of Methods for Photochemical Water Decomposition[92]

C1 $X + H_2O \xrightarrow{h\nu} X^+ + H + OH^-$

$\dfrac{X^+ + \frac{1}{2}H_2O \rightarrow X + H^+ + \frac{1}{4}O_2}{\frac{1}{2}H_2O \xrightarrow{h\nu} H + \frac{1}{4}O_2}$

C2 $Y + H_2O \xrightarrow{h\nu} Y^- + H^+ + OH$

$\dfrac{Y^- + H_2O \rightarrow Y + OH^- + \frac{1}{2}H_2}{H_2O \xrightarrow{h\nu} \frac{1}{2}H_2 + OH}$

C3 $Z + H_2O \xrightarrow{h\nu} ZO + H_2$

$\dfrac{ZO \rightarrow Z + \frac{1}{2}O_2}{H_2O \xrightarrow{h\nu} H_2 + \frac{1}{2}O_2}$

or $cis\text{-}ML_4H_2^{2+} \xrightarrow{h\nu} ML_4^{2+} + H_2$ $\quad$ (i)

$ML_4^{2+} + H_2O \rightarrow ML_4 + 2H^+ + \frac{1}{2}O_2$ $\quad$ (ii)

$\dfrac{ML_4 + 2H^+ \rightarrow cis\text{-}ML_4H_2^{2+}}{H_2O \xrightarrow{h\nu} H_2 + \frac{1}{2}O_2}$ $\quad$ (iii)

C4 $ML_n + H^+ \xrightarrow{h\nu} (ML_n - H)^+$ $\quad$ (iv)

$(ML_n - H)^+ \rightarrow \frac{1}{2}(ML_n)_2^{2+} + \frac{1}{2}H_2$ $\quad$ (v)

$\dfrac{\frac{1}{2}(ML_n)_2^{2+} + \frac{1}{2}H_2O \rightarrow ML_n + H^+ + \frac{1}{4}O_2}{\frac{1}{2}H_2O \rightarrow \frac{1}{2}H_2 + \frac{1}{4}O_2}$ $\quad$ (vi)

phosphine. If a single ligand or readily interchangeable combination of ligands could be used to run reactions (i)–(iii) consecutively, reaction (i) being endergonic, then water would be photodecomposed without the energetically wasteful production of intermediate H or OH. The threshold energy requirement for C3 is 420 nm.

C4 involves the photochemical formation of a monohydrido metal complex, followed by dimer formation and reductive cleavage. In this and the previous cases, it is possible and would be advantageous for more than one of the reactions to be photochemical. The $HIr(PF_3)_4$ complex has been shown to undergo reaction (v) on irradiation, and reaction (vi) has been observed for several complexes.[87,92] Two photons are used to split one water molecule in cycle C4, and that is the reason why the threshold energy is so low (840 nm). In order to have a cycle of this type, either two photons must be absorbed simultaneously by the catalyst (and this is impossible for a low density photon source such as solar radiation), or intermediate compounds must be formed which are sufficiently stable to react with each other, as in (iv)–(vi), or two coupled consecutive reactions must be driven by light against a gradient of chemical potential, as in green plant photosynthesis.

7.3. Woodward–Hoffmann Reactions

Certain types of photochemical transformations of unsaturated organic molecules in principle offer an elegant solution to the troublesome problem of rapid back reactions.[87] The ideal reactant A in a photochemical energy storage scheme should undergo the endergonic transformation to B in a single step, without the intermediate production of highly active intermediates which tend to react indiscriminately with molecules in their vicinity, and whose subsequent reactions must anyway be spontaneous—that is, wasteful of the light energy originally absorbed by A. B, although thermodynamically unstable with respect to A, should be ideally prevented from reforming A by some sort of selection rule. Exactly such selection rules operate in several organic reactions, of which three prototypes are shown in Fig. 20.

These three A ⇌ B reactions are all photochemical transformations which observe certain symmetry requirements known as the Woodward–Hoffmann rules. Woodward and Hoffmann, working together at Harvard, proposed these sweeping new rules in 1965 to rationalize a mass of data on the stereochemical course of photochemical rearrangements of the type shown. These reactions are 'concerted' processes—that is, all the valence electrons concerned in the reaction rearrange simultaneously, and no intermediates are formed. In such a case, the electronic orbitals of the reactants are transformed to those of the products in a continuous manner, and the symmetry classification of each molecular orbital must be conserved throughout the conversion. In a photo-

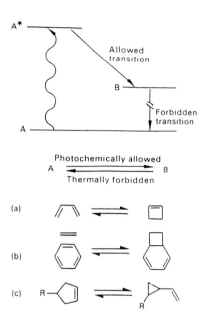

Figure 20. Three electrocyclic rearrangements which obey the Woodward–Hoffmann rules. Reaction (a), valence isomerization; reaction (b), cycloaddition; reaction (c), sigmatropic shift.

chemical transformation, the crucial orbital is the excited high energy orbital into which absorption of light promotes an electron. In the case of an ordinary, thermally activated reaction, the crucial orbital is the highest occupied orbital in the ground state. Since these two orbitals are generally of different symmetry, a transformation which is photochemically allowed may be thermally forbidden, and *vice versa*. In accordance with the principle of microscopic reversibility, the same rules apply to the reverse process.

In the transformations shown, the Woodward–Hoffmann rules act to 'allow' the forward photochemical transformation but 'forbid' the reverse thermal process. Thus reactant A is readily converted to B on irradiation. Each B contains a small 3- or 4-membered ring of carbon atoms in which there is considerable thermal strain, which ensures that the formation of B from A is endergonic. Each B is forbidden by the Woodward–Hoffmann rules from back reacting thermally to form A, but B can be pushed into reforming A by raising the temperature or by adding a catalyst. (Raising the temperature of course requires the input of a certain amount of energy, but once B is at the right temperature, the heat released in the back reaction causes it to be self-sustaining.)

The heat storage capacity in these photochemical transformations is good compared with conventional methods of solar heat storage in water or rocks.

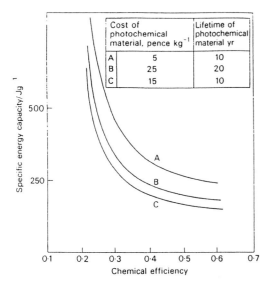

Figure 21. Plot of the specific energy capacity and chemical efficiency required of a photochemical system to match the winter operation of a typical (35% efficient) solar water heating system in the southern United States. The specific energy capacity is the specific thermal capacity of the reaction A $\xrightarrow{h\nu}$ B. The chemical efficiency is the function of the incident solar energy that is stored as chemical energy, η_{pc}. Adapted from Ref. 94.

Moreover B, once it is formed, does not dissipate as does heat, and it can be stored for a long time at room temperature, so that its energy can be released at will. Unfortunately, as is so often the case with photochemical reactions, u.v. rather than visible light is required in the three reactions shown in Fig. 20. Moreover, most of these photochemical electrocyclic reactions can lead to more than one allowed product, which is likely to be disadvantageous. It is, however, possible that the addition of chromophoric groups to the reactants to render them visibly coloured and photochemically sensitive to visible light, and the blocking of molecular sites susceptible to secondary reactions, may produce a more practical photochemical energy storage system.[87]

Photochemical storage of heat by a reaction of the type discussed in this section has been compared with a conventional hot water solar energy system by a team at Battelle Columbus.[94] Conversion efficiency, energy storage capacity and life-cycle costs were the primary bases of comparison. Among the potential advantages of photochemical systems that this research identified were cloudy day effectiveness, a smaller storage tank and uniform energy levels in the stored fluid, irrespective of season. Other conclusions are summarized in Fig. 21, which is a plot of the specific energy capacity and efficiency (η_{pc}) required of a photochemical system to match the winter opera-

tion of a hot water thermal system. The area to the upper right of these curves represents conditions where a photochemical system will have the advantages of lower cost or better performance.

The curves in Fig. 21 indicate that in developing a photochemical fluid to be competitive with a hot water thermal system, chemical efficiency can be traded off against energy storage capability, to some extent. However, there are limits. Even with the highest reasonable storage capacity, a minimum η_{pc} of 0.2 is necessary, and even with very high efficiency, a minimum thermal capacity of about $150\,J\,g^{-1}$ is necessary. The latter condition is not unduly hard to meet, but the first certainly is.

7.4. The Reductive Fixation of Carbon Dioxide

As reduced carbon compounds are convenient fuels, the possibility of achieving the reductive fixation of CO_2 *in vitro* is appealing, if remote. There are very few data on photochemical reactions of carbon dioxide in simple systems, for CO_2 has no low-lying excited states, and has not historically been of much interest to the photochemist. The first step in a photochemical reaction leading to reduced carbon compounds would probably be the formation of the radical anion CO_2^- from CO_2. The only simple photochemical reaction involving formation of CO_2^- from CO_2 appears to be the quenching of tertiary amine fluorescence by exciplex formation[95] followed by electron transfer, according to the scheme of (22)

$$R_3N \xrightarrow{h\nu} R_3N^* \xrightarrow{CO_2} (R_3N\cdots CO_2)^* \rightarrow (R_3N + CO_2) \nearrow\searrow R_3N^{\cdot+} + CO_2^- \qquad (22)$$

This mechanism has been confirmed by the correlation of fluorescence quenching rates with the ionization potential of the amine and the observation of the spectrum of the cation radical $R_3N^{\cdot+}$ in flashed amine/CO_2 systems.[96]

What appears to be the first abiological example of the photofixation of CO_2, by phenanthrene and other aromatic molecules in the presence of various amines, has recently been reported.[97]

$$\text{(structure)} \xrightarrow[\text{in organic solvent}]{CO_2, \text{amines}} \text{(structure)} \qquad (23)$$

The mechanism appears to be attack of CO_2^-, formed as in (22), on the 9 position of phenanthrene.

The reductive fixation of carbon dioxide may be achieved indirectly. For example, in photosynthetic species, the fixation of CO_2 from the atmosphere is a dark, not a photochemical process, although the energy that is necessary to drive the sequence is derived from solar energy by the photochemical production of $NADPH_2$ and ATP. In plants, the production of fructose requires, in the case of the Calvin reduction cycle, no fewer than 12 enzymically catalysed steps. The structure of the carboxylating enzymes involved, which permit the addition of $-CO_2^-$ to a carbon chain, are of considerable interest in connection with attempts at the abiological fixation of carbon dioxide. Modern methods of enzyme structure determination will undoubtedly increase our present rather sketchy knowledge of the structure of the active centres of these enzymes.

Finally, it may be remarked that the photochemical reductive fixation of atmospheric nitrogen would also be of potential value to agriculture, although this presents even more technical problems than does carbon dioxide reduction.

8. CONCLUDING REMARKS

Some developments in those parts of chemistry which have direct or indirect relevance to solar energy conversion have been outlined above. All the research described is fundamental in character, and in view of the many scientific questions that await an answer, estimates of the eventual feasibility of photochemical energy conversion can only be highly speculative. The most optimistic viewpoint would be that a cheap, durable, easily managed method of, say, achieving the decomposition of water with visible light will undubitably be discovered one day, if a sufficient number of laboratory man-hours are spent on the subject. Even a pessimist would probably cede the technical feasibility of such a process, but would conclude, from results so far achieved, that degradation of systems in continuous use and the associated high materials costs, will always render such processes uneconomic with respect to conventional energy sources. It is at the moment impossible to say which view is correct, because insufficient work has been done. Most chemists would therefore agree that a vigorous fundamental research programme should be initiated and maintained, so that a more informed assessment can be made in the future. In fact, an increasing number of photochemists throughout the scientific world are becoming interested in solar energy conversion, and are initiating research programmes on the subject.

The major question then becomes how long will it take before any breakthrough is made? Many of the problems delineated are of considerable fundamental interest and are likely to be solved eventually in the course of normal so-called pure research, though whether this increased knowledge will lead on to the fabrication of economical photochemical systems for harnessing solar energy is another matter. However, those who on these grounds now question the economic wisdom of carrying out solar/photochemical research would do well to ponder the history of the silicon solar cell. This was developed largely within 10 years from the early 1950s to meet the problem of generating electric power on unmanned satellites which were intended to spend a number of years in space, and therefore could not carry sufficient fuel to keep fuel cells operational. At the time, the silicon cells were developed virtually without regard to cost: they simply represented the only feasible way of converting solar energy to electric energy continuously for periods of years by a lightweight maintenance-free device. They have now reached such a stage of development that their use on earth to generate electric power, on a small or even a medium scale, is a definite possibility. Yet had one asked, in 1950, whether a research programme on the use of the new silicon photovoltaic cell for terrestrial power generation, was justifiable, one might have been tempted to answer that it was not. Indeed, that is exactly what the Bullard Report on new forms of energy for the United Kingdom did say in 1952.

Fundamental academic research in chemistry is, on the whole, cheap to carry out. Very much more money has to be committed to develop and to scale-up any device shown to work on a laboratory scale. Therefore, the right course, as far as solar/photochemical work is concerned, must be to initiate and maintain a good number of fundamental projects, so that as many systems as possible are developed in the laboratory to the point at which they work reasonably. Then, and only then, can an informed assessment of the economic and technical feasibility of photochemical solar energy conversion be made.

The ever present challenge to those who seek a useful method for photosynthesis *in vitro* is that abundantly successful solar energy conversion process—photosynthesis *in vivo*.

The manuscript was received on 15 February 1976.
© 1976. Heyden and Son Ltd.

LITERATURE CITED

1. K. L. Coulson, *Solar and Terrestrial Radiation*. Academic Press, New York (1975).
2. N. Robinson, *Solar Radiation*. Elsevier, Amsterdam (1966).
3. J. P. Cooper (editor), *Photosynthesis and Productivity in Different Environments*. Cambridge University Press, Cambridge (1975).
4. *Solar Electromagnetic Radiation*. NASA Space Vehicle Design Criteria, Monograph NASA SP-8005.
5. S. T. Henderson and D. Hodgkiss, *Br. J. Appl. Phys.* **14**, 125 (1963).
6. P. T. Landsberg, *Proc. Roy. Soc. (London)* **74**, 486 (1959).
7. M. Planck, *Theory of Heat Radiation*. Dover Publications, New York (1913).
8. M. A. Weinstein, *J. Opt. Sci. Am.* **50**, 597 (1960).
9. R. G. Mortimer and R. M. Mazo, *J. Chem. Phys.* **35**, 1013 (1961).
10. R. T. Ross, *J. Chem. Phys.* **45**, 1 (1966); **46**, 4590 (1967).

11. G. C. Dousmanis, C. W. Mueller, H. Nelson and K. G. Petzinger, *Phys. Rev. A* **133**, 316 (1964).
12. P. T. Landsberg and D. A. Evans, *Phys. Rev.* **166**, 242 (1968).
13. W. Feist and G. Wade, *J. Appl. Phys.* **41**, 1799 (1970).
14. D. C. Spanner, Introduction to Thermodynamics. Academic Press, New York (1964).
15. *Solar Cells: Outlook for Improved Efficiency.* Report of Ad Hoc Panel on solar cell efficiency, National Academy of Sciences, U.S.A. (1972). (Available from: Space Science Board, 2101 Constitutional Ave., Washington, D.C. 20418, U.S.A.)
16. Govindjee (editor), *Bioenergetics of Photosynthesis*, pp. 3–4. Academic Press, New York (1975).
17. J. J. Loferski, *J. Appl. Phys.* **27**, 777 (1956).
18. P. T. Landsberg and J. R. Mallinson, XIth IEEE Photovoltaic Specialist Conference, Phoenix, Arizona, U.S.A. (May 1975); J. R. Mallinson and P. T. Landsberg, Proceedings of the 3rd European Physical Society International Conference on Energy and Physics, Bucharest, Romania (September 1975).
19. W. W. Parson, *Annu. Rev. Microbiol.* **28**, 41 (1974).
20. W. W. Parson and R. J. Cogdell, *Biochim. Biophys. Acta* **416**, 105 (1975).
21. J. M. Anderson, *Biochim. Biophys. Acta* **416**, 191 (1975).
22. J. M. Anderson, *Nature* **253**, 536 (1975).
23. R. E. Fenna and B. W. Matthews, *Nature* **258**, 573 (1975).
24. G. Porter and G. S. Beddard, *Nature* in press.
25. R. S. Knox, *Biogenetics of Photosynthesis*, pp. 183–221. Academic Press, New York (1975).
26. W. Yu, P. P. Ho, R. R. Alfano and M. Seibert, *Biochim. Biophys. Acta* **387**, 159 (1975).
27. S. L. Shapiro, V. H. Kollman and A. J. Campillo, *FEBS Lett.* **54**, 358 (1975).
28. G. S. Beddard, G. Porter, C. J. Tredwell and J. Barber, *Nature* **258**, 167 (1975).
29. J. R. Bolton and K. Cost, *Photochem. Photobiol.* **18**, 417 (1973).
30. P. L. Dutton, J. S. Leigh and D. W. Reed, *Biochim. Biophys. Acta* **292**, 654 (1973).
31. G. Feher, R. A. Isaacson, J. D. McElroy, L. C. Ackerson and M. W. Okamura, *Biochim. Biophys. Acta* **368**, 135 (1974).
32. M. C. W. Evans, C. K. Silva, J. R. Bolton and R. Cammack, *Nature* **256**, 668, 1975.
33. M. G. Rockley, M. W. Windsor, R. J. Cogdell and W. W. Parson, *Proc. Nat. Acad. Sci. U.S.A.* **72**, 2251 (1975).
34. W. W. Parson, R. K. Clayton and R. J. Cogdell, *Biochim. Biophys. Acta* **387**, 265 (1975).
35. K. J. Kaufmann, P. L. Dutton, T. L. Netzel, J. S. Leigh and P. M. Rentzepis, *Science* **188**, 1301 (1975).
36. J. Fajer, D. C. Brune, M. S. Davis, A. Forman and L. D. Spaulding, *Proc. Nat. Acad. Sci. U.S.A.* **72**, 4956 (1975).
37. Govindjee and R. Govindjee, pp. 31–34, ref. 16.
38. C. J. Arntzen and J.-M. Briantais, Ch. 2, ref. 16.
39. C. F. Fowler and B. Kok, *Biochim. Biophys. Acta* **357**, 308 (1974).
40. G. S. Beddard, G. Porter and G. M. Weese, *Proc. Roy. Soc. London Ser. A* **342**, 317 (1975).
41. H. Ti Tien and S. R. Verma, *Nature* **227**, 1232 (1970).
42. M. Calvin, *Science* **184**, 375 (1974).
43. T. Yoshida, S. Ogura and M. Okuyama, *Bull. Chem. Soc. Jpn* **48**, 2775 (1975).
44. S. J. Valenty, paper presented to VIII International Conference on Photochemistry, Edmonton, Canada (August 1975).
45. H. Gaffron and J. Rubin, *J. Gen. Physiol.* **26**, 219 (1942).
46. D. T. Sawyer and M. E. Bodini, *J. Am. Chem. Soc.* **97**, 6588 (1975).
47. S. A. Alkaitis, M. Grätzel and A. Henglein, *Ber. Bunsenges. Phys. Chem.* **79**, 541 (1975).
48. M. Grätzel, A. Henglein and E. Janata, *Ber. Bunsenges. Phys. Chem.* **79**, 475 (1975).
49. M. Grätzel, J. K. Thomas and L. K. Patterson, *Chem. Phys. Lett.* **29**, 393 (1974).
50. M. Grätzel, J. J. Kozak and J. K. Thomas, *J. Chem. Phys.* **62**, 1632 (1975).
51. K. Kano and T. Matsuo, *Bull. Chem. Soc. Jpn* **47**, 2836 (1974).
52. M. Wolf, *Energy Convers.* **11**, 63 (1971).
53. B. Chalmers, *et al.*, NSF Report No. NSF/RANN/SE/GI-37067X/FR/75/1.
54. J. S. Escher and D. Redfield, *Appl. Phys. Lett.* **25**, 702 (1974).
55. D. Redfield, *Appl. Phys. Lett.* **25**, 647 (1975).
56. V. K. Jain and C. S. Jain, *Phys. Status Solidi* **30**, K69 (1975).
57. S. Wagner, J. L. Shay, K. J. Bachmann and E. Buehler, *Appl. Phys. Lett.* **26**, 229 (1975).
58. K. Ho and T. Ohsawa, *Jpn J. Appl. Phys.* **14**, 1259 (1975).
59. S. Wagner, J. L. Shay and P. Migliorato, *Appl. Phys. Lett.* **25**, 434 (1974).
60. J. L. Shay, S. Wagner and H. M. Kasper, *Appl. Phys. Lett.* **27**, 89 (1975).
61. A. L. Fahrenbuch, V. Vasilchenko, F. Buch, K. Mitchell and R. H. Bube, *Appl. Phys. Lett.* **25**, 605 (1974).
62. D. L. Pulfrey and R. F. McOuat, *Appl. Phys. Lett.* **24**, 167 (1974).
63. W. A. Anderson, A. E. Delahay and R. A. Milano, *J. Appl. Phys.* **45**, 3913 (1974).
64. W. A. Anderson and R. A. Milano, *Proc. IEEE.* **63**, 206 (1975).
65. M. D. Archer, *J. Appl. Electrochem.* **5**, 17 (1975).
66. W. J. Albery and M. D. Archer, *Electrochim. Acta*, in press.
67. H. Gerischer, *J. Electroanal. Chem. Interfacial Electrochem.* **58**, 263 (1975).
68. H. Gerischer and J. Gobrecht, in press.
69. J. H. Wang, *Proc. Nat. Acad. Sci. U.S.A.* **62**, 653 (1969).
70. A. Fujishima and K. Honda, *Bull. Chem. Soc. Jpn* **44**, 1148 (1971).
71. A. Fujishima and K. Honda, *Nature* **238**, 37, 1972.
72. T. Ohnishi, Y. Nakato and H. Tsubomura, *Ber. Bunsenges. Phys. Chem.* **79**, 523 (1975).
73. T. Watanabe, A. Fujishima and K. Honda, *Chem. Lett.* **8**, 897 (1974).
74. A. Fujishima, K. Kohayakawa and K. Honda, *Bull. Chem. Soc. Jpn* **48**, 1041 (1975).
75. A. Fujishima, K. Kohayakawa and K. Honda, *J. Electrochem. Soc.* **122**, 1487 (1975).
76. M. S. Wrighton, D. S. Grinley, P. T. Wolczanski, A. B. Ellis, D. L. Morse and A. Linz, *Proc. Nat. Acad. Sci. U.S.A.* **72**, 4 (1975).
77. H. Yoneyama, H. Sakamoto and H. Tamura, *Electrochim. Acta* **20**, 341 (1975).
78. K. L. Hardee and A. J. Bard, *J. Electrochem. Soc.* **122**, 739 (1975).
79. J. Keeney, D. H. Weinstein and G. M. Haas, *Nature* **253**, 719 (1975).
80. A. J. Nozik, *Nature* **257**, 383 (1975).
81. E. Rabinowitch, *J. Chem. Phys.* **8**, 551, 560 (1940).
82. A. E. Potter and L. H. Thaler, *Sol. Energy* **3**, 1 (1957).
83. L. J. Miller, *A Feasibility Study of a Thionine Photogalvanic Power Generation System*. Final Report, Contract No. AF33(616)-7911, Sunstrand Aviation, ASTIA Document No. 282870 (1962).
84. R. A. Hann, C. Read, D. R, Rosseinsky and P. Wassell, *Nature (London) Phys. Sci.* **244**, 126 (1973).
85. W. D. K. Clark and J. A. Eckert, *Sol. Energy,* **17**, 147 (1975).
86. C. Creutz and N. Sutin, *Proc. Nat. Acad. Sci. U.S.A.* **72**, 2858 (1975).
87. *The Current State of Knowledge of Photochemical Formation of Fuel.* Report of NSF Sponsored Workshop, Osgood Hill, Massachusetts, U.S.A. (23–24 September 1974). (Report available from Professor N. N. Lichtin, Department of Chemistry, Boston University, 685 Commonwealth Ave., Boston, Massachusetts 02215, U.S.A.)
88. J. G. Calvert, in *Introduction to the Utilization of Solar Energy*, A. M. Zarem and D. D. Erway (editors), McGraw-Hill, New York (1963).
89. S. Levine, H. Halter and F. Mannis, *Sol. Energy*, **2**, 11 (1958).
90. R. J. Marcus and H. C. Wohlers, *Sol. Energy*, **5**, 44 (1961).
91. W. E. McKee, E. Findl, J. D. Margerum and W. B. Lee, U.S. Department of Commerce Office of Technical Services, Report No. AD267060 (1961).
92. V. Balzani, L. Moggi, M. F. Manfrin, F. Bolletta and M. Gleria, *Science* **189**, 852 (1975).
93. L. J. Heidt and A. P. McMillan, *J. Am. Chem. Soc.* **76**, 2135 (1954).
94. S. G. Talbot, D. H. Frieling, J. A. Eibling and R. A. Nathan, *Sol. Energy* **17**, 367 (1975).
95. R. J. Macdonald and B. K. Selinger, *Aust. J. Chem.* **26**, 2715 (1973).
96. A. Harriman and G. Porter, unpublished work.
97. S. Tazuke and H. Ozawa, *Chem. Commun.* **7**, 237 (1975).

J. Chem. Soc. Faraday Trans. 2 **80**, 867–876 (1984).

Reproduced by permission of the Royal Society of Chemistry

Photoredox Processes in Metalloporphyrin–Crown Ether Systems

By Georges Blondeel,† Anthony Harriman,* George Porter and
Aleksandra Wilowska

Davy Faraday Research Laboratory, The Royal Institution, 21 Albemarle Street,
London W1X 4BS

Received 20th January, 1984

Metalloporphyrins have been prepared with benzo-crown-ether groups attached at each of the four *meso* positions. Cations can be inserted into the crown-ether void, and the photophysical properties of the resultant complexes depend upon the nature of the cation. With Eu^{3+} as cation, intramolecular electron transfer occurs from the triplet excited state of the metalloporphyrin to Eu^{3+}, but the large Eu^{2+} ion so formed is displaced from the crown-ether void so that reverse electron transfer is bimolecular. In the presence of an electron donor, irradiation of the metalloporphyrin forms the π-radical anion which transfers its extra electron to the Eu^{3+} ion housed in the crown-ether void.

Natural photosynthetic processes employ electron-transferring agents to impart vectorial charge separation and to ensure that the yield of redox ion products from the primary photoreaction is maximal.[1] In such environments light harvested by the ancillary pigments is transported to a specialised (bacterio)chlorophyll molecule housed in the reaction centre complex, which is in close contact with an electron acceptor (A). Upon excitation this (bacterio)chlorophyll donates an electron to the acceptor (which is either a chlorophyll or a pheophytin molecule) with unit quantum efficiency and, rather than reverse electron transfer, the electron is transported from $A^{·-}$ to secondary acceptors. By a series of thermal steps the electron is transported to a pool of plastoquinone, where it can be used to produce a fuel. Concomitantly, the oxidising equivalent remaining on the (bacterio)chlorophyll molecule is transferred to secondary donors (*e.g.* cytochromes), and in green plants it can be used to oxidise water to oxygen. This sequence of rapid electron transfers is rendered possible by the unique structure of the natural systems, which use proteins to hold the various reactants at optimal sites. Consequently, electrons can be transferred over quite large distances with little degradation of the optical energy.

Model systems for the conversion of solar energy into chemical potential have not really exploited the use of electron-transport chains, although several groups have reported on the photosensitisation of redox reactions across interfaces.[2–5] Instead, most models have employed temporal charge separation *via* the use of irreversible redox couples;[6–9] however, such systems are not practical for solar-energy storage devices. The difficulty of achieving sequential electron transfer lies with the need to site the various redox couples at well defined positions so that electron transfer becomes directional. Our manipulation of proteins has not advanced sufficiently for us to try to mimic the complex electron-transport chains found in nature, although work in this direction is proceeding.[10,11] More realistically, it should be possible to use rigid organic groups to space out the redox couples,

† Present address: Laboratorium voor Organische Chemie, Rijksuniversiteit Gent, B-9000 Gent, Krijgslaan 281-S4, Belgium.

and some progress has been made recently in this field.[12,13] So far, this work has been restricted to small organic groups so that the photochemistry requires u.v. excitation. In this paper we have attempted to extend the field by using metalloporphyrins, which absorb in the visible region, as the chromophores. In order to attach a second redox couple adjacent to the porphyrin we have covalently bonded benzo-15-crown-5 groups at the *meso* positions, as first reported by Thanabal and Krishnan.[14] Cations can be inserted into the crown-ether void so that intramolecular electron transfer should be possible.

EXPERIMENTAL

Meso-5,10,15,20-tetra(benzo-15-crown-5)porphine was prepared by the method of Thanabal and Krishnan[14] and purified by repeated chromatography on alumina using a mixture of chloroform, tetrahydrofuran and methanol in the ratio 90:8:2 as eluant. The final material gave a single spot on analytical-scale thin-layer chromatography. The corresponding benzo-18-crown-6 porphyrin was prepared and purified by an analogous procedure. Metal ions were inserted into the porphyrin ring by conventional methods[15] and the isolated products were rechromatographed on alumina. Cations were inserted into the crown void by adding a solution of the cation in methanol or acetone to a solution of the metalloporphyrin in chloroform or acetone and stirring in the dark to achieve equilibrium. Stability constants for the complexes formed between the various cations and the porphyrin–crown-ether system were not determined, and it is known that substituents on the benzene ring have a marked effect on stability constants for the simple benzo-15-crown-5–cation complexes.[16] These stability constants also depend upon solvent and the nature of the counterion.[17] In our work, ZnPCE(5) was dissolved in a 1:1 chloroform+methanol mixture or acetone to give a final concentration of $(1–5) \times 10^{-5}$ mol dm^{-3}. Small aliquots of a stock solution of NaSCN, KSCN, CuCl$_2$, Ni(NO$_3$)$_2$ or Eu(NO$_3$)$_3$ in methanol or acetone were added *via* a microsyringe and the course of reaction was followed by absorption spectroscopy. The ratio of cation to porphyrin required for complete complexation was 2, 10, $>10^3$, $>10^3$ and 4, respectively, for the cations K$^+$, Na$^+$, Cu^{2+}, Ni^{2+} and Eu^{3+}. The complex formed between Eu^{3+} and ZnPCE(5) is believed to exist in the form (Eu^{3+}, crown)$^{3+}$ (NO$_3^-$)$_3$.[18] Transition-metal ions tend to form unstable complexes with crown ethers,[19] and for Cu^{2+} the resultant complexes are often in the form (Cu$_2$, crown)$^{4+}$. Monomeric complexes are formed with Ni^{2+} but the stability constants are relatively low.

Absorption spectra were recorded with a Perkin-Elmer 554 spectrophotometer and emission spectra were recorded with a Perkin-Elmer MPF 4 spectrofluorimeter. Fluorescence quantum yields and excited singlet-state lifetimes were determined in outgassed solution as reported previously.[20] Flash-photolysis studies were made with outgassed solutions using conventional methods. The instruments used were an Applied Photophysics 200J conventional instrument (pulse duration 10 μs) with the photolysis lamps filtered to remove light of wavelength <500 nm and a frequency-doubled Nd^{3+} laser (pulse duration 15 ns) system.

ABBREVIATIONS

MPCE(5)X

$M = 2H^+, Zn^{2+}, Mn^{3+}, Sn^{4+}$

$X = Na^+, K^+, Cu^{2+}, Ni^{2+}, Eu^{3+}$

Table 1. Photophysical properties of the various cation complexes
of ZnPCE(5) in acetone solution

compound	cation	ϕ_F	τ_S/ns	τ_T/ms
ZnTPP	—	0.040	1.9	0.94
ZnPCE(5)	—	0.038	1.7	0.95
ZnPCE(5)	Na^+	0.038	1.7	0.96
ZnPCE(5)	K^+	0.0008	<0.2	—
ZnPCE(5)	Eu^{3+}	0.029	1.3	0.38

RESULTS AND DISCUSSION

SPECTROSCOPIC STUDIES

The absorption and emission spectra of $H_2PCE(5)$ and $H_2PCE(6)$ and their Zn^{II}, Mn^{III} and Sn^{IV} derivatives are essentially identical to the corresponding *meso*-tetraphenylporphyrins. For the Zn^{II} complexes there are minor changes in the fluorescence quantum yield (ϕ_F), excited singlet-state lifetime (τ_S) and excited triplet-state lifetime (τ_T), as shown by the data in table 1. The crown-ether cavity of ZnPCE(5) has a radius of 1.0 ± 0.1 Å and it will complex small cations, as was first demonstrated by Thanabal and Krishnan.[14] Titration of ZnPCE(5) in acetone or a 1:1 chloroform+methanol mixture with NaSCN solution causes only a slight increase in the absorption bands of the metalloporphyrin. There is a wavelength shift of only *ca.* 1 nm and a slight increase in the oscillator strength for both Q and B transitions. Thus Na^+ ions (radius 0.95 Å) fit comfortably into the crown-ether void without perturbing the porphyrin ring. Similar titrations with KSCN solution cause large changes in the absorption spectrum (fig. 1) which are consistent with dimerisation. Under these conditions the large ion K^+ (radius 1.33 Å) does not fit properly into the crown-ether void, and consequently a 2:4 ZnPCE(5):K^+ complex is formed.

A typical titration curve for addition of K^+ to ZnPCE(5) is shown as an insert to fig. 1; it is clear that absorption-spectral changes are essentially complete at a molar ratio ZnPCE(5):K^+ of 1:2. Therefore, we can write a simple expression for the equilibrium constant for complexation:

$$K = \frac{[ZnPCE(5)K^+]}{[ZnPCE(5)]^2 [K^+]^4}. \tag{1}$$

The calculated equilibrium constant has values around 10^{22}–10^{25} $dm^5 mol^{-5}$. As shown in fig. 1, complexation with K^+ causes the metalloporphyrin B band ($\lambda = 426$ nm) to undergo a fairly large blue shift ($\Delta = 740$ cm^{-1}). This shift corresponds to an exciton coupling energy of *ca.* 1750 cm^{-1} and an approximate porphyrin face-to-face separation distance of 4.1 Å.[21] The magnitude of this exciton coupling is sufficient to promote internal conversion from the singlet excited state of the dimer,[22] so that complexation with K^+ causes severe fluorescence quenching whereas complexation with Na^+ has no effect upon the fluorescence yield (table 1). In fact, the dimer does fluoresce weakly with a red shift ($\Delta = 270$ cm^{-1}) relative to ZnPCE(5).

Replacement of Zn^{II} in the centre of the porphyrin ring by Mn^{III} or Sn^{IV} ions leads to a substantial reduction in the ease of dimer formation. For ZnPCE(5) dimerisation is complete at a molar ratio K:ZnPCE(5) of 2:1, whereas with MnPCE(5) complete dimerisation requires a molar ratio K^+:MnPCE(5) of *ca.*

870   METALLOPORPHYRIN–CROWN ETHER SYSTEMS

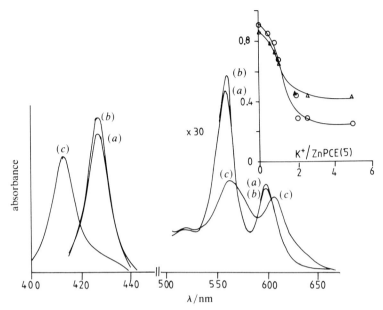

Fig. 1. Absorption spectra of ZnPCE(5) recorded in chloroform + methanol solution (a) and in the presence of (b) Na$^+$ and (c) K$^+$. The insert shows the change in absorbance at $\bigcirc$, 426 and $\triangle$, 555 nm as a function of the molar ratio of K$^+$ to ZnPCE(5).

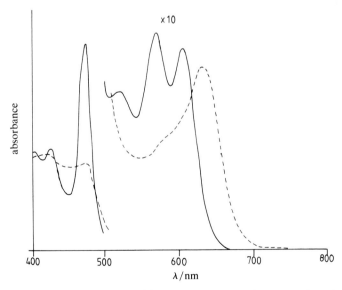

Fig. 2. Absorption spectra of MnPCE(5) recorded in chloroform + methanol solution (——) and in the presence of K$^+$ (– – – –).

40:1. Dimerisation was not observed with SnPCE(5), even with a very large excess of K^+. Since the Mn^{III} and Sn^{IV} porphyrins contain one and two axially coordinated chloride ions, respectively, the relative efficiency of dimerisation presumably reflects the increasing amount of steric hindrance imposed by the axial ligands. Incidently, dimeric manganese complexes are frequently proposed as the active catalysts for O_2 liberation in natural photosynthetic organisms,[23] and the dimer observed here is a possible candidate for use in model systems. For this reason we have included its absorption spectrum (fig. 2).

Other cations can be inserted into the crown-ether cavity. Large cations such as NH_4^+ (radius 1.48 Å) and Ba^{2+} (radius 1.35 Å) cause dimerisation of the metalloporphyrin, as found with K^+ ions. Several smaller transition-metal ions, such as Cu^{2+} (radius 0.72 Å), form very weak complexes with ZnPCE(5) and quench the excited states of the metalloporphyrin by the paramagnetic effect.[24] Similarly, Ni^{2+} (radius 0.72 Å) forms a complex with ZnPCE(5) in which there is the possibility for intramolecular energy transfer from metalloporphyrin excited states to low-energy dd states of the cation.[25] However, with all these transition-metal ions the stability constants for complexation are so low that it is not possible to distinguish between intra- and inter-molecular quenching. Much more efficient complexation occurs with Eu^{3+} ions (radius 1.03 Å), and here there is clear evidence for intramolecular quenching of both singlet and triplet excited states of the metalloporphyrin. Since Eu^{3+} ions cannot quench the excited-state porphyrin by energy transfer or paramagnetic effects and there is no indication of dimerisation, the most probable quenching mechanism is *via* electron transfer.

INTRAMOLECULAR PHOTOREDOX PROCESSES

Of the various cations that can be inserted into the crown-ether void, Eu^{3+} is the most interesting because it can be reduced easily to Eu^{2+}. Flash-photolysis studies performed with ZnTPP in outgassed acetone solution showed that both free Eu^{3+} ions and Eu^{3+} complexed with benzo-15-crown-5 quenched the triplet excited state:

$$ZnTPP^* + Eu^{3+} \rightarrow ZnTPP^{\cdot+} + Eu^{2+} \qquad (2)$$
$$k_T = (5 \pm 2) \times 10^5 \text{ dm}^3 \text{ mol}^{-1} \text{ s}^{-1}$$

$$ZnTPP^* + CEEu^{3+} \rightarrow ZnTPP^{\cdot+} + CEEu^{2+} \qquad (3)$$
$$k_T = (3 \pm 1) \times 10^6 \text{ dm}^3 \text{ mol}^{-1} \text{ s}^{-1}.$$

The bimolecular triplet quenching rate constant (k_T) can be written in the form

$$k_T = Kk_{ET} \qquad (4)$$

where

$$K = k_{DIFF} / k_{DISS} \qquad (5)$$

and the diffusion-controlled rate limit (k_{DIFF}) has a value of $(2.1 \pm 0.3) \times 10^{10} \text{ dm}^3 \text{ mol}^{-1} \text{ s}^{-1}$. The rate constant for dissociation of the encounter complex (k_{DISS}) can be estimated from the molar volumes of the reactants[26] as described previously,[27] so that the rate constant for the electron-transfer step (k_{ET}) can be calculated for each system; the calculated values lie in the range $2 \times 10^6 < k_{ET}/s^{-1} < 3 \times 10^7$. This slow rate of electron transfer is in keeping with the known properties of Eu^{3+}, which invariably undergoes non-adiabatic electron-transfer reactions.[28] Neither free Eu^{3+} ions nor $CEEu^{3+}$ quenched the singlet excited state of ZnTPP in acetone solution ($k_S < 10^8 \text{ dm}^3 \text{ mol}^{-1} \text{ s}^{-1}$).[29]

872   METALLOPORPHYRIN–CROWN ETHER SYSTEMS

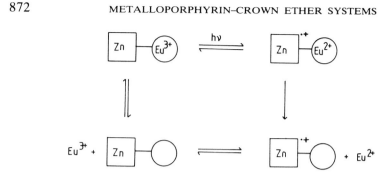

$$\boxed{Zn} \equiv ZnP \qquad \bigcirc \equiv CE$$

Fig. 3. Scheme for intramolecular electron transfer with ZnPCE(5)Eu^{3+} in acetone solution.

Flash-photolysis studies of ZnPCE(5)Eu^{3+} in outgassed acetone solution showed that, in addition to the triplet excited state, a long-lived transient species was formed upon excitation. The absorption spectrum of this transient was in good agreement with the known absorption spectrum of ZnTPP π-radical cation (ZnTPP$^{\cdot+}$),[30] showing that electron transfer occurs (fig. 3):

$$ZnPCE(5)Eu^{3+} \xrightarrow{h\nu} {}^{\cdot+}ZnPCE(5)Eu^{2+}. \qquad (6)$$

With metalloporphyrins, redox ion products are normally observed only from triplet-state photoreactions, and the triplet lifetime of the ZnPCE(5)Eu^{3+} complex can be expressed in the form

$$\tau_T = 1/(k_D + k_{ET} + k_T[Eu^{3+}]) \qquad (7)$$

where k_D refers to the rate constant for non-radiative decay of the triplet state of ZnPCE(5) in the absence of Eu^{3+} [$k_D = (1.05 \pm 0.08) \times 10^3 \, s^{-1}$], k_{ET} refers to the rate constant for intramolecular electron transfer and $k_T[Eu^{3+}]$ refers to the pseudo-first-order rate constant for quenching the triplet state by uncomplexed Eu^{3+} ions ($k_T[Eu^{3+}] = 20 \pm 4 \, s^{-1}$). Thus for the intramolecular system k_{ET} has a value around $1560 \pm 160 \, s^{-1}$. This is a much slower rate of electron transfer than found for the intermolecular case and it might be indicative of the different geometries of the intermolecular encounter complex and the intramolecular system. In the latter case the Eu^{3+} ions are held some distance away from the porphyrin π-system and at a particular orientation, whereas the intermolecular complex favours closer approach of the two reactants and probably has no preferred orientation.

By an analogous procedure, k_{ET} for the corresponding excited singlet-state reaction was found to be $(1.6 \pm 0.4) \times 10^8 \, s^{-1}$ for ZnPCE(5)Eu^{3+}. Based upon previous findings,[31] the consequence of intramolecular singlet-state quenching is probably enhanced internal conversion, so that the quantum yield for formation of the triplet excited state will be lower for ZnPCE(5)Eu^{3+} than for the uncomplexed ZnPCE(5) by a factor of 25%.

Although the triplet-quenching rate constants are low, the efficiency for charge separation into redox ion products is high (>70%) for the intermolecular systems in acetone solution. This is true also for the intramolecular system; based upon the fraction of the triplet state that is quenched ($= k_{ET} \times \tau_T$), the efficiency for formation of the porphyrin π-radical cation is ca. $60 \pm 10\%$. However, because only 60% of the triplet state is quenced under such conditions and because the triplet quantum yield is only ca. 66%, the quantum yield for formation of redox ion products is fairly low (ca. 0.3).

Decay of the porphyrin π-radical cation, as formed by reaction (6), followed second-order kinetics, at least over three half-lives. This is inconsistent with intramolecular electron transfer and suggests that the large Eu^{2+} ion (radius 1.12 Å) is displaced from the crown-ether void. An alternative explanation based upon self-exchange between complexed Eu^{2+} and free Eu^{3+} ions seems unlikely, since the self-exchange rate constant for the $Eu^{3+/2+}$ couple is very low (in aqueous solution).[32] Thus, reduction of Eu^{3+} is probably followed by physical displacement, and hence separation of the charges, and reverse electron transfer becomes a bimolecular process. The rate constant for the reverse step was found to be $(4.0 \pm 0.3) \times 10^9 \, dm^3 \, mol^{-1} \, s^{-1}$, which is essentially diffusion controlled when Coulombic forces are taken into account. The overall scheme for $ZnPCE(5)Eu^{3+}$ is given in fig. 3.

Analogous flash-photolysis studies carried out with the $ZnPCE(6)Eu^{3+}$ system showed an identical intramolecular photoredox step, with $k_{ET} = 1300 \pm 150 \, s^{-1}$. However, despite the larger size of the crown-ether cavity, the resultant Eu^{2+} ion was dislodged and reverse electron transfer was again a diffusion-controlled process [$k = (3.5 \pm 0.5) \times 10^9 \, dm^3 \, mol^{-1} \, s^{-1}$]. Note that these bimolecular rate constants for reverse electron transfer were calculated on the basis of the extinction coefficients determined[30] for $ZnTPP^{\cdot+}$ and are therefore subject to some measure of uncertainty.

SEQUENTIAL ELECTRON TRANSFER

In fluid solution the triplet excited state of ZnTPP is reduced by many electron donors (e.g. ascorbate, cysteine, benzoin,[33] NADH and EDTA[34]) and the primary photoproduct is the metalloporphyrin π-radical anion which has a characteristic absorption spectrum in the near-infrared region:[35]

$$ZnTPP^* + D \rightarrow ZnTPP^{\cdot-} + D^{\cdot+}. \tag{8}$$

As monitored by flash-photolysis techniques, in outgassed acetone solution $ZnTPP^{\cdot-}$ decays by a diffusion-controlled bimolecular process which resulted in irreversible formation of the corresponding chlorin. Presumably this decay process involves a disproportionation step, as has been shown by pulse-radiolysis studies:[35]

$$2ZnTPP^{\cdot-} + 2H^+ \rightleftharpoons ZnTPP + ZnTPPH_2. \tag{9}$$

In fact the metalloporphyrin π-radical anions are quite powerful reductants, and in aqueous solution they have been used to generate H_2 from water.[36] Similarly, in acetone solution $ZnTPP^{\cdot-}$ reduced both Eu^{3+} ions and $CEEu^{3+}$:

$$ZnTPP^{\cdot-} + Eu^{3+} \rightarrow ZnTPP + Eu^{2+}. \tag{10}$$

The bimolecular rate constant for reaction (10) was found to be $(1.1 \pm 0.2) \times 10^5 \, dm^3 \, mol^{-1} \, s^{-1}$, which is very much less than the diffusion-controlled limit and again shows the inherent non-adiabaticity of Eu^{3+} in electron-transfer reactions.

Flash photolysis of $ZnPCE(5)Eu^{3+}$ in outgassed acetone solution containing an electron donor (e.g. EDTA or triethanolamine) resulted in formation of the metalloporphyrin π-radical anion (fig. 4) but it decayed via a first-order process. The

874   METALLOPORPHYRIN–CROWN ETHER SYSTEMS

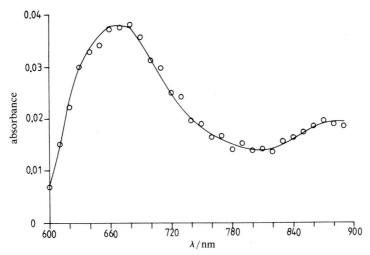

Fig. 4. Transient difference spectrum of the metalloporphyrin π-radical anion observed 100 μs after flash excitation of ZnPCE(5)Eu^{3+} in acetone solution containing triethanolamine (10^{-3} mol dm^{-3}).

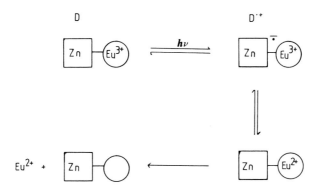

Fig. 5. Scheme for photoinduced electron-transfer processes for ZnPCE(5)Eu^{3+} in acetone solution containing triethanolamine (10^{-3} mol dm^{-3}).

rate of decay was quite slow [$k = (2.8 \pm 0.2) \times 10^3$ s^{-1}] and was independent of the concentrations of porphyrin or donor. Under these conditions the pseudo-first-order rate constant for oxidation of the metalloporphyrin π-radical anion by uncomplexed Eu^{3+} ions is only 5 ± 1 s^{-1}. Upon steady-state irradiation, no chlorin was formed during the early part of the irradiation, but after an inhibition period (*ca.* 30 min) the chlorin began to accumulate. In the absence of Eu^{3+}, chlorin formation commenced without any detectable inhibition period. These findings are consistent with the π-radical anion transferring its extra electron to the Eu^{3+} ion housed in the crown-ether cavity in preference to undergoing disproportionation. The thermodynamic driving force for this intramolecular electron transfer is quite large ($\Delta G° \approx -65$ kJ mol^{-1}) and the slow rate is a consequence of the poor orbital overlap between the reactants. Again, Eu^{2+} ions will be displaced from the crown-ether cavity as shown in fig. 5.

In most experiments the oxidised form of the donor ($D^{\cdot+}$) was a good reductant and was able to reduce Eu^{3+} in a dark reaction:[37]

$$D^{\cdot+} + Eu^{3+} \rightarrow D^{2+} + Eu^{2+} \tag{11}$$

$$^{\cdot}D^{2+} \rightarrow \text{irreversible products.} \tag{12}$$

Thus upon steady-state irradiation the concentration of Eu^{3+} is quickly depleted. At this point photoreduction of the metalloporphyrin begins with subsequent formation of the chlorin. Prolonged steady-state irradiation of $ZnPCE(5)Eu^{3+}$ in a $1:4$ water+acetone mixture containing triethanolamine (10^{-3} mol dm^{-3}) and colloidal platinum did not result in H_2 formation. This is surprising, because under similar experimental conditions reduced methyl viologen reduces water to H_2 at an appreciable rate.[38] Perhaps the absence of H_2 found with the Eu^{2+} system is related to thermodynamic properties of the metal ion in the mixed solvent medium; however, we cannot expand this argument at the present time.

CONCLUSIONS

We have demonstrated that an electron can be transferred from a metalloporphyrin π-radical anion to a covalently linked acceptor group. In our case the rate of intramolecular electron transfer was slow ($k \approx 3 \times 10^3$ s^{-1}) because of the nature of the acceptor, but it should be possible to make use of a more adiabatic redox couple as acceptor. The work is preliminary; before it has any real application it must be extended in several important directions. First, electron transfer from $ZnP^{\cdot-}$ to Eu^{3+} is essentially irreversible owing to the large thermodynamic driving force for the forward reaction. This wastes most of the optical energy and gives a final product (i.e. Eu^{2+}) which has limited potential for forming a chemical fuel. Secondly, the rate of intramolecular electron transfer is much too slow for the reaction to compete with geminate recombination of the primary photoproducts. Thirdly, we are restricted to a single electron transfer whereas we would like to have several different redox couples sited on a single chromophore. Finally, we must immobilise the reagents to remove the problems of diffusional recombination, which will always be a limiting factor in fluid solution.

We thank the S.E.R.C., the E.E.C. and G.E. (Schenectady) for financial support of this work.

1 *Photosynthesis*, ed. Govindjee (Academic Press, New York, 1982).
2 J. J. Grimaldi, S. Boileau and J. M. Lehn, *Nature (London)*, 1977, **265**, 229.
3 W. E. Ford, J. W. Otvos and M. Calvin, *Nature (London)*, 1978, **274**, 507.
4 T. Sugimoto, J. Miyazaki, T. Kokubo, S. Tanimoto, M. Okano and M. Matsumoto, *J. Chem. Soc., Chem. Commun.*, 1981, 210.
5 T. Nagamura, N. Takeyama and T. Matsuo, *Chem. Lett.*, 1983, 1341.
6 B. V. Koryakin, T. S. Dzhabiev and A. E. Shilov, *Dokl. Akad. Nauk SSSR*, 1977, **233**, 359.
7 J. M. Lehn and J. P. Sauvage, *Nouv. J. Chim.*, 1977, **1**, 547.
8 K. Kalyanasundaram, J. Kiwi and M. Gratzel, *Helv. Chim. Acta*, 1978, **61**, 2720.
9 A. Harriman, G. Porter and M. C. Richoux, *J. Chem. Soc., Faraday Trans. 2*, 1981, **77**, 833.
10 J. J. Hopfield, *Proc. Natl Acad. Sci. USA*, 1974, **71**, 3640.
11 J. L. McGourty, N. V. Blough and B. M. Hoffman, *J. Am. Chem. Soc.*, 1983, **105**, 4470.
12 P. Pasman, N. W. Koper and J. W. Verhoeven, *Rec. J. R. Neth. Chem. Soc.*, 1982, **101**, 363.
13 L. T. Calcaterra, G. L. Closs and J. R. Miller, *J. Am. Chem. Soc.*, 1983, **105**, 670.
14 V. Thanabal and V. Krishnan, *J. Am. Chem. Soc.*, 1982, **104**, 3643.
15 J. H. Fuhrhop, in *The Porphyrins*, ed. D. Dolphin (Academic Press, New York, 1978), vol. 2, part B, chap. 5.

876   METALLOPORPHYRIN–CROWN ETHER SYSTEMS

[16] R. Ungaro, B. E. Haj and R. Smid, *J. Am. Chem. Soc.*, 1976, **98**, 5198.

[17] M. Hiraoka, *Crown Compounds, Their Characteristics and Applications* (Elsevier, New York, 1982), p. 95.

[18] R. B. King and P. R. Heckley, *J. Am. Chem. Soc.*, 1974, **96**, 3118.

[19] D. De Vos, J. Van Daalen, A. C. Knegt, Th. C. Van Heyningen, L. P. Otto, M. V. Vonk, A. J. M. Wijsman and W. L. Driessen, *J. Inorg. Nucl. Chem.*, 1975, **37**, 1319.

[20] A. Harriman, *J. Chem. Soc., Faraday Trans. 1*, 1980, **76**, 1978.

[21] R. L. Fulton and M. Gouterman, *J. Chem. Phys.*, 1964, **41**, 2280.

[22] A. Harriman and A. D. Osborne, *J. Chem. Soc., Faraday Trans. 1*, 1983, **79**, 765.

[23] A. Harriman, *Coord. Chem. Rev.*, 1979, **28**, 147.

[24] G. Porter and M. R. Wright, *Discuss. Faraday Soc.*, 1959, **27**, 18.

[25] K. C. Marshall and F. Wilkinson, *Z. Phys. Chem. (Frankfurt)*, 1976, **101**, 67.

[26] T. W. Swaddle, *Inorg. Chem.*, 1983, **22**, 2663.

[27] A. Harriman, G. Porter and A. Wilowska, *J. Chem. Soc., Faraday Trans. 2*, 1983, **79**, 807.

[28] E. L. Yee, J. T. Hupp and M. J. Weaver, *Inorg. Chem.*, 1983; **22**, 3465.

[29] W. Potter and G. Levin, *Photochem. Photobiol.*, 1979, **30**, 225.

[30] N. Carnieri and A. Harriman, *Inorg. Chim. Acta*, 1982, **62**, 103.

[31] M. Gouterman and D. Holten, *Photochem. Photobiol.*, 1977, **25**, 85.

[32] M. Chou, C. Creutz and N. Sutin, *J. Am. Chem. Soc.*, 1977, **99**, 5615.

[33] G. R. Seely and M. Calvin, *J. Chem. Phys.*, 1955, **23**, 1068.

[34] A. Harriman, G. Porter and M. C. Richoux, *J. Chem. Soc., Faraday Trans. 2*, 1981, **77**, 1939.

[35] S. Baral, P. Neta and P. Hambright, *J. Phys. Chem.*, in press.

[36] A. Harriman and M. C. Richoux, *J. Photochem.*, 1981, **15**, 335.

[37] M. Kirsch, J. M. Lehn and J. P. Sauvage, *Helv. Chim. Acta*, 1979, **62**, 1345.

[38] A. Harriman, G. Porter and M. C. Richoux, *J. Chem. Soc., Faraday Trans. 2*, 1981, **77**, 833.

(PAPER 4/111)

Reprinted from The Dimbleby Lecture, BBC books (1988)

Knowledge Itself is Power

It is an honour to be invited to give this lecture commemorating that great broadcaster and communicator, Richard Dimbleby. He is remembered particularly as one who brought us all together, on important and often majestic national occasions, to remember and celebrate our history.

No scientist can enter this theatre without feeling a sense of our history as he remembers the immortals who have worked here. And its history has many lessons for us at this time, when there is such pressure to foresake our great scientific tradition for something more immediately profitable.

This royal institution was founded by Count Rumford in 1799 explicitly for "The application of science to the common purposes of life". As such, it was almost an instant failure but within a few years the direction was changed and Humphry Davy, then Michael Faraday, by pursuing knowledge itself, produced more exploitable discoveries here than have come from any other building of its size, before or since.

When Mrs Thatcher was interviewed recently in the television programme "Favourite Things", she put

7

Michael Faraday at the head of her list. When she went to number 10, one of the first changes she made was to add a statue of Faraday to the numerous generals and politicians portrayed there. One can understand her admiration of this great individualist, largely self-taught and self-supporting in this private institution; but there is another lesson to be learnt from his life. His survival to do his greatest work was a close-run thing. It was very nearly lost to the world because he had to fund his laboratory by doing short-term applied work which was not in any way of comparable importance. One episode of his life illustrates so well our difficulties in this country today that it is worth recalling.

Faraday was asked, by the Royal Society no less, to try to improve optical glass and he felt that he couldn't refuse. After nearly half a decade of rather fruitless work he asked permission, in 1831, to "lay the glass work aside for a while, that I may enjoy the pleasure of working out my own thoughts on other subjects", adding that he could never promise success with the glass because "to perfect a manufacture, not being a manufacturer, is what I am not bold enough to promise". That was written on July 4th 1831 and within two months, on August 29th 1831, he discovered

8

electromagnetic induction.

It was already known that electricity flowing in a wire or coil of wire produced magnetism but a stationary magnet in a coil like this produced no electric current; the needle remains motionless. But Faraday found that electricity _was_ produced when the magnet was moved in the coil, and again in the opposite direction when the movement is the other way. This is the actual coil that he used to make that historical discovery, he wound it and made the insulated wire himself. This was the beginning of electromagnetic field theory and it was also the first dynamo. The next year, in France and, a little later, in England, small dynamos like this were made, in which the coil or the magnet are moved continuously and generate a continuous electric current (though the electric light bulb came rather later). With some further development, here is Battersea Power Station!

There is a well known anecdote that when Faraday showed the effect to the Chancellor of the Exchequer, Mr Gladstone asked "but what is the use of your discovery?" to which Faraday replied "I know not, sir, but I'll warrant one day you'll tax it." They did. And now they are planning to sell part of it for over twenty billion pounds.

9

This lesson hardly needs any elaboration. A highly
qualified committee perceived optical glass to be an
urgent problem, which to some extent it was, and they
persuaded Faraday to follow their ideas rather than his
own. He did so for a reason not unfamiliar today -
money. The money was not for himself but for the
laboratory of this institution which, although it was
the greatest in England, was almost destitute.

When I was asked to give this lecture the producer
suggested that the title might be "the death of British
science", but this would have been an exaggeration, like
that reported of Mark Twain. If, nevertheless, you would
like a few words about the death of British science, how
about this?

"(Since the war) in the rivalry of skills, England
alone has hesitated to take part. Elevated by her
war-time triumphs she seems to have looked with contempt
on the less dazzling achievements of her philosophers
.... Her artisans have quitted her service, her
machinery has been exported to distant markets, the
inventions of her philosophers, slighted at home, have
been eagerly introduced abroad, her scientific
institutions have been discouraged and even abolished,
the articles which she supplied to other states have

10

been gradually manufactured by and transferred
themselves to other nations. Enough we trust, has
been said to satisfy every lover of his country that
the sciences and the arts of England are in a wretched
state of depression, and their decline is mainly owing
to the ignorance and supineness of the Government."

Those words were written by Sir David Brewster just
after the Napoleonic Wars and, in 1830, Charles
Babbage, the inventor of the first computer, wrote his
pamphlet "Reflections on the Decline of Science in
England". One year later Faraday discovered
electromagnetic induction. Science is a great
survivor, and it would be premature to read an obituary
on British science. But this is not to say that all is
well.

How has it come about that, at the height of our
scientic prowess, when science is the jewel in our
national crown, its future seems so uncertain? In this
perceived hour of need, scientists are being drafted
into the short-term exploitation of science, even if it
means that ultimately we shall have no science to
exploit. For over three centuries Britain played a
leading role in the scientific revolution that changed
our lives. The revolution will continue, but we shall

11

not be part of it if, by pursuing only short-term goals, we forfeit the science base that makes it all possible.

Government support of science in England and, to a lesser extent, in the rest of the United Kingdom, has usually lagged behind the rest of Europe, except in times of war or imminent war. It was Germany's rapid rise in both the industrial and military spheres during the later part of the nineteenth century that eventually forced the Government to introduce university reforms and subsidise large research programmes in various fields of science. Britain slowly caught up with her continental rivals and by the end of the nineteenth century she had developed a method of scientific research that was to prove second to none up to the end of World War II. In that war our very survival depended on those who developed radar, the jet engine, the mathematics of code breaking and penicillin, and we must be ever grateful for the prejudices of Hitler against "Jewish physics" which ensured that Germany's expertise in atomic research before the war was not used.

In the period immediately afterwards, the war-time momentum of science carried it forward at an unprecedented pace. At that time, science, measured by people involved or money spent, was growing at an

12

annual rate of more than 10% and the discovery of new

Knowledge advanced as never before.

It was a golden era for British science, too good to

last and something had to give or, more correctly, not

to give. In 1971, Mrs Shirley Williams rang the bell

and announced that for scientists the party was over.

Governments have been ringing bells ever since. There

has been a 60% reduction in funding for astronomy, base

and particle physics research since the mid-1970s and,

for the core sciences, the roughly level funding of the

research councils has been accompanied by cuts in

university provision. These core sciences of

mainstream physics, chemistry and biology, along with

mathematics form the knowledge base for most other

sciences as well as medicine, agriculture, engineering

and technology. Now the 10% growth has declined to less

than zero growth in real terms. There now seems to be a

deliberate policy of downgrading the pursuit of

knowledge in deference to the pursuit of affluence.

This change in attitudes followed the rather sudden

realisation that our industry had fallen behind that

of our main competitors. Then followed the most

remarkable non-sequitur, "our science is very good yet

our industry is not competitive. If our scientists'

ideas are developed elsewhere but not in this country

13

it must be our scientists who are to blame". As the
Daily Telegraph put it "Britain appears devoted to
destroying its science base to save the embarrassment
caused by its inability to exploit ideas". But our
industrial decline can hardly have happened because our
science was too good!

In giving evidence to the House of Lords Select
Committee on Science and Technology the chief civil
servant in the Treasury gave it as his view that "The
national source of science and technology is less
important than the abliity to assimilate and apply
scientific and technological ideas whatever their
origin". Any professional scientist or technologist
would tell the mandarins of the Treasury that only
those actually engaged at the frontiers of research
have the ability to assimilate and apply the ideas of
others and that they continually do so.

They will also point out that "once a base of knowledge
and skill has been removed, the cost and time factor
involved in its restoration is such that it is
virtually impossible to catch up with the uninterrupted
competition". Today the competition is so fierce that
a lead of a few months or even weeks can keep one
company ahead of another. And the knowledge base is
needed at short notice for other purposes than

14

industrial competition. New problems, even crises, are continually occuring even in peace-time. Industrial accidents, natural disasters, new diseases and epidemics, large scale pollution - these cannot wait months or years while we learn our basic science.

Things would be more tolerable if governments indicated that, although they had to take care of the pennies, they did recognise the importance of the scientific enterprise. To say that it was paid lip service in the recent election manifestos of the main parties would be an exaggeration and, whilst we hear frequent speeches emphasising the importance of _exploiting_ science, support for science itself is usually given with the enthusiasm of an atheist supporting the church. I have been told in turn that there is too much science and that the importance of Nobel Prizes went out with Harold Wilson.

There are _some_ signs of improvement; in the House of Commons debate last month, tributes were paid by the Secretary of State and other speakers to Britain's scientific achievements, and scientists are now cautiously looking over the parapet to see whether the shooting is over.

I hope so because the disagreement arises out of lack of understanding about how science works. There are

15

two arms to the science-technology endeavour, first the pursuit of knowledge and understanding for its own sake, and second the application of that knowledge to useful purposes. These operate with different motivations and in different environments. They derive their strength from different sources: the first from individuals free to explore the world in their own way and the second from teamwork in institutions directed from above towards a clear goal. One works from the bottom up, the other from the top down. Both are essential, and they must meet if we are to get the full material benefit from science.

 Basic science, or the "improvement of natural knowledge" for which the Royal Society was founded, has always been done best by young people, or in close company with them, and has flourished most spectacularly in the universities. The motivation of those who do it is curiosity. It is here that today one finds so much discouragement of the talented young scientist at the beginning of his career. Unless it is understood that science is not the same as technology, that a science research programme is different from a "Tomorrow's World" programme, we shall lose many of our most capable young scientists to other countries or to more trivial pursuits.

16

It should always be possible for a promising young
scientist to pursue his own ideas by applying in his
own name for a research grant. This so called
"responsive mode funding", this belief in people rather
than institutions, is being increasingly directed to
areas that a committee has decided are exploitable.
This was not the way of discovery and originality in
the past, in this country or in any other, and it
never will be. It is true that large numbers of
scientists respond to such offers of directed funding
but "they would, wouldn't they" if this is their only
entrance into a well-found laboratory. Some have even
managed to explore their own ideas secretly behind the
fume cupboard; and I am told that one acknowledged this
at the end of his paper with the words "the work
described in this paper was unwittingly supported by
several research contracts".

It is doubtful whether there is such a thing as useless
knowledge. Ignorance, the only alternative to
knowledge, can be no more than temporary bliss. But
for knowledge to be made useful the second arm is
essential, the arm of technology, development and the
engineering of products. The motivation may again be
creative but that is no longer enough, the cost and the
profit now become all important. This is a competitive

17

free market and only those engaged in the market know
how to operate it successfully. The research is the
confidential property of those who carry it out and, if
it is successful, they are the ones who profit from it.
There should therefore be little need for government to
interfere with or finance this research.

 A closer link between basic science and applied
research is essential but this does not mean that we
should expect our scientists to do industry's
development work or our industrialists to run the basic
research. In my local pub there used to be a notice over
the bar that read "We have agreed with the banks that we
will not cash cheques. They for their part have agreed not
to sell beer." Likewise we would like to agree with our
industrial friends that if they will not plan our search
for new knowledge, we for our part will undertake not to
manufacture motor cars or run a bank (though we would
like to cooperate by cashing a cheque from time to
time).

We must stop agonising about whether basic science is
exploitable. The link-up between basic research in the
universities and industrial exploitation needs pull as
well as push. Industry will have the full cooperation
of the university research worker; the days of ivory
tower snobbishness about "pure" research are long since

18

past. But a young research scientist, however
brilliant he may be in his science, is usually
uneducated in the affairs of finance, patent law and
market forces, nor has he the resources to take risks.
As Monsieur Georges Pompidou said "There are three ways
to ruin yourself: gambling, women and technology.
Gambling is the fastest. Women are the most
pleasurable. Technology is the most certain."

It is the scientist's duty to make his discoveries
widely known, as quickly as possible. From then on
industry must be in the saddle; it is industry's job,
after close consultation with the research scientist,
perhaps through bodies like the new Centre for
Exploitation of Science and Technology, to decide what
are the best areas for the industrial exploitation of
research.

I have said very little so far about funding, and I
shall not inflict upon you the armoury of statistics
fired from all sides. All agree that, in comparison
with our main competitors, the total spending in the
UK by both industry and government, on civil research
and development as a proportion of GDP is low. The
deficiency is particularly marked in the research
carried out and paid for by industry, it is 30% lower
than in the USA and 60% lower than in Germany or Japan.

19

Moreover, this year's expenditure white paper shows
that government spending on civil science and
technology is planned to fall, in proportion to total
government spending, from 1985 through to 1990. But
the trend which gives most concern is that every new
directive needs new money which is rarely made
available and the meagre funds allotted to research in
the core sciences are the first to be plundered.

 To give one example; of the £317 million made available to
the Science and Engineering Council in 1986/7, less
than 10% was used to fund applications for research in
the core sciences of physics, chemistry, biology and
mathematics, in all our universities. Last year,
nearly 80% of applications, many of them from
scientists of the highest ability and internationally
known, could not be met. Some have just given up
applying, some leave the country. Of the 28 United
Kingdom nationals elected to the Fellowship of the
Royal Society this year, one quarter are resident
overseas.

 The matter is too urgent to wait for restructuring of
the universities and research councils; we have trained
these first-class young scientists for a research
career and it is wasteful of talent, as well as unjust,
to deny them reasonable facilities for their work.

20

The minimum that is immediately needed is a positive gesture of support for the basic sciences by allotting 1% of the government civil research and development spend, or £25 million, to fund the top quarter of those individuals whose applications for equipment have been refused. In addition, £2 million is needed annually to provide small grants of up to about £20,000 to individual research workers for equipment and consumables. It is increasingly difficult to get these from the University Grants Committee and the Science and Engineering Research Council is no longer interested in applications of less than £25,000.

If this support for the science base were made the first priority for government funding the cost would be very small and it would begin to restore confidence in the Government's intentions towards basic science. The areas of "big science" like particle physics, astronomy and space research, which make contributions to basic knowledge of the greatest importance, are generally "one-off" projects and they have to be judged individually, especially since they are increasingly international enterprises.

Who should fund the science base? Although industry should take more responsibility for applied research and development in the future, almost the only source

21

of funding for basic science, in the United Kingdom at
the present time, is the taxpayer. Basic research is
freely given to the public and therefore has to be
publicly funded. Those who do research are no happier
about this than the taxpayer because it leads to a
government monopoly of science and no single authority
should be trusted with the power of knowledge itself.
But how can the pursuit of knowledge be freed from the
purse strings of government?

Already 43% of research contracts and grants to
universities are derived from non-government sources
but these are mainly directed to specified projects.
For the pursuit of knowledge by the individual it seems
the only alternative to funding from taxes is to seek
the support of the taxpayer voluntarily. I must be
joking! How can we convince the public that it should
pay for basic research? A demonstration by our
much-admired nurses is a powerful public message, a
demonstration by university dons doesn't have quite the
same pull. At best we might expect the reaction of the
American taxpayer who said "I am proud to pay taxes in
my country. The only thing is that I could be just as
proud for half the money."

But our universities are being increasingly brought
under a central nationalised direction at a time when in

22

most other things privatisation is the order of the
day. Unfortunately a university doesn't own its
student products; if it did the value added would
make it an immensely profitable enterprise. Yet the
great American universities, like Harvard, Princeton,
Yale, Chicago, MIT, Cal. Tech. and Stanford, are
private institutions which depend for their huge
programmes of basic research, as well as their teaching,
to a large extent on endowments and donations from
industry, alumni and individuals. Almost every
building on their campus is adorned with the name of a
benefactor. With recent changes to less penal taxation
may we not expect the wealthy British citizens to be as
generous in their support of education and the
advancement of knowledge as their American cousins?

Having strayed into the realm of education let me stay
there for a moment. Knowledge is little use if it is
not made available to others. Our country has an
enviable record in the highest achievements of
scientific research, and its education at the highest
levels in the universities is unsurpassed. But the
majority of our compatriorts, including those in
positions of administrative power, are among the most
scientifically illiterate of the civilised world.

The problem seems endemic in this country. Michael

23

Faraday once wrote:

"I have had occasion to go over to France with a Deputy
Board to look at their lighthouses, and we find
intelligent men there whom we cannot get here. In
regard to the electric light, which you may have heard
of, we have had to displace keeper after keeper for the
purpose of getting those who could attend to it
intelligently.

I trace everything to the ignorance of the learned in
literature as often as the unlearned, and their want of
judgement in natural things, where often there is a
fine intellect in other things."

The problem is early specialisation in our schools.
This is intellectual deprivation whatever the
speciality, but it is ruinous when it occurs to the
exclusion of science and numeracy. Without some
introduction to science, and particularly mathematics,
at an early age, the way for the late developer is hard
indeed.

The one recent government initiative that may do more
than anything else for the long-term prosperity of this
country is the introduction of a core curriculum in which
science, along with mathematics and English, is taught

24

to all children up to school-leaving age. Many
previous attempts to broaden and improve our
educational system have been frustrated and once again
these good intentions are being obstructed by parts of
the educational establishment. But in this section of
his Bill the Secretary of State will have the support
of all who wish to see Britain catch up with its
neighbours as we move into the twentieth century.

In a Dimbleby Lecture we must not forget television
which is, for some children, a more powerful influence
than books, their home or their school. A recent study
by the Department of Education and Science found that
children in the United Kingdom in the age range 6-14
spend, on average, 23 hours a week watching the box.
What an enormous power this is, what a responsibility
for those who produce programmes, and what an
opportunity for promoting civilisation and an
understanding of science.

But why is it necessary for the layman to understand
science? If for no other reason, because, in a
democracy, his opinion ultimately determines the course
of a future that will be increasingly based on science
and technology. As an example let me refer briefly to a
problem of considerable scientific intricacy and importance
that is to be decided for us by our elected Members of

25

Parliament. Later this year, they will have a free
vote on whether experiments on the human embryo up to
the age of 14 days after fertilisation, under
strict control of a licensing authority, shall or shall
not be permitted. The possibility has arisen because
of the rapid development in new medical procedures such
as in vitro fertilisation - "test-tube babies". The
embryos under discussion are surplus to those needed in
the treatment of infertility and are destroyed anyway.
 Much of the debate has been about the exact stage at
which life may be said to begin. A distinguished
medical friend of mine, when asked "But when does human
life really begin"? , replied "When the children have
left the home and you have got rid of the dog".
 But there are good biological reasons for choosing
14 days - up to this time it is not even possible
to say whether the embryo will develop into one person
or two. There is no question of pain since, at the
fourteen day stage, there is no development of even the
most primitive nervous system. The embyros are the
property of the parents whose permission for their use
in medical experiments would be obligatory by law. The
research offers the possibility of understanding better
and diagnosing earlier such defects as Down's Syndrome
and the causes of infertility.

26

Here again, in the opponents of the bill, one sees opposition to the pursuit of knowledge but not to its application, because, if the vote goes against research, operations on the embryos which are to be replaced in the womb will be permitted but operations on those that are to be destroyed will be totally forbidden.

Ethics are a personal matter which, some would say, transcend mere logic, but they need not exclude logic altogether. It is our responsibility to improve the health and happiness of our fellow human beings by seeking this knowledge. The work has few attractions in itself and we should be grateful to those small number who engage in it on our behalf. To suppress this knowledge is to outlaw learning and to license ignorance, a strange interpretation of what is ethical. I hope that Members of Parliament by their vote will reaffirm their belief in the goodness as well as the power of knowledge.

Over the last 200 years, pure and applied science have transformed the way we live, and each decade the changes in our lives have been more remarkable than the last. The majority of us live in far greater material comfort than any of our countrymen of two centuries ago and our expectation of life is twice as long as theirs.

27

High living standards no longer require the servitude
of others - Michael Faraday and James Watt freed more
men and women from slavery even than Abraham Lincoln.

There are no obvious limits to the advancement of
knowledge or to the practical applications of this
knowledge to improving our health, wealth and
happiness. In health, medical research has still much
to do to reduce the suffering still with us. In
wealth, rich as most of us are in material things
compared with our ancestors, we have seen only the
beginning of what will be available to us. And happiness?
Nothing can guarentee this, not even science. But science
can, and has already, eliminated many causes of great
<u>unhappiness</u>. In the words of Horace, "Knowledge of that
which   underlines   everything, gives   true   happiness,
unshakeable peace of mind, by eliminating the wonder at our
fate."

That is the ultimate value of knowledge. It is not
only a material power, it is a spiritual power and a
part of the human purpose. A country that lets it
languish becomes impoverished in more than its
Treasury.

28

Reproduced from *Proc. Am. Philos. Soc.* **136**, 521–525 (1992)
with permission from the American Philosophical Society

Can Science Policy Be Left to the Scientists?*

LORD PORTER

*Chairman, Centre for Photomolecular Sciences
Imperial College*

It is a great personal pleasure to take part in this joint meeting of two societies, each the most ancient philosophical society of its country. The question that I have asked myself and you is "Can Science policy be left to the scientists?" In trying to answer this let me begin with two case histories that bear on the problem and then, quite unjustifiably of course, argue from the particular to the general.

The first involves the founder of the American Philosophical Society and an active member of both of our societies—Benjamin Franklin. The story is rather well known how, when the Royal Society was asked by the ordnance board to advise on lightning conductors, or rods as they were called, for the arsenal at Purfleet, the society set up a committee and naturally made the inventor, Franklin, a member of it. The committee had to decide between pointed conductors, favoured by Franklin, and blunted ones. With one dissenter they recommended pointed ones. But the dissenting fellow was not satisfied and began a long correspondence including extensive scientific papers in the Transactions of the Royal Society.

This case would seem an unpromising one for politicians but Franklin and the supporters of pointed conductors became associated with "the insurgent colonists across the sea" and "those who objected to blunt points were considered disaffected subjects." George III naturally supported the home side and had the sharp points on his palace conductors fitted with knobs. He interviewed the president of the Royal Society and asked him to change the Society's views in favour of knobs. The president, Sir John Pringle, rightly refused, as respectfully as he could, and was replaced by Sir Joseph Banks at the next election.

The moral of the story is clear: science policy will not be left to the scientists if the scientists disagree with each other. When that happens a

---

* Read at a joint meeting of the Royal Society and the American Philosophical Society, May 1991.

PROCEEDINGS OF THE AMERICAN PHILOSOPHICAL SOCIETY, VOL. 136, NO. 4, 1992

522   LORD PORTER

vacuum is created which tends to be filled by politicians, who abhor a vacuum.

This year is the bicentenary of the birth of the greatest individual experimental scientist who ever lived—Michael Faraday, a fellow of the Royal Society and director of our sister the Royal Institution. As a second case history I would like to recall an incident in his life that throws some light on the question before us.

Faraday's survival to do his greatest work was a close-run thing. He was asked, by the Royal Society no less, to try to improve optical glass and he felt that he couldn't refuse. His masterpiece was very nearly lost to the world because he had to fund his laboratory by doing short-term applied work which was not in any way of comparable importance. After nearly half a decade of rather fruitless work he asked permission, in 1831, to "lay the glass work aside for a while, that I may enjoy the pleasure of working out my own thoughts on other subjects," adding that he could never promise success with the glass because "to perfect a manufacture, not being a manufacturer is what I am not bold enough to promise." That was written on 4 July 1831 and within two months, on 29 August 1831 he discovered electromagnetic induction. That decision was one of the most important science policy decisions ever made, and it was made by an individual against the wishes of a most prestigious committee of "peers."

This lesson hardly needs elaboration. This highly qualified committee perceived optical glass to be an urgent problem, which to some extent it was, and they persuaded Faraday to follow their ideas rather than his own. He did so for a reason all too familiar today—money. The money was not for himself but for the laboratory of his institution which, although it was the greatest in England at that time, was almost destitute.

There is a well-known anecdote that when Faraday showed his electricity generation to the Chancellor of the Exchequer, Mr. Gladstone asked "But what is the use of your discovery?" to which Faraday replied "I know not Sir, but I'll warrant one day you'll tax it." They did and now they have sold part of it for over twenty billion pounds.

Now from the particular to the general. Let me try to give my own answer to the question "Can science policy be left to the scientist," and then attempt to justify it. There are two answers, which not surprisingly, are "yes" and "no."

Yes, when the purpose of the scientist is the pursuit of knowledge itself, without himself applying that knowledge and always providing that his own experiments do not themselves endanger others. The scientist himself is the best judge of what is timely and promising for him and must be left to decide his own science policy.

No, when the purpose of the work is to apply knowledge in a way that will affect the lives of others. In this case the others or, at least in a democracy, their elected representatives, must decide policy, even though they will usually wish to consult the scientists as to what the options are.

But now come the complications. There is just one caveat to the rule

LEAVE SCIENCE POLICY TO SCIENTISTS?   523

of leaving *pure* science to the scientist. What if the scientist needs money—which is not infrequently the case? Indeed this is the big difference between the independent, free, individual scientist of earlier centuries and the same scientist today. In earlier times the cost of research was relatively small and many of the scientists, even if not wealthy, could afford most of their needs from their own pocket.

The income of the scientist and the cost of his research in most subjects lost sight of each other some years ago. In the past, scientists who needed big money, for telescopes perhaps, found a wealthy sponsor who asked nothing more than the privilege of sharing, as an amateur, in the joy of discovery. There are few of these today. Governments, the wisest ones anyway, decide how much funding shall be available overall for basic science and then leave the rest of the policy decisions to some sort of advisory committee, which includes scientists appointed by the government. There is not, and never will be, enough money for all, so choices have to be made—and there's the rub.

The choices that have to be made can only be made by peer review and, for basic science, the only peers are other scientists. But scientists in committee need to be watched. Remember Faraday and the Royal Society. The Royal Society committee making science policy for Faraday was about as well informed as they come, then or today, but, not to put too fine a point on it, it used its money to bribe Faraday to do work that it, the committee, thought more important than the ideas of Faraday.

Scientists are active in demanding more funds—but not very successful in distributing them. They are intensely parochial—whoever heard an organic chemist support a particle physics project in preference to any kind of organic chemistry—or a particle physicist do the same in reverse?

Nevertheless, in the relative privacy of the funding councils, decisions are made somehow or other, largely I suspect on the basis of a share of the spoils based roughly on what has happened before and, although I may seem to have been unenthusiastic about scientists' behaviour in this game, we have to conclude that the alternatives are so much worse as to be beyond consideration.

One has to pray therefore that, since peer review is inevitable, these peers will pay more attention to the scientist, his track record, his potential and his needs, than to their own views, often very ill informed about the value of the work that the scientist wishes to do. In the dreadful jargon of today's policy debates, they should favour responsive-mode-funding over dirigism! Also, since picking winners is so uncertain there are bound to be losers who ought to be winners—and this is a strong argument for a plurality of funding bodies and of selection methods.

Such bodies should not themselves be both donor and acceptor, as are those research councils who have to choose between funding research in one of their own establishments or in an independent laboratory such as a university. The decision is almost inevitable and is one of the reasons for the decline of research in the universities, especially in countries with a centrally planned economy. Whenever a good project is not funded it is safer to blame the government than to close down a not very effective establishment.

Given that basic science must be left to the scientist while the overall sum available has to be determined by the donor, how does the donor, the government in reality, determine the absolute sum. The government of this country has accepted in recent years that it is responsible for basic science, although not for near-market research. But how responsible? There is probably no logical way of deciding what is the right sum to be taken from the taxpayer but there is much to be said for making a decision, such as some proportion of the GDP, and sticking to it for some years because science is a long-term activity. In any case this decision cannot be left to the scientist or we shall hear again the accusation levelled by one U.S. Senator who said, "The trouble with the scientific pork-barrel is that the pigs themselves are running it."

Now we must ask the same question about applied research, where the answer will be quite different. The first new factor that arises here is that the purpose of industrial R and D is to make profits. Most would agree that the best people to decide the science policy for this purpose are those who make the profits, if they get it right, and the losses if they don't. A young research scientist, however brilliant he may be in his science, is usually uneducated in the affairs of finance, patent law and market forces, nor has he the resources to take risks.

It is the scientist's duty to make his discoveries widely known, as quickly as possible. From then on industry must be in the saddle, it is industry's job, after close consultation with the research scientist, to decide what are the best areas for the industrial exploitation of research.

Finally, we must turn from fiscal and economic problems to science policy decisions that have ethical associations.

Ethics are a personal matter which, some would say, transcend mere logic, but they need not exclude logic altogether. First, let it be said that there are, to my mind, no possible ethical problems that arise in connection with the pursuit of knowledge for its own sake within the reservations, such as safety, that I have already mentioned. Knowledge is value free, except in respect of its probability of being true. The only alternative is ignorance and one needs considerable knowledge of a matter before one can chose to remain ignorant of it. (To remain ignorant of Antarctica, for example, one would have to avoid running into it, and therefore one would need to know where it is, and to keep an eye on it in case it moves!)

It is in this application of knowledge that a science policy has to embrace ethics, and it is obvious that this cannot be left to the scientist alone, though it should also be evident that little progress can be made without his advice on the present state of knowledge and the likely developments.

The recent debate about research on the human embryo was a paradigm of collaboration of this kind—would that the problems of animal experiments were likewise. But in these cases which, like most of the difficult ethical/science areas at present, are in the medical field, the system of setting up a statutory body composed of laymen and scientists, with a majority of the former, from which a licence must be obtained for work in one of the sensitive areas, works well. Having got some sort of agreement on when life begins, perhaps we can now face the medical problems of euthanasia and death, and the relative value of, say, prolonging senility and organ transplants for the young.

Somewhat surprisingly, these ethical areas, involving our religious views and deepest prejudices, seem to find some sort of solution or, at least a compromise, while the more practical science-policy problems, like the greenhouse effect, seem to offer few routes to agreement. Often this is because the mechanisms for this are no more in place for science than for other international problems.

But it has to be said that, even if such mechanisms were available and even if the scientists were called to give advice, as of course they must, the real problems would then begin, because scientists would inevitably disagree, mainly because the knowledge base is so incomplete but also because value judgments would have to be made. On the overall advantages and disadvantages of nuclear and renewable energies, for example, scientists from different disciplines tend to have quite different views, views that are not always based wholly on thermodynamic or other scientific considerations.

So, to summarise: the science policy to be pursued in his own research by the scientist who seeks only to "improve natural knowledge" can and must, within the overall funding available, be left to the scientist.

The policy for overall scientific expenditure, and for the applications of science and technology, whether they involve ethical considerations or just material ones, must be chosen, not by scientists, but by elected representatives of the people as a whole, on the best scientific and other advice available.

This will, I know, be seen by some as a scientist ducking his responsibility for the consequences of modern technology. Far from it; I for one am proud of the overall balance-sheet of technology which gave us ploughshares as well as swords and where it is for every man to determine which he will have. Today there are few things that science cannot do to improve the health, wealth and happiness of our fellow human beings. It is the duty of the scientist to explain to all people what these things are and it is the duty of every person to try to understand science and the natural world so as to be able to decide what kind of a future he or she wants. Once that is agreed, science policy will be a relatively easy matter.

Presidential Farewell Address to The Royal Society (1990)

THE MOST IMPORTANT TASK

On 30 November 1990, the retiring President of the Royal Society, Professor Lord Porter, OM, FRS, delivered his final presidential address to the Society. He used the occasion to review the five distinguished years of his presidency, and to anticipate the tasks that the Society would face under his successor as it moved, for the fourth time, towards a new century. In the course of his address he had important things to say about the place of science in our educational system and in our national life. With his permission and good wishes, *Headlines* now publishes what he had to say about this, the most important task.

Lord Porter

P erhaps the most important task before the Society over the next few years will be to finish the job that has been begun in bringing science education to all our citizens.

There are scholars, a few of them fellows of the Society, who, having had no formal education in science, have acquired in later life a real interest and understanding of natural philosophy: but they are quite exceptional. The Society has therefore pressed for over a century that all our citizens should be taught science up to school-leaving age and none of them should be taught science alone. The National Curriculum, with 20% science, was a great step in this direction, and the retrograde step of allowing those who probably need science most to opt for science in only 12% of their time needs to be rectified as soon as possible.

I personally do not worry too much about whether science is taught as co-ordinated double award science (which the Society advocates) or as three subjects – physics, chemistry and biology – as long as all three core sciences are taken by everybody and subjects like astronomy, earth sciences and

human biology are incorporated into them.

The worst aspect of our educational system

The worst aspect of our educational system is the early specialisation at A-level. There are many ways of ensuring that all boys and girls take some science and, I hope, at least one foreign language. Committees and commissions, not least those of the Royal Society, have discussed them for over a century and have generally agreed in principle. Further enquiries in this area can serve little purpose; the matter is urgent and we must now take the decision to broaden the country's educational system and extend it to all. Five A-levels, taken at either principal or subsidiary level, including at least one science and one language, would be generally acceptable and there would be- no need, for the present, to tinker with the content or standard of our well- respected A-level syllabuses. I say this cautiously because as a grammar school boy I saw the school which gave me a start in life ravaged, during the 1950s, in a political war.

Some of the teachers in universities will demand a sweetener if they are to co-operate in any change, and the common ploy here is to say that entry standards in their own particular subject will fall if students waste time on other subjects. So, the argument goes, four-year courses at universities will be

necessary, in *their* subject anyway, and will have to be paid for.

This is, of course, justified and already applies to those who are taking a professional qualification in such subjects as medicine, and sometimes in engineering or law. For those who are going to become research workers the need for longer undergraduate courses is questionable because there is a lot to be said for starting research as soon as possible and taking more courses throughout one's career as the need becomes apparent. And the case does not apply at all to the much larger number who at present do not get to university, and do not seek an academic career. Many of these would be satisfied with, and even prefer, a two-year pass degree, so that the financial problem of supporting the academically gifted scholars for an extra year could be balanced by a more flexible system with a two-year pass, a three-year honour's and a four-year master's degree.

It is difficult for ministers, some with little or no secondary education in science, to appreciate the anger and frustration which scientists have long felt at a system which is controlled and guided by those who have little understanding of what makes scientists tick or appreciation of what science has done and will do for mankind. The politicians themselves are not to blame any more than are the editors who fill our newspapers, radio and television with every trivial crumb of literary gossip, but often cannot themselves supply a correctly worded headline for a scientific paragraph, even when their science writer has done an excellent job on it. They are all victims of a system that will not be changed until, like most other European countries, we have enough teachers of science, properly trained in their subjects, teaching all children in school from the age of 5 to 18, so that never again are we all half-educated, with most of our leaders chosen from the other half.

Why support science?

There is no avoiding the fact that in the modern world, basic science depends on the public for its funding and support. Many of the public are either ignorant of science or, worse, they are against it and more interested in mysticism, astrology, 'alternative' medicine and similar mumbo-jumbo. The nature of the transformation in the way we live and the role of science and technology in it are poorly understood by the great majority of people. This is why COPUS (the Committee on the Public Understanding of Science) was set up by the Society five years ago. It has been quite active and its work will be of increasing importance in the future, although, as I have already argued, the real trouble begins much earlier in the schools.

We know that the funding of science is inadequate and always will be because there are so many worthwhile things to be done. But what shall we tell our fellow men and women, almost wholly ignorant of the scientific endeavour, about why the support of science should be one of the highest priorities of mankind?

The material argument for the pursuit of science is the one that has the best hope of being understood by governments and also up to a point by the public, although not all of them are certain that they *want* further material progress.

But the pursuit of science is more than the mere pursuit of affluence. We are all born scientists; as children we constantly seek an understanding of our environment and ourselves, and carry out experiments on everything around us, until we are restrained from destroying all the apparatus in the house by our philistine parents.

The message has got through, and has little appeal

One of the most worrying trends of the last year or two has been the turning away from science of the boys and girls in our schools. A-level entries in chemistry, for example, are falling faster than the population of A-level age. Why is this happening in our schools when the advances of science are so exciting? We hear that it is the poor teaching or the poor funding or the poor financial rewards of a scientific career, and scientists themselves do little to encourage the young to follow their career when they spread gloom among the young about the state of British science. But I have yet another cause to

suggest; I believe the ubiquitous message that the purpose of science is to make things to sell has got through to the young people in our schools and has little appeal for them.

I became a scientist because I was fascinated by the world around me and I wanted to know what made it work. The burning desire for wealth creation never entered my soul.

Not all young scientists are like me, fortunately, but I believe a lot of them are being turned off by the identification of science with materialism. Science, engineering and technology are not as much fun as they used to be when those of my generation were young scientists.

The true scientist is he or she who has retained into adult life that innate human need for understanding, in spite of all the discouragements that may have been encountered during the education process. This is driven by a deep instinct within us for something that gives meaning and purpose to our human existence.

Much of our anxiety and unhappiness today stems from a lack of purpose which was rare a century ago. Science has destroyed many old beliefs but has

not yet helped us to find new ones. Some argue that science has nothing to do with belief or purpose but, if this were true, science would not have changed our fundamental beliefs in the way it has. Any force that has an influence on man or matter is open to experimental investigation and if it has no such influence it is of no consequence.

The ultimate purpose of science

It is the pursuit of knowledge that will lead us to better understanding of man's condition and purpose. Of course, it is quite possible that we can never understand, never discover a purpose, but we shall not succeed if we do not try. There is absolutely no evidence that the great intellectual power with which mankind is endowed has any limitations and, until evidence to the contrary is produced, we shall be wise not to give up the search . . .

There is one great purpose for mankind, which is to try to *discover* man's purpose by every means in our power. That is the ultimate purpose of science and of the Royal Society of London for Improving Natural Knowledge.   ■

The Athenæum Lecture (1999)

THE ARTS AND SCIENCES

THE TWO CULTURES 40 YEARS ON

T HE ROYAL SOCIETY, our neighbour across the road, has an ancient dining club which meets here regularly and drinks a toast to 'the arts and sciences'. This is very similar to the title of the lecture suggested for me by the Talk Dinner Committee and since it follows so closely the title of the memorable first Athenæum lecture of Lord Habgood, on *Theology and the Sciences,* I hope it is an appropriate sequel, though I cannot hope to match the Archbishop's eloquence and knowledge of his subject.

Your Committee's invitation to me referred to 'the two cultures, forty years on'. C.P. Snow, though often derided and misunderstood, was describing a real division of minds when he introduced his concept of the two cultures. Speaking in his Rede lectures he said, 'constantly I felt that I was moving among two groups – comparable in intelligence, identical in race, not grossly different in social origin, earning about the same incomes, who had almost ceased to communicate at all, who in intellectual, moral and psychological climate had so little in common that instead of going from Burlington House or South Kensington to Chelsea one might have crossed an ocean. In fact, one had travelled much further than across an ocean – because after a few thousand Atlantic miles, one found Greenwich Village talking precisely the same language as Chelsea, and both having about as much communication with M.I.T. as though the scientists spoke nothing but Tibetan.'

In the past, most educated people liked to be called philosophers; they made little distinction between the cultures and wanted to know something of both the sciences and the arts, of natural philosophy as well as moral philosophy. The great civilisations have always developed the sciences alongside the arts. Ancient cave drawings show an

· 1 ·

THE ATHENÆUM LECTURE

interest in anatomy as well as art and Neolithic man recognised that agriculture demanded some knowledge of the lunar and solar cycles. In Egypt, the Old Kingdom (2778–2263 BC), when the great pyramids were built, was the most fruitful period of all Egyptian history and it was also in that period that the basic discoveries of Egyptian mathematics, astronomy and medicine were made.

Greek scientists inherited this, but Hellenic science itself made huge advances by valuing science for its own sake. Greek mathematics developed rapidly over the two centuries (6th and 5th BC) from Pythagoras to Euclid and music theory developed along with it. Hippocrates was the greatest medical scientist of the ancient world but Plato and Homer made their contributions too. Over a few centuries all this flourished along with the incomparable architecture, sculpture and technology of classical Greece.

In the middle ages, science almost ceased and, consequently, so did development. Science and technology, as we know them today, began with the Renaissance and rose spectacularly in the 17th century. Francis Bacon's *New Atlantis* set a pattern for the Royal Society whose coat of arms declares emphatically, in the words of Horace, *Nullius in Verba*, its determination to have done with received wisdom and the dogma of authority.

When the Renaissance came, it was again a development of knowledge as a whole – it involved Galileo and Leonardo as well as Michelangelo. Indeed, one culture is impossible without the other. Literature only became generally available with the development of printing. It is arguable that the great development of music in the golden age of the 18th century owed as much to technological as to artistic genius. Some instruments essential for a Beethoven symphony orchestra had not been developed in the time of Bach.

The peripatetic scientists of the 18th and 19th centuries travelled widely and knew each other personally. They had a single international language – Latin. At the beginning of the 19th century the Athenæum appointed as its first Chairman the most brilliant young chemist, Humphry Davy, also a notable poet and friend of Coleridge, and of Wordsworth whom Davy helped with his Lyrical Ballads,

THE ATHENÆUM LECTURE

correcting the English at which Wordsworth said, 'I am no adept'. They, like many of their contemporaries, were polymaths.

Then, about the latter part of the last century, all this changed. Giving evidence to the public schools commission in 1862, Michael Faraday, Davy's assistant and the first Secretary of the Athenæum, a deeply religious man, said, 'The sciences make up life. The highly educated man fails to understand the simplest things of science and has no particular aptitude for grasping them. Persons who have had the discipline of classical instruction, persons who have been educated by the present system, are ignorant of their ignorance at the end of all that education. I trace everything to the ignorance of the learned in literature as often as the unlearned, and their want of judgement in natural things, where often there is a fine intellect in other things.'

A contemporary of Faraday, Charles Babbage, the father of the computer, wrote, 'As there exists with us no peculiar class professedly devoted to science, it frequently happens that, when a situation requiring for the proper fulfilment of its duties considerable scientific attainments is vacant, it becomes necessary to select from amateurs … This is amongst the causes why it so rarely happens that men in public situations are at all conversant even with the commonest branches of scientific knowledge … In other countries it has been found, and is admitted, that a knowledge of science is a recommendation to public appointments and that a man does not make a worse ambassador because he has directed an observatory or has added by his discoveries to the extent of our knowledge of animated nature.'

It was the literary critic Lionel Trilling who wrote, 'The exclusion of most of us from the mode of thought which is habitually said to be the characteristic achievement of the modern age is bound to be experienced as a wound to our self-esteem.'

George Orwell, whose concern for the future was quite considerable, said, in *The Road to Wigan Pier*, 'Every sensitive person has moments when he is suspicious of machinery and to some extent of physical science. But it is important to sort out the various motives, which have differed greatly at different times, from hostility to science

· 3 ·

and machinery, and to disregard the jealousy of the modern literary gent who hates science because science has stolen literature's thunder.'

Incomprehensible as it may be to those of us who love science and see it as one of the few redeeming features of this rather unhappy world, science has become unpopular with some sections of the public. Its detractors see little connection between the great thoughts of the philosophers and political historians on which our future civil servants and politicians are reared, and the mundane matters of heat, light, electricity and stinks which are, to many, a quite different matter, associated more with plumbing than with the higher planes of intellectual, social and artistic life.

There are some signs, as we approach the next millennium, that this cultural vacuum in British education may be replaced by a more liberal system, such as the International Baccalaureate, embracing the arts and sciences in both their academic and practical aspects.

Most of the public's concern is not so much about the advance of knowledge but about its social consequences such as technology, nuclear war, dark satanic mills and so forth. The euphoria which greeted the advances of scientific knowledge and the resulting technological progress is now accompanied by doubts. It is not that people have lost respect for science; on the contrary they now have that extreme form of respect which is called fear. What will scientists do next ... where are they taking us?

It is the natural reaction of the layman to call, somewhat helplessly, for social responsibility in the scientist. But neither layman nor scientist is very clear what this means. It is quite impossible to predict all the consequences of research and discovery, or the future applications of new knowledge. There is nothing in this world, whether natural or man-made, which is wholly good or wholly bad. Any object can be used to inflict pain on others. The discovery of iron gave us swords as well as ploughshares.

Most members of the public, and those who present science to them in the media, see the purpose of science as wealth creation with the improvement of health running second. We need to tell of science as a great odyssey, a search for truth and understanding of ourselves

· 4 ·

THE ATHENÆUM LECTURE

and our universe. These two aspects, the pure and applied aspects of the pursuit of knowledge are inseparable. It has been well said that there are only two kinds of science, applied science and science not yet applied.

We must face the fact that the improvement of knowledge is making all things possible, and unfortunately that knowledge is advancing faster than we can adapt to it. One possible solution would be to put a moratorium on the advancement of knowledge and research until we have had time to catch up. But this is not possible without draconian restrictions on human freedom; how can one outlaw knowledge and legalise ignorance which can be no more than temporary bliss? Many of our problems stem from too little natural knowledge, not too much.

Given the will, a modest Utopia is not beyond the power of modern technology or of world resources. Food supply is no longer a scientific problem. Fossil fuels and nuclear power should be adequate to give us the breathing space to develop renewable sources of energy. Two or three generations ahead, we may look forward to a stable or decreasing population adequately supplied with the material resources it needs for the good life – whatever, by that time, the good life is thought to be.

Most people would agree that there are things yet to be done by science in this imperfect world but, some will maintain, we must eventually reach a point where even the underdeveloped countries have everything that technology can provide for their needs. 'Then,' they may say, 'you scientists will just be doing science for your own amusement. I have no objection to this as long as it's safe, and I understand that it's fun and compulsive, like playing chess, but why should I pay for your game?'

One answer is that the sciences are part of the good life we seek, an intellectual activity, like the arts, and should be supported for the same reasons.

The scientist and the artist have much in common; both strive for originality through imagination; each tries to make a new statement and each hopes that the statement will be in some way acceptable to others. The fundamental difference between them is in the type of statement that is made.

The scientist's statement must be in a form which can be tested by anyone who is prepared to learn the necessary skills, and it is a statement about things which are common to all. The artist's statement is always partly about himself. The artist does not say, 'This is how it is' but, 'This is how I see it.' He makes a statement which can never be tested, never proved right or wrong. Thus, while science strives for a consensus in the belief that there is an ultimate truth which is common to all, the arts strive for individuality in the belief that each person's experience of truth, of nature, is different from that of others. These two views are not incompatible; it is quite consistent to believe that there is an external objective world, common to all, which can be discovered by science, and that there are also personal subjective impressions made by this world on our minds, which differ from one person to another.

Of course, the subjective experiences of different men and women may have much in common, there may be some sort of consensus; but there is little logical basis for judging the relative merits of Beethoven and the Beatles, of Rembrandt and Picasso, of Michaelangelo and Henry Moore, or indeed of claret and burgundy. On the other hand, the statements of Newton and Einstein can be compared in a meaningful way until a consensus of scientists is reached.

The artist and the scientist can nevertheless help each other in some ways. Much of modern art draws its ideas from recently discovered patterns of nature and the artistic imagination is drawn upon by scientists. It is often told how Kekulé conceived the ring structure for benzene molecules when, whilst dreaming on top of a bus, he saw a monkey grasp its tail and complete the circle. In Leonardo's anatomical drawings, art and science become indistinguishable. The mathematician, Dirac, once wrote, 'It is more important to have beauty in one's equations that to have them fit reality.'

It may be that in the forthcoming search for understanding of the brain and its workings the artists' and scientists' conceptions of things will converge. Imagination followed by reason is the way of discovery.

THE ATHENÆUM LECTURE

According to Popper, a scientific generalisation can never be proved, but may readily be disproved. But the statement becomes more and more probable as more and more unsuccessful attempts are made to disprove it. In the arts, the statement, being subjective, cannot even be disproved, it can only stimulate thought.

On these grounds, science may be considered superior to the arts. But to most people, music and painting, the novel and the theatre, are not primarily about knowledge. The artist's purpose, for them, is to give pleasure and stimulation to them rather than to express himself. Here, the arts have an enormous advantage over the sciences. People are not often as interested in others as in themselves; they care more about their own impressions than those of the artist or scientist. When they enjoy music, they enjoy the impressions it makes on them, not what it meant to Beethoven. This is readily verified by the fact that computer produced music can sometimes be enjoyed. The significance of this to science is that, if it is necessary to participate to some extent in order to derive enjoyment, only the scientist himself, or another scientist already closely involved in the same field, is likely to derive great intellectual pleasure from a new discovery. Others will derive a similar pleasure only if they rediscover the phenomenon for themselves.

This, fortunately, is the way we learn science and most people who have made a reasonably serious study, even of very elementary science, will have experienced the joy and excitement of understanding for the first time some great law of nature, or of proving for themselves some theorem of geometry. Some of the greatest ideas of science can continue to give pleasure throughout life, as one constantly meets new and ever more beautiful examples of their universal truth; the second law of thermodynamics is my favourite in this respect.

But again, if I examine the pleasure I derive from this great law, it always involves participation, rediscovery of some things I had forgotten or discovery of a new facet or application of the law. There is little pleasure to be gained from merely listening to a record which recites the second law of thermodynamics continuously (even with the help of Flanders and Swann). Yet one can listen to the same piece of music many times before becoming tired of it because, each time, the listener participates in different ways according to his own different moods. Those who get fun and intellectual satisfaction from science are, largely, the scientists.

If this is true, the last question of the layman remains to be answered: 'Why should I pay for your game?'

I believe that there is a very good reason, though I don't expect it to appeal to everyone. So far, we have answered the layman entirely in material terms. This has less and less appeal as material needs are satisfied and spiritual needs assume greater importance. Science has increased our health and wealth; now, what about our happiness?

Although our spiritual development is generally associated with religion or the arts rather than with science, our material development, our standard of living, our health and wealth, are almost wholly dependent on developments in science and medicine. The health and wealth of most people have developed out of recognition. In the so-called developed countries, the expectation of life has more than doubled in this century. The great majority of people today have material wealth and comfort of which kings and queens of the past would have been envious. James Watt and Michael Faraday released more men and women from slavery even than Abraham Lincoln. The memory of how people suffered without science is soon lost, and younger people never knew. They see only the remaining problems and human errors, some of which we can solve and prevent but with some of which we shall always have to live.

What we see now is a rapid transfer of responsibility for future evolution into the hands of one species, *Homo sapiens*. We are no longer pawns in the game, we are not even the kings and queens; we are the players.

And there is no going back – it is too late. We have interfered with out destiny to such an extent that we must continue to interfere, even to survive. Natural processes alone are no longer capable of feeding the multitudes of today.

There are no obvious limits to the advancement of knowledge or to the practical applications of this knowledge to improving our

THE ATHENÆUM LECTURE

health and wealth. In health, the magic bullets of chemistry and phar-macy have already worked miracles. In wealth, rich as most of us are in material things compared with our ancestors, we have seen only the beginning of what will be available to us.

And happiness? Science has increased our health and wealth; now what about our happiness?

To answer this question we have to ask deeper ones which are at the basis of our philosophy, our religion, and our ethics. What is it that we want of ourselves, of man, of our earth, of the universe? Human reason does not permit us to think happy thoughts which are irrational, and blind faith is no more acceptable today than it was to the twelfth century scientist:

> Then to the rolling Heav'n itself I cried,
> Asking 'What Lamp had Destiny to guide
> Her little Children stumbling in the Dark?'
> And – 'A blind understanding!' Heav'n replied

Omar's great dilemma, and ours, is that science has not yet helped man to find a new religion which replaces the old ones. There are philosophies of life, such as humanism, which provide a *modus viven-di* but do little to solve the basic questions answered so confidently by religion.

The discoveries of Copernicus, Darwin and the molecular biolo-gists have irrevocably changed our beliefs about our place in the world, but the new understanding has been negative in the sense of invalidating many old conceptions and faiths without providing a new, positive philosophy and purpose.

Is it not possible that our way to a new understanding, a new pur-pose for life, is through further knowledge and understanding of nature?

It is, of course, quite possible that we can never understand, never discover a purpose, but we shall not succeed if we do not try. Time and time again in science some artificial barrier has been proposed beyond which science could not pass, and many of those barriers are now behind us. The synthesis of organic substances, for example, was said to require a vital force until Wohler, in 1828, destroyed the idea in the only convincing way, by synthesising one *without* a vital force. There is absolutely no evidence that the great intellectual power with which we are endowed has any limitations and, until evidence to the contrary is produced, we shall be wise not to give up the search. The situation we face is the same as that faced by Pascal in his wager on the existence of God. 'A game is played at the extremity of this infinite distance where heads or tails will turn up. What will you wager? If you gain, you gain all, if you lose, you lose nothing.'

It might be argued that it is impossible for us to imagine any conceivable purpose in the universe, and therefore what we pursue is a mirage. But not many years ago it was impossible to imagine any solution to the chicken-and-egg problem of the origin of life; yet a simple solution, understandable to all, has been found. When the earth was thought to be flat, the problems about its ends seemed as insuperable as the problems of cosmic space and time seen to us today, but a spherical earth is now so obvious that it is hardly necessary to employ imagination to understand it. Could it be that our purpose will one day be as obvious as the spherical earth?

If only we could ask the right questions we would be well on the way to a solution. Perhaps we should rephrase the question, 'What is the human purpose?' in terms of the more elementary questions which might be asked by a child: 'Where am I?' and 'What am I?' Only later come the questions: 'Why am I here?' and 'Why am I made like I am?' 'Now that I'm here, what am I supposed to do?'

Until we know how to ask the right question, these simple, if imperfect, formulations of it have a heuristic value. Questions asking 'why?' are often satisfactorily answered, or dissolve away when we know the facts of how and where. 'Why don't Australians fall off the bottom of the earth?' is such a question. To the question, 'Why am I here? some logical positivists would say – and have said – that the only answer that can be given is, 'Because your father and mother produced you by the well-established method.' That used to be the only answer but, today, there is another answer, involving the story of how life started on earth and how the evolution of the species occurred. To

THE ATHENÆUM LECTURE

me that answer is a very significant step towards answering what is in my mind when I ask, 'Why am I here?' To know more of the 'how' brings us nearer to understanding the 'why', and to know all is to understand all.

The success which has attended the scientific approach is due to the supreme importance which science attached to observation and experiment, whereas the philosopher places his trust in reason – an essential but unreliable tool. Experiments with bits of glass and falling weights may seem trivial when compared with the great thoughts of the theologian and the philosopher but they have proved to be more reliable, and the best route to sure understanding. Most of the great scientific thinkers of the past were also experimenters and, of course, careful observers. The scientist-philosopher, the logician and the theologian must work together towards a common understanding. Newton was a noted theologian and saw his science as seeking the works of the Almighty. Zeno said that 'what moves matter may equally well be called providence or nature.' Can knowing nature and knowing God be very different?

If our problems seem insuperable, and the route interminably long, we should remind ourselves that modern science started only 400 years ago and has already transformed our lives and understanding. In this endeavour it is earlier than we think. What may we not achieve in the four billion years which remain before the earth becomes uninhabitable?

What is it that we want to achieve? Is it merely the greatest happiness of the greatest number? How many people do we want on earth anyway and what sort of people should they be? Until we have more understanding, all our ambitions for the world are, at best, short term and, at worst, may be quite wrongly conceived. Our ethics and morals must ultimately be decided in the light of this understanding.

Vanevar Bush has written, eloquently as always, 'Science has a simple faith which transcends utility. Nearly all men of science, all men of learning for that matter, and men of simple ways too, have it in some form and in some degree. It is the faith that it is the privilege of man to learn to understand and that this is his mission. Why does the

· 11 ·

shepherd at night ponder the stars? Not so that he can better tend his sheep. Knowledge for the sake of understanding, not merely to prevail: that is the essence of our being. None can define its limits or set its ultimate boundaries.'

There is, then, one great purpose for man, and for us today, and that is to try to discover man's purpose by every means in our power. That is the ultimate aspiration of science, and not only of science, but of every branch of learning that can improve our understanding.

Contribution from

GRAHAM R. FLEMING

University California, Berkeley
USA

Born 3rd December 1949. Educated University of Bristol, BSc (Honours) Chemistry (1971), PhD Physical Chemistry (1974). Research Fellow, California Institute of Technology, USA (1974–1975), University of Melbourne, Australia (1975). ARGC Research Assistant, University of Melbourne, Australia, Leverhulme Fellow, Royal Institution, UK (1977–1979). Professor, The University of Chicago (1985–1987). Arthur Holly Compton Distinguished Service Professor, University of Chicago (1987–1997), Professor, University of California Berkeley (1997–present), Melvin Calvin Distinguished Professor of Chemistry (2002–present), Director, Physical Biosciences Division, Lawrence Berkeley National Laboratory (1997–2005), Associate Laboratory Director for Physical Sciences (2002–2005), Deputy Laboratory Director (2005–present), Director (Berkeley), California Institute for Quantitative Biomedical Research, UCB, UCSF, UCSC (2000–present).

Winner of Marlow Medal, Royal Society of Chemistry 1981; Alfred P. Sloan Foundation Fellow 1981; Camile and Henry Dreyfus Teacher-Scholar 1982; Coblentz Award (Coblentz Society) 1985; John Simon Guggenheim Fellowship 1987; Tilden Medal, Royal Society of Chemistry 1991; Fellow, American Academy of Arts and Science 1991; Fellow, Royal Society of London 1994; Nobel Laureate Signature Award for Graduate Education in Chemistry, American Chemical Society 1995; Inter-American Photochemical Society Award 1996; Centenary Lecture and Medal, Royal Society of Chemistry; Peter Debye Award in Physical Chemistry, American Chemical Society 1998; Harrison Howe Award, American Chemical Society 1999; Earle K. Plyler Prize for Molecular Spectroscopy, George C. Crouch Foundation, American Physical Society 2002; Sierra Nevada Section Distinguished Chemist, American Chemical Society 2003; George Porter Medal, Photochemistry Societies of Europe, Asia/Oceania and the Americas 2004.

Royal Institution personnel in 1973. Graham Fleming is third from right, back row. George Porter is centre, front row, flanked on his left by Professor Ronald King and Bill Coates, and on his right by David Miller, Bursar, and Mike West.

Graham R. Fleming on
"Flash Photolysis and Spectroscopy: A New Method for the Study of Free Radical Reactions"
G. Porter

Proc. Roy. Soc. A **200**, 284 (1950)

The technique of flash photolysis emerged whole and fully formed in the first paper to describe it. That paper has a single author, a postgraduate student at Emmanuel College Cambridge, working under the supervision of Professor R. G. W. Norrish who was well known for his work in photochemistry. The student was George Porter, who had gone to Cambridge in 1946, having been demobilised from the Royal Navy after World War II. George's original problem was concerned with the methylene radical using a technique based on the time to destroy a metal mirror at a variable distance from the source of gaseous free radicals (the Paneth mirror technique). George's own view of his first paper is not charitable: "Paper published on this in 1947 (1). Not very good." In fact the paper was for the 1947 Discussion of the Faraday Society on "The Labile Molecule". The meeting was entirely concerned with the study of short-lived chemical species such as free radicals, but did not consider direct methods of observation at all. As Harry Melville said in his introductory remarks, in reference to the low concentrations of radicals obtained, "The direct physical methods of measurement simply cannot reach these magnitudes, far less make accurate measurements in a limited period of time, for example 10^{-3} s." George had, in fact, already begun on the development of flash photolysis, the method that was going to revolutionise the study of chemical kinetics, and over the course of the next fifty years would lead to the creation of the new fields of physical, chemical and biological dynamics of molecules down to timescales of hundreds of attoseconds (10^{-18} s).

In papers he left in the Royal Society archives George described how he got the idea for the flash photolysis method and it seems best to use his own words:

"I was sent to the Siemens lamp works in Preston to collect a large mercury arc which I wanted for my research and took the opportunity to look around. Siemens was using electronic flash lamps and talked of milliseconds duration and 10,000 J energy. I had been trying to obtain the spectrum of methylene which I knew lived for a millisecond and could be produced by light. My experience in the Navy made me familiar with rapid recording techniques. I went back to my rather sordid hotel room and in the space of about half an hour I knew that here was a new technique for photochemisty and free radicals. I wrote out the idea in detail and Norrish was enthusiastic immediately. I begged and borrowed condensers from the Navy and Air Force. But it was nearly another year before I had the idea of using a second flash to record the spectrum having previously wasted much time trying to build a rapid recording grating spectrograph."

Thus the impetus for the technique was the spectroscopic detection and study of free radicals in the gas phase. The first experiment designed to produce organic free radicals certainly showed that the concentration of free radicals would be adequate for detection: photolysis of acetone with a 4000 J flash resulted in total destruction of the acetone and the deposition of filaments of carbon throughout the 1 m long sample cell! It is interesting to speculate if C_{60} or carbon nanotubes were present in these filaments.

The second paper[2] written with Norrish describes the effect of such high intensity flashes on photochemical reactions and is clearly aimed at creating high concentrations of radicals. Today it seems a fairly standard paper and gives no hint that three days later (August 9, 1949) George would submit to the Proceedings a single-author paper describing the fully-fledged method of flash photolysis containing spectra as a function of time delay, for several radicals and transient species, including new spectra for the ClO and CH_3CO radicals.[3]

The method involving an intense "photolysis" flash and a second, delayed "spectro flash", to record the spectrum on a photographic plate at the image plane of a spectrograph, was revolutionary and set in train attempts, in which George was to play a leading role, to record the fastest processes in chemistry that continues to this day.

There are many remarkable aspects of this paper on "Flash photolysis and spectroscopy: A new method for the study of free radical reactions".[3] First, as George notes above, his original intention was to record the full two-dimensional surface of time and wavelength (frequency) information for the experimental system, something that is rarely attempted today, and then only with very high repetition rates, ultra-stable, laser sources. Second, George was concerned that the very intense photolysis flash would produce unmanageable quantities of scattered light, and as a result did not use an electronic delay between the two flashes, but instead a rotating wheel which synchronised a shutter to block the photolysis flash, the firing of the photolysis and spectroflash, and an oscilloscope time base. Timing was controlled by the gearing to the wheel for coarse delays and by altering the separation of the two contacts on the wheel which produce the trigger pulses to the two lamps. The condensers (capacitors) required for the discharge were given free by the Royal Navy, who even paid the chemistry department £20 for returning their packing cases!

Equally striking were the long term consequences of the true "basic" research in that first paper more than 30 years after the recording of the ClO spectrum and subsequent careful study of its reactions by George and F. J. Wright; ClO was found to play an important role in the catalytic destruction of stratospheric ozone (the "ozone hole" above Antarctica) via the reactions:

$$Cl + O_3 \rightarrow ClO + O_2$$
$$\underline{Cl + O \rightarrow Cl + O_2}$$
$$O_3 + O \rightarrow 2O_2$$

with the chlorine atoms being generated by photochemical decomposition of chlorofluorocarbons.

A remarkable burst of activity followed with studies ranging from the explosive reaction of hydrogen and oxygen and the kinetics of HO radicals, fragmentation of organic molecules, the explosive combustion of hydrocarbons to the study of such phenomena as knock, carbon formation and atom recombination reactions.

References

1. G. Porter and R. G. W. Norrish, *Disc. Faraday Society* **2**, 97 (1947).
2. G. Porter, R. G. W. Norrish, *Nature* **164**, 658 (1949).
3. G. Porter, *Proc. Roy. Soc. A* **200**, 284 (1950).

Reprinted from *J. Photochem. Photobiol. A: Chem.* **142**, 107–119 (2001) with permission from Elsevier

ELSEVIER   Journal of Photochemistry and Photobiology A: Chemistry 142 (2001) 107–119

Journal of
Photochemistry
and
Photobiology
A:Chemistry

www.elsevier.com/locate/jphotochem

The mechanism of energy transfer in the antenna of photosynthetic purple bacteria

Mino Yang*, Ritesh Agarwal, Graham R. Fleming

Department of Chemistry, Lawrence Berkeley National Laboratory, Physical Biosciences Division, University of California, Berkeley, CA 94720, USA

This paper is dedicated to Lord Porter on the occasion of his 80th birthday.

Abstract

The mechanism of energy transfer in the antenna system of purple bacteria is investigated by combination of photon echo spectroscopy and disordered exciton theory. In the B800 component of light harvesting complex 2 (LH2), a picture of incoherent hopping between monomers provides an excellent description of the photon echo data recorded as a function of excitation wavelength. In the B850 pigments of LH2, and to a somewhat greater extent in the B875 pigments of light harvesting complex 1 (LH1), the excitation is delocalized over several pigments. The observed dynamics correspond to relaxation between exciton states as a result of exciton–phonon coupling. Nonetheless, a picture of "hopping" between small groups of molecules provides a crude description of the motion of the excitation in B850, and B875. The electronic coupling required to simulate the experimental absorption spectrum and photon echo data is larger in LH1 than in LH2 (B850). © 2001 Published by Elsevier Science B.V.

Keywords: Purple bacteria; Antenna; Photosynthesis

1. Introduction

The process of light harvesting in which several hundred chlorophyll or bacteriochlorophyll molecules act as an antenna to collect and direct solar energy to a reaction center is central to the efficiency of terrestrial photosynthesis. Often, the overall efficiency of energy transfer from the antenna to the reaction center is above 95%, an astonishing fact given that the excitation must be transferred hundreds of times, and that the energy transfer efficiency of a spatially random solution of chlorophyll at the concentration (i.e. number density) of the chloroplast is vanishingly small.

The work of Lord Porter over the past half century has stimulated and inspired much of modern research on photosynthetic light harvesting. During this period, methods of temporal and spatial resolution have steadily improved along with the theoretical tools necessary to interpret and even predict the dynamical behavior of photosynthetic systems. Porter et al. [1] (and independently Alfano et al. [2,3] and Shapiro et al. [4]) were the first to use the picosecond time resolved spectroscopy to study energy migration in the antenna systems of plants and algae. Perhaps the culmination of this pioneering work was the beautiful demonstration of the energy funnel in phycobilisomes with the clear sequen-

tial population of progressively lower energy species as the energy flows through the phycobilisome to the core reaction center complex [5]. In parallel with these elegant experiments, Porter's group was pursuing some of the puzzling theoretical issues in light harvesting. Of particular significance was the demonstration by Beddard and Porter [6] of the role of concentration quenching in spatially random arrangements of chlorophyll molecules, and the consequent suggestion that chlorophyll molecules must be held in fixed positions in close (say 12 Å) proximity, but not allowed to come into full electronic contact. This was an extremely important suggestion at the time because it was not universally accepted (especially in the USA) that the antenna chlorophylls were bound exclusively in protein complexes rather than being in the lipid membrane. Of course, subsequent X-ray crystallographic studies have shown that all antenna molecules are bound in pigment–protein complexes in fixed and generally in highly organized arrangements [7–13].

The ability to obtain high resolution structures of membrane proteins, first of the reaction center [14–17] followed by the peripheral light harvesting complex of purple bacteria LH2 [7–9] has revolutionized our understanding of the primary steps of photosynthesis. At the time of writing, high resolution structures of photosystems I and II (PSI and PSII) of cyanobacteria and green plants have been announced and will be published shortly. These structures allow questions of mechanism and structure/function/dynamics relationships to

---
* Corresponding author.
E-mail address: minoyang@uclink4.berkeley.edu (M. Yang).

1010-6030/01/$ – see front matter © 2001 Published by Elsevier Science B.V.
PII: S1010-6030(01)00504-4

108   *M. Yang et al. / Journal of Photochemistry and Photobiology A: Chemistry 142 (2001) 107–119*

be brought into sharp relief. In parallel with the advances in structure determination, advances in ultrafast spectroscopy and in theoretical chemistry allow many of these fundamental questions to be addressed at a quantitative level.

Two such fundamental questions can be simply stated. (1) What is the interaction mechanism between the chromophores in such densely packed molecular aggregates? Is it the standard point dipole–dipole interaction responsible for long range energy transfer between strongly allowed transitions or do different mechanisms, perhaps arising from orbital overlap contribute or even dominate when donors and acceptors are spaced by distances less than their size? A related question concerns the role of 'dark' states, i.e. states with forbidden transitions in the energy migration process. Can such states play a role and if so, what is the mechanism? (2) Given that electronic couplings between individual molecules may be large compared to their reorganization energies and static variations in their energy levels, what is the mechanism by which excitation energy moves in the antenna? Is it appropriate to describe the process as incoherent hopping between individual molecules, or is the excitation at least partially delocalized over several or many molecules, moving in a coherent fashion? To elaborate the question a little, what states does the excitation process prepare and how do they evolve? Does the excitation become localized over time and if so, what is the timescale compared to energy migration and overall trapping?

Answers to these two questions should help to elucidate the design principles underlying the efficiency of photosynthetic light harvesting and inform the design and synthesis of synthetic antenna devices, such as dendrimeric light harvesting systems ([18] and references therein). In this paper, we

describe calculations and experiments designed to answer both questions in the context of the purple bacterial antenna complexes LH2 and LH1. The structure of LH2 is shown in Fig. 1, while the structure of LH1 is believed to be very similar to LH2 but with a larger ring containing 16 dimeric subunits (32 bacteriochlorophyll (BChl) molecules) rather than the 8 [9] or 9 [7] (16 or 18 BChls) dimeric subunits in LH2. In addition, LH1 lacks the ring of nine (in the *Rps. acidophila* structure) apparently monomeric BChl molecules present in LH2. Both structures also contain carotenoid molecules (see Fig. 1) acting in both protective and light harvesting roles [19]. The excited states of carotenoids, in common with other long chain polyenes, are unusual. The first excited singlet-state is of nominal A_g symmetry and thus, optical transitions from the ground state are one-photon forbidden. The transition to the second singlet state is strongly allowed and can couple to the BChl molecules via the standard Coulombic mechanism. However, the location and effectiveness in energy transfer of the first singlet state has been, until very recently, quite unclear. With the notable exceptions, [20–22] it has generally been supposed by most previous workers that the coupling of the S_1 state to the BChl molecules occurs via the Dexter exchange mechanism [23]. As we have shown elsewhere, this is incorrect and the Coulombic coupling mechanism also mediates very rapid (∼1 ps) energy transfer from the carotenoid S_1 state to the BChls, even though no absorption band for these states is detectable in the one-photon spectrum [24,25].

In this paper, we focus on the electronic interactions and energy transfer mechanisms in the three distinct sets of BChl molecules in the antenna system of purple bacteria: B800 and B850 of LH2 and B875 of LH1. Because the

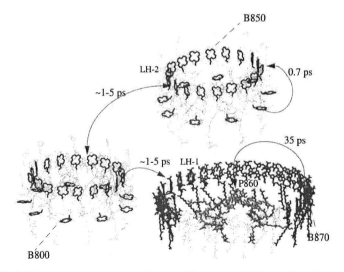

Fig. 1. Schematic view of the light harvesting antenna system of purple bacteria. The structure of LH2 is from [7] while that of LH1 is based on [10,66]. LH1 is shown cut away to show the reaction center.

M. Yang et al./Journal of Photochemistry and Photobiology A: Chemistry 142 (2001) 107–119   109

electronic interactions between the chromophores are relatively weak, energetic disorder, which is an inevitable feature of self-assembled systems, plays a major role in determining the electronic structure and dynamics. Study of such systems is greatly aided by experimental techniques that are sensitive to distributions of optical transition frequencies. Photon echo spectroscopy in general and the three-pulse photon echo peak shift technique in particular give incisive information about dynamics in inhomogeneous systems. In contrast, more traditional approaches, such as spontaneous fluorescence or transient absorption spectroscopy are quite insensitive to the nature of inhomogeneous dynamics. We, thus, focus our paper on photon echo measurements and their analysis via theoretical models.

2. The measurement of photosynthetic energy transfer via photon echo spectroscopy

While the pioneering picosecond measurements of Porter, Treadwell, Beddard, and coworkers gave access to overall energy trapping timescales, the timescale of energy transfer between individual molecules was inaccessible before the development of sub-100 fs laser sources. Although in reality, a continuum of possibilities exist, for the purpose of organizing this discussion we will distinguish two types of energy transfer: intra- and inter-band transfer. By the former, we mean transfer to molecules of similar spectral character whose absorption band lies within the spectrum of the laser used for excitation. An example of this, intraband transfer is the transfer within the B800 or B850 rings of LH2 (see Fig. 1) or the B875 ring of LH1. In interband transfer, the excitation moves to a species absorbing outside the excitation laser spectrum, examples being B800 to B850 transfer, or carotenoid to either B800 or B850 transfer in LH2. Clearly, the interband transfers, given adequate time resolution, are accessible to the standard methods of pump-probe and fluorescence upconversion spectroscopy. However, the intraband cases are just those where electronic delocalization effects might be most important since the energy gaps between molecules are small, allowing the possibility of strong coupling.

Over the past 5 years, we have been developing a type of photon echo measurement, the three pulse echo peak shift method (3PEPS) which seems specially suited to the measurement of intraband energy transfer [26–47]. The 3PEPS technique has been described in detail elsewhere [27,48,49], and here we will confine ourselves to a brief description of the information content of the technique and some recent developments. A peak shift, $\tau^*(T)$, plot, such as the one shown in Fig. 2 is obtained by determining the position of the three pulse echo profile maximum with respect to zero delay for the coherence period (τ) as a function of the population period, T. Fig. 2 shows peak shift data for a dilute solution of a dye molecule IR144 in a polymer glass PMMA [30,36]. The long-time value of the peak shift is finite,

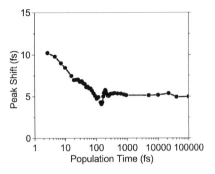

Fig. 2. Plot of the three pulse echo peak shift for a dilute solution of the dye IR144 in PMMA at room temperature [30]. The oscillatory features result from vibrational wavepackets. Note that the peak shift ($\tau^*(T)$) becomes consistent after about 200 fs, whereas in fluid solution, the peak shift decays to zero on a picosecond timescale.

unlike the case of a fluid solution, [48] and reflects the presence of static inhomogeneous broadening in the transition frequencies of the individual chromophores.

Now consider an ensemble of molecular aggregates, each aggregate consisting of a set of chemically identical chromophores. Fig. 3 shows two possible scenarios for the distribution of site energies (and optical transition frequencies) for the chromophores. In the first picture, the mean energy of each aggregate is the same, in other words, the energies of each member of the aggregate are chosen at random from the full distribution and have no correlation with each other. In the second case, there is some correlation between individual members of a particular ensemble and as a consequence, the means of the individual aggregates are also distributed. In this context, we can write the static energy (or transition energy) of an individual chromophore, i, in aggregate 'a' as

$$\varepsilon_a^i = \langle \varepsilon \rangle + E_a + \delta\varepsilon_a^i \tag{1}$$

where $\langle \varepsilon \rangle$ is the mean energy of the monomers over the whole ensemble, E_a the static offset of the mean energy of the monomers in the aggregate a from $\langle \varepsilon \rangle$ and $\delta\varepsilon_a^i$ is the static offset of the energy of the chromophore, i, from E_a. The standard deviations of the distributions of E_a and $\delta\varepsilon_a^i$

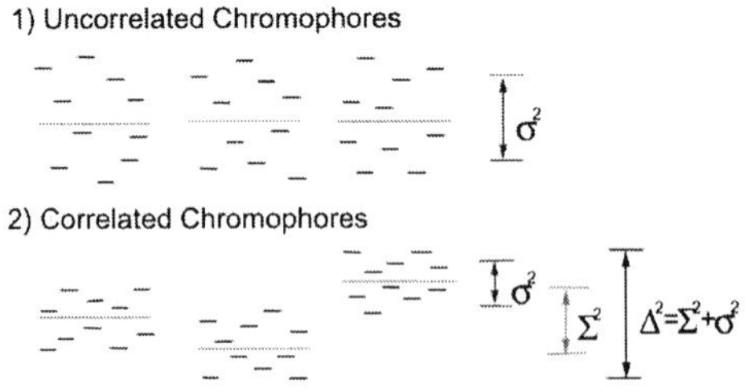

1) Uncorrelated Chromophores

2) Correlated Chromophores

Fig. 3. An illustration of the two types of disorder discussed in the text.

are defined by Σ and σ, respectively, and the total inhomogeneous width is given by

$$\Delta^2 = \Sigma^2 + \sigma^2 \qquad (2)$$

The energy transfer time within an aggregate or the size of the delocalization of an exciton state depends only on σ. On the other hand, the energy transfer time scale between aggregates will depend mainly upon Σ. The bandwidth of the absorption spectrum will depend on both quantities. Thus, the characterization of these quantities is important.

In photon echo experiments, the presence of static inhomogeneous broadening allows rephasing because the memory of a given molecule's transition frequency is retained (totally in the absence of homogeneous dephasing, partially in the presence of both inhomogeneous and homogeneous dephasing). In the peak shift measurement, this leads to a finite long-time value of the peak shift, $\tau^*(\infty)$. This discussion applies to a dilute system in which excitation remains confined to the molecule that was initially excited. Now in a molecular aggregate in which energy transfer occurs between individual monomers, the excitation can explore the σ distribution, but the Σ distribution either not at all, or on a much longer timescale. Thus, the energy transfer process itself leads to a loss of transition frequency memory and, thus, a decay of the peak shift. The timescale of the energy transfer is represented directly in the dependence of the τ^* on T. If $\Sigma = 0$, the energy transfer will lead to complete loss of memory and $\tau^* = 0$ at longer time than energy transfer time scale. If $\Sigma \neq 0$, this is not so and the form of $\tau^*(T)$ can be used to determine both Σ and σ [45,50]. Fig. 4 shows the remarkable sensitivity, in the presence of intraband energy transfer, to the site energy distributions of the 3PEPS method. By contrast, both transient absorption and transient grating methods are quite insensitive to the presence or absence of intraband energy transfer in such a system; still less the nature of the ensemble disorder (Fig. 4).

Before describing a specific light harvesting system, it is worthwhile exploring the origins of the σ and Σ distributions a little further. The individual sites of particular chromophores can give rise to variations in the transition frequency for a variety of reasons. For example, side chain conformational disorder undoubtedly exists in the polypeptides that bind the chromophores. Ionizable side chains may be close to their pK and thus distributed between charged and neutral forms. The transition frequency of chlorophyll molecules is known to be quite sensitive to distortions of the macrocycle from planarity [51]. Any or all of these effects (originating from local inhomogeneities around a chromophore) will contribute to σ.

There are two types of contribution to Σ: the first has been previously discussed in the context of single molecule spectroscopy [52–54] and arises from global distortions of the whole complex, thus shifting the mean transition frequency of the entire set of chromophores in the complex. Several research groups have introduced this variable based on the physical heterogeneity among the aggregates and the

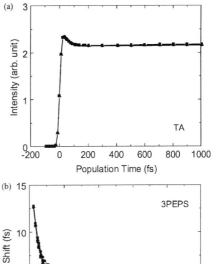

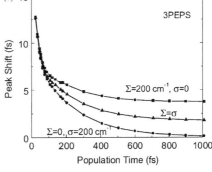

Fig. 4. Calculated transient absorption (TA) (a) and peak shift (3PEPS) (b) signals for three different energy transfer systems with (i) completely uncorrelated disorder ($\Sigma = 0, \sigma = 200\,\text{cm}^{-1}$); (ii) partially correlated $\Sigma = \sigma = 141\,\text{cm}^{-1}$; (iii) completely correlated ($\Sigma = 200\,\text{cm}^{-1}, \sigma = 0$). All three curves are essentially identical for TA, but give very different results in 3PEPS.

mechanisms producing the fluctuations in E_a and $\delta\varepsilon_a^i$ have often been implicitly assumed to be independent. In this case, the aggregate energy will be different from one aggregate to another and, thus, the introduction of a stochastic variable, E_a, for the aggregate energy is required. For example, Freiberg et al. attribute the fluctuation of E_a to random solvent shifts and the fluctuations of $\delta\varepsilon_a^i$ to different environments surrounding the monomers [52]. van Oijin et al. also introduced intra- and inter-complex heterogeneity but without any specific mechanism [53]. Mostovoy and Knoester successfully analyzed recent single molecule spectra by employing inter-ring disorder [54]. They suggested that the sources of the inter-ring disorder could be structural deformation or finite correlation between monomers with a ring. Agarwal et al. also used a two-level disorder model to explain their 3PEPS data and linear absorption spectrum of B800 in LH2 [50].

The second contribution to Σ is purely statistical and does not seem to have been discussed earlier. This effect arises from the fact that each individual aggregate samples

M. Yang et al./Journal of Photochemistry and Photobiology A: Chemistry 142 (2001) 107–119   111

a rather small portion of the total distribution (originating from the local inhomogeneities around a chromophore). The fact that the number of monomers in a given aggregate is not sufficient to fully sample the distribution, necessarily produces a distribution in the means of the individual complexes. We will refer to this finite sampling contribution to Σ as Σ_{loc} whereas the global distortion contribution we denote as Σ_{glob}.

Assuming these two are independent, we have

$$\Sigma^2 = \Sigma_{glob}^2 + \Sigma_{loc}^2 \tag{3}$$

and, equivalently, the total inhomogeneous width

$$\Delta^2 = \Sigma_{glob}^2 + \Sigma_{loc}^2 + \sigma^2 \equiv \Sigma_{glob}^2 + \Delta_{loc}^2 \tag{4}$$

One can easily realize from simple Gaussian statistics that

$$\Sigma_{loc} = \frac{\Delta_{loc}}{\sqrt{N}} \tag{5}$$

which means the aggregate energy fluctuates with Gaussian statistics characterized by Σ_{loc} (unless $N \rightarrow \infty$) even in the absence of any global fluctuation of a whole complex. In this case, the standard deviation of the intra-aggregate fluctuation σ is written in terms of the total fluctuation width of the local inhomogeneities, Δ_{loc}, as

$$\sigma = \sqrt{\Delta_{loc}^2 - \Sigma_{loc}^2} = \Delta_{loc}\sqrt{\frac{N-1}{N}} \tag{6}$$

Then if the inter-aggregate disorder originates only from this kind of statistical effect, i.e. $\Sigma = \Sigma_{loc}$, we arrive at a simple relation between Σ and σ

$$\frac{\sigma}{\Sigma} = \sqrt{N-1} \tag{7}$$

and the effect of inter-aggregate disorder will relatively increase when the number of monomers is small as in the B800 band of LH2.

Fig. 5 illustrates the statistical contribution to Σ. In the upper panel, the open circles show the distribution of transition frequencies of individual monomers randomly chosen from a Gaussian distribution. The solid bars show the means of individual complexes containing nine monomers, and are clearly characterized by a distribution of their own. The lower panel quantifies this relationship showing the standard deviations within a complex (intra, open circles) and between complexes (inter, filled squares) as a function of the number of monomers. As an example, consider the B800 component of LH2 in *Rps. acidophila*, which contains nine monomers per LH2. If the total inhomogeneous width (Δ_{loc}) is 100 cm^{-1}, from Eqs. (5) and (6), we expect $\sigma = 94$ cm^{-1} and $\Sigma_{loc} = 34$ cm^{-1}.

We can compare the predictions of Eq. (7) with some literature data. Freiberg et al. simulated the linear and nonlinear spectra of both the B800 and B850 bands of LH2 [52]. They concluded that $\sigma^{B850} = 216$ cm^{-1}, $\Sigma^{B850} = 54$ cm^{-1}; $\sigma^{B800} = 46$ cm^{-1}, $\Sigma^{B800} = 46$ cm^{-1} which give ratios

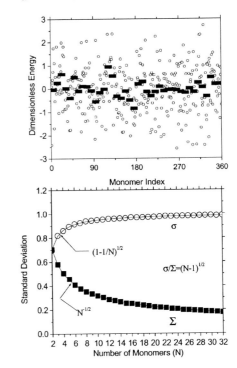

Fig. 5. Upper panel: plot of distribution of monomer energies (open circles) for 360 monomers chosen from a Gaussian distribution of standard deviation, 1. The bars indicate the mean energies of sets of nine chromophores chosen from the distribution. Lower panel: plot of standard deviation within a complex (open circles) and between complexes (filled squares) as a function of the number of monomers in a complex.

of $\sigma^{B850}/\Sigma^{B850} = 4$ and $\sigma^{B800}/\Sigma^{B800} = 1$. We can see the ratio for the B850 surprisingly agrees well with Eq. (7) which gives 4.1, but the ratio for the B800 is different from our estimate of 2.8 and from Agarwal et al.'s value of 1.8 (see Section 3.1). However, the qualitative finding that the ratio decreases with decreasing numbers of monomers is in accordance with Eq. (7) and, thus, we conclude that a significant portion of the inter-aggregate disorder they observed arises from statistical fluctuations.

Since $\sigma \approx \Delta_{loc}$, as can be seen in Eq. (6), for typical LH complexes, the influence of different levels of local inhomogeneity may be difficult to observe in studies concentrating on the electronic structure and dynamics occurring only within one aggregate [55]. However, when the intra-aggregate properties are combined with optical properties affected by overall distribution of energy (in most ensemble-averaged experiments and even in some single molecule spectroscopy studies [52–54]), one should carefully describe this second level of static disorder.

112   M. Yang et al. / Journal of Photochemistry and Photobiology A: Chemistry 142 (2001) 107–119

3. Nonlinear response of weakly coupled systems

In order to model an actual photosynthetic complex, a detailed model for the population dynamics is required. Since the rate of energy transfer between chromophores i and j will depend on their energies in disordered systems, it is necessary to properly account for the ensemble effect. Noting the optical responses from energy donors (excited by laser) and acceptors (excited by energy transfer) have different characteristics, Yang and Fleming showed that, for weakly coupled systems, to a reasonable approximation the nonlinear optical signal can be written as [45]

$$\text{Signal} \approx \langle R_D \rangle_{\text{disorder}} \cdot \langle P_D \rangle_{\text{disorder}}$$
$$+ \langle R_A \rangle_{\text{disorder}} \cdot \langle P_A \rangle_{\text{disorder}} \qquad (8)$$

where $\langle R_{D(A)} \rangle_{\text{disorder}}$ and $\langle P_{D(A)} \rangle_{\text{disorder}}$ are the ensemble-averaged optical response function and population, respectively, of energy donors (acceptors). The factorization in Eq. (8) neglects any correlation between inhomogeneous optical responses and inhomogeneous population kinetics and is a mean kinetics approximation. With this factorization, instead of full Monte Carlo computation in three dimensions (with respect to time variables in the third-order experiments), we can carry out a Monte Carlo procedure only for the population kinetics as described below and then combine those with analytic expressions for the optical response functions [45] to make the simulation much more efficient.

For a given configuration of static energies of chromophores, we calculate the time dependent population using the Pauli master equation

$$\frac{\mathrm{d}}{\mathrm{d}t} P_i(t) = -\tau_i^{-1} P_i(t) - \sum_{\substack{j=1 \\ j \neq i}}^{N} [F_{ji} P_i(t) - F_{ij} P_j(t)] \qquad (9)$$

Here F_{ij} is the rate constant for transfer from site j to site i, $P_i(t)$ the probability of occupation of site i, τ_i the lifetime of the excited state, which in this care corresponds to the B800 to B850 energy transfer time. Downhill rates are calculated from the overlaps of site-energy-dependent donor emission and acceptor absorption spectra, while uphill rates are calculated via the detailed balance equation

$$F_{ji} = F_{ij} \exp\left(\frac{-\Delta E_{ji}}{k_B T}\right) \qquad (10)$$

where ΔE_{ji} is the difference between the peak frequencies of the donor and acceptor absorption bands. The effect of disorder on energy transfer rate appears directly through the overlap integral between two chromophores with their own site energies. The Pauli master equation is solved using the Green function solution of the coupled equations [56]. Each chromophore is assigned a site energy randomly chosen from a Gaussian distribution by a Monte Carlo sampling procedure and then the averaged populations are inserted into Eq. (8). In order to model the data of Fig. 6, we input

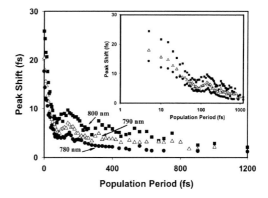

Fig. 6. Peak shift data for the LH2 complex of *Rps. acidophila* at three different wavelengths within the B800 band. Inset shows the same data on a logarithmic timescale.

the homogeneous lineshape, the coupling strength and the laser spectrum. The energy transfer rates are calculated from these parameters within the model.

3.1. Application to the B800 band of LH2

The approach described above has been applied to photon echo data for the B800 band of LH2 [50]. In this work, pulses of 25 fs were used. These pulses have a bandwidth which covers the entire absorption band of the B800 chromophores. The intraband energy transfer was shown to occur with an average timescale of 500–600 fs [50,57,58]. Further, it was not possible to describe the data using only a value for σ. The data were best fit by $\sigma = 90\,\text{cm}^{-1}$ and $\Sigma = 50\,\text{cm}^{-1}$ ($\Delta = 103\,\text{cm}^{-1}$). The values are extremely close to those expected if $\Sigma = \Sigma_{\text{loc}}$, in other words, the variations in the mean energies of the B800 sets in individual LH2 complexes appears to arise almost entirely from the limited sampling effect illustrated in Fig. 5.

The coupling between the B800 molecules in LH2 is calculated to be $30\,\text{cm}^{-1}$, [57,59,60] which is small compared to the homogeneous linewidth, disorder, and thermal energy ($k_B T$). We, therefore, expect a model based on incoherent hopping in an inhomogeneous ensemble to accurately describe the energy transfer dynamics. To test this idea in detail, we carried out excitation wavelength dependent 3PEPS studies with excitation pulses centered at 800, 790, and 780 nm. The results of these experiments are shown in Fig. 6. All three peak shift curves show the presence of $165\,\text{cm}^{-1}$ vibrational wavepackets persisting up to at least 1 ps at 800 nm. The amplitude of the quantum beats decreases with excitation wavelength as does the initial value of the peak shift. Both of these effects can be understood by constructing an accurate vibronic model for the absorption spectrum and have recently been described in detail for dye molecules in solution [46,47]. Of more interest for the present discussion is the finding that the decay timescale of

M. Yang et al. / Journal of Photochemistry and Photobiology A: Chemistry 142 (2001) 107–119   113

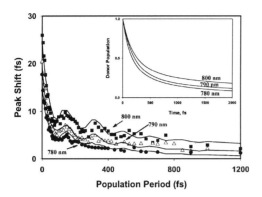

Fig. 7. Calculated peak shifts at 800, 790, and 780 nm based on the inhomogeneous hopping model described in the text. Inset shows the population kinetics for the three different excitation wavelengths. For clarity, the B800 to B850 energy transfer is not included in this plot, it is, however, properly included in the calculated peak shifts of the main figure.

the slower component in the peak shift (representing B800 to B800 energy transfer) decreases from about 600 fs at 800 nm to 500 fs at 790 nm and 400 fs at 780 nm.

Such an effect is characteristic of energy transfer in an inhomogeneous energy transfer system, because the pigments in the low energy position of the distribution have a much higher likelihood of making an 'uphill' energy transfer step. Fig. 7 shows the result of such calculations using a value of the inhomogeneous width $\Delta_{\text{loc}} = 100\,\text{cm}^{-1}$. Clearly, the experimental data are reproduced extremely well. The inset shows the decay of an initially populated donor at three different wavelengths. The wavelength dependent kinetics is obtained by filtering the Monte Carlo procedure by the laser spectrum. The more rapid decay at the higher energy side of the distribution is apparent. For clarity, the B800 to B850 transfer process is not included in the calculations shown in the inset, it is, however, properly included in the simulations of the experimental data.

To sum up, the energy transfer within the B800 ring of LH2 can be well understood as incoherent hopping on a timescale of $\sim$500 fs via the Förster mechanism in an inhomogeneous ensemble. The inhomogeneity does not appear to have any correlation in individual aggregates beyond that expected from the limited sampling of the full distribution by the individual groups of B800 molecules. In other words, the energy transfer dynamics are dominated by the local disorder and the spatial correlation length of the factors responsible for the site energy variation is, therefore, not much larger than the spacing between molecules.

4. Nonlinear response of excitonically coupled systems

In the case of the B850 molecules of LH2 or the B875 molecules of LH1, the situation is more complicated and

more interesting. Now the excitonic interactions are at least similar to the site energy disorder and to $k_{\text{B}}T$ and the electronic states will be at least partially delocalized. This delocalization leads to the phenomenon of exchange narrowing [61–63], whereby the distribution of site energies is apparently narrowed by the averaging effect of the delocalized electronic states. Thus, the intra-complex disorder (and, therefore, the total disorder) associated with the delocalized states become dependent on J, the electronic coupling. The inter-complex disorder Σ, however, remains independent of J.

The third-order response function governing the nonlinear signals is described by a density matrix in the representation of the static-energy-dependent exciton basis which is, as usual, obtained by numerical diagonalization of Hamiltonian [64,65]. Zhang et al. formulated a third-order response function in this context based on the projection operator technique and applied it to photon echo and pump-probe studies of the B850 systems of LH2 [65]. The evolution equations for those density matrix elements are given by

$$\frac{\text{d}}{\text{d}t}\rho(t) = -i(L_0 + L')\rho(t) \qquad (11)$$

where L_0 is a diagonal Liouville operator governing exciton dynamics in the absence of any exciton transfer process. The off-diagonal term, L', in Eq. (11) is responsible for the population transfer process between the exciton states and we employ a course-grained description for those population transfer processes

$$\frac{\text{d}}{\text{d}t}\rho(t) \approx iL_0\rho(t) - R\rho(t) \qquad (12)$$

where the matrix R is given by the Redfield tensor which is based on a second-order approximation with respect to the off-diagonal coupling. Solution of Eq. (12) is combined with the expression for the polarization to give the third-order signal arriving from a given configuration of the static nuclear component. Finally, we use a take Monte Carlo sampling procedure in order to get the macroscopic polarization averaged over the random configuration of the static energies.

4.1. Application to LH1

The most crucial parameters determining the mechanism of the excitation energy transfer within LH1 are the Coulombic interaction and dynamic and static disorder of transition energies. Two-dimensional (2D) electron diffraction of LH1 reveals a larger ring very similar to B850 with 32 BChl, possibly segregated into pairs [10]. However, the resolution of this study is not sufficient to specify precisely microscopic parameters, such as the distances between pigments and the orientation of the transition dipole moments. Despite the importance of the LH1 complexes in the sequence of energy transfer from the LH2 to the reaction center, less extensive efforts have been made on the study of the LH1 complexes because of the lack of structural information.

114   M. Yang et al. / Journal of Photochemistry and Photobiology A: Chemistry 142 (2001) 107–119

Recently, Hu and Schulten suggested a model for the structure of LH1 of *Rb. sphaeroids* as a hexadecamer of αβ-heterodimers [66]. The characteristic structure of the basic unit of the complex is the same as in the B850 band of the LH2 complex. The model structure shows that the LH1 complex contains a ring of 32 BChls (18 BChls in the B850 band of the LH2) referred to as B875 BChls according to their main absorption band. The Mg–Mg distance between neighboring B875 BChls is 9.2 Å within the αβ-heterodimer and 9.3 Å between neighboring αβ-heterodimers which are surprisingly similar to those in the LH2 complex. Since the number of heterodimers in the LH1 and LH2 complexes differ by a factor of 2, the overall size and angle between the monomeric BChls must be different. Even though this difference could yield different values of the Coulombic coupling between the BChls in the LH1 compared to those of LH2, the similar intermolecular distances in the two complexes leads us to assume that the Coulombic coupling strength should be in the same order of magnitude. We will assume the circular structure suggested by Hu and Schulten and construct model Hamiltonians for LH1 and, for the purpose of comparison, the B850 band of LH2 and calculate the peak shift and absorption spectrum of the two complexes.

From the well-resolved structural information, the Coulombic coupling strengths in the LH2 complexes have been studied experimentally [67–72] and theoretically [59,60,73–78] to give a range of 200–800 cm^{-1}. With those relatively well established values of the Coulombic coupling, various experimental techniques have been used to determine the dynamic and static disorder [26,31,33]. Among those, we will take the parameter values which have been employed by Scholes and Fleming [79]. They determined the intra- and inter-dimer nearest neighbor Coulombic coupling strength as 320 and 255 cm^{-1}, respectively. The monomeric transition energies of two different types of BChl molecules within a dimer were suggested to be different by 530 cm^{-1}. By introducing a line broadening function (obtained by an analysis of 3PEPS data [31] with a simple two-level model), they obtained absorption and CD spectra in excellent accord with the experimental data. Since the CD spectrum is very sensitive to the variation of the parameters mentioned above [80], we assume the values they determined are quite reliable.

Fig. 8a shows the experimental peak shift behavior of the LH2 of *Rb. sphaeroids* at room temperature measured by Jimenez et al. [31]. The pulse width and center wavelength are 35 fs and 850 nm, respectively. By use of our model described in Section 4, we simulated the experimental data in Fig. 8a. Although one can see minor oscillatory features in the peak shift data which presumably arise from coherent vibrational motion, we ignore this effect in the present paper. The effect of nuclear motion is described only by one Gaussian component in the transition frequency correlation function, $M(t)$, which represents fast fluctuations of the chromophore and protein molecules: $M(t) = e^{-(t/\tau_g)^2}$ with $\tau_g = 150$ fs. The reorganization energy for BChl-a monomer in

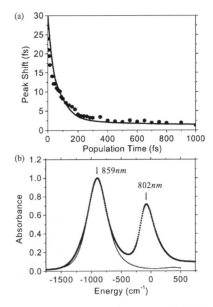

Fig. 8. Peak shift data (a) and absorption spectrum (b) of LH2 from *Rb. sphaeroids* (solid dots). The experimental data are from [31], the solid lines calculated using the theory outlined in the text.

the light harvesting complex is assumed to be 120 cm^{-1}. The effect of local static disorder is incorporated by a Gaussian distribution of transition energies of monomers ($\Delta_{loc} = 150$ cm^{-1}) and the global static disorder is assumed not to exist ($\Sigma_{glob} = 0$). The current model calculation fits the peak shift rather well. With this set of parameters, the average of delocalization length (defined by the inverse participation ratio [57,81–83]) weighted by oscillator strength is obtained as 4.

With the same parameter set, we simulated the absorption spectrum as shown in Fig. 8b. Exchange narrowing and lifetime broadening of the line shape (resulting from dephasing process due to population transfer) as well as pure dephasing process are properly incorporated within the present model. The simulated absorption spectrum almost completely agrees with the experimental data with a minor deviation on the red side of the B850 band. The disagreement on the blue side around −500 cm^{-1} mainly results from our neglect of high frequency modes in the simulation in order to save calculation time. Also, the blue side of the absorption spectrum should have a small contribution from the B800 band, which is not taken into account in our simulation. From the comparison of simulations and experiments of the peak shift and absorption spectrum, we believe that spectral information associated with the BChl molecules in the B850 band of the LH2 complex are reasonably well quantified and will be described in detail elsewhere [84].

M. Yang et al. / Journal of Photochemistry and Photobiology A: Chemistry 142 (2001) 107–119

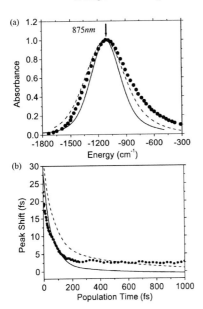

Fig. 9. Absorption spectrum (a) and peak shift data (b) for LH1 from *Rb. sphaeroids* (solid dots). The data are from [31], the solid lines from calculations using the LH2 parameter, the dashed line uses an inhomogeneous width (Δ_{loc}) of $250\,\mathrm{cm}^{-1}$ rather than $150\,\mathrm{cm}^{-1}$.

Now we extend our calculations to the LH1 complex. For our initial calculations we employ the same values of parameters as used for LH2. In this case, the only difference between LH1 and LH2 is in the number of monomers included in each complex. Fig. 9a and b show the resulting peak shift and absorption spectrum, respectively, along with the experimental data measured by Jimenez et al. [31]. Neither experiment could be well reproduced with the LH2 parameter set.

The major differences between the simulation and the experiment lie in the long time value of the peak shift and line width of the absorption spectrum. The simulated peak shift for LH1 at long time is smaller than that for LH2. There are two possible origins for this effect and to obtain a physical picture, we base our discussion on previous results for an isolated two-level system. The delocalization scale of the exciton states in LH complexes is greater than one monomer and, thus, the contribution of transitions from one-exciton to two-exciton states will yield a somewhat different picture of the peak shift from that of an isolated two-level system. However, we believe such an approach will allow us to qualitatively understand the peak shift behavior of the exciton systems without relying on a complicated calculation. For a two-level system with moderate static disorder, Cho et al. [28] derived the long time value, $\tau^*(\infty)$, of the peak shift as:

$$\tau^*(\infty) \sim \frac{\Delta^2}{\Delta_{\mathrm{el-ph}}^3} \tag{13}$$

where Δ and $\Delta_{\mathrm{el-ph}}$ are the static disorder of the monomer and the electron–phonon coupling strength, respectively. Assuming the delocalized exciton states to be effective two-level systems, one could, in the absence of any exciton transfer process, approximately extend the applicability of Eq. (13) to the exciton system

$$\tau^*(\infty) \sim \frac{\Delta_{\mathrm{ex,in}}^2}{\Delta_{\mathrm{ex-ph}}^3} \tag{14}$$

where $\Delta_{\mathrm{ex,in}}$ and $\Delta_{\mathrm{ex-ph}}$ are the static disorder associated with the effective two-level exciton system and its coupling strength to the phonon. If the size of delocalization length is smaller than the size of whole aggregate, the former is approximated by

$$\Delta_{\mathrm{ex,in}}^2 \approx \Sigma^2 + \frac{\sigma^2}{N_{\mathrm{del}}} \tag{15}$$

where Σ and σ are introduced in Section 2. Here N_{del} is the delocalization length of the exciton state. The first term in Eq. (15) describes the fluctuations of the mean energy of the aggregate and the second term is the exchange-narrowed intra-aggregate disorder of the exciton state. The exciton–phonon coupling strength is also exchange-narrowed and thus, if we neglect any exciton transfer process, we get $\Delta_{\mathrm{ex-ph}} = \Delta_{\mathrm{el-ph}}/\sqrt{N_{\mathrm{del}}}$ and the long-time peak shift is given by

$$\tau^*(\infty) \sim \frac{\Sigma^2 + \sigma^2/N_{\mathrm{del}}}{\Delta_{\mathrm{el-ph}}^3/N_{\mathrm{del}}^{3/2}} \tag{16}$$

This is an approximate expression for long time value of peak shift of the exciton system with a delocalization length N_{del} in the absence of any exciton transfer process. The exciton transfer can be approximately incorporated into Eq. (16) by using our previous studies on the peak shift of energy transfer systems [40,45]. Using these results, we infer that the peak shift in the presence of exciton transfer is given by

$$\tau^*(\infty) \sim \frac{\Sigma^2 + (\sigma^2/N_{\mathrm{del}})f_i}{\Delta_{\mathrm{el-ph}}^3/N_{\mathrm{del}}^{3/2}} \tag{17}$$

where f_i is the fraction of the exciton population remaining on the initial states out of the total population of optically active exciton states. Physically the factor f_i describes how many optically active exciton states have a rephasing capability (associated with the intra-aggregate disorder) to give a finite peak shift. If the excitons stay in their initial states forever, $f_i = 1$ and if all the excitons have moved to different states, $f_i = 0$. In Eq. (17), we did not include a change in the effective exciton–phonon coupling strength due to lifetime broadening based on the assumption that the latter is smaller than the former. Inserting Eqs. (3), (5) and (6) into

Eq. (17), we finally get

$$\tau^*(\infty) \sim \frac{N_{del}^{3/2}}{\Delta_{el-ph}^3} \Sigma_{glob}^2 + \sqrt{N_{del}} \frac{\Delta_{loc}^2}{\Delta_{el-ph}^3} \left[\frac{N_{del}}{N} + \frac{N-1}{N} f_i \right]$$

$$\sim \frac{N_{del}^{3/2}}{\Delta_{el-ph}^3} \Sigma_{glob}^2 + \sqrt{N_{del}} \frac{\Delta_{loc}^2}{\Delta_{el-ph}^3} \left[\frac{N_{del}}{N} + f_i \right]$$

when $N \gg 1$ (18)

The accuracy of this expression needs to be checked by comparison with the exact calculation discussed in Section 4 in future work. Looking at Eq. (18), we can see why the simulated 3PEPS data in the LH1 and LH2 have different offsets. Since we have used the same set of parameters, the delocalization length should be very similar in the calculations for the two complexes. Then only the differences come from Σ_{glob}, the number of monomers N and the factor f_i in Eq. (18). The N^{-1}-dependence on the second term comes from the effect of inter-aggregate fluctuation with a width Σ_{loc}. As N increases (LH1), the value of Σ_{loc} become reduced as shown by Eq. (5) to lower the rephasing capability associated with it. The factor f_i will also be inversely proportional to the value of N. As the number of monomers becomes larger, the number of excitons which have optical activity will also increase unless the exciton size change. Thus, the fraction of excitons staying on the initial state out of the full set of optically active states will be reduced as the number of monomers increases. As a result, we will get a reduced peak shift at long times. These mechanisms will not affect the absorption spectrum significantly (as shown in the simulation) since the total magnitude of disorder is not different. Since both mechanisms arise only from the difference in the number of monomers, they are related to each other.

Now, based on the intuition obtained from Eq. (18), we will try to fit the experimental data using exact simulations. From our studies on isolated two-level systems and Eq. (18), we know that the long time value of the peak shift can be increased by increasing the magnitude of the static disorder. Of course, this also makes the bandwidth of the absorption spectrum wider. Thus, in order to improve the agreement of simulation and the experimental data, one could try to employ a larger value of static disorder. The dashed lines in Fig. 9 represent the case of $\Delta_{loc} = 250\,cm^{-1}$. As expected, the bandwidth of the absorption spectrum is increased and the overall peak shift is moved up. However, we can clearly see that a slowly decaying component in the simulated peak shift data became more apparent. This decay results from energy hopping between exciton units (delocalized over a few monomers). As we increase the static disorder, exciton states become more localized and the energy gap between adjacent exciton states becomes larger. As a result, we see energy hopping process occurring on a longer time scale. In contrast to the simulated data, the experimental data show almost no decrease in the peak shift after about

400 fs. This means equilibration of the excitation within a circular ring is almost complete on this timescale. Increase of electron–phonon coupling strength will make the rate of exciton equilibration faster but also bring down the overall peak shift, which does not improve the fit of the simulated data to the experimental data. Since the bandwidth of the simulated absorption spectrum for $\Delta_{loc} = 250\,cm^{-1}$ is already broad enough, it would make no sense to increase the static disorder and electron–phonon coupling strength any more. The second possibility suggested by Eq. (18) to increase the long time value of peak shift is to reduce the electron–phonon coupling strength. However, this will significantly increase the initial peak shift and give a much larger disagreement with the experimental initial peak shift.

One solution to this problem is to introduce a non-zero value of Σ_{glob}. Since we perform a Monte Carlo simulation based on a Gaussian distribution of monomer energies, the inter-aggregate disorder originating from the statistical fluctuation is automatically incorporated in our previous simulation. Therefore, extra degrees of inter-aggregate disorder resulting from physical heterogeneity may be necessary in order to get a reasonable fit. In Fig. 10, we present the absorption spectrum and the 3PEPS simulated with the same parameters as in Fig. 9 except for introducing a Σ_{glob} of $50\,cm^{-1}$ and the two vibrational modes previously captured in [31]. Even though a little fine-tuning in the simulated

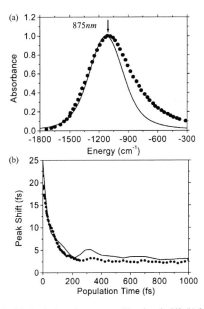

Fig. 10. Calculated absorption spectrum (a) and peak shift (b) for LH1 (lines) compared with experimental data (solid dots) for a model including global heterogeneity between complexes ($\Sigma_{glob} = 50\,cm^{-1}$). The two vibrational modes (with the frequencies 110 and 190 cm^{-1}) captured in [31] are included. The other parameters remain the same as in Fig. 9.

3PEPS is required, we could get much improved fits of both the absorption and the 3PEPS data.

Another insight from Eq. (18) is that the long time value of the peak shift is proportional to the delocalization length. Thus, in order to increase the simulated peak shift in Fig. 9, we could try to increase the delocalization length. Increase of the electronic coupling strength only, will make the absorption spectrum too narrow due to strong exchange narrowing. Instead, therefore, we increase the static disorder, Δ_{loc}, as well as the electronic coupling strength. Fig. 11 shows a simulation with three parameter sets that give $N_{\text{del}} = 8$ ($\Delta_{\text{loc}} = 105\,\text{cm}^{-1}$, $J = 400\,\text{cm}^{-1}$), ($\Delta_{\text{loc}} = 185\,\text{cm}^{-1}$, $J = 600\,\text{cm}^{-1}$), ($\Delta_{\text{loc}} = 225\,\text{cm}^{-1}$, $J = 700\,\text{cm}^{-1}$), while the reorganization energy and Σ_{glob} are fixed to $240\,\text{cm}^{-1}$ and 0, respectively. The last set of parameters gives the best fit to the experimental data.

If we accept the increased coupling model, the calculations shown in Fig. 11 suggest that the excitonic states are somewhat more delocalized in LH1 than those of LH2. The

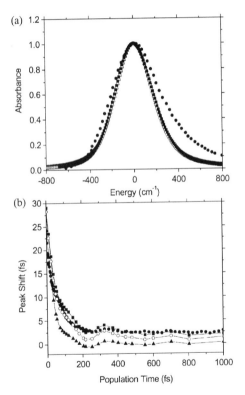

Fig. 11. Effect of increasing the electronic coupling (J) and the disorder (Δ_{loc}) on the calculated LH1 absorption spectrum (a) and peak shift (b). Again the experimental data are shown with solid dots. $\Delta_{\text{loc}} = 105\,\text{cm}^{-1}$, $J = 400\,\text{cm}^{-1}$ (solid triangle), $\Delta_{\text{loc}} = 185\,\text{cm}^{-1}$, $J = 600\,\text{cm}^{-1}$ (open circle), $\Delta_{\text{loc}} = 225\,\text{cm}^{-1}$, $J = 700\,\text{cm}^{-1}$ (solid square). Σ_{glob} is fixed at zero and the reorganization energy is $240\,\text{cm}^{-1}$ in all cases.

superradiance studies of Monshouwer et al. [83] find emitting dipole strengths (relative to bacteriochlorophyll a) of 2.8 for LH2 and 3.8 for LH1. To avoid confusion, we note that the dipole strengths refer to the equilibrated (long time) state, whereas our N_{del} refers to the initially prepared exciton states. The two pairs of number cannot be compared directly, but their ratios are similar for LH1:LH2, and we would clearly expect the delocalization to be less in the equilibrium state, thus, the two measures are consistent. However, to definitively decide between the two possible interpretations of our LH1 data: increased coupling, increased Δ_{loc} versus introduction of Σ_{glob} requires more modeling-for example of the CD spectrum.

In either case, the dynamical behavior in both LH1 and LH2 (B850) is clearly rather different from that in B800 of LH2. In our model, the relaxation process occurring on the 100 fs timescale corresponds to relaxation between partially delocalized exciton states, with a timescale controlled by the exciton–phonon coupling strength. However, investigation of the wavefunctions of the exciton states of individual complexes, in the context of the model outlined in Section 4, suggests a relatively simple picture for the spatial motion of the excitation. The relaxation within the exciton manifold of a given complex is controlled by the energy gap between the exciton states and the spatial overlap of their wavefunctions. On the average, using the parameters of Figs. 8 and 11, the individual exciton states are initially delocalized over four and eight monomers in LH2 and LH1, respectively. The overlap requirement then implies a picture of the excitation moving between groups of monomers, since to a good approximation in LH2, and a reasonable approximation in LH1, the non-zero amplitudes of the wavefunctions are concentrated on neighboring molecules. In other words, a reasonable crude picture of the spatial motion of excitation in LH2 (B850) and LH1 (as a result of exciton relaxation) is of hopping between groups of molecules, rather than hopping between monomers as in B800 of LH2.

5. Concluding remarks

The light harvesting antenna of purple bacteria is based on an energy funnel beginning with B800 and proceeding via B850 and B875 to the reaction center. There are two obvious ways to construct such a funnel: (1) select different chemical species that absorb at the required wavelengths or (2) use exciton (and solvation) interactions to progressively red shift the spectra of the same chemical species. In the purple bacteria, nature has adopted the second strategy and it is tempting to speculate on whether there is an intrinsic advantage to this solution (ignoring completely any biosynthetic implications of the first solution). To effectively and efficiently transfer energy between many identical molecules, it is important to avoid losing too much energy per step. This, in turn, implies that both homogeneous and inhomogeneous contributions to the lineshape should be relatively small. In

118   *M. Yang et al. / Journal of Photochemistry and Photobiology A: Chemistry 142 (2001) 107–119*

the excitonic system, the phenomenon of exchange narrowing reduces the intrinsic values of both broadening contributions. Relatively weak electron–phonon coupling can induce rapid spatial motion of the exciton, without a large loss in energy (and consequent decrease in Franck–Condon factor) because of the exchange narrowing. Similarly, the influence of static variations of site energies, which are presumably inevitable in natural systems, is substantially decreased by the exchange narrowing. Both effects are somewhat reduced by the lifetime broadening effect of the exciton relaxation, but overall there is a significant reduction in linewidth in both B850 and B875. In contrast, in B800, up to a reasonable level, disorder is not a major problem since the transfer to B850 is significantly downhill.

With our present knowledge, it is not possible to say definitively whether the delocalization is simply a consequence of the excitonic red shift or an important design feature. Such questions are likely to provoke spectroscopic studies of antenna systems for many years to come.

Acknowledgements

G.R.F. thanks Lord Porter for support and encouragement over the past 30 years, and for inspiring our studies of photosynthetic energy transfer. This work was supported by the Director, Office of Science, Office of Basic Energy Sciences, Chemical Sciences Division, of the US Department of Energy under Contract no. DE-AC03-76SF0098.

References

[1] G.S. Beddard, G. Porter, C.J. Tredwell, Nature 258 (1975) 166.
[2] M. Seibert, R.R. Alfano, S.L. Shapiro, Biochim. Biophys. Acta 292 (1973) 493.
[3] M. Seibert, R.R. Alfano, Biophys. J. 14 (1974) 269.
[4] S.L. Shapiro, V.H. Kollman, A.J. Campillo, FEBS Lett. 54 (1975) 358.
[5] G. Porter, C.J. Tredwell, G.F.W. Searle, J. Barber, Biochim. Biophys. Acta 501 (1978) 232.
[6] G.S. Beddard, G. Porter, Nature 260 (1976) 366.
[7] G. McDermott, S.M. Prince, A.A. Freer, A.M. Hawthornthwaite-Lawless, M.Z. Papiz, R.J. Codgell, N.W. Isaacs, Nature 374 (1995) 517.
[8] A.A. Freer, S. Prince, K. Sauer, M.Z. Papiz, A.M. Hawthornthwaite-Lawless, G. McDermott, R.J. Cogdell, N.W. Isaacs, Structure 4 (1996) 449.
[9] J. Koepke, X. Hu, C. Muenke, K. Schulten, H. Michel, Structure 4 (1996) 581.
[10] S. Karrasch, P.A. Bullough, R. Ghosh, EMBO J. 14 (1995) 631.
[11] D.E. Tronrud, M.F. Schmidt, B.E. Matthews, J. Mol. Biol. 188 (1986) 443.
[12] Y.F. Li, W. Zhou, R.E. Blankenship, J.P. Allen, J. Mol. Biol. 271 (1997) 456.
[13] N. Krauss, W.-D. Schubert, O. Klukas, P. Fromme, H.T. Witt, W. Saenger, Nature Struct. Biol. 3 (1996) 965.
[14] J. Deisenhofer, O. Epp, K. Miki, R. Huber, H. Michel, Nature 318 (1985) 618.
[15] H. Michel, O. Epp, J. Deisenhofer, EMBO J. 5 (1986) 2445.
[16] J.P. Allen, G. Feher, T.O. Yeates, H. Komiya, D.C. Rees, Proc. Natl. Acad. Sci. U.S.A. 84 (1987) 5730.
[17] J.P. Allen, G. Feher, T.O. Yeates, H. Komiya, D.C. Rees, Proc. Natl. Acad. Sci. U.S.A. 84 (1987) 6162.
[18] A. Adronov, S.L. Gilat, J.M.J. Fréchet, K. Ohta, F.V.R. Neuwahl, G.R. Fleming, J. Am. Chem. Soc. 122 (2000) 1175.
[19] H.A. Frank, R.J. Cogdell, in: A. Young, G. Britton (Eds.), Carotenoids in Photosynthesis, Chapman & Hall, London, 1993, p. 252.
[20] H. Nagae, T. Kakitani, T. Katoh, M. Mimuro, J. Chem. Phys. 98 (1993) 8012.
[21] A. Damjanovic, T. Ritz, K. Schulten, Phys. Rev. E 59 (1999) 3293.
[22] J.-P. Zhang, R. Fujii, P. Qian, T. Inaba, T. Mizoguchi, Y. Koyama, K. Onaka, Y. Watanabe, H. Nagae, J. Phys. Chem. B 104 (2000) 3683.
[23] D.L. Dexter, J. Chem. Phys. 21 (1953) 836.
[24] P.J. Walla, P.A. Linden, C.-P. Hsu, G.D. Scholes, G.R. Fleming, Proc. Natl. Acad. Sci. U.S.A. 97 (2000) 10808.
[25] C.-P. Hsu, P.J. Walla, G.R. Fleming, M. Head-Gordon, J. Phys. Chem. B, in press.
[26] T. Joo, Y.W. Jia, J.-Y. Yu, D.M. Jonas, G.R. Fleming, J. Phys. Chem. 100 (1996) 2399.
[27] T. Joo, Y.W. Jia, J.-Y. Yu, M.J. Lang, G.R. Fleming, J. Chem. Phys. 104 (1996) 6089.
[28] M. Cho, J.-Y. Yu, T. Joo, Y. Nagasawa, S.A. Passino, G.R. Fleming, J. Phys. Chem. 100 (1996) 11944.
[29] S.A. Passino, Y. Nagasawa, T. Joo, G.R. Fleming, J. Phys. Chem. A 101 (1997) 725.
[30] Y. Nagasawa, S.A. Passino, T. Joo, G.R. Fleming, J. Chem. Phys. 106 (1997) 4840.
[31] R. Jimenez, F. van Mourik, J.-Y. Yu, G.R. Fleming, J. Phys. Chem. B 101 (1997) 7350.
[32] S.A. Passino, Y. Nagasawa, G.R. Fleming, J. Chem. Phys. 107 (1997) 6094.
[33] J.-Y. Yu, Y. Nagasawa, R. van Grondelle, G.R. Fleming, Chem. Phys. Lett. 280 (1997) 404.
[34] Y. Nagasawa, J.-Y. Yu, M. Cho, G.R. Fleming, Faraday Discussions 108 (1997) 23.
[35] M.L. Groot, J.-Y. Yu, R. Agarwal, J.R. Norris, G.R. Fleming, J. Phys. Chem. B 102 (1998) 5923.
[36] Y. Nagasawa, Y.-Y. Yu, G.R. Fleming, J. Chem. Phys. 109 (1998) 6175.
[37] M. Yang, G.R. Fleming, J. Chem. Phys. 110 (1999) 2983.
[38] M.J. Lang, X.J. Jordanides, X. Song, G.R. Fleming, J. Chem. Phys. 110 (1999) 5884.
[39] M. Yang, K. Ohta, G.R. Fleming, J. Chem. Phys. 110 (1999) 10243.
[40] M. Yang, G.R. Fleming, J. Chem. Phys. 111 (1999) 27.
[41] X.J. Jordanides, M.J. Lang, X. Song, G.R. Fleming, J. Phys. Chem. B 103 (1999) 7995.
[42] D.S. Larsen, K. Ohta, G.R. Fleming, J. Chem. Phys. 111 (1999) 8970.
[43] Q.-H. Xu, G.D. Scholes, M. Yang, G.R. Fleming, J. Phys. Chem. A 103 (1999) 10348.
[44] R. Agarwal, B.P. Krueger, G.D. Scholes, M. Yang, J. Yom, L. Mets, G.R. Fleming, J. Phys. Chem. B 104 (2000) 2908.
[45] M. Yang, G.R. Fleming, J. Chem. Phys. 113 (2000) 2823.
[46] D.S. Larsen, K. Ohta, Q.-H. Xu, G.R. Fleming, J. Chem. Phys., submitted for publication.
[47] K. Ohta, D.S. Larsen, M. Yang, G.R. Fleming, J. Chem. Phys., submitted for publication.
[48] W.P. de Boeij, M.S. Pshenichnikov, D.A. Wiersma, J. Phys. Chem. 100 (1996) 11806.
[49] W.P. de Boeij, M.S. Pshenichnikov, D.A. Wiersma, Ann. Rev. Phys. Chem. 49 (1998) 99.
[50] R. Agarwal, M. Yang, Q.-H. Xu, G.R. Fleming, J. Phys. Chem. B 105 (2001) 1887.
[51] E. Gudowska-Nowak, M.D. Newton, J. Fajer, J. Phys. Chem. 94 (1990) 5795.
[52] A. Freiberg, K. Timpmann, R. Ruus, N.W. Woodbury, J. Phys. Chem. B 103 (1999) 10032.

M. Yang et al. / Journal of Photochemistry and Photobiology A: Chemistry 142 (2001) 107–119   119

[53] A.M. van Oijen, M. Ketelaars, J. Köller, T.J. Aartsma, J. Schmidt, Biophys. J. 78 (2000) 1570.

[54] M.V. Mostovoy, J. Knoester, J. Phys. Chem. B 104 (2000) 12355.

[55] S.E. Dempster, S. Jang, R. Silbey, J. Chem. Phys., submitted for publication.

[56] J.M. Jean, C.-K. Chan, G.R. Fleming, Israel J. Chem. 28 (1988) 169.

[57] V. Sundström, T. Pullerits, R. van Grondelle, J. Phys. Chem. B 103 (1999) 2327.

[58] Y.-Z. Ma, R.J. Cogdell, T. Gillbro, J. Phys. Chem. B 102 (1998) 881.

[59] B.P. Krueger, G.D. Scholes, G.R. Fleming, J. Phys. Chem. B 102 (1998) 5378.

[60] G.D. Scholes, I.R. Gould, R.J. Cogdell, G.R. Fleming, J. Phys. Chem. B 103 (1999) 2543.

[61] E.W. Knapp, Chem. Phys. 85 (1984) 73.

[62] J. Knoester, J. Chem. Phys. 99 (1993) 8466.

[63] M. Wubs, J. Knoester, Chem. Phys. Lett. 284 (1998) 63.

[64] T. Meier, V. Chernyak, S. Mukamel, J. Chem. Phys. 107 (1997) 8759.

[65] W.M. Zhang, T. Meier, V. Chernyak, S. Mukamel, J. Chem. Phys. 108 (1998) 7763.

[66] X. Hu, K. Schulten, Biophys. J. 75 (1998) 683.

[67] T. Pullerits, V. Sundström, Acc. Chem. Res. 29 (1996) 381.

[68] T. Pullerits, M. Chachisvilis, V. Sundström, J. Phys. Chem. 100 (1996) 10787.

[69] R. van Grondelle, J.P. Dekker, T. Gillbro, V. Sundström, Biochim. Biophys. Acta 1187 (1994) 1.

[70] A. Freiberg, J.A. Jackson, S. Lin, N.W. Woodbury, J. Phys. Chem. A 102 (1998) 4372.

[71] M.H.C. Koolhaas, G. van der Zwan, R.N. Frese, R. van Grondelle, J. Phys. Chem. B 101 (1997) 7262.

[72] M.H.C. Koolhaas, G. van der Zwan, R. van Grondelle, J. Phys. Chem. B 104 (2000) 4489.

[73] K. Sauer, R.J. Cogdell, S.M. Prince, A. Freer, N.W. Issacs, H. Scheer, Photochem. Photobiol. 64 (1996) 564.

[74] X. Hu, T. Ritz, A. Demjanovic, K. Schulten, J. Phys. Chem. B 101 (1997) 3854.

[75] R.G. Alden, E. Johnson, V. Nagarajan, W.W. Parson, C.J. Law, R.G. Cogdell, J. Phys. Chem. B 101 (1997) 4667.

[76] J. Linnanto, J.E.I. Korppi-Tommola, V.M. Helenius, J. Phys. Chem. B 103 (1999) 8739.

[77] S. Tretiak, C. Middleton, V. Chernyak, S. Mukamel, J. Phys. Chem. B 104 (2000) 4519.

[78] S. Tretiak, C. Middleton, V. Chernyak, S. Mukamel, J. Phys. Chem. B 104 (2000) 9540.

[79] G. Scholes, G.R. Fleming, J. Phys. Chem. B 104 (2000) 1854.

[80] M.H.C. Koolhaas, G. van der Zwan, R.N. Frese, R. van Grondelle, J. Phys. Chem. B 101 (1997) 7262.

[81] H. Fidder, J. Knoester, D.A. Wiersma, J. Chem. Phys. 95 (1991) 7880.

[82] R. Jimenez, S.N. Dikshit, S.E. Bradforth, G.R. Fleming, J. Phys. Chem. 100 (1996) 6825.

[83] R. Monshouwer, M. Abrahamsson, F. van Mourik, R. van Grondelle, J. Phys. Chem. B 101 (1997) 7241.

[84] M. Yang, G.R. Fleming, in preparation.

Contribution from

BRIAN ARTHUR THRUSH

University of Cambridge
UK

B orn in Hampstead Garden Suburb, London on 23 July 1928. Only child of the late Arthur Albert Thrush (a publisher and book writer) and the late Dorothy Charlotte Thrush (née Money). In 1958, married Rosemary Catherine Terry, daughter of the late George and Gertrude Terry of Ottawa, Canada, and has one son and one daughter.

Educated at the Haberdashers' Aske's Hampstead School (now at Elstree) from 1939 to 1946 and from which he won a Major Open Scholarship to Emmanuel College, Cambridge. Achieved First Class Honours in both parts of the Natural Sciences Tripos, and from 1949 to 1950 undertook research on ionisation in flames with Morris Sugden as a Bachelor Scholar of Emmanuel College. Took his Cambridge BA in 1949, MA and PhD in 1953, and ScD in 1965.

Appointed an Assistant in Research in Physical Chemistry in 1950 and constructed the first flash photolysis apparatus to use electronic timing. Became a University Demonstrator in 1953, an Assistant Director of Research in 1959 and a University Lecturer in 1964. Appointed to a Readership in 1969 and made an *ad hominem* Professor at Cambridge in 1978. Was Head of Department of Physical Chemistry from 1986 to 1988, and first Head of the (united) Department of Chemistry from 1988 to 1993. Became an Emeritus Professor in 1995. Was elected to a Fellowship of Emmanuel College in 1960, being a Tutor from 1963 to 1967, Vice-Master from 1986 to 1990, and Acting Master from 1986 to 1987.

After spending a sabbatical year as a Consultant Physicist at the US National Bureau of Standards in Washington DC in 1957–1958, his research interests moved from flash photolysis to elementary reactions in discharge flow systems, particularly chemiluminescence (the subject of his 1965 Tilden Lecture), spectroscopy from the far infrared to vacuum ultraviolet (Rank Prize for Opto-electronics 1992) and reactions of atmospheric importance (M. Polanyi Medal 1980).

Elected a Fellow of the Royal Society in 1976 and served on its Council from 1989 to 1991. Was a Visiting Professor of the Chinese Academy of Sciences in the 1980s. Served on the US National Academy of Sciences Panel on Atmospheric Chemistry from 1975 to 1980, the Lawes Agricultural Trust Committee from 1979 to 1989, the Council of NERC from 1985 to 1990. Was President of the Chemistry Section of BAAS in 1986 and was elected to the Academia Europaea in 1990, serving on its Council from 1992 to 1998.

Brian A. Thrush on
"Flash Photolysis and Spectroscopy: A New Method for the Study of Free Radical Reactions"
G. Porter
Proc. Roy. Soc. A **200**, 284 (1950)

The Department of Physical Chemistry at Cambridge which George Poter entered as a Research Student after demobilisation in 1945 strongly reflected Professor R. G. W. Norrish's interest in gas phase reactions, notably photochemistry and chain reactions. Wartime studies had included the suppression of gun flash and measurement of ionisation in flames (where a rocket's exhaust might affect its radio control).

At the end of the war, it was clear that progress in reaction kinetics and photochemistry was limited by the lack of techniques for observing and studying the free radicals and other short-lived intermediates which product analysis and overall kinetic studies showed must be present in these systems. Wartime developments had led to the production of much specialised equipment which was now surplus to Government requirements. Initially, Norrish and Porter used a 7.5 kW mercury arc mounted in a searchlight to generate methylene radicals by the photolysis of ketene or diazomethane in a gas flow, but their attempts to detect CH_2 radicals by the removal of a tellurium mirror using the Paneth mirror technique had limited success.[1]

The high intensity flash tubes which were developed during the war for night-time aerial photography were orders of magnitude more intense than a mercury arc. Furthermore, they had the potential to record the absorption spectrum of transient products within their lifetime. George Porter, Morris Sugden and Tony Harding discussed the possibilities of this approach, and Porter decided to try it. The Royal Navy gave a large number of 1 and 2 μF condensers free and actually paid the Department £20 for returning the empty packing cases. The condensers were installed in a small basement room and connected in parallel yielding up to 2000 μF which could be charged to 4 kV and then discharged through a 1 m long quartz flash tube in about 2 ms. This and the reaction vessel were surrounded by a magnesium oxide coated reflector. The absorption spectra of the products were recorded using a much smaller flash of duration of 50 μs, the timing of which was determined by contacts on a rotating wheel.[2] Unfortunately the absorption spectra of important polyatomic species such as HCO, CH_2, CH_3, NH_2 and HNO lay outside the range where prism spectrographs provided reasonable resolution. These spectra were, however, obtained by flash photolysis and analysed by Herzberg and Ramsay in Ottawa using large grating spectrographs.[3] In addition to being used by Porter and Frank Wright to make a pioneering study of the kinetics of the ClO radical,[4] Porter used this original flash photolysis apparatus to study the kinetic behaviour of the HO radical in the hydrogen–oxygen reaction photosensitised by NO$_2$.[5]

The highlighted paper[6] begins with a brief description of a more compact flash photolysis apparatus using a much shorter photolysis flash and electronic timing to bring the time resolution down from 1 ms to 20 μs, an approach which lasted until overtaken by laser technology. It describes the use of this apparatus to observe the absorption spectra of OH, CH, C$_2$, CN and NH in acetylene/oxygen explosions photosensitised by NO$_2$. It raised the interesting question as to whether C$_2$ had a $^3\Pi$ or $^1\Sigma$ ground state, now settled in favour of the latter. It was also observed that sometimes the concentration of the radicals, notably CN, had changed greatly during the the duration of the spectroscopic flash, i.e. much less than 50 μs. Rather than attempting to reduce greatly the duration of the spectroscopic flash, it was decided to use photoelectic observation of the absorption spectra present. A xenon arc was the source and two photomultiplier cells were mounted behind adjustable slits in the focal surface of the Littrow spectrograph.[7] In addition to measuring the transient absorption by the free radicals present, these studies exhibited very sharp emission peaks of chemiluminescent origin. Scattered light from the initiating flash made it necessary to study systems where there was an induction period greater than 20 μs before ignition and it was difficult to ensure that ignition was simultaneous throughout the 50 cm reaction vessel. It was established in the paper reproduced here[8] that the sharp emission peak was due to the impact of a detonation wave on the window of the reaction vessel closest to the spectrograph slit.

As compared with studying combustion processes using flames,[9] flash photolysis provided long path absorption measurements which were easier to translate into concentrations than were emission studies. The latter could also be affected by chemiluminescence.The ability to detect trace species, particularly in the induction period, was used by Erhard and Norrish[10] to identify the reactive species present when anti-knock agents such as lead tetraethyl are added to explosions initiated photochemically.

However it was clear that flash photolysis had great potential for the study of elementary gas phase reactions in isolation. Combustion systems were too complex overall. This account began with the flash photolysis of NO$_2$, and this process on its own was shown by Lipscomb, Norrish and Thrush[11] to yield highly vibrationally excited oxygen molecules:

$$NO_2 + h\nu = NO + O$$
$$O + NO_2 = NO + O_2^{*}$$

A similar phenomenon was observed in the flash photolysis of ClO$_2$ and of O$_3$.

In historical terms, it should be noted that Porter left Cambridge in 1954 to become briefly Assistant Director of the British Rayon Research Association and then Professor of Physical Chemistry at Sheffield. With him went his interests in condensed phase systems, notably the triplet states of aromatic molecules and their free radicals. At Cambridge, Norrish and his colleagues continued to use flash photolysis to study gas phase systems, particularly those related to combustion and the formation of excited species.

References

1. R. G. W. Norrish and G. Porter, *Disc. Faraday Soc.* **2**, 97 (1947).
2. G. Porter, *Proc. Roy. Soc. A* **200**, 284 (1950).
3. G. Herzberg, *Disc. Faraday Soc.* **35**, 7 (1963).
4. G. Porter and F. J. Wright, *Disc. Faraday Soc.* **14**, 23 (1953).
5. R. G. W. Norrish and G. Porter, *Proc. Roy. Soc. A* **210**, 439 (1952).
6. R. G. W. Norrish, G. Porter and B. A. Thrush, *Proc. Roy. Soc. A* **216**, 165 (1953).
7. R. G. W. Norrish, G. Porter and B. A. Thrush, *Eighth Symposium on Combustion*, (Reinhold, New York), p. 651.
8. B. A. Thrush, *Proc. Roy. Soc. A* **233**, 147 (1955).
9. A. G. Gaydon and H. G. Wolfhard, *Flames* (Chapman and Hall, London, 1953).
10. K. H. L. Erhard and R. G. W. Norrish, Proc. Roy. Soc. A **234**, 179 (1956).
11. F. J. Lipscomb, R. G. W. Norrish and B. A. Thrush, *Proc. Roy. Soc. A* **233**, 455 (1956)

Reprinted from *Proc. Roy Soc.* A **233**, 147–151 (1956) with permission from The Royal Society

The homogeneity of explosions initiated by flash photolysis

By B. A. Thrush

Department of Physical Chemistry, University of Cambridge

(*Communicated by R. G. W. Norrish, F.R.S.—Received* 14 *July* 1955)

[Plate 4]

Photo-electric studies have been made of the light emitted from flash-initiated explosions of acetylene and oxygen photo-sensitized by nitrogen dioxide in a long quartz tube. These confirmed the presence of a very short-lived intense emission from the diatomic radicals which had previously been found by Norrish, Porter & Thrush (1953a). It is shown that, whilst the main part of the explosion is homogeneous, two detonation waves travel a short distance to the ends of the reaction vessel. The sharp peak is caused by the impact of one of these detonation waves on the vessel window nearer the spectrograph.

Introduction

The narrowness of the reaction zone and the steep temperature gradients make it very difficult to study the reactions occurring in flames (Gaydon & Wolfhard 1953), although some separation can be achieved by working at low pressure (Gaydon & Wolfhard 1950). A homogeneous system would appear to offer many advantages, as it would then be possible to study the kinetics of the reaction rather than its propagation through space. It is, therefore, necessary to produce a large number of reaction centres in a very short time if this is to be achieved. With adiabatic compression, mechanical difficulties make it almost impossible to obtain the very rapid pressure and temperature rise needed for a truly homogeneous reaction (Taylor, Taylor, Livengood, Russell & Leary 1950).

The use of a high intensity flash discharge to initiate an explosion photo-chemically provides a simple method of producing a relatively homogeneous explosion in a vessel which is sufficiently long to give good sensitivity for absorption spectroscopy. This type of apparatus has been described by Norrish, Porter & Thrush (1953b, 1955a), who investigated explosions of the lower hydrocarbons, particularly acetylene, with oxygen using nitrogen dioxide as photo-sensitizer. They found that the duration of the flash discharge which they were using to photograph the absorption spectrum was too great for the more rapid processes to be studied. Owing to this and to variations in the initial period before the appearance of the free radicals, these authors used photo-electric methods to follow the change of light emission or absorption of selected wavelength ranges with time (1953a, 1955b). They showed that the diatomic radicals, but not the continuous spectra, gave an intense emission with a duration of about 3 μs superimposed on a radiation curve which was mainly thermal in origin. It was shown that this sharp peak was due to a sudden excitation and not to a sharp change in radical concentration since it could not be reversed by a 4000° K source. It will be shown that this radiation is due to an inhomogeneity in the system.

[147]

148   B. A. Thrush

EXPERIMENTAL

The general form of the apparatus and techniques used have been described elsewhere (Norrish *et al.* 1953 *b*, 1955 *b*), and figure 1 is a simplified diagram of the modifications needed for these experiments. To check the homogeneity of the explosions, the light emitted in a narrow wavelength range has been observed by both looking lengthwise down the reaction vessel and by viewing a 5 mm length of the vessel perpendicular to its length. For this purpose a photomultiplier cell housing with collimating slits has been fitted on a sliding carrier vertically above the reaction vessel, which is viewed through a slit cut along the length of the reflector. As it was not practicable to mount a spectrograph in this position, an Ilford 603 filter was placed in front of the photocell, this transmits in the region

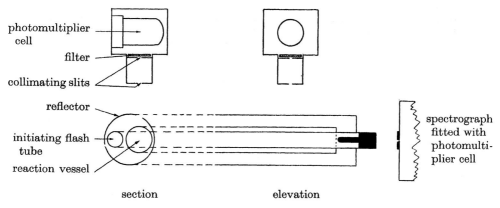

FIGURE 1. Simplified diagram of the apparatus.

of the C_2 Swan (0, 0) sequence, and the mixture studied was chosen so that it radiated these bands strongly without much continuous radiation. The second photocell, which was mounted on the spectrograph and observed the vessel lengthwise, was adjusted to receive the same spectral region as the first photocell. As some of the light from the initiating flash was reflected directly on to the first photomultiplier cell, the initial period of the explosion had to be at least one millisecond to avoid the long 'tail' of the initiating flash, also the bleeder current in the voltage divider which supplied the multiplying electrodes of the photocell was restricted to 100 μA to avoid possible damage by overloading.

The output of each photomultiplier cell was fed through a cathode follower and a specially constructed d.c. amplifier on to a Y plate of a Cossor 1049 double-beam oscillograph operated at 4 kV, a time resolution of 1 μs was obtained with negligible trace interaction. A linear time-base was built to give a scan which covered the full diameter of the cathode-ray tube; this was triggered by one amplifier output when the initiating flash was fired.

RESULTS AND DISCUSSION

Figure 2, plate 4, consists of a series of pairs of light emission-time curves for the same mixture (13 mm C_2H_2, 10 mm O_2, 2 mm NO_2); in each case the duration of the trace shown is 1 ms with time increasing to the right. In these experiments,

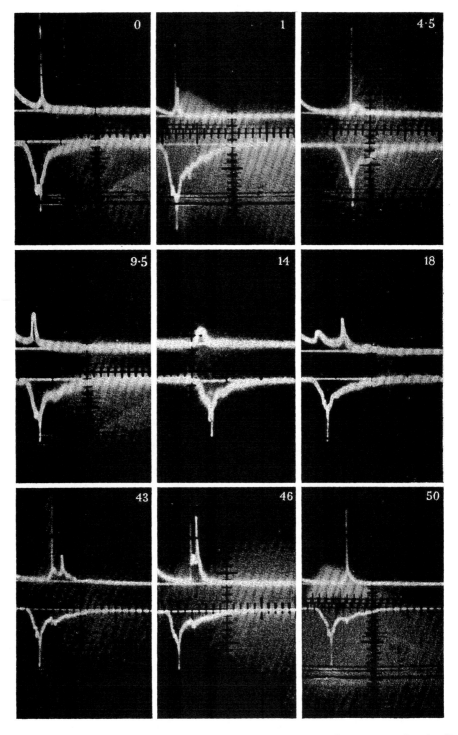

FIGURE 2. Light emission-time curves; upper traces transverse, lower traces longitudinal. The numbers shown on each photograph are the distances in cm of the second photocell from the window nearer the spectrograph.

The homogeneity of explosions initiated by flash photolysis   149

the efficiency of the 15 cm of the reflector farthest from the spectrograph was reduced slightly so that the effect of non-uniform illumination could be observed. The upper trace shows the transverse emission of a 5 mm length of the reaction vessel, the numbers given being the distances in centimetres of the centre of the zone observed from the end of the reaction vessel nearer to the spectrograph. The left-hand side of this trace sometimes shows the last part of the decay of the flash used to initiate the explosion, but this can easily be distinguished from the light emitted by the explosion itself. The lower trace (for which light emission increases downward) gives the emission observed looking lengthwise down the reaction vessel, this is the same in every case and provides a convenient time reference for the events observed on the other trace. For ease of presentation, an interpretation of the lower trace will be given first, and it will be shown that this is consistent with the light emission observed sideways.

The lower trace can be divided into three parts: a steady rise lasting about 100 μs, a sharp peak with a duration of about 5 μs and a slow decay in which it is possible to detect a much smaller second peak some 100 μs after the main one. During the first part of the rise, an approximately homogeneous explosion occurs in the central part of the vessel, the reaction is then propagated outward by two detonation waves which are established about 7 and 20 cm from the ends nearer to and farther from the spectrograph respectively. When the detonation wave which is travelling towards the spectrograph hits the window at that end, it produces the intense emission peak observed in the lower trace. The treatment of the reflector mentioned above results in a lower light intensity at the far end of the vessel, causing the gas there to react more slowly than in the centre. The second detonation wave is therefore established farther from that end, and reaches it after the first detonation wave has hit the window nearer the spectrograph. This second impact produces a small peak in the emission observed through the spectrograph about 100 μs after the main peak. The great relative reduction in the height of the second peak is due to absorption of the radiation in passing through 50 cm of products containing the same species as emit the radiation (C_2, CH and CN). The difference in the distance of the two impacts from the spectrograph can have very little effect on the apparent intensity owing to the small aperture of the instrument. After the passage of these detonation waves, the pressure and temperature of the reaction products will vary considerably along the vessel, and compression and rarefaction waves will travel along the tube and affect the radiation by heating or cooling the products.

Considering now the emission observed sideways, that at 9·5, 14 and 18 cm is characteristic of the homogeneous reaction which occurs in the centre of the vessel. Apart from a second emission peak at 18 cm, which is presumably due to reheating of the burnt gas by a shock wave, the radiation from this part of the vessel shows that a substantially homogeneous explosion occurs in this region during the rise of the emission observed lengthwise. The light observed sideways from the region nearer the spectrograph (4·5, 1 and 0 cm) shows intense pulses of short duration, although that observed at the end window (0 cm) lasts longer than the others. On close comparison of the pairs of traces (for instance, with a set square) it can be

B. A. Thrush

seen that, whilst the peak seen sideways at the end of the vessel coincides in time with the one in the lower trace, those observed in the upper trace at 1 and 4·5 cm occur at very small but increasing intervals before the peaks observed lengthwise. The pulses in the upper traces at 4·5 and 1 cm are due to the light received from a detonation front as it passes the observation points travelling towards the spectrograph; the impact of this wave on the end window of the reaction vessel produces the emission peak observed lengthwise on the lower trace in all cases.

As mentioned previously, comparison of the traces at 50 cm shows that the impact of the detonation wave produces only a very small peak in the light observed lengthwise. The emission from this second detonation wave as it passes the photocell slits appears as a narrow pulse in the upper traces at 43 and 46 cm; the sensitivity of the amplifier for this sideways emission trace has been increased in these two cases to show a second peak with a much slower decay; this is due to reheating of the reaction products by a shock wave which travels back from the end window after the impact of the detonation wave. Using the lower trace as reference, it is possible to compare the times at which these waves pass various points, and hence measure both the detonation velocity and that of the reflected shock wave; these measurements also confirm the directions of propagation stated.

The detonation wave travels at about 2×10^5 cm/s as is to be expected, but little information can be obtained from the accurate measurement of this velocity, since the wave is propagated through gas which is at a slightly earlier phase of the reaction. These experiments show clearly how easily detonation waves can be established in an explosive system even where there is a large and almost uniform concentration of reaction centres. A clear analogy exists between the appearance of detonation waves in these systems and 'knock' in internal combustion engines, as it has been shown (Miller 1947; Male 1949) that this phenomenon is caused by the onset of detonation ahead of the main flame front in highly compressed gas which is already reacting rapidly. This technique provides a relatively simple method of studying the transition from rapid burning to detonation under conditions similar to those encountered in engines, without the attendant complexities.

As these emission peaks are confined to the diatomic radical spectra in all mixtures which have been investigated (Norrish et al. 1955b), it is clear that these species are present in the detonation front, whilst those responsible for the continuous spectra are formed at a later stage. In the most acetylene-rich mixtures part of this continuous spectrum is due to solid carbon, but these systems and ones which are not quite rich enough to deposit carbon show also a distinctive continuum, the origin of which will be discussed elsewhere.

The author would like to express his gratitude to Professor R. G. W. Norrish, F.R.S., for many helpful discussions.

The homogeneity of explosions initiated by flash photolysis 151

REFERENCES

Gaydon, A. G. & Wolfhard, H. G. 1950 *Fuel*, **29**, 15.

Gaydon, A. G. & Wolfhard, H. G. 1953 *Flames*. London: Chapman and Hall.

Male, T. 1949 *Third symposium on combustion*, p. 721. Baltimore: Williams and Wilkins.

Miller, C. D. 1947 *S.A.E. Quart. Trans.* **1**, 98.

Norrish, R. G. W., Porter, G. & Thrush, B. A. 1953a *Nature, Lond.*, **172**, 71.

Norrish, R. G. W., Porter, G. & Thrush, B. A. 1953b *Proc. Roy. Soc.* A, **216**, 153.

Norrish, R. G. W., Porter, G. & Thrush, B. A. 1955a *Proc. Roy. Soc.* A, **227**, 423.

Norrish, R. G. W., Porter, G. & Thrush, B. A. 1955b *Fifth symposium on combustion*, p. 651. New York: Reinhold.

Taylor, C. F., Taylor, E. S., Livengood, J. C., Russell, W. A. & Leary, W. A. 1950 *S.A.E. Quart. Trans.* **4**, 232.

Contribution from

AHMED ZEWAIL

California Institute of Technology
USA

Born 26 February 1946 in Damanhur, the "City of Horus", 60 km from Alexandria, the only son in a family of three sisters and two loving parents. Married Dema in 1989, and has two daughters, Maha and Amani, and two sons, Nabeel and Hani.

Academic Positions

Director, Physical Biology Center for Ultrafast Science & Technology (2005–)
Director, NSF Laboratory for Molecular Sciences (LMS), Caltech (1996–)
Linus Pauling Professor of Chemistry and Professor of Physics, Caltech (1995–)
Linus Pauling Professor of Chemical Physics, Caltech (1990–1994)
Professor of Chemical Physics, Caltech (1982–1989)
Associate Professor of Chemical Physics, Caltech (1978–1982)
Assistant Professor of Chemical Physics, Caltech (1976–1978)
IBM Postdoctoral Fellow, University of California, Berkeley (1974–1976)
Predoctoral Research Fellow, University of Pennsylvania (1970–1974)
Teaching Assistant, University of Pennsylvania (1969–1970)
Instructor and Researcher, Alexandria University (1967–1969)
Undergraduate Trainee, Shell Corporation, Alexandria (1966)

Academic Degrees

Alexandria University, Egypt (1967): BS, First Class Honours
Alexandria University, Egypt (1969): MS
University of Pennsylvania, Philadelphia, USA (1974): PhD

Honours, Prizes

King Faisal International Prize in Science (1989)
First Linus Pauling Chair, Caltech (1990)
Wolf Prize in Chemistry (1993)
Robert A. Welch Award in Chemistry (1997)
Benjamin Franklin Medal, The Franklin Institute, USA (1998)
Egypt Postage Stamps, with Portrait (1998); "The Fourth Pyramid" (1999)
Nobel Prize in Chemistry (1999)
Order of the Grand Collar of the Nile, highest honour of Egypt, conferred by President Mubarak (1999)
Ahmed Zewail Prizes and Awards; American Chemical Society, University of Pennsylvania, AUC (2000–)

Also holds 20 Honorary professorships, 30 Honorary degrees from universities worldwide, and 29 honorary fellowships, including Foreign Membership of the Royal Society of London, Royal Swedish Academy of Sciences, Russian Academy of Sciences, and 50 named Medals and Awards.

Professional Activities

Member of Advisory and Editorial Boards; Editor of scientific journals (Chemical Physics Letters at present) and book series; Chairman and Member of Organizing Committees of national and international conferences; Member of Boards including the following:

World Scientific Advisory Board (1994–)
Max Planck Institute, Board of Advisors (1994–)
American University in Cairo, Board of Trustees (1999–)
Bibliotheca Alexandria (The Alexandria Library), Board of Trustees (2001–)
The Welch Foundation, Scientific Advisory Board (2002–)
Multilateral Initiative on Malaria, Patron (2003–)
Qatar Foundation, Board of Directors (2003–)
Chalmers University, Scientific Advisory Board, Sweden (2003–)
TIAA-CREF, Board of Trustees (2004–)
NTU, Advisory Board, Singapore (2005–)
École Normale Supérieure, Paris, France (2005–)

Publications and Presentations

Some 450 articles have been published in the fields of science (authored and co-authored with members of the research group), education, and world affairs. Some three hundred named, plenary, and keynote lectures have been given and include the following: Bernstein, Berson, Bodenstein, Celsius, Condon, Aimé Cotton, Coulson, Debye, Einstein, Eyring, Faraday, Franklin (Benjamin), Gandhi, Helmholtz, Hinshelwood, Karrer, Kirkwood, Kistiakowsky, Lawrence, London, Nobel, Novartis, Noyes, Onassis, Pauling, Perrin, Pimentel, Max Planck, Polanyi, Raman, Roberts, Röntgen, Schrödinger, U Thant (United Nations), Thomson (J. J.), Tolman, Watson, and Wilson.

Research, Public Education and World Affairs

Current research is devoted to dynamical chemistry and biology, with a focus on the physics of elementary processes in complex systems. In the Laboratory for Molecular Sciences (LMS) Center, collaborative multidisciplinary research has been established to address the role of complexity in the primary function of real systems including enzyme catalysis, protein-RNA transcription, electron transport in DNA, and the role of water in protein and DNA recognitions. At the Center for Physical Biology, the focus is on Ultrafast Science and Technology (UST) with techniques which include ultrafast diffraction, crystallography, and microscopy for the imaging of transient structures in space and time with atomic-scale resolution.

A significant effort is also devoted to giving public lectures to enhance awareness of the value of knowledge gained from fundamental research, and helping the population of developing countries through the promotion of science and technology for the betterment of society.

Ahmed Zewail giving an impromptu address at the Banquet of the International Conference on Photochemistry, August 1995. Edward Schlag, Frank Wilkinson, George and Stella Porter are identifiable at the table.

Ahmed Zewail on
"Flash Photolysis and Spectroscopy: A New Method for the Study of Free Radical Reactions"
G. Porter
Proc. Roy. Soc. A **200**, 284 (1950)

"Nanosecond Flash Photolysis"
G. Porter and M. R. Topp
Proc. Roy. Soc. London A **315**, 163–184 (1970)

"Time-resolved Fluorescence in the Picosecond Region"
G. Porter, E. S. Reid, and C. J. Tredwell
Chemical Physics Letters **29**, 469–472 (1974)

George Porter made significant contributions to the advancement of science, particularly in the field of flash photolysis. He also contributed to the popularisation of science through his lucid lectures and demonstrations at the Royal Institution and elsewhere. The importance of flash photolysis stems from the ability, through intense milli-to-microsecond flashes, to generate high concentrations of intermediates; e.g., free radicals, which live on this timescale and can be probed optically using another beam of light. The applications of flash photolysis became evident in many photochemical studies. At about the same time the "Relaxation Method" of Manfred Eigen was developed and for this contribution Eigen shared the 1967 Nobel Prize in Chemistry with R. G. W. Norrish and George Porter "for their studies of extremely fast chemical reactions, effected by disturbing the equilibrium by means of very short pulses of energy."

The origins of two-pulses optical "pump-probe" techniques go back to the 19th century. It was known then that electrical sparks and Kerr cell shutters can be used to record transient phenomena. Abraham and Lemoine in 1899,[1] in an ingenious experiment, demonstrated that the Kerr response of carbon disulphide was faster than ten nanoseconds [it is in fact about two picoseconds]. They used an electrical pulse that produced a spark and simultaneously activated a Kerr shutter. Light from the spark was collimated through a variable delay path, and through the Kerr cell, consisting of crossed polarisers, and a carbon disulphide cell. The rotation of the analyser indicated the presence of birefringence in the cell for short optical delays, but this disappeared for pathlengths greater than several meters, reflecting an overall optical/electrical response time of 2.5 ns. This simple but elegant experiment provides the genesis of the methodology for using the difference in light path to set the desired time delay — 1 foot difference corresponds to 1 ns delay. The approach was used to such effect by Porter as reported in the 1950 paper and with Topp in the 1970 paper. The use of the

Kerr cell to provide a shutter in combination with an optical delay was used by Porter in the 1974 paper.

The invention of lasers changed the landscape. Flash lamps did not provide the versatility and intensity of pulsed lasers and George and his group were among the first to use these pulsed lasers to improve the time resolution. But George was concerned about the limitation imposed by the "uncertainty principle" on future advances in the picosecond and especially the femtosecond time regime. What was not obvious at the time was the fundamental role of "coherence".[2-4] Even in isolated molecules, coherence can be induced among quantum states, in fact by exploiting the uncertainty principle. What is relevant is the energy uncertainty compared to the total energy of the system under consideration. These realisations led to the advances made in studies of ultrafast molecular phenomena at much shorter timescales, in 1980 with picosecond and in 1987 with femtosecond time resolution.[4] As importantly, it was possible to distinguish between "kinetics of ensembles" and "dynamics of single-molecule trajectories". Without clear understanding of the meaning of coherence, these conceptual limitations would have impeded, or perhaps delayed, the impact of ultrashort time resolution in chemistry and biology.

There is no doubt that George's efforts (with Norrish) in developing flash photolysis for spectroscopic and kinetic applications, especially on the millisecond to nanosecond scale, will be remembered as flash photolysis has already made it into the lexicon of knowledge.

References

1. H. Abraham and J. Lemoine, *C. R. Hebd. Seances Acad. Sci.* **129**, 206–208 (1899).
2. A. H. Zewail, *Nature (London)* **412**, 279 (2001)
3. D. Phillips, Ultrafast processes, in *100 Years of Physical Chemistry* (Black Bear Press, Cambridge, UK, 2003), p. 105.
4. A. H. Zewail, *Les Prix Nobel Article*, referenced in the accompanying paper which gives a summary of the concepts involved.

Used with permission from *J. Chem. Ed.* **78**(6), 737–751 (2001).

Nobel Prize Report

Freezing Atoms in Motion: Principles of Femtochemistry and Demonstration by Laser Stroboscopy

J. Spencer Baskin and Ahmed H. Zewail*
Arthur Amos Noyes Laboratory of Chemical Physics, California Institute of Technology, Pasadena CA 91125

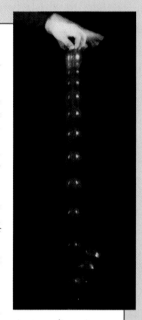

I. Introduction

The concept of the atom, proposed 24 centuries ago and rejected by Aristotle, was born on a purely philosophical basis, surely without anticipating some of the 20th century's most triumphant scientific discoveries. Atoms can now be seen, observed in motion, and manipulated (*1*). These discoveries have brought the microscopic world and its language into a new age, and they cover domains of length, time, and number. The *length* (spatial) resolution, down to the scale of atomic distance (angstrom, 1 Å = 10^{-8} cm), and the *time* resolution, down to the scale of atomic motion (femtosecond, 1 fs = 10^{-15} s), have been achieved. The trapping and spectroscopy of a single ion or electron and the trapping and cooling of neutral atoms have also been achieved. All these achievements have been recognized by the awarding of the Nobel Prize to scanning tunneling microscopy (1986), to single-electron and -ion trapping and spectroscopy (1989), to laser trapping and cooling (1997), and to laser femtochemistry (1999).

Before femtochemistry, the actual atomic motions involved in chemical reactions had never been observed in real time despite the rich history of chemistry over two millennia. Chemical bonds break, form, or geometrically change with awesome rapidity. Whether in isolation or in any other phase, this ultrafast transformation is a dynamic process involving displacements of electrons and atomic nuclei. The speed of atomic motion is ~1 km/s (10^5 cm/s), and hence, to record atomic-scale dynamics over a distance of an angstrom, the average time required is ~100 fs. The very act of atoms within molecules (reactants) rearranging themselves to form new molecules (products) is the focus of the field of femtochemistry. With femtosecond time resolution we can freeze atoms in motion and study the evolution of molecular structures as reactions unfold and pass through their transition states (configurations intermediate between reactants and products), giving a motion picture of the transformation.

Note: The demonstration reported in this paper is now on display at the Nobel Museum in Stockholm.

Femtochemistry shares some common features with stop-motion photography using cameras with high-speed shutters and with flash stroboscopy, in which short light flashes illuminate an object under study. Capturing a well-defined image or impression of an object at a given point in its motion by short-duration illumination is the methodology common to all of these techniques. The technical requirements in each case are (i) attaining adequate spatial resolution to define the object of study and (ii) synchronization to a well-defined time-axis, permitting reconstruction of the dynamic process undergone.

This contribution discusses the basic principles of femtochemistry and the analogy to stop-motion photography. We describe an educational exhibit that highlights the experimental approach by demonstrating the use of light pulses in the study of a rapidly moving object. Some theoretical concepts unique to the molecular world of femtochemistry are discussed and illustrated with applications to real molecules.

II. Principles of Stop-Motion Photography

A. High-Speed Shutters

As a starting point, we may consider the historic experiments in stop-motion photography performed more than a hundred years ago by photographer Eadweard Muybridge. Some of those experiments, carried out under the patronage of Leland Stanford at Stanford's Palo Alto farm, are commemorated by a plaque on the campus of Stanford University (Fig. 1). Muybridge's work, which began in 1872, was inspired by a debate over the question of whether all four hooves of a trotting horse are simultaneously out of contact with the ground at any point in its stride. Muybridge believed he could answer this question by taking photographs with a camera shutter and film capable of capturing reasonably sharp images of a horse trotting at high speed. (The trot is a mode of locomotion, or gait, in which diagonal legs move together in pairs—it is a smooth, steady motion, in contrast to the more energetic bounding motion of the gallop, which is the gait used to attain ultimate speed for short distances. The details of each gait were eventually elucidated by Muybridge, but his written account clearly describes the seminal role of the trot in inciting his interest [2].)

We can estimate the duration of the shutter opening, Δt, that Muybridge needed for his camera from consideration of the necessary spatial resolution and the velocity υ of the horse. For a clearly defined image of a horse's legs, a resolution (Δx) of 1 cm is reasonable; that is, 1 cm is small compared to the relevant dimensions of the problem (the dimensions of the leg and its displacement during the course of a stride). Taking $\upsilon \approx 10$ m/s for the velocity of a horse (the legs will, in fact, at times be moving several times faster), and using the relation $\Delta x = \upsilon \Delta t$, leads to

$$\Delta t \approx \frac{1\ \mathrm{cm}}{10\ \mathrm{m/s}} = 10^{-3}\ \mathrm{s} = 1\ \mathrm{ms}$$

Indeed, Muybridge was able to achieve the necessary exposure times to capture an image of a trotting horse with all four feet in the air. Subsequent to this initial success, he devoted many years to the photographic study of animals and humans in motion, for some time in Palo Alto and later at the University of Pennsylvania (2).

Figure 1. Tablet erected on the Stanford University campus in 1929, commemorating "motion picture research" of Eadweard Muybridge at the Palo Alto farm of Leland Stanford in 1878 and 1879.

In these studies, Muybridge sought not only to provide the isolated images required to answer questions such as that which first attracted his interest, but also to document the entire sequence of an animal's leg motions during its stride. To establish the required absolute timing of the photographs, he initially set up a row of equally spaced cameras along a track at the Palo Alto farm. The shutter of each camera was activated by a thread stretched across the track in front of the camera. Thus a horse running down the track at speed υ recorded a series of photographs, and the point in time associated with the ith photo could be calculated as $\sim d_i/\upsilon$, where d_i was the distance from the starting gate to the ith camera. The separation in time between frames was $\tau = \Delta d/\upsilon$, where $\Delta d = d_{i+1} - d_i$, so the number of frames per second was $\upsilon/\Delta d$. Although the absolute timing of the frames of the series was imperfect, tied as it was to the velocity of the horse from camera to camera, the images nevertheless permitted detailed analysis of the motion. The imprecision of the chronology of images obtained in this manner was subjected to some criticism, and in his later studies, Muybridge used cameras with shutters triggered sequentially by a clockwork mechanism (2) to obtain photographs regularly spaced in time, such as those shown in Figure 2.

Etienne-Jules Marey, professor at the Collège de France and a contemporary of Muybridge, invented "chronophotography", a reference to the regular timing of a sequence of images (2, 3) recorded by a *single* camera using a rotating slotted-disk shutter (the analogue of shutter-stroboscopy, the concept of which was developed as early as 1832; the related flash-stroboscopy is discussed in what follows). The recording was made on either a single photographic plate or a film strip, the precursor to modern cinematography. Marey, like Muybridge, focused on investigations of humans and animals in motion, such as the motion in righting of cats.

Figure 2. A series of photographs by Eadweard Muybridge of a trotting horse, taken during studies at the University of Pennsylvania (1884–1885). The time interval between photos is regular, 0.052 s. Note that in the third photograph (top right), all four hooves are simultaneously off the ground in the trot motion.

Muybridge also developed an apparatus to project sequences of images obtained in his experiments to give "the appearance of motion" in demonstrations representing an early stage in the development of motion picture technology. His device, which he called a zoöpraxiscope, made use of a projecting lamp and counter-rotating image and shutter disks to produce a rapid succession of apparently stationary bright images and periods of intervening darkness. With image frames appearing at a rate greater than about 20 per second, the viewer's perception of each bright image persisted across the dark intervals, merging with its successor to form the desired continuous, but moving, picture. A simpler device for viewing the sequences, which was well known in Muybridge's time, is the zoetrope (from the Greek *zoe* 'life' and *tropos* 'turning'). It consists of a cylinder spinning on its axis with a series of slits equally spaced around its circumference. Through the slits, photos on the opposite inner surface are briefly glimpsed in rapid sequence, giving an impression of continuous motion.

B. Stroboscopy

An alternative approach to the study of rapid motions, which has also proved capable of reaching much shorter time scales than possible with fast shutters, is the use of short light flashes, which make an object moving in the dark visible to a detector (observer's eye or photographic plate, for example) only during the light pulse. Thus the pulse duration Δt plays the same role as the opening of a camera shutter and can be thought of in just the same way. An instrument that provides a series of short light pulses is a stroboscope (*strobos* from the Greek word for "whirling" and *scope* from the Greek for "look at", for its original use in viewing rotating objects). Combined with a camera with an open shutter and with an appropriately chosen Δt for the light pulses, a stroboscope can produce a well-resolved image of an object as fast as a bullet.

In the mid-19th century, spark photography had been demonstrated for stopping rapid motions. The development of stroboscopic photography in the mid-20th century was greatly advanced by Harold Edgerton, professor at MIT and cofounder of EG&G electronics, through the development of electronic flash equipment capable of producing reliable, repetitive, and microsecond-short flashes of light. (Edgerton and EG&G also developed camera shutters based on optical

principles with no moving parts that are far faster than any conventional mechanical shutter [3].)

An example of the use of a stroboscope is shown in Figure 3, a precisely timed sequence of images of a falling apple. With the apple's velocity limited to $\upsilon \leq 5$ m/s and a value of Δx of ~1 mm for a sharp image, Δt of the flash must be ~$(1 \text{ mm})/(5 \text{ m/s}) = 2 \times 10^{-4}$ s or less, well within the stroboscope's range. An absolute time axis is established here by electronic timing of the flashes. The picture shows the effect of gravity, which can be quantified by analyzing the successive positions in which the light illuminates the apple. According to the law of uniformly accelerated motion

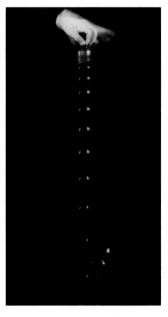

Figure 3. A falling apple photographed by stroboscopic illumination at intervals of ~1/25 s (3). The effect of gravity is clear.

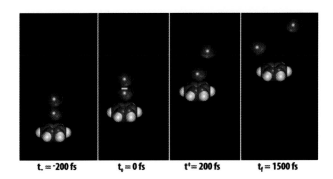

Figure 4. Molecular structures for a reaction in progress involving two molecules (bimolecular). The diatomic iodine molecule (I_2, top) is split by exchange of an electron with the ring molecule benzene (C_6H_6).

$t_- = $ -200 fs $t_0 = 0$ fs $t^+ = 200$ fs $t_f = 1500$ fs

$x = \frac{1}{2} a t^2$, for the position x of an object at rest at $t = 0$ and subjected to acceleration a. Therefore, flashes equally spaced in time record images of uniformly increasing separation (up until impact and rebound from the table); the slope of the plot of image separation versus time is equal to $g\tau$, where g is the acceleration of gravity (~9.8 m/s²) and τ is the spacing of the flashes. When an apple height of 10 cm is assumed in Figure 3, one finds that τ is ~0.04 s; that is, the strobe flash rate was ~25 per second.

C. Femtoscopy

If the above ideas can be carried over in a straightforward manner to the study of atoms in motion, then the requirements for a femtochemistry experiment are easily determined. For a molecular structure in which atomic motions of a few angstroms typically characterize chemical reactions, a detailed mapping of the reaction process will require a spatial resolution Δx of less than 1 Å, more than 8 orders of magnitude smaller than needed by Muybridge. Therefore, the Δt required to observe with high definition molecular transformations in which atoms move at speeds of the order of 1000 m/s is $(0.1 \text{ Å})/(1000 \text{ m/s}) = 10^{-14}$ s = 10 fs. While this time scale has been recognized theoretically as the time scale for chemical reactions since the 1930s, it became possible to directly see the detailed steps in molecular transformations for the first time in the 1980s (4) with the development of femtosecond lasers (see Appendix). Such minute times and distances mean that molecular-scale phenomena are governed by quantum mechanical principles. Quantum theoretical considerations that are fundamental to the observation of atoms in motion are discussed in Section IV.

Flashing a molecule with a femtosecond laser pulse can be compared to the effect of a stroboscope flash or the opening of a camera shutter. Thus a pulse from a femtosecond laser, combined with an appropriate detector, can produce a well-resolved "image" of a molecule as it passes through a specific configuration in a process of nuclear rearrangement, as Muybridge caught the horse with all four feet in the air. (The detection step is based on spectroscopic or diffraction techniques, and the measured signal can be analyzed to give information about the positions of the molecule's atoms; see Appendix.) The pulse that produces such an image is called a *probe* pulse, because it is used to probe the molecule's structure just as a shutter opening and a stroboscope flash probed the positions of the horse and apple in Figures 2 and 3, respec-

tively. The use of laser pulses to "stop the motion" of atoms and obtain instantaneous molecular structures may be called femtoscopy; femto, the prefix meaning 10^{-15}, is from *femton*, the Scandinavian word for "fifteen". Molecular structures determined at different stages of a reaction process can be treated as the frames of a motion picture, allowing the motion of the atoms to be clearly visualized, as illustrated in Figure 4; the number of frames in a molecular movie could then be as high as 10^{14} per second.

Probing is not the whole story. For the entire course of any motion to be recorded, that motion must be initiated so that it takes place in the time span accessible to a sequence of probe snapshots. In photographing the horse and apple, the processes are initiated by opening a starting gate for the horse and by releasing the apple, and the respective probing sequences are arranged to coincide closely in time to those actions. For femtochemistry, the analogous operation is realized by launching the molecule on its path using a femtosecond initiation pulse (the *pump* pulse) passing the sample. This establishes a temporal reference point (time zero) for the changes that occur in the molecular motion or reaction. The timing relative to the probe pulses is accomplished by generating the pump and probe pulses from a common source and sending either the pump or the probe along an adjustable optical path to the sample (Fig. 5A). The difference between pump and probe path lengths divided by the constant speed of light of about 300,000 km/s (actually 299,792 km/s) precisely fixes each probe image on the time axis established by the pump.

Such use of optical path differences to measure transient phenomena dates back at least to 1899, when Abraham and Lemoine reported their measurement in France of the Kerr response of carbon disulfide (Fig. 5B) (5). They used the breakdown of air at a spark gap to *simultaneously* ($t = 0$) create a probe light pulse (spark) and discharge a parallel plate capacitor immersed in carbon disulfide. Before the discharge of the capacitor, the carbon disulfide molecules between the plates retained a net alignment with the applied electric field, resulting in optical birefringence, a difference in refractive index for light polarized parallel or perpendicular to the plates (Kerr effect). Upon discharge, the spark-produced probe light monitored the Kerr effect as a function of time by traversing a variable-length optical delay before passing between the plates of the capacitor. Polarization analysis showed that the substantial Kerr effect to which the probe light was subject at the shortest measurable delays was progressively reduced and

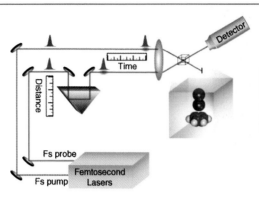

(A) The concept of femtosecond (pump-probe) experiments. After the probe pulse has been delayed by diversion through a variable-length optical path, femtosecond pump and probe pulses are focused into a volume containing the molecules to be studied. The detector responds to the probe pulse by any of a variety of schemes (see Appendix).

(B) Kerr-cell response measurement of Abraham and Lemoine (1899). The high voltage applied across the plates of the capacitor induce a Kerr effect in the carbon disulfide. The breakdown of air at the spark gap discharges the capacitor and creates a pulse of light, which follows the path indicated (reflecting from mirrors M_2, M_3, M_4, and M_1; L_2, L_3, and L_1 are collimating lenses) to probe the Kerr effect. The time dependence of the effect is determined by moving mirrors M_2 and M_3 as indicated, to vary the time delay for passage of the probe pulse through the Kerr cell. The light is polarized before the cell by the polarizer P, and the Kerr

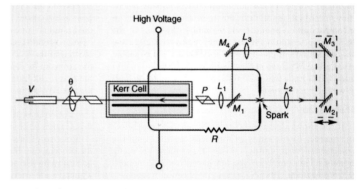

effect is quantified by adjusting the rotation angle θ of an analyzing polarizer to measure the polarization properties of the light after passage through the Kerr cell, as observed at the viewing telescope V.

Figure 5. Experimental layouts using changes in optical path length to establish timing.

ultimately disappeared as the delay path length was increased. From this observation, a half-life of 2.5 nanoseconds, corresponding to a path length of 75 cm, was determined for the combined electrical and Kerr (alignment) response time of the system. We now know that the Kerr response due to reorientation of the molecules occurs in ~2 ps (1 ps = 10^{-12} s), with a femtosecond component associated with electronic polarization effects.

A fundamental difference between femtoscopy and the horse and apple analogies is that, in femtochemistry experiments, one probes typically millions to trillions of molecules for each initiation pulse, or repeats an experiment many times, to provide a signal strong enough for adequate analysis. A comparable situation would arise in stroboscopy of the apple if capture of a distinct photographic image could only be accomplished by using many different apples or repeated exposures. It is clear that success in such a case would require (i) precise *synchronization* of the strobe (probe) pulse sequence with the release of the apple, to less than or about the strobe pulse duration, for optimum resolution; and (ii) a precisely defined launching configuration of each apple, to a fraction of an apple diameter.

By the same reasoning, to synchronize the motion of many independent *molecules* so that all have reached a similar point in the course of their structural evolution when the probe pulse arrives to capture the desired impression, the relative timing of pump and probe pulses must be of femtosecond precision, and the launch configuration must be defined to subangstrom resolution. It is only by means of such synchronization that the signals from many molecules may be added together without hopelessly blurring the molecular structure derived from the measurement. The use of pulses generated from the same source and an optical path delay as described above provides the required high degree of timing precision. A typical optical path accuracy of 1 μm corresponds to absolute timing of the molecular snapshots of 3.3 fs (see Fig. 5A).

Of equal importance is the required definition of the launch configuration. This definition is naturally realized because the femtosecond pump pulse snaps all members of the molecular ensemble from their ground states, which have a single well-defined structure. Moreover, on the femtosecond time scale, moving atoms are coherent or particle-like in their trajectories. The quantum description of how the localization of nuclei in motion is achieved in the molecule and the ensemble is discussed in Section IV.

III. Demonstration

Because the scales of distance and time of the motions that are the subject of femtochemistry research are almost unimaginably small and the measured signals require a sophisticated apparatus and analysis of data, we have, for educational purposes, designed a simple exhibit capable of highlighting the basic concepts of femtochemistry and stop-motion stroboscopy. The exhibit gives a student or visitor a concrete and visually interesting illustration of the use of short light pulses in the study of a rapidly moving object, in this case a molecular model.

The exhibit occupies a 3×5-ft table (Fig. 6, top). The output of a continuous-wave, diode-pumped solid-state laser (532 nm, 15 mW; appropriate safety precautions should be observed in handling such laser powers) is chopped by a rotating opaque disk 6 cm in radius provided with two openings (4.8 cm and 5.1 cm from the rotation axis), of tangential lengths 2 mm and 8 mm, respectively. The chopper is rotated by a dc motor (driving voltage 0–12 V, 2600 rpm at 4.5 V) and is mounted on a translation stage allowing either of the two openings to be positioned in the optical path.

The chopped laser beam passes through three lenses at ~3-inch spacing: two strongly diverging lenses (ca. -25 mm focal length each) and a 2-inch diameter 300-mm focal length converging lens. These lenses expand the 1-mm diameter beam into a gently diverging light cone (~6° divergence) of 7-cm diameter at a distance of 40 cm from the last lens. At that point the light illuminates a molecular model (here, methane, 5-cm diameter) mounted on the axle of a second dc motor. After an additional 40-cm path length the light is reflected by a plane mirror toward a screen at a distance of 60 cm from the mirror. The shadow of the model cast on the screen by the laser light is about 3 times the size of the model. The camera in the exhibit is representative of the detection step in a real experiment.

When the molecular model (Fig. 6, bottom, far left) is set in rapid rotation (frequency $\nu > 10$ Hz), it appears only as a blur to the eye if viewed under continuous illumination (Fig. 6, bottom, second from left), and a blurred shadow is cast on the screen if the laser is not chopped. With pulsed laser illumination, we can freeze the motion for study in the manner discussed in the previous section, if the spatial resolution and timing requirements can be satisfied. First, the model atoms are about 1 cm in dimension, so the spatial resolution needed to produce a sharply defined image is $\Delta x \approx 1$ mm. The linear speed υ of an atom moving across the laser beam

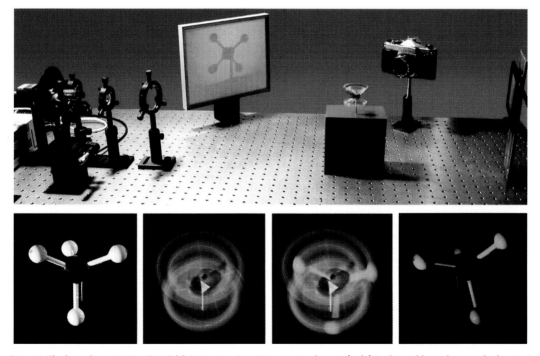

Figure 6. The laser demonstration. Top: Exhibit overview. A continuous green laser at far left is chopped by a chopper wheel, passes through three lenses, illuminates a spinning molecular model, and is reflected by the mirror at far right to the screen at center left. Bottom: Four views of the molecular model are. From left to right, the model is (1) stationary under room light only; (2) spinning under room light only; (3) spinning under room light and pulsed laser illumination; and (4) spinning under pulsed laser illumination only.

may be calculated from the relation $\upsilon = \omega\, r$ for a point at distance r from the rotation axis of a body rotating at angular velocity ω. For r_{max} = 2.5 cm and ν = 43 Hz (~2600 rpm), $\omega = 2\pi\nu$ = 270 rad/s and υ_{max} = 675 cm/s. Therefore, the first requirement of the exhibit laser pulses is that their duration Δt be $\leq (0.1 \text{ cm})/(675 \text{ cm/s}) = 1.5 \times 10^{-4}$ s = 150 μs to produce a sharp image or cast a distinct shadow.

The synchronization by the pump pulse in femto-chemistry is necessary to allow a well-defined image consti-tuted from many independent molecules, as discussed above. Similarly, in the exhibit, the observer can study a persistent image or shadow of the object only when the model is rap-idly and repetitively exposed to the laser flashes in precisely the same spatial configuration. Otherwise, even when each pulse is short enough to produce an isolated sharp image, the rapid sequence of different and fleeting images will leave a blurred or chaotic impression on the eye. Since the rotating model undergoes a periodic motion, it will always be at the same point in its trajectory as the probe pulse arrives when the frequency (repetition rate) of the probe pulses created by the chopper wheel is adjusted to match the frequency of rotation of the model. The resulting sequence of synchronized, identical images can then be viewed *as though the molecule were stationary*.

The two right-hand photos of the spinning model at the bottom of Figure 6 show this effect with and without room lights. The enlarged shadow cast on the screen appears likewise as the shadow of a stationary object (Fig. 6, top). Note that stationary *multiple* images of the model can also be produced by many other chopper frequencies not equal to the model rotation frequency. For example, when the chopper frequency is 3/4 of the rotation frequency, the model is illuminated every 4/3 rotation, giving three stationary overlapping images separated from each other by 120° rotations.

By tuning the voltage driving the motor of the chopper and observing the model or its shadow, the synchronization condition to observe a single stationary image can be found. The pulse duration Δt is then fixed by the length of the aperture. To illustrate the role of pulse duration in determin-ing image definition, there are apertures of different lengths in the wheel. With the chopper wheel rotating at 2600 rpm, $\Delta t \approx 150$ μs is obtained when the laser passes through the 2-mm aperture. As determined in the calculation above, these 150-μs pulses give a sharp image of the molecule. Translation of the chopper to let the laser pass the 8-mm opening at slightly larger radius on the chopper disk results in a train of 580-μs pulses. With these pulses, the structure of the molecule is still easily distinguishable, but it is somewhat blurred rather than sharply defined.

We note that with the setup as described, only ~0.38 mW (2.5% of the laser light) and ~0.1 mW pass the 8-mm and the 2-mm aperture, respectively. These power levels of 532 nm light allow the stroboscopic effect to be visible in ambient illumination. For comparable visibility, a 60-mW red helium–neon laser would be required, since the relative visibility of light is ~3.8 times higher at 532 nm than at 633 nm. If desired, a laser of lower visibility can be used for the exhibit in an appropriately darkened environment. Indeed, the use of a laser is not critical, and any sufficiently bright light source, either a conventional continuous-wave light with chopper or a pulsed light (strobe) with adjustable pulse lengths, may be substituted.

IV. Femtochemistry—Principles and Applications

Up to this point, in applying the principles of strobos-copy to the molecular realm, we have treated the motion of atoms no differently from those of macroscopic objects like horses or apples. These latter objects of everyday experience are governed by what are known as the classical mechanical laws of motion, due to Newton, in which each body has a well-defined position and velocity at any time. However, when dealing with the actual molecular scale of motions in a femtochemistry experiment, quantum mechanical concepts become of paramount importance. Specifically, at the scale of atomic masses and energies, the quantum mechanical particle–wave duality of matter comes into play, and the notions of positions and velocities common to everyday life must be applied cautiously. In this section, we will first examine how the principles of quantum mechanics and uncertainty apply in the world of molecules, then show how on the femtosecond time scale it is nevertheless appropriate to describe atomic motion classically—the atoms move like localized particles with well-defined velocities. Examples of the behavior of real molecular systems are given.

A. Matter Particle–Wave Duality and Concept of Coherence

One of the profound findings of quantum mechanics is the particle–wave duality of matter expressed by the remarkable relationship of de Broglie in 1924: $\lambda = h/p$, where λ is the de Broglie matter wavelength, h is the Planck constant, and p is the momentum of the particle/wave, $p = m\upsilon$. In wave mechanics, the state of any material system is totally defined by a spatially varying wave function from which it is possible to determine only *probabilities* of the system's having a certain spatial configuration or having a certain momentum, rather than specific values for those quantities. For example, the probability that a particle whose wave function is $\Psi(r,t)$ will be found at position r is $|\Psi(r,t)|^2$, the square modulus of the wave function.

The form of wave functions representing atomic motion is determined by the energy landscape of the system, called the potential energy surface, which in a molecule depends on the specific arrangement of electrons around the nuclei. The total energy of an isolated system is constant, and the difference between the total energy and the potential energy is energy of motion, or kinetic energy (E_{kin}), of the constituent particles. The classically allowed region of the potential energy landscape of a system consists of all regions in which the total energy is greater than the potential energy; that is, where E_{kin} is nonnegative. According to classical mechanics, motion of the system is only possible in this region. If the classically allowed region is limited in extent (i.e., the system is confined or bound), only certain specific values of the energy are pos-sible.

When the energy of a system is exactly defined (energy uncertainty ΔE = 0), its wave function spreads over (and even somewhat beyond) the entire classically allowed region of space, and the position probability density (ppd) is independent of time. Mathematically, the spatial and temporal dependence of the wave function are separable, and $\Psi(r,t)$ may be written as $\Psi(r,t) = \psi(r) \exp(-iEt/\hbar)$, where the position-independent temporal phase factor, $\exp(-iEt/\hbar)$, is a complex number of

Nobel Prize Report

modulus 1. Thus

$$|\Psi(r,t)|^2 = \psi(r)\exp(-iEt/\hbar)\,\psi^*(r)\exp(iEt/\hbar) = |\psi(r)|^2$$

with no time dependence.

To illustrate these concepts, we present in Figure 7 the simple case of a diatomic molecule vibrating in a parabolic, or harmonic, potential energy well, which is an approximate representation of the vibrating iodine molecule. The horizontal axis represents the separation, r, of atomic nuclei. The possible energy values of the system are represented by the horizontal lines labeled by the quantum number n. Examples of wave functions $\psi_n(r)$ corresponding to $n = 0$ and 16–20 are plotted, with the zero of the wave function amplitude axis set equal in each case to the corresponding energy value. The sign of the wave function alternates from positive to negative with a spatial period determined by the kinetic energy, which is zero at the walls of the well and has its maximum value at the center.

When the physical state of a molecule is described by one of these wave functions of definite energy, a measurement of atomic position will reflect only the stationary probability distribution, with no nuclear motion. Taking $n = 20$ of Figure 7, for example, there are 21 peaks in the ppd (corresponding to both positive and negative peaks in the wave function), distributed from one wall of the potential to the other. As another example, the ppd and the momentum probability distribution for a stationary state of a more realistic potential, one modeled on the quasi-bound excited state of sodium iodide (NaI) (see below), are shown in Figure 8. The ppd fills the classically allowed region of the potential well, from 2.75 Å to 10.19 Å, and the momentum probability has peaks that correspond to the kinetic energy ($\sim$5300 cm^{-1}; 1 cm^{-1} = 1.98648×10^{-16} ergs) near the center of the well.

Motion is observed in ultrafast laser experiments when the system is prepared in a *coherent (superposition) state*, making the state function a coherent sum of stationary wave functions for different energies E_i:

$$\Psi(r,t) = \sum_i c_i\,\psi_i(r)\exp(-iE_it/\hbar) \qquad (1)$$

The coefficients c_i determine the relative contributions of the wave functions to the superposition. The sum is said to be coherent when well-defined phase relations exist between the constituent wave functions (see below). The ppd in this case is simply

$$\Psi(r,t)\,\Psi^*(r,t) =$$
$$\left[\sum_i c_i\,\psi_i(r)\exp(-iE_it/\hbar)\right]\left[\sum_j c_j^*\,\psi_j^*(r)\exp(+iE_jt/\hbar)\right] \qquad (2)$$

One can see that cross-terms ($i \neq j$) in this product are *not* time independent (the phase factors do not cancel), but oscillate at the frequencies determined by the energy differences ($E_i - E_j$), so the expected outcome of a position measurement changes with time—*the system is in motion!*

B. Analogy with Light

The above is an example of wave superposition and *interference*, a more familiar example of which is the Young double-slit light interference experiment, illustrated at the bottom of Figure 7. As the wave function is a probability wave, so light is a wave of electric and magnetic fields whose amplitudes change at a regular interval, the wavelength, from positive to negative to positive along the direction of propa-

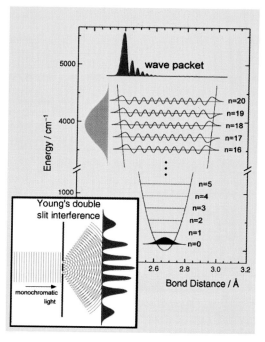

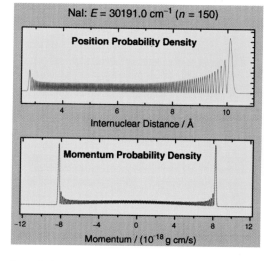

Figure 7. Diatomic molecule in a harmonic oscillator potential: stationary wave functions and formation of a localized wave packet. Inset: Thomas Young's experiment (1801) on the interference of light.

Figure 8. Stationary state of the theoretical bound potential of NaI (modeled on the real quasi-bound potential that is covalent in nature at short distances and ionic at long distance)—position and momentum probability distributions.

gation. Like the ppd, the intensity of light at a point in space is given by the square of the field amplitude. When light from two or more sources overlaps in space, the instantaneous field amplitudes (not intensities) from each source must be added together to produce the resultant light field.

In Thomas Young's two-slit experiment, reported in 1801, light from a single source passes through two parallel slits cut in an opaque screen to produce, in the space beyond, two *phase-coherent* fields of equal wavelength and amplitude. At points for which the distances to the two slits differ by $n + \frac{1}{2}$ wavelengths (for integer n) the amplitudes of the two superposed fields are opposite in sign and add to zero at all times; no light is detected. Elsewhere, the amplitudes do not cancel. Thus, a stationary pattern of light and dark interference fringes is produced. If, instead, *incoherent* (without a well-defined phase relation) fields of the same amplitudes were superposed, the intensities observed from each slit individually would simply add together and no interference fringes would be seen.

C. Wave Packets and Uncertainty—the Classical Limit

In a manner analogous to the interference of light waves in the Young experiment, the constituent matter wave functions of a coherent superposition state interfere constructively and destructively, and the resultant total wave function may have a large amplitude in only a limited area of the classical region at any given time. This constitutes a *wave packet*. At the top of Figure 7 is shown a specific example of the ppd of a wave packet formed from a superposition of the plotted wave functions $n = 16$ to 20 with weighting according to the distribution curve at the left of the figure. As time advances ($t > 0$ in $\exp(-iE_it/\hbar)$), the phase relationships between wave functions change, causing the wave packet to move. As long as the ppd remains sufficiently localized on the scale of the total classical region, as it is in the figure, a discussion in terms of the classical concepts of particle position and momentum is entirely appropriate. Such is the case for the wave packet of Figure 7, which will oscillate back and forth across the potential well indefinitely as would a classical particle at the same energy in that potential.

For a laser pulse to prepare a system in a localized wave packet, the deposited energy cannot be precisely defined, as wave functions at a broad range of energies E_i must be included in the superposition state. Given that the sizes of the energy quanta of a radiation field are determined by the constituent wavelengths, this translates into a requirement for a broad range of wavelengths in the field, which is also consistent with the nature of ultrafast pulsed laser sources. (Ultrashort light pulses are produced much as wave packets are formed, by coherently combining a large number of different wavelengths of light, each corresponding to a characteristic mode of the laser cavity; see Appendix.) Thus localization in time and in space are simultaneously achievable by virtue of the energy uncertainty!

When the extent of the classically allowed region is unbounded, the energy may take on all values rather than being restricted to quantized levels, but the same principle of superposition of wave functions applies to building wave packets; the sum in eq 1 becomes an integral. One example of unbound motion is motion at constant potential energy—that is, for a *free particle*. It is possible in this case to create a

wave packet in one dimension with a very simple bell-shaped spatial profile defined by the Gaussian equation

$$\Psi(x, t) = N(t) \exp\left(-\frac{(x - x_0 - \langle \upsilon \rangle t)^2}{4[\Delta x(t)]^2}\right) \quad (3)$$

where $\Delta x(t)$ can be shown to be the root mean square (rms) deviation in the ppd, $|\Psi(x,t)|^2$, and $N(t)$ is a normalization constant. (The rms deviation of a probability distribution $P(x)$ is given by

$$\left[\int_{-\infty}^{+\infty} (x - \bar{x})^2 P(x) \mathrm{d}x\right]^{1/2}$$

where $\bar{x}$ is the mean value $\int_{-\infty}^{+\infty} xP(x)\mathrm{d}x$.) The time dependence of Δx is considered below. The maximum of the wave function or ppd is at $x_0 + \langle \upsilon \rangle t$ and thus moves at velocity $\langle \upsilon \rangle$. Note that the full width at half maximum (FWHM) of a Gaussian function is $2(2 \ln 2)^{1/2} = 2.3548$ times its rms deviation, so the full width, $\Delta x^{\mathrm{FW}}(t)$, of the ppd is equal to $2.3548 \times \Delta x(t)$.

The well-known Heisenberg uncertainty relation states that the degree of localization in space of a quantum system is inversely related to the minimum uncertainty in the system's momentum. That is, the more limited the extent of the ppd $|\Psi(x,t)|^2$, the larger the uncertainty in the momentum, and thus the less well defined the future trajectory can be. When the uncertainties Δx and Δp are defined as rms deviations of the probability distributions, then the uncertainty relation is $\Delta x \, \Delta p \geq \hbar/2$, where $\hbar = h/2\pi$.

For a Gaussian free particle wave packet having the minimum value of the uncertainty product, $\Delta x \, \Delta p = \hbar/2$, the momentum distribution is also Gaussian, and the contribution of the momentum uncertainty to the widening of the packet as it moves is expressed in the relationship

$$\Delta x(t) = [\Delta x^2(t=0) + (\Delta \upsilon \times t)^2]^{1/2} \quad (4)$$

where $t = 0$ is the time of minimum uncertainty and $\Delta \upsilon = \Delta p/m$. Because Δp is inversely related to $\Delta x(0)$, it is only possible to slow the spreading by an increase in the initial Δx, thereby allowing a smaller $\Delta \upsilon$. Note that spreading is inevitable for the minimum-uncertainty, free-particle wave packet, but not for wave packets in general, which can be shaped by their potentials. Equation 4 can be used to give a feeling of the time needed for an appreciable spreading to take place—for $\Delta x(t)$ to increase by a factor of $\sqrt{2}$, the product $\Delta \upsilon \times t$ must be equal to $\Delta x(t=0)$. From this equality and the uncertainty relation for rms deviations, we can express t_s, the time for spreading by ~40%, as

$$t_s = \Delta x(0)/\Delta \upsilon = m\Delta x(0)/\Delta p = 2m\Delta x^2(0)/\hbar \quad (5)$$

When a free particle wave packet is initiated by an ultrashort light pulse so that it has the minimum value of the $\Delta x \, \Delta p$ uncertainty product, one can calculate the relation between the pulse duration and the wave-packet spatial width. It turns out to be quite simple: $\Delta x = \langle \upsilon \rangle \Delta t$.

The size of $\hbar$, 1.05457×10^{-27} erg-second, means that the fuzziness required by the uncertainty principle is imperceptible on the normal scales of size and momentum, but becomes important at atomic scales. For example, if the position of a stationary 100-g apple is initially determined

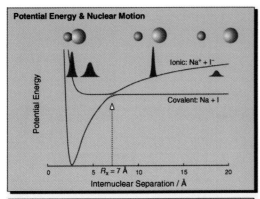

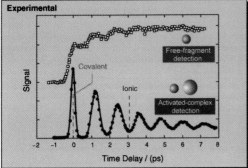

Figure 9. Femtoscopy of the bond breakage and bond reformation of sodium iodide. Top: The two potential energy functions, the two channels of wave packet evolution (bound and free), and schematic of wave packet motion. Bottom: Time sequence of measured populations of free fragments (open squares) and of quasi-bound complex transition-state configurations (see Appendix).

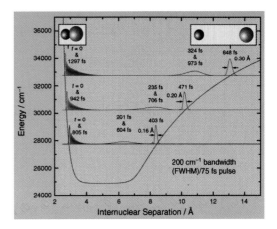

Figure 10. Evolution of a 200-cm^{-1} FWHM bandwidth wave packet, corresponding to pulses of 73.6-fs (~75-fs) FWHM, on the theoretical bound potential of NaI (see legend to Figure 8) at three excess energies: 2862, 5362, and 7862 cm^{-1}.

a classical *single-molecule trajectory* have been met. The key to achieving localization of the molecular ensemble is generally provided by the naturally well-defined initial equilibrium configuration of the studied molecules before excitation by the pump pulse. For example, most iodine molecules at room temperature or below are in their $n = 0$ vibrational state, the wave function of which, shown in Figure 7, yields a width of the ppd of only 0.08 Å FWHM. This spatial confinement establishes the phase relations between the excited wave functions that produce and define the localized wave packets, so the entire sample of molecules are launched from essentially the same starting point—a single-molecule trajectory is observed. The femtosecond time resolution is critical to exploit this localized launch configuration before the dispersion or propagation of the wave packet.

D. A Paradigm Case Study—Sodium Iodide

Observation of the motion of nuclei as wave packets was realized in experiments conducted at Caltech in the 1980s. A prime example is represented by experiments on the wave packet motion of sodium iodide, and Figure 9 depicts the potential energy for the motion and the experimental observations. The quasi-bound upper potential curve is composed of a covalent inner part and an ionic long-range part created by the crossing of the covalent and ionic curves, as shown. As discussed below (see Appendix), the measurements show a large fraction of the excited wave packet oscillating in this well, between covalent and ionic bonding, while with each stretch a smaller fraction remains on the covalent surface and separates as free atoms. The role of coherence is critical in preparing the wave packet, and a single-molecule trajectory of the ensemble is assured when the pump pulse synchronizes all molecules near the 2.67-Å bond length of the ground state minimum at time zero. This system has been studied in great detail, highlighting these and other concepts, and it represents a paradigm case of molecular dynamics with atomic-scale

to a small fraction of a wavelength of light, say $\Delta x = 10$ nm, a momentum uncertainty greater than

$$\frac{\hbar/2}{\Delta x} = \frac{0.5 \times 10^{-27} \text{ erg-s}}{1 \times 10^{-6} \text{ cm}} = 5 \times 10^{-22} \text{ g cm/s}$$

will be imposed, or $\Delta \upsilon \geq 5 \times 10^{-24}$ cm/s. Under the minimum uncertainty assumption, eq 5 can be used to calculate that the apple's position uncertainty will spread by ~40% only after 2×10^{17} s, or 6 billion years! On the other hand, an electron with a mass 29 orders of magnitude smaller would spread by 40% from an initial 1 Å localization after only 0.2 fs. Clearly, the much greater masses of atomic nuclei make their wave packets spread orders of magnitude slower than the wave packet of the electron (see below).

As mentioned in Section II C, atomic-scale localization is critical for femtoscopy: not only localization of the wave packets formed in each molecule, but of the total spread in position among wave packets formed in the large number of molecules on which the measurement is performed. When both types of spread are small with respect to the relevant extent of the classical region, the conditions for observation of

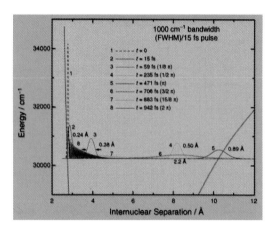
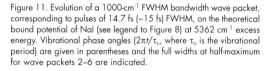

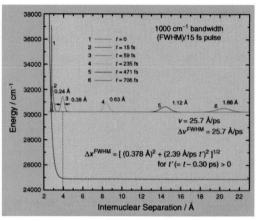

Figure 11. Evolution of a 1000-cm⁻¹ FWHM bandwidth wave packet, corresponding to pulses of 14.7 fs (~15 fs) FWHM, on the theoretical bound potential of NaI (see legend to Figure 8) at 5362 cm⁻¹ excess energy. Vibrational phase angles ($2\pi t/\tau_v$, where τ_v is the vibrational period) are given in parentheses and the full widths at half-maximum for wave packets 2–6 are indicated.

Figure 12. Evolution of a 1000-cm⁻¹ FWHM bandwidth wave packet, corresponding to pulses of 14.7 fs (~15 fs) FWHM, for unbound NaI at 5362 cm⁻¹ excess energy.

resolution (see ref *4* for more details). It is natural therefore to illustrate the properties and time evolution of wave packets as developed above with calculations based on the NaI system.

First, we give some examples of wave packet propagation on the theoretical bound state potential, which is modeled on the quasi-bound NaI state. This potential is shown in Figure 10, with an energy axis referenced to the NaI ground-state potential minimum. Five snapshots, representing one full vibrational cycle, of the ppd's of wave packets formed at three different center energies by coherent superpositions with 200-cm⁻¹ FWHM Gaussian population bandwidths are also shown. (This bandwidth corresponds to ~75-fs FWHM laser pulses.) Note that the wave function whose ppd is shown in Figure 8 is one of the $\psi_i(r)$ contributing to the sum in eq 1 for the superposition at the middle energy, 30,250 cm⁻¹ or 5362 cm⁻¹ excess vibrational energy. At time zero, the constituent wave functions are all taken as real and positive at the inner turning point of the potential. As time advances, the phase factors, $\exp(-iE_it/\hbar)$, cause the wave packets to move, in each case broadening in the center of the well and sharpening at each turning point.

These wave packets are seen to remain well localized in comparison to the extent of the classically allowed region throughout their first vibrational periods. In fact, at this initial Δx, 10 or more vibrational cycles can be completed without the progressive spreading of the wave packet destroying the usefulness of a classical description of the motion. The femto-chemistry experiments referred to above (Fig. 9), with pulses of similar duration, have shown this long-time particle-like behavior. In contrast, a calculation for a 1000-cm⁻¹ FWHM bandwidth superposition, or ~15-fs pulse, at 5362 cm⁻¹ excess energy, shown in Figure 11, is initially much more localized (smaller $\Delta x(0)$) but spreads much more quickly (larger Δp). Now the spreading of the wave packet in even one vibrational period is substantial.

As shown in Figure 9, the real NaI system involves two crossing potentials, one of which allows the Na and I to separate totally as free atoms. If we use the relative motion of Na and I at a constant potential as an example of free particle motion, for an average (kinetic) energy of 5362 cm⁻¹, the mean velocity of separation $\langle v \rangle$ is 25.7 Å/ps or 2.57×10^5 cm/s ($E_{kin} = \frac{1}{2}\mu\langle v\rangle^2$, where $\mu = 19.46$ amu $= 3.232 \times 10^{-23}$ g, the reduced mass of Na and I). A 1000-cm⁻¹ Gaussian energy bandwidth corresponds to a 14.72-fs FWHM Gaussian pulse. Such a pulse will create a minimum uncertainty wave packet of 0.378 Å FWHM. Restating the minimum uncertainty relation in terms of the full widths at half maximum as $\Delta x^{FW} \times \Delta p^{FW} \geq 8 \ln 2 \times \hbar/2 = 2.7726\hbar$, we can calculate a momentum spread Δp^{FW} of 7.73×10^{-19} g cm/s or $\Delta v^{FW} = \Delta p^{FW}/\mu = 2.39$ Å/ps.

The growth with time of the wave packet spatial width can then be calculated from eq 4. Using FWHM values for the present case leads to

$$\Delta x^{FW}(t) = [(0.378 \text{ Å})^2 + (2.39 \text{ Å/ps} \times t)^2]^{1/2}$$

This gives a width of 2.4 Å after 1 ps. If the initial wave packet is twice as broad (0.756 Å), Δv^{FW} will be half as large and the wave packet will spread more slowly, to only 1.4 Å after 1 ps. Twice the initial wave packet width will be produced for the same $\langle v \rangle$ by a pulse twice as long, or for the same pulse width by an average energy four times higher. Similarly, eq 5 can be used to calculate the time for the packet to spread by 40%: if $\Delta x^{FW}(0) = 1$ Å, that is, $\Delta x(0) = (1/2.3548)$ Å , then $t_s = 1.1$ ps; the wave packet in that time travels over 28 Å. If $\Delta x(0)$ doubles, the wave packet takes 4.4 ps, four times as long, to spread by the same factor.

Figure 12 shows the evolution of an unbound NaI wave packet with energy spread corresponding to a 14.72-fs laser pulse. The wave packet is formed on the steep repulsive potential wall of the bound well, but evolves after the initial

---

Correction: In Figure 12, the Δv^{FWHM} value should be 2.3 Å/ps, and not 25.7 Å/ps.

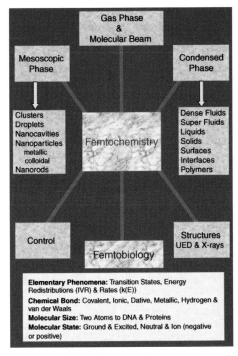

Figure 13. Scope of femtochemistry applications.

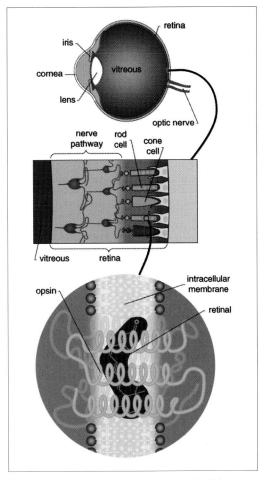

Figure 14. The molecular basis of vision. Within the light-sensitive rod and cone cells in the retina are membranes in which are embedded photoreceptive molecules such as rhodopsin, which comprises a light-absorbing part, retinal, and a protein, opsin. Top: simplified diagram of cross-section of the eye. Middle: enlargement of a portion of the retina, showing rod and cone cells. Bottom: schematic of part of a membrane within a rod cell. (Drawing by Randall J. Wildman.)

repulsion on the unbound, constant-energy surface. The initial packet is quite narrow and virtually identical to the bound packet of Figure 11; that is, its form depends only on the local potential surface. While the potential is changing sharply with distance (t < 30 fs) the above calculations concerning width and spreading of the wave packet for a constant potential energy are not applicable. However, after the packet enters the force-free region, it reproduces quite well the behavior predicted above for a free particle wave packet of minimum uncertainty at t = 30 fs.

These calculations for NaI show how, despite the important role of quantum uncertainty and wave mechanical behavior, it can be entirely appropriate to treat the nuclear motion of reacting molecular systems as particle-like, as we did in our presentation of the technique of femtoscopy.

E. World of Complexity

Femtochemistry is by no means limited to studies of small systems such as the two-atom system used as an example above. Femtochemistry has found applications in all phases of matter (Fig. 13) and in sibling fields including femtobiology of complex systems. For example, vision is the result of the conversion of light energy to an electrochemical impulse. The impulse is transmitted through neurons to the brain, where signals from all the visual receptors are interpreted. One of the initial receptors is a pigment called *rhodopsin*, which is located in the rods of the retina. The pigment consists of an organic molecule, retinal, in association with a protein named

opsin (Fig. 14). A change in shape of retinal, which involves twisting of a double bond, apparently gives the signal to opsin to undergo a sequence of dark (thermal) reactions involved in triggering neural excitation.

The primary process of twisting takes 200 fs, similar to the time seen in prototypical chemical systems (stilbene); and coherent motion along the reaction coordinate continues after the reaction, in the photoproduct (*6*). The speed of the reaction and the product coherence indicate that the energy is not first absorbed, then redistributed to eventually find the reaction path; instead, the entire process proceeds in a coherent manner—that is, as a wave packet. This coherence is credited for making possible the high (70% or more) efficiency of the initial step, despite the large size of the rhodopsin molecule and the many channels for dissipation of energy. These dynamics are similar to those of NaI, discussed above, even though the contrast in size is huge, from two atoms to a protein.

V. Conclusion

This paper presents an elementary description of the concepts basic to femtochemistry—the ability to freeze atoms in motion with femtosecond (10^{-15} s) time resolution. Unlike macroscopic motions, motion at the molecular scale requires new concepts to be addressed, and we elucidate the importance of coherence and the relationship to matter particle–wave duality and to the uncertainty principle. We provide a demonstration using a pulsed laser and molecular model to capture the fundamental idea behind the experimental technique. The exhibit is simple and can be made easily for educational and conceptual purposes. The analogy with stop-motion photography was highlighted over a century of development, from high-speed shutters to stroboscopes and on to femtoscopy; future directions include the use of ultrafast diffraction methods (7) to probe the totality of molecular structural changes in complex systems such as biological assemblies.

Acknowledgments

We thank the National Science Foundation for the support of this educational research. We wish also to thank the Nobel Museum Project (Stockholm) for triggering our interest in this effort, and Carsten Kötting and Dongping Zhong for help with three of the figures.

Literature Cited

1. von Baeyer, H. C. *Taming the Atom*; Random House: New York, 1992.
2. Muybridge, E. *Animals in Motion*; Dover: New York, 1957. See also Haas, R. B. *Muybridge*; University of California Press: Berkeley, 1976.
3. Jussim, E.; Kayafas, G. *Stopping Time*; H. N. Abrams: New York, 1987. See also Frizot, M. *La Chronophotographie*; Exposition á la Chapelle de l'Oratoire: Beaune, France, 1984.
4. Zewail, A. H. In *Les Prix Nobel (The Nobel Prizes 1999)*; Frängsmyr, T., Ed.; Almqvist and Wiksell: Stockholm, Sweden, 2000; pp 110–203 and references therein. A shorter version was published in *J. Phys. Chem., A* **2000**, *104*, 5660–5694.
5. Abraham, H.; Lemoine, J. *C. R. Hebd. Seances Acad. Sci.* **1899**, *129*, 206–208.
6. See: Wang, Q.; Schoenlein, R. W.; Peteanu, L. A.; Mathies, R. A.; Shank, C. V. *Science* **1994**, *266*, 422–424. For stilbene, see Pedersen, S.; Bañares, L.; Zewail, A. H. *J. Chem. Phys.* **1992**, *97*, 8801–8804.
7. Ihee, H.; Lobastov, V. A.; Gomez, U. M.; Goodson, B. M.; Srinivasan, R.; Ruan, C.-T.; Zewail, A. H. *Science* **2001**, *291*, 458–462.
8. Hopkins, J.-M.; Sibbett, W. *Sci. Am.* **2000**, *283* (3), 73–79, and references therein. References 18 and 19 in *Les Prix Nobel* (*4*) give more details.

Appendix: Ultrashort Laser Pulses and Experimental Techniques

Techniques of femtochemistry involve the use of ultrashort pulses and sensitive methods of detection, and the generation, amplification, and characterization of ultrashort pulses become part of the experiments. The generation of such laser pulses has a rich history, beginning with the realization of the first (ruby) laser in 1960 and the dye laser in 1966. Pulses of picosecond and femtosecond duration are obtained using what is referred to as mode-locking, and in combination with pulse compression methods, a 6-fs pulse was obtained in 1987 from a dye laser at Bell Laboratories, USA. The current state of the art in pulse generation involves self-mode-locking of solid-state lasers (titanium–doped sapphire or Ti:sapphire), which can be operated more routinely and with great reliability; pulses as short as 4 fs have been produced. This self-mode-locking in Ti:sapphire crystal was first observed by researchers at the University of St. Andrews in Scotland. Below, we briefly describe the concepts involved in generation and use of ultrashort pulses, following two recent review articles (*4*, *8*).

The laser, light amplification by stimulated emission of radiation, operates on two concepts: amplification and feedback. Thus one must have a gain medium (e.g. Ti:sapphire crystal) for amplification and a cavity (e.g. two mirrors) for the feedback. The excited Ti:sapphire crystal amplifies the light, and with the feedback of the cavity, the bouncing of light back and forth, an intense laser beam is produced from one of the mirrors, which is partially transmitting (Fig. 15). The cavity (of length L) has modes, or light of certain wavelengths, which are separated in frequency by $c/(2L)$. If one mode is isolated, then lasing occurs in a narrow-band frequency in what is referred to as CW or continuous-wave operation. For generation of an ultrashort pulse, light of a large number (possibly millions) of modes (wavelengths) is emitted together; the modes are said to be locked together. Mode-locking is a superposition of light waves in phase (Fig. 15), and as such it is similar to the creation of wave packets of matter waves as described in Section IV. In Figure 15 we also display the shape of a femtosecond pulse at 800 nm; the variation of the electric field at a point may be expressed as

$$E(t) = E_0 \exp\left(-\frac{t^2}{2\sigma^2}\right) \text{Re}\left[\exp\left(i\omega_0 t\right)\right] \quad \text{(A1)}$$

a packet with a carrier frequency $\omega_0/2\pi$ and rms width σ of the field amplitude envelope.

Three essential elements are contained in the cavity of a femtosecond laser: a gain medium, a mode-locking element, and an element that compensates for group velocity dispersion. The gain medium has a broad spectrum to ensure a large number of modes sufficient for femtosecond pulse generation; a 10-fs pulse requires a spectral width of about 100 nm at an 800-nm center wavelength. The group-velocity-disper-

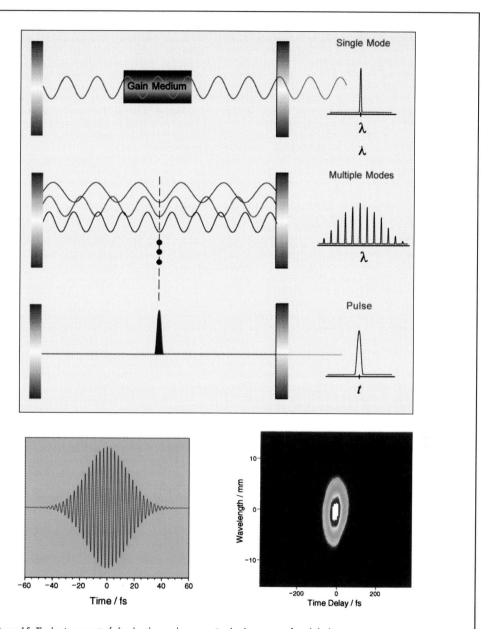

Figure 15. The basic concept of ultrashort-laser-pulse generation by the process of mode-locking.

The upper figure shows (top) continuous-wave lasing in single-mode operation; (middle) the superposition of several modes, which, if locked together in large numbers as indicated by the three vertical dots, gives (bottom) pulsed operation such as that depicted here. Note the analogy with matter's waves superposition in Figure 7.

The lower figures show (left) field of a femtosecond pulse at 800 nm (eq A1, σ = 15 fs), together with (right) an experimental display of a pulse in wavelength (nm)–time (fs) domain.

sion element usually consists of a combination of prisms arranged so that the broad range of spectral components in the pulse can all stay in step while circulating in the cavity. There are two forms of mode-locking, active and passive; the idea is to provide a time-dependent modulation of loss in (or) gain in the cavity. Active mode-locking uses an externally driven modulator, whereas in passive mode-locking an element that lets the pulse itself vary the gain/loss is included in the cavity. One method of passive mode-locking uses an organic molecule as a "saturable absorber" to enhance the propagation of a pulse over that of continuous lasing in the cavity. When the light is so intense that the saturable absorber's rate of absorption exceeds its rate of ground-state population recovery, it becomes temporarily transparent—it is saturated. This transparency occurs at the peak intensity levels of pulsed operation, allowing light to pass and circulate in the cavity to ultimately produce ultrashort pulses; the lower light intensities achievable in continuous lasing are blocked.

For Ti:sapphire lasers, the favoring of pulsed operation is achieved using self-mode-locking. The lasing material has a refractive index that changes depending on the light intensity. Because the laser beam in the cavity has a transverse spatial distribution with the largest intensity in the middle (Gaussian profile), the speed of light propagation for the intense part decreases relative to that of the lower intensity part, resulting in a lens-type focusing behavior (optical Kerr effect). This effect is pronounced only under the high intensities possible with short pulses. With a narrow aperture in the cavity where the pulsed beam focuses, much of the beam from any lower intensity mode of operation, which does not focus, can be blocked, while the pulses pass unimpeded. Thus the favored pulsed lasing will be the natural mode of operation.

To carry out femtosecond experiments, we need both a pump and a probe pulse as discussed in Section IIC. These are usually created from the output of a single femtosecond laser by a variety of nonlinear optical mixing processes chosen to produce the specific wavelengths of light appropriate for the molecule under study. The generated femtosecond pump and probe pulses are characterized by measuring the intensity, duration, spectral bandwidth, and polarization of the fields. The pump pulse sets the $t = 0$ by initiating the molecular process under study, and the relative timing of pump and probe passage through the sample is controlled by an optical delay. The probe pulse is usually absorbed, and the absorption varies as a function of the probe delay time, in response to the spectral changes of atoms or molecules as the reaction proceeds. The amount of probe absorption is reflected in a signal acquired by one of a variety of detection schemes. Two of these are laser-induced fluorescence and mass spectrometry.

In laser-induced fluorescence, the absorption of probe light energy creates an electronically excited species, which then emits light characteristic of the species; that is, it fluoresces. The fluorescence intensity is measured at each delay time to determine the amount of absorption. For example, for NaI (discussed in Section IVD), the absorption spectra of various configurations of the transition-state complex Na···I are very different from those of free Na and I atoms. Thus by tuning the probe to the wavelength absorbed by Na···I with Na and I close together we can observe a periodic fall and rise of the fluorescence signal with delay as the population vibrates back and forth in the quasi-bound well, passing repeatedly through the strongly absorbing configuration (Fig. 9). Since some of the molecules break apart with every bounce, the peak of every return is also smaller than the last. On the other hand, when the probe is tuned to a wavelength absorbed by free Na atoms, the fluorescence increases in stepwise fashion (Fig. 9) as, at each bounce, a new batch of free atoms is added to the existing pool; for the entire process all molecules move coherently.

In mass spectrometric detection, the probe creates a population of ions which can be sorted by mass. Thus, changes in mass as well as changes in absorption spectra can be monitored as bonds are broken and formed. There are variant methods of detection such as photoelectron spectroscopy (*4*). Ultrafast electron diffraction is being developed to observe the bond length change with time for more complex structures (*7*).

Contribution from

HARRY KROTO

University of Sussex, UK

and

BARRY WARD

UK

Harold Kroto

Born 1939 in Wisbech Cambridgeshire. Educated Bolton School. BSc (First class honours degree Chemistry, 1961) and a PhD (Molecular Spectroscopy, 1964) University of Sheffield. Postdoctoral work at the National Research Council (Ottawa, Canada 1964–1966) and Bell Telephone Laboratories (Murray Hill, NJ USA 1966–1967); Tutorial Fellow 1967, Lecturer 1968, Reader 1977, University of Sussex (Brighton) in 1967. Became a Professor in 1985 and a Royal Society Research Professor in 1991. In 1996, was knighted for his contributions to chemistry and later that year, together with Robert Curl and Richard Smalley (of Rice University, Houston, Texas), received the Nobel Prize for Chemistry for the discovery of C_{60} Buckminsterfullerene a new form of carbon.

Research Fields Cover Several Major Topics:

(1) (1961–1970) (i) Electronic spectroscopy of free radicals and unstable intermediates in the gas phase; (ii) Raman spectroscopy of intermolecular interactions in the liquid phase, and (iii) theoretical studies of electronic properties ground and excited states of small molecules and free radicals.

(2) (1970–1980) Research focused on the creation of new molecules with multiple bonds between carbon and elements, mainly of the second and third rows of the Periodic Table (S, Se and P), which were reluctant to form such a link. These studies showed that many of these previously-assumed impossible species could be produced, studied by spectroscopy and used as valuable synthons leading to a wide class of new phosphorus containing compounds. In particular the spectroscopic studies of molecules with carbon–phosphorus

355

multiple bonds (C=P and C≡P) were the pioneering studies that initiated the now prolific field of phosphaalkene/alkyne chemistry.

(3) (1975–1980) Laboratory and radioastronomy studies on long linear carbon chain molecules (the cyanopolyynes) led to the surprising discovery (by radioastronomy) that they existed in interstellar space and also in stars. Since these first observations the carbon chains have become a major area of modern research by molecular spectroscopists and astronomers interested in the chemistry of space.

(4) (1985–1990) The revelation (1975–1980) that long chain molecules existed in space could not be explained by the then accepted ideas on interstellar chemistry and it was during attempts to rationalise their abundance that C_{60} Buckminsterfullerene was discovered. Laboratory experiments at Rice University, which simulated the chemical reactions in the atmospheres of red giant carbon stars, serendipitously revealed the fact that the C_{60} molecule could self-assemble. This ability to self-assemble has completely changed our perspective on the nanoscale behaviour of graphite in particular and sheet materials in general. The molecule was subsequently isolated independently at Sussex and structurally characterised.

(5) (1990–) Present research focuses on Fullerene chemistry and the nanoscale structure of new materials, in particular nanotubes. This has led to a wide range of new nanostructured materials, the first insulated nanowires and new perspectives on the mechanism of nanotube formation.

Key Collaborations: With D. R. M. Walton (Sussex), T. Oka, L. Avery, N. Broten and J. MacLeod (NRC Ottawa) on carbon chain molecules in the laboratory and space; J. F. Nixon on phosphaalkene/alkyne chemistry (at Sussex); with J. P. Hare, P. R. Birkett, A. Darwish, M. Terrones, W. K. Hsu, N. Grobert, Y. Q. Zhu, R. Taylor and D. R. M. Walton on Fullerene chemistry and nanostructures (at Sussex); with R. F. Curl, J. R. Heath, S. C. O'Brien, Y. Liu and R. E. Smalley (at Rice University Texas) on the discovery of Buckminsterfullerene.

Education: Chairman of the board of the Vega Science Trust which produces science programmes for network television. 75 have been made and so far 55 have been broadcast on the BBC Learning Zone educational slot. Member of National Advisory Committee on Cultural and Creative Education.

Scientific Awards etc.: Tilden Lectureship of the RSC (1981); International Prize for New Materials by the American Physical Society (1992, shared with R. Curl and R. Smalley); Italgas Prize for Innovation in Chemistry (1992); Royal Society of Chemistry Longstaff Medal (1993); Hewlett Packard Europhysics Prize (1994, shared with W. Kraetschmer, D. Huffman and R. Smalley); Nobel Prize for Chemistry in 1996 (shared with R. Curl and R. Smalley); American Carbon

Society Medal for Achievement in Carbon Science (1997, shared with R. Curl and R. Smalley); Blackett Lecturship 1999 (Royal Society); Faraday Award and Lecture 2001 (Royal Society); Dalton Medal 1998 (Manchester Lit and Phil), Erasmus Medal of Academia Europaea, Ioannes Marcus Marci Medal 2000 (Prague) for contributions to molecular spectroscopy.

Fellowships etc.: Fellow of the Royal Society (1990); Fellow of the Royal Society of Chemistry; President of the Royal Society of Chemistry (2002–2004); Mexican Academy of Science; Member Academia Europaea (1993); Hon. Foreign Member Korean Academy of Science and Technology (KAST) (1997); Hon. Fellow of the Royal Microscopical Society (1998); Hon. Fellow of the Royal Society of Edinburgh (1998); Hon. Fellow of the RSC (2000).

Honorary Degrees: Université Libre (Bruxelles), Stockholm (Sweden), Limburg (Belgium), Sheffield, Kingston, Sussex, Helsinki (Finland), Nottingham, Yokohama City (Japan), Sheffield-Hallam, Aberdeen, Leicester, Aveiro (Portugal), Bielefeld (Germany), Hull, Manchester Metropolitan, Exeter, Hong Kong City (China), Gustavus Adolphus College (Minnesota, USA), University College London, Patras (Greece), Halifax (Nova Scotia, Canada), Strathclyde; Hon. Fellowship: Bolton Institute.

Graphic Design Work has resulted in numerous posters, letterheads, logos, book/ journal covers, medal design etc. Awards: Sunday Times Book Jacket Design competition (1964) and more recently the Moet Hennessy/Louis Vuitton Science pour l'Art Prize (1994). Citation in the international design annual "Modern Publicity" (1979) for the cover of "Chemistry at Sussex".

TV/Internet Science Programmes: Prix Leonardo Bronze Medal (2001); Chemical Industries Association (Presidents Prize short list, 1998 and 1999).

Harry Kroto inspiring young molecular modellers.

Barry Ward

I was born in 1939 in Co. Durham and attended St Aidan's Grammar School in Sunderland where my favourite teacher taught chemistry and coached football. The choice of career was not easy but I opted for science and graduated in Chemistry at the University of Sheffield in 1961. It was then my good fortune to be accepted by George Porter as a research student and after three tough but rewarding years I was awarded a PhD for a thesis titled "A Study of Some Aromatic Free Radicals by High Resolution Spectroscopy". During the studentship George, kind as always, arranged for me to spend several weeks in Paris, working in Sidney Leach's group on free radical emission spectra.

Sheffield University Chemical Society in 1963. Barry Ward (centre) was chairman, and George Porter (on his left), President. On Barry Ward's right is Professor R. N. Howarth, and on George Porter's left are "Doc" Lawrence and Richard Dixon.

In October 1964, my new wife and I joined the "brain drain" to the USA where I worked in Max Matheson's group applying pulse radiolysis to a variety of liquid and gaseous systems. Some two years later, against the shadow of a looming US draft, I opted for the UK rather than Vietnam and joined Shell Research near Chester where Maurice Sugden had been tempted from Cambridge to establish a Combustion Research Laboratory. Quite coincidentally, Maurice had been the external examiner for my thesis. There, while working on free radicals using ESR and a rotating cryostat, I became interested in the wider context of industrial research and between 1969 and 1971 worked in the office of Lord Rothschild (Head of Shell Research) in The Hague on Research Planning and Programming.

I then moved into Shell Chemicals. Much of my career for the next twenty years was involved with the commercialisation of new products emerging from the laboratories, the struggle to find the right balance between targeted and unprogrammed research and the development of new businesses. Challenging jobs in Delft, Melbourne, Toronto and London and a large family saw those years pass quickly and in 1995 I retired from Shell.

There was now more time to spend on voluntary work, tennis, skiing and trekking in wild places but, perhaps most importantly, there was the opportunity to pursue an interest in Art which had been the victim in the grammar school of a Sophie's Choice between Football and Art on a Wednesday afternoon. In 2001 my wife and I graduated together from the University of Southampton with BA Degrees in Visual Art. The title of my dissertation was "The Importance of the Image in Contemporary Science: a Contribution to the Debate on the Two Cultures". It reflected my special interest in the links between Art and Science. However, nothing I have ever produced which could be labelled as "art" has matched the purity and beauty of the phenyl radical spectra as they emerged in the darkroom in Sheffield in 1964. But the struggle continues.

FLASHES OF MEMORY
Harry Kroto and Barry Ward on
"The High Resolution Absorption Spectroscopy of Aromatic Free Radicals"
G. Porter and B. Ward
J. Chim. Phys. 1517–1522 (1964)

Preamble

It is an honour to be asked to write about one of George's papers as well as a great pleasure. George planted so many indelibly pleasant memories that this task is a good excuse to recall some of them and relive some of the best moments of our student days in Sheffield. We are writing of a time, which in retrospect was the Golden Age of British University Science. The mood of the period was embodied in the 1963 Robbins Report on Higher Education which recognised the economic value of higher education but also stressed that the salaries, ultimately achieved by graduates, were not the only measure of its value to society. It stated that the aim of a university was to produce not mere specialists but rather cultivated men and women. These views were not seriously challenged and the report resulted in a threefold expansion of the University system. The prestige of science in society was at a high, research was deemed a priority and undergraduate places on science degree courses were in demand and highly valued. Research students were quite well supported with DSIR grants and academic openings and job opportunities in the science and engineering industries seemed to be unlimited. At least that is how it seemed at the time.

Our relationships with George over many years were multi-faceted and so we have decided to split the writing task into two personal accounts:

Little Flashes (by Harry Kroto)

There was one paper in particular that I admired and it was a study by Barry Ward, who was my exact contemporary at Sheffield and close friend. Barry carried out the work with George between 1961–1964. It was with some prescience that David Phillips asked me to write about this particular paper and it has given me a great opportunity to collaborate with Barry (really for the first time). During this period we worked on the same apparatus — a flash photolysis apparatus, of course — but one designed specifically for the study of unstable species in the gas phase using a high-resolution spectrometer. All the other flash photolysis systems at Sheffield were attached to relatively low-resolution instruments and focused mainly on reaction kinetics. This one, built by my supervisor Richard Dixon, was designed to detect and analyse the electronic spectra of new small free radicals using the high-resolution spectroscopic approaches developed at the National Research Council in Ottawa, Canada. Although I was a research student working with Richard Dixon, and Barry was working with George and we shared the apparatus, I do not remember a single moment when there was any problem sharing time on the equipment although we both had extremely time-consuming experimental projects. I think this was as much a tribute to the friendly atmosphere that George and Richard engendered as it was to the mutual respect and friendship that Barry and I developed for each other. Of course it also may have helped that Barry played football for Sheffield during the winter months whereas I played tennis for Sheffield during the summer. Our "beloved(!)" flash apparatus catalysed the friendship and it is appropriate that this request has catalysed our first paper — or article — together. It seemed appropriate to split the task of writing this article as Barry can describe the intimate experiences of working directly with George as well as the more specific details associated with their outstanding scientific breakthrough — a breakthrough that revealed, for the first time, detailed spectra of aromatic free radicals at high resolution in the gas phase.

Chemistry Undergraduate Days in Sheffield

Classical thermodynamics had always seemed to me to be a somewhat abstract field, whose mysterious ways were cloaked in the macroscopic view of matter, which had been left behind in the 19th century. Maybe it was appropriate for the building of steam engines like those that I had spent many hours waiting for and watching during the avid train-spotting phase of my childhood, which in its rather fanatical form I guess probably lasted about a year. However there seemed little relationship with the quantum mechanical world of molecules and spectra with which I had become enamoured as I started to become acquainted with atoms, molecules and quantum mechanics. As a consequence I remember well the lectures on statistical mechanics that George gave. The world of thermodynamics, for which I had heretofore had so little feel, suddenly became quite clear as the plethora of thermodynamic terms fell out using straightforward application of statistical relations to energy level data. I also remember that some of these lectures started at 9 am on Saturday mornings — and I do mean Saturday. On one occasion George, having become irritated by latecomers, locked the door to the lecture theatre and the memory of the banging on the door to this day reminds me of the last scene of the film "The Heiress" based on Henry Irving's play "Washington Square".

Down in the Dungeons

Well below ground at the end of a long narrow passage was a darkroom inside which was the 21-foot Eagle high-resolution spectrometer mounted on a massive concrete block hanging on damping springs. The flash apparatus was on the outside and light flashes from the apparatus were focused into the darkroom through a slit in the intervening wall. The tension that built up over long runs lasting many hours (often 6 and sometimes even 18 hours — I think Barry may have done a 24-hour run) to see whether there was anything on the photographic plate was nerve-racking. There is a certain type of mindset with its own vocabulary associated with the moment, when after an 8 or more hour run one opens up the plate holder in the dark and one's fingers discover an empty plateholder revealing the awful fact that you have forgotten to put a plate in the plateholder. It may be because of such experiences that I can hardly bear to watch an infamous scene in "Faulty Towers" in which John Cleese lifts off the cover of a dish only to discover that instead of the cordon bleu roast duck he had promised his clients, it is a blancmange. The scene in which his hands dive into the blancmange in desperate hope of finding an incarcerated duck does evoke the frantic feeling in the dark for the missing plate. It has always seemed to me that it's by learning how to cope with these sorts of exasperating eventualities with equanimity that researchers are created. The plateholder was so incredibly heavy that I think that Richard had built it so he could do a bit of weight lifting on the side or perhaps because he expected the Terminator to do a PhD with him rather than take the softer option of becoming Governor of California. Sometimes long runs of ca 1000 flashes (taking about a minute a flash) were needed. In fact the apparatus that one finds in modern workout clubs is for wimps compared with the repetitive physical exercise machine that George and Richard had set up for Barry and me. It was fortunate that we both played sport fairly seriously. At one point I decided to do a rudimentary ergonomic analysis of the operation of the flash apparatus. Operation of the apparatus required our bending down each time to switch off the main charger unit when the condensers were charged up. This was not much of a problem if only 10 flashes were required for a run but if it was 1000 ...! I decided to rig up a mirror angled so I could remain in a sitting position on a high stool, see the meter under the bench, trip off the charger switch with the big toe of my outstretched left leg while carrying out operations above the bench often far to the right. Barry and I often ended up after long runs paralysed in an extended ballet position. I guess training for Swan Lake was better than training to be stooped over models for Millet's "the Gleaners". One night at about 2 am during one of these long runs I got a pain in my abdomen. I was alone in the building — no health and safety regulations in "them thar days lad" — and I remember crawling on my hands and knees along the corridor and staggering out to a bus stop. I made it home on a late night bus. This sort of eventuality was a necessary component of any PhD in those days. The doctor who came reckoned it was a grumbling appendix and said if it was still bad the next day he would take it out — assuming that I was still alive, that is— "Ahh them wer't days — an'tell students that nowadays and they won't believe yer."

The Paper

When I arrived at Sheffield from Bolton School I had already learned all about benzyne intermediates from my schoolteacher Harry Heaney who had given me a copy of a Quarterly Review paper he had written. Harry had also recommended that I come to read chemistry at Sheffield — so it is all his fault! Harry soon left the school to return to academia and later became Professor of Organic Chemistry at Loughborough University. I remember that it was a 50–50 decision for me as to whether I would do Organic Chemistry research with Professor R. D. Haworth or spectroscopy with Richard for my PhD. I also remember occasionally feeling slightly envious of Barry whose project seemed to combine so beautifully both the organic free radical chemistry that attracted me with the quantum mechanics that had started to fascinate me. It seemed that George had an unerring feel for the key chemical intermediates to detect using the new high-resolution apparatus. I remember many times that George would come in to the upstairs lab where we had desks to see how Barry was getting on and if Barry happened to be downstairs in the Flash Lab, George always appeared equally happy to chat to me about the smaller molecules that Richard and I were working on. I remember one time during a period in which I was trying to detect the ion (NH_3^+) with a hollow cathode lamp; we started to discuss the possibility of detecting negative ions. George went out and came back with a fascinating paper — which I may still have somewhere — entitled something like: "Gibt es ein OH^- radikal?". I know he must have been thrilled to see Richard Saykally's brilliant microwave studies many years later on negative ions and the recent detection of the phenyl radical by microwave spectroscopy.

Later

In 1967 after two years in Canada and a year in the US (Bell Labs, Murray Hill) I returned to an academic position in the UK (University of Sussex). From then on our paths crossed numerous times sometimes at conferences and on every occasion — no matter with whom he might have been speaking — George always greeted me warmly. For a young scientist starting an academic career it was a tremendous boost in confidence to be treated in such a fashion by someone who had since become a Nobel Laureate (an award I used to be somewhat more in awe of than now!). It was with some regret that having asked me to give my first Friday evening RI discourse in 1985 he had just left as Director and was unable to be present — especially as it was one of the first presentations — indeed the first major presentation — on C_{60} Buckminsterfullerene.

Aromatic Flashes (by Barry Ward)

I was incredibly fortunate — not just in having George as an inspiring supervisor but also by being able to draw freely on the help of the brilliant scientists, both staff and students, in his department. George was an excellent research manager and visionary teacher who provided his people with resources and ideas, a firm steer and imaginative support. In this respect Richard Dixon, the eminent spectroscopist and John Murrell, the noted theoretical chemist, and their students were of enormous help at various stages of the work. Harry and I were always friends and never competitors, except on the squash court. We enjoyed our collaboration, although occasionally we did argue about religion without ever resolving the problem of existence. For me it was natural to like and admire a Renaissance man with so many talents who was happy to trade his old guitar in exchange for the volume of the *Transactions of the Faraday Society* containing his first publication, which was on the electronic spectrum of CBr.

The plan for my PhD project was to take the study of aromatic free radicals by absorption spectroscopy into a new dimension using high resolution equipment. These radicals had been a research interest of George's for some ten years during which he had identified strong UV absorptions of the benzyl, phenoxy and anilino radicals. He suggested I start by looking for a weak, long wavelength transition of the benzyl radical and correlating it with emission spectra from electrical discharges. I quickly realised how thorough a scientist he was when at the outset he gave me a list of some 30 papers, many in German, to be read and digested before his return from a trip to the USA. It was clear that there were to be no short cuts. Generally, he supervised with a light touch and regular meetings and reports were not his style. Nevertheless, a message from Judith, his secretary, that the "Prof." would like to see me always raised the pulse.

The experimental side was demanding and, by today's standards, not that safe. Working alone and often overnight in a basement dungeon surrounded by discharging high voltage condensers irradiating some rather noxious compounds can best be described as testing. Our only protection in this hostile environment was a pair of ear-plugs. I liked to think George understood the effort invested in those beautiful spectra but the statement in his Nobel Lecture that "the experiments do not take very long to perform" suggests he had forgotten that some of my (and Harry's) experiments required hundreds of flashes to build up sufficient exposure of the plates! Indeed, I almost missed the Cuban Missile Crisis because of a long-running experiment.

He was a patient and humane man who believed that "young people should always be given the benefit of the doubt" and undoubtedly his patience was stretched by the length of time it took to generate the first results. The transition from undergraduate to research student was not easy and it was more than a year before the benzyl spectrum emerged as a faint line spotted on a photographic plate by the expert eye of Sydney Leach visiting from Paris. From then on progress was rapid and it truly became the exploratory investigation that George had envisaged. The assignments of spectra to particular species relied in large measure on comparisons of results from a range of substituted precursors together with a slice of Occam's Razor. Experiments examining the diffuse spectra of anilino and phenoxy radicals detected new, transient, highly resolved spectra which were very difficult to assign. For a time we thought we had detected benzyne but eventually concluded that it was the cyclopentadienyl radical. Sadly, there was not the time to follow up our speculations about the mechanism of its formation from phenols and anilines as our real interest was then in finding the phenyl radical. After photolysing simple halogenated benzenes we found the phenyl radical absorption at a wavelength where George felt it should be, rather than on the basis of theoretical calculations — for such was not his style. I can still remember his excitement (not a common occurrence as he was usually very composed) when he rushed into the upstairs lab to look at the first plate showing an exquisitely-resolved spectrum. He particularly liked the analysis of the spectra of the halogen-substituted radicals using MO theory, which supported the idea of a $\pi \rightarrow n$ transition from a $\pi^6 n$ ground state of the radical. The vibrational structure could be analysed in terms of benzene-like modes of vibration. These results were first presented at a conference in Bordeaux, where he initiated my appreciation of St. Emilion wines, and at a Royal Society Conversazione. He was always prepared to share the limelight and give credit while attaching great importance to the establishment of priority.

He had a quiet sense of humour: his statement of the Inverse Boyle's Law (the greater the volume, the greater the pressure) was almost a ritual at student functions in Sheffield and in his undergraduate lectures he clearly enjoyed describing the third body in iodine atom recombination as a chaperone. The department's weekly "Coffee and Kinetics" research seminars, at which attendance was expected, were a significant part of our training. In his genial fashion however, he once agreed to move a talk to another day so as to accommodate one of my more important football games, provided I made the coffee as a trade-off.

He was a keen sailor and very competitive. He could not resist a challenge from his department's yachtsmen but unfortunately his Drake succumbed to the student Armada on a cold autumn day in the Peak District. The grace and humour with which a becalmed "Prof." accepted defeat was typical of him. He was a consultant to General Electric (USA) during my time at Sheffield and he valued the contact with industry and thought highly of GE's efforts to apply fundamental scientific insights. His attitude possibly stemmed from his time at the Rayon Research Association.

The results from the project were certainly interesting but limited by the technology available at that time. Perhaps the real merit of the work was that it opened up a new field of study which generated more fascinating questions than answers, questions that would encourage the development of new technological approaches.

Postscript

It was a highly formative experience for us as students and embryonic researchers to have been closely associated with George. Against the background of our youthful existential debates we can look back and say that perhaps his lasting gift to us was a belief in the intrinsic value of Science in our Culture. He was very clear about this when he wrote: "There is, then, one great purpose for man and for us today, and that is to try to discover man's purpose by every means in our power. That is the ultimate relevance of science, and not only of science, but of every branch of learning which can improve our understanding".[1]

Reference

1. A. L. Mackay, *A Dictionary of Scientific Quotations* (Adam Hilger, Bristol, 1991), p. 201.

Reprinted from *Science*, **242**, 1139–1145 (1988)

Space, Stars, C$_{60}$, and Soot

Harold Kroto

Although carbon has been subjected to far more study than all other elements put together, the buckminsterfullerene hollow-cage structure, recently proposed to account for the exceptional stability of the C$_{60}$ cluster, has shed a totally new and revealing light on several important aspects of carbon's chemical and physical properties that were quite unsuspected and others that were not previously well understood. Most significant is the discovery that C$_{60}$ appears to form spontaneously, and this has particularly important implications for particle formation in combustion and in space as well as for the chemistry of polyaromatic compounds. The intriguing revelation that 12 pentagonal "defects" convert a planar hexagonal array of any size into a quasi-icosahedral cage explains why some intrinsically planar materials form quasi-crystalline particles, as appears to occur in the case of soot. Although the novel structural proposal has still to be unequivocally confirmed, this article pays particular attention to the way in which it provides convincing explanations of puzzling observations in several fields, so lending credence to the structure proposed for C$_{60}$.

I N THIS ARTICLE SOME EXCITING NEW AVENUES IN THE chemistry of one element, carbon, which are a consequence of the premise that the C$_{60}$ molecule has the high symmetry of a truncated icosahedron, are explored. We shall see that this novel proposal allows many pieces of the carbon chemistry jigsaw puzzle to fall neatly into place.

When David Walton, some years ago, introduced me to some polyyne ($\cdots$C≡C–C≡C–C≡C–C≡C$\cdots$) chain molecules that

he and his colleagues had made (*1, 2*), they called to mind the problems that a microscopic baton twirler would have in catching a baton that was flexing wildly as it spun in the air. To study this process quantum mechanically we, with Anthony Alexander, made H–C≡C–C≡C–C≡N (HC$_5$N) and measured its microwave rota-

The author is at the School of Chemistry and Molecular Sciences, University of Sussex, Brighton, BN1 9QJ, United Kingdom.

tional frequencies (3). These measurements, which coincided with advances in the detection of molecules in space by radio astronomy pioneered by Townes and co-workers (4), led to a successful search for HC_5N in space with Takeshi Oka, Lorne Avery, Norman Broten, and John MacLeod (5). This detection was surprising at the time and it instigated further work, with Colin Kirby, on HC_7N (6) (Fig. 1a), which led to the even more surprising radio detection of it (Fig. 1b) (7) as well as the even longer HC_9N (8) and $HC_{11}N$ (9). Many molecules appear to form in the cold interstellar medium by ion-molecule reactions as discussed by Herbst and Klemperer (10) and Dalgarno and Black (11). The chains were, however, peculiar, as

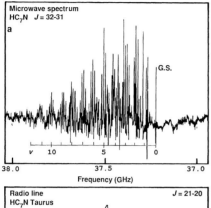

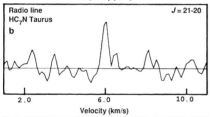

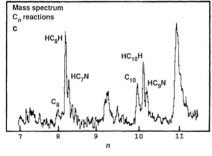

Fig. 1. Data on H–C≡C–C≡C–C≡C–C≡N. (**a**) A microwave rotational transition; J, rotational quantum number; v, lowest binding vibrational quantum number. The "vibrationless" ground-state line is at the far right, G.S. ($v = 0$). Most of the molecules are, however, flexing violently as indicated by the multitude of bending vibrational satellites (all the other lines with $v > 0$). This is the quantum picture of the rotation/flexing process. (**b**) A rotational line detected by radio astronomy from a molecular cloud with a Doppler velocity of 6 km/s in the constellation of Taurus. (**c**) The mass spectrum of the C/H/N reaction. The bare C_8 and C_{10} peaks (at 8×12 and 10×12 amu positions, respectively) develop satellites at +2 amu for HC_8H and $HC_{10}H$ and at +3 amu for HC_7N and HC_9N, respectively.

cool, carbon-rich, red giant stars, which are constantly puffing out copious quantities of dust, seemed more plausible sources (12). In such stars the possibility that some symbiotic chain/dust chemistry, perhaps related to soot formation (13), seemed worth considering and also susceptible to study on the basis of work by Hintenburger, Franzen, and Schuy (14) who had detected 33-atom species in carbon arcs in the 1960s. The possibility of detecting 33-atom polyyne chains in space seemed a wildly exciting prospect.

In a meeting with Robert Curl and Richard Smalley the possibility of simulating carbon star chemistry was proposed. Smalley and co-workers (15) had developed a powerful technique in which a laser vaporizes atoms of a refractory material, such as a metal, in a carrier gas (usually helium) where they reaggregate forming clusters which are then cooled by supersonic expansion, skimmed into a beam, and detected by laser photoionization mass spectrometry. The technique seemed ideal for the study of carbon cluster reactions and spectroscopy. The first carbon cluster-beam study was carried out by Rohlfing, Cox, and Kaldor who found clusters much larger than C_{33} but, most curiously, with only even numbers of atoms (16). Our project, to simulate carbon-star chemistry, was initiated some time later with graduate students Jim Heath and Sean O'Brien, and fascinating new results poured out from day one. Many details are given by Curl and Smalley (17) though in the interests of a self-consistent narrative and an insight into this author's perspective there is some overlap.

Carbon Cluster Results

Initially, cluster reactions were probed which showed that C_n ($n < 30$) clusters did indeed react with H and N to form polyynes, such as HC_7N and HC_9N (Fig. 1c) (18, 19), which had been detected in space, a result satisfyingly consistent with the idea of a stellar source of interstellar chains. The larger clusters were totally inert, and as the experiments progressed it became impossible to ignore the antics of the C_{60} peak which varied from relative insignificance to total dominance depending on the clustering conditions (Fig. 2, a through d).

After much discussion we conjectured that the bizarre behavior, particularly the dominance of C_{60} in Fig. 2c, could be the result of stabilization by closure of a graphitic net into a hollow chicken-wire cage similar to the geodesic domes of Buckminster Fuller (20) [Fig. 3 (21)]. Such closure would eliminate all 20 or so reactive edge bonds of a 60-atom flat sheet. This led to the realization that there was a most elegant and, at the time, overwhelming solution—the truncated icosahedron cage:

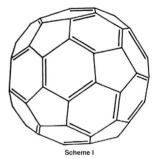

Scheme I

This structure necessitated the throwing of all caution to the wind (the Greek icosahedron) and it was proposed immediately (22); after all, it was surely too perfect a solution to be wrong. We named C_{60}

(23) after Buckminster Fuller, which has turned out to be a highly appropriate name. Subsequently Martyn Poliakoff, at Nottingham, drew our attention to an idea of David Jones who had, in a delightfully inventive article under the pseudonym of Daedalus in the *New Scientist*, already proposed such cages *(24, 25)*. Jones showed that Euler's network closure requirement,

$$12 = 3n_3 + 2n_4 + 1n_5 + 0n_6 - 1n_7 - 2n_8 - \ldots$$

(where n_k is the number of k-sided faces), applies. For carbon only $k = 5$ or 6 are likely (though $k = 7$ should not be overlooked) so 12 pentagonal faces are necessary though the number of hexagonal ones is arbitrary. Jones was influenced by Thompson who had pointed out, in his elegant book *(26)*, that though a sheet of hexagons may

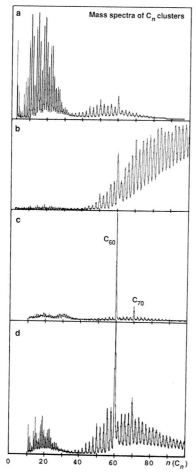

Fig. 2. Mass spectra under various conditions. (**a**) Moderate clustering: both polyyne chain precursors ($n < 36$) and fullerenes ($n > 20$) observed. (**b**) Significant clustering: only even clusters (all fullerenes?) detected. (**c**) Extensive clustering: C_{60} and C_{70} remain and appear to be dominant because the very large clusters are well above the detection range. (**d**) Clustering similar to (c) but at about 100 times the ionizing laser flux: extensive fragmentation of large clusters (particles) into fullerenes and chain precursors occurs.

warp and bend, it can never close. In fact C_{60} itself had been predicted, first in 1970 by Osawa and Yoshida *(27, 28)* and later by Bochvar and Gal'pern *(29, 30)*, Davidson *(31)*, and (coincidentally with our discovery) Haymet *(32)*.

Closed Fullerene Cages

In addition to the basic proviso that only even clusters can close perfectly, some simple physicochemical principles apply to closed 5/6-ring (fullerene) cages *(33, 34)*. They predict stability if (i) curvature-related strain is symmetrically (geodesically) distributed, (ii) there is aromaticity, (iii) the electronic shell is closed, and (iv) pentagons are isolated as much as possible by hexagons to avoid the instability inherent in fused-pentagon configurations. These principles predict unique stability for C_{60} because it is the only 5/6-ring cage for which all atoms are equivalent (so strain is perfectly distributed); has 12,500 resonance forms *(35)*—slightly more than the archetypal aromatic benzene; should have a closed shell of electrons *(36)*; and is, almost obviously, the smallest fullerene for which all pentagons can be isolated. General consideration of principle (iv) reveals that the next fullerene able to avoid abutting pentagons is, surprisingly, C_{70} *(33, 34)* (Fig. 4). This result, together with Fig. 2c, offers strong support for closure as a general phenomenon. Generalization of principle (iv) to the feasibility of isolating fused-pentagon multiplets suggests that C_{50} (Fig. 4) is relatively stable as it is the smallest cage that can avoid fused triplets *(33, 34)*. Further analysis indicates that C_{20} is the least stable, C_{22} cannot exist, C_{24} is the first stabilized cage, and C_{28} is the first cage able to avoid directly fused quartets and be geodesically stabilized (Fig. 4). Thus closure predicts the rather unusual magic number sequence: 20, 24, 28, 32, 36, 50, 60, 70, with 22 totally anti-magic *(33)*. This sequence is compellingly consistent with experiment; see, for instance, Fig. 5 *(37)*. This leads to the inescapable conclusion that the whole set are fullerenes, and perhaps also are all the large even clusters in Fig. 2b as well as those off-scale. Note that odd-numbered clusters cannot close perfectly.

The Icospiral Particle Nucleation Scheme

How could we have inadvertently synthesized C_{60} in 100 μs when Paquette and co-workers had taken somewhat longer to make dodecahedrane, $C_{20}H_{20}$ *(38)*? How could the entropy factor inherent in the spontaneous creation of so symmetric an object from a chaotic plasma have been overcome? Herein lies the most important aspect of the discovery, and the answer appears to explain many puzzling features of the mysterious process that occurs whenever wood burns to form soot. It also seems to present a plausible mechanism for primordial particle formation in space. In addition to polyynes and particles, a new character, C_{60}, has emerged, whose shadowy role, like that of the Third Man, has only now come to light.

The answer itself is incredibly simple. Although the stability of graphite indicates that carbon must form hexagonal networks rather readily, we also know (now) that such networks cannot close perfectly. A moment's thought, however, suggests that, for a small sheet, energetics should favor closure owing to the bond energy released by eliminating the reactive edge. Indeed, triple (benzyne) bonds will make the edges curl anyway *(39)*. To curve toward closure pentagons are necessary, and from the appearance of C_{60} we conclude that they must occur rather frequently. After all, saucer-shaped corannulene ($C_{20}H_{10}$, a pentagon surrounded by five hexagons) is stable *(40)* and pentagons are legion in tars *(41)*. The

Fig. 3. View inside the U.S. pavilion at Expo 67 in Montreal (*21*). One pentagon is evident. Note that the hexagons are distorted. [With permission of Graphis Press, Zurich]

original nucleation scheme (*42*) has been refined by introducing epitaxial control along with the energetic considerations, and this has enabled a study, with Kenneth McKay, of the detailed structures of the resulting particles to be made (*43*). The initial hypothetical steps involve highly reactive open spiral shell (nautilus-like) embryos (Fig. 6). Smaller accreting carbon fragments are mopped up by adsorption on the surface of such shells and rapidly knit into the advancing edge. Polyyne cross-linking is explosive so adsorption is probably rate-limiting, resulting in kinetics proportional to the surface area of the nucleus. Once the trailing edge has been bypassed (Fig. 6c), closure should be impossible (unless annealing takes place) and new network will form under epitaxial control so that the skin lies as closely as possible to 3.4 Å (the graphite interlayer spacing) above the lower layer (Fig. 6d). Now the peculiar 12- and only 12-pentagon closure requirement, together with epitaxy, result in a particle with a fascinating shape and internal structure.

Before describing this, we should recall that the aim of the mechanism was to account for the spontaneous creation of C_{60}. This is readily explained by the occasional, statistical closure during nucleation of a network with the correct disposition of pentagons. On closure C_{60}, along with other fullerenes, is unable to grow and is left behind as the rest "snowball" into giant particles. Ultimately we are left with C_{60} (and C_{70}) as the lone survivors, surrounded by large particles. That large particles are present is demonstrated by the crucial results in Fig. 2d, which show that they can be photofragmented into smaller, detectable species by high-power laser radiation. Probably the less stable fullerenes are attacked and devoured by

a host of voracious carbon particles. This scenario suggests that early C_{60} signals might include contributions from closed isomers that react to form large clusters or rearrange to the ultrastable C_{60} itself.

To predict the structures of the large particles, symmetric giant fullerenes with up to 6000 atoms have been constructed as a preliminary to computer study. This yielded an unexpected result in that, as the fullerenes evolve to ever larger structures, the more or less round shape of the small ones metamorphoses into a more or less icosahedral shape (Fig. 7). The truncation remains, but at microscopic dimensions, forming 12 effective cusps in a smoothly warping single continuous plane of hexagons. The giant fullerenes are essentially the structures of Daedalus (*24, 25*), but they are not round. A puzzle in the analogy with the geodesic domes was herewith resolved: In the domes the lengths of the struts near the pentagons are very unequal (Fig. 3) (*21*), and this is necessary to achieve a smooth round shape. Carbon bonds, on the other hand, will not "conveniently" extend and contract and so the bulk of the curvature is focused at the 12 cusps. An intriguing way to consider this surface is as a single sheet of hexagonal graphite with 12 pentagonal defects (only 12, whatever the size) systematically inserted during growth to form an essentially continuous icosahedral monosurface. The growing onion-like particle would have a distorted icosahedral shape, and, as epitaxy causes the cusps to lie more or less above each other, the resulting polyhedral shells will be effectively similar in the geometric sense (*43*).

This result enabled a further puzzling result to fall neatly into place. Sumio Iijima had drawn our attention to his beautiful electron microscope pictures of carbon particles taken in 1980 (*44*), which showed concentric graphitic shell internal structure with polyhedral shapes (Fig. 8). The scheme explains such particles as quasi-icosahedral spirals, which are quasi-crystals of graphite composed entirely of sp^2 hybridized carbon. There are 20 pyramidal graphite microcrystallite wedges, the layers of which curve into the layers of those that abut on the three internal faces. C_{60} should not lie at the center of the particle. It is intriguing that the icospiral shell, which does not appear to have been discussed previously, combines the two most bioemotive shapes, the helix and the icosahedron.

Soot

The scheme predicts C_{60} to be a by-product of soot formation (*42*), and it is most satisfying that significant support for the C_{60} proposal is provided by the work of Gerhardt, Löffler, and Homann who recently found that C_{60}^+ is indeed a dominant ion in a sooting flame (*45*). As there is much hydrogen in soot, the question arises as to where it is. The answer may lie in the fact that there is no necessity for pentagons to form stable more or less closed structures in the

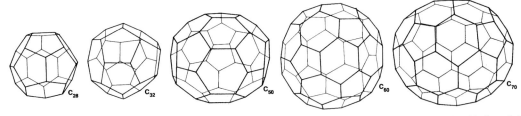

Fig. 4. Five "magic" fullerenes. C_{28} is a very interesting tetrahedral molecule with a striking family resemblance to Gomberg's stable free radical triphenylmethyl, $\cdot C(C_6H_5)_3$, the first organic free radical and the forerunner of free radical chemistry. C_{32}, here viewed more or less along its threefold axis, is handed. One of two feasible C_{50} semistable isomers is shown and the one and only C_{60} buckminsterfullerene. C_{70} is formed by separating two halves of C_{60} by a ring of ten extra atoms to form a U.S. football.

Fig. 5. A particularly interesting spectrum taken from the data of Cox *et al.* (*37*). The correspondence with the magic fullerene sequence, 24, 28, 32, 50, 60, and 70 (*33*), is compelling evidence for closure. Note that only C_{60} and C_{70} are ultrastable (Fig. 2c). The sharp cutoff at C_{24} is particularly satisfying as no C_{22} fullerene is possible.

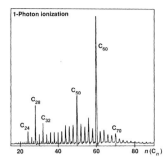

presence of hydrogen (*43, 46*); stability can be achieved simply with hexagonal nets and internal C–H bonds (Fig. 9). Soot forms under conditions where the balance between the C–C network and C–H bond formation is continually oscillating and thus gaps in the network must occur as an icospiral grows. This will spoil the particle's polyhedral angularity as epitaxial control is lost and gaps—perhaps large ones—are covered over, trapping C–H and dangling bonds in a three-dimensional spiral with a smoother, more spheroidal shape (*46*).

Cosmic C_{60}

When polyynes and particles form in the laboratory, the fullerenes always turn up as well (Fig. 2). As this must also be true in space, it is worth considering the possible astrophysical consequences of C_{60}, the other fullerenes, and icospiral particles (*47–50*). It should be stressed that the consequences are based on the experimental observations and are for the most part totally independent of the structure proposal, which must still be considered circumstantial though it does tie the whole picture together in a most neat and satisfying way.

It seems likely that the cosmic particles associated with polyynes will have basic icospiral structure with varying degrees of hydrogenation. Such species will have infrared signatures similar to those of polyaromatic hydrocarbons (PAH). Indeed, that vital link in planet formation, the primordial solid particle, may well have been carbonaceous and if so it too could have had icospiral structure. The infrared emission from interstellar material near some stars has features consistent with PAH-like material (*51, 52*) and thus also the particles discussed here. Indeed, the presence of internal C–H bonds such as those in Fig. 9 gives rise to bands that allow an even more satisfactory fit to the interstellar data (*53*).

Rieu, Winnberg, and Bujarrabal (*54*) recently found HC_7N outside the Egg Nebula, which is a dust cloud (*55*) surrounding a star (Fig. 10). This suggests that polyynes are produced by ambient starlight photofragmentation of dust (*56*). As chains and fullerenes also form during particle nucleation (Fig. 2, a through c) and again during particle photofragmentation (Fig. 2d), we may conclude that C_{60} and other fullerenes are also produced and can survive together with polyynes in the Egg's extended envelope as well as other interstellar regions where chains and dust occur. The results presented in Fig. 2 show that a family of carbon species with 30 to 100 atoms is formed on carbon nucleation and on carbon photofragmentation, suggesting some degree of resilience. The dimensions of these species must be ~8 ± 2 Å, a result independent of the structure. This experimental observation ties in intriguingly well (*56*) with the interesting suggestion by Sellgren that certain interstellar infrared features are consistent with emission from particles with dimensions of ~10 Å (*57*).

Fig. 6. The icospiral nucleation process starts with a reactive saucer-shaped (corannulene) C_{20} (**a**) and grows by accretion via the half-shell (**b**) to (**c**), where edge bypass has occurred, and then on to (**d**), where the second growth shell is forming under epitaxial control. Statistical closure at stage (c) is proposed as the explanation for fullerene formation.

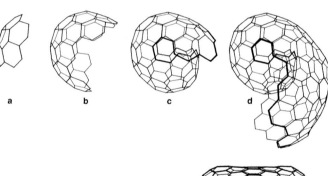

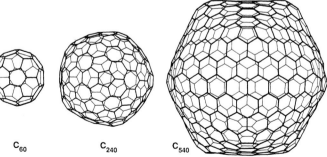

Fig. 7. C_{60}, C_{240}, and C_{540} with relative diameters 1:2:3. The rapid shift toward icosahedral shape is dramatic. The objects are really truncated but the truncation remains at microscopic dimensions as the cluster grows to macroscopic size. The surface thus becomes a smoothly curving net with 20 more or less flat triangular segments connecting the 12 cusps as in an icosahedron.

The ubiquitously distributed molecular material responsible for the Diffuse Interstellar Bands (58) has remained a mystery for decades, and there have been numerous conjectures as to the identity of the carrier. The most important aspect of the problem is that there are not many spectroscopic features, about 40, and so very few carriers can be responsible and they must be very stable to starlight. The carbon nucleation scheme is unique in that it alone offers a scenario for the production of relatively few stable large species from a chaotic interstellar mix. Starlight will ionize much of the interstellar C_{60} present and so C_{60}^+ should be ubiquitously distributed in space, making it a viable candidate for the carrier (47, 49, 50). Again, this is an experimental conclusion quite independent of the structure proposed for C_{60}. Heath et al. (59) have detected a weak absorption feature for neutral C_{60} which does not coincide with any known diffuse feature. Similar arguments to the above hold for another puzzling interstellar feature, the 2170 Å band. The ease of formation and the resilience of metallofullerene caged-metal complexes (60) indicate that they also must be abundant in space. Optical spectra of such species as $C_{60}Na$, $C_{60}K$, and $C_{60}Ca$, most probably ionized, are also worthy of consideration (47, 49).

Some interesting isotope anomalies occur in some meteorites (61), and it is possible that the metallofullerenes could play a role (62) in, for instance, the case of the high abundance of ^{22}Ne (63). Clayton has pointed out that as large amounts of ^{22}Na, which decays via β^+ emission to ^{22}Ne with a 2.6-year half-life, are ejected from the helium shells of supernovae and the surface explosions of novae, this element should precipitate in intimate chemical association with any surrounding dust as ejected material expands and rapidly cools (64). Such conditions seem ideal for the formation of C_nNa metallofullerenes in which ^{22}Na decay produces caged ^{22}Ne which cannot escape (62). Such species, occluded in carbonaceous chondrites, offer

support for the suggestion that the ^{22}Ne is an extinct radioactive anomaly bearing witness to the meteorite's presolar origin (64). Alkali and alkaline-earth metals readily intercalate in carbon particles, an observation consistent with the icospiral graphite model. The aggregation of a mix of metallofullerenes and particles with intercalated atoms suggests entrapment sites of at least two kinds for which there is some evidence (61). The icospiral particle might also provide a convenient nucleus for efficient pressure- or shockwave-induced metamorphosis to diamond (62), perhaps bearing on the recent exciting discovery of diamond-like domains in meteorites (65). In this case a third possible trapping site is suggested, as such a scenario leads to diamond zones with significant impurity loads.

Discussion

There is a serious note to the story relating to the Chernobyl disaster. The icospiral is a perfect vehicle for windborne dispersal of

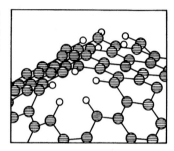

Fig. 9. Hydrofullerenes have hydrogen atoms (open circles) replacing carbon atoms (hatched circles) at the pentagonal cusps.

Fig. 8. Particles photographed by Sumio Iijima (44) which show clear evidence of polyhedral concentric shell structure consistent with the icospiral graphite particle model. At the right the contours are delineated.

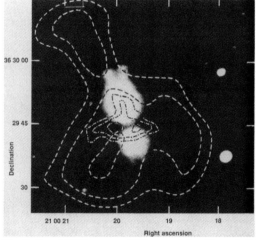

Fig. 10. Map of HC_7N (– – –) and NH_3(– · – · –) from Rieu et al. (54) superimposed on a photograph of the Egg Nebula scattering (55). The dust emanates in two lobes from a central star. These data suggest that HC_7N results from fragmentation of carbonaceous dust by starlight; compare with Fig. 2d. Laboratory data (Fig. 2) indicate that even carbon clusters, in particular C_{60} (most likely ionized as C_{60}^+), must be present along with chains.

trapped (radioactive) atoms. This recalls the Minotaur imprisoned in a maze designed (coincidentally) by the original Daedalus, the great craftsman of antiquity. Unfortunately in this case the prisoner need not escape to cause havoc.

A rather esoteric interest in quantum molecular dynamics led to the discovery of long chains in space and subsequently the discovery of C_{60}, the icospiral carbon particle, and the associated nucleation scheme. The associated prediction that C_{60} should be a by-product of combustion and a key to the soot formation process is a prime example of the way in which an interest in fundamental problems for their own intrinsic sake, irrespective of their predicted use, can yield results of applied significance.

If the spontaneous C_{60} formation proposal is correct, and as we have seen all the observations augur well, then a new direction in organic chemistry is presaged. Under the correct conditions, presumably those which favor sp^2 carbon network formation (perhaps under epitaxial control) at the expense of other pathways, puckered nets and closed cages should be susceptible to synthesis. Indeed, there is some evidence in the elegant corannulene synthesis of Barth and Lawton (40) that the step in which saucer shape was achieved may be more favored than some flat-earth chemists might think.

For me the most intriguing revelation of the whole C_{60} story has been the realization that the inclusion of pentagonal, or effectively pentagonal, defects in the propagating front of a growing layer results in cusps; and with but 12 cusps an icosahedral shell results, whatever the size. It seems likely that this is the solution to a wide range of problems such as, for instance, the formation of suspensions of materials that tend to form layers and the microscopic structures of intrinsically layered materials. There does not now seem to be any reason why any molecular or atomic array should grow as a perfectly flat plane unless epitaxially restricted to do so. In fact, energetics at the microscopic level suggests that materials should form under the influence of the mysterious, ubiquitous, and all-powerful icosahedral template. It is thus very intriguing to find

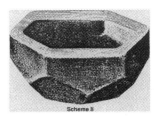

Scheme II

hidden in the carbon literature this schematic drawing by Heydenreich, Hess, and Ban of a graphitized carbon black particle, 500 million pounds of which were produced in the United States in 1968 (66).

REFERENCES AND NOTES

1. R. Eastmond, T. R. Johnson, D. R. M. Walton, *Tetrahedron* **28**, 4601 (1972).
2. B. F. Coles, P. B. Hitchcock, D. R. M. Walton, *J. Chem. Soc. Dalton Trans.* **1975**, 442 (1975).
3. A. J. Alexander, H. W. Kroto, D. R. M. Walton, *J. Mol. Spectrosc.* **62**, 175 (1976).
4. A. C. Cheung, D. M. Rank, C. H. Townes, D. C. Thornton, W. J. Welch, *Phys. Rev. Lett.* **21**, 1701 (1968).
5. L. W. Avery, N. W. Broten, J. M. MacLeod, T. Oka, H. W. Kroto, *Astrophys. J.* **205**, L173 (1976).
6. C. Kirby, H. W. Kroto, D. R. M. Walton, *J. Mol. Spectrosc.* **83**, 26 (1980).
7. H. W. Kroto *et al.*, *Astrophys. J.* **219**, L133 (1978).
8. N. W. Broten, T. Oka, L. W. Avery, J. M. MacLeod, H. W. Kroto, *ibid.* **223**, L105 (1978).
9. M. B. Bell, S. Kwok, P. A. Feldman, H. E. Matthews, *Nature* **295**, 389 (1982).
10. E. Herbst and W. Klemperer, *Astrophys. J.* **185**, 505 (1973).
11. A. Dalgarno and J. H. Black, *Rep. Prog. Phys.* **39**, 573 (1976).
12. H. W. Kroto, *Int. Rev. Phys. Chem.* **1**, 309 (1981).
13. ———, *Chem. Soc. Rev.* **11**, 435 (1982).
14. V. H. Hintenberger, J. Franzen, K. D. Schuy, *Z. Naturforsch. Teil A* **18**, 1236 (1963).
15. J. B. Hopkins, P. R. R. Langridge-Smith, M. D. Morse, R. E. Smalley, *J. Chem. Phys.* **78**, 1627 (1983).
16. E. A. Rohlfing, D. M. Cox, A. Kaldor, *ibid.* **81**, 3322 (1984).
17. R. F. Curl and R. E. Smalley, *Science* **242**, 1017 (1988).
18. J. R. Heath, Q. Zhang, S. C. O'Brien, R. F. Curl, H. W. Kroto, R. E. Smalley, *J. Am. Chem. Soc.* **109**, 359 (1987).
19. H. W. Kroto, J. R. Heath, S. C. O'Brien, R. F. Curl, R. E. Smalley, *Astrophys. J.* **314**, 352 (1987).
20. R. B. Fuller, *Inventions—The Patented Works of Buckminster Fuller* (St. Martin's Press, New York, 1983).
21. M. Proulx (photographer), *Graphis* **132**, 378 (1967) (Graphis Press, Zurich).
22. H. W. Kroto, J. R. Heath, S. C. O'Brien, R. F. Curl, R. E. Smalley, *Nature* **318**, 162 (1985).
23. A. Nickon and E. F. Silversmith, *Organic Chemistry—The Name Game: Modern Coined Terms and Their Origins* (Pergamon, New York, 1987).
24. D. E. H. Jones, *New Sci.* (3 November 1966), p. 245.
25. ———, *The Inventions of Daedalus* (Freeman, Oxford, 1982), pp. 118–119.
26. D. W. Thompson, *On Growth and Form* (Cambridge Univ. Press, Cambridge, 1942).
27. E. Osawa, *Kagaku (Kyoto)* **25**, 854 (1970) (in Japanese).
28. Z. Yoshida and E. Osawa, *Aromaticity* (Kagakudojin, Kyoto, 1971) (in Japanese).
29. D. A. Bochvar and E. G. Gal'pern, *Dokl. Akad. Nauk SSSR* **209**, 610 (1973) [English translation, **209**, 239 (1973)].
30. I. V. Stankevich, M. V. Nikerov, D. A. Bochvar, *Russ. Chem. Rev.* **53**, 640 (1984).
31. R. A. Davidson, *Theor. Chim. Acta* **58**, 193 (1981).
32. A. D. J. Haymet, *J. Am. Chem. Soc.* **108**, 319 (1986).
33. H. W. Kroto, *Nature* **329**, 529 (1987).
34. T. G. Schmalz, W. A. Seitz, D. J. Klein, G. E. Hite, *J. Am. Chem. Soc.* **110**, 1113 (1988).
35. D. J. Klein, T. G. Schmalz, G. E. Hite, W. A. Seitz, *ibid.* **108**, 1301 (1986).
36. P. W. Fowler and J. I. Steer, *J. Chem. Soc. Chem. Commun.* **1987**, 1403 (1987).
37. D. M. Cox, K. C. Reichmann, A. Kaldor, *J. Am. Chem. Soc.* **110**, 1588 (1988).
38. L. A. Paquette, R. J. Ternansky, D. W. Balogh, G. Kentgen, *ibid.* **105**, 5446 (1983).
39. J. Almlöf, in *Carbon in the Galaxy*, S. Chang, Ed. (National Aeronautics and Space Administration, Washington, DC, in press).
40. W. E. Barth and R. G. Lawton, *J. Am. Chem. Soc.* **93**, 1730 (1971).
41. E. Clar, *Polycyclic Hydrocarbons* (Academic Press, New York, 1964), vols. 1 and 2.
42. Q. L. Zhang *et al.*, *J. Phys. Chem.* **90**, 525 (1986).
43. H. W. Kroto and K. G. McKay, *Nature* **331**, 328 (1988).
44. S. Iijima, *J. Cryst. Growth* **50**, 675 (1980).
45. P. H. Gerhardt, S. Löffler, K. H. Homann, *Chem. Phys. Lett.* **137**, 306 (1987).
46. H. W. Kroto, K. G. McKay, D. R. M. Walton, S. G. Wood, in preparation.
47. H. W. Kroto, in *Polycyclic Aromatic Hydrocarbons and Astrophysics*, A. Leger *et al.*, Eds. (Reidel, Dordrecht, 1987), p. 197.
48. ———, *Proc. R. Inst. G.B.* **58**, 45 (1986).
49. ———, *Philos. Trans. R. Soc. London Ser. A* **325**, 405 (1988).
50. ———, in *Carbon in the Galaxy*, S. Chang, Ed. (National Aeronautics and Space Administration, Washington, DC, in press).
51. W. W. Duley and D. A. Williams, *Mon. Not. R. Astron. Soc.* **196**, 269 (1981).
52. A. Leger and J. L. Puget, *Astron. Astrophys.* **137**, L5 (1984).
53. H. W. Kroto, in preparation.
54. Nguyen-Q-Rieu, A. Winnberg, V. Bujarrabal, *Astron. Astrophys.* **165**, 204 (1986).
55. E. P. Ney, K. M. Merrill, E. E. Becklin, G. Neugebauer, C. G. Wynn-Williams, *Astrophys. J.* **198**, L129 (1975).
56. H. W. Kroto and M. Jura, in preparation.
57. K. Sellgren, *Astrophys. J.* **277**, 623 (1984).
58. G. H. Herbig, *ibid.* **196**, 129 (1975).
59. J. R. Heath, R. F. Curl, R. E. Smalley, *J. Chem. Phys.* **87**, 4236 (1987).
60. J. R. Heath *et al.*, *J. Am. Chem. Soc.* **107**, 7779 (1985).
61. E. Anders, in *Meteorites and the Early Solar System*, J. F. Kerridge, Ed. (Univ. of Arizona Press, Tucson, 1987), chap. 13.1.
62. K. McKay, L. Dunne, H. W. Kroto, in preparation.
63. D. C. Black and R. O. Pepin, *Earth Planet. Sci. Lett.* **6**, 395 (1969).
64. D. D. Clayton, *Nature* **257**, 36 (1975).
65. R. S. Lewis, T. Ming, J. F. Wacker, E. Anders, E. Steel, *ibid.* **326**, 160 (1987).
66. R. D. Heydenreich, W. M. Hess, L. L. Ban, *J. Appl. Crystallogr.* **1**, 1 (1968). Scheme II is published here with the permission of the *Journal of Applied Crystallography*.
67. I am very grateful to many colleagues, in particular: A. J. Alexander, E. Anders, D. D. Clayton, J. D. Cornforth, R. F. Curl, F. Diederich, L. Dunne, P. A. Fowler, M. Jura, J. R. Heath, D. E. H. Jones, C. Kirby, W. Klemperer, K. G. McKay, A. McKay, S. O'Brien, T. Oka, G. Ozin, A. Parsonage, J. Pethica, T. Schmalz, R. E. Smalley, S. J. Wood, and B. Zuckerman. I also thank S. Iijima for sending the images from which Fig. 8 was produced. Finally, I am particularly grateful for the help and encouragement of D. R. M. Walton during many crucial phases of this work.

Contribution from

F. Sherwood Rowland

University of California, Irvine
USA

Born 28th June 1927. Educated at Ohio Wesleyan University (BA, 1948), and University of Chicago (MS, 1951 and PhD, 1952). Was on the Faculty at Princeton from 1952–1963, before becoming Professor of Chemistry at the University of Kansas, 1964–1970. Since then, has been Professor of Chemistry at University of California, Irvine, where he is currently Bren Research Professor, Earth System Science, School of Physical Sciences. Honours include the Tyler Prize in Ecology (1983), Japan Prize in Environment and Energy (1989), American Chemical Society Peter Debye Award (1993), American Geophysical Union Roger Revelle Medal (1994), and the Nobel Prize in Chemistry (1995) joint with M. Molina for work on the depletion of the ozone layer by chlorofluorocarbon molecules. Is a member of the National Academy of Sciences since 1978, and has been Foreign Secretary since 1994.

Sherry Rowland meets the Pope in 1983.

Sampling the Vatican atmosphere.

F. Sherwood Rowland on
"Studies of Free Radical Activity by the Methods of Flash Photolysis. The Reaction Between Chlorine and Oxygen"
G. Porter and Franklin J. Wright
Faraday Society Discussion **14**, 23–34 (1953)

George Porter's introduction of flash photolysis into chemical kinetics in the immediate aftermath of World War II opened the field of gas phase free radical chemistry. Direct observations of concentrations on a microsecond time were now available. The scientific pressures during the war had brought orders of magnitude increases in financial support for instrumental development, and much of this equipment became readily available as surplus when fighting had ended. A very significant aspect of the flash photolysis frontier was the creation for the first time in the laboratory free radical concentrations so high that (a) their spectra were available for study, and as identification for kinetic purposes, and (b) that radical–radical reactions often became highly competitive with the much more familiar reaction involving radical attack on stable molecules. As described in the first sentence of the Porter/Wright paper under *Experimental*, the Porter group had already identified 20, and rapidly growing in number, free radicals — half of them previously unknown. The flash photodecomposition of Cl_2 into Cl atoms in the presence of O_2 ran through such radicals as ClOO into ClO, and then the dimer of the latter, $(ClO)_2$. These chlorinated species were only a small part of the flash revolution because each of these species could now react not only with itself, but with all of the other new radicals as well. But the ClO radical was now available for study, and the $(ClO)_2$ dimer had entered the literature.

I entered graduate school in chemistry at the University of Chicago in autumn 1948. The wartime efforts in nuclear chemistry toward the development of nuclear bombs had exposed in passing innumerable opportunities for the peacetime application of the newly discovered radioisotopes of most of the elements in the Periodic Table. Although the world's first nuclear reactor, constructed under the football stands of Stagg Field at the University of Chicago had already been dismantled, the University had managed to attract to the post-war Departments of Chemistry and Physics a stellar group of scientists. I was fortunate enough to be able to attend lectures in physical chemistry by Harold Urey, inorganic chemistry by Henry Taube, nuclear physics by Willard Libby — past or future Nobel Prize winners. Edward Teller taught the course in chemical physics and Nobel Laureate James Franck was present as well.

Professor Libby's initial post-war research effort had concerned the postulated production of ^{14}C in the lower stratosphere by cosmic ray bombardment, and had grown into full bloom as the basis of carbon-14 dating of archeological and geological artefacts with ages going backward in time to about 25,000 years before the present. His research proposal to me in 1949 as a new research student was the study of the reactions of radioactive bromine atoms freshly created with energies minuscule on the nuclear scale, but with kinetic energies per atom of hundreds of electron volts, enormous on the chemical scale. These bromine atoms included three different radioisotopes with half-lives of 18 minutes, 4.5 hours, and 36 hours, and each of these was explosively released from its existing chemical bonds (as $C_5H_{11}Br$, for instance). However after losing most of its excess kinetic energy to glancing collisions with its surrounding, a substantial portion of the radiobromine atoms re-entered organic combination. This kinetic arena became known as "hot area chemistry", and was avidly pursued by many radiochemists around the world for the next couple of decades.

My own research group worked extensively on the reactions of energetic tritium atoms (^3H = T) formed by the nuclear reactions $^3He(n,p)T$ [gas phase} and $^6Li(n, ^4He)T$ {condensed phase]. We also utilised the photolysis of TBr with UV radiation to produce T atoms with initial energies of 2.8 electron volts at 184.9 nanometers, or less energy with longer UV wavelengths. Hydrogen atoms including the isotopes D and T, do not react with methane at room temperature, and then only by abstraction of H to form molecular hydrogen as a product at higher temperatures. However with photochemically-induced energies of 3 electron volts available, the substitution of T for H in methane is readily observed with a threshold of 1.5 electron volts. Similarly, highly-energetic radioactive ^{18}F atoms were observed to react with the usually inert molecule CF_4 to form $CF_3^{18}F$, and even the radical products $CF_2^{18}F$ and $CF^{18}F$. The formation of $CF^{18}F$ is almost 10 electron volts endothermic, and serves as a useful reminder that "inert" polyatomic molecules are not necessarily inert when abundant amounts of energy are available within the molecule.

I will not go into detail as to how and why Dr Mario Molina and I began in October 1973 to explore the behaviour of the "inert" molecules, CCl_3F and CCl_2F_2. Suffice it to say, we concluded that each of these molecules could absorb solar ultraviolet radiation when the wavelengths became shorter than 220 nanometers, and would decompose as given in reaction (1) with the release of atomic chlorine:

$$CCl_xF_y + h\nu \rightarrow CCl_{x-1}F_y + Cl \qquad (1)$$

However, solar radiation of such short wavelengths is not available in the lower atmosphere because of absorption at higher altitudes by O_2 or O_3. Only when the chlorofluorocarbon molecules have risen above most of the O_2 or O_3 do they encounter this energetic ultraviolet radiation and undergo decomposition. The Cl atoms then react with O_3 by reaction (2) to produce ClO and O_2. Frequently, the follow-on reaction to (2) is the interception of ClO by O in (3), with the

$$Cl + O_3 \rightarrow ClO + O_2 \tag{2}$$

$$ClO + O \rightarrow Cl + O_2 \tag{3}$$

$$O + O_3 \rightarrow O_2 + O_2 \tag{4}$$

consequence that the sequence (2) plus (3) provides a catalytic chain reaction to convert one atom — which usually would have combined with an O_2 molecule to form and O_3 molecule — and a bona fide O_3 molecule back to two molecules of O_2. The system is then ready to repeat the process many thousands of times — thereby causing stratospheric ozone depletion.

The situation in the stratosphere has some analogies to the laboratory conditions first encountered in Porter's flash photolysis experiments. The stratosphere, composed primarily of such chemically unreactive molecules as N_2, O_2, Ar, CO and H_2O, is such that substantial fractions of atoms such as Cl, H and N (other than N_2) exist as free radicals in ClO_x (Cl, ClO), HO_x (HO, HO_2) and NO_x (NO, NO_2, NO_3) series. Furthermore, the cross-product reactions among these radical chains form important molecules in the stratosphere, including HOCl, H_2O_2, $HONO_2$ (nitric acid, but written to demonstrate H bonding to O, not N), $ClONO_2$, HONO, ClONO, N_2O_5, and Porter and Wright's $(ClO)_2$.

Reaction (2) occurs so much faster than (3) in the stratosphere because of the much greater concentration of O_3 than O, that detection of this ClO_x chain quickly centred on the radical ClO, which became the "smoking gun" symbol of ozone depletion occurring in the stratosphere, successfully carried out in 1975. Our realisation that the photolysis rate of the cross-product molecule chlorine nitrate ($ClONO_2$) had been overestimated in the original Rowland/Molina paper brought strong emphasis on detection of this molecule in the stratosphere in 1976. Laboratory work on chlorine nitrate is difficult because it reacts on contact with water molecules on almost any surface, but spectroscopic detection in the stratosphere was successful in 1977.

Laboratory experiments by Haruo Sato demonstrated in 1984 that the reactions of HCl and H_2O with $ClONO_2$ were extremely fast with the formation of HNO_3 and the release of Cl_2 and HOCl, respectively. In the following year, the heavy loss of ozone over Antarctica every springtime was reported by the British Antarctica Survey, and quickly confirmed by NASA's Nimbus-7 satellite. This loss was attributed to a chemical mechanism involving these two reactions, especially the one with HCl, and involving an entirely new ClO_x chain reaction. In the lower stratosphere with the sun near the horizon in Antarctica, O atoms for reaction (3) are scarce, and reaction (2) is now followed by the dimer-forming reaction (4) of Porter and Wright to make $(ClO)_2$, and its subsequent photolysis (5) plus (6) returns the Cl to the atomic form without an O atomic reaction.

$$2(Cl + O_3) \qquad \rightarrow 2ClO + 2O_2 \qquad\qquad (2) + (2)$$
$$ClO + ClO + M \rightarrow (ClO)_2 + M \qquad\qquad (4)$$
$$(ClO)_2 + h\nu \qquad \rightarrow Cl + ClOO \qquad\qquad (5)$$
$$ClOO \qquad\qquad \rightarrow Cl + O_2 \qquad\qquad\qquad (6)$$

$$O_3 + O_3 \qquad\qquad \rightarrow O_2 + O_2 + O_2$$

A new wrinkle in the radical–radical reactions has been raised with the consideration that a radical such as ClO might bond at either the Cl or O end of the molecule, and NO_2 might bond either to N or to O. *Ab initio* calculations have moved forward sufficiently that good energetic calculations have been made of the expected energies of the possible isomers, including *cis* and *trans* varieties where they occur. These calculations were carried out with the cooperation of Professor Warren Hehre and his research group. When ClO bonds to itself, the molecule might hypothetically have the structure ClOOCl or ClOClO, or even ClClO$_2$, but the winner is the peroxy form, ClOOCl. The cross-product of ClO with NO_2 could similarly result in ClOONO, ClONO$_2$, OClNO$_2$ or OClONO. The calculations indicate that only the ClONO$_2$ isomer is energetically accessible when ClO reacts with NO_2, and no evidence exists for the transient formation. Finally the HO radical might react with NO_2 by attack on either the N atom to form nitric acid HONO$_2$, or the peroxy form HOONO. In this case small amounts of HOONO were reported in matrix isolation experiments at very low temperatures, and more recently this isomer has been detected in the stratosphere.

George Porter lived to see the importance of ClO, ClONO$_2$ and ClOOCl in the stratosphere, and would, I believe, have been very pleased to see how his creation of free radical spectroscopy has become so important in the exploration of the atmosphere.

Review of Geophysics and Spacephysics, **13** (1975). Copyright 1975 American Geophysical Union.
Reproduced by permission of American Geophysical Union

Chlorofluoromethanes in the Environment

F. S. Rowland and Mario J. Molina

Department of Chemistry, University of California, Irvine, California   92664

The molecules CF_2Cl_2 and $CFCl_3$ are released into the atmosphere following their extensive use as propellants for aerosol spray cans and in refrigeration. Since they are chemically inert and have low solubility in water, these chlorofluoromethanes have very long atmospheric residence times and can be detected throughout the troposphere in amounts roughly corresponding to the integrated world industrial production to date. The most important sink for atmospheric $CFCl_3$ and CF_2Cl_2 appears to be photolytic dissociation in the stratosphere by ultraviolet radiation around 2000 Å. Upon photolysis the chlorofluoromethanes release active chlorine atoms, which initiate an extensive catalytic chain reaction

$$Cl + O_3 \rightarrow ClO + O_2 \qquad ClO + O \rightarrow Cl + O_2$$

leading to the net destruction of O_3 in the stratosphere. This chain reaction can be diverted through reaction of ClO with NO, which interconnects the NO_x and ClO_x catalytic cycles. The Cl—ClO chain is interrupted by the reaction of Cl with methane or other hydrogenous species to form HCl, and it is renewed by reaction of OH with HCl. One-dimensional diffusion calculations show that present O_3 depletion levels resulting from the presence of the chlorofluoromethanes are of the order of 1%. This depletion would increase up to 15 or 20% if the chlorofluoromethane injection were to continue indefinitely at present rates. Furthermore, the calculations show that the full stratospheric effect of the photodissociation of CF_2Cl_2 and $CFCl_3$ is not immediately felt after their introduction at ground level because of the delay required for upward diffusion to the 25- to 30-km level. If the atmospheric injection of these compounds were to terminate only after causing an observable depletion of stratospheric ozone, the depletion would intensify for sometime thereafter and would remain significant for a period of a century or more.

Contents

The ubiquitous observation in the earth's troposphere of the molecule $CFCl_3$ in amounts estimated to be comparable to the total industrial production integrated to that time prompted us to investigate the eventual terrestrial chemical sinks for this and similar molecules [*Molina and Rowland, 1974a, b*]. One obvious eventual sink was always solar photodissociation at high altitudes and short wavelengths, and we have been unable as yet to uncover any other sinks as important as this photochemical one. We therefore estimated the lifetimes of such molecules against solar photodissociation processes and then compared the rates of other possible processes with these. Our work has concentrated on the photoprocesses involved with CF_2Cl_2 and $CFCl_3$, both of which are now readily detectable in the atmosphere in mole fractions exceeding 1×10^{-10} and 8×10^{-11}, respectively [*Lovelock, 1971; Lovelock et al., 1973; Wilkniss et al., 1973, 1974; Su and Goldberg, 1973*]. We have then followed the chemistry of the photodecomposition products to their ultimate sinks as well. Since the most important chemical reactions of one important photoproduct, atomic chlorine, involve a chain reaction resulting in a net combination of $O + O_3 \rightarrow 2O_2$, we have also estimated the magnitude of this chlorine atom catalyzed decomposition of atmospheric ozone. Two brief summaries of these processes and conclusions have already been published [*Molina and Rowland, 1974a, b*]. Quantitative estimates of the magnitude and time scales of these effects have now appeared from several laboratories and are in general agreement that the potential future effects on the ozone level of the atmosphere are sufficiently severe that steps should be taken to avoid them [*Crutzen, 1974a; Cicerone et al., 1974; Wofsy et al., 1975; Rowland and Molina, 1974*]. Much of the material of this paper has not been published before except in the form of an Atomic Energy Commission report, no. 1974-1, from the University of California, Irvine [*Rowland and Molina,* 1974]. The present version involves some revision and includes additional material published during the two months since the issuance of the Atomic Energy Commission report.

Both CF_2Cl_2 and $CFCl_3$ are transparent at ultraviolet wavelengths longer than about 2300 Å. The important wavelengths for solar photodissociation of both CF_2Cl_2 and $CFCl_3$ thus fall in a window between O_2 and O_3 absorption at 1850–2270 Å. Since these wavelengths lie at the very edge of the existing measurements of the absorption cross sections for these molecules, we have measured them again, extending them to longer wavelengths, and have obtained accurate values over this entire range. These cross sections can then be combined with estimates of the solar flux penetrating to each altitude (for each wavelength and each solar zenith angle) to calculate rate constants for photodissociation for each molecule at each altitude.

With appropriate models for the average vertical diffusion of trace atmospheric components, the relative rates of diffusion and photodissociation can be combined in a steady state calculation to determine the equilibrium concentrations of both $CFCl_3$ and CF_2Cl_2 versus altitude, a constant rate of introduction of each in the troposphere being assumed. Since very little radiation in this wavelength range penetrates even to the tropopause, the photodissociation lifetimes for these molecules in the troposphere are much longer than the rates for diffusion into the stratosphere, and the actual photolysis by solar radiation occurs almost entirely in the stratosphere. Photolysis is relatively rapid above about 25-km altitude, and not many molecules of either compound survive to diffuse as high as 40 km above the surface. Circulation, especially vertical, is much slower in the stratosphere than in the troposphere, and these molecules and their photodissociation products can spend months or even years in the processes of diffusing from tropopause to dissociation altitudes and back down again.

Integration over all altitudes of these molecular distributions and the actual rates of photodissociation provides an estimate of the average lifetime of these molecules in the atmosphere. For convenience these distributions versus altitude have been normalized to a yearly input flux of CF_2Cl_2 and $CFCl_3$ equal to the 1973 industrial input of each species.

The photolysis of each of these molecules involves the loss of one Cl atom with the formation of a free radical, as in (1) and (2).

$$h\nu + CF_2Cl_2 \rightarrow CF_2Cl + Cl \qquad (1)$$

$$h\nu + CFCl_3 \rightarrow CFCl_2 + Cl \qquad (2)$$

The immediate reaction of these radicals with O_2 releases another Cl atom (perhaps as ClO), leaving CF_2O or $CFClO$, respectively. The further decomposition of these residual molecules depends upon photolysis rates for each and has not yet been considered in detail. The subsequent effects of the photolysis processes are therefore largely concerned with the fate of the Cl atoms released in the photolysis and the following reaction of the initial free radical with O_2. These atoms are unable to remain permanently at these altitudes, either in the form of atoms or in some other chemical form, and are eventually removed from the stratosphere by diffusion to the tropopause, below which they can be removed by the weather processes occurring in the troposphere. On the basis of present knowledge, this process of diffusion to the tropopause is presumed to

occur with the various chlorine-containing species existing almost entirely as gaseous species.

At altitudes of 20–50 km the most probable reaction of Cl in the stratosphere occurs with O_3, as in (3).

$$Cl + O_3 \rightarrow ClO + O_2 \qquad (3)$$

At the higher altitudes the most probable reaction of ClO occurs with O, as in (4).

$$ClO + O \rightarrow Cl + O_2 \qquad (4)$$

The net reaction of (3) plus (4) is the removal of one molecule of O and O_3, the Cl atom being returned to its initial chemical state, and thus the formation of a chain reaction for the removal of odd oxygen (O and O_3).

$$\text{Net: } O + O_3 \rightarrow O_2 + O_2$$

Through this chain process, one Cl atom can catalytically convert hundreds or even thousands of molecules of odd oxygen into molecular oxygen before it diffuses down to the tropopause and is itself removed from the atmosphere by tropospheric processes. The potential importance of such atmospheric reactions is thus intimately tied in with this catalytic multiplication of the ability of Cl atoms for removing odd oxygen.

The chain reaction of (3) plus (4) can be diverted in the lower stratosphere through reaction of ClO with NO, as in (5).

$$ClO + NO \rightarrow Cl + NO_2 \qquad (5)$$

Sometimes this reaction is followed by (6),

$$NO_2 + O \rightarrow NO + O_2 \qquad (6)$$

and the net of (3) + (5) + (6) is the same as for (3) + (4). However, in most instances, the NO_2 formed in (5) is photolyzed, as in (7),

$$h\nu + NO_2 \rightarrow NO + O \qquad (7)$$

and the overall result of (3) + (5) + (7) is no net chemical change in the odd oxygen concentration of the atmosphere. The ClX (Cl, ClO, and HCl) chain reaction in the lower stratosphere is thus thoroughly interconnected with the NO_x chain catalysis of odd oxygen removal, which involves (6), (7), and (8),

$$NO + O_3 \rightarrow NO_2 + O_2 \qquad (8)$$

and some less important reactions. In the atmosphere now the most important processes for removal of odd oxygen are the NO_x-catalyzed reactions. However, an increase in the concentration of ClX species in the atmosphere by 10–20 times might well make the ClX-catalyzed processes the dominant ones. The continued injection of CF_2Cl_2 and $CFCl_3$ into the stratosphere at current rates would be sufficient to raise the stratospheric concentrations by a factor of 15–30 within about 50 years.

The Cl-ClO chain can also be interrupted by the reaction of Cl with some of the hydrogenous species in the stratosphere, including CH_4, H_2, HO_2, H_2O_2, and HNO_3, as is shown in (9)–(13).

$$Cl + CH_4 \rightarrow HCl + CH_3 \qquad (9)$$

$$Cl + H_2 \rightarrow HCl + H \qquad (10)$$

$$Cl + HO_2 \rightarrow HCl + O_2 \qquad (11)$$

$$Cl + HOOH \rightarrow HCl + HO_2 \qquad (12)$$

$$Cl + HONO_2 \rightarrow HCl + ONO_2 \qquad (13)$$

The HCl formed in these reactions is not catalytically active in odd oxygen removal and is a temporary sink for Cl atoms until they are released again and the Cl atom chain is restarted through the attack of OH on HCl, as in (14).

$$OH + HCl \rightarrow H_2O + Cl \qquad (14)$$

The distribution of ClX among the species HCl, ClO, and Cl is thus controlled by reactions involving O_3, O, NO, CH_4, H_2, HO_2, H_2O_2, HNO_3, and OH, as summarized in the diagram:

$$HCl \underset{RH}{\overset{OH}{\rightleftarrows}} Cl \underset{O, NO}{\overset{O_3}{\rightleftarrows}} ClO$$

A precise calculation of the expected perturbation of trace atmospheric composition through the inclusion of increasing amounts of CF_2Cl_2 and $CFCl_3$, and thereby ClO, Cl, and HCl, is exceedingly complicated. All of the components listed above interact with each other and with a few other species (H, $O(^1D)$, etc.) by at least 60 separate chemical reactions, many of them exhibiting temperature dependence over the $210°-275°K$ range characteristic of stratospheric temperatures. In addition, all of these components are subject to meteorological transport as well as chemical reaction. Further, changes in O_3 concentration affect the radiation balance of the stratosphere and its temperature profile versus altitude. Calculations embodying most of these complexities have been made by *Crutzen* [1974a] and by *Wofsy et al.* [1975]. Partial calculations have also been performed by *Cicerone et al.* [1974] and by *Molina and Rowland* [1974b].

Several other sinks have been considered, but none so far appear to represent more than a minor correction to stratospheric ultraviolet photolysis. Since these molecules have no absorption beyond about 2200–2300 Å and radiation at these wavelengths (and shorter) is severely attenuated by absorption at higher altitudes, photodissociation in the troposphere is negligibly slow. The reaction of $O(^1D)$ with CF_2Cl_2 and $CFCl_3$ is quite rapid, but even with a reaction occurring approximately on every collision, the amounts of $O(^1D)$ (primarily from ozone photolysis) are negligibly small in the troposphere, and the reaction is negligible. In the stratosphere, reaction on every collision with chlorofluoromethanes makes $O(^1D)$ about 100 times less important than UV photolysis in removing $CFCl_3$ and about 10 times less important than UV photolysis in removing CF_2Cl_2.

The attack of OH on CH_4 is believed to be important in the troposphere, but the corresponding Cl atom abstraction reaction (e.g., $OH + CFCl_3 \rightarrow HOCl + CFCl_2$) is endothermic and negligible under all atmospheric conditions. Neither $CFCl_3$ nor CF_2Cl_2 is very soluble in water, and they are not removed by rainout in the troposphere, as evidenced by the long residence times already observed. Details on possible biological involvement of the chlorofluoromethanes are very scarce, since these are not natural materials (except possibly for minute amounts from volcanic activity), and rapid biological removal seems highly unlikely.

The relative insolubility indicates that dissolution in the ocean should not be an important sink. The measurements of *Lovelock et al.* [1973] and *Wilkniss et al.* [1974] have indicated equilibrium between the atmosphere and the surface layers of the ocean, further implying that the oceanic sink is not very important. The rates of hydrolysis are quite slow but show $CFCl_3$ as considerably more reactive than CF_2Cl_2. Thus even if there were removal of $CFCl_3$ by this route (and there is no

evidence for it), the removal of CF_2Cl_2 would be very much slower and therefore negligible. Again, since electron capture by $CFCl_3$ is about 100 times more probable than for CF_2Cl_2, $CFCl_3$ would be preferentially removed by such a process, and the removal of CF_2Cl_2 by e^- capture must be negligible by analogy with the observed long atmospheric lifetime for $CFCl_3$.

Current Inventories of Atmospheric $CFCl_3$ and CF_2Cl_2

At least four sets of measurements have been published for the $CFCl_3$ content of tropospheric air samples and two sets for CF_2Cl_2. Additional measurements are now being carried out by several groups. The mixing ratios of $CFCl_3$ observed by *Lovelock et al.* [1973] for Atlantic Ocean surface air near the end of 1971 (sampled during a voyage from the United Kingdom to Antarctica and return) varied from 4 to 8×10^{-11}, with a marked northern/southern hemisphere variation. Except for the variation with latitude, the scatter in the data was relatively small. The mean aerial concentration from these measurements was given as $(5.0 \pm 0.7) \times 10^{-11}$.

The mixing ratios for $CFCl_3$ measured by *Su and Goldberg* [1973] for southern California samples in early 1973 varied widely, from 1.0 to 220×10^{-11}, as did their CF_2Cl_2 mixing ratios, $30-800 \times 10^{-11}$. The highest values can be attributed to the collection of the samples in coastal cities in which anthropogenic surface sources can readily add fresh material to the background average. However, the measurement of some concentrations markedly lower than the consistent southern hemisphere values of Lovelock earlier in time suggests that a systematic discrepancy exists between these sets of measurements.

The measurements of Su and Goldberg for tropospheric air in the Anza Borrego desert 100 km northeast of San Diego showed concentrations in early 1973 of 9.7×10^{-11} and 70×10^{-11} for $CFCl_3$ and CF_2Cl_2, respectively. The concentration of CF_2Cl_2 seems too high to be representative of thoroughly mixed tropospheric air. Su and Goldberg estimated a residence time of around 30 years from this concentration together with an estimated 0.54 Mt/year (metric) world production rate. Such an estimate, however, must be based on an assumed constancy of world production of CF_2Cl_2 for a period approximately as long as the estimated residence time, an assumption which is clearly unsatisfactory for CF_2Cl_2 over the past 30 years. A mixing ratio of 70×10^{-11} for CF_2Cl_2, if applicable to the atmosphere of the entire earth, implies the presence in the atmosphere of about 14 Mt of CF_2Cl_2. This amount is several times larger than the entire integrated world production to date of CF_2Cl_2, much of which has been used as a refrigerant and may not yet have been released to the atmosphere. Consequently, we conclude that the concentrations of CF_2Cl_2 reported for these desert samples are not valid as estimates of the concentration characteristic of a thoroughly mixed troposphere in 1973.

Wilkniss et al. [1973] measured mixing ratios for $CFCl_3$ in November and December 1972 in Pacific surface air from 15°N to 74°S latitude. These values showed a mean of $(6.1 \pm 1.3) \times 10^{-11}$ and were consistently higher than those of Lovelock et al. at comparable latitudes. *Hester et al.* [1974] have obtained values consistent with about 10×10^{-11} for CF_2Cl_2 and 5×10^{-11} for $CFCl_3$ for samples collected in 1973 in southern California.

Since measurement of these molecules at atmospheric levels has only been carried out in the last 3 years, efforts are just beginning to intercalibrate equipment and to prepare appropriate standard samples. Further, the occasional high measurement in the results of Wilkniss et al., the discrepancy between Lovelock et al. and Wilkniss et al. over the presence or absence of CH_3I in such samples, and the fourfold or more difference between the Su-Goldberg minimum values and those of the other three groups all suggest that routine measurement of $CFCl_3$ in atmospheric samples has not yet been attained. Appreciable problems of sample contamination and of calibration must be solved before accurate estimates can be made of the current atmospheric burden of each of these molecules. Reports of measurements on more recently collected samples indicate that the atmospheric burden of $CFCl_3$ has increased in early 1974 over the late 1972 values by about 36%, from $(6.1 \pm 1.3) \times 10^{-11}$ to $(8.3 \pm 0.6) \times 10^{-11}$ vol/vol [*Wilkniss et al.,* 1974]. Both sets of measurements were carried out with essentially the same equipment by the same research group and therefore can be fairly compared with one another for an estimate of percentage increase in concentration of $CFCl_3$ over that 15-month period.

Neither CF_2Cl_2 nor $CFCl_3$ is a naturally occurring molecule, with the possible exception of trace amounts formed in volcanic eruptions [*Stoiber et al.,* 1971]. The overwhelming source of these compounds in the terrestrial atmosphere is the anthropogenic use of both molecules as propellant gases in aerosol sprays and of CF_2Cl_2 as a refrigerant.

WORLDWIDE INDUSTRIAL PRODUCTION OF CHLOROFLUOROMETHANES

Estimates of the amount of chlorofluoromethanes produced per year have been made by R. L. McCarthy (private communication, 1974), whose data are summarized in Table 1. The U.S. production data are regularly given in U.S. Tariff Commission reports, but the data for the rest of the world are of uncertain accuracy; McCarthy gives the accuracy of his more recent estimates as ±25%. A 1972 estimate of the growth of the U.S. fluorocarbon industry as a whole, of which the chlorofluoromethanes are the most important component, is also given in Table 1. The molecule CF_2Cl_2 is the most widely used of the fluorocarbons. Although production was originally confined to the United States, the chlorofluoromethanes are now being manufactured throughout the world. About 8% is currently produced in Japan with about 45–50% each in the United States and Europe. A list of trade names for these molecules as manufactured in 20 different countries is given in Table 2.

Estimates of the present atmospheric burden of $CFCl_3$ can be made from these production figures if the following additional data are known: the fraction of $CFCl_3$ released to the atmosphere, the average delay time between production and release to the atmosphere, and the average lifetime in the atmosphere. An estimated 85% of current $CFCl_3$ production is used in aerosol sprays, and another 5–8% may be released to the atmosphere in processes involving the foaming of polyurethane. A negligible amount of $CFCl_3$ is used in refrigeration. The average delay time between production and release has been estimated by McCarthy as 6 months for $CFCl_3$ used as aerosol spray.

ATMOSPHERIC RESIDENCE TIME FOR $CFCl_3$

We have attempted to estimate the residence time of $CFCl_3$ in the atmosphere from the total production to date as compared with the amount now found to be in the atmosphere. In

TABLE 1. Estimated World Production of CF_2Cl_2 and $CFCl_3$ in Kilotons (10^9 Grams)

| Year | CFCl₃ | | CF₂Cl₂ | |
	United States	World	United States	World
1950–1955	54	54	...	...
1956–1957	32	32	...	...
1958	23	23	59	59
1959	27	27	71	71
1960	33	40	75	87
1961	41	52	78	94
1962	56	72	94	117
1963	64	83	98	129
1964	67	93	103	143
1965	77	112	123	175
1966	77	122	130	196
1967	83	139	141	225
1968	93	165	148	256
1969	108	197	167	300
1970	111	217	170	329
1971	117	241	177	363
1972	136	285	191	422
1973	147	313	221	469
Cumulative to November 1, 1974		2576		3830

The data are converted from original data by R. L. McCarthy (private communication, 1974) given in millions of pounds.

The chemical profile [*Chemical Marketing Reporter,* 1972] for fluorocarbons in the United States has a history for the period 1961–1971 of 8.7% growth per year, the largest volume fluorocarbon being CF_2Cl_2.

Also during 1948–1972, 50% of the total fluorocarbons were used as aerosol propellants, 28% as refrigerants, 10% as plastics, 5% as solvents, and 7% as blowing agents, exports, and other miscellany.

For 1971 the demand was 340×10^6 kg (750×10^6 lb); for 1972, 374×10^6 kg (825×10^6 lb); and for 1976 the estimated demand is 454×10^6 kg (1000×10^6 lb).

this calculation we have not specified the nature of the processes removing the $CFCl_3$ and have assumed that there is no delay period before each injected molecule becomes accessible to this unspecified sink. This 'no delay period' hypothesis should be approximately true for any tropospheric sinks, at least for lifetimes longer than one year, but would not be correct for the stratospheric sink, which we believe to be important for these molecules. In the calculation we have used the $CFCl_3$ production data of Table 1 and have assumed that a constant fraction (0.90) of this production has been released to the atmosphere after a delay period of 6 months and have then estimated the amount still to be expected in the atmosphere in early 1974 for various average atmospheric lifetimes.

Calculations for various hypothetical lifetimes and an injection pattern parallel to the production pattern of Table 1 indicate that about 75% of the $CFCl_3$ would still be in the atmosphere if its atmospheric lifetime were 20 years, about 85% for a 30-year lifetime, and 90% for a 50-year lifetime. The integrated amount of $CFCl_3$ manufactured to the end of 1973 is 2.2 Mt from Table 1, of which we assume 10% or 0.22 Mt was used in processes not involving release to the atmosphere. If a 6-month time lag is invoked, approximately 0.16 Mt had not yet been released to the atmosphere at the end of 1973, and the potential integrated atmospheric burden at the end of 1973 would then be 1.8 Mt if the atmospheric lifetime were infinite. The corresponding amounts in the atmosphere would be about 1.6, 1.5, and 1.4 Mt for average lifetimes of 50, 30, and 20 years, respectively.

Rowland and Molina: Chlorofluoromethanes in the Environment

TABLE 2. List of Trade Names

Country	Company	Trade Name
Argentina	Fluoder S.A.	Algeon
	I.R.A., S.A.	Frateon
Czechoslovakia	Slovek Pro Chemickov A Hutni Vyobu, Ustianad Cabem	Ledon
England	Imperial Chemical Industries, Ltd.	Arcton
	Imperial Smelting Corp., Ltd.	Isceon
France	Products Chimiques Penchiney-Saint-Gobain	Flugene
	Société d'Electro-Chimie d'Electro-Metallurgie et des Acières Electrique d'Ugine	Forane
East Germany	V.E.B. Alcid Fluorwerk Dohna	Frigedohn
West Germany	Chemische Fabrik von Heyden AG	Heydogen
	Farbwerke Hoechst AG	Frigen
	Kali-Chemie AG	Kaltron
India	Everest Refrigerants, Ltd.	Everkalt
	Navin Fluorine Industries	Mafron
Italy	Montecatini-Edison	Algofrene
		Edifren
Japan	Daikin Kogyl Co., Ltd.	Daiflon
	Asahi Glass Co., Ltd.	Asahiflon
Netherlands	Uniechemie N.V.	Fresane
	Noury van Der Lande N.V.	FCC
Russia		Eskimon
United States	E. I. du Pont de Nemours and Co.	Freon
	Allied Chemical and Dye Corp.	Genetron
	Kaiser Chemicals	Kaiser
	Pennwalt Chemical Co.	Isotron
	Racon, Inc.	Racon
	Union Carbide Corp.	Ucon

The chlorofluoromethanes are frequently described by trademark names plus identifying number, such as Freon-11 for $CFCl_3$ and Freon-12 for CF_2Cl_2 for these compounds when manufactured by E. I. duPont de Nemours and Company. The numbering system used for these molecules is the following: the hundreds digit is the number of carbon atoms in the molecule minus 1, the tens digit is the number of hydrogen atoms in the molecule plus 1, and the units digit is the number of fluorine atoms in the molecule. With $CFCl_3$, these numbers are 011, and the zero is dropped in the description.

This class of compounds was developed in 1930 by the General Motors Research Laboratory during a search for a nontoxic nonflammable refrigerant.

The entire atmosphere contains approximately 1.1×10^{44} molecules (5.1×10^{18} cm^2 of earth's surface $\times 2.1 \times 10^{25}$ molecules cm^{-2}), of which about 1.0×10^{44} are in the troposphere. If the integrated atmospheric burdens are converted to predicted mole fractions when distributed over the entire troposphere, the estimated levels for January 1974 are about 7.9×10^{-11} for an infinite average lifetime, 6.7×10^{-11} for 30 years, and 5.9×10^{-11} for 20 years. If all of the estimated production to the end of 1973, without any correction for delay time or nonatmospheric release, were distributed through the troposphere, the calculated concentration would be 9.5×10^{-11}.

The experimental $CFCl_3$ mole fraction of $(8.3 \pm 0.6) \times 10^{-11}$ measured by Wilkniss et al. in early 1974 is thus consistent with any estimated lifetime of 30 years or longer [Machta, 1973], whereas there is too much $CFCl_3$ present to be consistent with any shorter lifetimes. However, the quoted uncertainty of 25% in the knowledge of recent industrial production rates for $CFCl_3$ makes precise comparisons impossible.

For the stratospheric sink, which we believe to be highly important for $CFCl_3$, the atmospheric burden should not be divided by the total atmosphere but only by that below about 20 km (~90% of the atmosphere). There is also an appreciable induction period between introduction at ground level and exposure to the stratospheric sink. The basic conclusion remains unchanged for a stratospheric sink; the present atmospheric burden of $CFCl_3$ appears to correspond to that expected for an

atmospheric lifetime of >20 years. Furthermore, there are uncertainties in amount manufactured, in fractional release to the atmosphere, in delay times in storage before use, and in estimates of the present atmospheric levels. Since, in addition, most of the $CFCl_3$ has only recently entered the atmosphere and has therefore not been exposed to removal processes for very long, a substantial improvement in the average lifetime estimate as obtained from direct observation of the terrestrial concentrations cannot be expected until a number of years have elapsed.

Similar estimates can also be attempted for CF_2Cl_2 but with the severe added complication that appreciable fractions of CF_2Cl_2 are used as a refrigerant cooling fluid. Although most refrigerant fluid is eventually released to the atmosphere, the average delay time is difficult to estimate, presumably a few years. The amount now present in the atmosphere is harder to measure because of the much lower sensitivity for CF_2Cl_2 compared with $CFCl_3$ for the electron capture detectors used in gas chromatographic assay and has been much less widely measured. Until much more detailed information is available about the worldwide distribution of CF_2Cl_2 no comparisons of any accuracy can be made. Most of the uncertainties are much larger here than with $CFCl_3$, but the measurements of Hester et al. [1974] indicate concentrations of the order of 10×10^{-11} vol/vol, which correspond to atmospheric lifetimes comparable to or longer than that of $CFCl_3$.

ADDITIONAL CHLORINE-CONTAINING MOLECULES

Many other volatile chlorine-containing molecules are manufactured for various technological purposes, and some of these materials are undoubtedly released to the atmosphere. The 1972 United States production and sales of the most important of these are summarized in Table 3. The discrepancy between production and sales figures frequently is an indication that a particular molecule is used without sale for the manufacture of another (for example, CH_2ClCH_2Cl in the production of $CH_2{=}CHCl$). The usage patterns for these molecules is quite varied, and it is difficult to estimate what fraction of each may be released to the atmosphere. Molecules such as CCl_4, CH_3CCl_3, $CCl_2{=}CCl_2$, and $CHCl{=}CCl_2$ have been detected, especially in urban atmospheres [Lovelock et al., 1973; Lovelock, 1974].

Of all the molecules listed in Table 3 only CCl_4 should have any appreciable current stratospheric significance. This molecule has been in widespread technological use since about 1910 and was long employed as a cleaning agent, involving eventual loss to the atmosphere. Its current major use, however, is in the manufacture of $CFCl_3$ and CF_2Cl_2, and most of the amount listed in Table 3 was not released to the atmosphere.

The tropospheric concentration of CCl_4 is currently approximately one part in 10^{10}, quite comparable to that of CF_2Cl_2 and $CFCl_3$. Its stratospheric behavior should be similar to that of the chlorofluoromethanes, and its photodissociation will also initiate Cl atom chains. If CCl_4 were released to the atmosphere in yearly amounts comparable to those of CF_2Cl_2 and $CFCl_3$, it would be of similar importance as a stratospheric hazard.

The average atmospheric residence times for CCl_4 with photodissociation as the only sink are about 35–50 years [Molina and Rowland, 1974c], somewhat shorter than those of $CFCl_3$ with comparable models for vertical diffusion, as given later in Table 7. The present tropospheric concentrations of CCl_4 probably represent an accumulation over several decades and thus are much less likely to increase rapidly with time. Nevertheless, the origin of CCl_4 is not firmly established and the suggestion has been made that it may even be of natural origin [Lovelock et al., 1973]. Further work will be required to establish the sources and source strengths for CCl_4.

The cumulative effects of ClO_x catalysis in the stratosphere represent, of course, a sum of contributions from all of the chlorine-containing species, and at the present time the three molecules $CFCl_3$, CF_2Cl_2, and CCl_4 are roughly equivalent in importance. The contribution from the first two, however, is expected to increase rapidly with time, whereas that of CCl_4 is not.

The other molecules in Table 3 all contain either C=C double bonds or C–H bonds. In each case the molecule is susceptible to attack by OH radicals in the troposphere, either by addition or by abstraction as in (15) and (16),

$$OH + \ \mathord{>}C{=}C\mathord{<} \ \rightarrow \ \mathord{>}COH{-}C\mathord{<} \ \rightarrow \ decomposition \qquad (15)$$

$$OH + RH \rightarrow H_2O + R \qquad (16)$$

and should have a relatively short (perhaps 1–5 years) atmospheric residence time. Furthermore, the C=C double bond is susceptible to attack by additional reactive species such as ozone. In these cases, then, stratospheric reactions should involve only a small fraction of the total, and corresponding larger atmospheric releases would be required to produce a given level of stratospheric effect.

ABSORPTION CROSS SECTIONS FOR PHOTODISSOCIATION OF CHLOROFLUOROMETHANES

The chlorofluoromethanes all absorb electromagnetic radiation only in the far ultraviolet, as shown in Figure 1, which shows the data of Doucet et al. [1973] for CF_2Cl_2. Measurements in the near ultraviolet and visible regions have been reported only for CCl_4, which has absorption cross sections of less than 10^{-25} cm^2 throughout the range from 5000 Å down to 3000 Å and which is therefore frequently used as a spectroscopic solvent [Hampel, 1971]. Infrared absorption is of course observed but is incapable of causing photodissociation.

While chlorofluoromethanes at the outer edge of the atmosphere would be exposed to, and could be dissociated by, radiation over the entire range of wavelengths shorter than about 2200 Å, such molecules diffusing upward from the surface of the earth will be exposed to this full range of radiation wavelengths only if diffusion upward is much more rapid than the photodissociation process itself. Otherwise, the upwardly diffusing molecules are selectively exposed to those wavelengths of radiation that are most successful in penetrating the various partially absorbing components in the

TABLE 3.   Production of Halogenated Hydrocarbons in the United States for 1972

	Production		Sales	
Chemical	Kilotons	Millions of Pounds	Kilotons	Millions of Pounds
CH_2ClCH_2Cl	3542	7809	656	1447
$CH_2{=}CHCl$	2308	5088	1516	3343
CCl_4	452	997	422	930
$CCl_2{=}CCl_2$	333	734	328	723
CH_3CH_2Cl	261	576	88	194
CH_2Cl_2	214	471	201	443
CH_3Cl	206	454	94	208
CH_3CCl_3	200	441	176	389
CF_2Cl_2	199	439	190	419
$CHCl{=}CCl_2$	194	427	200	441
CH_2BrCH_2Br	143	315	78	173
$CFCl_3$	136	300	130	286
$CHCl_3$	107	235	92	203
All others	782	1723	82	180

From U.S. Tariff Commission [1973].

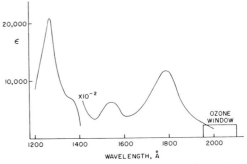

Fig. 1.   Absorption spectrum of CF_2Cl_2, 1200–2000 Å [from Doucet et al., 1973].

layers at higher altitudes. Since stratospheric diffusion processes are relatively slow, the chlorofluoromethanes actually will be decomposed only by radiation in the relatively narrow band between 1840 and 2250 Å, a region of lesser atmospheric absorption between the more intense absorptions of O_2 at shorter wavelengths and O_3 at longer wavelengths [Ackerman, 1971]. Figure 2 illustrates this 'window' in the absorption of the solar radiation flux. Since these wavelengths are just at the onset of absorption by $CFCl_3$ and CF_2Cl_2, they were not of primary interest to Doucet et al., nor did their published measurements extend to the absorption edges. We have remeasured the absorption cross sections for these molecules in the wavelength range from 1850 to 2270 Å with a Cary 14 UV spectrometer, using a path length of 10 cm and a pressure of 5–50 torrs. These cross sections are listed in Table 4, together with the earlier measurements of Doucet et al., with which the agreement is not particularly good. We have also remeasured the absorption cross sections for CCl_4 in this wavelength range, as summarized in Table 4.

The C−Cl bond in the chlorofluoromethanes has a bond

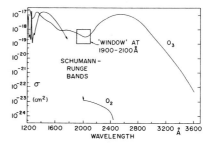

Fig. 2. Absorption cross sections for O_2 and O_3, 1200–3600 Å.

dissociation energy of about 70–75 kcal/mol [Curran, 1961; Kerr, 1966] and energetically could be dissociated with absorbed light at any wavelength shorter than about 5000 Å. However, absorption is observed only in the far ultraviolet for which the energy is far in excess of the minimum required for breaking the C−Cl bond. The photochemical process has been

TABLE 4. Absorption Cross Sections for CF_2Cl_2, $CFCl_3$, and CCl_4 in the Wavelength Range 1850–2272 Å

Midpoint of Interval λ, Å	Interval Boundaries ν, 10^3 cm^{-1}	CF_2Cl_2 Cross Section, 10^{-20} cm²		$CFCl_3$ Cross Section, 10^{-20} cm²		CCl_4 Cross Section, This Work 10^{-20} cm²
		This Work	Doucet et al. [1973]	This Work	Doucet et al. [1973]	
	44.0					
2260		...	*	0.9	0.7†	6.8
	44.5					
2235		...	*	1.3	1.2†	9.6
	45.0					
2210		...	*	1.8	*	14.4
	45.5					
2186		...	*	3.4	*	21.2
	46.0					
2162		<0.2	*	5.1	*	27.6
	46.5					
2139		0.30	*	8.0	9.1, 7.0‡	32.9
	47.0					
2116		0.70	*	12.5	12.9	42.5
	47.5					
2094		1.36	*	18.5	18.3	48.6
	48.0					
2073		1.81	*	25.0	26.0	56.0
	48.5					
2051		3.11	*	34.5	37.0	62.5
	49.0					
2030		4.9	*	46.0	51.8	66.0
	49.5					
2010		7.2	5.2	59.0	72.6	68.4
	50.0					
1990		11.3	6.6	73.0	101	70.0
	50.5					
1970		16.4	8.7	88.5	137	71.2
	51.0					
1951		24.6	11.0	113	175	73.3
	51.5					
1932		34.5	13.0	137	214	86.0
	52.0					
1914		49.1	15.6	164	255	113
	52.5					
1896		65.5	18.2	197	299	164
	53.0					
1878		86.5	22.8	227	349	240
	53.5					
1860		91.2	26.0	241	~410	336
	54.0					

* Not measured.
† Lacher et al. [1950].
‡ Marsh and Heicklen [1965].

thoroughly studied and is interpreted as a transition ($n \rightarrow \sigma^*$) involving excitation to a repulsive electronic state that immediately dissociates by breaking the carbon-chlorine bond [*Majer and Simons*, 1964]. The C−F bond in these molecules is much stronger (110–130 kcal/mol) [*Curran*, 1961; *Kerr*, 1966], but the energy available for some of the wavelengths in the absorption region is actually sufficient for dissociation of the C−F bonds. However, there is no information from any photolysis measurements to suggest that the C−F bonds are actually broken in competition with C−Cl bond breakage, and we have assumed that the only important photochemical dissociation processes are (1) and (2), with direct loss of one Cl atom.

MECHANISMS AND QUANTUM YIELDS FOR PHOTODISSOCIATION OF CF_2CL_2 AND $CFCL_3$

Very few experiments on the photolysis of the mixed chlorofluoromethanes have been mentioned in the published literature, although an extensive series of experiments has been carried out involving the radicals CF_3 and CCl_3 [e.g., *Heicklen*, 1969]. The initial step in all of these photodissociation experiments has been assumed to be the breaking of the weakest carbon-halogen bond, and the subsequent reaction mechanism studies have been largely concerned with the ultimate fate of the residual radical, i.e., largely with the chemistry of CF_3 and CCl_3. In the presence of O_2 these two radicals under a wide variety of conditions lead eventually to the formation of CF_2O and CCl_2O, respectively, as the only observed carbon-containing products, with quantum yields of unity. By analogy, the expected reactions of $CFCl_2$ and CF_2Cl with O_2 would lead to the formation of CFClO and CF_2O, respectively, if the reacting radical always were to lose the most weakly bonded halogen atom during its path to the phosgene-type molecule. Heicklen's experiments on the oxidation of CF_3 with O_2 led him to conclude that the reaction mechanism probably involved the direct reaction (17) in one step,

$$CF_3 + O_2 \rightarrow CF_2O + OF \qquad (17)$$

although he did not completely rule out the possibility of reactions involving the stabilized intermediate CF_3O_2 [*Heicklen*, 1966].

Marsh and Heicklen [1965] separately photolyzed $CFCl_3$ with the 2138-Å line of a zinc resonance lamp in the presence of NO, in the presence of O_2, and with no scavenger added. In the first two cases, substantial amounts of product were spectroscopically observed and assigned as $CFCl_2NO$ and CFClO, respectively.

The yields of CFClO and $CFCl_2NO$ from $CFCl_3/O_2$ and $CFCl_3/NO$ mixtures were unaffected ($\pm 10\%$) by the total pressure or by the $CFCl_3$/additive ratio ($CFCl_3/O_2$ was varied from 0.1 to 10), the suggestion being that the observed product was being formed by an essentially exclusive pathway. From this evidence, Marsh and Heicklen assumed in analogy with the CF_3 and CCl_3 experiments that the quantum yield for CFClO formation was unity. Since the wavelength region of interest in solar photodissociation is relatively narrow and includes the zinc resonance line at 2138 Å, we conclude that the best estimate for solar ultraviolet photodissociation of $CFCl_3$ is $\phi = 1.0 \pm 0.1$ for reaction (2), with no evidence for any other contributing pathway.

The observation of CFClO with a quantum yield of unity in photolyzed $CFCl_3/O_2$ mixtures implies the immediate formation of two ClO_x entities per photolyzed molecule, one from the initial photodissociation step (Cl atom) and one from the

reaction of $CFCl_2$ with O_2 (either Cl or ClO). The chemical fates of the third Cl atom and of the F atom from $CFCl_3$ are then determined by the subsequent stratospheric reactions of CFClO, including the possibility of direct solar photolysis of CFClO, chemical attack by various radical species, hydrolysis by H_2O, or downward diffusion to the troposphere with subsequent rainout.

Photodissociation experiments with CF_2Cl_2 in O_2 have demonstrated that the quantum yield for photooxidation is $\Phi = 1.0 \pm 0.1$ at 1849 Å and therefore that the number of ClO_x chain carriers released is 2 per absorption [*Milstein and Rowland*, 1975].

In the stratosphere then each initial photodissociation of one molecule of CF_2Cl_2 is expected to form two ClO_x entities, one from the initial photolysis (Cl atom) and one from reaction of CF_2Cl with O_2 (either Cl or ClO). The fate of the two F atoms from CF_2Cl_2 is then determined by the subsequent reactions in the stratosphere of CF_2O.

The two lowest states of the chlorine atom, the ground state $^2P_{3/2}$ and the 2.5-kcal/mol higher $^2P_{1/2}$ excited state, are both present in almost all Cl atom experiments. We have assumed an equilibrium mixture throughout this work. None of the higher excited states of the Cl atom are energetically accessible in any of the reactions discussed for the atmosphere.

SOLAR IRRADIATION FLUX VERSUS WAVELENGTH AND ALTITUDE

The absorption by O_2 of radiation with wavelengths shorter than 2400 Å is the initial source of odd oxygen in the terrestrial atmosphere through the formation of two $O(^3P)$ atoms, each of which can then react with O_2 to form O_3. Ozone in turn strongly absorbs, also with dissociation, longer wavelength radiation out to 3100 Å, as well as, less strongly, wavelengths in the visible and infrared regions [e.g., *Nicolet*, 1971]. Solar ultraviolet radiation in the 1800- to 2400-Å region is strongly absorbed in the stratosphere by both O_2 and O_3, and its depth of penetration is a function chiefly of the incoming solar flux, the cross sections of O_2 and O_3, the O_3 distribution versus altitude, and the zenith angle of the sun [*Ackerman*, 1971].

In the 1800- to 2000-Å region, absorption in O_2 can occur either in the very narrow intense lines of the Schumann-Runge bands or in the underlying continuum [e.g., *Kockarts*, 1971]. In this band region the depth of UV penetration varies sharply with wavelength changes of 0.1 Å, and exact calculations must be performed with a very fine wavelength grid. Various approximation processes have been developed to permit estimation of the diminution of solar intensity with altitude in this band region, and we have used a procedure modeled after those of *Kockarts* [1971] and of *Brinkmann* [1971]. Beyond about 2000 Å the absorption is essentially controlled by continuums in O_2 and O_3, and the calculations are relatively simpler. When the solar fluxes calculated to penetrate below 50 km are combined with the absorption cross sections of CF_2Cl_2 and $CFCl_3$, absorption in the Schumann-Runge band region is found to be small relative to that at longer wavelengths, and errors introduced by the approximations are therefore not important.

The measured solar flux in the wavelength region around 1800–2300 Å is somewhat uncertain, perhaps by as much as a factor of 2. It is not yet clear how much of this uncertainty is attributable to natural variation in the solar flux with the 11-year solar cycle and how much arises from discrepancies among the various experimental techniques for measuring the solar flux in this range [*Ackerman*, 1974]. For our calculations, we have adopted the solar fluxes given by *Ackerman* [1971]. The very recent measurements by *Simon* [1974] average about 20% lower than those of Ackerman in the important region between 1960 and 2100 Å.

The solar fluxes, attenuated with decreasing altitude, and the absorption cross sections can be used to calculate a partial rate of removal for each molecule versus wavelength for overhead sun, a quantum yield of unity being assumed as discussed above. The rates summed over all wavelengths then provide a total rate of photodissociation J for each molecule for overhead sun and at various altitudes, shown as the 0° line in Figures 3 and 4 for CF_2Cl_2 and $CFCl_3$, respectively. The rates of photodissociation have been similarly calculated for different zenith angles and altitudes, as shown in Figures 3 and 4. A combined global average photodissociation rate, including both diurnal and latitudinal effects, is indicated by the dotted line on each of these figures.

The cross products of averaged solar flux and absorption cross section are shown in Figure 5 for CF_2Cl_2 at altitudes from 10 to 50 km and indicate that >90% of the actual photolysis processes occur in the relatively narrow band of wavelengths between 1950 and 2150 Å. The absorbing band becomes narrower at lower altitudes, but the summed photodissociation coefficients become so small that photodissociation in the real atmosphere is negligible below about 25 km. Over the altitudes in which dissociation is calculated to occur, the effective absorption band does not vary much in width or in location of the maximum with altitude. A similar graph for $CFCl_3$ at 40 km is also included in Figure 5. The absorption in $CFCl_3$ is stronger at all wavelengths than for CF_2Cl_2, and the location of the wavelength band shifts slightly as a consequence; no difference of practical importance arises from this shift in the wavelengths of absorption. The higher-absorption cross sections always translate to more rapid photodissociation, and the lifetime of $CFCl_3$ molecules in the atmosphere is therefore shorter than that of CF_2Cl_2 molecules.

The calculated average lifetimes for photodissociation at a fixed altitude are summarized in Table 5 for several altitudes. Below 20 km these photochemical lifetimes are so long that atmospheric diffusion processes are much more rapid, and little photodissociation takes place before diffusion away from these altitudes occurs.

ATMOSPHERIC DIFFUSION MODELS

The rate at which molecules of CF_2Cl_2 and $CFCl_3$ released at ground level arrive at stratospheric altitudes is determined by

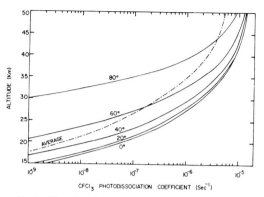

Fig. 4. Photodissociation rate of $CFCl_3$ in the stratosphere.

upward diffusion involving a very complex set of atmospheric motions. All atmospheric observations are consistent with the general assumption that there is no gravitational separation of atmospheric components of different molecular weights below about 100-km altitude. In a sample collected between 44 and 62 km by rocket and analyzed in the laboratory, the ratios to N_2 of ^{20}Ne, ^{40}Ar, and ^{84}Kr are all within 1.005 of the ratios found in air samples collected at the surface of the earth [*Martell*, 1973]. Consequently, we assume that heavier molecules such as CF_2Cl_2 and $CFCl_3$ (molecular weights 121 and 137.5), as well as lighter species such as Cl, HCl and ClO, should all mix completely without gravitational separation to altitudes of ≥60 km. The rate at which the products from photodissociation are eventually removed from the stratospheric ozone layer is determined by downward diffusion involving another complex set of motions.

At the present time the effects of these stratospheric motions in mixing trace atmospheric components are being modeled by progressively more complicated systems of one dimension [e.g., *Crutzen*, 1972; *McElroy*, 1972], two dimensions [e.g., *Hesstvedt*, 1973, 1974; *MacCracken*, 1973], and three dimensions [e.g., *Mahlman*, 1972, 1973; *London and Park*, 1974], introducing vertical, longitudinal, and latitudinal motions in that order. The current three-dimensional models are sufficiently difficult in terms of computer time and capacity that only limited sets of chemical reactions can be included in studies of the variability of chemical composition with altitude, latitude, and longitude. Most of the published

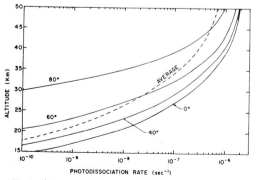

Fig. 3. Photodissociation rate of CF_2Cl_2 in the stratosphere.

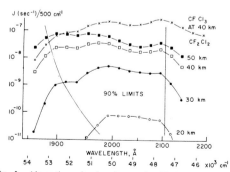

Fig. 5. Absorption of solar photons by CF_2Cl_2 and $CFCl_3$ for wavelength bands between 1850 and 2200 Å.

TABLE 5. Calculated Average Photodissociation Lifetimes for CF_2Cl_2 and $CFCl_3$ at Various Altitudes

Altitude, km	CF_2Cl_2 Lifetime, s	$CFCl_3$ Lifetime, s
10	1.2×10^{14} (4×10^6 yr)	8.9×10^{12} (3×10^5 yr)
20	2.1×10^9 (66 yr)	2.1×10^8 (6.6 yr)
30	2.9×10^7 (11 months)	3.4×10^6 (1.3 months)
40	3.3×10^6 (1.3 months)	4.0×10^5 (4.7 days)
50	1.4×10^6 (16 days)	1.8×10^5 (2.1 days)
Top of atmosphere	3.0×10^5 (3.5 days)	5.3×10^4 (15 hours)

treatments of the variation of trace components with altitude have involved one-dimensional diffusion models in which vertical mixing of components is treated through the concept of a global average vertical 'eddy' diffusion coefficient, and solutions have been sought to the diffusion equation

$$n \frac{\partial f_i}{\partial t} + \frac{\partial \phi_i}{\partial z} = P_i - L_i$$

in which P_i and L_i are production and loss terms, f_i is the volume-mixing ratio of the ith component, n is the total atmospheric number density, ϕ_i is the particle flux in the vertical direction,

$$\phi_i = -nK_z \, \partial f_i / \partial z$$

and K_z is the eddy diffusion coefficient.

The magnitude of the eddy diffusion coefficient is used in such calculations as an adjustable parameter to be fitted to some appropriate set or sets of data concerning average mixing patterns versus altitude in the real atmosphere. The same eddy diffusion coefficients are used for all gaseous species.

The one-dimensional models have had considerable success in correlating calculated properties with the actual observed average values for minor trace components in the atmosphere and have been successfully applied to the calculation of the rates of NO_x and HO_x reactions with O_3 and O during the past several years. Such models, of course, cannot predict latitudinal variations because the latitude is not a variable parameter of the system. This simple diffusion concept thus ignores the observation that diffusion upward into the stratosphere is more prominent near the equator, whereas downward diffusion is favored at higher latitudes. The photodissociation of the chlorofluoromethanes is of course more important for the overhead sun conditions prevailing near the equator, and the meteorological transport processes will tend to carry the photodissociation products poleward during the course of downward diffusion.

We have carried out several one-dimensional vertical diffu-

sion calculations with a single component and no chemical reactions except photolysis for evaluation of the upward mixing of the chlorofluoromethanes and the altitude distribution of the photodissociation reactions. We have then calculated the downward mixing of their ultimate photodissociation products to the tropopause using one component, ClX, for the sum of HCl, ClO, and Cl. The eddy diffusion coefficients in our calculations have been fixed at values similar to those recently used by others in their published model calculations. All have been solved for a steady state condition in which a constant flux of CF_2Cl_2 or $CFCl_3$ is introduced at ground level. Calculations with an eddy diffusion coefficient of 3×10^5 cm^2 s^{-1} for the troposphere showed an essentially imperceptible gradient of concentration from the ground to the tropopause. Accordingly, we have simply adopted the approximation of rapid complete mixing in the troposphere and carried out the calculations only above the tropopause. The chlorofluoromethanes then diffuse upward until solar photodissociation competes successfully, and both upward and downward diffusion of the photodissociation products can then occur. The only sink for dissociation products in these models is diffusion across the tropopause in a downward direction, after which they are assumed to be immediately removed by tropospheric processes, primarily through rainout.

Two separate calculations are thus carried out for each of the chlorofluoromethanes: a steady state solution for the concentrations of CF_2Cl_2 (or $CFCl_3$) required in order to produce a removal rate by photodissociation equal to the constant rate of introduction by diffusion upward through the tropopause and a steady state solution for the concentration of ClX to be found for a source given by the photodissociation rate at each altitude and for a sink at the tropopause of equal magnitude. The boundary conditions for the first calculation are that the upward flux of CF_2Cl_2 through the tropopause is fixed and the mole ratio of CF_2Cl_2 in the atmosphere goes to zero at infinite altitude. For the second calculation, the downward flux through the tropopause is fixed, and the concentration goes to zero at the tropopause. The step size has typically been fixed at 0.1 km, but little change was observed with a step size of 0.01 km.

In each of these calculations the tropopause is fixed at a single altitude as an approximation to the fluctuations of tropopause with latitude and season in the real atmosphere. The eddy diffusion coefficients have been those chosen by other modelers, and the location of the fixed tropopause at 10, 15, and 16 km conforms to the assumptions of their individual models. The model atmosphere of *Banks and Kockarts* [1973] has been used for temperature and density versus altitude (Table 6). The constant fluxes upward at the tropopause have

TABLE 6. Atmospheric Model

Altitude, km	Scale Height, km	Temp., °K	Pressure, torrs	Density, g cm^{-3}	Concentration, cm^{-3}
15	6.20	210.8	84.8	1.87×10^{-4}	3.89×10^{18}
20	6.45	218.9	38.5	8.16×10^{-5}	1.70×10^{18}
25	6.70	227.1	18.0	3.68×10^{-5}	7.65×10^{17}
30	6.95	235.2	8.64	1.71×10^{-5}	3.55×10^{17}
35	7.45	251.7	4.31	7.96×10^{-6}	1.66×10^{17}
40	7.95	268.2	2.25	3.90×10^{-6}	8.11×10^{16}
45	8.15	274.5	1.21	2.05×10^{-6}	4.26×10^{16}
50	8.15	274.0	0.656	1.11×10^{-6}	2.31×10^{16}
55	8.15	273.6	0.355	6.02×10^{-7}	1.25×10^{16}
60	7.54	252.8	0.188	3.45×10^{-7}	7.17×10^{15}

been fixed at 0.5 and 0.3 Mt per year for CF_2Cl_2 and $CFCl_3$, respectively (CF_2Cl_2, 1.54×10^7 molecules cm^{-2} s^{-1}; $CFCl_3$, 8.2×10^6 molecules cm^{-2} s^{-1}), corresponding generally to production levels characteristic of about 1973 or 1974. These time-independent solutions of the diffusion equation thus provide an estimate of the levels of the various chlorinated species to be expected if a steady state condition were to be reached at the current injection rates. The average residence time for CF_2Cl_2 and $CFCl_3$ can be estimated for each of these models simply by integrating the total amount present and dividing by the upward flux at the tropopause.

Calculations have been carried out with several sets of eddy diffusion coefficients as graphed in Figure 6. The choice of values above 60 km has a negligible effect on all of our calculations, since the chlorofluoromethane and ozone concentrations are negligible at these altitudes and the total amount of ClX contained in the atmosphere above 60 km is a very small fraction of the total.

The distributions of $CFCl_3$ with altitude calculated for diffusion models B, C, and D are illustrated in Figure 7, and the distribution of CF_2Cl_2 with altitude for model D is shown in Figure 8. In both Figures 7 and 8 the production rate of Cl atoms versus altitude is also shown. We have assumed 2 Cl atoms per CF_2Cl_2 molecule photolyzed and 2.5 Cl atoms per $CFCl_3$. In each case, all Cl atoms are assumed to be formed at the same altitude as that of the initial photolysis.

In the absence of good spectra for CFClO, we have assumed that 0.5 of the Cl atoms in CFClO are released and that 0.5 diffuse to the tropopause without photodissociation. The downward fluxes of Cl at the tropopause as ClX (and CFClO) are set equal to the upward fluxes of Cl as CF_2Cl_2 and $CFCl_3$.

In subsequent calculations involving reaction rates, several trace components have been fixed at average concentrations similar to those obtained from measurements and from models of the photochemical and transport processes affecting these species in the stratosphere. The fixed concentrations of six of these species are shown in Figure 9. The concentrations of CH_4 and H_2 are shown in Figures 10 and 11.

These concentrations are all average quantities, and specifically are intended as 24-hour averages. Some species (O, NO, OH, etc.) show substantial variations between night and day, consistent with their formation rather directly from solar irradiation and with their high reactivity such that they are removed by chemical reaction within a short time following the setting of the sun. In the real atmosphere then the concentrations for these particular species must vary from near zero at night to twice the listed concentrations during midday.

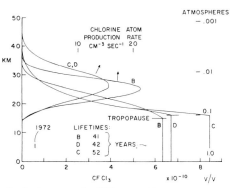

Fig. 7. Stationary state for $CFCl_3$ versus altitude with eddy diffusion coefficients for three different models. Models are defined in Figure 6.

Similarly, these average concentrations may also vary for a particular species with latitude and season. Calculations with diurnal and seasonal variations built into them are very much more complex and are now being carried out by other atmospheric modeling groups. The general expectation from early results is that these refinements will not alter appreciably the semiquantitative evaluation of global average effects of trace concentration measurements. These improvements in modeling will, of course, be very important in attempts to understand the diurnal variations in concentration themselves, measurements at sunrise or sunset, etc.

The calculated lifetimes for CF_2Cl_2 and $CFCl_3$ with each of these eddy diffusion coefficient models are summarized in Table 7. The 'best' model for the eddy diffusion coefficients versus altitude is still subject to wide variation in choice, with consequent substantial uncertainties in lifetimes, steady state concentrations, etc., for the chlorofluoromethanes. Conclusive choice among these possibilities will require extensive additional stratospheric data.

Since an uncertainty of as much as a factor of 2 has been discussed for the stratospheric solar flux around 2000 Å, we have calculated lifetimes for photodissociation rates greater and less than the current best estimate by a factor of $(10)^{1/2}$ and have listed these in Table 8. For convenience in estimation of relative lifetimes against photodissociation for other molecules, we have extended these calculations to photodissociation rates ranging from 0.01 to $100J$. These

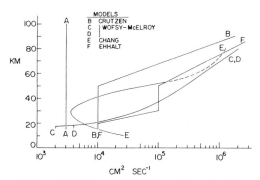

Fig. 6. Eddy diffusion coefficients for six different models.

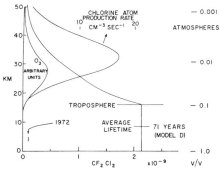

Fig. 8. Stationary state for CF_2Cl_2 versus altitude for eddy diffusion coefficients for model D, defined in Figure 6.

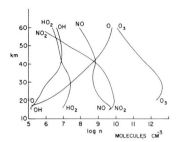

Fig. 9. Assumed global 24-hour average concentrations for six species of importance in stratospheric chlorine chemistry (modeled after those of *McConnell and McElroy* [1973] and *McElroy et al.* [1974]).

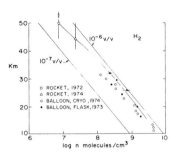

Fig. 11. Molecular hydrogen concentration versus altitude (adapted from T. M. Hard, private communication, 1974). The squares are data from *Ehhalt et al.* [1972], the triangles are data from *Ehhalt* [1974], the lines are the adopted profile for H_2, the closed circles are data from *Ehhalt and Heidt* [1973], and the open circles are data from *Ehhalt et al.* [1974].

values could be used, for example, to estimate photochemical lifetimes (and altitudes of photodissociation) for other chlorofluorocarbon molecules as soon as the absorption cross sections in the 1850- to 2300-Å range have been measured.

The generally shorter lifetimes for model D than B reflect the more rapid upward diffusion above about 25 km for the former, thereby exposing the molecules to the larger values of J at higher altitudes, with consequently more rapid photodissociation.

It is obvious from Table 8 that an uncertainty of a factor of 2 in solar flux means an error of about 20%, i.e., 71 ± 15 years in the estimated lifetime for CF_2Cl_2 with diffusion model D. Until the solar flux is more firmly established, and that might require observations with one specific instrument over an entire 11-year solar cycle, the estimated lifetimes for photochemical dissociation must have a substantial error in them from this source alone. At the present time, moreover, the uncertainties in the choice of eddy diffusion model introduce substantial spread into the calculated lifetimes, as was summarized earlier in Table 7.

CHAIN REACTIONS IN THE OZONE LAYER

All of the chief stable components and most of the minor components of the earth's atmosphere have even numbers of electrons per molecule, and with the exception of the oxygen species O and O_2, all of the electrons are fully paired. The only stable odd electron molecules ('free radicals') in the at-

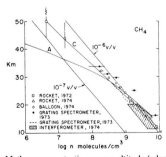

Fig. 10. Methane concentration versus altitude (adapted from T. M. Hard, private communication, 1974). The squares are data from *Ehhalt et al.* [1972], the triangles are data from *Ehhalt et al.* [1974], the open circles are data from *Ehhalt* [1974], the closed circles are data from *Ackerman and Muller* [1973], the dotted line is data from *Cumming and Lowe* [1973], the hatching is data from *Farmer et al.* [1974], A is adopted CH_4 profile A, and C is adopted CH_4 profile C.

mosphere are NO and NO_2. The so-called odd oxygen species, O and O_3, also contain even numbers of electrons but are reactive toward each other and toward most odd electron chemical species. For many atmospheric considerations, it is convenient to calculate variations in the odd oxygen concentration, i.e., the sum of the concentrations of O and O_3. The only important reaction in the atmosphere that increases the odd oxygen concentration is the photolysis of molecular O_2 by ultraviolet radiation with wavelength shorter than 2420 Å, as in (18).

$$h\nu + O_2 \rightarrow O + O \tag{18}$$

One prime mechanism for removal of odd oxygen in the stratosphere is the direct reaction of O with O_3, as in (19).

$$O + O_3 \rightarrow O_2 + O_2 \tag{19}$$

In addition, a variety of reaction sequences have been discovered involving various odd electron species, including the HO_x(H, OH, HO_2) series, the NO_x(NO, NO_2) series, and the ClO_x(Cl, ClO, ClOO, OClO) series, which accomplish either reaction (19) or (20) by a catalytic chain sequence that sums to one of these reactions. Reaction (20) does not occur directly.

$$O_3 + O_3 \rightarrow O_2 + O_2 + O_2 \tag{20}$$

Similar reactions can also be written for the FO_x series (F, OF, FO_2), and presumably for many other free radical species. These other possible catalytic chains of course only become important in the stratosphere when some process exists by which these free radicals can diffuse into or be injected into the stratosphere.

The absorption of ultraviolet radiation by O_3 with the formation of O atoms, as in (21),

$$h\nu + O_3 \rightarrow O + O_2 \tag{21}$$

is very important in the atmosphere in removing both ultraviolet and visible radiation from the incoming solar flux. The combination of O atoms with molecular O_2, as in (22),

$$O + O_2 + M \rightarrow O_3 + M \tag{22}$$

then releases this energy into the stratosphere as kinetic or internal energy of either the ozone or of the other molecule M and is responsible for the substantial rise in stratospheric temperatures (210° to 274°K) occurring as the altitude increases from 15 to 50 km. However, neither (21) nor (22) results in a net change in the total odd oxygen concentration of

TABLE 7.   Calculated Average Atmospheric Residence Times for CF_2Cl_2 and $CFCl_3$ With Different Eddy Diffusion Coefficient Models

Diffusion Model	Average Residence Times, years	
	$CFCl_3$	CF_2Cl_2
A	85	205
B	41	102
C	52	82
D	42	71
E	60	160
F	29	53

the atmosphere, and both are omitted from calculations involving changes in the odd oxygen concentration. The four reactions (18), (19), (21), and (22) are those originally suggested by *Chapman* [1930] for estimating ozone levels in the atmosphere.

The reactions of NO_x that result in the greatest net removal of odd oxygen from the atmosphere are the pair of reactions given earlier in (6) and (8) [e.g., *Johnston*, 1974]. (At lower altitudes, for which reaction (6) is slow because of the low concentration of O atoms, reaction (7) competes as a sink for NO_2, and the combination of (8) + (7) does not result in any change in the total amount of odd oxygen present.) As summarized below, several similar sequences of catalytic ozone removal involving reactions (23)–(27) have also been identified for HO_x.

$$H + O_3 \rightarrow OH + O_2 \qquad (23)$$

$$OH + O \rightarrow H + O_2 \qquad (24)$$

$$OH + O_3 \rightarrow HO_2 + O_2 \qquad (25)$$

$$HO_2 + O \rightarrow OH + O_2 \qquad (26)$$

$$OH + O_3 \rightarrow H + O_2 + O_2 \qquad (27)$$

However, the net contributions in the stratosphere of the HO_x cycles are considerably less important than those of the natural NO_x chains when considering total removal of odd oxygen, as was revealed by a detailed study of the individual reaction rate constants [e.g., *McConnell and McElroy*, 1973; *Nicolet*, 1974].

Since all of the reactions in these chains are between a free radical (odd electron species) and an even electron species, one of the products must always be an odd electron species, and the total number of free radicals is not changed by the occurrence of the reaction. The propagation of these free radical

chains can thus only be stopped by formation of a nonreactive free radical or by reaction of two free radicals with one another (including reaction with particulate matter on which radicals might hypothetically accumulate). The radical HO_2 is relatively nonreactive, and the HO_x chain removal of odd oxygen is slowed down by the formation of HO_2. The cross termination of two chains is illustrated by reaction (28),

$$OH + NO_2 \rightarrow HONO_2 \qquad (28)$$

which removes one OH radical and one odd electron NO_2 with the formation of nitric acid, HNO_3, an even electron species. Almost all even electron species formed by the reaction of two free radicals are very unreactive toward O_3 and O, and the chain processes are thus effectively shut off by such dual termination processes, unless the terminating molecule is forced into the reaction sequence again, most often by its photolysis, as illustrated by the photolysis of HNO_3 to reform $OH + NO_2$.

CLO_x CHAINS FOR DECOMPOSITION OF OZONE

The existence of a chain reaction involving Cl and ClO reactions with ozone and oxygen atoms was considered by several groups during 1973, especially after the presentation by Stolarski and Cicerone at the IAGA meeting in Kyoto in September 1973. The ClO_x chain has been treated in several recent publications with respect to the possible effects of natural chlorine-containing compounds, HCl introduced into the troposphere from industrial processes, and HCl introduced into both the troposphere and stratosphere by the proposed U.S. space shuttle (from NH_4ClO_4 used as the oxidant) [*Hoshizaki et al.,* 1973; *Stolarski and Cicerone,* 1974; *Crutzen,* 1974b; *Wofsy* 1974]. The importance of the chain in connection with the photolysis of chlorofluoromethanes in the stratosphere has also been indicated earlier [*Molina and Rowland,* 1974a, b]. The data pertinent to chlorofluoromethane reactions are considered in the following section.

Chlorine atoms introduced into the stratosphere can be viewed simply as an additional group of free radicals, and sets of reactions with O_3 and O corresponding to those of HO_x and NO_x need to be considered for each of the possible ClO_x species. Of the several ClO_x species, the chloroperoxy radical ClOO is least important in the stratosphere because of its weak Cl—O bond and the accompanying probability of rapid dissociation into $Cl + O_2$. From the heats of formation [*Watson,* 1974], the Cl—O bond energy in ClOO is only 9 kcal/mol, and the activation energy for dissociation is probably only slightly higher. Correspondingly, the reaction of Cl with O_2 to form ClOO has a very low activation energy, and Cl atoms should

TABLE 8.   Variation of Average Atmospheric Residence Time With Different Values of J for Photodissociation of CF_2Cl_2 and $CFCl_3$

	CF_2Cl_2 Average Residence Time, years		$CFCl_3$ Average Residence Time, years	
	Model B	Model D	Model B	Model D
100J	17.2 (19–24)	23.9 (17–25)	7.3 (16–21)	10.8 (16–19)
10J	40.0 (21–29)	41.4 (21–32)	17.0 (18–24)	23.7 (17–25)
3.1J	62.7 (23–31)	53.5 (23–36)	26.1 (20–26)	31.9 (19–28)
1J	102 (24–34)	71.2 (25–41)	40.5 (21–29)	41.7 (21–32)
0.31J	173 (26–37)	102 (27–45)	64.4 (22–32)	54.5 (23–36)
0.1J	311 (27–41)	168 (29–49)	106 (24–35)	73.6 (25–41)
0.01J	1330 (29–48)	537 (30–52)	337 (27–42)	186 (29–49)

The values in parentheses are photodissociation altitudes in kilometers. The range of altitudes is the maximum and minimum altitudes for which the rate of formation of Cl atoms exceeds 0.5 times the maximum Cl formation rate.

react readily with O_2 in the atmosphere. Since both reactions (29) and (30) occur rapidly in the stratosphere,

$$Cl + O_2 + M \rightarrow ClOO + M \tag{29}$$

$$ClOO + M \rightarrow Cl + O_2 + M \tag{30}$$

the relative concentrations of ClOO and Cl can be estimated from the free energy of the reaction, for which the value is ΔG = 1.4 kcal/mol at 298°K and is almost independent of temperature [Watson, 1974]. The equilibrium ratio for (Cl/ClOO) is about 450 at 298°K and 0.1 atm, increasing to about 10^5 at 55 km. Consequently, the ClOO concentration represents essentially a negligible fraction of the total ClO_x concentrations at all stratospheric temperatures and pressures.

The role of the symmetrical chlorine dioxide, OClO, in the stratosphere is undetermined at the present time because of the uncertainty as to whether it is formed by reaction (31). Reaction has been observed marginally in laboratory experiments for the two reactants, ClO and O_3 [Clyne and Watson, 1974], but may not be real; even if it is real,

$$ClO + O_3 \rightarrow OClO + O_2 \tag{31}$$

$$ClO + O_3 \rightarrow ClOO + O_2 \tag{32}$$

$$ClO + O_3 \rightarrow Cl + O_2 + O_2 \tag{33}$$

the mechanism of reaction has not been distinguished from among the three possibilities given in (31) to (33). The stratospheric consequences if the mechanism of the observed reaction is the formation of OClO as in (31) are discussed later. Symmetrical OClO is highly unstable and has only 5 kcal/mol of stability relative to decomposition to Cl + O_2. However, this decomposition process requires rearrangement during reaction and probably would proceed first throuh isomerization to ClOO. No activation energy is known for this reaction, but it is probably large enough to prevent isomerization under stratospheric conditions.

The most important chain process in the stratosphere for ClO_x species involves the Cl/ClO chain working through reactions (3) and (4), with the net conversion of O + O_3 to $2O_2$. Both of these reactions take place essentially without an activation energy and with rather high collision efficiencies. The other most important stratospheric reactions for Cl-containing species include the reactions of ClO with NO, Cl with several hydrogen-containing species, OH with HCl, and possibly ClO with O_3. Current estimates for the values of these rate constants are listed in Table 9 and discussed below. The earlier values for many of these reactions and a brief critical discussion of the best current choices for these rate constants have been issued by Watson [1974].

One other reaction that would be of importance in the

stratosphere if its rate were fast would be the reaction of ClO with CO, as in (34).

$$ClO + CO \rightarrow Cl + CO_2 \tag{34}$$

However, the rate constant of 1.4×10^{-15} at 300°K [Clyne et al., 1974; Harker, 1972], when taken with the relative abundances of CO in the stratosphere [Wofsy et al., 1972], indicates that reaction (34) is too slow to be a significant sink for ClO.

ABSTRACTION REACTIONS BY THERMAL CL ATOMS

The termination, at least temporarily, of Cl atom chains in the stratosphere can be accomplished by the reaction of Cl atoms with hydrogenous materials, including both stable molecules (CH_4, H_2, H_2O_2, HNO_3, etc.) and free radicals (HO_2). The attack of Cl upon H_2O is endothermic by 10 kcal/mol, which eliminates this reaction from any importance in the stratosphere. Under most circumstances the combination of atmospheric abundance and rate constants for reaction makes abstraction from CH_4 the dominant process among the stable molecular reactants, as illustrated in Table 10, and the others can be effectively ignored except under conditions of unusually low CH_4 mole fraction, expected somewhere in the stratosphere above 50 km. Some recent discussions have considered NH_3 as a tropospheric reactant [Wofsy et al., 1972], but its solubility in rainwater and consequent loss by rainout keep Cl atom reactions with it of negligible importance relative to abstraction from CH_4 and H_2.

The actual stratospheric reaction rate of Cl with H_2O_2 is quite uncertain, both because the laboratory reaction rate has not been measured and because H_2O_2 has not yet been detected in the stratosphere. The situation is somewhat better for Cl + HNO_3, for the latter has been measured in the stratosphere. An estimate of the Cl atom reaction rate with these molecules can be made through comparison of the analogous reactions of OH with the same molecules. The reaction rate constants for Cl with CH_4 and H_2 are, respectively, about 16 and 5 times as rapid as for OH with the same molecules at stratospheric temperatures. Since, in addition, Cl and OH are isoelectronic and rather similar in thermochemistry, approximate rate constants can be estimated by assuming that the Cl atom reactions are 10 times faster than the corresponding OH reactions. These values are summarized in Table 9.

Combination of the Cl + HNO_3 rate constant from Table 9 with measured HNO_3 concentrations indicates that this reaction is considerably less important than Cl + CH_4 even at 25 km. The possible contribution to HCl formation becomes negligible for HNO_3 (less than 1%) above 40 km.

The H_2O_2 level in the stratosphere has been estimated from different models of the photochemical and meteorological processes, with variations in predicted levels of a factor of 5 or

TABLE 9.   Rate Constants for Reactions Involved in the ClO_x Cycles in the Stratosphere

Reaction	Reactants	k, cm^3 molecule^{-1} s^{-1}	Reference
3	Cl + O_3	$(1.85 \pm 0.36) \times 10^{-11}$	Clyne and Watson [1974]
4	ClO + O	$(5.3 \pm 0.8) \times 10^{-11}$	Bemand et al. [1973]
5	ClO + NO	$(1.7 \pm 0.2) \times 10^{-11}$	Clyne and Watson [1974]
9	Cl + CH_4	$5.1 \times 10^{-11} \exp(-1790/T)$	Clyne and Walker [1973]
10	Cl + H_2	$5.7 \times 10^{-11} \exp(-2260/T)$	Clyne and Walker [1973]
11	Cl + HO_2	2×10^{-11}	Estimated
12	Cl + H_2O_2	$1.7 \times 10^{-10} \exp(-910/T)$	Estimated
13	Cl + $HONO_2$	$6 \times 10^{-12} \exp(-400/T)$	Estimated
14	OH + HCl	$2.0 \times 10^{-12} \exp(-313/T)$	Zahniser et al. [1974]
31–33	ClO + O_3	$<5 \times 10^{-15}$ at 298°K	Clyne et al. [1974]

TABLE 10. Calculated Reaction Rates for Chlorine Atoms at Various Stratospheric Altitudes

	A = 25 km T = 227.1°K D = 7.65 × 10¹⁷	A = 30 km T = 235.2°K D = 3.55 × 10¹⁷	A = 35 km T = 251.7°K D = 1.66 × 10¹⁷	A = 40 km T = 268.2°K D = 8.11 × 10¹⁶	A = 45 km T = 274.5°K D = 4.26 × 10¹⁶	A = 50 km T = 274.0°K D = 2.31 × 10¹⁶	A = 55 km T = 273.6°K D = 1.25 × 10¹⁶
Concentrations, cm⁻³ (from Figures 9–11)							
O_3	4.0×10^{12}	3.8×10^{12}	1.8×10^{12}	5.8×10^{11}	2.0×10^{11}	7.1×10^{10}	2.7×10^{10}
CH_4(C, Fig. 10)	5.4×10^{11}	2.1×10^{11}	7.6×10^{10}	3.1×10^{10}	1.3×10^{10}	6.0×10^{9}	3.0×10^{9}
H_2	6.1×10^{11}	2.8×10^{11}	1.2×10^{11}	4.9×10^{10}	2.1×10^{10}	9.2×10^{9}	3.8×10^{9}
HO_2	2.4×10^{7}	2.6×10^{7}	1.7×10^{7}	8.9×10^{6}	6.3×10^{6}	4.6×10^{6}	3.3×10^{6}
HNO_3	4.0×10^{9}	6.0×10^{8}	1.5×10^{8}	2.5×10^{7}	3.0×10^{6}	$<10^{6}$	$\cdots$
H_2O_2*	1.8×10^{9}	1.2×10^{9}	2.8×10^{8}	4.0×10^{7}	1.2×10^{7}	4.4×10^{6}	1.8×10^{6}
H_2O_2†	3.2×10^{8}	3.5×10^{8}	1.6×10^{8}	4.0×10^{7}	1.5×10^{7}	5.3×10^{6}	1.9×10^{6}
Rates of removal‡ of Cl atoms, s⁻¹							
$k_3[O_3]$	74	70	33	10.7	3.7	1.3	0.50
$k_9[CH_4]$	1.0×10^{-2}	5.2×10^{-3}	3.2×10^{-3}	2.0×10^{-3}	9.8×10^{-4}	4.4×10^{-4}	2.2×10^{-4}
$k_{10}[H_2]$	1.7×10^{-3}	1.1×10^{-3}	0.8×10^{-3}	0.6×10^{-3}	3.2×10^{-4}	1.4×10^{-4}	0.5×10^{-4}
$k_{11}[HO_2]$	0.5×10^{-3}	0.5×10^{-3}	0.3×10^{-3}	0.2×10^{-3}	1.3×10^{-4}	0.9×10^{-4}	0.7×10^{-4}
$k_{13}[HNO_3]$	4.1×10^{-3}	0.6×10^{-3}	0.2×10^{-3}	0.3×10^{-4}	0.5×10^{-5}	$\cdots$	$\cdots$
$k_{12}[H_2O_2]$*	5.6×10^{-3}	4.3×10^{-3}	1.3×10^{-3}	0.2×10^{-3}	0.7×10^{-4}	0.3×10^{-4}	0.1×10^{-4}
$k_{12}[H_2O_2]$†	1.0×10^{-3}	1.2×10^{-3}	0.7×10^{-3}	0.2×10^{-3}	0.9×10^{-4}	0.3×10^{-4}	0.1×10^{-4}
$\sum k_x[HX]$* $x = 9\text{–}13$	2.0×10^{-2}	1.2×10^{-2}	5.8×10^{-3}	3.0×10^{-3}	1.5×10^{-3}	7.0×10^{-4}	3.5×10^{-4}
$\sum k_x[HX]$†	1.7×10^{-2}	8.6×10^{-3}	5.2×10^{-3}	3.0×10^{-3}	1.5×10^{-3}	7.0×10^{-4}	3.5×10^{-4}
Ratio of $k_3[O_3]$ to sum of HCl- producing reactions							
Case b*	3.7×10^{3}	5.8×10^{3}	5.7×10^{3}	3.5×10^{3}	2.5×10^{3}	1.9×10^{3}	1.4×10^{3}
Case c†	5.3×10^{3}	8.1×10^{3}	6.3×10^{3}	3.5×10^{3}	2.5×10^{3}	1.9×10^{3}	1.4×10^{3}

A stands for altitude, T for temperature, and D for number density.
* H_2O_2 concentrations from *McConnell and McElroy* [1973].
† H_2O_2 concentrations from P. J. Crutzen (private communication, 1974).
‡ Rate constants from Table 9.

more at a given altitude. If the high values in estimated concentrations are combined with the estimated rate constant of Table 9 for reaction (12), HCl formation from H_2O_2 is almost equal to that from CH_4 at 25–30 km, falling to less than 10% as large as CH_4 at 40 km, as is shown in Table 10. Since the estimated rate constant for abstraction of H from H_2O_2 by Cl is only a factor of 3 slower than Cl reaction with O_3 at 275°K, it is unlikely that this estimate of the rate constant is too low, and it may well be high by a factor of 2–4. If the lower estimates of H_2O_2 concentrations are used, then reaction (12) represents only a minor correction to the dominant CH_4 reaction throughout the stratosphere [*Molina and Rowland*, 1975].

The most important potential H atom source among the free radicals is apparently HO_2, whose abundance is calculated to rise as high as 10^7 radicals cm⁻³ at altitudes of 30–60 km. However, the rate coefficient for the reaction of Cl with HO_2 has not yet been measured. If the value is assumed to be 2×10^{-10} molecule cm⁻³ s⁻¹ and is used with typical values for HO_2 concentration from recent atmospheric modeling calculations, then this reaction is approximately as important as abstraction from CH_4 in removing Cl atoms with the formation of HCl. However, if this reaction rate is only 2×10^{-11} molecule cm⁻³ s⁻¹, then it is less important than reaction with CH_4 (but not negligible) over the entire altitude range from 30 to 50 km. We believe that a rate constant of 2×10^{-10} is unlikely, since geometrical considerations alone (e.g., collision of Cl with the O—O end of HO_2) should preclude collision efficiencies near unity, as required for a rate constant of about 2×10^{-10}. A recent measurement for the similar reaction of H + HO_2 to form H_2 + O_2 indicates a rate constant of 2.8×10^{-11} cm³ molecule⁻¹ s⁻¹ [*Baldwin et al.*, 1974].

Among all of the important atmospheric rate constants concerned in the various sets of reactions with O_3 and O, only the reactions of $O(^1D)$ with CH_4, H_2, and H_2O have rate constants ~3×10^{-10} [*Heidner and Husain*, 1973; *Heidner et al.*, 1973], while that for OH + HO_2 has been quoted with a value approaching 2×10^{-10} for its preexponential factor [*Demore and Tschuikow-Roux*, 1974; *Hochanadel et al.*, 1972], and the value of 2×10^{-10} without an activation energy has been used for OH + HO_2. The latter rate constant is the subject for continuing investigation, and a spread of rate constants from 2×10^{-11} to 2×10^{-10} for (35) has been recommended for consideration by atmospheric modelers in the most recent compilations of rate constants. Reaction (35)

$$OH + HO_2 \rightarrow H_2O + O_2 \qquad (35)$$

is intimately involved in estimates of the OH and HO_2 concentrations at 30–60 km in the stratosphere, since reaction with each other is a major component in the removal rates for each. No direct measurement of the HO_2 concentration in the atmosphere has yet been reported.

The rate constant for the reaction of Cl with OH does not seem to be known, but the activation energies for the reverse of reaction (36)

$$Cl + OH \rightarrow HCl + O \qquad (36)$$

have been measured as 4 and 7 kcal/mol in two separate experiments [*Balakhnin et al.*, 1971; *Wong and Belles*, 1972]. Although the agreement between the two sets of data is poor, both are in accord that the activation energy is at least 4 kcal/mol. Since the forward reaction is about 0.8 kcal/mol exothermic, it must have an activation energy of more than 3 kcal/mol in the forward direction, i.e., sufficient to make the rate constant for reaction (36) $<5 \times 10^{-13}$ and the rate therefore negligible relative to abstraction of H atom from either CH_4 or HO_2.

The rate constant for the reaction of Cl with CH_4 has been measured near room temperature by two separate groups with good agreement: $(1.5 \pm 0.1) \times 10^{-13}$ cm³ molecule⁻¹ s⁻¹ by *Davis et al.* [1970] at 298°K; and $(1.3 \pm 0.1) \times 10^{-13}$ cm³ molecule⁻¹ s⁻¹ by *Clyne and Walker* [1973]. Clyne and Walker also measured these rate constants over the temperature range from 300° to 686°K and have derived the equation $k = 5.1 \times 10^{-11}$ exp $(-1790/T)$ cm³ molecule⁻¹ s⁻¹ as the best fit to their data. We have used the rate constants versus temperature obtained from Clyne and Walker's results in all of our calculations. Extrapolation of these values to stratospheric temperatures leads, for example, to a rate constant of 1.0×10^{-14} cm³ molecule⁻¹ s⁻¹ at 210°K, a factor of 13 slower than the slowest actually measured rate in their experimental determinations. However, in the most important 35- to 55-km region the extrapolated rate constant has a value between 3 and 7×10^{-14}.

The rate constant for the reaction of Cl with H_2 has been measured several times, with some discrepancy in the data [*Fettis and Knox*, 1964; *Benson et al.*, 1969; *Davis et al.*, 1970; *Galante and Gislason*, 1973]. In our calculations we have used the recent critical estimate of *Clyne and Walker* [1973], which gives the value $k = 5.6 \times 10^{-11}$ exp $(-2260/T)$ cm³ molecule⁻¹ s⁻¹ and which is stated by the authors to give the better representation of the activation energy of the reaction. In this case, however, the experimental temperature range (195°–496°K) includes all of the temperatures pertinent to the stratosphere, so that no extrapolations are required.

Comparison of the relative rates of reaction, as in Table 10, shows that the removal of Cl by reaction with H_2 is only about 20–30% as important as the CH_4 reaction. With increasing altitude above about 20 km the CH_4 mixing ratio decreases because of high-altitude sinks, such as solar photodissociation and attack by $O(^1D)$ atoms. The H_2 mixing ratio, on the other hand, shows a small increase from 5×10^{-7} to 8×10^{-7} vol/vol at about 35 km and then begins to decrease again [*Ehhalt et al.*, 1972, 1974]. The reactions of Cl with H_2 and with HO_2 thus serve as a base level for removal of Cl atoms by abstraction, even if the CH_4 reaction were to decrease sharply because of lowered CH_4 mole fraction at some higher altitudes.

At stratospheric temperatures the reaction efficiency per collision of Cl with CH_4 is thus $\sim 10^{-4}$, and the mixing ratio of CH_4 is $\sim 10^{-6}$. Consequently, even with unit collision efficiencies, only species with mixing ratios of $\sim 10^{-10}$ in the atmosphere could have an effect comparable to that of CH_4 in

terminating Cl atom chains. Except for the free radical HO_2, most species either fall below that concentration level or have substantial activation energies for reaction, or both. Since even molecules with a relatively weak C—H bond, such as $CHCl_3$, have collision efficiencies of $<10^{-3}$ for the abstraction reaction at these temperatures [*Clyne and Walker*, 1973], the possibility of high-efficiency removal of Cl atoms by reaction with undetected trace hydrogenous compounds can be effectively dismissed.

The relative rates of reaction of Cl with O_3, CH_4, H_2, HO_2, H_2O_2, and HNO_3 have been calculated and are shown in Table 10 for several altitudes by using typical concentrations consistent with both measured and estimated properties of the natural atmosphere. The distribution of most of these components with altitude was illustrated earlier in Figures 9–11. Over the entire range of altitudes of Table 10 the overall rate of removal of Cl by reaction with O_3 is from 1400 to 8000 times as rapid as by all HCl-forming reactions summed together, and the average lifetime of a free Cl atom varies from about 2 s at 55 km to less than 0.02 s at 20 km.

If the rate constant for k_{11} is estimated as 2×10^{-10} cm³ molecule⁻¹ s⁻¹, then all of the estimates in the row for $k_{11}[HO_2]$ are increased by a factor of 10 and would then vary from being about equal to $k_9[CH_4]$ at 30 km to being 3 times as large at 55 km, as shown in Table 11. These numbers are significantly different from those of Table 10 but have no effect on the lifetimes of Cl atoms, which are still dominated entirely by $k_3[O_3]$, nor are changes made in any of the qualitative conclusions about overall stratospheric effects. In our subsequent calculations we have usually used the data of Table 10 with the rate constant $k_{11} = 2 \times 10^{-11}$ cm³ molecule⁻¹ s⁻¹ unless it is specifically mentioned that some variation from these data has been introduced.

RELEASE OF CL ATOMS BY OH ATTACK ON HCL

The chief reaction in the atmosphere that can release Cl atoms from HCl is the attack by OH radicals, as in (14). The rate constant for this reaction was first measured indirectly at 1940°K from the structure of CH_4—O_2 flame fronts inhibited by HCl [*Wilson et al.*, 1969]. A measurement at 295°K when combined with the earlier high-temperature data led to an estimate of 2.6 kcal/mol as the activation energy for (14) [*Takacs and Glass*, 1973]. However, two additional measurements, each over a range of temperatures, indicate a very much lower activation energy while agreeing very closely with the Takacs-Glass value at 295°K. We have used the rate equation $k_{14} = 2.0 \times 10^{-12}$ (exp $-313/T$) from the work of *Zahniser et al.* [1974] in our calculations. The equation $k_{14} = 4.1 \times 10^{-12}$ (exp

TABLE 11.   Calculated Reaction Rates for Chlorine Atoms at Various Altitudes With Different Assumptions for Rate Constants and Concentration

	A = 25 km	A = 30 km	A = 35 km	A = 40 km	A = 45 km	A = 50 km	A = 55 km
Rates of removal of Cl atoms							
$k_{11}[HO_2]$	4.8×10^{-3}	5.2×10^{-3}	3.4×10^{-3}	1.8×10^{-3}	1.3×10^{-3}	9.2×10^{-4}	6.6×10^{-4}
$\sum k_x[HX]$*	2.6×10^{-2}	1.6×10^{-2}	8.9×10^{-3}	4.6×10^{-3}	2.7×10^{-3}	1.5×10^{-3}	9.4×10^{-4}
$x = 9-13$							
Ratio of $k_3[O_3]$ to sum of HCl-producing reactions	2.8×10^3	4.3×10^3	3.7×10^3	2.3×10^3	1.4×10^3	8.5×10^2	5.3×10^2

Rate constant k_{11} is assumed to be 2×10^{-10} throughout. Other rate constants are as given in Table 10. A is altitude.
* Using H_2O_2 concentrations from *McConnell and McElroy* [1973].

$-530/T$) given by *Smith and Zellner* [1974] differs by 15% or less throughout the important altitude region from 35 to 50 km.

METHANE CONCENTRATIONS AT 50 KM

Only two actual sample measurements of CH_4 at altitudes above about 30 km have been made. These were the rocket-collected samples of *Ehhalt et al.* [1972, 1974], which indicated an average concentration of 0.2×10^{-6} by volume for a sample collected in a liquid neon–cooled Dewar opened for gas collection on the ascending rocket between 44 and 62 km and 0.38×10^{-6} for one collected between 40 and 50 km. Prior to these measurements, most estimates of CH_4 concentrations showed much more rapid falloff with increasing altitude above 30 km, corresponding to much lower general estimates of the eddy diffusion coefficients appropriate to altitudes between 30 and 50 km.

The measurements of CH_4 at 40–62 km are probably satisfactory but might be misleadingly high because both samples were collected at the same location and could by chance have had a fortuitously large CH_4 content from local atmospheric circulation patterns. We have no reason to believe that such events occurred but wished to examine the possible consequences of too great a reliance on these CH_4 concentration measurements through the critical region. Consequently, we have also carried out calculations similar to those of Table 10, substituting a literature CH_4 profile curve from 1972 (line A in Figure 10) prior to the publication of the rocket measurement of Ehhalt et al.

The chief conclusion that can be drawn from these estimates is that if the rocket measurements of CH_4 are seriously in error on the high side for whatever reason then the control of the Cl atom chain through HCl formation would depend upon the reactions with H_2 and HO_2 and would permit appreciably longer chains. The error, however, would only be of the order of a factor of 2 or 3 at the important altitudes and not qualitatively different.

RATE CONSTANTS FOR CLO$_x$ REACTIONS WITH OZONE AND OXYGEN ATOMS

The attack of Cl atoms on O_3 is very rapid, and the rate constant has recently been measured to have a value of $(1.85 \pm 0.36) \times 10^{-11}$ cm^3 molecule^{-1} s^{-1} [*Clyne and Watson*, 1974]. No temperature studies have been made for this reaction, but the high value of k insures that the activation energy is minimal and can be neglected. Accordingly, this value of k has

been used at all temperatures in our calculations. In contrast, the reaction of NO with O_3 is very much slower, 1.8×10^{-14} cm^3 molecule^{-1} s^{-1} at 300°K, with an activation energy of about 2 kcal/mol [*Niki*, 1974].

The rate constant for reaction of ClO with O has also been recently measured to have the still faster rate constant of $(5.3 \pm 0.8) \times 10^{-11}$ cm^3 molecule^{-1} s^{-1} [*Bemand et al.*, 1973]. Again, no temperature dependence has been measured for this reaction, but the activation energy must be negligibly small. This reaction is about 6 times as fast as the reaction of NO_2 with O atoms [*Davis et al.*, 1973].

The most important reactions in the chlorine atom-catalyzed decomposition of ozone are very rapid in the higher stratosphere, as summarized in Tables 10 and 12. At 40 km, for example, with O_3 and O atom concentrations of about 5.8×10^{11} and 3.9×10^8 molecules cm^{-3}, respectively, the average time required for attack of Cl on O_3 is 0.1 s and for ClO on O is <1 min. This chain reaction can be diverted from this path of one complete cycle every minute by two competitive reactions: ClO can also react with NO, and the fraction of ClO that reacts with O is determined by the O/NO ratio; and a very small fraction of Cl atoms react with CH_4 or other species to form HCl. From the reaction rate constants of Tables 10 to 12 and the concentrations of Figure 9, some typical fractions of reaction by each path have been calculated in Table 12.

With the rate of removal of Cl atoms determined essentially entirely by reaction with O_3 and that of ClO by reaction with either O or NO, the ratio of Cl/ClO can be readily calculated as being equal to the ratio of

$$\frac{(Cl)}{(ClO)} = \frac{k_4[O] + k_5[NO]}{k_3[O_3]} \quad (37)$$

Since all of these reactions have negligible activation energies, the calculated ratio is independent of temperature and dependent only on the relative concentrations of O, NO, and O_3, plus the rate constants given in Table 9.

The Cl/ClO ratio has been evaluated for the typical concentration profile of Figure 9 in Table 12. The rates of removal of ClO by reaction with O and NO are also given in this table. The combined rate of removal of ClO corresponds to lifetimes varying from 100 s at 30 km to less than 5 s at 55 km. The corresponding values for the actual rates of removal of Cl atoms have been given earlier in Table 10.

The ultimate consequence of the reaction of ClO with NO for the chain removal of odd oxygen depends upon the fate of the NO_2 formed in the process. If the NO_2 in turn reacts with

TABLE 12. Stratospheric Reaction Rates of ClO With O and NO

	A = 25 km	A = 30 km	A = 35 km	A = 40 km	A = 45 km	A = 50 km	A = 55 km
Concentrations, cm^{-3}							
O	6.8×10^6	3.2×10^7	1.3×10^8	3.9×10^8	1.2×10^9	2.7×10^9	4.2×10^9
NO	7.5×10^8	4.9×10^8	5.1×10^8	7.0×10^8	5.9×10^8	3.4×10^8	1.8×10^8
Rate of removal of ClO, s^{-1}							
Reaction with O	3.6×10^{-4}	1.7×10^{-3}	6.9×10^{-3}	2.1×10^{-2}	6.4×10^{-2}	1.4×10^{-1}	2.2×10^{-1}
Reaction with NO	1.3×10^{-2}	8.3×10^{-3}	8.7×10^{-3}	1.2×10^{-2}	1.0×10^{-2}	0.6×10^{-2}	0.3×10^{-2}
Total	1.3×10^{-2}	1.0×10^{-2}	1.6×10^{-2}	3.3×10^{-2}	7.4×10^{-2}	1.5×10^{-1}	2.3×10^{-1}
Fraction of ClO reaction with O	0.03	0.17	0.44	0.63	0.86	0.96	0.99
Cl/ClO Ratio	1.8×10^{-4}	1.4×10^{-4}	4.7×10^{-4}	3.1×10^{-3}	2.0×10^{-2}	1.1×10^{-1}	4.5×10^{-1}

A is altitude.

O, the overall three-step reaction leads to the removal of two equivalents of odd oxygen. In the course of this reaction, with large amounts of ClO_x present, the NO/NO_2 ratio can be markedly shifted.

However, the increasingly successful competition of NO with O for ClO at lower altitudes is chiefly the result of decreasing O atom concentration. Consequently, the rate of reaction of NO_2 with O is also slowed down, and the primary fate of NO_2 in the 25- to 30-km range is photolysis back to NO + O. For this overall sequence, the initial attack of Cl on O_3 leads eventually back to an O atom, with no overall net change in the odd oxygen concentration. The bulk of the odd oxygen depletion by ClO_x thus occurs at 30 km or higher. At 45 km, for example, Cl reacts with O_3 in 0.3 s, ClO with O or NO (86% O) every 14 s, and about 4 cycles (8 odd oxygen equivalents) are completed every minute until the chain is interrupted by one of the side reaction pathways forming HCl. Table 10 shows that the frequency of this side reaction at 45 km is about 1 in 2500 cycles, indicating that the chain proceeds for an average of 10 hours with 4 cycles per minute before interruption. In the next section the rate of restarting of ClO_x chains at 45 km by attack of OH on HCl is estimated as 4.9×10^{-6} s^{-1}, or about 2.4 days. At this altitude then each ClX entity removes about 5000 odd oxygen species every 3 days and will continue to do this until it diffuses downward into less reactive regions and eventually is permanently removed at the tropopause. Rough estimates suggest that the Cl atoms from each molecule of CF_2Cl_2 remove on the average about 10^5 molecules of odd oxygen before permanent removal of the Cl species from the stratosphere.

The fraction of time during which ClO_x species are actively involved in removing O_3 or O by chain processes is basically determined by the time spent as Cl/ClO versus time spent as HCl. During the interim following abstraction of H from CH_4, H_2, etc., the chain process is in abeyance until initiated again by the attack of OH upon HCl. The lifetime of HCl in the 25- to 55-km-altitude range is essentially determined by the concentration of OH (and the temperature), whereas the time required to re-form the HCl is given by the lifetime of Cl multiplied by the inverse of the fraction of Cl atoms reacting by the H abstraction pathway. Typical lifetimes for HCl, given in Table 13, are in the range of a few times 10^5 s (i.e., a few days) and are usually longer than the sum of the lifetimes of the Cl and ClO species (almost entirely lifetime of ClO).

The (HCl/ClO) ratio has been calculated from the ratio given in (38), in which the $\sum k_x[HX]$ is the sum of reaction rates for abstraction of H from CH_4, H_2, and HO_2:

$$\frac{(HCl)}{(ClO)} = \left[\frac{\sum k_x[HX]}{k_8[O_3]} \right] \times \left[\frac{k_4[O] + k_5[NO]}{k_{14}[OH]} \right] \quad (38)$$

The calculated (HCl/ClO) ratios are summarized in Table 13 and are combined with the (Cl/ClO) ratios from Table 12 to give the fractional time spent by Cl species as HCl, ClO, and Cl at each altitude. These fractional times vary with changes in adopted concentration profiles and with the assumed rate constant for k_{11}, but the general conclusions are similar: HCl is the dominant ClX species at all altitudes for almost all calculations (exception: with CH_4 profile A, ClO exceeds HCl at 40–45 km); ClO ranges from 5 to 50% in the 30- to 50-km range; the concentration of Cl is always <2% of the total ClX concentration in the stratosphere.

CHAIN-TERMINATING REACTIONS

The introduction of new free radical species into the stratospheric chemistry requires consideration of the complete set of termination reactions involving each of these species with each of the free radical species always present in the atmosphere. The possible dual chain termination steps for the ClO_x species thus include cross products of Cl, ClO, and OClO with each of the following species: H, OH, HO_2, NO, NO_2, Cl, ClO, OClO. The possible compounds formed by such cross-termination processes are illustrated in Table 14. These cross-radical reactions can be classified into three categories: (1) combination to form a stable molecule, (2) reaction to form two stable molecules, (3) reaction to form two different free radicals. Only (1) and (2) are actually termination processes. The most important example of (3) is the reaction of ClO with NO, as was discussed earlier.

An important example of (2) is the reaction of Cl with HO_2 to form HCl + O_2, both products being stable molecules with low cross sections for photodissociation at the altitude of formation. Such a reaction can thus terminate two chains, one HO_x and one ClO_x, in such a way that neither is likely to be restarted in the stratosphere without a considerable time delay.

Several of the chemical species in Table 14 are not known and may well not be stable. Several of the others are well known but have strong absorption spectra in the near ultraviolet and/or visible regions and would have very short photodissociation lifetimes if formed in the stratosphere. Molecular Cl_2, for example, has been estimated to have a photodissociation lifetime of only a few minutes even in the troposphere. The molecule HNO_3 is a stable molecule formed by the cross termination of OH and NO_2 radicals, and its rate of formation and photodissociation lifetime are very impor-

TABLE 13. Estimated Rates of Reaction of OH With HCl in the Stratosphere

	A = 25 km	A = 30 km	A = 35 km	A = 40 km	A = 45 km	A = 50 km	A = 55 km
Concentration of [OH], cm^{-3}	6.0×10^5	1.3×10^6	3.2×10^6	6.7×10^6	8.0×10^6	6.8×10^6	5.2×10^6
Rate constant k_{14} for OH + HCl	4.7×10^{-13}	4.9×10^{-13}	5.4×10^{-13}	5.9×10^{-13}	6.1×10^{-13}	6.1×10^{-13}	6.1×10^{-13}
Rate $k_{14}[OH]$ of removal of HCl, s^{-1}	2.8×10^{-7}	6.4×10^{-7}	1.7×10^{-6}	4.0×10^{-6}	4.9×10^{-6}	4.1×10^{-6}	3.2×10^{-6}
HCl/ClO	7.8	1.5	1.2	2.1	5.9	19	49
Fraction of time* spent by Cl as HCl	0.89	0.60	0.54	0.68	0.85	0.94	0.97
Fraction of time spent by Cl as ClO	0.11	0.40	0.46	0.32	0.15	5.0×10^{-2}	2.0×10^{-2}
Fraction of time spent by Cl as Cl	2.0×10^{-5}	5.7×10^{-5}	2.1×10^{-4}	9.8×10^{-4}	2.9×10^{-3}	5.8×10^{-3}	9.0×10^{-3}

A is altitude.
* Rate constants and concentrations of Table 10.

TABLE 14.   Possible Products From Chain Termination Reactions of ClX Species With Other Radicals

	Cl	ClO	OClO	NO	NO$_2$	H	OH	HO$_2$
Cl	Cl$_2$	Cl$_2$O	ClO + ClO	NOCl	(ClO + NO) ONOCl	HCl	HCl + O HOCl	HCl + O$_2$ HOClO
ClO	...	Cl$_2$ + O$_2$	Cl$_2$O$_3$	Cl + NO$_2$ ONOCl	NOClO$_2$	(HCl + O) HOCl	HCl + O$_2$(?) HOClO	HOCl + O$_2$
OClO	...	...	?	NOClO$_2$	ONOClO$_2$	HCl + O$_2$(?) HOClO	HOClO$_2$	HOClO$_3$

Endothermic reactions are listed in parentheses and are not important in the stratosphere.

tant for calculations concerning the effects of the NO$_x$ cycle on natural ozone levels. Aside from the reaction forming HCl + O$_2$, none of the cross-termination reactions appear likely to form a molecule that would be less photoreactive than HNO$_3$. Consequently, our calculations concerning possible termination steps for the Cl/ClO species in the stratosphere have indicated importance only for those reactions leading to the formation of HCl.

IMPORTANCE OF SYMMETRICAL OClO AND HIGHER OXIDES IN THE STRATOSPHERE

Laboratory experiments have shown that the reaction of ClO with O$_3$ is not rapid at 298°K, with no indication that reaction occurs at all [Clyne and Watson, 1974]. If reaction did occur to form OClO, then the stratospheric chemistry of this additional species must be considered. If OClO were to have a long stratospheric lifetime, then the calculated concentrations of ClX would have to be expanded to include OClO together with HCl, ClO, and Cl.

The quantum yield for photodissociation of OClO is not too well characterized, and calculation of its rate of photolytic disappearance cannot be done with as much accuracy as with other molecules, such as CF$_2$Cl$_2$, CFCl$_3$, and HCl. The existence of substantial absorption bands in the near ultraviolet, essentially unshielded by O$_3$, O$_2$, or other major atmospheric components, strongly indicates that the photochemical lifetime of OClO will be relatively short at all altitudes. (The upper electronic state characterized by these bands is very probably predissociated [Watson, 1974].) Consequently, although OClO itself would not react rapidly with O$_3$ or O, its photolysis would rapidly reestablish the Cl/ClO chain. A more accurate measurement of the OClO photodecomposition is desirable, but the overall effect on the Cl/ClO chains is probably going to be small in comparison with other uncertainties at the present time.

Since the upper limit for the rate constant for ClO reaction with O$_3$ is about 10^4 times slower than for ClO with O and 3000 times slower than for ClO with NO (Table 9), reaction of ClO with O$_3$ will be of negligible importance at altitudes of 40 km or above under any circumstances. Since most of the destruction of odd oxygen by the Cl/ClO chain occurs above 35 km, we have not attempted any calculations in which estimates of the effect of OClO formation is included.

The reaction of OClO with O$_3$ to form ClO$_3$ has been reported [Schumacher, 1957], and other higher oxides of chlorine have also been reported occasionally from laboratory experiments. High concentrations of these higher oxides of chlorine seem rather improbable in the stratosphere because of their appreciable photochemical cross sections in the near ultraviolet and visible parts of the spectrum [Goodeve and Richardson, 1937]. Solar photolysis of such oxides should be quite rapid, in close analogy to the behavior of some of the ox-

ides of nitrogen, with the result that negligible fractions of Cl should be found as higher oxides. The important sinks should all be species that are relatively stable toward solar photolysis, i.e., HCl, ClO, and Cl. Even ClO has a nonnegligible absorption cross section for solar photons, and a few percent of the ClO radicals will actually be dissociated photochemically before reaction with O or NO can occur.

FLUORINE ATOM CHAINS

Both CFCl$_3$ and CF$_2$Cl$_2$ carry not only Cl but also F atoms into the stratosphere, and some consideration needs also to be given to the potential reactions of the FO$_x$ series. Although the same types of reactions are anticipated in the stratosphere for F atoms as for Cl atoms, the rates of reaction and the energetics of these reactions are quite different for these two halogen series. The chain attack of F on O$_3$ and FO on O is possible, as in (39) and (40),

$$F + O_3 \rightarrow FO + O_2 \tag{39}$$

$$FO + O \rightarrow F + O_2 \tag{40}$$

with rate constants comparable to the corresponding Cl atom chain. However, the abstraction reactions of F are very much more rapid with CH$_4$ (and with H$_2$) than the corresponding reactions of Cl atoms, with the result that relatively few ozone molecules will be removed before the FO$_x$ chain is diverted into HF. The rate constants for reactions (41) and (42)

$$F + CH_4 \rightarrow HF + CH_3 \tag{41}$$

$$F + H_2 \rightarrow HF + H \tag{42}$$

have been given as 6.0 × 10^{-11} and 2.5 × 10^{-11} cm^3 molecules^{-1} s^{-1}, respectively [Clyne et al., 1973], corresponding to collision efficiencies of ~0.1 and activation eneroies of ≤2 kcal/mol. Thus even at stratospheric temperatures and concentrations, F atoms will on the average be removed before 100 chain cycles have been completed. Once formed, the molecules of HF are verystrongly bonded (bond dissociation energy equal to 135 kcal/mol) and are not attacked by OH because of the endothermicity of reaction (43):

$$OH + HF \rightarrow H_2O + F \qquad \Delta H = +17 \text{ kcal/mol} \tag{43}$$

Photodissociation of HF is also difficult, exhibiting a low σ in the crucial 1800- to 2200-Å region. Consequently, in contrast to the situation with HCl, once HF is formed, neither attack by OH nor photodissociation has a high probability for restarting the FO$_x$ chain. The FO$_x$ chains from photodissociation of chlorofluoromethanes can thus be disregarded relative to the ClO$_x$ chains from the same molecules, insofar as the removal of ozone equivalents is concerned. Nevertheless, the molecule HF is itself a reactive gas, and its introduction at steady state levels approaching 10^{-9} mole fraction might have other consequences as yet unexplored.

The total concentrations of F atoms in all chemical forms following photodissociation of CF_2Cl_2 is of course the same as the total concentrations of ClX. Consequently, all of the calculations for ClX versus altitude for CF_2Cl_2 and its photodecomposition can be immediately read as predictions of the FX concentrations with the same numerical values at particular altitudes. The chief components of FX are CF_2O and HF, and the corresponding components for the photodissociation of $CFCl_3$ are $CFClO$ and HF. Assessment of the distribution of FX among these possible products requires careful measurement of the absorption spectra of CF_2O and $CFClO$ and evaluation of the relative rates of photolysis, hydrolysis, and diffusion to the tropopause. Once CF_2O or $CFClO$ has been photolyzed, the chemical path leading the F atoms into HF requires only a very short time period.

POSSIBLE REMOVAL OF CHLOROFLUOROMETHANES BY ELECTRON CAPTURE AND/OR IONIC REACTION

The more chlorinated chlorofluoromethanes have very high electron capture cross sections, the suggestion being that consideration needs to be given to the possibility that such processes as (44) and (45)

$$e^- + CFCl_3 \rightarrow Cl^- + CFCl_2 \qquad (44)$$

$$e^- + CF_2Cl_2 \rightarrow Cl^- + CF_2Cl \qquad (45)$$

might constitute important atmospheric sinks for these molecules. In the ionized layers of the high atmosphere designated as E and F_2, high concentrations of electrons are present from high-energy photodetachment processes at altitudes of about 120 and 300 km, respectively. However, these altitudes are so far above those at which solar photodissociation is important that electron capture by $CFCl_3$ and CF_2Cl_2 will be of negligible importance for molecules introduced initially as gases in the troposphere and arriving in these ionization layers only by vertical diffusion.

The peak cross sections for electron capture with dissociative electron attachment have been measured to be 1.6×10^{-14} cm²/molecule for CCl_4, 9.5×10^{-15} cm²/molecule for $CFCl_3$, and 1.1×10^{-16} cm²/molecule for CF_2Cl_2 [Christophorou and Stockdale, 1968]. The first two peaks occur with thermal energy electrons, but the peak cross section for electron capture by CF_2Cl_2 occurs for electrons at 0.15-eV energy. For a thermal electron concentration of 10 cm⁻³ the removal of CF_2Cl_2 by electron capture in the stratosphere would occur at a rate corresponding to an average lifetime of 10 years. The corresponding concentration required for a 10-year average lifetime for $CFCl_3$ is about 0.05 cm⁻³. The concentration of electrons in the E and F_2 layers has been measured to rise as high as about 10^5 and 10^6 el/cm³, respectively. At these altitudes, if chlorofluoromethanes were injected directly (e.g., from a rocket), their lifetimes against electron capture would be minutes or less. However, in the troposphere and lower stratosphere the concentration of electrons is kept very low by the immediate trapping in reaction (46):

$$e^- + O_2 + M \rightarrow O_2^- + M \qquad (46)$$

If dissociative electron capture were to occur with $CFCl_3$ and CF_2Cl_2, the initial immediate result would be the formation of Cl^- and either $CFCl_2$ or CF_2Cl. If such a reaction were to occur in the stratosphere or above, the Cl^- species would be sub-ject to photoelectron detachment with radiation in the near ultraviolet and would be dissociated to $e^- + Cl$ atom unless first attached to some particulate matter. The stratospheric or mesospheric consequences of dissociative electron capture would thus usually lead to the same fragments as those resulting from direct photodissociation, i.e., a Cl atom and a CF_nCl_{3-n} radical.

The 100-fold difference in electron capture rates for $CFCl_3$ and CF_2Cl_2, taken together with the measured amounts of $CFCl_3$ in the atmosphere, demonstrates that electron capture in the natural environment cannot be an important sink for CF_2Cl_2. If electron capture were the sole sink for $CFCl_3$ and if it has an atmospheric lifetime of 20 years or more, then the atmospheric lifetime for CF_2Cl_2 would be ∼100 times longer because of the difference in capture cross sections for the two species and hence would have a lifetime of $>10^3$ years against dissociative electron capture. Since this lifetime is much longer than those postulated for CF_2Cl_2 from photodissociation in the ultraviolet, dissociative electron capture must be negligible relative to the solar photodissociation sink for CF_2Cl_2. The estimate of a 20-year lifetime for $CFCl_3$ in the atmosphere is of course a lower limit, and there is no evidence that electron capture even contributes to its removal, let alone being entirely responsible for it. Consequently, the estimated lifetime of $>10^3$ years for CF_2Cl_2 against dissociative electron capture is a very conservative evaluation, and the lifetime is probably very much longer.

In the stratosphere and troposphere, electrons are primarily released by the passage of cosmic radiation and are immediately converted to O_2^-. This species is also very reactive, and a complex series of rapid reactions can ensue, leading finally to the formation of NO_3^-. As far as is known, NO_3^- is a 'terminal' ion in both the stratosphere and troposphere in the sense that it does not undergo further charge or atom transfer reactions [Mohnen, 1971; Ferguson, 1974]. The ultimate fate of NO_3^- is either reaction with a positive ion, electron photodetachment to $NO_3 + e^-$, or condensation reactions with H_2O to form a cluster ion $NO_3(H_2O)_x^-$. The negative ion concentration peaks at about 10^3 ions/cm³ in the stratosphere with lower values in the troposphere. The cluster ions become increasingly important at lower altitudes.

If ionic reaction were to occur with the chlorofluoromethanes very rapidly (e.g., $k \sim 3 \times 10^{-10}$ cm³ molecule⁻¹ s⁻¹), then negative ions at 10^3 ions/cm³ would remove molecules at a rate of 3×10^{-7} s⁻¹, which would be competitive with photodissociation. However, the NO_3^- ion has a high electron affinity [Berkowitz et al., 1971] and cannot undergo dissociative charge transfer with CF_2Cl_2 or $CFCl_3$. Thus the only energetically feasible reactions are displacement reactions such as (47):

$$NO_3^- + CFCl_3 \rightarrow CFCl_2ONO_2 + Cl^- \qquad (47)$$

Relatively few experiments seem to have been performed with NO_3^- ions present, and none with NO_3^- plus CF_2Cl_2 or $CFCl_3$. There is no indication that reactions such as (47) occur, let alone that they are exceptionally fast. The displacement of Cl^- by NO_2^- in reaction with CH_3Cl is a fast, but not extremely fast, reaction ($k = 1.6 \times 10^{-11}$ cm³ molecule⁻¹ s⁻¹ [McIver, 1971]) and that of NO_3^- should be slower, whereas $NO_3(H_2O)_x^-$ ions would react much more slowly, if at all, with chlorofluoromethanes. No evidence was found for reaction with C_3H_7Cl of NO_3^- formed in gaseous ethyl nitrate; reaction would have probably been observable only for reactions occurring with $k > 10^{-10}$ cm^3 molecule^{-1} s^{-1} [*Kriemler and Buttrill*, 1973].

It seems doubtful that ionic reactions are more than minor sinks at most for chlorofluoromethane removal, but direct experimental measurement of the rate constants for some possible reactions would be desirable.

Possible Removal of Chlorofluoromethanes by Reaction With O(^1D) or With OH

Singlet oxygen atoms, O(^1D), react rapidly with CH$_4$, H$_2$O, and H$_2$ in the atmosphere with a rate constant of $(3 \pm 1) \times 10^{-10}$ cm^3 molecule^{-1} s^{-1} [*Heidner et al.*, 1973; *Heidner and Husain*, 1973; *Cvetanović*, 1974]. Similarly, the rate constants for O(^1D) reaction with CF$_2$Cl$_2$ and CFCl$_3$ are also very fast and have been measured to be 5.3×10^{-10} and 5.8×10^{-10} cm^3 molecule^{-1} s^{-1}, respectively [*Pitts et al.*, 1974; *Sandoval et al.*, 1974]. However, the concentration of O(^1D) atoms, chiefly from the photolysis of ozone at wavelengths shorter than 3100 Å, is quite low except at very high altitudes and does not compete well with the rates for photolysis for chlorofluoromethanes at the same altitudes. The average concentrations of O(^1D) for overhead sun conditions as estimated by *Nicolet* [1972] are given in Table 15, together with the calculated rates of reaction with CF$_2$Cl$_2$ and CFCl$_3$. Even with reaction on every collision, the removal of CF$_2$Cl$_2$ by O(^1D), as in reaction (48),

$$O(^1D) + CF_2Cl_2 \rightarrow CF_2Cl + ClO \qquad (48)$$

is 10% or less of the rate of solar photolysis in the critical altitudes around 30 km, at which most of the CF$_2$Cl$_2$ actually is decomposed (Table 8 and Figure 8). At altitudes of 15 km and below the rate constants for reaction of CF$_2$Cl$_2$ with O(^1D) are competitive with its photolysis, but both rates are negligible. The lifetime for O(^1D) removal of CF$_2$Cl$_2$ at 15 km is 200–600 years, far longer than times for diffusion away from that level.

Since the net result of O(^1D) reaction with CF$_2$Cl$_2$ is the formation of CF$_2$Cl + ClO, the atmospheric result on ozone is indistinguishable in the 25- to 50-km range from photolysis, except that the rate constant for removal of CF$_2$Cl$_2$ might be increased by a factor of ~1.1, a factor well within the error of knowledge of the solar flux at any altitude. Calculations carried out with decomposition rates summing the photolysis rates of Figure 3 for CF$_2$Cl$_2$ and the O(^1D) reaction rates of Table 15 give average lifetimes about 4% shorter than those given in Table 8 for photolysis alone.

The photolysis rates of CFCl$_3$ are always about a factor of 10 higher than those of CF$_2$Cl$_2$ at each altitude, and the re-

moval of CFCl$_3$ by O(^1D) is thus no more than 1% of the rate of removal by photolysis in the 25- to 30-km region.

The attack of OH on CH$_4$ by abstraction of H is believed to be important in the troposphere and therefore is a prime factor in maintaining the average atmospheric lifetime of CH$_4$ of <10 years [*Wofsy et al.*, 1972; *Levy*, 1973]. However, the corresponding Cl atom abstraction reaction (49) is highly endothermic

$$OH + CFCl_3 \rightarrow HOCl + CFCl_2 \qquad (49)$$

and will not occur in either the troposphere or stratosphere. Although a displacement reaction such as (50)

$$OH + CFCl_3 \rightarrow CFCl_2OH + Cl \qquad (50)$$

is presumably exothermic, the reaction would be expected to have a very high activation energy and is unknown even in laboratory experiments.

Hydrolysis of Chlorofluoromethanes; Lack of Biological Interactions

Neither CFCl$_3$ nor CF$_2$Cl$_2$ is very soluble in water, and neither is removed by rainout in the troposphere, as evidenced by the long atmospheric residence times already observed for both of the chlorofluoromethanes. The relative insolubility in water together with their chemical stability (especially toward hydrolysis) [*Hudlicky*, 1962; *Sanders*, 1974] indicates that these molecules will not be rapidly removed by dissolution in the ocean, and the few measurements made so far indicate equilibrium between surface ocean and air and therefore no major oceanic sink [*Lovelock et al.*, 1973; *Liss and Slater*, 1974]. Mixing into the deep ocean is a process requiring hundreds of years, and so a stable equilibrium will exist between troposphere and stratosphere for a long time before any substantial amount of CFCl$_3$ or CF$_2$Cl$_2$ will penetrate well below the thermocline. The amount of chlorofluoromethane in surface waters at equilibrium with the atmosphere is a negligible fraction of the amount in the atmosphere.

The rates of hydrolysis in containers with mixtures of chlorofluoromethanes and water or aqueous solutions are reported to be very slow for both CFCl$_3$ and CF$_2$Cl$_2$, with the latter less reactive than the former [*Sanders*, 1974; *Johnsen et al.*, 1972]. Nevertheless, the homolytic cleavage of the C—Cl bond in CFCl$_3$ mixed with alcohol has been reported in metal containers to lead to its decomposition, as in (51)

$$CFCl_3 + CH_3CH_2OH \rightarrow CH_3CHO + HCl + CHFCl_2 \qquad (51)$$

[*Sanders*, 1960, 1965; *Minford*, 1964; *Bohac*, 1968]. A similar

TABLE 15.   Calculated Relative Rates of Atmospheric Removal of Chlorofluoromethanes by Reaction With O(^1D) Atoms and by Solar Ultraviolet Photolysis

Altitude, km	O(^1D) atoms/cm^3 *Nicolet* [1972]	*McElroy et al.*[*] [1974]	Removal Rate[†] by O(^1D), s^{-1}	$h\nu$[‡]/O(^1D) for CF$_2$Cl$_2$	$h\nu$/O(^1D) for CFCl$_3$
10	≤0.1	0.01	≤5 × 10^{-11}	≥10^{-3}	≥0.1
15	0.1–0.3	0.04	5–16 × 10^{-11}	1–4	13–40
20	0.8–2.2	0.5	4–12 × 10^{-10}	5–14	50–130
25	5–12	3	3–6 × 10^{-9}	8–19	70–170
30	25–35	12	1–2 × 10^{-8}	12–17	100–140
40	350	90	2 × 10^{-7}	7	45
50	600	180	3 × 10^{-7}	6	40

[*] 30°N latitude, solar declination +12°.

[†] Rates for O(^1D) reaction in cm^3 molecule^{-1} s^{-1}: CF$_2$Cl$_2$, 5.3×10^{-10}; CFCl$_3$, 5.8×10^{-10} [*Pitts et al.*, 1974].

[‡] Overhead sun conditions for both ultraviolet photolysis ($h\nu$) and O(^1D).

process has also been suggested in the atmosphere [Su and Goldberg, 1973], with the combination of atmosphere plus metal container reaction accounting for the smaller concentration of $CFCl_3$ in the atmosphere than that of CF_2Cl_2. Since the low values found by Su and Goldberg for the mixing ratio of $CFCl_3$ in the atmosphere have not been confirmed by any other group, the conclusion that the atmospheric $CFCl_3/CF_2Cl_2$ ratio is appreciably lower than the ratios of industrial production ($CFCl_3/CF_2Cl_2 \sim$ ⅔) is probably incorrect.

Details about possible biological interactions of these molecules in the environment are very scarce. Since they are almost entirely of recent anthropogenic origin and are chemically very inert, the chlorofluoromethanes might be expected to be almost totally unreactive biologically. The experiments carried out so far have indicated that soils and plants do not act as sinks for either CF_2Cl_2 or $CFCl_3$ (experiments by O. C. Taylor, N. E. Hester, and E. A. Cardiff; N. E. Hester, private communication, 1974). Rapid biological removal of these molecules from the atmosphere thus seems highly unlikely.

POSSIBLE REACTIONS OF CHLOROFLUOROMETHANES WITH AEROSOL PARTICLES IN THE STRATOSPHERE

Lidar experiments and collection experiments with filter papers have both indicated the presence of stratospheric aerosol particles, especially in the 15- to 20-km range a few kilometers above the tropopause [e.g., Junge et al., 1961; Hoffman, 1974; Bricard and Vigla, 1974]. The presence of these condensed phase particles raises the possibility of heterogeneous surface reactions in the stratosphere, either for the chlorofluoromethanes themselves or for their photodecomposition products. The possible absorption without chemical reaction of $CFCl_3$ or CF_2Cl_2 on the surfaces of these particles cannot be an important sink for them because it is not permanent. If the chlorofluoromethanes were to be carried into the troposphere on the surface of the sulfate aerosol, the progressively warmer conditions at lower altitudes would eventually release the chlorofluoromethanes into the atmosphere chemically unchanged.

The sources of the sulfate aerosols are themselves still quite uncertain, although the volcanic eruption of Mount Agung (Bali) in 1963 certainly caused an increase in the amount of stratospheric aerosol. An apparent sharp falloff in their concentration between 1969 and 1971–1972 has not been satisfactorily explained [Cadle, 1972]. The most important negative ion positively identified in these aerosols is SO_4^-, while NH_4^+ is found in much smaller quantities, and NO_3^- is found with some filter collections [Cadle, 1972; Hirono et al., 1974; Lazrus and Gandrud, 1974]. The absence of NO_3^- with polystyrene filters is explained through the hypothesis that the NO_3^- ion is not a constituent of the aerosols themselves but is retained in the actual stratospheric experiments by chemical reaction of gaseous HNO_3 with the cellulose component of IPC filters [Lazrus and Gandrud, 1974]. The postulated compositions for the aerosol include concentrated sulfuric acid, dilute sulfuric acid, and ammonium sulfate, depending on altitude and some other parameters.

The amount of SO_4^- collected by cellulose-filter experiments over the central United States in 1970–1971 was in the range of 2.8–20 × 10^{-14} g cm^{-3} at about 18-km altitude. A series of polystyrene filters exposed to stratospheric air at 16–18 km in October 1970 showed 0.4–5.2 × 10^{-14} g SO_4^- cm^{-3}. Neutron activation analysis of the latter indicated chlorine concentrations of 0.2–1.2 × 10^{-14} g cm^{-3}, and Cl/Br

ratios (by weight) of 12 to 19 [Cadle, 1972]. For comparison, the troposphere in 1972 contained about 10 × 10^{-11}, 6 × 10^{-11}, and 7 × 10^{-11} vol/vol of CF_2Cl_2, $CFCl_3$, and CCl_4, respectively, with somewhat smaller amounts of the first two presumably present in 1970–1971. At 10-km altitude the mixing ratios for these three chlorinated species would correspond to approximately 10^{-13} g cm^{-3} of Cl in each of the three compounds in gaseous form. The precise chemical identity and amount of the chlorinated species to be expected at 18 km from the presence of these organic compounds in the troposphere are not certain, and the concentration of all gases is about a factor of 4 less at 18 km than at 10 km, but the total mole fraction of chlorine so near the tropopause should be similar to the average ratio found in the troposphere, or about 10×10^{-14} g cm^{-3} at 18 km from the abundance data given above. This amount is comparable (0.5–25 times as large) to the amount of SO_4^- and substantially larger (8–50 times as much) than the chlorine content measured for these aerosols.

The aerosol particles appear to remain for a long time in the stratosphere, the lifetimes lengthening rapidly with altitude above the tropopause. Under these circumstances, major removal of chlorinated species from the stratosphere by the sulfate aerosols would require much higher chlorine contents than are observed. The most recent estimate of SO_4^- flux from the stratosphere is 2.1 × 10^5 tons/year, with Na^+ and Cl^- concentrations described as 'vanishingly small and erratic' [Lazrus and Gandrud, 1974]. This flux is substantially smaller than the current combined upward chlorine fluxes as CF_2Cl_2, $CFCl_3$, and CCl_4.

The reactions of ClO and/or Cl with aerosol particles may lead to the removal of some Cl atoms from the gas phase. However, reaction at 18 km would have very little effect on the concentrations of these species at 35–55 km. The termination of some of these chains on aerosol particles in the lower stratosphere might only serve as an alternate sink for our present assumption of rainout at the tropopause, without having any material effect on the calculations in the upper stratosphere. Calculations from the known thermodynamics of HCl-H_2O mixtures (S. C. Wofsy, private communication, 1974) indicate that HCl should not be retained in the condensed phase even at the low stratospheric temperatures.

The sulfate aerosols might become involved in the chlorofluoromethane cycle in another manner if the particle surfaces served as catalytic sites for the decomposition of the molecules, even if they did not act to remove the decomposition species. Many collisions will occur in the stratosphere between gaseous chlorine-containing species and the surface of such aerosol particles at temperatures as low as 210°K. Furthermore, recent laboratory experiments with concentrated sulfuric acid at room temperature have demonstrated that $CFCl_3$, at least, can react at these temperatures [Siegemund, 1973]. However, the recent work of Hester et al. [1975] shows that $CFCl_3$ and CF_2Cl_2 are stable in the presence of SO_3 and H_2SO_4 aerosol particles produced under simulated heavy smog conditions. In addition, laboratory comparison experiments of hydrolysis rates have rather consistently shown that neither $CFCl_3$ nor CF_2Cl_2 hydrolyzes easily or rapidly but that the former is more reactive when such reactions do occur [Sanders, 1974]. Since $CFCl_3$ is present in the troposphere in amounts roughly comparable to the integrated industrial production to date, there is no indication that there is any rapid selective removal of $CFCl_3$ versus CF_2Cl_2, as might be anticipated were hydrolysis by H_2SO_4 an important stratospheric sink.

The Cl/Br ratios found in marine aerosols close to the sea surface near Hawaii reflect the Cl/Br ratio in seawater of about 1000 when collected over water, but only about 100 when collected over land [*Duce et al., 1965*]. These ratios are considerably richer in chlorine (or poorer in bromine) than those found in the stratospheric aerosols, and the differences have been suggested as possibly arising from worldwide atmospheric pollution by bromine compounds used in gasoline for automobile engines [*Cadle, 1972*]. Much more work will be necessary to determine the source of these aerosols and of the halogens found in them. Although the stratospheric aerosols are obviously not very well understood at present, we conclude that there is now no available information to suggest that they interfere in a major way with the model involving solar photolysis of chlorofluoromethanes and subsequent gas phase reactions by their photodecomposition products. Similarly, atmospheric calculations of the interactions of NO and NO_2 with the odd oxygen of the stratosphere give very good results with the basic assumption that the NO_x species and HNO_3 are not removed by heterogeneous reaction with the aerosol layers. The absence of HNO_3 from the polystyrene filters collected at 18 km provides confirmation that these aerosol particles are not an important sink for NO_x or HNO_3 [*Cadle, 1972*].

CALCULATED OZONE DEPLETION BY CL*X* SPECIES

The atmospheric diffusion calculations described earlier have been carried out with CF_2Cl_2 and $CFCl_3$ independently. With a constant flux of each crossing the tropopause (fixed at approximately the 1973 industrial production levels), a simple time-independent steady state solution was sought first. In such calculations, as illustrated in Figures 7 and 8, the rate of disappearance by photolysis is defined to be equal to the average rate of introduction at the tropopause (assumed equal to the rate of introduction at ground level), and the concentrations versus altitude of CF_2Cl_2 and $CFCl_3$ required to sustain such photodecomposition rates are obtained from the calculation. Integration of these concentration-altitude graphs provides an estimate of the total atmospheric content of each. The atmospheric content divided by the rate of photolysis provides the average atmospheric lifetime of each species.

In our diffusion calculations, the chlorine-containing compounds formed subsequent to photodissociation are all cumulatively considered as 'Cl*X*,' and a distribution of these summed concentrations versus altitude is also obtained from the calculations. In such a steady state situation with CF_2Cl_2, the total Cl content (i.e., $2 \times [CF_2Cl_2] + [ClX]$) has a fixed ratio to the other components of the atmosphere at all altitudes. The downward flux of Cl atoms as [Cl*X*] across the tropopause in the steady state calculation must, of course, be equal to the upward flux of Cl atoms as CF_2Cl_2. The distribution of Cl*X* versus altitude at steady state is shown in Figure 12 for CF_2Cl_2 with eddy diffusion models A, B, C, and D. The distribution for $CFCl_3$ at steady state is shown in Figure 13.

In a 'complete' one-dimensional calculation, each of the possible Cl*X* species would be considered separately with its own mechanisms for formation and removal involving interchange with many of the other Cl*X* species and for transport in the vertical direction. Above about 25–30 km the rates of chemical formation and removal are very much larger than the transport rates involving diffusion upward and downward to and from adjacent layers of the atmosphere. (The time required to diffuse 5 km vertically is about 1 year for a diffusion coefficient of 10^4 cm² s⁻¹.) Consequently, the atmosphere can

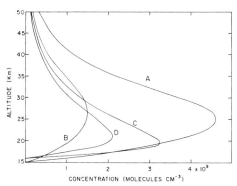

Fig. 12. Steady state concentrations of Cl*X* from CF_2Cl_2 versus altitude for four different eddy diffusion coefficient models. Models are defined in Figure 6.

be assumed at these altitudes to be approximately in local chemical equilibrium (i.e., photochemical stationary state), and reasonable estimates of the individual concentrations of HCl, ClO, Cl, etc., can be obtained by applying (37) and (38) to the summed concentrations [Cl*X*]. At altitudes of 25 km and below the rates of transport are no longer negligible, and local chemical equilibrium is no longer a safe assumption. However, as is indicated below, the depletion of O_3 by Cl/ClO chains occurs almost entirely above 30 km and therefore in the atmospheric regions for which local chemical equilibrium is a reasonable approximation. From the reaction rate constants given earlier, it is apparent that the chief Cl*X* constituents are HCl and ClO, with Cl atoms of lesser concentration throughout. The fractional concentrations of HCl, ClO, and Cl with altitude for the average trace concentrations of Figures 9–11 were given in Table 13. The individual concentrations versus altitude at steady state with the eddy diffusion coefficients of model D are shown in Figure 14 for CF_2Cl_2 and Figure 15 for $CFCl_3$.

Some of the Cl atoms from $CFCl_3$ are found after photolysis in the form of CFClO, and the total Cl content is then given by $3 \times [CFCl_3] + [CFClO] + [ClX]$. We have assumed for con-

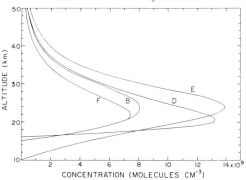

Fig. 13. Steady state concentrations of Cl*X* from $CFCl_3$ versus altitude for four eddy diffusion coefficient models. Models are defined in Figure 6.

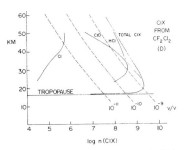

Fig. 14.   Concentrations of ClO, Cl, HCl, and total ClX from CF₂Cl₂ versus altitude with the eddy diffusion coefficient of model D.

venience in calculation that half of the CFClO molecules are photolyzed at the same altitude as the initial CFCl₃ photolysis and that half are not photolyzed at all. In our calculation then, the same relative distribution with altitude applies for 0.5 molecule of CFClO as for 2.5 molecules of ClX. This circumstance is obviously incorrect, for CFClO has a different sensitivity than ClX to photodissociation, and CFClO will certainly photolyze more rapidly at higher altitudes. The more correct calculational procedure, however, would involve the much more complicated solution of the diffusion equation with CFClO injected into the stratosphere with a pattern corresponding to the initial photolysis pattern for CFCl₃. In our calculation, 80% of the Cl atoms (2.0 per CFCl₃ molecule) are produced at the initial photolysis altitude, and the assumption that the additional 20% (0.5 atom per CFCl₃ molecule) are formed at the altitude of CFCl₃ photolysis instead of that of CFClO photolysis is a minor error. Calculation of the behavior of CFClO and CF₂O (from CF₂Cl₂) in atmospheric photolysis is now being carried out.

The most important potential environmental effects of ClX in the stratosphere are all concerned with the possible depletion or redistribution of the ozone layer as a consequence of the chain reaction sequence involving (3) and (4). An initial estimate of possible ozone depletion can be calculated through a 'perturbation' technique in which these ClX concentrations of Figures 14 and 15 are used in conjunction with the concentrations given in Figures 9–11 and are compared with the present rates of removal of O₃ by the reactions now believed to be rate controlling in the atmosphere. We describe this as a 'per-

turbation' technique because the concentrations of Figure 9 have been obtained by others as solutions of complete sets of the possible formation and removal mechanisms for each in the stratosphere. These concentrations are not necessarily the same as those that would be appropriate to the solutions of still larger sets of chemical reactions including in addition the formation and removal of the ClX species. In practice then the 'perturbation' calculation is likely to be accurate only if its results indicate that the overall effects of inclusion of these species are minor. As shown below, the effects are major, and a more detailed calculation is required.

In one recent calculation, the relative rates of removal of odd oxygen by the four processes listed below are estimated to account for 60–74%, 10–20%, 15–20%, and 1%, respectively, of the present atmospheric removal: NOₓ catalysis; HOₓ catalysis; direct reaction, i.e., O₃ + O; transport to the troposphere [*Johnston*, 1974]. A 'perturbation' calculation of the relative rates of odd oxygen removal for reactions (3) and (4) and for NO₂ + O, as in reaction (6), is illustrated in Figure 16 for diffusion models B and D at steady state. The most important conclusion to be drawn from Figure 16 is the simple qualitative conclusion that the rates of removal of O atoms by reactions of NO₂ and ClO are roughly comparable, i.e., that the removal of odd oxygen by ClX under the postulated steady state conditions is of approximately equal importance with the NOₓ reaction sequence, which is now the dominant ozone control mechanism in the stratosphere. An immediate semiquantitative corollary can be stated: in the 30- to 55-km-altitude range, the inclusion of such quantities of ClX would result in appreciable diminution of the ozone concentrations. The comparison of Figure 16 also clearly shows that the 'perturbation' calculation cannot be used for quantitative estimates since the concentrations of some species, e.g., O₃, would be sufficiently affected as to require a full solution of the original calculation with the ClX reactions included in detail.

Some specific remarks can be made about the details of Figure 16. The removal of odd oxygen by the NOₓ catalytic system can be simply calculated as equal to twice the observed rate for NO₂ + O, allowing for the additional removal of O₃ by reaction with NO in the chain. However, direct calculation of O₃ removal by the reaction of NO + O₃ would not match the NO₂ + O curve because many of the NO + O₃ reactions are followed, not by NO₂ + O, but by the photolysis of NO₂ to NO and O, with no net change in the odd oxygen concentration for the sum of the two reactions. A similar situation pertains to the rates calculated for reactions (3) and (4) in Figure 16, for which it can be seen that the rate of removal of O₃ is

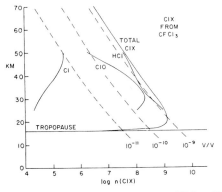

Fig. 15.   Concentrations of ClO, Cl, HCl, and total ClX from CFCl₃ versus altitude with the eddy diffusion coefficient of model D.

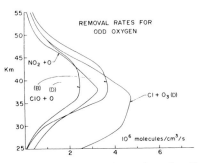

Fig. 16.   'Perturbation' calculation of removal rates for odd oxygen by ClX and by NO₂.

greater than that for removal of O. Here the difference arises because many of the ClO reactions occur, especially at altitudes below 35 km, with NO and not with O. As with the NO_x chain itself, many of the NO_2 molecules formed by this reaction of ClO with NO are subsequently photolyzed, and the sum of reactions (3) + (5) + photolysis of NO_2 also leads to no net change in the odd oxygen concentration. Thus, in addition to the marked changes in O_3 concentration, the inclusion of steady state amounts of ClX in the stratosphere will also affect the NO/NO_2 ratio, shifting it toward NO_2.

FURTHER CALCULATIONS

The extension of such 'perturbation' calculations to obtain more realistic quantitative solutions is a complicated problem involving several considerations: the time delays between manufacture and injection and between injection and the appearance of the ozone depletion effects; feedback in ozone depletion calculations; the possibility of uneven distribution of effects with latitude or longitude such that global average calculations give a misleading impression; and the applicability of a steady state assumption.

The steady state calculation involves the assumption that the rate of industrial production of CF_2Cl_2 and $CFCl_3$ will remain for the next 50–100 years at the present level. Such a projection for the years in the twenty-first century is obviously without a rational basis and depends upon a large number of essentially unknown factors. The steady state assumption is indeed an extremely conservative prediction of the probable rates of industrial production in the next 10–15 years. A more rational prediction is that the production of CF_2Cl_2 and $CFCl_3$ will continue to increase in 1975–1980 at approximately the same rate that it has increased during the years 1964–1974, roughly 13% per year on a worldwide basis. The estimated world and U.S. productions for $CFCl_3$ and the U.S. production for CF_2Cl_2 are graphed versus time in Figure 17. Over the last few years, an increase in the 12–15% range is consistent with the world growth pattern. As a rough approximation then the magnitude of the 'perturbation' shown in Figure 16 should be multiplied by a factor of 2 for every 6 years in which the industrial production rate continues to increase before settling down to a constant level. If the industrial production were to increase by 8–15% per year through the 1980's, as suggested by a recent prediction [*Reid,* 1974], and then to level off, the steady state calculation would then leave the NO_2 + O reaction as a minor perturbation on the overwhelmingly dominant ClO + O reaction. Accordingly, in the time-dependent calculations described below, we have adopted several models for the industrial production of CF_2Rl_2 and $CFCl_3$ versus time during the next two decades.

MULTIDIMENSIONAL CALCULATIONS

The current distribution of ozone in the atmosphere varies widely with the day, the month, and the season at a given location, and with the latitude and longitude of the measurement site [*Dütsch,* 1974]. In general, the thinnest ozone layers are found at the equator, and the thickest layers occur in the polar regions during winter. The ozone layer also tends to be found at higher altitudes in tropical as compared with polar regions. Insofar as the concern with ozone depletion problems is ultimately directed toward possible increases in the penetration of ultraviolet radiation to the earth's surface, the polar regions are of lesser importance because the thicker ozone shields and the oblique angles of entrance of solar rays insure that the ground ultraviolet levels are considerably lower than

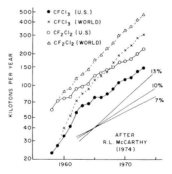

Fig. 17. Chlorofluoromethane production versus year.

those characteristic of the equator. Two-dimensional and three-dimensional models of the behavior of the circulation patterns of the stratosphere incorporate a general tendency for upward diffusion into the stratosphere near the equator, circulation poleward, and downward diffusion to the troposphere nearer the poles. The long atmospheric lifetimes of the chlorofluoromethanes will tend strongly toward equalized concentrations throughout the troposphere, reducing latitudinal and longitudinal variations in concentration. However, the stratospheric photolysis of the chlorofluoromethanes takes place prevalently in the regions with the most overhead sun, i.e., near the equator, and the subsequent motions should then tend to carry the ClX poleward. Since the ozone layers are generally the thinnest and at the highest altitudes near the equator, the possibility certainly exists that the ozone removal processes will be more strongly augmented by ClX in the tropical regions than nearer the poles. The average global depletions of ozone calculated with one-dimensional models may well be grouping together substantial depletions near the equator with relatively small effects at higher latitudes. If this particular pattern were to be substantiated by two- or three-dimensional calculations, then estimates of the increased levels of ultraviolet radiation at the earth's surface made on the basis of average depletion effects might be grossly misleading. The increase in ground level ultraviolet radiation from a 10% ozone reduction between 0° and 30° latitude would be very much greater than a 10% reduction between 30° and the pole, and the one-dimensional calculation provides no information about such possible latitudinal variations from the overall average.

FEEDBACK IN THE OZONE DEPLETION CALCULATION

There are several successive stages of more sophisticated calculations that can be applied if an initial calculation suggests the possibility of net ozone depletion. First, the partial depletion of O_3 can result in a net increase in the rate of formation of odd oxygen [*Johnston,* 1974; *Wofsy,* 1974]. Second, the broad temperature maximum at about 50 km is basically the result of radiative heating caused by the absorption of radiation by O_3, and both the amount of radiation absorbed and the initial distribution with altitude of this absorbed energy are dependent upon the distribution with altitude of the O_3 absorber. Any change in the temperature profile of the stratosphere immediately can be fed back into the calculation of all competitive chemical rates of reaction involving activation energy barriers to reaction. Ultimately, too, the stratospheric circulation patterns must be somewhat dependent upon the temperature profile of the stratosphere. Since no generally satisfactory understanding yet exists of the degree of interaction between temperature profile and circulation patterns, no satisfactory predictions can be made of the magnitude of possible circulation changes that might result from a specific pattern of ozone depletion and/or redistribution [Dickinson, 1974]. The feedback possibilities can then be extended further by noting that the eddy diffusion coefficients used in the initial stages of calculation, modeling the vertical circulatory motions in an average way, might also be affected by changes in the stratospheric circulation pattern accompanying a shift in the average temperature profile.

INCREASED ODD OXYGEN PRODUCTION WITH OZONE DEPLETION

The initial source of odd oxygen in the atmosphere is the photolysis of $O_2 \rightarrow 2O$ with ultraviolet radiation of ≤ 2420 Å. As discussed earlier, the subsequent addition reaction of O to O_2 and the photolysis of $O_3 \rightarrow O_2 + O$ each result in no net change in the total odd oxygen present. Hence the absorption by O_3 of radiation with $\lambda > 2420$ Å has no effect on total odd oxygen. Similarly, radiation with 1400 Å $< \lambda <$ 1800 Å is absorbed essentially entirely by O_2 and hence is unaffected by any changes in the thickness or altitude distribution of the O_3 component. However, between 1800 and 2300 Å, and especially 2000–2300 Å, the higher σ for O_3 roughly counterbalances its lower concentration, and both O_3 and O_2 absorb comparable fractions of the incoming radiation. In this wavelength region, depletion of O_3 in the higher levels permits the penetration of increased fluxes of 2000- to 2300-Å radiation to lower levels, some of which will then be absorbed by O_2 and not O_3, and the total production rate of odd oxygen species will be correspondingly increased. An illustrative calculation of such a shift in absorption from O_3 to O_2 with diminished O_3 is shown in Table 16 for overhead sun conditions. In this case, the maximum change in flux (bottom line) occurs in the 2100- to 2200-Å band. The complete estimate of changes in total odd oxygen requires such a calculation for all zenith angles, and with

an ozone distribution consistent with the ClX, NO_x, etc., concentrations for the specific system at each altitude.

The number of quanta per 20-Å band around 2100–2300 Å is comparable to the sum of the numbers of photons for all wavelengths below 1900 Å (Figure 18), thus insuring a rather substantial feedback of increased odd oxygen formation for depletion of the ozone layer. Detailed calculations of the importance of this feedback mechanism require integration over a wavelength/altitude grid and are routinely included in some complex atmospheric modeling. Even if the total amount of ozone per square centimeter is not varied, general shifting of the existing ozone to lower altitudes will result in increased odd oxygen production. In general, the greater the distortion in ozone distribution, the more feedback into total odd oxygen production, while the calculated distortion is itself strongly dependent upon the model chosen for eddy diffusion coefficients.

Two calculations with different eddy diffusion models have been carried out by Crutzen [1974a] and have included this odd oxygen feedback into the estimates of overall ozone depletion and redistribution at steady state. In each case, there has been a net ozone depletion plus a substantial alteration in the relative distribution of ozone versus altitude in the resulting equilibrium calculation. The calculated changes at steady state are shown in Figure 19 for model B and sum to a net depletion in O_3 of 7% for a constant input flux of CF_2Cl_2 and $CFCl_3$ characteristic of the 1972 world production rates. In this calculation then both a net O_3 reduction and a substantial redistribution with altitude have been calculated. The further feedbacks involving absorbed radiation or changed temperature profile versus altitude were not incorported into these calculations. The maximum ozone depletion at steady state with model B is about 42% at 40-km altitude. The other diffusion model used by Crutzen was that of Ehhalt (model F). In this case the net O_3 depletion is 6.5%, and the higher diffusion coefficients between 30 and 50 km result in a net depletion at all altitudes from 20 to 50 km with a maximum decrease of 17% at 40 km.

A separate estimate of the steady state ozone depletion to be

TABLE 16.   Calculated Fractional Absorption in O_3 and O_2 for Overhead Sun Conditions in the 2020- to 2247-Å Range

Ozone Level	WB = 2020–2041 Å F = 1.74	WB = 2062–2083 Å F = 4.20	WB = 2128–2150 Å F = 10.6	WB = 2174–2198 Å F = 13.2	WB = 2222–2247 Å F = 18.0
Normal (see Fig. 9)					
Fraction absorbed by O_2	0.49	0.39	0.20	0.11	0.06
Fraction absorbed by O_3	0.51	0.61	0.80	0.89	0.94
Fraction absorbed above 30 km	0.74	0.78	0.92	0.98	1.00
Median altitude for absorption, km	35	36	38	41	43
0.5 × normal above 30 km; normal below 30 km					
Fraction absorbed by O_2	0.61	0.52	0.30	0.17	0.10
Fraction absorbed by O_3	0.39	0.48	0.70	0.83	0.90
0.5 × normal above 15 km					
Fraction absorbed by O_2	0.67	0.59	0.35	0.19	0.10
Fraction absorbed by O_3	0.33	0.41	0.65	0.81	0.90
Fraction absorbed above 30 km	0.62	0.63	0.77	0.89	0.96
Median altitude for absorption, km	32	33	36	38	39
Normal					
Flux × 10^{12} absorbed by O_2	0.85	1.66	2.1	1.4	1.1
0.5 × normal					
Flux × 10^{12} absorbed by O_2	1.17	2.47	3.7	2.5	1.8
Increase from normal to 0.5 × normal					
Flux × 10^{12} absorbed by O_2	0.3	0.8	1.6	1.1	0.7

WB denotes wavelength band. Each range corresponds to a 500 cm^{-1} band pass, i.e., 2020–2041 Å represents the band from 49,000 to 49,500 cm^{-1}.

F denotes flux × 10^{12} solar photons cm^2 s^{-1}. Total flux for all wavelengths below 1900 Å is 10.9 × 10^{12} cm^{-2} s^{-1}. Flux data and σ for O_2 and O_3 from Ackerman [1971].

expected in the future can be made by combining our calculation of the steady state level of ClX with the calculation of *Wofsy and McElroy* [1974] for fixed arbitrary constant vol/vol ratios of Cl at all altitudes. Those calculations were concerned primarily with Cl in the form of HCl, as for example from the proposed U.S. space shuttle, and assumed Cl in the usual ClX forms (HCl, ClO, Cl) without including the possibility of its being present as CF$_2$Cl$_2$, CFCl$_3$, or other organic forms. In our steady state calculation with eddy diffusion model D for CF$_2$Cl$_2$, we obtained a limiting vol/vol ratio for ClX of 4.2 × 10^{-9} at high altitude, with ClX ratios of 3.8 × 10^{-9} at 40 km, 3.4 × 10^{-9} at 35 km, and 2.9 × 10^{-9} at 30 km. From Figure 16 we estimate that the median altitude for the Cl/ClO destruction of odd oxygen occurs at about 40 km and therefore adopt 3.7 × 10^{-9} vol/vol of ClX as a reasonable approximation to a weighted average for use with the Wofsy-McElroy ClX calculation. The corresponding steady state calculation for CFCl$_3$ gives a limiting value of 1.68 × 10^{-9} vol/vol for ClX from this source. Since CFCl$_3$ photolyzes at lower altitudes than CF$_2$Cl$_2$ does (Table 8), the 40 km vol/vol value for ClX is almost the limiting value: 1.67 × 10^{-9}. Together then at steady state for CF$_2$Cl$_2$ and CFCl$_3$ with diffusion model D, we estimate a total ClX concentration at 40 km of 5.4 × 10^{-9} vol/vol. The corresponding ozone depletion [*Wofsy and McElroy*, 1974, Figure 6] is calculated to be 13%. Since the concentrations adopted in Figure 9 rather closely follow those of typical Wofsy-McElroy calculations and diffusion model D is taken directly from their work, this estimate of 13% depletion is essentially a complete calculation from a consistent set of data. The earlier ClX calculation of *Wofsy and McElroy* [1974] has now been redone [*Wofsy et al.*, 1975] with the rate constant for OH + HCl changed to that of *Zahniser et al.* [1974]. This correction increases the O$_3$ depletion effect from 13% to 18%.

TIME DELAY EFFECTS IN THE APPEARANCE OF OZONE DEPLETION

Six separate calculations have now been made of the time dependence of ozone depletion by chlorofluoromethane release to the atmosphere. The calculations by *Cicerone et al.* [1974] (Figure 20) and *Rowland and Molina* [1974] show the time dependence of the chlorofluoromethane and of the ClX concentrations but do not give independent estimates of the ozone depletion expected at each stage. The calculation of *Wofsy et al.* [1975] is complete in itself, giving predicted ozone depletion with time (Figure 21). Complete time-dependent calculations have now also been carried out by *Crutzen* [1974c], *Turco and Whitten* [1974], and Chang, Wuebbles, Duewer, Molina, and Rowland (unpublished work, 1975).

These calculations all have made various assumptions about

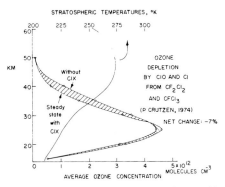

Fig. 19.   Ozone depletion by ClO and Cl in steady state with eddy diffusion model B [from *Crutzen*, 1974a].

the possible usage patterns for chlorofluoromethanes in the future. Line A of Figure 21 illustrates the growth in ozone depletion toward an asymptote near the middle of the twenty-first century at a constant level of release of CF$_2$Cl$_2$ and CFCl$_3$. The extrapolated technological pattern (Figure 17) indicates that a 13% increase per year is a more valid estimate of industrial plans, at least for the near future. Lines B and C of Figure 21 both represent the effects of 10% growth rates with sudden cessation in 1978 and 1995, respectively. Line D extrapolates this 10% increase indefinitely; the release rate 40 years from now would then be about 45 times the present release rate. Lines E and F provide corresponding estimates for a more rapid exponential increase of 22% per year.

Figure 20 illustrates a time-dependent calculation for immediate cessation (line 3), constant release from now (line 2), and continued exponential growth with a 3.5-year doubling time. The eddy diffusion coefficients for these calculations are quite similar to those of model D.

In the following section we describe some of the effects of the choice of different eddy diffusion parameters on the time development of ClX patterns in the stratosphere. In these calculations, we have also used models involving (1) cessation after 1975; (2) constant production at present rates through 1975, ceasing abruptly in 1990; and (3) production increasing at 10% per year, until 1990, and then ceasing abruptly. Although exponential increase of 20% per year certainly oc-

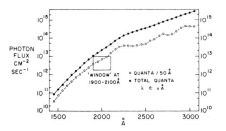

Fig. 18.   Differential and integral solar photon fluxes for wavelength range 1450–3100 Å.

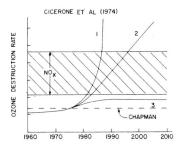

Fig. 20.   Calculated ozone destruction rates versus calendar year for ClX [from *Cicerone et al.*, 1974]. The dashed line labeled Chapman is the calculated level for Chapman reactions, the shaded region labeled NO$_x$ is the calculated level for NO$_x$ reactions, line 1 is the calculated ClX effect for continued growth of production model, line 2 is the calculated ClX effect for constant production model at current rates, and line 3 is the calculated ClX effect with production of chlorofluoromethanes stopped immediately.

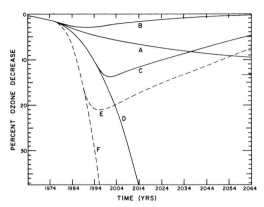

Fig. 21.   Calculated ozone destruction rates versus calendar year for six models of fluorocarbon use [from *Wofsy et al.*, 1975]. Line A is constant production at 1972 levels; line B, emissions increase at 10% per year and cease abruptly in 1978; line C, emissions increase at 10% per year and cease abruptly in 1995; line D, emissions increase continuously at 10% per year; line E, emissions increase at 22% per year and cease abruptly in 1987; line F, emissions increase continuously at 22% per year.

curred in the 1950's, the prospect for an increase at more than 13% per year seems quite unlikely, and we have not used any predictive models involving 20% growth rates.

All of these calculations involve a time-dependent solution of the diffusion equation. In our calculations below, the mixing in the troposphere is again assumed to be very fast and is approximated as instantaneous on the time scales appropriate to stratospheric diffusion. The concentration of ClX is fixed at zero at the tropopause, i.e., the troposphere is a perfect sink. The numerical solution has been obtained through the explicit finite difference technique. The time steps were 0.001 year for eddy diffusion model D and 0.004 year for models B and E. The altitude steps were variable, from about 0.3 km near the tropopause to >2 km above 50 km. The numerical accuracy was confirmed by sample calculations with smaller time steps and/or altitude steps. The calculations were also tested with initial steady state conditions, for which the time-dependent numerical solution preserves the steady state condition, i.e., no spurious losses or sources of particles.

TIME-CONCENTRATION PATTERNS WITH 1-YEAR STEP FUNCTION INJECTIONS OF CF_2Cl_2

We have carried out such calculations for several of the eddy diffusion models described earlier. With diffusion model D the

introduction (hypothetical) of CF_2Cl_2 at a constant rate for a 1-year-long step function (with zero rates both before and after the 1-year period) leads to the development of the CF_2Cl_2 and ClX altitude profiles versus time shown in Figure 22. The CF_2Cl_2 distribution with altitude reaches near equilibrium in about 4 years; the concentration at the 16-km level (tropopause) reaches a maximum just at the end of the injection year and then begins to decrease as photodecomposition gradually removes CF_2Cl_2 without any replacement.

On the other hand, the ClX distribution versus altitude continues to increase for about a decade after introduction (Figure 23) and is about 50% of the final ClX concentrations within approximately 3 years after the beginning of the injection. For this model of eddy diffusion then the effect of a given year's injection of CF_2Cl_2 has reached half of the maximum effect about 3 years later, whereas the maximum actually occurs after about 10 years. This general type of time response is characteristic for eddy diffusion models with relatively rapid diffusion between 30 and 60 km, as shown earlier in Figure 6.

Similar 1-year step injections have also been simulated for the Crutzen (model B) and Chang (model E) choices of eddy diffusion coefficients. With model B the diffusion of CF_2Cl_2 into near equilibrium with the troposphere takes somewhat longer than with model D illustrated earlier. This tendency is even more marked with model E, for which the maximum CF_2Cl_2 concentration at 30 km is reached about 7 years after the beginning of the injection.

In Crutzen's model B, ClX requires considerably longer to reach maximum values at the higher elevations than with the Wofsy-McElroy model D, as shown in Figure 24. The maximum concentration for 35 km is reached in Figure 24 at about 13 years, for 40 km is reached at about 15 years, and for 45 km at about 19 years. The corresponding times to reach half of the maximum concentrations are about 4–5 years.

With model E [*Chang*, 1974] the minimum in eddy diffusion coefficient (Figure 6) near 30 km slows down still more the appearance of ClX at the higher altitudes. The maximum ClX concentrations at the 40- to 45-km level are reached about 30–35 years later, and the times for reaching half maximum concentrations are 8–10 years, approximately twice as long as in the Crutzen model and 3 times as long as for the Wofsy-

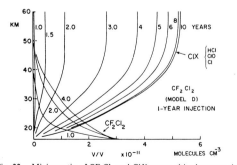

Fig. 22.   Mixing ratio of CF_2Cl_2 and ClX versus altitude versus time for 1-year step injection, eddy diffusion model D.

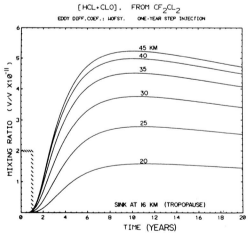

Fig. 23.   Mixing ratio of ClX versus time for various altitudes; 1-year step injection, eddy diffusion model D.

McElroy model. The distribution of ClX with time for a 1-year injection with model E is shown in Figure 25. The progressively slower diffusion in the 30-km range for models D, B, and E also leads to successively longer estimated lifetimes of 71, 102, and 160 years, as given earlier in Table 7.

The relative rates of removal of O atoms by reaction with ClO were shown earlier in Figure 16 to occur predominantly in the 35- to 45-km range of altitudes. If the ClX concentration at the 40-km level is used as a measure of odd oxygen removal, then the Chang model shows essentially negligible odd oxygen removal by ClX from CF$_2$Cl$_2$ until 4 or 5 years after the initial release of CF$_2$Cl$_2$ into the atmosphere. This delay, or 'induction period,' prior to observation of any effect on odd oxygen removal is about 2 years for model B and 1½ years for model D. The length of any such induction period is quite important in estimation of the expected growth pattern for an ozone-depletion effect. Since presently available stratospheric data do not permit an unequivocal choice among such widely divergent models as D and E, the length of any such induction period in the real atmosphere is very much open to question.

TIME-CONCENTRATION PATTERNS FOR SEVERAL HYPOTHETICAL MODELS OF FUTURE RELEASE OF CHLOROFLUOROMETHANES

Time-dependent calculations of the concentrations of CF$_2$Cl$_2$ and ClX versus altitude have also been carried out with three projected models for the future course of CF$_2$Cl$_2$ industrial production. In all three the growth pattern to the end of 1975 has been modeled after the exponential growth illustrated in Figure 17. The model is fitted to a 1973 CF$_2$Cl$_2$ production of 454 kt arbitrarily divided into two exponential segments, 18.5% per year growth from 1955 to 1970 and 10.5% per year growth from 1971 to 1975. These segments were chosen to fit an earlier version of Table 1 [*Rowland and Molina*, 1974]. The revised data in Table 1 of this article are better represented by 15.5% growth per year prior to 1964 and 12.3% since 1964 with a 1973 value of 470 kt.

Three models were chosen for the manufacture and release of CF$_2$Cl$_2$ subsequent to 1975: (1) immediate termination of all CF$_2$Cl$_2$ production; (2) termination after 1975 of all growth of production, with constant production at the 1975 rate through 1990, followed by complete termination at that time; (3) continued growth in production and release until 1990 at 10.5% per year, followed by complete termination of production in 1990. In all three models, no time delay between manufacture and release to the atmosphere has been assumed, i.e., additional time delays between manufacture and release would modify the predictions made by these models. The injection patterns are illustrated by plus signs on the figures.

In addition to these three growth/termination patterns the long-term limits of concentrations to be expected for any constant level of production can be estimated directly from the steady state models described earlier.

STRATOSPHERIC CONCENTRATIONS FROM CF$_2$CL$_2$ INJECTED BY 1975

With termination after 1975 and eddy diffusion coefficient model D the concentrations of CF$_2$Cl$_2$ in the stratosphere between 25 and 45 km reach a maximum within 2–3 years after injection ceases. The ClX concentrations with model D reach maximum values for all altitudes between 1980 and 1985, as shown in Figure 26, and then slowly decline. The concentrations during the 1980's are about 60% higher than those observed in 1975 and again fall to the 1975 level about 40 years after complete termination of the introduction of CF$_2$Cl$_2$.

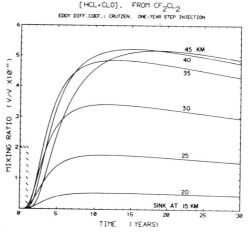

Fig. 24.   Mixing ratio of ClX versus time for various altitudes; 1-year step injection, eddy diffusion model B.

With the Crutzen model for eddy diffusion coefficients, the 40-km concentration of ClX maximizes about 1985, 10 years after termination of injection, at a level about 80% higher than the 1975 level and has not yet fallen quite to the 1975 level again 60 years later in 2035. With the Chang model E the concentrations are again much slower to reach a maximum, and again slower to decay. With this model the maximum concentrations are reached at lower altitudes by about 1990 and are about twice the 1975 concentrations. At the higher altitudes, however, the maximum concentrations are attained about the year 2000, 25 years after termination of injection, and range from 3 to 5 times as large as the 1975 ClX concentrations at these altitudes.

These three calculations for growth pattern 1 essentially present the estimates of the inevitable, i.e., the consequences in terms of stratospheric CF$_2$Cl$_2$ and ClX concentrations for material whose manufacture and use has either already occurred or which will be committed by the end of 1975. Each

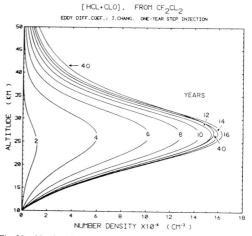

Fig. 25.   Number density of ClX versus altitude for various times; 1-year step injection, eddy diffusion model E.

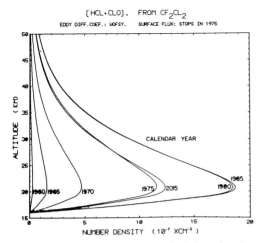

Fig. 26.   Number density of ClX versus altitude for various times. Flux stops in 1975, eddy diffusion model D.

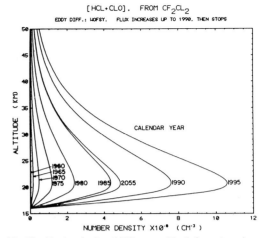

Fig. 28.   Number density of ClX versus altitude for various times. Flux increases to 1990, then stops; eddy diffusion model D.

of these models of eddy diffusion processes in the stratosphere is reasonably satisfactory for explaining the existing measurements of various trace constituents and cannot yet be eliminated as possibly the best model.

ESTIMATES OF CONSEQUENCES FOR TWO DIFFERENT MANUFACTURING PATTERNS FOR CF$_2$CL$_2$ BETWEEN 1975 AND 1990

The assumption of a constant rate of production from 1975 to 1990 at the 1975 rate is a conservative estimate for the probable chlorofluoromethane production pattern in the next decade or two. The assumption of continued exponential growth at 10.5% per year is probably a more accurate estimate of worldwide current industrial planning. Accordingly, we have carried out calculations for both of these patterns to 1990 and then have arbitrarily assumed complete termination at that point. The mixing ratios versus time are shown in Figure 27 for several altitudes with diffuse model B. The maximum concentrations at the 40- to 45-km level occur about the year 2000 for termination in 1990. The steady state concentrations for these hypothetical 1975 and 1990 production rates are

about 1.2 and 5.2 times the values calculated earlier for a rate of 0.5 Mt/year.

The ClX concentrations versus time and altitude are shown in Figures 28 and 29 for diffusion models D and E with a pattern of 10% increase until 1990, followed by complete termination of further injection. In both models the ClX concentration at 40 km can be seen to be very high in the year 2055 and would then be falling with the lifetimes given in Table 7. With each diffusion model the tropospheric concentrations of CF$_2$Cl$_2$ in 1990 are about 6×10^{-10} and 12×10^{-10}, respectively, for the constant pattern (2) and for the increasing pattern (3) models of 1975–1990 production.

All of these projected models have several attributes in common, although differing in quantitative detail: the predicted levels of CF$_2$Cl$_2$ and ClX are all much larger than the present levels; the cessation of further atmospheric injection is followed by a predicted delayed maximum in ClX concentrations in the stratosphere, and therefore in calculated depletions of

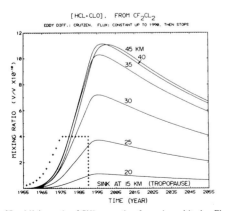

Fig. 27.   Mixing ratio of ClX versus time for various altitudes. Flux is constant to 1990, then stops; eddy diffusion model B.

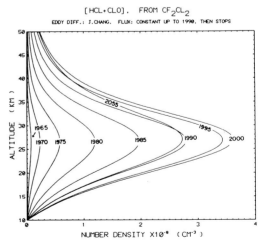

Fig. 29.   Number density of ClX versus altitude for various times. Flux increases to 1990, then stops; eddy diffusion model E.

the ozone concentrations in the stratosphere; and the predicted concentrations decrease slowly from these maximums, lasting at high levels for the entire twenty-first century.

The models discussed above have all considered only the time delays involved in atmospheric diffusion processes. These estimates can then be modified by consideration of the highly uncertain time delays between manufacture of CF_2Cl_2 and its release to the atmosphere. Approximately half of the CF_2Cl_2 manufactured is used in aerosol sprays for which the estimate has been made of a 6-month delay between production and atmospheric release. A comparable amount of CF_2Cl_2 is used as a refrigerant coolant gas, and the time delay appropriate to refrigerant use is difficult to evaluate. General experience suggests that most refrigerant coolant gases are eventually liberated to the atmosphere with delay times varying from a few years to 15–20 years, depending on the relative timing of leaks in the system and obsolescence of the equipment. These two processes for CF_2Cl_2 atmospheric release can be roughly modeled with a 3- to 4-year average delay between manufacture and release. For these circumstances, then, one would calculate that immediate cessation of CF_2Cl_2 injection into the atmosphere is not a physically real possibility; the aerosol contribution would continue for about 6–12 months and the refrigerant contribution for the better part of a decade. The maximum ozone depletion effect for diffusion model D might thus be delayed until about 1990 for immediate cessation of the manufacture of CF_2Cl_2 and a gradual cessation of its atmospheric release.

Summary of Ozone Depletion Effects

A general summary of the conclusions obtained by combining the diffusion calculations of model D with the ozone depletion calculation of *Wofsy and McElroy* [1974] is shown in Figure 30, which displays selected ClX profiles from CF_2Cl_2 versus altitude. The profiles from left to right are as follows: (1) the ClX profile at the end of 1975, assuming (*a*) no delays between production and atmospheric release and (*b*) continued 10.5% yearly increase in production for the years 1974 and 1975; (2) the profile in 1980 if release to the atmosphere were to cease abruptly at the end of 1975; (3) the profile in 1995 if production were held constant at the 1975 rate through 1990 and then abruptly stopped, the maximum effect being calculated for the period 1995–2000; (4) the profile in 1995 if production were to continue to double every 7 years (10.5% per year) from 1975 to 1990 and then to cease abruptly; (5) the steady state profile for the 1973 production rate; (6), (7), and (8) the steady state profiles for the 1980, 1985, and 1990 production rates, estimated on the basis of rising production at 10.5% yearly, followed by leveling off after 1980, 1985, and 1990, respectively. These steady state rates are roughly characteristic of the year 2050 in these calculations.

Earlier we estimated from our steady-state calculations for CF_2Cl_2 plus $CFCl_3$, together with the calculations of *Wofsy and McElroy* [1974] for ClX of HCl origin, that the 1973 rate of introduction of these chlorofluoromethanes would lead to a 13% depletion of ozone (18% with their revised calculation [*Wofsy et al.,* 1975]) if continued until a steady state is reached. Since the contribution of CF_2Cl_2 is the major part of this estimate, rough estimates of percentage ozone depletions accompanying these varying profiles can be made by assuming that ozone depletions are approximately linear with ClX concentrations.

The present calculated O_3 depletion level is roughly 1% as compared with the prechlorofluoromethane atmosphere, with

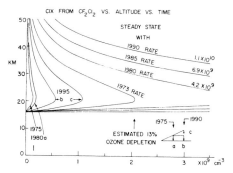

Fig. 30.   Number density of ClX versus altitude for various assumed flux models; eddy diffusion model D.

an inevitable rise to about 1.5–2.0% even if chlorofluoromethane release to the atmosphere were to be immediately terminated. Profiles 3 and 4 correspond to peak ozone depletions of about 4% and 7% in the period 1995–2000, following termination of atmospheric injection in 1990.

Current Stratospheric Levels of ClX

Collection experiments in the troposphere have clearly established the presence of chlorine, presumably in the form of HCl, in variable amounts ranging into the 10^{-9} vol/vol range. Such tropospheric HCl is subject to removal by rainout, washout, heterogeneous adsorption, and reaction with NH_3 and has a very short survival time, estimated to be at most a week or two [*Wofsy,* 1974]. Among the sources for tropospheric HCl are volcanic activity, industrial production, and marine sea spray [*Stolarski and Cicerone,* 1974; *Crutzen,* 1974*b*; *Wofsy and McElroy,* 1974]. An upper limit of 1.8×10^{-10} vol/vol has been placed on the HCl concentration of the stratosphere above 15 km through measurements in infrared absorption from high flying aircraft [*Farmer,* 1974]. Stratospheric measurements of other Cl species are very scarce; Lovelock has reported CCl_4 at about 18 km [*Lovelock et al.,* 1973]. Additional measurements have indicated the presence of $CFCl_3$ in the stratosphere at 10 km [*Lovelock,* 1974] and at altitudes from 12 to 19 km (P. C. Krey, private communication, 1974).

The photolysis of CF_2Cl_2, $CFCl_3$, and CCl_4 all can lead to the eventual formation of HCl by the reaction of Cl atoms with CH_4, H_2, etc. The transport of chlorine in these organic forms to 20- to 30-km altitudes provides a source for stratospherically introduced Cl atoms and therefore for the formation of HCl at altitudes well above the tropopause. Estimates of predicted levels of ClX, with HCl the major component, can thus be obtained from the time-dependent calculations given earlier. Since CCl_4 has been in industrial use for a much longer time than CF_2Cl_2 or $CFCl_3$, the stratospheric concentrations of ClX should be reasonably near equilibrium with CCl_4 itself; this is certainly the case if CCl_4 were to be of natural origin [*Molina and Rowland,* 1974*c*].

The integrated amount of ClX above 15 km from Figure 26 corresponds to an average concentration of 6×10^{-11} vol/vol in 1975, eddy diffusion model D being used. Although the mixing ratio of ClX above 30 km is about 1.7×10^{-10} vol/vol, it falls off rapidly below 20 km, the result being the much lower value averaged from 15 km up. When allowance is made for additional ClX from CCl_4 and from $CFCl_3$ and then for the fact that some of the ClX is not HCl, the predicted vol/vol ratio in 1975 for HCl above 15 km is less than the upper limit of detection of 1.8×10^{-10} given by *Farmer* [1974]. (In mid-1973, at the time of Farmer's measurements, the predicted ClX concentrations in the stratosphere are only about two thirds as large as the predicted values for the end of 1975.)

With diffusion models B and E the integrated amounts of ClX from CF_2Cl_2 correspond to average concentrations in 1975 of about 4×10^{-11}, somewhat less than those predicted by diffusion model D and therefore well below the detection limits.

With either of the hypothetical growth rates for 1975–1990 and any of the diffusion models, stratospheric HCl should be directly detectable in 1985–1990 by instruments operating with the presently stated sensitivity.

POSSIBLE BIOLOGICAL AND CLIMATOLOGICAL EFFECTS OF STRATOSPHERIC OZONE DEPLETION

This review has not been directed toward scientific evaluation of any of the environmental consequences of ozone depletion. However, the possible biological consequences of ozone depletion have been evaluated by a special committee of the National Academy of Sciences and reported in detail in 1973 [*Smith*, 1973]. The most readily evaluated effect quantitatively is the increase in skin cancer to be anticipated from the increased levels of ultraviolet radiation in the 2900- to 3100-Å region, and this report suggests a minimum of 8000 additional cases of skin cancer per year in the white population of the United States for a 5% decrease in average ozone level over the United States. This calculation is based essentially on the approximately threefold higher incidence of such skin cancers in the southwestern United States as compared to rates in the northern and eastern sections, an incidence rate that parallels rather closely the increased ultraviolet exposure under natural circumstances in the southwest. The most recent evaluation of this risk indicates that the actual incidence of skin cancer is about 5 times the minimum given in the Academy report, and the incremental increase from O_3 depletion would therefore also be 5 times larger, about 8000 cases per year per percent of O_3 depletion [*Urbach et al.*, 1974]. The increased ultraviolet radiation from depletions in ozone level would of course be worldwide in influence, but the appropriate background medical data required for quantitative estimates of the effects are generally not available for most population groups.

The other possible major environmental effects are climatological and are very difficult to evaluate with any precision. *Dickinson* [1974] has recently discussed these interactions in detail, without any firm conclusions concerning the onset levels for any such phenomena. He has stated that 'a 1% reduction in solar radiation reaching the lower troposphere would produce essentially a 1°C drop in global mean temperature if surface albedo, water vapor distribution, and cloudiness were to remain fixed.' As he has pointed out, however, an increase in ice cover would increase the albedo, thereby reflecting still more radiation and producing still more temperature drop. Similarly, if relative humidity, rather than amount of water vapor, were conserved, a further decrease in temperature could be expected, and the combination of increased albedo and decreased water vapor could magnify the 1°C drop into a 6°C drop for 1% reduction in solar radiation entering the lower troposphere. From this, he has concluded that 'there is consequently the possibility of an ice age with as little as 1% reduction in solar radiation if it were to persist long enough for ocean surface temperatures to come into equilibrium.' The effect of ozone depletion and redistribution on total radiation

penetrating to the lower troposphere is not easily calculated, even though the penetration of some wavelengths will obviously be enhanced. The reinforcing feedback mechanisms should work in either direction, and an increase in solar radiation into the lower troposphere would tend toward melting of the ice caps. All of the calculations of ClX effects on O_3 indicate that any effects once observed will persist for 50 to 100 years, presumably long enough for equilibrium in surface ocean temperatures.

Further study of the interactions between possible alterations in the stratospheric temperature and radiation profile and climatological effects is obviously needed before any definite statements can be made about the magnitude and onset levels of any global scale changes of climate to be expected from possible ozone depletion.

OBSERVABLE TRENDS IN THE AVERAGE OZONE CONCENTRATION OF THE ATMOSPHERE?

In an earlier section we estimated that the present ClX levels in the atmosphere from the photodecomposition of CF_2Cl_2 and $CFCl_3$ are sufficient to cause a 1% decrease in the average O_3 concentration of the atmosphere. An additional 1% depletion from CCl_4 can also be calculated, but this figure is presumably near steady state [*Molina and Rowland*, 1974c] and is not likely to increase sharply in the coming decades. A 1% change is completely undetectable with present instrumentation and with any that might be available in the foreseeable future because the wide natural fluctuations of ozone level at a given recording station vary with day and season as well as with the latitude and longitude of the recording station. The long-term trends in average ozone concentration measured at Tromsø in Norway show a distinct 11-year cycle, which has been correlated with the 11-year solar cycle [*Angell and Korshover*, 1973; *Ruderman and Chamberlain*, 1975]. Variations in the amount of incoming ionizing radiation can lead to alterations in the amount of NO_x formed as the consequence of the passage of this radiation into the atmosphere, and the NO_x variations can in turn influence the amount of ozone present. The strength of such ionizing radiation is always more intense in polar regions than in equatorial regions because of the effects of the earth's magnetic field in shielding the latter, and the effects on ozone content are expected to be greater for recording stations nearer the pole. The 11-year cycle can also be seen to a lesser extent in the measurements at Arosa, Switzerland, much further from the pole than Tromsø.

There appears to be general agreement now that the average ozone content of the northern hemisphere rose by 5–7% during the period from 1961 to 1970 [*Komhyr et al.*, 1971; *Johnston et al.*, 1973; *London and Kelley*, 1974], and there is one recent report that the average ozone concentration at one station in the southern hemisphere (Aspendale, Australia, 38°S) has decreased by about 3 ± 1% overall during the period 1965–1974 [*Pittock*, 1974]. One proposed explanation for the increase in the northern hemisphere is that the ozone level there was reduced by extensive atmospheric testing of nuclear weapons during the period 1960–1963 with substantial formation of NO_x through heating of air in the nuclear fireballs and that the atmosphere has been recovering from this depletion since the ban on atmospheric testing was established in 1963. Since such weapons testing was carried out predominantly in the northern hemisphere, with much of it in the arctic regions, the southern hemisphere ozone concentration would have been much less affected by any NO_x formed in this manner. Unfortunately, only three stations have been recording ozone data long enough to include several 11-year solar cycles, so that possible natural cyclic variations in ozone content are simply not a matter of extensive experimental record.

During the past 3 years the Climatic Impact Assessment Program has been concerned with possible impacts of projected supersonic transport operations on the world's climate and specifically with the potential effects of NO_x from jet engine exhaust on O_3 content. The consensus from this program seems to be that the magnitude of ozone change required before it could be detected falls someplace between the optimistic hope for detection of a 5% change lasting for 5 years and the pessimistic viewpoint that the minimum detectable change would be a 10% change lasting for 10 years. On the various calculated production profiles for the chlorofluoromethanes, except immediate termination, the calculated ozone depletion reaches the measurable level sometime in the 1985-1990 period with present ozone-measuring techniques.

Acknowledgments. Part of this paper originally appeared as Atomic Energy Commission Report 1974-1 under contract AT(04-3)-34, P.A. 126.

REFERENCES

Ackerman, M., Ultraviolet solar radiation related to mesospheric processes, in *Mesospheric Models and Related Experiments,* edited by G. Fiocco, pp. 149-159, D. Reidel, Dordrecht, Netherlands, 1971.

Ackerman, M., Solar ultraviolet flux below 50 kilometers, *Can. J. Chem., 52,* 1505-1509, 1974.

Ackerman, M., and C. Muller, Stratospheric methane and nitrogen dioxide from infrared spectra, *Pure Appl. Geophys., 106-108,* 1325-1340, 1973.

Angell, J. K., and J. Korshover, Quasi-biennial and long-term fluctuations in total ozone, *Mon. Weather Rev., 101,* 426-443, 1973.

Balakhnin, V. P., V. I. Egorov, and E. I. Intezarova, Kinetic investigation of elemental reactions of oxygen atoms in the gas phase by EPR, 2, The reaction O + HCl = OH + Cl, *Kinet. Catal. USSR, 12,* 258-262, 1971.

Baldwin, R. R., M. E. Fuller, J. S. Hillman, D. Jackson, and R. W. Walker, Second limit of hydrogen + oxygen mixtures: The reaction H + HO_2, *J. Chem. Soc. Faraday Trans. 1, 70,* 635-641, 1974.

Banks, P. M., and G. Kockarts, *Aeronomy, Part A,* p. 39, Academic, New York, 1973.

Bemand, P. P., M. A. A. Clyne, and R. T. Watson, Reactions of chlorine oxide radicals, 4, Rate constants for the reactions Cl + OClO, O + OClO, H + OClO, NO + OClO and O + ClO, *J. Chem. Soc. Faraday Trans. 1, 69,* 1356-1374, 1973.

Benson, S. W., F. R. Cruickshank, and R. Shaw, Iodine monochloride as a thermal source of chlorine atoms: The reaction of chlorine atoms with hydrogen, *Int. J. Chem. Kinet., 1,* 29-43, 1969.

Berkowitz, J., W. A. Chupka, and D. Gutman, Electron affinities of O_2, O_3, NO, NO_2, NO_3 by endothermic charge transfer, *J. Chem. Phys., 55,* 2733-2745, 1971.

Bohac, S., Conductometric testing and corrosion study of non-anhydrous ethanol systems for hair spray, *J. Soc. Cosmet. Chem., 19,* 149-158, 1968.

Bricard, J., and D. Vigla, Stratospheric aerosols physicochemistry, *Can. J. Chem., 52,* 1479-1490, 1974.

Brinkmann, R. T., Photochemistry and the escape efficiency of terrestrial hydrogen, in *Mesospheric Models and Other Related Experiments,* edited by G. Fiocco, pp. 89-102, D. Reidel, Dordrecht, Netherlands, 1971.

Cadle, R. D., Composition of the stratospheric sulfate layer, CIAP Proceedings of the Survey Conference, *DOT-TSC-OST-72-13,* Dep. of Transp., Washington, D. C., 1972.

Chang, J. S., Simulations, perturbations, and interpretations, Proceedings of the Third Conference on Climatic Impact Assessment Program, *DOT-TSC-OST-74-15,* pp. 330-341, U.S. Dep. of Transp., Cambridge, Mass., February 26-March 1, 1974.

Chapman, S., A theory of upper atmospheric ozone, *Quart. J. Roy. Meteorol. Soc., 3,* 103, 1930.

Chemical Marketing Reporter, Chemical profile: Fluorocarbons, *202*(8), 9, 1972.

Christophorou, L. G., and J. A. D. Stockdale, Dissociative electron attachment to molecules, *J. Chem. Phys., 48,* 1956-1960, 1968.

Cicerone, R. J., R. S. Stolarski, and S. Walters, Stratospheric ozone destruction by man-made chlorofluoromethanes, *Science, 185,* 1165-1167, 1974.

Clyne, M. A. A., and R. F. Walker, Absolute rate constants for elementary reactions in the chlorination of CH_4, CD_4, CH_3Cl, CH_2Cl_2, $CHCl_3$, $CDCl_3$ and $CBrCl_3$, *J. Chem. Soc. Faraday Trans. 1, 69,* 1547-1567, 1973.

Clyne, M. A. A., D. J. McKenney, and R. F. Walker, Reaction kinetics of ground state fluorine, F (^2P), atoms, 1, Measurements of fluorine atom concentrations and the rates of reactions F + CHF_3 and F + Cl_2 using mass spectrometry, *Can. J. Chem., 51,* 3596-3604, 1973.

Clyne, M. A. A., and R. T. Watson, Kinetic studies of diatomic free radicals using mass spectrometry, 2, Rapid bimolecular reactions involving ClO($X^2\pi$) radicals, *J. Chem. Soc. Faraday Trans. 1, 70,* 2250-2259, 1974.

Clyne, M. A. A., D. J. McKenney, and R. T. Watson, Reactions of chlorine oxide radicals, 5, The reaction 2 ClO($X^2\pi$) → products, *J. Chem. Soc. Faraday Trans. 2,* in press, 1974.

Crutzen, P. J., The photochemistry of the stratosphere with special attention given to the effect of NO_x emitted by supersonic aircraft, CIAP Proceedings of the Survey Conference, *DOT-TSC-OST-72-13,* pp. 80-89, Dep. of Transp., Washington, D. C., 1972.

Crutzen P. J., Estimate of possible future ozone reductions from continued use of fluorochloromethanes, *Geophys. Res. Lett., 1,* 205-208, 1974a.

Crutzen, P. J., A review of upper atmospheric photochemistry, *Can. J. Chem., 52,* 1569-1581, 1974b.

Crutzen, P. J., The future effects of the continued use of chlorofluoromethanes on stratospheric ozone (abstract), *Eos Trans. AGU, 56*(12), 1153, 1974c.

Cumming, C., and R. P. Lowe, Balloon-borne spectroscopic measurement of stratospheric methane, *J. Geophys. Res., 78,* 5259-5263, 1973.

Curran, R. K., Positive and negative ion formation in CCl_3F, *J. Chem. Phys., 34,* 2007-2010, 1961.

Cvetanović, R. J., Excited state chemistry in the stratosphere, *Can. J. Chem., 52,* 1452-1464, 1974.

Davis, D. D., W. Braun, and A. M. Bass, Reactions of Cl($^3P_{3/2}$): Absolute rate constants for reaction with H_2, CH_4, CH_2Cl_2, C_2Cl_4, C_2H_6 and c-C_6H_{12}, *Int. J. Chem. Kinet., 2,* 101-114, 1970.

Davis, D. D., J. T. Herron, and R. E. Huie, Absolute rate constants for the reaction $O(^3P)$ + NO_2 → NO + O_2 over the temperature range 230-339°K, *J. Chem. Phys., 58,* 530-535, 1973.

Demore, W. B., and E. Tschuikow-Roux, Temperature dependence of the reactions of OH and HO_2 with O_3, *J. Phys. Chem., 78,* 1447-1451, 1974.

Dickinson, R. E., Climatic effects of stratospheric chemistry, *Can. J. Chem., 52,* 1616-1624, 1974.

Doucet, J., P. Sauvageau, and C. Sandorfy, Vacuum ultraviolet and photoelectron spectra of fluoro-chloro derivatives of methane, *J. Chem. Phys., 58,* 3708-3716, 1973.

Duce, R. A., J. W. Winchester, and T. W. Van Nahl, Iodine, bromine, and chlorine in the Hawaiian marine atmosphere, *J. Geophys. Res., 70,* 1775-1799, 1965.

Dütsch, H. U., The ozone distribution in the atmosphere, *Can. J. Chem., 52,* 1491-1504, 1974.

Ehhalt, D. H., Sampling of stratospheric trace constituents, *Can. J. Chem., 52,* 1510-1518, 1974.

Ehhalt, D. H., and L. E. Heidt, Vertical profiles of molecular H_2 and CH_4 in the stratosphere, paper presented at the AIAA/AMS International Conference on the Environmental Impact of Aerospace Operations in the High Atmosphere, Amer. Meteorol. Soc., Denver, Colo., June 11-13, 1973.

Ehhalt, D. H., L. E. Heidt, and E. A. Martell, The concentration of atmospheric methane between 44- and 62-km altitude, *J. Geophys. Res., 77,* 2193-2196, 1972.

Ehhalt, D. H., L. E. Heidt, R. H. Lueb, and N. Roper, Vertical profiles of CH_4, H_2, CO, N_2O, and CO_2 in the stratosphere, Proceedings of the Third Conference on Climatic Impact Assessment Program, *DOT-TSC-OST-74-15,* pp. 153-160, U.S. Dep. of Transp., Washington, D. C., 1974.

Farmer, C. B., Infrared measurements of stratospheric composition, *Can. J. Chem., 52,* 1544-1559, 1974.

Farmer, C. B., R. A. Toth, O. F. Raper, and R. A. Schindler, Recent results of aircraft infrared observations of the stratosphere,

34   ROWLAND AND MOLINA: CHLOROFLUOROMETHANES IN THE ENVIRONMENT

Proceedings of the Third Conference on Climatic Impact Assessment Program, *DOT-TSC-OST-74-15,* pp. 234–245, U.S. Dep. of Transp., Washington, D.C., 1974.

Ferguson, E. E., Ion molecule reactions, Chemical Kinetics Data Survey VII, Tables of Rate and Photochemical Data for Modelling of the Stratosphere, *NBSIR 74-430,* pp. 82–101, Nat. Bur. of Stand., Washington, D. C., 1974.

Fettis, G. C., and J. H. Knox, The rate constants of halogen atom reactions, *Prog. Reaction Kinetics,* vol. 2, edited by G. Porter, pp. 1–38, Macmillan, New York, 1964.

Galante, J. J., and E. A. Gislason, Critical re-examination of the reaction Cl + H₂ = HCl + H, *Chem. Phys. Lett., 18,* 231–234, 1973.

Goodeve, C. F., and F. D. Richardson, The absorption spectrum of chlorine trioxide and chlorine hexoxide, *J. Chem. Soc. Faraday Trans. 1, 33,* 453–457, 1937.

Hampel, B., *DMS-UV-Atlas of Organic Compounds,* vol. 5, edited by H. H. Perkampus, I. Sandeman, and C. J. Timmons, p. M/10, Plenum, New York, 1971.

Harker, A. B., Reaction kinetics of the photolysis of NO₂ and the spectra and reaction kinetics of the free radicals in the photolysis of CO₂–O₂–Cl systems, Ph.D. thesis, 195 pp., Univ. of Calif., Berkeley, 1972.

Heicklen, J., Photolysis of trifluoroiodomethane in the presence of oxygen and nitric oxide, *J. Phys. Chem., 70,* 112–118, 1966.

Heicklen, J., Gas phase oxidation of perhalocarbons, in *Advances in Photochemistry,* vol. 7, edited by J. Pitts, G. Hammond, and W. A. Noyes, pp. 57–148, Interscience, New York, 1969.

Heidner, R. F., and D. Husain, Electronically excited oxygen atoms, O(2¹D₂): Time-resolved study of the collisional quenching by the gases H₂, D₂, NO, N₂O, NO₂, CH₄, and C₃O₂ using atomic absorption spectroscopy in the vacuum ultra-violet, *Int. J. Chem. Kinet., 5,* 819–831, 1973.

Heidner, R. F., D. Husain, and J. R. Wiesenfeld, Kinetic investigation of electronically excited oxygen atoms, O(2¹D₂), by time resolved attenuation of atomic resonance radiation in the vacuum ultra-violet, *J. Chem. Soc. Faraday Trans. 2, 69,* 927–938, 1973.

Hesstvedt, E., Comments on photochemical modeling, Proceedings of the Second Conference on CIAP, *DOT-TSC-OST-73-4,* pp. 285–290, Dep. of Transp., Washington, D. C., 1973.

Hesstvedt, E., Reduction of stratospheric ozone from high-flying aircraft, studied in a two-dimensional photochemical model with transport, *Can. J. Chem., 52,* 1592–1598, 1974.

Hester, N. E., E. R. Stephens, and O. C. Taylor, Fluorocarbons in the Los Angeles basin, *J. Air Pollut. Control Ass., 24,* 591–595, 1974.

Hester, N. E., E. R. Stephens, and O. C. Taylor, Fluorocarbon air pollutants, 2, *Atmos. Environ.,* in press, 1975.

Hirono, M., M. Fujiwara, O. Uchino, and T. Itabe, Observations of stratospheric aerosol layers by optical radar, *Can. J. Chem., 52,* 1560–1568, 1974.

Hochanadel, C. J., J. A. Ghormley, and P. J. Ogren, Absorption spectrum and reaction kinetics of the HO₂ radical in the gas phase, *J. Chem. Phys., 56,* 4426–4432, 1972.

Hoffman, D. J., Stratospheric aerosol determinations, *Can. J. Chem., 52,* 1519–1526, 1974.

Hoshizaki, H., J. Y. Myer, and K. O. Redler, Potential destruction of ozone by HCl in rocket exhausts, *Rep. LMC-D-354204,* Lockheed Missiles and Space, Palo Alto, Calif., 1973.

Hudlicky, M., *Chemistry of Organic Fluorine Compounds,* pp. 340–341, Macmillan, New York, 1962.

Johnsen, M. A., W. E. Dorland, and E. K. Dorland, *The Aerosol Handbook,* pp. 251–286, W. E. Dorland Co., Caldwell, N. J., 1972.

Johnston, H. S., Photochemistry in the stratosphere, *Acta Astronaut., 1,* 135–156, 1974.

Johnston, H. S., G. Whitten, and J. Birks, Effects of nuclear explosions on stratospheric nitric oxide and ozone, *J. Geophys. Res., 78,* 6107–6135, 1973.

Junge, C. E., C. W. Chagnon, and J. E. Manson, Stratospheric aerosols, *J. Meteorol., 18,* 81–108, 1961.

Kerr, J. A., Bond dissociation energies by kinetic methods, *Chem. Rev., 66,* 465–500, 1966.

Kockarts, G., Penetration of solar radiation in the Schumann-Runge bands of molecular oxygen, in *Mesospheric Models and Related Experiments,* edited by G. Fiocco, pp. 160–176, D. Reidel, Dordrecht, Netherlands, 1971.

Komhyr, W. D., E. W. Barret, G. Slocum, and H. K. Weickmann, Atmospheric total ozone increase during the 1960's, *Nature, 232,* 390–391, 1971.

Kriemler, P., and S. E. Buttrill, Positive and negative ion-molecule

reactions and the proton affinity of ethyl nitrate, *J. Amer. Chem. Soc., 92,* 1123–1128, 1973.

Lacher, J. R., L. E. Hummel, E. F. Bohmfalk, and J. D. Park, The new ultraviolet absorption spectra of some fluorinated derivatives of methane and ethylene, *J. Amer. Chem. Soc., 72,* 5486–5489, 1950.

Lazrus, A. L., and B. W. Gandrud, Stratospheric sulfate aerosol, *J. Geophys. Res., 79,* 3424–3431, 1974.

Levy, H., Photochemistry of minor constituents in the troposphere, *Planet. Space Sci., 21,* 575–591, 1973.

Liss, P. S., and P. G. Slater, Flux of gases across the air-sea interface, *Nature, 247,* 181–184, 1974.

London, J., and J. Kelley, Global trends in total atmospheric ozone, *Science, 184,* 987–989, 1974.

London, J., and J. H. Park, The interaction of ozone photochemistry and dynamics in the stratosphere: A three-dimensional atmospheric model, *Can. J. Chem., 52,* 1599–1609, 1974.

Lovelock, J. E., Atmospheric fluorine compounds as indicators of air movements, *Nature, 230,* 379, 1971.

Lovelock, J. E., R. J. Maggs, and R. J. Wade, Halogenated hydrocarbons in and over the Atlantic, *Nature, 241,* 194–196, 1973.

Lovelock, J. E., Atmospheric halocarbons and stratospheric ozone, *Nature, 252,* 292–294, 1974.

MacCracken, M. C., Zonal atmospheric model ZAM2, Proceedings of the Second Conference on CIAP, *DOT-TSC-OST-73-4,* pp. 298–320, Dep. of Transp., Washington, D. C., 1973.

Machta, L., Global scale atmospheric mixing, paper presented at the Second International Union of Theoretical and Applied Mechanics and International Union of Geodesy and Geophysics Symposium on Turbulence in Environmental Pollution, Int. Union of Geod. and Geophys., Charlottesville, Va., April 8–14, 1973.

Mahlman, J. D., Preliminary results from a three-dimensional general circulation/tracer model, Proceedings of the Second Conference on CIAP, *DOT-TSC-OST-73-4,* pp. 321–337, Dep. of Transp., Washington, D. C., 1973.

Mahlman, J. D., and S. Manabe, Numerical simulation of the stratosphere: Implications for related climatic change problems, CIAP Proceedings of the Survey Conference, *DOT-TSC-OST-72-13,* pp. 186–193, Dep. of Transp., Washington, D. C., 1972.

Majer, J. R., and J. P. Simons, Photochemical processes in halogenated compounds, in *Advances in Photochemistry,* vol. 2, edited by J. Pitts, G. Hammond, and W. A. Noyes, pp. 137–181, Interscience, New York, 1964.

Marsh, D., and J. Heicklen, Photolysis of fluorotrichloromethane, *J. Phys. Chem., 69,* 4410–4415, 1965.

Martell, E. A., The distribution of minor constituents in the stratosphere and lower mesosphere, in *Physics and Chemistry of the Upper Atmosphere,* edited by B. M. McCormac, pp. 24–33, D. Reidel, Dordrecht, Netherlands, 1973.

McConnell, J. C., and M. B. McElroy, Odd nitrogen in the atmosphere, *J. Atmos. Sci., 30,* 1465–1480, 1973.

McElroy, M. B., Conference summary, CIAP Proceedings of the Survey Conference, *DOT-TSC-OST-72-13,* pp. 260–269, Dep. of Transp., Washington, D. C., 1972.

McElroy, M. B., S. C. Wofsy, J. E. Penner, and J. C. McConnell, Atmospheric ozone: Possible impact of stratospheric aviation, *J. Atmos. Sci., 31,* 287–303, 1974.

McIver, R. T., Kinetic and equilibrium studies of gas phase ionic reactions using pulsed ion cyclotron resonance spectroscopy, Ph.D. thesis, 198 pp., Stanford Univ., Stanford, Calif., 1971.

Milstein, R., and F. S. Rowland, Quantum yield for the photolysis of CF₂Cl₂ in O₂, *J. Phys. Chem.,* in press, 1975.

Minford, J. D., Compatibility studies of aluminum with propellant and solvents for use in aerosols, *J. Soc. Cosmet. Chem., 15,* 311–326, 1964.

Mohnen, V. A., Discussion of the formation of major positive and negative ions up to 50 km level, in *Mesospheric Models and Related Experiments,* edited by G. Fiocco, pp. 210–219, D. Reidel, Dordrecht, Netherlands, 1971.

Molina, M. J., and F. S. Rowland, Stratospheric sink for chlorofluoromethanes: Chlorine atom catalysed destruction of ozone, *Nature, 249,* 810–812, 1974a.

Molina, M. J., and F. S. Rowland, Stratospheric sink for chlorofluoromethanes—Chlorine atom catalyzed destruction of ozone, paper presented at the Second International Conference on the Environmental Impact of Aerospace Operations in the High Atmosphere, Amer. Meteorol. Soc., San Diego, Calif., July 8–10, 1974b.

Molina, M. J., and F. S. Rowland, Predicted present stratospheric abundances of chlorine species from photodissociation of carbon tetrachloride, *Geophys. Res. Lett.*, *1*, 309–312, 1974c.

Molina, M. J., and F. S. Rowland, Some unmeasured chlorine atom reaction rates important for stratospheric modeling of chlorine atom catalyzed removal of ozone, *J. Phys. Chem.*, in press, 1975.

Nicolet, M., Aeronomic reactions of hydrogen and ozone, in *Mesospheric Models and Related Experiments*, edited by G. Fiocco, pp. 1–51, D. Reidel, Dordrecht, Netherlands, 1971.

Nicolet, M., Aeronomic chemistry of the stratosphere, CIAP Proceedings of the Survey Conference, *DOT-TSC-OST-72-13*, pp. 40–70, Dep. of Transp., Washington, D. C., 1972.

Nicolet, M., An overview of aeronomic processes in the stratosphere and mesosphere, *Can. J. Chem.*, *52*, 1381–1396, 1974.

Niki, H., Reaction kinetics involving O and N compounds, *Can. J. Chem.*, *52*, 1397–1404, 1974.

Pittock, A. B., Trends in the vertical distribution of ozone over Australia, *Nature*, *249*, 641 643, 1974.

Pitts, J. N., H. L. Sandoval, and R. Atkinson, Relative rate constants for the reaction of $O(^1D)$ atoms with fluorocarbons and N_2O, *Chem. Phys. Lett.*, *29*, 31–34, 1974.

Reid, J. B., The aerosol industry tomorrow, *Aerosol Age*, *19*(7), 21–24, 1974.

Rowland, F. S., and M. J. Molina, Chlorofluoromethanes in the enviroment, *AEC Rep. 1974-1*, Univ. of Calif., Irvine, September 1974.

Ruderman, M. A., and J. W. Chamberlain, Origin of the sunspot modulation of ozone: Its implications for stratospheric NO injec-tion, *Planet. Space Sci.*, *23*(1), in press, 1975.

Sanders, P. A., Corrosion of aerosol cans, *Soap Chem. Spec.*, *36*, 95–103, 1960.

Sanders, P. A., Reaction of propellant 11 with water, *Soap Chem. Spec.*, *41*, 117–124, 1965.

Sanders, P. A., Propellants and solvents, in *The Science and Technology of Aerosol Packaging*, edited by J. J. Sciarra, B. S. Pharm, and L. Stoller, pp. 97–149, John Wiley, New York, 1974.

Sandoval, H. L., R. Atkinson, and J. N. Pitts, Reactions of elec-tronically excited $O(^1D)$ atoms with fluorocarbons, *J. Photochem.*, *3*, 325–327, 1974.

Schumacher, H. J., Der Mechanismus der thermischen Reaktion zwischen Chlor und Ozon, *Z. Phys. Chem. Frankfurt am Main*, *13*, 353–367, 1957.

Siegemund, G., Einfache Synthese von Carbonylchloridfluorid und Carbonylbromidfluorid, *Angew. Chem.*, *85*, 982, 1973.

Simon, P., Balloon measurements of solar fluxes between 1960 Å and 2300 Å, Proceedings of the Third Conference on Climatic Impact Assessment Program, *DOT-TSC-OST-74-15*, pp. 137–142, U.S. Dep. of Transp., Washington, D. C., 1974.

Smith, K. C., Biological impacts of increased intensities of solar ultraviolet radiation, Report to the Environmental Studies Board, 44 pp., Nat. Acad. of Sci., Washington, D. C., 1973.

Smith, I. W. M., and R. Zellner, Rate measurements of reactions of OH by resonance absorption, 3, Reactions of OH with H_2, D_2, and hydrogen and deuterium halides, *J. Chem. Soc. Faraday Trans. 2*, *70*, 1045–1056, 1974.

Stoiber, R. E., D. C. Leggett, T. F. Jenkins, R. P. Murrmann, and W. I. Rose, Organic compounds in volcanic gas from Santiaguito volcano, Guatemala, *Geol. Soc. Amer. Bull.*, *82*, 2299–2302, 1971.

Stolarski, R. S., and R. J. Cicerone, Stratospheric chlorine: A possible sink for ozone, *Can. J. Chem.*, *52*, 1610–1615, 1974.

Su, C. W., and E. D. Goldberg, Chlorofluorocarbons in the atmosphere, *Nature*, *245*, 27, 1973.

Takacs, G. A., and G. P. Glass, Reactions of hydroxyl radicals with some hydrogen halides, *J. Phys. Chem.*, *77*, 1948–1951, 1973.

Turco, R. P., and R. C. Whitten, A study of the possible long-term effects on the ozone layer of the continuing production of freon gases (abstract), *Eos Trans. AGU*, *55*(12), 1153, 1974.

Urbach, F., D. Berger, and R. E. Davies, Field measurements of biologically effective ultraviolet radiation and its relation to skin cancer in man, Proceedings of the Third Conference on Climatic Impact Assessment Program, *DOT-TSC-OST-74-15*, pp. 523, U.S. Dep. of Transp., Washington, D. C., 1974.

U.S. Tariff Commission, Synthetic organic chemicals, U.S. production and sales for 1972, *TC-681*, Washington D. C., 1973.

Watson, R. J., Chemical kinetics data survey 8, rate constants of C10, of atmospheric interest, *NBSIR 74-516*, Nat. Bur. Standards, Washington D. C., 1974.

Wilkniss, P. E., R. A. Lamontagne, R. E. Larson, J. W. Swinnerton, C. R. Dickson, and T. Thompson, Atmospheric trace gases in the southern hemisphere, *Nature Phys. Sci.*, *245*, 45–47, 1973

Wilkniss, P. E., D. J. Bressan, R. A. Carr, R. A. Lamontagne, and W. Swinnerton, Occurrence and removal of gaseous and particulate contaminants in the marine atmosphere, paper presented at the 168th American Chemical Society Meeting, Amer. Chem. Soc., Atlantic City, N. J., September 1974.

Wilson, W. E., J. T. O'Donovan, and R. M. Fristrom, Flame inhibition by halogen compounds, *12th Symposium (International) on Combustion*, pp. 929–942, The Combustion Institute, Pittsburg, Pa., 1969.

Wofsy, S. C., Atmospheric photochemistry of N-, H-, and Cl-containing radicals, Proceedings of the Third Conference on Climatic Impact Assessment Program, *DOT-TSC-OST-74-15*, pp. 359–375, Dep. of Transp., Washington, D. C., 1974.

Wofsy, S. C., and M. B. McElroy, HO_x, NO_x, and ClO_x: Their role in atmospheric photochemistry, *Can. J. Chem.*, *52*, 1582–1591, 1974.

Wofsy, S. C., J. C. McConnell, and M. B. McElroy, Atmospheric CH_4, CO, and CO_2, *J. Geophys. Res.*, *77*, 4477–4493, 1972.

Wofsy, S. C., M. B. McElroy, and N. D. Sze, Freon consumption: Implications for atmospheric ozone, *Science*, in press, 1975.

Wong, E. L., and E. F. Belles, Rate measurements for the reaction of hydrogen chloride and deuterium chloride with atomic oxygen, *NASA Tech. Note TM D-6495*, 21 pp., 1972.

Zahniser, M. S., F. Kaufman, and J. G. Anderson, Kinetics of the reaction of OH with HCl, *Chem. Phys. Lett.*, *27*, 507–510, 1974.

(Received October 25, 1974;
accepted October 29, 1974.)

Contribution from

FRANK WILKINSON

Loughborough University of Technology
UK

orn 3rd June 1936, in Tanfield, County Durham, England. Attended Stanley Grammar School, from 1947 till 1954 and then moved to Sheffield University where the then Professor George Porter lectured to him on Physical Chemistry. In 1957 having obtained a BSc (First Class Honours in Chemistry) he started research under the supervision of Lord Porter obtaining his PhD degree in 1960. His thesis was entitled "The Transfer of Electronic Excitation Energy from the Triplet State in Solution".

Academic Career

1960–1964   Research Fellow and College Tutor, Pembroke College, Oxford. He built the first flash photolysis apparatus in Oxford and carried out Photochemical Research in the Physical Chemistry Laboratory in Oxford.

1964–1967   Lecturer in Physical Chemistry at University of East Anglia.
1967–1979   Reader in Physical Chemistry at University of East Anglia.

1979–2001   Professor of Physical Chemistry, Loughborough University.
1996–1999   Head of Chemistry Department, Loughborough University.
2001–   Professor Emeritus, Loughborough University.
2001–   Editor in Chief, Photochemical & Photobiological Sciences.

Other Activities

President of the European Photochemical Association (1984–1988).

Chairman of the XIth IUPAC Symposium on Photochemistry, Lisbon (1986).

Visiting Professor at the Radiation Laboratory Data Centre, University of Notre
    Dame, USA (1978, 1989 and 1991) and at the Universities of Bordeaux
    (1983), of Chile (1994), of Fribourg (1988), of Montpellier (1989) and of
    Berne (1991).

Consulted for Ciba-Geigy, Kodak, Pilkingtons, Tissuemed Ltd and Dupont.

Course Assessor on Reaction Kinetics, Open University.

Author of textbook *Chemical Kinetics and Reaction Mechanisms* (Van Nostrand
    Reinhold Ltd).

Published over 320 papers in International Journals on Excited State Properties
    and Laser Flash Photolysis. Recent examples include a study using Diffuse
    Reflectance Laser Flash Photolysis: "Quantitative rate constants for radical
    reactions in the nanopores of cotton", P. Hunt, D. R. Worrall, F. Wilkinson,
    S. N. Batchelor, *J. Amer. Chem. Soc.* **124**, 8532–8533 (2002); and on singlet
    oxygen formed by electronic energy transfer: "Electronic to vibrational
    energy conversion and charge transfer contributions during quenching by
    molecular oxygen of electronically excited triplet states", A. A. Abdel-Shafi
    and F. Wilkinson, *Phys. Chem. Chem. Phys.* **4**, 248–255 (2002).

Songsters at the 10th International Conference on the Photochemical Conversion and Storage
of Solar Energy (Interlaken, 1994). Frank Wilkinson is far left, George Porter far right and Mary
Archer centre.

Frank Wilkinson on
"Energy Transfer from the Triplet State"
G. Porter and F. Wilkinson
Proc. Roy. Soc. **264**, 284 (1961)

The research work described in this paper was carried out between October 1957 and September 1960 in the Chemistry Department of Sheffield University where I had the privilege to study for my PhD degree under the supervision of the then Professor George Porter. In December 1958, he included, with due acknowledgment, some of our work on energy transfer from triplet benzophenone to naphthalene in his Tilden Lecture to the Chemical Society at Burlington House, London. He also delivered this prize lecture[1] entitled "The Triplet in Chemistry" at Aberdeen, Leicester, Newcastle and Southampton Universities early in 1959. I made an oral presentation[2] of some of our results at the Faraday Society Discussion on "Energy Transfer with Special Reference to Biological Systems" held in Nottingham in 1959, but typical of Porter, he waited until we had final results from 20 different combinations of donor, acceptor and solvent before submitting our manuscript to the Royal Society.

The results given in our paper established that triplet energy transfer, i.e. the reaction

$$D^* \text{ (triplet)} + A \text{ (singlet)} \rightarrow D \text{ (singlet)} + A^* \text{ (triplet)}$$

occurs as a spin-allowed[5] collisional process in fluid solution independent of the nature of the triplet states of the donor D, or the singlet states of the acceptor A, and is diffusion controlled provided the triplet level of the donor is well above that of the acceptor. While this work was being carried out, Backstrom and Sandros[3] independently showed that triplet energy transfer to and from biacetyl, one of the few molecules which phosphoresces with relatively high yield in room temperature fluid solution, was a highly-efficient process. Prior to this in 1956, Terenin and Ermolaev[4] had studied energy transfer in rigid glasses from the triplet states of benzophenone and other carbonyl containing compounds as donors to produce the triplet states of naphthalene and some of its derivatives as acceptors by demonstrating quenching of the phosphorescence of the donors and sensitised production of phosphorescence from the acceptors.

The impact of Porter's paper on studies of primary photochemical studies was considerable. This was due in part to the fact that it was possible to use triplet–triplet energy transfer to inhibit "unwanted" triplet state photoreactions by the use of additives, which could act as energy acceptors efficiently accepting the energy and dissipating it as heat. The process could also be used to sensitise "wanted" triplet state reactions of acceptors in the absence of singlet reactions by using sources of light which the acceptor absorbs only weakly or not at all. The paper was a citation classic and although it is now cited in almost all photochemical textbooks, it is still cited 2 or 3 times a year in the current primary literature. Triplet energy transfer can be used to establish whether singlet or triplet states are responsible for observed photochemical reactions. Porter was already interested in the photoreactions of ketones and in 1960 we published the first demonstration of "photochemical protection" to prove that photoreduction of benzophenone in alcoholic solution was due to hydrogen atom abstraction by the triplet state of benzophenone.[5] Porter went on to make a major contribution concerning the photochemical reactions of many ketones[6] while, inspired by him, I applied the principles I had learnt with him to show the photoreactive states in the photoreactions of anthraquinone,[7] duroquinone[8] and other quinones were also due to the reaction of the triplet states of these quinones.

The editors have asked contributors to this volume to include a paper of their own work to follow one of Porter's classic papers. My co-author and I are happy to dedicate the paper entitled "Diffuse Reflectance Triplet–Triplet Absorption Spectroscopy of Aromatic Hydrocarbons Chemisorbed on γ-Alumina" to the memory of Lord George Porter who taught me how to carry out photochemical studies and was an inspiration to me throughout my academic career. Porter was awarded the Nobel Prize jointly for his development of flash photolysis,[9] one of the most powerful techniques for investigating photoinduced transient intermediates following pulsed excitation by directly observing, using transmission spectroscopy, absorption by the intermediates. Our paper[10] is the first example of the extension, by using diffuse reflectance spectroscopy, of the flash photolysis technique to opaque and to highly-scattering media thus enabling the study of photoinduced intermediates of molecules adsorbed at the surface of catalytic metal oxide surfaces, within highly-scattering microcrystals and dyes adsorbed on fabrics and chemically bound to polymers etc.[11]

References

1. G. Porter, *Proc. Chem. Soc.* 291–302 (1959).
2. F. Wilkinson, *Faraday Soc. Disc.* 96–97 (1959).
3. H. L. J. Backstrom and K. Sandros, *Acta Chem. Scand.* **12**, 823 (1958) and **14**, 48 (1960).
4. A. Terenin and V. Ermolaev, *Trans. Faraday Soc.* **52**, 1042 (1956).
5. G. Porter and F. Wilkinson, *Trans. Faraday Soc.* **52**, 1686 (1961).
6. G. Porter and P. Suppan, *Trans. Faraday Soc.* **61**, 1664 (1965).
7. K. Tickle and F. Wilkinson, *Trans. Faraday Soc.* **61**, 1981 (1965).
8. J. Nafisi-Movaghar and F. Wilkinson, *Trans. Faraday Soc.* **66**, 2257 (1970).
9. G. Porter, *Proc. Roy. Soc. A* **200**, 284 (1950).
10. R. K. Kessler and F. Wilkinson, *J. Chem. Soc. Faraday Trans.* **77**, 309 (1981).
11. F. Wilkinson, *J. Chem. Soc. Faraday Trans.* **82**, 2073 (1986).

J. C. S. Faraday Trans 1, 77, 309–320 (1981).
Reproduced by permission of the Royal Society of Chemistry

Diffuse Reflectance Triplet–Triplet Absorption Spectroscopy of Aromatic Hydrocarbons Chemisorbed on γ-Alumina

By Rudolf W. Kessler† and Francis Wilkinson*‡

School of Chemical Sciences, University of East Anglia, Norwich NR4 7TJ

Received 25th January, 1980

An apparatus developed to enable diffuse reflectance flash photolysis measurements to be made is described. Transient spectra observed from five aromatic hydrocarbons adsorbed at less than monolayer levels on activated γ-alumina powder are shown to be due to triplet–triplet absorption. These spectra are strikingly different from the triplet–triplet spectra of these same hydrocarbons in homogeneous media and these differences are interpreted as arising mainly from transitions to charge-transfer states of the surface–adsorbate systems. The potential of this method for investigating photocatalytic reactions and excited states properties on surfaces is demonstrated.

The increasing number of publications in recent years concerned with the properties of adsorbed species demonstrates the considerable and widespread interest which exists in this field. However, the photochemistry of adsorbed species is still a relatively underdeveloped area worthy of much more intensive study. Thus although there have been several investigations concerning 'heterogeneous photocatalysis' and photosensitation of molecules adsorbed on metal oxides[1, 2] very little is known about the properties of electronically excited states of chemisorbed species. Emission[3-6] and diffuse reflectance[7, 8] absorption studies have established that there are often considerable changes in the energy and the nature of electronically excited singlet and triplet states upon adsorption. Consequently the probabilities of various photochemical[9-11] and photophysical processes[3, 4] are altered.

One of the most powerful techniques for investigation of primary photochemical and photophysical processes is flash photolysis which enables one to monitor absorption by transient intermediates, including electronically excited states with extremely short lifetimes. We have therefore developed a new method of diffuse reflectance flash photolysis for examining transient absorption spectra and decay kinetics from molecules adsorbed on highly light-scattering surfaces. During the course of this work the flash photolysis of pyrene adsorbed on Vycor glass has been reported by making use of the transparency, large surface area and large total porosity of Vycor glass.[12] Since our method is not restricted to transparent media it should have wider applications, *e.g.* to many powdered samples.

In order to convincingly demonstrate the validity of the method we have chosen to investigate aromatic hydrocarbons adsorbed on γ-alumina powder since absorption and luminescence studies have already been made on such systems.[3-8] Furthermore, phosphorescence is often easy to detect and when the diffuse reflectance transient absorption and the phosphorescence have identical kinetic behaviour, this confirms

† Present address: Institut für Physikalische und Theoretische Chemie der Universität, D-7400, Tubingen, West Germany.
‡ Present address: Department of Chemistry, University of Technology, Loughborough, Leicestershire LE11 3TU.

the assignment of the transient as the emitting triplet state. Chemisorbed aromatic hydrocarbons on γ-alumina constitute favourable systems for testing the method since (i) the surface and its dehydration procedure is well established,[13, 14] (ii) the chemisorbed aromatic hydrocarbons are relatively photochemically stable (in the absence of oxygen), (iii) controllable variations in the pretreatment procedure of the surface causes decisive changes in the properties of the excited states of the aromatic hydrocarbons, allowing studies to be made as a function of the surface activity, and (iv) several adsorbed aromatic hydrocarbons give intense phosphorescence even at room temperature[3, 4] so that the decay of the triplet–triplet absorption and of the phosphorescence often can be measured for the same samples. In addition, γ-alumina gives no detectable transient absorption with our apparatus.

In this paper we report transient spectra in the range 330-1000 nm which we were able to unambiguously assign to the triplet–triplet absorption spectra of aromatic molecules adsorbed on γ-alumina. This is the first time that absorption spectra of excited states of chemisorbed species have been obtained from highly reflecting catalytic surfaces. Since these spectra arise from excited states of adsorbed molecules at a gas–solid interface with coverages of far less than a monolayer, variations in the spectra and decay are attributable to the interactions of the molecule with the surface. It appears likely, therefore, that application of this method should yield results capable of bringing a new insight and understanding of adsorbed species as well as a new means for probing their photochemistry.

EXPERIMENTAL

SAMPLE PREPARATION

γ-alumina (Fa, Merck, active, neutral) with a specific surface area of 98 m² g⁻¹ was used as the adsorbent and was heated for 4 days under high vacuum (5×10^{-5} Torr) at the specified temperature (T_a). After this pretreatment of the surface typically 0.2-1 mg of the compound per g of γ-alumina was adsorbed onto the surface by slow sublimation under vacuum to give a coverage of far less than a monolayer. The long pretreatment procedure and the slow sublimation ensures a high reproducibility of the adsorbed system. The adsorption technique is described in detail in ref. (5). Naphthalene, biphenyl, triphenylene and pyrene were zone refined and used without further purification. Phenanthrene was purified as described in ref. (15).

APPARATUS

A schematic diagram of the apparatus is given in fig. 1. The sample contained in an evacuated quartz spectroscopic cell is continuously irradiated with light from a quartz iodine lamp (60 W). The diffuse reflected light was collected and dispersed by a Hilger and Watts D285 prism monochromator, typically the bandpass was 3-9 nm. Transient spectra were measured point by point at wavelength intervals no greater than 5 nm. The photomultipliers used were an EMI 6255 (S-13) for detection in the ultraviolet and visible regions and an RCA 4832 photomultiplier which could be used for the complete range investigated, i.e. between 330 and 1000 nm. Excitation was either with a ca. 20 mJ, ca. 20 nanosecond laser pulse at 265 nm (J, K Lasers, Nd frequency quadrupled) or with a xenon flash lamp with an f.w.h.m. of ca. 3 ms and ca. 60 J output. The xenon flash lamp which was used for investigation of all the longer-lived transients was filtered so that excitation of the sample covered the range 300-380 nm. The transient absorption signal observed immediately following the flash excitation was usually 3-8% of the diffuse reflected light. In order to record the absorbed signal effectively, a current offset was applied to the output of the photomultiplier. The signal was stored in either a transient recorder Datalab D921 or in a Northern Instruments multichannel analyser used in MCS mode. The MCA was used for storing and accumulating the transients with longer lifetimes.

R. W. KESSLER AND F. WILKINSON

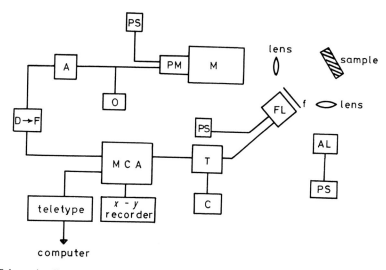

FIG. 1.—Schematic diagram of the apparatus. A = amplifier, AL = analysing light, C = counter, D → F = d.c. to frequency converter, f = filter, FL = flash lamp, M = monochromator, O = current offset, PM = photomultiplier, PS = power supply and T = trigger.

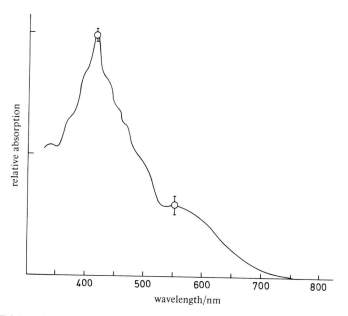

FIG. 2.—Triplet–triplet absorption spectrum of naphthalene chemisorbed on γ-alumina 0.15 s after excitation. ($T_a = 500\,°C$, 0.8 mg, naphthalene per g γ-alumina, $\tau_1 = 0.55$ s).

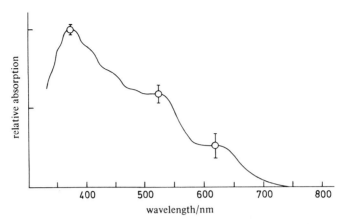

FIG. 3.—Triplet–triplet absorption spectrum of biphenyl chemisorbed on γ-alumina 0.15 s after excitation. ($T_a = 500\ °C$, 0.8 mg biphenyl per g γ-alumina, $\tau_1 = 1.0$ s).

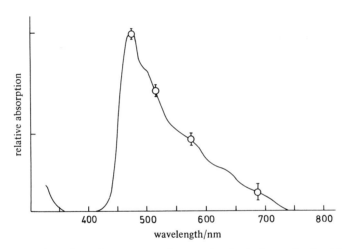

FIG. 4.—Triplet–triplet absorption spectrum of pyrene chemisorbed on γ-alumina 0.1 s after excitation. ($T_a = 500\ °C$, 0.67 mg pyrene per g γ-alumina, $\tau_1 = 0.4$ s).

The noise to signal (N/S) ratio was usually *ca.* 10% and in order to improve the S/N ratio for decay measurements of the longer-lived transients the transient absorption data were accumulated (60-120 flashes).

RESULTS

The emission spectra observed were similar to those observed by Oelkrug *et al.* [3-5] The triplet–triplet absorption spectra of naphthalene, biphenyl, pyrene, triphenylene and phenanthrene adsorbed on γ-alumina, activated at the temperature, T_a, are shown in fig. 2-6. The error bars on the spectra correspond to the mean error derived from repeated measurements at that wavelength. Whenever necessary the triplet–triplet absorption spectra were corrected for overlapping phosphorescence.

R. W. KESSLER AND F. WILKINSON

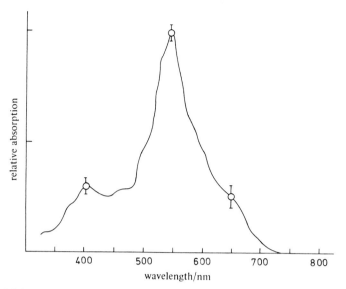

FIG. 5.—Triplet–triplet absorption spectrum of triphenylene chemisorbed on γ-alumina 0.2 s after excitation. ($T_a = 500\,^{\circ}$C, 0.8 mg triphenylene per g γ-alumina, $\tau_1 = 11$ s).

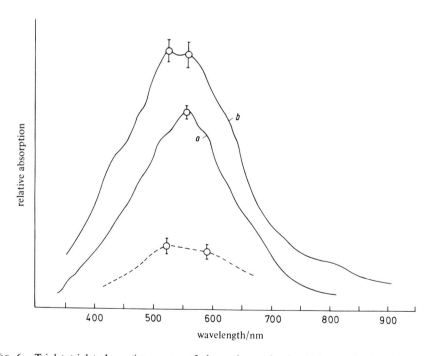

FIG. 6.—Triplet–triplet absorption spectra of phenanthrene chemisorbed on γ-alumina (0.5 mg phenanthrene per g γ-alumina). Continuous lines show spectra recorded 0.15 s after excitation with $T_a = (a)$ 500 and (b) 660 °C. The broken line spectrum is with $T_a = 500\,^{\circ}$C, 0.8 s after excitation. (With $T_a = 500\,^{\circ}$C, $\tau_1 = 0.8$ s and for perdeuterated phenanthrene $\tau_1 = 1.5$ s).

Detailed kinetic analysis was possible when many traces were accumulated and this showed that the slopes of first-order plots of the logarithm of either the phosphorescence intensity or the transient intensity of absorption as a function of time invariably decreased with increasing time to give a limiting value. As the traces are non-exponential, only the limiting lifetime value is given (τ_1). In each case, except for pyrene which phosphoresces in a spectral region where our apparatus has a low sensitivity, the triplet–triplet absorption decay time and the phosphorescence lifetime were found to be specific for each adsorbed molecule and were equal within experimental error. No significant dependence on analysing wavelength could be detected for any sample. In the case of phenanthrene the coincidence of triplet–triplet-absorption decay and phosphorescence decay was also found with the perdeuterated hydrocarbon. A triplet–triplet-absorption decay trace and a phosphorescence decay trace of phenanthrene adsorbed on γ-alumina is shown in fig. 7. A full discussion of diffuse reflectance of inhomogeneously absorbing samples will be published elsewhere.[16]

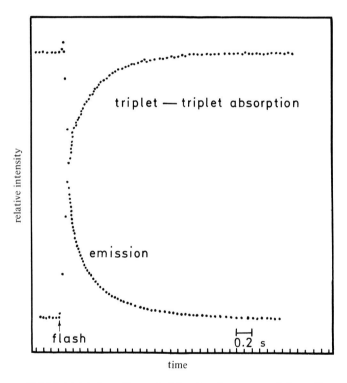

Fig. 7.—Decay of triplet–triplet absorption and of the phosphorescence of phenanthrene chemisorbed on γ-alumina ($T_a = 500\,°C$, 0.5 mg phenanthrene per g γ-alumina).

Many experiments were made to exclude the possibility of irreversible photochemistry. For samples with highly activated γ-alumina many flashes (ca. 600-1000) were possible before one could detect a growing-in of a new transient, e.g. in the case of phenanthrene at 600 nm. However, with phenanthrene adsorbed on non-activated, γ-alumina and therefore with the surface covered with water, changes arose rapidly after exposure to only 5 laser excitation pulses and the samples became coloured.

DISCUSSION

On the basis of the similar kinetic behaviour observed for the decay of the transient absorption and of the phosphorescence for four out of five of the aromatic hydrocarbons investigated, we confidently assign the spectra shown in fig. 2-6 to triplet–triplet absorption spectra of the adsorbed hydrocarbons. The lack of vibrational fine structure was established by narrowing the monochromator slits to increase the spectral resolution when no further fine structure was detected. However, it should be recognised that the noise/signal ratio was much greater for the measurements made with the higher spectral resolution. Apart from being broad the transient absorption spectra show many differences from the triplet–triplet absorption spectra of the free molecules; for example, as reported for these same aromatic hydrocarbons in rigid media [e.g. see ref. (17)].

The phosphorescence spectra of the adsorbed hydrocarbons[3, 4, 16] show that the lowest triplet state is similar in nature and energy to that observed in rigid glasses. The relatively long lifetimes and the intense phosphorescence of the adsorbed species on highly activated surfaces indicate low mobility. Taking into account the detailed investigations concerning adsorption by aromatic hydrocarbons on γ-alumina made by Oelkrug et al.[5] it seems unlikely that, at the low coverages used in this work, there will be excimer formation and thus other reasons have to be found to explain the observed spectral differences.

The multi-exponential decay of the triplet state of the adsorbed hydrocarbons indicates a variety of environment (sites) for the adsorbed hydrocarbons. Naphthalene and biphenylene give almost single exponential decay while with triphenylene the decay traces are multi-exponential indicating very inhomogeneous adsorption. This suggests the larger the molecule the less probable each molecule is to find a similar environment. This is not unexpected since even the number of surface atoms or groups with which each adsorbate interacts will increase with its size. This kind of heterogeneity of adsorption probably contributes to the broadness of the spectra observed.

Adsorption of a molecule on a surface results in a lowering of the symmetry which can lift degeneracies of electronic states as well as modifying optical selection rules.[18] It is therefore necessary to consider whether the observed spectra can be attributed to shifts in band positions and to changes in the intensities of transitions from those found in the free molecules. The lack of information on and agreement concerning the assignments of the triplet–triplet absorption bands for the free molecule makes such a treatment difficult. A very good discussion concerning triplet–triplet absorption assignments which demonstrates the alarming diversity of theoretical estimates of states likely to give rise to weak triplet–triplet transitions in the 'free' aromatic hydrocarbons is given in ref. (17).

An alternative approach recognizes that the surface sites on a γ-alumina surface can act as either electron donors or acceptors.[19] Indeed, the occurrence of singlet charge-transfer states of aromatic hydrocarbons adsorbed on γ-alumina is firmly established from fluorescence studies.[3-5] Transitions from the lowest triplet state (T_1) which is essentially molecular in nature to higher triplet charge-transfer states of the adsorbent–adsorbate systems may account for the differences in the triplet–triplet spectrum observed for the adsorbed species. Before discussing this possibility we detail below the differences which need explaining for each aromatic hydrocarbon investigated.

NAPHTHALENE

Triplet–triplet absorption spectra of naphthalene have been reported by many authors following the pioneering work of McClure,[20] Craig and Ross[21] and Porter and Windsor.[22] Meyer $et\ al.$[23] determined a detailed spectrum with band positions, λ, extinction coefficients, ε, and assignments as given below:

$$\lambda = 629 \text{ nm } (14430 \text{ cm}^{-1}), \varepsilon = 250 \text{ dm}^3 \text{ mol}^{-1} \text{ cm}^{-1}, {}^3A_g^+ \leftarrow {}^3B_{2u}^+;$$
$$\lambda = 489 \text{ nm } (20300 \text{ cm}^{-1}), \varepsilon = 300 \text{ dm}^3 \text{ mol}^{-1} \text{ cm}^{-1}, {}^3B_{1u}^+ \leftarrow {}^3B_{2u}^+;$$
$$\lambda = 414 \text{ nm } (24150 \text{ cm}^{-1}), \varepsilon = 33000 \text{ dm}^3 \text{ mol}^{-1} \text{ cm}^{-1}, {}^3B_{1g}^- \leftarrow {}^3B_{2u}^+;$$
$$\lambda = 400 \text{ nm } (25000 \text{ cm}^{-1}), \varepsilon = 3000 \text{ dm}^3 \text{ mol}^{-1} \text{ cm}^{-1}, {}^3A_g^+ \leftarrow {}^3B_{2u}^+.$$

A further two bands were reported outside the region we investigated at 275 and 236 nm.[23] The assignments given here agree fairly well with the theoretical estimates of De Groot and Hoytink[24] and Orloff[25] although the calculated energies are different [see ref. (17)].

In the case of adsorbed naphthalene we find a strong triplet–triplet absorption band at 418 nm (23900 cm^{-1}), a shoulder at $ca.$ 500 nm and a broad band at $ca.$ 550 nm ($ca.$ 18000 cm^{-1}). Since the lowest triplet state ($T_1 \equiv {}^3B_{2u}^+$) is red shifted by $ca.$ 1000 cm^{-1}, assignment of the band at 418 nm as being due to the same transition as that observed for the free molecule which is almost universally agreed to be due to the transition to the ${}^3B_{1g}^-$ state suggests that this state is red shifted by a similar amount to that observed for T_1, $i.e.$ $ca.$ 1000 cm^{-1}. The major observed difference is the loss of vibrational structure. However, to explain the broader band observed at $ca.$ 550 nm one would need to consider that the upper triplet state for the transition was red shifted by > 3000 cm^{-1}. In addition, the relative intensities of the two bands in the different environments are very different, being $> 10^2$ in the case of the 'free' molecule but only $ca.$ 3 for the adsorbed species. Such large changes for transitions to some but not other excited triplet states seems unlikely.

BIPHENYL

Many of the same workers who studied naphthalene have also studied triplet–triplet spectra in the case of biphenyl.[22, 23] In addition, more recent work has been reported by Ota $et\ al.$[26] The band maxima and oscillator strengths, f, reported by Meyer $et\ al.$[23] are given here with their assignments:

$$\lambda = 370 \text{ nm } (27000 \text{ cm}^{-1}), f = 0.5, {}^3A_g^- \leftarrow {}^3B_{1u}^+;$$

$$\lambda = 484 \text{ nm } (20650 \text{ cm}^{-1}), f = 0.01, {}^3B_{3g}^- \leftarrow {}^3B_{1u}^+;$$

$$\lambda = 685 \text{ nm } (14600 \text{ cm}^{-1}), f = 0.01, {}^3A_g^+ \leftarrow {}^3B_{1u}^+.$$

Biphenyl adsorbed on γ-alumina shows a very broad triplet–triplet absorption band at $ca.$ 370 nm ($ca.$ 27000 cm^1) and at least two other bands at $ca.$ 530 and $ca.$ 640 nm (see fig. 3). As in the case of naphthalene the strongest transition of the adsorbed species in the same spectral region as the strongest transition in the free molecule and can be assigned to the same transition. However, the relative intensities and the positions of the weaker transitions do not correspond very well. To assign them to the two weaker transitions given above requires considerable intensification over that of the free molecule transitions ($>$ tenfold) and considerable spectral shifts to the red and to the blue.

PYRENE

In the region of interest there are many transitions in the triplet–triplet absorption spectrum of pyrene.[17, 27, 28] Reported values include

$$\lambda = 370 \text{ nm } (27000 \text{ cm}^{-1}), \varepsilon = 5500 \text{ dm}^3 \text{ mol}^{-1} \text{ cm}^{-1}, {}^3B_{1g}^- \leftarrow {}^3B_{2u}^+;$$

$$\lambda = 413 \text{ nm } (24200 \text{ cm}^{-1}), \varepsilon = 48000 \text{ dm}^3 \text{ mol}^{-1} \text{ cm}^{-1}, {}^3A_g^- \leftarrow {}^3B_{2u}^+;$$

$\lambda = 520$ nm (19 200 cm^{-1}), $\varepsilon = 12\,500$ dm^3 mol^{-1} cm^{-1}, $^3B_{1g}^- \leftarrow {}^3B_{2u}^+$;

$\lambda = 775$ nm (12 900 cm^{-1}), $\varepsilon \approx 126$ dm^3 mol^{-1} cm^{-1}, $^3B_{1g}^+ \leftarrow {}^3B_{2u}^+$;

$\lambda = 870$ nm (11 500 cm^{-1}), $\varepsilon \approx 165$ dm^3 mol^{-1} cm^{-1}, $^3A_g^+ \leftarrow {}^3B_{2u}^+$.

The assignments of the weaker transitions must be regarded as speculative.

As mentioned earlier in the introduction Piculo and Sutherland[12] have observed a triplet–triplet spectrum for pyrene adsorbed on Vycor glass. They found no transient absorption from 380 to 460 nm and band maxima (with relative intensities) as follows: $\lambda = 480$ nm (0.35), $\lambda = 522$ nm (1.0) and $\lambda = 625$ nm (0.4). They point out their triplet–triplet absorption spectrum is substantially red shifted over that of the 'free' molecule (*ca.* 100 nm) but the singlet–singlet transitions arising from the ground state of adsorbed pyrene are only shifted by $\leqslant 2$ nm from 'free' molecule spectra in hydrocarbon solvents.[12]

In our experiments (see fig. 4) we observe almost no absorption in the region 360–410 nm, but there is a main peak at 475 nm (21 050 cm^{-1}), a shoulder at *ca.* 500 nm (20 000 cm^{-1}) and indications of broad bands at *ca.* 575 nm (17 400 cm^{-1}) and *ca.* 640 nm (15 600 cm^{-1}). If the main band is assigned to the same transition as in the free molecule this would indicate a considerable shift but not so large as in the case of the triplet–triplet spectrum on Vycor glass.[12] We are at present investigating triplet–triplet spectra of pyrene adsorbed on other active surfaces and this will be discussed in detail elsewhere.[16] However, as far as this work is concerned it is clear that pyrene is different from naphthalene and biphenyl in that there is no longer a correspondence of the main peak for adsorbed and 'free' pyrene.

TRIPHENYLENE

Only a few triplet–triplet absorption spectra of triphenylene have been published[17] and assignments are also rather scarce. There are at least two bands for free triphenylene at $\lambda = 415$ nm (23 700 cm^{-1}) and $\lambda \approx 665$ nm (*ca.* 15 000 cm^{-1}). Noucli[29] assigned these as $^3C_b \leftarrow {}^3L_a$ and $^3K_a \leftarrow {}^3L_a$ in Platts' Nomenclature (*i.e.* both $^3E' \leftarrow {}^3A_2$ in D_{3h} symmetry).

In the case of the adsorbed species there is a strong triplet–triplet absorption band at 550 nm, a shoulder at *ca.* 650 nm and a band at *ca.* 410 nm. These weaker transitions correspond fairly well to the position of triplet–triplet absorption for the free molecule. However, with free triphenylene no band has been previously observed or theoretically predicted at around 550 nm and we assign this very strong band as a charge-transfer transition.

PHENANTHRENE

Triplet–triplet absorption of 'free' phenanthrene[25, 27] consists of a structured band at 490 nm (20 400 cm^{-1}, $^3A_1^- \leftarrow {}^3B_2^+$) and a broad band at *ca.* 350 nm (28 600 cm^{-1}, $^3B_2^- \leftarrow {}^3B_2^+$). By adsorption of phenanthrene on γ-alumina pre-treated at $T_a = 500$ °C we obtain a maximum triplet–triplet absorption at *ca.* 550 nm (18 200 cm^{-1}) which could be the $^3A_1^- \leftarrow {}^3B_2^+$ transition red shifted *ca.* 2000 cm^{-1}. This band is extraordinarily broad which may indicate that it is composed of more than one broad transition.

A considerable difference in the spectrum is observed on increasing the pretreatment temperature T_a to 650 °C. In this case we observe two bands in the triplet–triplet spectrum, one at 550 nm and the other at 510 nm (19 600 cm^{-1}). It has been shown that an increase in the pretreatment temperature causes an increase in the Lewis acidity of the surface.[3] Singlet charge-transfer bands of aromatic hydrocarbons with ionisation potentials in the range 7.4–8.1 eV show red shifts from 2000-6000 cm^{-1} in comparison with the free molecule.[3-5,30] In zero-order approximation[31] singlet and triplet charge-transfer states have the same energy; we can therefore estimate the variation expected in the energy of a triplet charge-transfer state as the electron affinity of the surface is increased as a result of a higher pretreatment temperature. Fig. 8 shows schematically the variation expected for free molecule and charge-transfer triplet states with

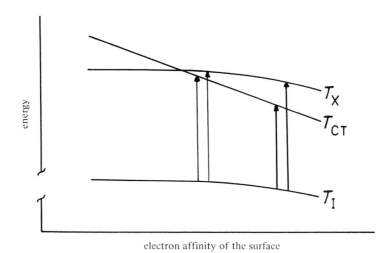

FIG. 8.—Illustration of how the dependence of triplet energy levels on the electron affinity of a surface and therefore on the pretreatment temperature can lead to spectral differences.

increasing electron affinity of the surface and how this could account for the appearance of two bands at the higher pretreatment temperature, although it has to be recognised that this is probably an oversimplification, since no account has been taken of possible changes in Franck–Condon factors for these transitions. Nevertheless, the sensitivity of the triplet–triplet spectra to the surface activity is supporting evidence that some of the observed spectral changes are due to transitions involving charge-transfer states.

ENERGIES OF TRIPLET STATES OF CHEMISORBED SPECIES

The energies of the lowest triplet states are obtained from phosphorescence spectra and these, combined with the observed triplet–triplet spectra, give estimates of the energies of the higher triplet states (based on band maxima). Fig. 9 plots these energies against the ionisation potential of the aromatic hydrocarbons. Lines with slopes of unity can be drawn through several of the states for the different adsorbed hydrocarbons. Recently Oelkrug et al.[30] have shown that the transition energy of the singlet charge-transfer states of aromatic hydrocarbons adsorbed on γ-alumina plotted against ionisation potential give a slope of 0.93, i.e. close to unity, and fig. 9 is again supporting evidence that triplet–triplet absorption spectra of these adsorbed aromatic hydrocarbons differ from those of free molecules, mainly as a result of transitions to states which are predominantly charge-transfer in nature. However, from the data available at present, it is impossible to be certain which transitions are localised on the hydrocarbon and why, in the case of pyrene, no localised transitions are observed below 400 nm where the 'free' molecule has a very strong transition.

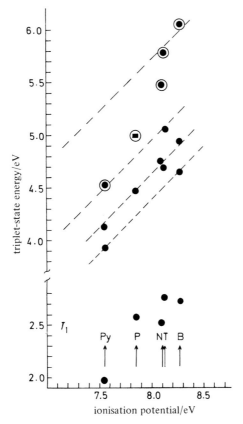

Fig. 9.—Triplet energies of chemisorbed aromatic hydrocarbons plotted against the ionisation potentials[32] of these donors. Triplet-state energies of chemisorbed molecules on γ-alumina ●, $T_a = 500$ °C and ■, $T_a = 650$ °C. ◉, ▣ indicate approximate coincidence with states arising from strong transitions of the free molecules. Dotted lines have been drawn with slopes of unity. B = biphenyl, N = naphthalene, T = triphenylene, P = phenanthrene and Py = pyrene.

CONCLUSIONS

A technique of diffuse reflectance flash photolysis has been developed which has enabled triplet–triplet absorption spectra of adsorbed species to be unambiguously detected. Many of the differences of these spectra from those observed for the same molecules in the gas phase, or in dilute homogeneous solutions, have been attributed to the presence of charge-transfer states of the adsorbent–adsorbate systems and the occurrence of transitions to these triplet charge-transfer states. Further work is needed which should allow studies of the mobility of triplet states on surfaces and many other applications are possible, *e.g.* energy transfer, electron transfer, radical formation, *etc.* It has been demonstrated that this technique is capable of direct observation of primary photophysical and photochemical processes of adsorbed species on highly reflecting surfaces of catalytic interest.

We thank the S.R.C. and the British Council for grants which have supported this work.

320   TRIPLET–TRIPLET ABSORPTION SPECTROSCOPY

1   M. Formenti and S. J. Teichner, *Catalysis* (Specialist Periodical Report, The Chemical Society, London, 1978), vol. 2, pp. 87–106.
2   F. Steinbach and R. Harborth, *Faraday Discuss. Chem. Soc.*, 1974, **58**, 143.
3   D. Oelkrug, M. Plauschinat and R. W. Kessler, *J. Lumin.*, 1979, **18/19**, 434.
4   D. Oelkrug, H. Erbse and M. Plauschinat, *Z. Phys. Chem. N.F.*, 1975, **96**, 283.
5   D. Oelkrug, M. Radjaipour and H. Erbse, *Z. Phys. Chem. N.F.*, 1974, **88**, 23.
6   V. A. Fenin, V. A. Shvets and V. B. Kazanski, *Kinet. Katal.*, 1978, **19**, 1289.
7   G. N. Asmdov and O. V. Krylov, *Kinet. Katal.*, 1978, **19**, 1004.
8   G. N. Asmdov and O. V. Krylov, *Kinet. Katal.*, 1978, **19**, 1208.
9   H. G. Hecht and J. L. Jensen, *J. Photochem.*, 1978, **9**, 33.
10 H. Gerischer, *Faraday Discuss. Chem. Soc.*, 1974, **58**, 219.
11 H. Moesta, *Faraday Discuss. Chem. Soc.*, 1974, **58**, 244.
12 P. L. Piculo and J. W. Sutherland, *J. Am. Chem. Soc.*, 1979, **101**, 3123.
13 J. B. Peri, *J. Phys. Chem.*, 1965, **69**, 220.
14 M. Zamora and A. Cordoba, *J. Phys. Chem.*, 1978, **82**, 584.
15 E. C. Kooyman and E. Farenhurst, *Trans. Faraday Soc.*, 1953, **43**, 58.
16 D. Oelkrug, M. Plauschinat, R. W. Kessler and F. Wilkinson, to be published.
17 H. Labhardt and W. Heinzelmann, *Organic Molecular Photophysics*, ed. J. B. Birks (J. Wiley, Chichester, 1973), vol. 1, p. 297 and references therein.
18 A. M. Bradshaw, *Z. Phys. Chem. N.F.*, 1978, **112**, 33.
19 G. M. Muha, *J. Catal.*, 1979, **58**, 470.
20 D. S. McClure, *J. Chem. Phys.*, 1951, **19**, 670.
20 D. P. Craig and I. G. Ross, *J. Chem. Soc.*, 1954, 1589.
22 G. Porter and M. Windsor, *Proc. R. Soc. London, Ser. A*, 1958, **245**, 235.
23 Y. H. Meyer, R. Astier and J. M. Leclerq, *Chem. Phys. Lett.*, 1969, **4**, 587.
24 R. L. De Groot and G. J. Hoytink, *J. Chem. Phys.*, 1967, **46**, 4523.
25 M. K. Orloff, *J. Chem. Phys.*, 1967, **47**, 235.
26 K. Ota, K. Murofushi, T. Hoshi, E. Shibuya and Y. Yoshino, *Z. Phys. Chem. N.F.*, 1979, **104**, 181.
27 M. Windsor and J. R. Novak, in *The Triplet State*, Proceedings of an International Symposium held at the American University of Beirut, Lebanon, ed. A. B. Zahlan (Cambridge University Press, 1967), p. 229.
28 D. Lavalette, *J. Chim. Phys.*, 1969, **66**, 1845.
29 G. Noucli, *J. Chim. Phys.*, 1969, **66**, 555.
30 D. Oelkrug and M. Radjaipour, *Z. Phys. Chem. N.F.*, to be published.
31 H. Beens and A. Weller, in *Molecular Luminescence*, ed. E. C. Lim (W. A. Benjamin, New York, 1969), p. 203.
32 J. B. Birks and M. A. Slifkin, *Nature (London)*, 1961, **191**, 761; G. Briegleb and J. Czekalla, *Z. Elektrochem.*, 1959, **63**, 6.

(PAPER 0/138)

Contribution from

GEORGE TRUSCOTT

and

EDWARD LAND

Keele University, UK

George Truscott

Born 1939 in South Wales. Obtained his BSc and PhD from the University of Wales in 1961 and 1964 and was awarded the DSc in 1983. Elected a Fellow of the Royal Society of Chemistry in 1974 and Fellow of the Royal Society of Edinburgh in 1988. In 1993, was the Lee Visiting Fellow in the Sciences at Christ Church College, Oxford University.

Has written two and edited three textbooks and published about 200 papers. Amongst his academic distinctions, was the founding President of the European Society of Photobiology, and, for 5 years, the Senior Editor of the *Journal of Photochemistry and Photobiology B (Biology)*. For about 10 years, was also Editor of *Photochemistry and Photobiology*. In 1999 was the Medallist of the European Society of Photobiology.

Has worked with many granting agencies, charities and major industries both in the UK and overseas including EPSRC, MRC, Cancer Research Campaign, the Wellcome Trust, the Psoriasis Association, the Leverhulme Trust, the Parkinson's Disease Society, American Institute for Cancer Research, the Association for International Cancer Research (UK), World Cancer Research Fund (UK), NATO, EC, ICI, Hoffman–La Roche (Basle), Givaudan Roure (Geneva), and Wella (Darmstadt, Germany).

Was Head of Department of Chemistry at Paisley University for 16 years and is currently Research Professor of Medicinal Chemistry at Keele University having earlier been Head of Department of Chemistry and Research Dean of Sciences at Keele.

Current major research interests concern molecular mechanisms involved with the beneficial and deleterious effects of dietary carotenoids (Leverhulme Trust) and mechanisms of hair bleaching and damage (Wella, Germany).

Edward Land

Born in 1937 in Birmingham. A graduate of Sheffield University, received his PhD in 1962, supervised by George Porter for "Flash Photolysis Studies of Aromatic Free Radicals Derived from Phenols and Anilines". From 1963 to 1997, at the Paterson Institute for Cancer Research, Manchester, he extended his studies of the interaction of radiation with biological materials, using flash photolysis and pulse radiolysis. His main contributions have been in the fields of:

(1) measuring photophysical and photochemical parameters of organic molecules in condensed media,
(2) primary molecular mechanisms involved in electron transport in mitochondria and chloroplasts,
(3) electron migration within peptides and proteins,
(4) aspects of the photochemotherapy of skin diseases by furocoumarins and of solid tumours by porphyrins,
(5) the chemistry of melanogenesis.

In 1998, became a Senior Research Fellow in Chemistry at Keele University and is currently an Honorary Professor in the School of Chemistry and Physics where his interests include the action of carotenoid anti-oxidants, and the early steps of melanogenesis.

Was a visiting scientist in the Chemistry Division at Argonne National Laboratory, Illinois in 1970, Professor Associe in the Department of Biophysique, Museum National d'Histoire Naturelle, Paris (1978–1979), and an Associate Editor of *Photochemistry and Photobiology* (1974–1976), and of the *International Journal of Radiation Biology* (1983–1988).

Was awarded the Weiss Medal of the Association for Radiation Research in 1982 and has authored or co-authored over 250 publications including two textbooks.

Edward Land and George Truscott on
"Primary Photochemical Processes in Aromatic Molecules. Part 3 – Absorption Spectra of Benzyl, Anilino, Phenoxy, and Related Free Radicals"

G. Porter and F. J. Wright

Trans. Faraday Soc. **51**, 1469–1474 (1955)

The discovery of the UV absorption of the relatively simple isoelectronic benzyl ($\phi CH_2^\bullet$), anilino ($\phi NH^\bullet$) and phenoxyl ($\phi O^\bullet$) radicals in the gas phase, followed 2 years later by the detection by Porter and Windsor of benzyl and the phenoxyl-related benzosemiquinone radical in liquid paraffin solution,[1] has been followed by many flash spectroscopic and kinetic studies of related radicals.

From 1959 Porter used flash photolysis to study aromatic radicals derived from phenols and anilines in fluid solution. The UV and visible bands of a number of phenoxyl and anilino radicals were identified and alterations of the acidity of their aqueous solutions resulted in changes in absorption spectra corresponding to protonation, phenoxyls being very strong acids (pK ~ −1) and anilinos much weaker (pK ~ 7).[2-4] 2,4,6-Tri-tertiary-butyl phenol, and the corresponding aniline, lead on photolysis to stable phenoxyl and anilinos with UV-VIS spectra very similar to their unsubstituted counterparts. The ESR spectra[5] of those stable radicals provided unequivocal assignments of their structures, previously deduced only indirectly from flash spectroscopic studies of series of closely-related molecules. Flash photolysis of toluene vapour enabled Porter and Ward[6] to identify the weak, sharp-banded visible spectrum of the benzyl radical, the maximum being at 447.7 nm, Porter and Strachan[7] having previously observed the visible absorption of benzyl at low temperatures in an isopentane + methyl cyclohexane rigid glass.

About a decade after flash photolysis was invented, the principle was applied to high energy radiation (pulse radiolysis), and this technique has also been used to study benzyl, phenoxyl and anilino radicals and derivatives. In 1960, the benzyl radical was, in fact, one of the first organic radicals to be studied using pulse radiolysis, by McCarthy and MacLachlan,[8] and their paper was the subject of a "Coffee and Kinetics" talk at Sheffield University by Porter in the same year (coffee price 1*d* per cup, served in 250 ml glass beakers — before the days of COSHH!). In radiolysis, unlike photolysis, the solvent mainly absorbs the radiation, so in the case of aqueous solutions, the most reactive primary products are predominantly $OH^\bullet$ radicals and the hydrated electron e^-_{aq}. For aqueous phenol solutions, the phenoxyl radical can be formed by initial addition of $OH^\bullet$ to phenol, leading to dihydroxycyclohexadienyl radicals which spontaneously eliminate water to give $\phi O^\bullet$, the driving force behind this process being the higher resonance stabilisation of phenoxyl.[9] The anilino radical can also be formed via $OH^\bullet$ addition to aniline, followed by H_2O elimination.[10]

There are many biologically important radicals related to phenoxyl and anilino. The common amino acids tyrosine (TyrOH) and tryptophan (TrpH) give rise, on either photolytic or radiolytic one-electron oxidation, to the substituted phenoxyl TyrO•, and the substituted anilino Trp•. The enzyme ribonucleotide reductase, which catalyses the reduction of ribonucleotides to their corresponding deoxynucleotides, contains a tyrosyl free radical TyrO•, which is necessary for its enzyme activity.[11] As the two radicals TyrO• and Trp• have prominent and distinct absorption spectra, and differing reduction potentials, there have been many studies of electron transfer from tyrosine to Trp• in peptides and proteins (see, for example, Ref. 12) which can lead to important structural information about these molecules.

The one-electron reduction potentials of tryptophan radicals are now well known as a function of pH, and there is current interest in using this radical as a standard for the determination of carotenoid radical cation reduction potentials.[13]

Semiquinones are involved in photosynthesis, one of Porter's later scientific interests. Thus, the reduced primary electron exceptor of Photosystem II in algae and higher plants is the photosemiquinone anion.[14] Similarly ubisemiquinone anion[15] results from such electron transfer in bacterial photosynthesis.

The propensity of semiquinones to rapidly disproportionate to quinones has been used to prepare high concentrations, otherwise unobtainable, of short-lived o-quinones involved in the formation of melanin pigments.[16,17] Recent interest has focussed on (a) how the various reaction rate constants affect the balance between eumelanogenesis and phaeomelanogenesis,[18] and (b) measurements of the rate constants of the first two chemical steps of eumelanogenesis: cyclisation of dopaquinone to cyclodopa, and redox exchange between dopaquinone and cyclodopa yielding dopachrome.[19]

The application of flash photolysis to the study of phenoxyl, anilino and benzyl radicals has been of enormous value to chemists for the past four decades, The emphasis has already swung towards biological and medical chemistry with the study of amino acids, vitamin E, electron transfer in proteins and molecules involved in vision. Without doubt, medical/biological researchers will use this pioneering work of Porter for decades to come.

References

1. G. Porter and M. W. Windsor, *Nature* **180**, 187 (1957).
2. E. J. Land, G. Porter and E. Strachan, *Trans. Faraday Soc.* **57**, 1885 (1961).
3. E. J. Land and G. Porter, *Trans. Faraday Soc.* **59**, 2016 (1963).
4. E. J. Land and G. Porter, *Trans. Faraday Soc.* **59**, 2027 (1963).
5. N. M. Atherton, E. J. Land and G. Porter, *Trans. Faraday Soc.* **59**, 818 (1963).
6. G. Porter and B. Ward, *J. Chim. Physique* **61**, 1517 (1964).
7. G. Porter and E. Strachan, *Spectrochim Acta* **12**, 299 (1958).
8. R. L. McCarthy and A. McLachlan, *Trans. Faraday Soc.* **56**, 1187 (1960).

9. E. J. Land and M. Ebert, *Trans. Faraday Soc.* **63**, 1181 (1967).

10. A. Wigger, W. Grunbein, A. Henglein and E. J. Land, *Z. Naturforschg* **24b**, 1267 (1969).

11. M. Sahlin, A. Graslund, A. Ehrenberg and B.-M. Sjoberg, *J. Biol. Chem.* **257**, 366 (1982).

12. R. V. Bensasson, E. J. Land and T. G. Truscott, *Excited States & Free Radicals in Biology & Medicine* (OUP, 1993).

13. M. Burke, R. Edge, E. J. Land, D. J. McGarvey and T. G. Truscott, *FEBS Lett.* **500**, 132 (2001).

14. H. J. Van Gorkom, *Biochim. Biophys. Acta* **347**, 439 (1974).

15. L. Slooten, *Biochim. Biophys. Acta* **275**, 208 (1972).

16. C. Lambert, J. N. Chacon, M. R. Chedekel, E. J. Land, P. A. Riley, A. Thompson and T. G. Truscott, *Biochim. Biophys. Acta* **993**, 12 (1989).

17. C. Lambert, E. J. Land, P. A. Riley and T. G. Truscott, *Biochim. Biophys. Acta* **1035**, 319 (1990).

18. E. J. Land and P. A. Riley, *Pigment Cell Res.* **13**, 273 (2000).

19. E. J. Land, S. Ito, K. Wakamatsu and P. A. Riley, *Pigment Cell Res.* **16**, 487 (2003).

Contributed Article
PULSED RADIATION STUDIES OF XANTHOPHYLLS
A. Cantrell, E. J Land, T. G Truscott

Introduction

A major interest of George Porter in his later research concerned the photochemistry of plant pigments. This emphasised the roles of chlorophyll in the reaction centres. The authors often benefited from discussions of these and other systems at visits to the Royal Institution, conferences and around external examining of PhD students.

The role of carotenoids (xanthophylls are oxygen-containing carotenoids) is of particular interest to photosynthesis and vision. In photosynthesis their role includes protection of the reaction centres via singlet oxygen and triplet chlorophyll quenching and in the antenna complex via the so-called xanthophyll cycle. In vision the xanthophylls protect the macula from light-induced damage via a simple blue light filtering mechanism and probably also via quenching of active oxygen species such as oxy-radicals and singlet oxygen.

In recent years much interest has centred on the beneficial and possibly deleterious effect of using carotenoids as food supplements against diseases such as cancer and, for xanthophylls, against age-related macular degeneration, the major cause of blindness in older people in the western world.

Carotenoids are one of the most common pigments in nature and are responsible for the red/yellow colours of many plant leaves, fruits, and fish. Recently, the epidemiological work which has associated carotenoid dietary intake with reduced risk of degenerative disease has caused much debate — this preventative activity being associated with both the antioxidant effects of carotenoids including xanthophylls, and via cell signalling mechanisms. In this paper we review the quenching of singlet oxygen by xanthophylls, and report new measurements for β-cryptoxanthin. We also review the properties of xanthophyll radical cations, again including new data for β-cryptoxanthin, and comment upon the reaction of the radical cations with water soluble reductants — so-called "repair reactions".

The structures of several xanthophylls are given in Fig. 1 together, for comparison, with that of β-carotene and lycopene, two hydrocarbon carotenoids much used as food colourants and claimed to have both beneficial and deleterious effects on humans.

Experimental

The major techniques used are laser flash photolysis and pulse radiolysis (the details of the equipment used have been described previously.[1,2] The carotenoids

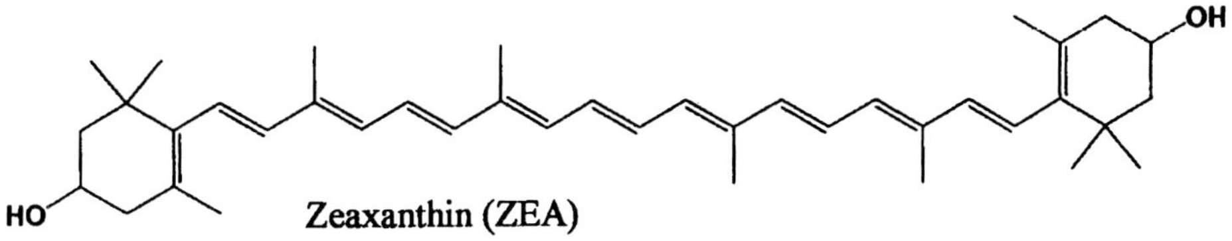

Zeaxanthin (ZEA)

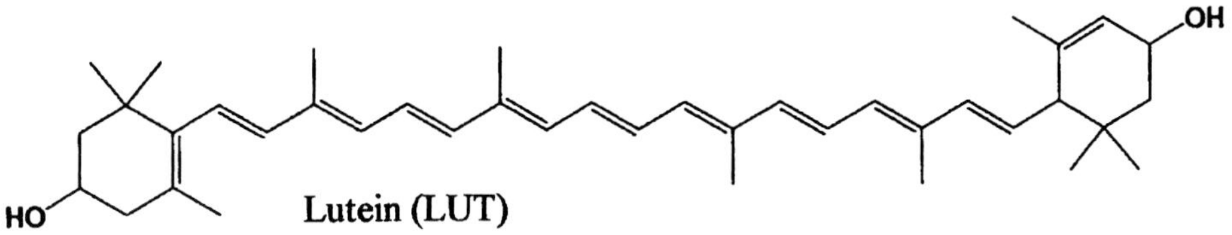

Lutein (LUT)

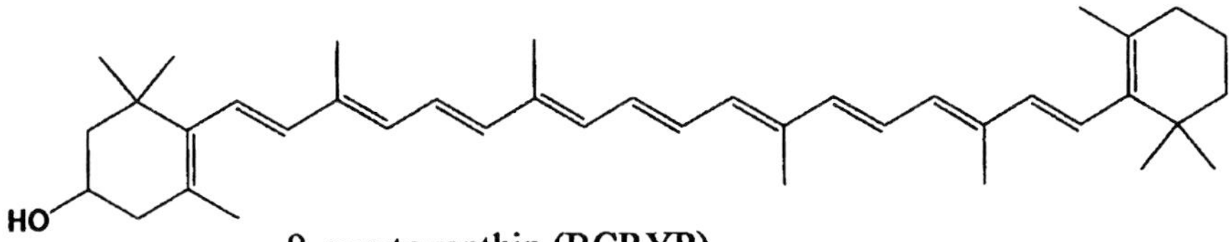

β-cryptoxanthin (BCRYP)

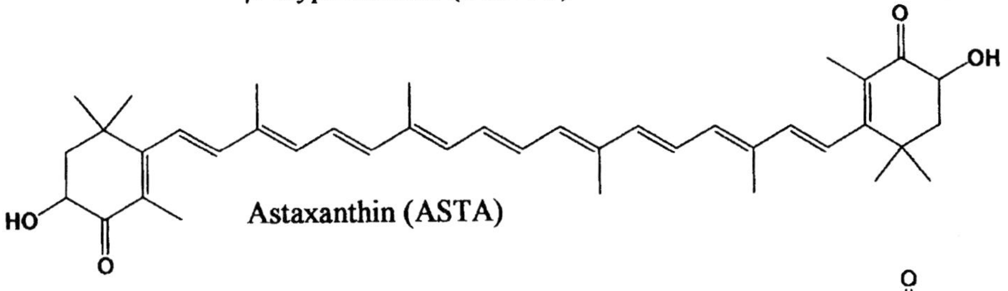

Astaxanthin (ASTA)

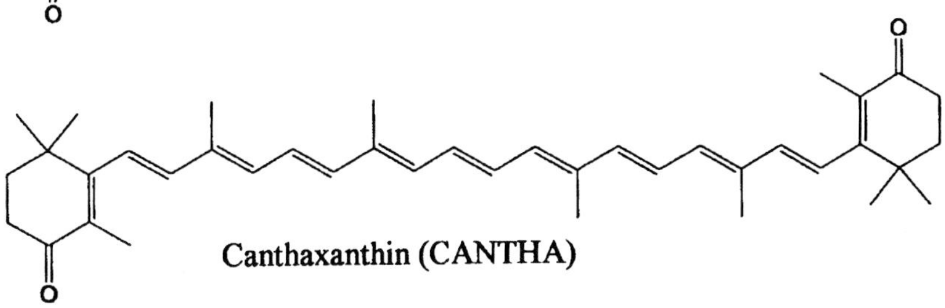

Canthaxanthin (CANTHA)

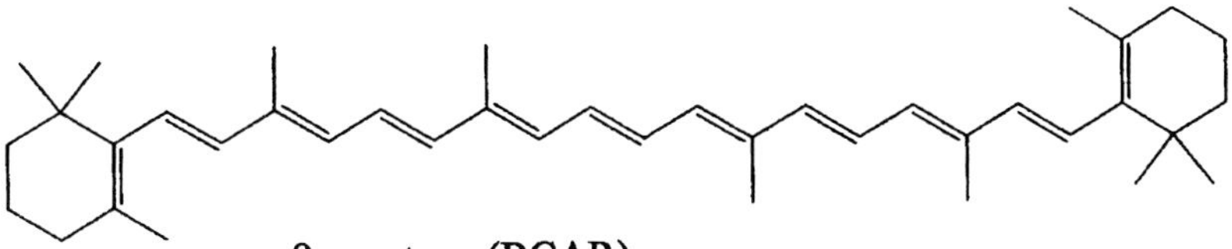

β-carotene (BCAR)

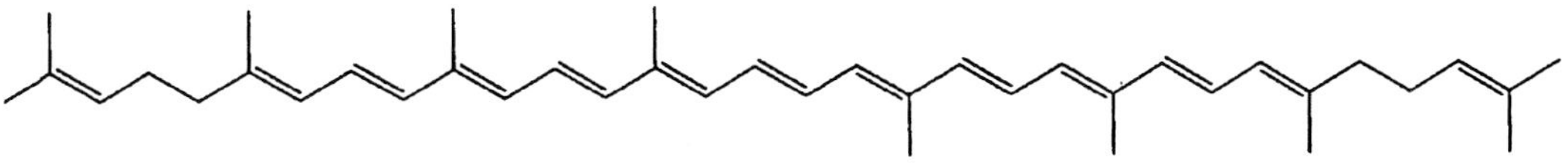

Lycopene (LYC)

Fig. 1. Structures of the xanthophylls and hydrocarbon carotenoids, β-carotene and lycopene.

were all supplied by Hoffmann–La Roche and were used without further purification. All solvents and other materials used were of the highest purity commercially available. The method for the preparation of the liposomes has also been presented previously.[3]

Results and Discussion

Singlet Oxygen Studies

The efficiency of singlet oxygen quenching of the carotenoids in organic solvents such as benzene has been shown to be related to the number of conjugated double bonds, such that for example, dodecapreno-β-carotene (a synthetic carotenoid with 15 conjugated double bonds) is a much more effective quencher than septapreno-β-carotene with only 8 conjugated double bonds.[4] The solvent systems described in these papers are far removed from the real *in vivo* situation and little is known of the singlet oxygen quenching ability of carotenoids in lipid membranes. Incorporation of carotenoids into a membrane environment may lead to very different behaviour.

In recent work[5] we have reported on the capacity of 4 dietary xanthophylls (ZEA, ASTA, CANTHA and LUT) and BCAR and LYC to quench singlet oxygen in a model cell membrane, where singlet oxygen has been generated in either the aqueous or lipid phase. We have used the lipid, dipalmitoyl phosphatidylcholine (a C16 chain length lipid, DPPC) to generate unilamellar liposomes that contain each of the various carotenoids. We now report the extension of these studies to BCRYP, an asymmetric oxy-carotenoid with only one –OH group (see Fig. 1). A major objective is to compare the singlet oxygen quenching abilities of the main dietary carotenoids in a model cell membrane environment and determine any differences between quenching rate constants when singlet oxygen is generated in the aqueous or lipid phase.

(1) Singlet oxygen sensitisation using water-soluble rose bengal

Figure 2 gives a typical singlet oxygen trace in the presence and absence of carotenoid, in this case for BCRYP. Figure 3 gives the Stern–Volmer type plot for the quenching of singlet oxygen by BCRYP. From such data the second-order rate constants for the quenching of singlet oxygen by a range of xanthophylls were obtained and are given in Table 1.

As can be seen from the table, CANTHA shows the same rate constant (2.3×10^9 M^{-1} s^{-1}) to BCAR for singlet oxygen quenching using rose bengal, but most other xanthophylls have somewhat lower quenching rate constants ($1 - 2 \times 10^8$ M^{-1} s^{-1}), BCRYP having a rate constant of 1.8×10^8 M^{-1} s^{-1}; ASTA shows an intermediate value with a rate constant of 6×10^8 M^{-1} s^{-1}; LUT is the least effective at singlet oxygen quenching with a rate constant of

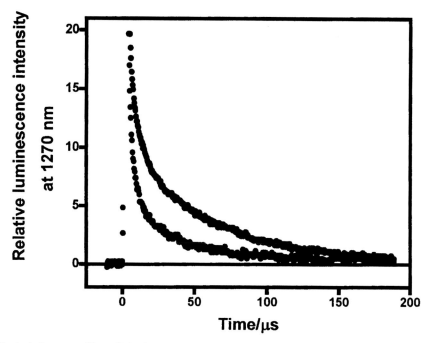

Fig. 2. Typical decay profiles of singlet oxygen sensitised by rose bengal (532 nm laser excitation) in the presence (lower trace) and absence (upper trace) of BCRYP in DPPC unilamellar liposomes.

Table 1. Second-order rate constants for the quenching of singlet oxygen by xanthophylls compared to β-carotene. RB refers to singlet oxygen sensitised from the triplet state of rose bengal in the aqueous phase. PBA refers to singlet oxygen sensitised from the triplet state of pyrene butyric acid in the lipid phase. ‡Value based on one concentration only. (Number of carbon–carbon conjugated double bonds is given in parentheses). (Data from *Ref. 4 and †Ref. 5).

$k/10^9$ M^{-1} s^{-1}		Benzene	RB in Liposomes	PBA in Liposomes
BCAR	(11)	*13.5	†2.3	†2.5
CANTHA	(11)	*13.2	†2.3	‡1.6
ASTA	(11)	*11.3	†0.59	—
ZEA	(11)	*12.6	†0.23	†0.17
BCRYP	(11)	13.0	0.18	0.14
LUT	(10)	*6.6	†0.11	†0.08

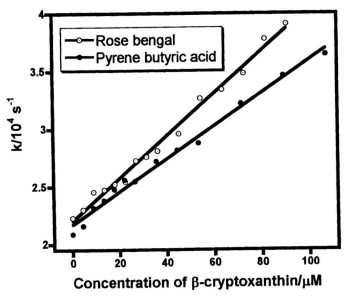

Fig. 3. Rate of decay of singlet oxygen against BCRYP concentration in air-saturated solutions of DPPC unilamellar liposomes using either rose bengal or pyrene butyric acid as singlet oxygen sensitiser.

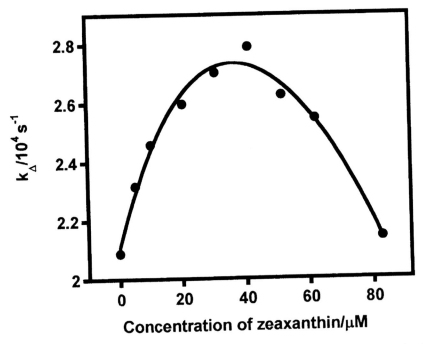

Fig. 4. Rate of decay of singlet oxygen against zeaxanthin concentration in air-saturated solutions of DPPC unilamellar liposomes using rose bengal as singlet oxygen sensitiser.

$1.1 \times 10^8 \ M^{-1} \ s^{-1}$ and ZEA shows a nonlinear plot (see Fig. 4). At concentrations up to ~15 μM ZEA, the plot is, within experimental error, linear and analysis of this portion of the curve gives a quenching rate constant of $2.3 \times 10^8 \ M^{-1} \ s^{-1}$, which is the value given in Table 1. As can be seen, at higher ZEA concentrations the rate begins to decrease and at 80 μM, ZEA shows essentially no singlet oxygen quenching at all. No such effect was observed for BCRYP as shown in Fig. 3.

(2) *Singlet oxygen sensitisation using lipid-soluble 4-(1-pyrene) butyric acid (PBA)*

Figure 5 shows a typical singlet oxygen decay trace sensitised from PBA in DPPC liposomes in the presence and absence of BCRYP.

From the corresponding Stern–Volmer type plots (Fig. 3), the singlet oxygen quenching rate constants were obtained and are given in Table 1. As for singlet oxygen generated in the aqueous phase, BCAR gives the highest singlet oxygen quenching rate constant. CANTHA gave a lower value of $1.6 \times 10^9 \ M^{-1} \ s^{-1}$, while LUT showed the lowest quenching ability once again. It was not possible to obtain a value for ASTA in unilamellar liposomes since, during the extrusion process, most of the ASTA was lost. From ZEA, a more linear plot is obtained than when the singlet oxygen was generated in the aqueous phase and the shape of the plot (not given, but see Ref. 5) appears to suggest that zero quenching would occur at much higher concentrations than for the water-soluble sensitisation

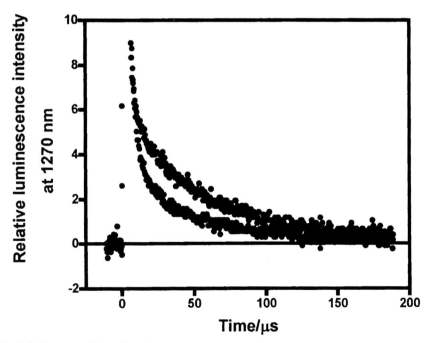

Fig. 5. Typical decay profiles of singlet oxygen sensitised by PBA (355 nm laser excitation) in the presence (lower trace) and absence (upper trace) of BCRYP in DPPC unilamellar liposomes.

experiments. For BCRYP, the quenching rate constant is apparently a little lower than when the singlet oxygen is generated in the aqueous phase, but this difference is probably not significant. The overall fact that the rate constants for singlet oxygen quenching are quite similar irrespective of whether the generation is in the aqueous phase or in the lipid phase is rather surprising and is discussed further below.

Radical Studies

The radical cation of BCRYP was produced via two methods; firstly by laser flash photolysis using electron transfer to 1-nitronaphthalene in methanol and secondly by pulse radiolysis in benzene and hexane solution. The spectra (not shown) gave single peaks at 910, 1020 and 1020 nm in methanol, benzene and hexane respectively. In general, these are quite similar to the previously reported radical cation spectra of other xanthophylls with 11 carbon–carbon double bonds.[6]

(1) One-electron reduction potentials

Determination of the efficiency of electron transfer between tryptophan radicals and xanthophylls at a specific pH allows the one-electron reduction potential of the carotenoids to be estimated:

$$TrpH^{\bullet+} + CAR \rightleftharpoons CAR^{\bullet+} + TrpH \quad (pH\ 4). \quad (1)$$

We have previously reported values[7] in the small range of 980–1060 mV for 5 carotenoids. In the present work we have extended this to BCRYP and obtained a value of 1028 mV (see Table 2) in micelles. This measurement shows that the radical cation of BCRYP, like other carotenoids, is itself a strong oxidising agent. For example, strong enough to oxidise tyrosine and other amino acids and lead to protein damage via cross-linking.

Table 2. One-electron reduction potentials for the xanthophylls and β-carotene and lycopene calculated from the equilibrium with TrpH[•+] in Triton-X 100 micelles ([*]values taken from Ref. 7, [†]measured in 4% Triton-X 100/405 mixed micelles).

Radical Cation	Reduction Potential/mV
[*]β-carotene	1060
[*]canthaxanthin	1040
[*]zeaxanthin	1031
[*]astaxanthin	1030
β-cryptoxanthin	1028
[*†]β-carotene	1028
[*†]lycopene	980

$$TyrOH + CAR^{\bullet+} \rightarrow CAR + TyrO^{\bullet} + H^{+} \qquad (2)$$

$$CySH + CAR^{\bullet+} \rightarrow CAR + CYS^{\bullet} + H^{+} \qquad (3)$$

$$2TyrO^{\bullet} \rightarrow \text{tyrosine dimers.} \qquad (4)$$

Clearly, as with BCAR, the possibility of damage [see, for example, Ref 8] arises from such oxidation and could be a drawback to dietary supplementation as we have suggested.[9] However, repair of these radical cations by water-soluble reductants such as vitamin C is a facile reaction that may preclude such damage (see below).

(2) Radical cation repair reactions

In previous work we have shown that water-soluble vitamin C can efficiently repair (i.e. convert to the parent molecule) carotenoid radical cations even though the carotenoids are all water insoluble[3]:

$$CAR^{\bullet+} + AscH_2 \rightarrow CAR + AscH^{\bullet} + H^{+} \qquad (5)$$

$$CAR^{\bullet+} + AscH^{-} \rightarrow CAR + AscH^{\bullet} \qquad (6)$$

We have suggested that the radical cation re-distributes itself within the membrane so as to make it more accessible to the vitamin C. The –OH group is thought to orientate zeaxanthin so that it spans the membrane, whereas lutein associates via the two –OH groups with the same water-lipid interface. In either case they are very accessible to water soluble components which may effect repair. Indeed, one may speculate that this is linked to the sole use by the eye of LUT and ZEA xanthophylls to protect from blue light sensitised reactions. The macular provides a very oxidative environment with plenty of oxygen, sensitisers and blue light, furthermore there is a good supply of vitamin C. We have extended these measurements of repair efficiency to BCRYP and to 4 water-soluble reductants,[9] vitamin C, uric acid, ferulic acid and trolox in micellar environments. The results are given in Table 3 and, as can be seen, all such reductants are efficient in the repair of the radical cation.

General Conclusions

A possibly surprising result of these investigations is that BCRYP, like the other carotenoids we have studied, quenches singlet oxygen with more-or-less the same efficiency irrespective of whether singlet oxygen is generated in the aqueous phase or in the lipid phase. We assume this means that the rate-determining step is the diffusion of singlet oxygen through the lipid membrane rather than in the aqueous phase to the lipid/water interface. As noted above ZEA does not show a linear dependence of singlet oxygen lifetime with ZEA concentration (Fig. 4), which has been attributed to ZEA aggregation at high concentrations.[5] This

Table 3. Second order rate constants ($/10^7$ M^{-1} s^{-1}) for repair of carotenoid radical cations by water-soluble biomolecules. [†]From Ref. 9 in Triton-X 100 micelles.

Carotenoid Radical Cation	Trolox	Ascorbic Acid	Ferulic Acid	Uric Acid
[†]β-carotene	19	1.0	0.1	1.1
[†]lycopene	19	1.8	0.1	1.0
[†]astaxanthin	52	5.5	0.6	12.1
[†]canthaxanthin	47	4.3	0.5	1.0
[†]lutein	31	1.8	0.2	1.5
[†]zeaxanthin	26	1.5	0.2	1.0
β-cryptoxanthin	19	2.7	0.12	1.97

effect was not observed for BCRYP. Furthermore, the spectroscopic behaviour of BCRYP is different from that of ZEA and LUT on aggregation. For the xanthophylls containing 2 –OH groups the carotenoids may well stack, i.e. align themselves next to each other spanning the membrane. This leads to a blue shift in their ground state absorption spectra and, for ZEA especially, a loss of efficiency in singlet oxygen quenching. For BCRYP with only one –OH group there will be no rigid spanning of the membrane and much less tendency to stack, and the result being a linear dependence of singlet oxygen lifetime with concentration. Furthermore, BCRYP exhibits a red shift on aggregation whereas the ZEA and LUT shift to the blue. Since protection against blue light is pivotal to the role of xanthophylls in the macular this may be related to the selectivity of the eye for LUT and ZEA rather than hydrocarbon or mono-hydroxy carotenoids — the aggregation of zeaxanthin and lutein possibly being protective for young children.[10]

Another aspect is the need for particularly efficient repair processes in which any carotenoid radical cation is rapidly reconverted to the parent molecule. The reduction potentials of all the radical cations, including our new data on BCRYP, show that such radical cations are very oxidative. In the highly oxidative environment of the eye very fast reduction of the radical cation is essential. The two –OH groups of LUT and ZEA may orientate these xanthophylls towards the water interface so that reduction by vitamin C, for example, is extremely effective.

Overall our new results with BCRYP suggest that, like other dietary carotenoids, their beneficial role in our diet depends on their interaction with other molecules such as vitamin C and excessive dietary supplementation with a single dietary component may not be appropriate.

Acknowledgments

The authors thank the World Cancer Research Fund and the National Lottery Charities Board for financial support, Roche Vitamins Ltd. for the carotenoids and Dr Regina Goralczyk for useful discussions. The new pulse radiolysis experiments were performed at the Free Radical Research Facility at Daresbury Synchrotron Laboratory. The FRRF is supported by the European Commission Human Potential Programme — Access to Infrastructure. We thank the Director of Daresbury Laboratory, for permission to use the facilities. In addition we thank Drs R. Edge and S. Navaratnam for assistance with some of the experiments.

References

1. J. H. Tinker, S. M. Tavender, A. W. Parker, D. J. McGarvey and T. G. Truscott, *J. Am. Chem. Soc.* **118**, 1756–1761 (1996).

2. J. Butler, B. W. Hodgson, B. M. Hoey, E. J. Land, J. S. Lea, E. J. Lindley, F. A. P. Rushton and A. J. Swallow, *Radiat. Phys. Chem.* **34**, 633–646 (1989).

3. M. Burke, R. Edge, E. J. Land and T. G. Truscott, *J. Photochem. Photobiol. B: Biol.* **60**, 1–6 (2001).

4. P. F. Conn, W. Schlach and T. G. Truscott, *J. Photochem. Photobiol. B: Biol.* **11**, 41–47 (1991).

5. A. Cantrell, D. J. McGarvey, T. G. Truscott, F. Rancan and F. Böhm, *Arch. Biochem. Biophys.* **412**, 47–54 (2003).

6. R. Edge, E. J. Land, D. J. McGarvey, L. Mulroy and T. G. Truscott, *J. Am. Chem. Soc.* **120**, 4087–4090 (1998).

7. M. Burke, R. Edge, E. J. Land, D. J. McGarvey and T. G. Truscott, *FEBS Lett.* **500**, 132–136 (2001).

8. The Alpha-Tocopherol Beta-Carotene Cancer Prevention Study Group, *New Eng. J. Med.* **330**, 1029–1035 (1994).

9. A. Cantrell and T. G. Truscott, in *Carotenoids in Health and Disease*, eds. N. I. Krinsky, S. T. Mayne and H. Sies (Marcel Dekker Inc. New York), pp. 31–52.

10. J. Roberts, in *Sun Protection in Man*, ed. P. Giacomoni (Elsevier Science B. V. Amsterdam, 2001), pp. 155–174.

Contribution from

DAVID PHILLIPS

Imperial College London
UK

Born 3 December 1939. Educated South Shields Grammar-Technical School, University of Birmingham, BSc Chemistry (1961), PhD Physical Chemistry (1964). Fulbright Scholar, University of Texas (1964–1966); Royal Society/Academy of Sciences of USSR Exchange Scientist (1966/1967). Lecturer/Senior Lecturer/Reader in Chemistry, University of Southampton (1967–1980). Wolfson Professor of Natural Philosophy (1980–1989), Acting Director (1986), Deputy Director (1986–89) The Royal Institution of Great Britain. Professor of Physical Chemistry (1989–present), Head of Department of Chemistry (1992–2002), Hofmann Professor of Chemistry 2000–present, Dean for the Faculties of Life Sciences and Physical Sciences, 2002–2005, Senior Dean, 2005–present, Imperial College London.

Council of the Royal Institution; Chairman of London Gifted and Talented; Chairman of RSC Education and Qualifications Board; Member of RSC Council, Fellow of Royal Society of Chemistry; Fellow of the NY Academy of Sciences; Winner of the Royal Society of Chemistry Nyholm Medal 1994; Royal Society Faraday Award, 1997; Sebetia-Ter Prize (Italy) 1999. OBE, June 1999 for services to science education. Delivered 1987 Royal Institution BBC TV Christmas Lectures jointly with J. M. Thomas. Author of children's books on science.

Members of the Porter and Phillips research groups, and staff of the Royal Institution, 1983.

David Phillips on
"Nanosecond Flash Photolysis"
G. Porter and M. R. Topp
Proc. Roy. Soc. Lond. A. **315**, 163–184 (1970)
and
"Time-resolved Fluorescence in the Picosecond Region"
G. Porter, E. S. Reid and C. J. Tredwell
Chem. Phys. Lett. **29**, 469–472 (1974)

The first paper describes the first use of a ruby laser to extend the time domain of flash photolysis experiments into the nanosecond regime. Hitherto, flash photolysis had been restricted to the microsecond region because it was impossible using conventional discharge lamps to produce both short pulses and high enough intensities to generate a significant concentration of transients such that they could be detected using absorption methods, although low intensity nanosecond discharges were used to excite fluorescence which was detected using time-correlated single photon counting. However, the advent of the laser permitted for the first time transient absorption techniques to be carried out .

As the paper points out, the essential requirements for any flash photolysis experiment are a spectral dispersive device (monochromator or spectrograph), an intense initiating photolysis pulsed light source, a monitoring pulsed light source, and a delay unit which can vary the time between excitation and detection, and of course, a detection system. The provision of these essentials in the laser experiment described in this paper was ingenious.

The primary pulsed light source was a home-built ruby laser delivering about 1.5 J (10^{18} photons) per pulse at 694.3 nm, and about 80 mJ per pulse when frequency-doubled to 347.1 nm. This was sufficient to generate transients at up to 5×10^{-4} mol. dm^{-3}. The interrogatory pulse and delay proved to be problematic initially. Attempts to use a commercially available nanosecond spark discharge as the spectroscopic flash were not successful. The discharge was triggered by focussing the red laser beam on to the spark gap, but the timing of the firing was irreproducible. Subsequent improvements still gave a timing jitter of a few tens of nanoseconds, but was reproducible only for 100 flashes or so. Moreover, the oxygen gas which was used in the spark source gave line emissions, which bedevilled the recording of transient spectra.

This problem was solved elegantly by using as the interrogatory light source fluorescence from a dye excited by a delayed fraction of the main laser pulse. Since the fluorescence could be chosen to cover a reasonably broad spectral range, and was structureless, it proved to be an ideal method for this timescale. Moreover, the delay between excitation and spectroscopic pulses was simply achieved by using the fact that light travels at 30 cm per nanosecond, and so a simple optical system was devised to vary the time delay from zero to 150 ns using mirrors. On this timescale, a metre rule sufficed to give the accuracy needed; however, when the principle was adopted in subsequent experiments on much shorter timescales, increasing accuracy of distance measurement was called for.

This simple, but very effective instrument was used by Porter and Topp to study the singlet and triplet state spectra of aromatic hydrocarbons, and their decay characteristics. It proved a model for subsequent studies by the Porter and other groups, and stands out as a classic development in the history of flash photolysis.

Our own work was influenced greatly by this first use of lasers to detect and record short-lived transients in photochemical systems. While extremely effective in the measurement of electronic absorption spectra of such transients, and in determining decay characteristics, the method does not in general generate much effective information on the structure of intermediates. From the 1980s, in a collaboration with, initially, Professor Ron Hester, York University, and colleagues (notably Drs. Tony Parker, Pavel Matousek, Mike Towrie, and Bill Toner) at the Laser Support Facility (now the Lasers for Science Facility) at the Central Laser Facility of the Rutherford Appleton Laboratories near Oxford, we set about developing a modification of the flash photolysis experiment in which the transients were detected using resonance Raman detection. The technique became known as "Time-resolved resonance Raman (TR^3)" spectroscopy. Since this generated in favourable cases the vibrational spectra of the transients, some structural information could be gleaned from the experiment. Initial experiments were on the triplet state of pairs of molecules involved in electron transfer reactions, on the nanosecond timescale.[1-4] We then progressed to the picosecond time-domain, and most recently, have combined with the group of Professor Mike George, Nottingham University, to construct a world-class apparatus on which both time-resolved infrared and TR^3 spectra can be measured.[5,6] A key development in the application of this instrumentation was the use of a Kerr cell to suppress fluorescence which in even weakly fluorescence molecules gives a signal six orders of magnitude greater than the Raman signal.[7] This is suppressed extremely efficiently by the use of the Kerr effect, in which the laser pulse is used to rotate the plane of polarisation of between crossed polarisers such that the optical shutter opens for a few picoseconds. This technique was utilised by Porter, Reid and Tredwell (accompanying paper) to time resolve fluorescence from organic molecules, where an optical delay line is used to delay the shutter opening such that a fluorescent decay profile could be measured. Our system can also be used this way, but for Raman suppression, the shutter is open coincident with the laser pulse, and is subsequently closed to suppress the fluorescence. As Zewail has pointed out,[8] the first use of the Kerr effect was due to Abraham and Lemoine in 1899,[9] the first recorded use of the pump and probe technique.

The Kerr effect method of time-resolving fluorescence was an alternative to the use of time-correlated single-photon counting, for which the time-resolution was much poorer when spark discharge lamps and photomultipliers were used, but improved to the 10 ps region with the subsequent use of mode-locked lasers and microchannel plates.[10]

The time-resolved vibrational spectroscopy apparatus is now in general use at RAL and forms an essential part of the investigative armoury available to UK and European users. This technique is described fully in the accompanying paper of ours which owes its inspiration to the Porter and Topp, and Porter, Reid and Tredwell articles.

References

1. J. N. Moore, G. H. Atkinson, D. Phillips, P. M. Killough and R. E. Hester, *Chem. Phys. Lett.* **107**, 381–384 (1984).
2. A. W. Parker, R. E. Hester, D. Phillips and S. Umapathy, *JCS. Faraday Trans.* **88**, 2649–2653 (1992).
3. E. Vauthey, D. Phillips and A. W. Parker, *J. Phys. Chem.* **96**, 7356–7360 (1992).
4. P. Matousek, R. E. Hester, A. J. Langley, J. N. Moore, A. W. Parker, D. Phillips, W. T. Toner, M. Towrie, I. C. E. Turcu and S. Umapathy, *Meas. Sci. Technol.* **4**, 1090–1095 (1993).
5. W. M. Kwok, C. Ma, M. W. George, D. C. Grills, P. Matousek, A. W. Parker, D. Phillips, W. T. Toner and M. Towrie, *Phys. Chem. Chem. Phys.* **5**, 1043–1050 (2003).
6. M. Towrie, D. C. Grills, J. Dyer, J. A. Weinstein, P. Matousek, R. Barton, P. D. Bailey, N. Subramaniam, W. M. Kwok, C. Ma, D. Phillips, A. W. Parker and M. W. George, *Appl. Spectrosc.* **57**, 367–380 (2003).
7. P. Matousek, M. Towrie, C. Ma, W. M. Kwok, D. Phillips, W. T. Toner and A. W. Parker, *J. Raman. Spectrosc.* **32**, 983–988 (2001).
8. J. S. Baskin and A. Zewail, *J. Chem. Ed.* **78**, 737–751 (2001).
9. H. Abraham and J. Lemoine, *C. R. Acad. Sci.* **129**, 206–208 (1899).
10. D. V. O'Connor and D. Phillips, *Time-Correlated Single-Photon Counting* (Academic Press, London, 1984), 288 pp.

Reprinted with permission from *Appl. Spectrosc.* 57(4), 367–380 (2003)

submitted papers

Development of a Broadband Picosecond Infrared Spectrometer and its Incorporation into an Existing Ultrafast Time-Resolved Resonance Raman, UV/Visible, and Fluorescence Spectroscopic Apparatus

MICHAEL TOWRIE, DAVID C. GRILLS, JOANNE DYER, JULIA A. WEINSTEIN, PAVEL MATOUSEK, ROBIN BARTON, PHILIP D. BAILEY, NARESH SUBRAMANIAM, WAI M. KWOK, CHENSHENG MA, DAVID PHILLIPS, ANTHONY W. PARKER, and MICHAEL W. GEORGE*

Central Laser Facility, CCLRC Rutherford Appleton Laboratory, Chilton, Didcot, Oxfordshire, OX11 0QX, United Kingdom (M.T., P.M., R.B., P.D.B., N.S., A.W.P.); School of Chemistry, University of Nottingham, University Park, Nottingham, NG7 2RD, United Kingdom (D.C.G., J.D., J.A.W., M.W.G.); and Department of Chemistry, Imperial College, Exhibition Road, London, SW7 2AY, United Kingdom (W.M.K., C.M., D.P.)

We have constructed a broadband ultrafast time-resolved infrared (TRIR) spectrometer and incorporated it into our existing time-resolved spectroscopy apparatus, thus creating a single instrument capable of performing the complementary techniques of femto-/picosecond time-resolved resonance Raman (TR³), fluorescence, and UV/visible/infrared transient absorption spectroscopy. The TRIR spectrometer employs broadband (150 fs, ~150 cm⁻¹ FWHM) mid-infrared probe and reference pulses (generated by difference frequency mixing of near-infrared pulses in type I AgGaS₂), which are dispersed over two 64-element linear infrared array detectors (HgCdTe). These are coupled via custom-built data acquisition electronics to a personal computer for data processing. This data acquisition system performs signal handling on a shot-by-shot basis at the 1 kHz repetition rate of the pulsed laser system. The combination of real-time signal processing and the ability to normalize each probe and reference pulse has enabled us to achieve a high sensitivity on the order of ΔOD ~ 10^{-4}–10^{-5} with 1 min of acquisition time. We present preliminary picosecond TRIR studies using this spectrometer and also demonstrate how a combination of TRIR and TR³ spectroscopy can provide key information for the full elucidation of a photochemical process.

Index Headings: Ultrafast time-resolved spectroscopy; Broadband; Infrared; Resonance Raman; Kerr gate; Fluorescence; Array detector; DNA probes; Photopolymerization initiators; Charge-transfer excited states.

INTRODUCTION

Time-resolved pump-probe spectroscopy is a widely applied technique for the direct detection of short-lived reactive intermediates and excited states. In early flash photolysis experiments, flash lamps were used to identify the transient species. Pulsed lasers rapidly superseded lamps and advances in technology have now made tran-

sient absorption studies possible on femtosecond time scales, extending from the UV to the mid-infrared (MIR) wavelength range. A key development in the emerging ultrafast techniques has been the advent of reliable and powerful commercially available femtosecond laser sources, generally based on titanium sapphire (Ti:Sapphire) oscillators and regenerative amplifiers. These provide intense kilohertz and multi-kilohertz repetition rate femtosecond and picosecond pulses at the energies required to drive optical parametric generators (OPGs) and amplifiers (OPAs). This enabled[1] the generation of independently tunable, synchronized pulses at micro-Joule levels and kilohertz repetition rates, providing the necessary flexibility for ultrafast pump-probe spectroscopic measurements.

Visible/near-UV transient absorption spectroscopy has benefited from such developments because ultrashort pulsed lasers are capable of generating a bright, stable femtosecond white-light continuum down to 10 fs.[2] The advent of techniques for the amplification of these spectrally broadband pulses in non-colinear OPAs has enabled sub 10 fs pump-probe visible absorption spectroscopy to be carried out.[3] Access to the UV wavelengths can be achieved through sum-frequency generation and optical rectification has also extended pump-probe techniques to the THz spectral region.[4,5] Time-resolved fluorescence spectroscopy has also been performed with a femtosecond time resolution using either up-conversion techniques[6–10] or a Kerr gate acting as a fast shutter[11,12] to monitor background-free fluorescence.

The last few years have seen many developments in femto-/picosecond vibrational spectroscopy. For example, time-resolved resonance Raman (TR³) spectroscopy is now a generally applicable analytical tool,[1,13–18] since the lasers can now be tuned to be resonant on the ground

Received 11 June 2002; accepted 25 October 2002.
* Author to whom correspondence should be sent.

0003-7028/03/5704-0367$2.00/0
© 2003 Society for Applied Spectroscopy

and intermediate state transitions at will[19,20] and intense fluorescence can efficiently be rejected from the Raman signals in the temporal domain using Kerr gate methodology.[15–18]

Ultrafast time-resolved infrared (TRIR) spectroscopy is also rapidly becoming a more widely used technique due to recent advances in IR laser and detector technology.[21] Early ultrafast TRIR experiments examined vibrational relaxation processes by both pumping and probing samples with narrowband MIR pulses.[22] Experiments combining UV/visible pump and IR probe pulses were subsequently performed.[23,24] In these measurements the sample was probed using narrowband picosecond MIR pulses and a single element IR detector. Spectra were built-up in a 'point-by-point' fashion by repeating the experiment at a series of IR frequencies. An alternative approach for ultrafast TRIR measurements used a continuous wave (cw) IR laser fixed at one particular IR frequency, which continuously probed the sample.[25] A picosecond UV pulse was used to initiate a photochemical reaction which resulted in a temporal change in the IR transmission. A second, delayed visible gating pulse was then combined with the transmitted IR beam in a LiIO₃ crystal, gener-ating light at the sum of the frequencies of the two beams. This up-converted visible pulse was detected using a photomultiplier tube at various gating pulse time delays. The apparatus was subsequently improved,[26] allowing TRIR experiments to be performed on the femtosecond timescale. Broadband ultrafast TRIR spectroscopy offers many advantages over narrowband TRIR since it allows complete transient IR bands to be simultaneously monitored, giving greater spectroscopic detail and much faster data acquisition times. A key development was the construction of a picosecond IR spectrometer that generated an MIR bandwidth of ~100 cm⁻¹ full width at half-maximum (FWHM), using the difference frequency mixing of two visible laser pulses in a nonlinear LiIO₃ crystal.[27] Up-conversion of the IR probe pulses to the visible permitted their detection using an optical multichannel analyzer and subsequently[28] a focal-plane charge-coupled device (CCD) detection system. The availability of multielement IR array detectors has now eliminated the need to up-convert the broadband IR probe pulses into the vis-'ble region, thus providing a significant increase in signal-to-noise ratio. In the first such experiments[29] the sample was probed with broadband (~65 cm⁻¹ FWHM) IR pulses, which were then detected by dispersion over 10-element HgCdTe (MCT) linear IR array detectors. The advantage of using small array detectors was that it allowed data acquisition to be performed at a high repetition rate (1 kHz), thus permitting extensive signal averaging. Heilweil and co-workers also replaced their up-conversion/CCD array detection system with 256 × 256 element InSb[30] and MCT[31] focal-plane IR array detectors and detected extremely broadband (~500 cm⁻¹ FWHM) IR pulses along linear tracks of the focal-plane arrays. The advantage of this approach is that transient IR spectra can be obtained over a much wider spectral range without the need to scan the probe laser wavelength. However, the repetition rate of the laser system was low (20 Hz) due to speed limitations in reading out the large amount of data generated by the array.

The early broadband TRIR spectrometers incorporated

dye lasers into their systems for the generation of the ultrashort IR pulses. Currently, ultrafast TRIR systems are mainly based around solid-state femtosecond Ti : Sapphire oscillators/regenerative amplifier systems, which are used to pump MIR OPAs. The broadband (~200 cm⁻¹) MIR pulses generated by these OPAs are then detected either along individual tracks of 256 × 256 element IR focal-plane array detectors at low repetition rate (e.g., 30 Hz)[32] or by 32-element linear IR array detectors at high repetition rate (1 kHz).[33,34] Advances in IR detector technology and computer processing power are now permitting the use of larger IR array detectors at high repetition rates. In this paper we describe the addition of a broadband ultrafast TRIR spectrometer to the existing[1] Rutherford Appleton Laboratory Laser Facility, thus producing the *first* single spectrometer capable of ultrafast fluorescence, UV/visible transient absorption, TRIR, and TR³ measurements. The apparatus described incorporates two 64-element MCT linear IR array detectors operating at 1 kHz, with the potential for scaling to data readout rates of 8 kHz (2 × 128 elements) or 1 kHz (2 × 1024 elements).

EXPERIMENTAL

Materials. All solvents were obtained from Aldrich and degassed with argon prior to use. Cr(CO)₆, (2,4,6-trimethylbenzoyl)diphenylphosphine oxide, and 4-(dimethylamino)benzonitrile (Aldrich) were used as received. *fac*-[(dppz)Re(CO)₃(py)]⁺ [PF₆]⁻ and *fac*-[(dppz)Re(CO)₃Cl] were synthesized according to literature procedures.[35] The synthesis of Pt(2,2'-bpy)(SC₆F₄CN)₂ will be described elsewhere.[36]

Laser Instrumentation. We have previously described in detail[1] our 1 kHz repetition rate femto-/picosecond dual OPA system for generating two independently tunable pulses in the range of ~205–2800 nm. This system is optimized for femto-/picosecond pump-probe TR³ and can be used for time-resolved UV/visible absorption, fluorescence, and reflectance spectroscopy. The system is centered around a Ti : Sapphire regenerative amplifier (Spectra Physics, Spitfire/Merlin) operating at a 1 kHz repetition rate at ~800 nm. This has recently been upgraded from ~0.7 to 2.5 mJ per pulse to permit the TRIR measurements (see below). The regenerative amplifier is seeded by an ~100 fs pulse from a mode-locked Ti : Sapphire laser (Spectra Physics, Tsunami). Two modes of operation are available with this system: (1) in the picosecond mode, spectral filtering of the Ti : Sapphire oscillator seed pulses produces 800 nm output of ~1–2 ps FWHM pulse duration and ~15–7 cm⁻¹ bandwidth respectively, and (2) the incorporation of a femtosecond stretcher/compressor arrangement provides an output of 150 fs and ~100 cm⁻¹ bandwidth. The 800 nm regenerative amplifier output is then frequency doubled (in a beta barium borate (BBO) crystal) and split to pump two OPAs. Spectral coverage of the OPAs when combined with up-conversion (including the idler tunability) is 205–2800 nm.

We have now extended the OPA tuning range to provide high brightness spectral coverage into the MIR region, down to ~10 μm (1000 cm⁻¹). There are three modes of infrared generation: narrowband MIR, broad-

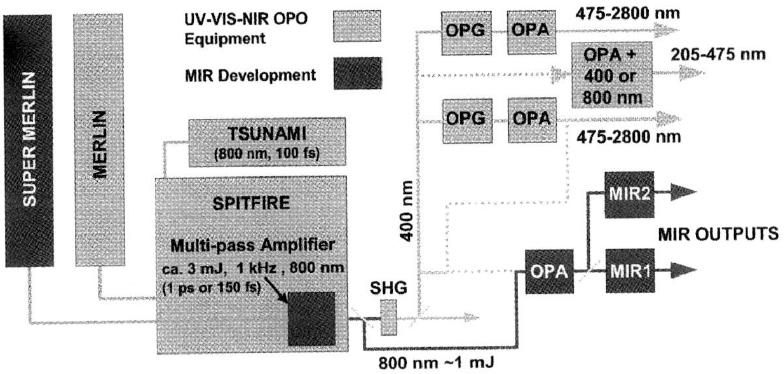

Fig. 1.   Schematic layout of the laser system and OPAs. Existing equipment is shown in light gray and the mid-infrared upgrade is shown in dark gray.

band MIR, and very broadband MIR. For the high brightness, narrow linewidth ($\sim$25 cm^{-1}) MIR light, the pump laser is operated in the picosecond mode. Since the laser system is required to drive a third, MIR, OPA, it was necessary to increase the energy of the 800 nm fundamental pulses from 0.7 mJ to $\sim$2–3 mJ. This was achieved by the addition of a second Nd : YLF pump laser (Spectra Physics, Super Merlin) and a two-pass Ti : Sapphire amplifier (Spectra Physics, Super Spitfire) as shown in the schematic of our laser system in Fig. 1. The narrowband OPA configuration enables a precise choice of IR laser frequency to specifically pump photochemical and photophysical processes in the MIR. When the pump laser is operated in the femtosecond mode ($\sim$150 fs pulse width) the broadband ($\sim$150–200 cm^{-1}) MIR pulses that are generated are used for spectrally dispersed TRIR spectroscopy using linear IR array detectors. In this case the laser system is operated without the final linear amplifier stage in order to provide higher stability MIR pulses.

NARROW BAND TYPE II AgGaS$_2$ OPTICAL PARAMETRIC AMPLIFIER

The narrowband MIR OPA is based on difference frequency generation between the signal and idler of an 800 nm pumped optical parametric amplifier, that in turn is seeded by the idler output of a narrowband 400 nm pumped picosecond OPA of the type described previously.[1] To summarize, $\sim$20 μJ pump energy at 400 nm is used to pump a two-pass optical parametric generator based on a 4 mm thick 27° cut type I BBO crystal. The signal output from the generator is collimated and then dispersed using a 2400 lines/mm grating to inject a narrow band ($<$10 cm^{-1}) spectral component of the OPG light as a seed into a second 4 mm type I BBO optical parametric amplifier pumped by $\sim$60 μJ of 400 nm light. This gives $\sim$5 μJ of narrowband, $\sim$20 cm^{-1}, low divergence signal and idler output. The idler output, $\sim$1 μJ, is then used as the signal in a single pass 800 nm pumped type II BBO OPA that uses two 5 mm long 31° cut type II crystals with orientation, 800.0(e) = 1400.0(e) + 1866.7(o). This orientation has the dual advantage that

angle walk-off can be compensated for and, unlike alternative orientations, group velocity mismatch delays the idler with respect to the signal by $\sim$100 fs. This helps compensate for group velocity mismatch in the silver gallium sulfide (AgGaS$_2$) crystal. The amplified signal and idler pass directly into a single 2 mm thick 41° cut type II AgGaS$_2$ OPA to create the MIR output by difference frequency generation (DFG). For narrow band operation, type II AgGaS$_2$ is favored over type I due to its 40% higher nonlinear coefficient (d_{eff} = 12.9 pm/V) and 50% lower acceptance bandwidth (25 cm^{-1}/cm).[37] Using 1.2 mJ of 800 nm pump energy, we obtained a maximum MIR output of 10 μJ ($\sim$4 μm wavelength), corresponding to 0.8% energy conversion efficiency and 4% photon conversion efficiency from the pump laser. However, the usual performance from the difference frequency mixing was $>$5 μJ from 2.5 to 6.0 μm and 1–5 μJ from 6 to 10 μm, with a typical linewidth of 25 cm^{-1}. The specifications of the IR output given above provide sufficient IR intensity for TRIR measurements and leave sufficient residual pump laser energy ($\sim$1.5 mJ) to drive at least two further 400 nm pumped OPAs for pump-probe experiments.

VERY BROAD BAND ZnGeP$_2$ OPTICAL PARAMETRIC AMPLIFIER

ZnGeP$_2$ has one of the highest nonlinearity coefficients of all crystals ($\sim$70 pm/V)[37] and a transparency in the range of 2–10 μm. We obtained broadband output by pumping a simple 48° cut type I ZnGeP$_2$ OPG/OPA at degeneracy with the 2–3 μm wavelength idler output of an 800 nm potassium titanyl phosphate (KTP) OPA. Previously, bandwidths of several hundred wavenumbers have been demonstrated using this method.[38] A schematic layout of the OPA is given in Fig. 2. The 2–3 μm pump beam ($\sim$30 μJ/pulse) is generated by replacing the BBO crystals in the 800 nm pumped OPA stage of the narrow linewidth OPA described above with two 41° cut type II KTP crystals. These have slightly higher transparency than BBO in the longer wavelength range.

The pump beam was passed through two 5 mm thick ZnGeP$_2$ crystals (anti-reflection coated) in a two-pass

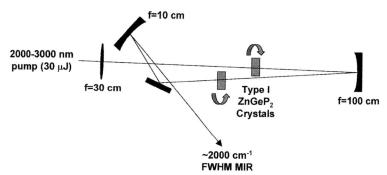

FIG. 2.   Schematic layout of the very broadband ZnGeP₂ mid-infrared OPA.

OPG and OPA configuration. On the first pass the pump beam was focused using a 30 cm focal length lens. On the second pass the pump beam and OPG light were focused into the second crystal by a 100 cm focal length gold mirror. Using a pump wavelength of ~2.5 μm we btained an MIR output extending from 3.2–8.3 μm (3100 to 1200 cm⁻¹). Using this configuration, the pulse energy was measured to be ~0.5 μJ.

While the bandwidth of this OPA is very large, it currently suffers from two limitations that make it inappropriate for high-sensitivity time-resolved IR measurements. Firstly, the output has large shot-to-shot spectral fluctuations that cannot easily be compensated for, even with the probe and reference design we have adopted, and secondly, the spectral brightness is too low to give good signal-to-noise performance from our detection apparatus.

BROADBAND TYPE I AgGaS₂ OPTICAL PARAMETRIC AMPLIFIER

To generate broadband MIR pulses more suitable for TRIR absorption spectroscopy we have adopted an OPA scheme similar to that described by Kaindl et al.[39] Stable MIR pulses are generated by the difference frequency mixing of the signal and idler outputs, obtained from a vhite light continuum seeded 800 nm pumped BBO OPA, in an uncoated type I, 36° cut 1.5 mm thick, AgGaS₂ crystal. The main feature of this design is that it provides MIR pulses that are both energetically and spectrally stable. This requires careful control of the amplifier gain characteristics and a highly stable pump source of good beam quality. With this OPA design, Kaindl et al.[39] reported that an extremely high stability of <1% shot-to-shot was attainable using the output of an ~80 fs pump laser derived directly from a regenerative amplifier, which in itself was stable to ~1%. Due to a number of factors, including longer pump pulse length (~150 fs), lower beam quality (due to the multi-pass amplifier), and energy instability in the regenerative amplifier trigger box, we found the IR output to be unstable at the ±10% level, with some variation in spectral profile and an occasional 'drop out' to less than 10% of the average value in about 2% of the laser shots. Therefore, for direct TRIR measurements at 1 kHz we could not expect to achieve better than ~3 × 10⁻³ ΔOD in one

second of accumulation. This is ten times lower in sensitivity than is possible with a system of 1% energy jitter. To compensate for these variations, a dual detector probe and reference scheme was adopted that succeeded in bringing the sensitivity to the ~3 × 10⁻⁴ ΔOD level in one second of accumulation. A typical MIR linewidth of ~150 cm⁻¹ has been achieved with this OPA, with a pulse energy of ~500 nJ.

Spectral Characterization of the Mid-infrared Pulses. Two methods have been employed for the spectral characterization of the MIR pulses. For the narrowband and very broadband pulses, a conventional FT-IR spectrometer (Thermo Nicolet, Avatar) was used. Typical FT-IR spectra of the narrowband and very broadband MIR pulses are shown in Fig. 3. The broadband (~150 cm⁻¹) MIR pulses, that have been used in all of the spectrally dispersed TRIR measurements described in this paper, were characterized from the output spectrum generated by the 64-element MCT array detectors. Calibration of the pixel numbers into wavenumbers was achieved using standard IR absorbing samples. A typical output of the broadband IR OPA is given in Fig. 4.

Experimental Setup for Broadband TRIR Spectroscopy. In order to minimize the effects of laser intensity fluctuations we have developed a highly sensitive multi-channel IR probe and reference approach to normalize out intensity variations in real time on a shot-by-shot basis at the 1 kHz laser repetition rate. A schematic diagram of the apparatus used for ultrafast broadband TRIR spectroscopy is shown in Fig. 5. The broadband (~150 cm⁻¹ FWHM) MIR probe beam generated by the OPA is split into probe and reference pulses using a 50% germanium beamsplitter. The probe beam is focused to a diameter of ~150 μm in the sample using an f = 30 cm gold-coated spherical mirror. The transmitted light from the sample is then imaged onto a 0.5 mm spectrograph input slit. A similar optical relay is used for the reference arm, but without a sample. The two ¼ m spectrographs (CVI Laser Corp., DKSP240) disperse the IR light across two 64-element MCT photoconductive IR array detectors (Infrared Associates Inc., MCT-13-64el) situated at the focal plane of the exit ports. The detectors are powered by pre-amplifiers (Infrared Systems Development Corp., MCT-64000) having a nominal gain of 100 and a 13 V maximum output. The detector elements are 2 mm high

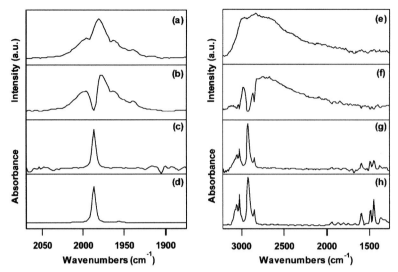

FIG. 3.   (a) A typical FT-IR spectrum of the narrowband MIR pulses. (b) FT-IR spectrum of these pulses after they have been transmitted through an IR cell containing Cr(CO)$_6$ in n-hexane. (c) Absorbance spectrum of Cr(CO)$_6$ obtained from the logarithm of the ratio of these spectra, and (d) a conventional FT-IR spectrum of Cr(CO)$_6$ in n-hexane for comparison. (e) A typical FT-IR spectrum of the very broadband MIR pulses. (f) FT-IR spectrum of these pulses after they have been transmitted through a polystyrene film. (g) Absorbance spectrum of polystyrene obtained from the logarithm of the ratio of these spectra, and (h) a conventional FT-IR spectrum of the polystyrene film for comparison. (a–d) were recorded at 4 cm^{-1} resolution and (e–h) at 16 cm^{-1} resolution. Small differences between spectra (g) and (h) are due to instabilities in the IR laser and the slightly poorer spectral resolution for (g) that is caused by the kHz laser repetition rate.

and 0.5 mm wide, with a 35 mm active length. They are liquid nitrogen cooled and protected by a 19 × 50 mm ZnSe window. The elements were operated with ~10 mA (1 V bias) and ~1500 V/W responsivity (before the pre-amps.) with a noise of 1 nV rms/Hz$^{1/2}$ (before the pre-amps.). Their spectral responsivity was specified as >30% over the wavelength range 2.5–15 μm with a maximum D* ~30 × 10^9 cm Hz$^{1/2}$ W^{-1}.

The spectrographs have two dispersion options with

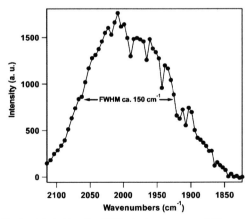

FIG. 4.   A typical broadband MIR output spectrum generated from one of the 64-element MCT linear array detectors.

gold gratings: 150 lines/mm (4000 nm blaze) or 300 lines/mm (2400 nm blaze), providing 25.6 and 12.8 nm/mm dispersion, respectively. They were adapted, by increasing the aperture size of the light baffle before the exit port and replacing the exit mirror with a larger rectangular aluminum mirror, to increase dispersion at the output (from 25 mm to 35 mm) in order to match the MCT array size. The pump pulse is sent through a variable delay line, reflected off an optical galvanometer, and focused into the sample to a spot size of ~200 μm. The galvanometer is synchronized to the laser and is used to deflect the pump beam onto and away from the sample on an alternate shot basis. This generates data in pump-on/pump-off pairs that can be compared to eliminate any long-term drifts of the IR laser intensity.

Data Acquisition and Processing. The outputs of both 64-element MCT arrays are fed simultaneously into an analog multiplexer board based on the custom-made Rutherford Appleton Laboratory HX2, originally designed for charge detection from multi-element X-ray detector arrays. Each HX2 integrated circuit integrates charge (up to 60 pC +ve charge and 20 pC −ve charge) over 16 parallel input signals with an electrical bandwidth of ~1 MHz and has an integration period determined by internal or external clock circuitry. Each board uses four HX2s to capture 64 channels simultaneously with all readouts multiplexed into a single output consisting of a sequence of 64 analog voltages. The buffered output generates an ~12 V full-scale voltage. The electronics circuit board (PC20001/5) incorporates pulse timing circuitry for charge integration times down to ~20 μs as set by the

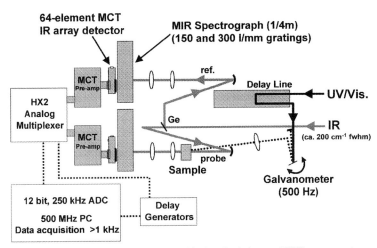

Fig. 5. Schematic diagram of the layout used for broadband picosecond TRIR measurements.

basic clocking structure that was tied to the readout rate. In order to match the detector response to the integration period, we provided an external clock pulse train to define an ~2 μs integration period followed by a 14 μs readout period. A second HX2 board, slaved to the first board, is used to acquire the signals from the second MCT detector array. The outputs from both boards are read out in simultaneous bursts of 64 analog signals, with each burst separated by 1 ms as defined by the 1 kHz laser repetition rate.

The charge injected from the detectors into the HX2 is adjusted to give ~12 V output for a 4 V MCT detector input signal. The multiplexed output has a full-scale range of about 12 V and baseline ripple of ~15 mV root mean square (rms) and noise of <6 mV rms. The noise is derived mostly from the MCT detectors at the ~2 mV level and the ripple from 50 Hz mains-related modulation. The mains ripple is correlated in all analog outputs and is largely removed in the subsequent data normalization process.

A low-pass radio frequency (RF) filter is used to reduce high-frequency noise prior to digitization. The HX2 board provides the trigger and clock pulses to synchronize outputs with the ADC process. The analog signals are passed into two channels of an 8-channel simultaneous sampling, 12-bit resolution ADC card (Datel 416J) housed in a standard 500 MHz PC. This card has the capacity to acquire at up to 250 kHz on each channel. The data is then streamed to the PC RAM via the ADC FIFO (FIFO = first in, first out) and PC direct memory access where data analysis involving signal discrimination, normalization, and averaging between detector channels is carried out on a shot-by-shot basis at the 1 kHz acquisition rate. A third channel of the ADC is used to monitor the pump-on/off status of the laser. Analysis of the pump-on/pump-off data pairs creates a rolling average for the change in transmission, ΔT_N using the following equation:

$$\Delta T_N = \frac{[(I_{probe}/I_{ref})_{pump\ on} - (I_{probe}/I_{ref})_{pump\ off}] + \Delta T_{N-1}(N-1)}{N}$$

The average change in absorbance, ΔA_N is then given by:

$$\Delta A_N = -\log\left(1 + \frac{I_R}{I_P}\Delta T_N\right)$$

where I_R and I_P are the final averages of the pump-off spectra on the reference and probe side, respectively, and N is the total number of acquisitions. Further discrimination by the software removes large fluctuations in the signal, such as laser 'drop outs', or other large variations, such as those associated with gas bubbles in the sample flow stream, on a shot-by-shot basis. This combination of probe and reference normalization and real-time pulse-to-pulse signal processing has enabled us to achieve, using a low-stability laser source (±10%), an extremely high sensitivity with our apparatus, on the order of ΔOD ~ 10^{-4}–10^{-5} with 1 min of acquisition time. This corresponds to an effective shot-to-shot stability of ~1%.

Sample Handling. We have used a variety of sample-handling techniques depending on the nature of the sample and solvent. For the majority of samples a conventional CaF_2 IR flow cell has been used (Harrick Scientific Corp.) with variable pathlength (6 μm–1 mm). In order to prevent sample degradation on the windows of the cell, it is randomly oscillated in the plane perpendicular to the direction of the laser beams using acentric motors attached to smooth ball slides mounted on an x–y stage. A magnetically coupled gear pump (Micropump) or a poly(tetrafluoroethylene) (PTFE) peristaltic pump (Masterflex) is used to flow the sample down to a minimum volume of ~10 cm^3 and for very low volume samples (down to ~50 μL) the flow can be stopped and the cell statically sealed.

Photosensitive samples are prone to pump-induced degradation. Even when using a rapidly oscillated flow

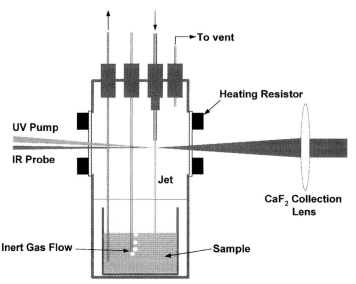

FIG. 6.   Diagram of the inert atmosphere sample chamber used for open-jet TRIR measurements on extremely photosensitive samples.

cell, this can lead to damage of the optical window sur-faces, which may interfere with the transmission of the beams through the sample. In such circumstances we use a sample chamber, which encloses a 1 mm diameter flow-ing sample jet situated ~1 cm away from the cell win-dows (see Fig. 6). The laser beams are focused into the jet and since the jet is not in contact with the windows, sample degradation problems at the interface are elimi-nated. Low f number CaF$_2$ lenses image the light from the sample jet onto the spectrograph slit. The cell also has the advantage of allowing an inert atmosphere to be used for air-sensitive samples. Resistive heaters on the windows are necessary to prevent condensation of sol-vent. This arrangement is suitable for use with solvents that are not strongly absorbing in the IR and also for UV/visible transient absorption measurements. For measure-nents in aqueous solution and below ~1700 cm^{-1}, where common organic solvents have strong absorptions, the thin enclosed flow cell is used.

All IR probe and reference beams and optics are en-closed within nitrogen-purged beam pipes and boxes to reduce absorption by atmospheric water vapor to an ac-ceptable level and to remove fluctuations in intensity caused by dust particles.

RESULTS AND DISCUSSION

The incorporation of the broadband femto-/picosecond TRIR spectroscopy apparatus, described in this paper, into the existing ultrafast spectroscopic facility at the Rutherford Appleton Laboratory has produced an instru-ment capable of any combination of Kerr gated TR^3, TRIR, UV/visible transient absorption, and fluorescence studies on time scales down to 400 fs (limited by the FWHM of the pump-probe cross-correlation function). A schematic of the apparatus is shown in Fig. 7. In addition

to conventional UV/visible pump experiments, IR pump experiments are also possible in combination with any of the probing techniques. Three-color experiments, such as UV pump–IR pump–TR^3 probe or UV pump–IR pump–UV/visible probe, can also be performed. We highlight the completion of our instrument by describing a series of preliminary picosecond TRIR and TR^3 investigations on organic and inorganic samples. In all of these exam-ples, the excitation density of the pump pulses was ~10 mJ cm^{-2} and to ensure that all signals were due to pop-ulation dynamics, the pump and probe pulse polarizations were set at the magic angle (54.7°).

The Development of New Infrared Probes for DNA. TRIR spectroscopy is a well-established technique for probing the excited states and elucidating the photochem-ical mechanisms of transition-metal carbonyl com-pounds.[40–43] This is due to the unique spectroscopic prop-erties of these compounds. For example, the ν(CO) stretching vibrations are virtually uncoupled from other molecular vibrations and they produce very intense and generally narrow IR bands in the 1750 to 2150 cm^{-1} re-gion, which can also be predicted from simple force-field models.[44] The positions and intensities of the ν(CO) bands are extremely sensitive to the electronic environ-ment and therefore act as direct reporters of the electronic density at the metal center.

There is currently a major research effort in trying to understand how charge is transported through DNA.[45–48] Much of the work to date has focused on probing the electron-transfer process using either intercalated or co-valently bound organic or transition-metal complexes, particularly complexes of Ru(II), Rh(III), and Cu(I).[49] For example, [Ru(bpy)$_2$(dppz)]$^{2+}$ (bpy = 2,2′-bipyridine, dppz = dipyrido[3,2-a:2′,3′-c]phenazine) acts as a 'mo-lecular light switch' for DNA.[50] In aqueous buffer solu-

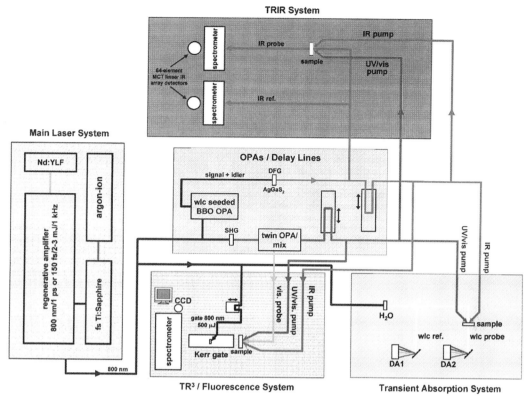

Fig. 7. Schematic layout of the ultrafast time-resolved spectroscopy apparatus showing how the laser pulses are integrated into the different spectroscopic techniques.

tion it is non-emissive; however, in the presence of DNA it displays an enhancement of luminescence. The majority of these DNA probes possess low-lying metal-to-ligand charge-transfer (MLCT) excited states, which serve ·o heighten their sensitivity to their surroundings. However, emission bands are generally broad and featureless, providing little structural information. Furthermore, the presence of more than one chromophoric ligand can introduce ambiguity in interpreting which ligand is associated with the emission.

We have focused our research on the development of a family of Re(I)dppz complexes of the general formula fac-[(dppz-R,R')Re(CO)$_3$(L)]$^{0,+}$ (L = ligand) as IR probes for DNA. These systems possess only one chromophoric ligand and also offer the advantage that they contain an inherent ν(CO) IR spectroscopic handle. However, incorporation of the strongly accepting dppz ligand into Re(I) complexes can complicate their photophysical properties by producing close-lying ligand-centered ($\pi \to \pi^*$) and MLCT lowest excited states.[35,51] Thus, the photophysical properties of all new DNA probes must be elucidated fully before an investigation of electron-transfer processes in DNA can be undertaken. It is hoped that the results of these studies will ultimately allow us to design Re–dppz complexes with tunable excited states that are suitable for DNA probing.

We present here a preliminary ps-TRIR investigation of the photophysical properties of two of these complexes, fac-[(dppz)Re(CO)$_3$(py)]$^+$ [PF$_6$]$^-$ (py = pyridine) (**1**) and fac-[(dppz)Re(CO)$_3$Cl] (**2**). The results demonstrate the potential of ps-TRIR spectroscopy for easily identifying the nature of the excited states of prospective transition-metal probes for DNA. The results·of our investigations on the whole family of complexes will be the subject of a forthcoming publication.[52]

(1) L = pyridine, n=1
(2) L = Cl⁻, n=0

The ps-TRIR spectrum in the ν(CO) region obtained 100 ps after 400 nm excitation of the charged complex

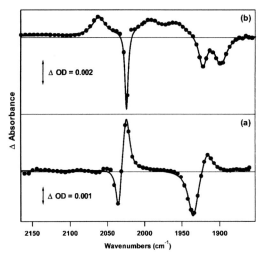

FIG. 8. TRIR spectra of (a) fac-[(dppz)Re(CO)$_3$(py)]$^+$ [PF$_6$]$^-$ in ace-tonitrile and (b) fac-[(dppz)Re(CO)$_3$Cl] in CH$_2$Cl$_2$, both recorded 100 ps after 400 nm excitation. The spectral resolution is ~8 cm^{-1}.

1 in acetonitrile is shown in Fig. 8a. The two negative bands correspond to a depletion of the ground state upon excitation. Two positive bands, corresponding to a new species, are observed at lower frequencies relative to the ground state. The positions of these bands are characteristic of a dppz-centered $\pi \rightarrow \pi^*$ excited state.[53] This is because the π^* orbitals of dppz are slightly more electron-donating than the π orbitals. Thus, population of the dppz π^* orbitals in the excited state causes a small in-crease in π back-bonding from the metal to the CO li-gands and a characteristic negative shift in the ν(CO) vibrations.

Replacing the pyridine ligand with Cl$^-$ yields the neu-tral complex **2**. The TRIR spectrum of **2** in CH$_2$Cl$_2$ re-corded 100 ps after 400 nm excitation is shown in Fig. 8b. The negative bands indicate depletion of the ground-state species, and three new bands are observed, shifted to higher frequencies relative to the ground-state bands. Such a positive shift of the ν(CO) bands is characteristic of a dπ(Re) $\rightarrow$ π^*(dppz) MLCT excited state.[53] This is because in an MLCT excited state there is a reduction of electron density at the metal center, causing a correspond-ing reduction in π back-donation and thus a strengthening of the CO bonds.

Although complex **2** was shown to possess an MLCT lowest excited state, this species is neutral. It is highly desirable for a DNA probe to be positively charged since this would enhance its ability to bind to the DNA base pairs in the helix and also increase its solubility in an aqueous environment. Work is currently in progress using the ps-TRIR system to study a large series of charged complexes in aqueous (D$_2$O) solution and in the presence of DNA. The experiments presented above demonstrate how ps-TRIR spectroscopy can be used for studying the nature of the lowest excited states of transition-metal car-bonyl complexes.

Investigation of the Reactivity of Photopolymeriza-tion Initiator Molecules. We have investigated the photochemistry of (2,4,6-trimethylbenzoyl)diphenylphos-phine oxide (**3**) using ps-TRIR. Complex **3** is one mem-ber of a large family of benzoyl compounds that have found extensive commercial use as photoinitiators in free radical polymerization processes.[54] Photolysis of **3** causes highly efficient α-cleavage ($\phi \sim 0.6$) from a short lived ($\tau < 1$ ns) triplet excited state and the formation of 2,4,6-trimethylbenzoyl (**4**) and diphenylphosphinoyl (**5**) radi-cals.[55]

The reactivity of **5** towards a number of substrates has been extensively studied using laser flash photolysis[56] and time-resolved ESR.[57] However, due to the low UV cross-section of benzoyl radicals in solution, the reactivity of **4** has not been as extensively investigated. Sluggett et al. were able to detect the presence of **4** in solution using nanosecond TRIR spectroscopy after UV photolysis.[55] Steady-state photolysis studies showed that a number of photoproducts are produced upon photolysis of **3** in the absence of any radical trapping agents. However, in the presence of BrCCl$_3$, in addition to the expected trapped products (2,4,6-trimethylbenzoyl bromide and diphenyl-phosphinic bromide), the rearrangement product, diphe-nyl[(2,4,6-trimethylbenzoyl)oxy]phosphine (**6**) was still formed.

This supported the conclusion obtained previously[58] from ^{31}P chemically induced dynamic nuclear polariza-tion (CIDNP) experiments that **6** is rapidly produced via an *in-cage* recombination of radicals **4** and **5**. An *in-cage* formation of **6** would compete with the generation of the polymerization-initiating radicals **4** and **5** and reduce the efficiency of the UV curing process. We have performed a ps-TRIR investigation of **3** in acetonitrile solution. Fig-ure 9a shows the ground-state FT-IR spectrum of **3** in

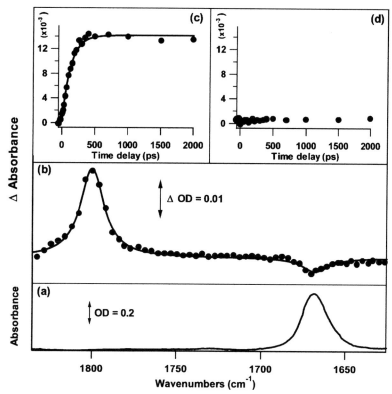

FIG. 9. (a) FT-IR spectrum of (2,4,6-trimethylbenzoyl)diphenylphosphine oxide in acetonitrile, showing the $\nu(CO)$ absorption. (b) TRIR spectrum obtained 500 ps after 400 nm excitation of this solution. Kinetic traces were recorded at (c) 1800 cm^{-1} and (d) 1738 cm^{-1}. The spectral resolution of the TRIR spectrum is ~8 cm^{-1}.

acetonitrile. Only the $\nu(CO)$ carbonyl band is observed in this region. Figure 9b shows the TRIR spectrum obtained using the open-jet configuration, 500 ps after 400 nm excitation. A small negative band is present, which is assigned to bleaching of the ground-state species upon photolysis. A large positive band centered at 1800 cm^{-1} is also formed, and this is assigned to the benzoyl radical 4 by comparison with previous ns-TRIR studies.[55] The kinetics of the formation of the benzoyl radical band at 1800 cm^{-1} are shown in Fig. 9c. This band grows-in, with a lifetime of 130 ps, and then remains constant over the observable timescale of the experiment (2 ns). These kinetics are in good agreement with the rate of formation of the phosphinoyl radical 5 (123 ps), previously obtained[59] using UV/visible transient absorption spectroscopy. This observation is consistent with the mechanism for α-cleavage of 3 proposed earlier,[59] whereby the rate-limiting step is intersystem crossing from the singlet excited state to the triplet state. Previous steady-state FT-IR studies[55] have shown that 6 has a strong IR absorption at 1738 cm^{-1}. Figure 9d shows the kinetic trace obtained at 1738 cm^{-1}. We find no evidence on the picosecond time scale for the generation of a transient species at 1738 cm^{-1}. These results appear at first to disagree with the

original conclusion that 6 is formed extremely rapidly via an *in-cage* recombination of 4 and 5. However, recent nanosecond TRIR power-dependence studies[60] suggest that 6 is formed via a multi-photon process and is therefore only observed under conditions of high UV photon flux nanosecond pulses. A more detailed study of this important class of photopolymerization initiators will be published shortly.[61]

Excited States of Inorganic Chromophores: High Spectral Resolution TRIR Measurements. The majority of ultrafast IR measurements are performed at relatively low spectral resolution (>8 cm^{-1}). However, there are many circumstances in which it would be advantageous to perform high spectral resolution TRIR measurements. We give one example here where there are two heavily overlapped bands in the TRIR spectrum. Recording such spectra at high resolution can often aid the interpretation of the results subsequent to an initial low-resolution scan of a large region. An optional 300 lines/mm grating is available in our spectrographs for such purposes and this provides spectral resolutions ranging from, e.g., ~5 cm^{-1} (at 2200 cm^{-1}) to ~2 cm^{-1} (at 1400 cm^{-1}).

Time-resolved infrared studies of the excited states of metal coordination compounds have so far mainly fo-

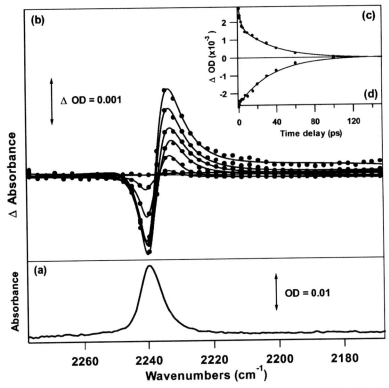

FIG. 10.   (a) FT-IR spectrum of Pt(2,2'-bpy)(4-CN-C₆F₄-S)₂ in dichloromethane. (b) TRIR spectra shown at 7 time delays between 1 and 500 ps after 400 nm excitation of this solution. (c) and (d) show kinetic traces obtained from the heights of the curve-fitted transient and bleach bands, respectively. The spectral resolution of the TRIR spectrum is ~5 cm⁻¹.

cused on octahedral d^6 chromophores, e.g., Re¹, Ru¹¹, and Os¹¹, in part due to their relatively long lifetimes (100 ns–1 μs).[42] There have been relatively few TRIR investigations of the excited states of d^8 chromophores.[62–64] We have therefore investigated the dynamics of the lowest electronic excited state of Pt(2,2'-bpy)(4-CN-C₆F₄-S)₂ (7) by following changes in the electronic density on the peripheral cyano reporter groups.

7

The FT-IR spectrum of **7** in CH_2Cl_2 (Fig. 10a) exhibits a single band at 2240 cm⁻¹ corresponding to the $\nu(CN)$ stretches of the peripheral cyano groups. Photolysis of this solution (400 nm, 150 fs) results in an instantaneous bleaching of the parent absorption and the formation of a new $\nu(CN)$ band at ~2233 cm⁻¹. This is shown in Fig. 10b at a series of selected time delays (1–500 ps) after excitation. With increasing time delay, the bleach and transient recover back to the baseline. The negative shift observed for the transient is very small and therefore the use of the high-resolution gratings greatly improves the accuracy of the subsequent spectral analysis. The shift of the $\nu(CN)$ band in the excited state reflects a depopulation of the bonding π-orbitals of the CN groups due to electron transfer from the thiolate moiety to the α-diimine ligand (as indicated by the arrow in the above structure). This indicates that the HOMO in Pt(2,2'-bpy)(4-CN-C₆F₄-S)₂ is mainly localized on the thiolate. The kinetics of this process were examined by plotting the heights of the Lorentzian curve fitted transient and bleach bands (obtained from custom-written curve-fitting software) versus pump-probe time delay (see Figs. 10c and 10d). The bleach fits well to a single exponential recovery giving a lifetime for the charge-transfer excited state of 33.0

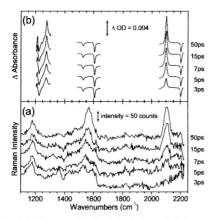

FIG. 11. (a) TR³ spectra of 4-dimethylaminobenzonitrile in acetonitrile recorded at 3, 5, 7, 15, and 50 ps with a 267/300 nm pump/probe configuration. Solvent peaks and background have been subtracted out. (b) TRIR spectra of this sample recorded at the same time delays after 267 nm excitation. These were obtained in three separate experiments, each probing in different regions of the infrared. Spectral resolutions of the TRIR spectra range from ~3–4 cm⁻¹ in the low wavenumber region to ~9–10 cm⁻¹ in the high wavenumber region.

± 3.0 ps. The transient decay fits to two exponential components (τ = 2.0 ± 0.5 and 35.0 ± 4.0 ps). The fast decay is assigned to an early relaxation process (cooling or solvation) associated with the decay of "hot" vibrational modes initially formed upon excitation. The slower component agrees well with the bleach recovery and represents the lifetime of the back electron transfer process. A complete investigation of the nature of the lowest excited state of 7, including resonance Raman and electrochemical studies, will be published elsewhere.[36]

Investigation of Intramolecular Charge-Transfer Dynamics in 4-Dimethylaminobenzonitrile. Although ps-TRIR spectroscopy is an extremely valuable tool for the elucidation of photochemical mechanisms, it is often necessary to combine it with another pump-probe technique in order to build a complete picture of the photochemistry involved. We describe below one such example, which demonstrates how a combination of ps-TRIR and ps-TR³ gives complementary results when investigating the intramolecular charge-transfer (ICT) process following electronic excitation of 4-dimethylaminobenzonitrile (DMABN). This molecule has received much attention over the last four decades[65] because the spectral profile of its solution-phase fluorescence is strongly dependent upon the nature of the solvent. In a non-polar solvent its fluorescence emission is 'normal', but in polar solvents it is deeply red-shifted. The large spectral shift is accepted to be due to an excited-state intramolecular electron-transfer reaction, but the structural changes that occur on the picosecond time scale during the course of this reaction remain controversial. Recent work surrounding the use of time-resolved vibrational spectroscopy to investigate this compound and its ICT reaction can be found in our[20,66–68] and other[69–71] papers and references therein. Figure 11a shows the ps-TR³ spectra of DMABN in acetonitrile using a 267 nm pump and 300 nm probe.

The accumulation time for each spectrum is about 600 s. Solvent peaks and background have been eliminated by subtracting the negative time-delay spectrum. In earlier work using other solvents we have assigned the 3 ps spectrum to the locally excited (LE) state of DMABN, which is formed in non-polar solvents and goes on to form the ICT state at later times in polar solvents.[72] Thus, the 50 ps spectrum corresponds to the relaxed ICT state. From these data we estimate the time-constant for the conversion from the LE to ICT state in this solvent to be 6.5 ps, which is consistent with the 6 ps time constant previously obtained using time-resolved fluorescence.[73]

Figure 11b shows ps-TRIR spectra of DMABN in acetonitrile, again pumping at 267 nm. The accumulation time for each spectral region is about 150 s. The negative peaks at ~1226, 1527, 1607, and 2202 cm⁻¹ are from bleaching of ground-state DMABN bands. The positive peaks at ~1214, 1277, and 2098 cm⁻¹ are assigned[69–71] to the ring C–CN stretch, ring C–N(Me₂), and the C≡N stretches, respectively, within the ICT state of DMABN. Interestingly, as with other workers,[70] we do not observe any IR bands associated with the LE state of DMABN at early time delays.

We now compare the time-resolved resonance Raman and infrared spectra and make the following comments. The TRIR spectra give a much better S/N ratio with just a quarter of the accumulation time used for the TR³ spectra. It is worth noting that the spectral resolution achievable for the Raman spectra is limited by the spectral linewidth of the probe laser, but for the IR spectra the resolution is only dependent on the spectrograph resolution. On the other hand, at present it is easier to obtain a wider spectral window for TR³ in a single accumulation than for the TRIR method because the ps-IR source can only cover ~150 cm⁻¹. This means that to accumulate a wide spectral range, several spectral windows need to be taken. Finally, at present it is very difficult for the ps-TRIR experiments to obtain spectral information below 1000 cm⁻¹ unlike for ps-TR³, which provides access down to ~100 cm⁻¹.[74,75]

The data obtained from both techniques show similarities and differences. That is, both spectra show the ~2100 cm⁻¹ band while the other bands are exclusive to each technique. Of particular note is the ICT band observed at ~1280 cm⁻¹ in the ps-TRIR spectrum. This band is assigned to a mode that is believed to play an important role in elucidating the structure of the ICT state. At present two main possibilities are debated. In the first, the dimethylamino group of DMABN changes from being near planar (slightly wagged) in the ground state to being perpendicular to the phenyl ring plane in the ICT state (the twisted-ICT or TICT model[76]). The second possibility is the amino group changes from the slightly wagged conformation to be in-plane with the ring, with the ring–N bond gaining some partial double-bond character (PICT model[77]). The ps-TRIR reported here for this mode shows a subtle peak shift with pump-probe time delay, moving from 1273 cm⁻¹ at 3 ps to 1277 cm⁻¹ at 50 ps. This would be difficult to observe using point-by-point IR methods[70] and has not been reported previously by Okamoto et al.,[71] who rely on the method of optically heterodyned detection of absorption anisotropy to observe such weak transients. The signal-to-noise ratio in Okamoto's experiments (S/N ~ 6 for the 1607 cm^{-1} band) is not as high as that obtained in the present experiments (S/N ~ 60 for the 1607 cm^{-1} band). Such temporal changes observed in vibrational spectroscopy are often related to intramolecular vibration relaxation (IVR), solvation changes of the newly created state, and to rapid structural modifications. Further work is presently underway to establish whether the observed spectral shifts can be related to the ICT process and to structurally define the nature of this state.

CONCLUSION

In summary, we have constructed a broadband (~150 cm^{-1} FWHM) picosecond TRIR spectrometer, based on OPA technology, which incorporates 64-element infrared array detectors and data-processing electronics. This has allowed us to operate at a high repetition rate of 1 kHz. Using a probe and reference normalization approach, we have achieved a high sensitivity on the order of ΔOD ~ 10^{-4}–10^{-5} after 1 min of acquisition time. The apparatus has the potential for scaling to data readout rates of 8 kHz (2 × 128 elements) or 1 kHz (2 × 1024 elements). n addition to operating in the broadband mode for spectrally dispersed TRIR spectroscopy, the IR pulses can be generated as narrowband (~25 cm^{-1} FWHM) picosecond IR pulses, which can be used for specifically pumping infrared transitions. We have shown how this spectrometer integrates into our existing ultrafast spectroscopy apparatus, providing a single system that allows us to access the complementary techniques of femto-/picosecond time-resolved resonance Raman, fluorescence, and UV/visible/infrared transient absorption spectroscopies. We have successfully demonstrated the potential and sensitivity of the TRIR spectrometer with a range of organic and inorganic samples and have shown one example where the use of both TR^3 and TRIR spectroscopy is vital for the elucidation of the photochemical processes involved.

ADDENDUM

Following submission, the shot-to-shot stability of the laser has been improved to <±2% and after probe/ref-'rence normalization it now has an effective stability of <±0.5%.

ACKNOWLEDGMENTS

This work was supported by EPSRC Research Grant GR/M40486. We also thank Professors J. M. Kelly, J. J. Turner, and M. Poliakoff for helpful discussions and Mr. J. Govans and Mr. K. Stanley for technical support. M.W.G. is grateful to the Royal Society of Chemistry for the award of the Sir Edward Franklin Fellowship.

1. M. Towrie, A. W. Parker, W. Shaikh, and P. Matousek, Meas. Sci. Technol. **9**, 816 (1998).
2. R. Huber, H. Satzger, W. Zinth, and J. Wachtveitl, Opt. Commun. **194**, 443 (2001).
3. A. Shirakawa, I. Sakane, M. Takasaka, and T. Kobayashi, Appl. Phys. Lett. **74**, 2268 (1999).
4. A. G. Markelz, A. Roitberg, and E. J. Heilweil, Chem. Phys. Lett. **320**, 42 (2000).
5. M. C. Beard, G. M. Turner, and C. A. Schmuttenmaer, Phys. Rev. B **62**, 15764 (2000).
6. R. Schanz, S. A. Kovalenko, V. Kharlanov, and N. P. Ernsting, Appl. Phys. Lett. **79**, 566 (2001).
7. H. Mahr and M. D. Hirsch, Opt. Commun. **13**, 96 (1975).
8. T. C. Damen and J. Shah, Appl. Phys. Lett. **52**, 1291 (1988).
9. A. Mokhtari, A. Chebira, and J. Chesnoy, J. Opt. Soc. Am. B **7**, 1551 (1990).
10. M. L. Horng, J. A. Gardecki, A. Papazyan, and M. Maroncelli, J. Phys. Chem. **99**, 17311 (1995).
11. Y. Kanematsu, H. Ozawa, I. Tanaka, and S. Kinoshita, J. Lumin. **87**, 917 (2000).
12. J. Takeda, K. Nakajima, S. Kurita, S. Tomimoto, S. Saito, and T. Suemoto, J. Lumin. **87**, 927 (2000).
13. W. L. Weaver, L. A. Huston, K. Iwata, and T. L. Gustafson, J. Phys. Chem. **96**, 8956 (1992).
14. K. Iwata and H. Hamaguchi, Chem. Phys. Lett. **196**, 462 (1992).
15. P. Matousek, M. Towrie, A. Stanley, and A. W. Parker, Appl. Spectrosc. **53**, 1485 (1999).
16. P. Matousek, M. Towrie, C. Ma, W. M. Kwok, D. Phillips, W. T. Toner, and A. W. Parker, J. Raman Spectrosc. **32**, 983 (2001).
17. N. Everall, T. Hahn, P. Matousek, A. W. Parker, and M. Towrie, Appl. Spectrosc. **55**, 1701 (2001).
18. P. Matousek, M. Towrie, and A. W. Parker, J. Raman Spectrosc. **33**, 238 (2002).
19. P. Matousek, G. Gaborel, A. W. Parker, D. Phillips, G. D. Scholes, W. T. Toner, and M. Towrie, Laser Chem. **19**, 97 (1999).
20. W. M. Kwok, C. Ma, D. Phillips, P. Matousek, A. W. Parker, and M. Towrie, J. Phys. Chem. A **104**, 4188 (2000).
21. M. D. Fayer, Ed., *Ultrafast Infrared and Raman Spectroscopy* (Marcel Dekker, New York, 2001).
22. E. J. Heilweil, M. P. Casassa, R. R. Cavanagh, and J. C. Stephenson, J. Chem. Phys. **81**, 2856 (1984).
23. T. Elsaesser and W. Kaiser, Chem. Phys. Lett. **128**, 231 (1986).
24. T. Elsaesser, W. Kaiser, and W. Luttke, J. Phys. Chem. **90**, 2901 (1986).
25. J. N. Moore, P. A. Hansen, and R. M. Hochstrasser, Chem. Phys. Lett. **138**, 110 (1987).
26. P. A. Anfinrud, C. H. Han, T. Q. Lian, and R. M. Hochstrasser, J. Phys. Chem. **95**, 574 (1991).
27. E. J. Heilweil, Opt. Lett. **14**, 551 (1989).
28. T. P. Dougherty and E. J. Heilweil, Opt. Lett. **19**, 129 (1994).
29. P. Hamm, S. Wiemann, M. Zurek, and W. Zinth, Opt. Lett. **19**, 1642 (1994).
30. S. M. Arrivo, V. D. Kleiman, T. P. Dougherty, and E. J. Heilweil, Opt. Lett. **22**, 1488 (1997).
31. T. A. Heimer and E. J. Heilweil, J. Phys. Chem. B **101**, 10990 (1997).
32. P. T. Snee, H. Yang, K. T. Kotz, C. K. Payne, and C. B. Harris, J. Phys. Chem. A **103**, 10426 (1999).
33. P. Hamm, M. Lim, and R. M. Hochstrasser, J. Chem. Phys. **107**, 10523 (1997).
34. Y. Wang, J. B. Asbury, and T. Lian, J. Phys. Chem. A **104**, 4291 (2000).
35. H. D. Stoeffler, N. B. Thornton, S. L. Temkin, and K. S. Schanze, J. Am. Chem. Soc. **117**, 7119 (1995).
36. J. A. Weinstein, A. J. Blake, E. S. Davies, A. L. Davis, M. W. George, D. C. Grills, I. V. Lileev, A. M. Maksimov, P. Matousek, M. Y. Mel'nikov, A. W. Parker, V. E. Platonov, M. Towrie, C. Wilson, and N. N. Zheligovskaya, Inorg. Chem., paper submitted (2002).
37. See the SNLO nonlinear optics code, available from A. V. Smith, Sandia National Laboratory, Albuquerque, NM 87185-1423.
38. K. L. Vodopyanov and V. G. Voevodin, Opt. Commun. **117**, 277 (1995).
39. R. A. Kaindl, M. Wurm, K. Reimann, P. Hamm, A. M. Weiner, and M. Woerner, J. Opt. Soc. Am. B **17**, 2086 (2000).
40. M. Poliakoff and E. Weitz, Adv. Organomet. Chem. **25**, 277 (1986).
41. M. W. George, M. Poliakoff, and J. J. Turner, Analyst (Cambridge, U.K.) **119**, 551 (1994).
42. M. W. George and J. J. Turner, Coord. Chem. Rev. **177**, 201 (1998).
43. K. McFarlane, B. Lee, J. Bridgewater, and P. C. Ford, J. Organomet. Chem. **554**, 49 (1998).
44. P. S. Braterman, *Metal Carbonyl Spectra* (Academic Press, London, 1975).
45. J. Jortner, M. Bixon, T. Langenbacher, and M. E. Michel-Beyerle, Proc. Natl. Acad. Sci. U.S.A. **95**, 12759 (1998).
46. M. Núñez, D. B. Hall, and J. K. Barton, Chem. Biol. **6**, 85 (1999).
47. P. O'Neill, A. W. Parker, M. A. Plumb, and L. D. A. Siebbeles, J. Phys. Chem. B **105**, 5283 (2001).
48. B. Giese and A. Biland, Chem. Commun., 667 (2002).

49. K. E. Erkkila, D. T. Odom, and J. K. Barton, Chem. Rev. **99**, 2777 (1999).

50. A. E. Friedman, J. C. Chambron, J. P. Sauvage, N. J. Turro, and J. K. Barton, J. Am. Chem. Soc. **112**, 4960 (1990).

51. V. W. W. Yam, K. K. W. Lo, K. K. Cheung, and R. Y. C. Kong, J. Chem. Soc., Dalton Trans., 2067 (1997).

52. J. Dyer, W. J. Blau, C. G. Coates, C. M. Creely, J. D. Gavey, D. C. Grills, S. Hudson, J. M. Kelly, P. Matousek, J. J. McGarvey, J. McMaster, A. W. Parker, M. Towrie, J. A. Weinstein, and M. W. George, Photochem. Photobiol. Sci., paper submitted (2002).

53. J. R. Schoonover, G. F. Strouse, R. B. Dyer, W. D. Bates, P. Chen, and T. J. Meyer, Inorg. Chem. **35**, 273 (1996).

54. R. M. Williams, I. V. Khudyakov, M. B. Purvis, B. J. Overton, and N. J. Turro, J. Phys. Chem. B **104**, 10437 (2000).

55. G. W. Sluggett, C. Turro, M. W. George, I. V. Koptyug, and N. J. Turro, J. Am. Chem. Soc. **117**, 5148 (1995).

56. T. Sumiyoshi and W. Schnabel, Makromol. Chem. **186**, 1811 (1985).

57. M. Kamachi, A. Kajiwara, K. Saegusa, and Y. Morishima, Macromolecules **26**, 7369 (1993).

58. W. Rutsch, K. Dietliker, D. Leppard, M. Koehler, L. Misev, U. Kolczak, and G. Rist, *Proceedings of the XXth International Conference in Organic Coatings Science and Technology* (1994), p. 467.

59. S. Jockusch, I. V. Koptyug, P. F. McGarry, G. W. Sluggett, N. J. Turro, and D. M. Watkins, J. Am. Chem. Soc. **119**, 11495 (1997).

60. C. S. Colley, Ph.D. Thesis, University of Nottingham, Nottingham, U.K. (2001).

61. C. S. Colley, D. C. Grills, N. A. Besley, S. Jockusch, P. Matousek, A. W. Parker, M. Towrie, N. J. Turro, P. M. W. Gill, and M. W. George, J. Am. Chem. Soc. **124**, 14952 (2002).

62. C. E. Whittle, J. A. Weinstein, M. W. George, and K. S. Schanze, Inorg. Chem. **40**, 4053 (2001).

63. G. D. Smith, M. S. Hutson, Y. Lu, M. T. Tierney, M. W. Grinstaff, and R. A. Palmer, Appl. Spectrosc. **55**, 637 (2001).

64. J. A. Weinstein, D. C. Grills, M. Towrie, P. Matousek, A. W. Parker, and M. W. George, Chem. Commun., 382 (2002).

65. E. Lippert, W. Rettig, V. Bonacickoutecky, F. Heisel, and J. A. Miehe, Adv. Chem. Phys. **68**, 1 (1987).

66. W. M. Kwok, C. Ma, P. Matousek, A. W. Parker, D. Phillips, W. T. Toner, and M. Towrie, Chem. Phys. Lett. **322**, 395 (2000).

67. C. Ma, W. M. Kwok, P. Matousek, A. W. Parker, D. Phillips, W. T. Toner, and M. Towrie, J. Phys. Chem. A **106**, 3294 (2002).

68. W. M. Kwok, C. Ma, P. Matousek, A. W. Parker, D. Phillips, W. T. Toner, M. Towrie, and S. Umapathy, J. Phys. Chem. A **105**, 984 (2001).

69. M. Hashimoto and H. Hamaguchi, J. Phys. Chem. **99**, 7875 (1995).

70. C. Chudoba, A. Kummrow, J. Dreyer, J. Stenger, E. T. J. Nibbering, T. Elsaesser, and K. A. Zachariasse, Chem. Phys. Lett. **309**, 357 (1999).

71. H. Okamoto, H. Inishi, Y. Nakamura, S. Kohtani, and R. Nakagaki, J. Phys. Chem. A **105**, 4182 (2001).

72. C. Ma, W. M. Kwok, D. Phillips, P. Matousek, A. W. Parker, M. Towrie, and W. T. Toner, Central Laser Facility, Rutherford Appleton Laboratory Annual Report 1999/2000 107 (2000).

73. P. Changenet, P. Plaza, M. M. Martin, and Y. H. Meyer, J. Phys. Chem. A **101**, 8186 (1997).

74. W. M. Kwok, C. Ma, A. W. Parker, D. Phillips, M. Towrie, P. Matousek, and D. L. Phillips, J. Chem. Phys. **113**, 7471 (2000).

75. W. M. Kwok, C. Ma, A. W. Parker, D. Phillips, M. Towrie, P. Matousek, X. M. Zheng, and D. L. Phillips, J. Chem. Phys. **114**, 7536 (2001).

76. K. Rotkiewicz, K. H. Grellmann, and Z. R. Grabowski, Chem. Phys. Lett. **19**, 315 (1973).

77. K. A. Zachariasse, M. Grobys, T. vonderHaar, A. Hebecker, Y. V. Ilichev, Y. B. Jiang, O. Morawski, and W. Kuhnle, J. Photochem. Photobiol., A **102**, 59 (1996).

Contribution from

ALEXANDER J. MACROBERT

University of College London
UK

Reader in Photochemistry and Photobiology at The Royal Free and University College Medical School, University College London. Researches on the development of photosensitisers for photodynamic therapy and new applications of optical spectroscopy in biomedicine.

Graduated from Cambridge University in Natural Science (1977) and took his PhD on free radical kinetics and spectroscopy under Michael Clyne's supervision at the Dept. of Chemistry, Queen Mary College, University of London. Carried out postdoctoral research at the Physical Chemistry Laboratory, Oxford University with Gus Hancock followed by two years in Germany on Royal Society research fellowship at Bielefeld University in the Faculty of Physics.

In 1985 took up an MRC research fellowship at The Royal Institution, Davy Faraday Laboratory, to work with David Phillips on phthalocyanine photosensitisers, before moving in 1989 to the Dept. of Chemistry, Imperial College, London. In 1992 took up present position at University College London.

MacRobert in a relaxed mood, Royal Institution.

Alexander J. MacRobert on
"Triplet State Quantum Yields for Some Aromatic Hydrocarbons and Xanthene Dyes in Dilute Solution"
P. G. Bowers and G. Porter
Proc. Roy. Soc. A. **299**, 348–353 (1967)

This work was one of a series of papers by George Porter and Peter G. Bowers which brought about a major advance for molecular photophysics and photochemistry by establishing a new method for determining the quantum yield for triplet state formation based on transient absorption spectroscopy following flash photolysis. The transient absorption technique had previously been developed for time-resolved studies of triplet state dynamics, and their simple yet elegant innovation was to reference the triplet–triplet absorption versus the known ground state depletion to provide a far more direct measurement of triplet state quantum yields compared to previous studies, which has proven so successful that it is now one of the standard techniques in the armamentarium of many photochemical laboratories. In the intervening years the main development has been the incorporation of pulsed laser excitation which itself was pioneered by Porter and co-workers.

The compounds investigated by Porter and Bowers included xanthene dyes which exhibit intense visible absorption bands and are thus amenable to flashlamp-induced depletion of the ground state. Among the dyes studied was eosin which was found to have a high triplet quantum yield of 0.7. The photosensitising properties of eosin have long been recognised, notably in the photobiological field, since eosin was one of the first photosensitisers used for photodynamic therapy (PDT) of tumours. Pioneering work was carried out by von Tappeiner and Jesionek in Germany and published in 1903 (H. von Tappeiner and A. Jesionek, *Muench. Med. Wochenschr*, **44**, 2024–2044) in which they demonstrated that the topical application of eosin followed by visible light illumination could be used to treat skin tumours. Their work was shortly followed by others showing that chlorophyll could act as a photosensitiser for the haemolysis of erythrocytes. This phenomenon, which relied on the presence of oxygen, was denoted as the "Photodynamische Wirkung" (or photodynamic effect) from which the present medical term is derived. Later studies demonstrated that the photodynamic effect relied upon the production of oxidising species, principally singlet oxygen through resonant energy transfer from the sensitiser triplet state to molecular oxygen. Their selection of eosin would have been serendipitous after much trial and error, and of course along with the advent of lasers and fibre-optic guidance, other dyes have supplanted these early attempts in photodynamic therapy. However, the same prerequisites still apply for photosensitiser selection in that a good triplet state yield and visible wavelength absorption are among the sought-after properties. Tetrapyrroles and their analogues fulfil these criteria well and the photochemistry of many of these compounds, which can also act as photosensitisers for solar energy conversion, were investigated by Porter and colleagues. Phthalocyanine dyes are among the leading second generation photosensitisers now being used for PDT and work carried out at the Royal Institution was instrumental in bringing these compounds forward. My own involvement in this field started there working with David Phillips on metallated sulphonated phthalocyanines which were chosen for their strong absorption near 700 nm and good triplet state yields, as measured using the transient absorption techniques originally inspired by Porter, which serves to highlight how fundamental advances may be translated into important practical applications.

Contributed Article

PHOTOSENSITISING PROPERTIES OF PROTOPORPHYRIN IX AND PHOTOPROTOPORPHYRIN IN RELATION TO PHOTODYNAMIC THERAPY

Alexander J. MacRobert

ABSTRACT: This article concerns the photosensitisation properties of porphyrin dyes which are among the most important sensitisers for photodynamic therapy (PDT). Some of the first sensitisers used for PDT were haematoporphyrin and its derivatives, but more recently protoporphyrin IX (PpIX) has attracted much interest since it is the photoactive metabolite of the prodrug, 5-aminolaevulinic acid, which offers some unique advantages over conventional methods. The photosensitiser triplet state plays a key role in the mechanism since it can be quenched by oxygen to produce the powerful oxidising species, singlet oxygen, under the type 2 process. However the sensitiser can in parallel undergo photodegradation as a result of singlet oxygen mediated reactions and type 1 processes which can ultimately limit its effectiveness. PpIX is exceptional in that it can also be photomodified to another photosensitiser, photoprotoporphyrin (Ppp), with longer peak red absorption at about 670 nm. This study compares the photosensitisation efficiency of PpIX and its photoproduct in cells using laser irradiation at the respective therapeutic absorption maxima (635 and 670 nm) and shows that Ppp is more effective than PpIX. A method is also presented for measuring the photodegradation quantum yields of PpIX and Ppp in cells using fluorescence spectroscopy together with the Ppp emission lifetime.

Introduction

The photophysical and photochemical properties of certain dyes such as porphyrins render them suitable for use as photosensitising agents in a variety of applications, including medicine. Photodynamic therapy (PDT) involves the administration of a photosensitising drug which is activated usually in the 600–700 nm range to induce the destruction of cancerous tumours and other lesions.[1] PDT-induced damage to cells is attributed to the oxidation of biomolecules (amino acids, lipids etc.) by reactive oxygen species, of which the most important is thought to be singlet oxygen produced via the type 2 mechanism:

$$PS\ (S_0) + h\nu \rightarrow PS\ (S_1) \rightarrow PS\ (T)$$
$$PS\ (T) + O_2 \rightarrow PS\ (S_0) + {}^1O_2.$$

An important feature of this process is that the recycled ground state photosensitiser (PS) can absorb more light thus enabling continuous generation of singlet oxygen, provided enough oxygen is present. Eventually the sensitiser will photodegrade and the oxygen supply may also decline but, owing to the efficiency of this cyclical mechanism, only small cellular concentrations of photosensitiser are required for PDT which is obviously important in any medical application. The initial criteria to be satisfied in selecting a photosensitiser for PDT are firstly, a good triplet yield which in turn should confer a good singlet oxygen quantum yield, and secondly, good absorption at red or near-infrared wavelengths where tissue is relatively transparent; although since the triplet state electronic energy should exceed that of singlet oxygen versus the ground state, the maximum dye absorption wavelength is to date limited to about 800 nm. In PDT, light is generally provided by a laser (typically a diode laser) which can be delivered via thin fibre optics for minimally invasive local destruction of tumours at virtually any site in the body. However the applications are not confined to tumour treatment and one of the most active fields is the treatment of age-related macular degeneration which is a common condition in later years ultimately leading to blindness.[2] PDT may also be used to destroy bacteria and viruses.[3] Gene therapy is now being combined with PDT in a number of ways, for example sub-lethal irradiation can be used to improve the intracellular delivery of therapeutic genes. Cells may also be genetically modified through luciferase transfection to self-destruct in the presence of a photosensitiser and luciferin. The resulting intracellular bioluminescence appears to lead to very efficient activation of the photosensitiser[4] despite the weak emission intensities and this method potentially offers a new therapy for disseminated disease since the light does not have to be delivered externally. In the present context it is perhaps appropriate to note we suggested that a Förster energy transfer mechanism may contribute to the efficiency of sensitiser activation in an analogous manner to photosynthetic reaction centres which Porter studied over many years.

The fact that tissue is relatively transparent at red and near-infrared wavelenths has lead to the investigation of many porphyrins and porphyrin analogues for PDT, intially with haematoporphyrin-based sensitisers and more recently with phthalocyanine dyes.[5] Some of these compounds have also been studied by Porter and co-workers[6] as photosensitising agents for solar energy conversion. In that work a palladium phthalocyanine was investigated and recently a palladium bacteriochlorophyll derivative has been shown to be an effective sensitiser for PDT.[7] The fact that it is phosphorescent makes this compound unique compared to other dyes used in PDT where fluorescence emission is the dominant radiative process. In a different approach to PDT, there has been much interest over the past few years in protoporphyrin IX (PpIX) since it is the photoactive metabolite of the prodrug, 5-aminolaevulinic acid (5-ALA), which offers some unique advantages over conventional methods. ALA is a naturally-occurring compound present in all nucleated mammalian cells which is metabolised to a protoporphyrin IX via the haem biosynthetic pathway. It is worth noting that PpIX and chlorophyll largely share the same biosynthetic pathway, which is probably no coincidence, and only diverge in the final steps with the insertion of magnesium to form chlorophyll. The conversion of PpIX to haem through the insertion of iron via ferrochelatase is a rate-limiting step which enables the intracellular accumulation of PpIX following exposure of cells to a large excess of ALA. The accumulation of the porphyrin can also be increased by inhibition of ferrochelatase using iron chelators. Clinical advantages of this approach include rapid clearance from tissue resulting in short-lived cutaneous photosensitivity and selective epithelial photosensitisation rendering it suitable for treatment of early tumours.[8] After administration of ALA, typically with topical application, a rapid build-up of PpIX takes place within a few hours and treatment is then carried out using red light, generally at 635 nm, to activate the PpIX.

PpIX absorbs light throughout the visible spectrum and emits characteristic red fluorescence over the 600–730 nm range. Upon irradiation in the presence of oxygen, PpIX in solution undergoes rapid photodegradation and new fluorescence bands are observed corresponding to a range of hydroxyaldehyde photoproducts resulting from singlet oxygen attack on the vinyl groups.[9–17] One of the main photoproducts produced is a chlorin moiety, photoprotoporphyrin (Ppp), with peak absorption around 670 nm and peak fluorescence emission around 674 nm. The photodegradation mechanisms, which can be strongly influenced by the solvent used, are presented and summarised in the literature.[18–20] The possibility that some of the photoproducts could also be effective photosensitisers has stimulated interest in their role, especially for Ppp, in the overall efficacy of ALA-PDT and has led to speculation that concomitant or sequential irradiation of ALA sensitised lesions with 635 and 670 nm light might be more effective in harnessing the therapeutic effects arising from both PpIX and Ppp.[21,22] Of course rapid photobleaching can be a disadvantage since the PDT effect becomes self-limiting. Recently two-photon excitation of PpIX has been used[23] in PDT which has the advantage that bleaching is confined to the focal plane of the laser although whether multiphoton irradiation will have any practical use remains to be seen.

Experimental Methods

Photosensitisers. PpIX dimethyl ester (HPLC > 95%) was obtained from Sigma while photoprotoporphyrin dimethyl ester (TLC > 95%) was obtained from Scientific Frontiers. Concentrated stock solutions were prepared in dimethyl sulphoxide, DMSO, for the cellular studies. The cell line used throughout was PAM 212 murine transformed keratinocytes grown in RPMI medium and provided by the Centre for Dermatology, University College London. A standard MTT assay from Sigma was used to assess the cell toxicity with a standard microplate reader (Dynatech MR 700). Comparative absorbance spectra of PpIX and Ppp were taken in DMSO using a Perkin-Elmer Lamda 40 spectrophotometer (Fig. 1). The spectra showed no evidence of aggregation.

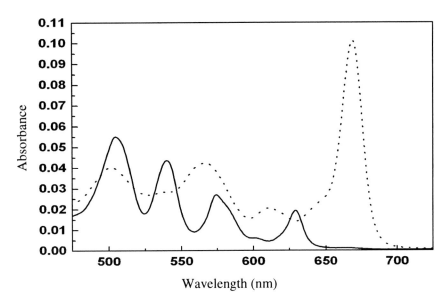

Fig. 1. Relative absorbance spectra of 20 μM PpIX and 20 μM Ppp in DMSO. The solid curve denotes the absorbance of PpIX and the dotted curve the absorbance of Ppp, shown in optical densities. The spectra were recorded using 1 mm path length glass cuvettes.

Fluorescence imaging. The fluorescence images of cells incubated with both Ppp and PpIX were taken with an Olympus IMT-2 microscope coupled to a liquid nitrogen cooled CCD (600 × 400 pixels, Mark 1 Wright Inst., Enfield, UK). The fluorescence excitation was carried out with a 543 nm He-Ne laser coupled via a liquid light guide to the microscope, and images were captured in greyscale format by the CCD camera via a long-pass filter.

Fluorescence lifetime measurements. The lifetime measurements used the same microscope as for the fluorescence imaging. Time-correlated single photon counting (TCSPC) was employed for the determination of lifetimes and a picosecond (40 ps FWHM) Picoquant PDL 800 diode laser operating at 635 nm and 5 MHz was used for the excitation of fluorescence. The output of this laser was weakly focused through the 20× microscope objective via a dichroic mirror in epi-fluorescence configuration onto a cell-plated microscope cover slip, giving a spot size of about 100 micron. The fluorescence from the illuminated spot, which sampled approximately 10 cells, was passed through the dichroic mirror to a trinocular port, where it was focused into a 1 mm fibre bundle positioned just above the exit port. The distal end of the fibre bundle was connected via an SMA connector to the entrance slit of a monochromator equipped with a long pass filter for extra rejection of the scattered light, and set to 670 nm which corresponds to the maximum of Ppp fluorescence emission. The monochromator output was then fed to a fast response photon-counting photomultiplier (PMT) module (Hamamatsu 5783-01), and the signal was acquired by a TCSPC photon counting card synchronised with the laser pulse (TimeHarp 100, Picoquant GmBh, Germany), installed on a dedicated PC. The lifetimes were obtained via the TimeHarp 100 interface and analysed together with the instrument response function (350 ns FWHM) using software (Fluofit) provided by Picoquant. This experimental set-up was optimised prior to the actual experiments with the use of a Ppp-DMSO solution between a coverslip and a microscope slide, identical to the ones used in the cell experiment. In the actual cell experiment the background was found to be virtually zero.

Fluorescence photobleaching experiments. Cells were irradiated at 670 nm (Hamamatsu LD670C diode laser) or at 635 nm (Diomed 635 nm diode laser). The irradiation doses were in the range of 0.1–15 Jcm^{-2}, at a laser fluence of 8.13 mW cm^{-2} for the 670 nm laser and the 635 nm irradiation of the Ppp loaded cells, and 40 mW cm^{-2} for the 635 nm irradiation of the PpIX loaded cells. A fluorescence spectrum was taken prior to irradiation and then after each irradiation dose with fluorescence excitation provided by a 543 nm He-Ne laser; no photobleaching effects were observed using this low power irradiation. The output of the He-Ne laser was coupled into a bifurcated probe with the single central 600 µM core fibre branch carrying the excitation light through the underside of the Petri dish. The second branch of the bifurcated probe consisting of a bundle of six fibres, concentrically arranged around the excitation fibre, was used to collect emission from the sample and deliver it to the entrance of a 600 l/mm diffraction grating spectrometer, equipped with a three-stage Peltier-cooled MCP-CCD (600 × 400 pixels, Mark II Wright instruments Ltd). An OG590 Schott long-pass filter was placed in front of the CCD array of the spectrometer to eliminate any spurious interference below 590 nm. The integration time of the spectrometer was set to 5 second/scan and data were acquired and analysed on a dedicated PC computer. The cells were washed with saline solution prior to irradiation. The output of the He-Ne laser was measured to be around 1 mW throughout the experiments and the spectra acquired were treated with background subtraction with respect to a control sample. The peak intensity value of each spectrum (at 636 nm for PpIX and 674 nm for Ppp, no deconvolution was necessary) was registered after each irradiation and the photobleaching curves were constructed from these points against the irradiation fluence.

Laser irradiation, uptake and survival assays. The cells (approximately 15,000/well) for uptake measurements were inoculated into 96 well plates until they were about 90% confluent. Following incubation with the dyes for set times, the plates were placed in the well plate reader of a LS50B Perkin Elmer spectrofluorimeter. For the Ppp incubated plate, an excitation wavelength of 430 nm was selected and the emission at 674 nm was detected. For the plate incubated with PpIX, the excitation wavelength was set to 405 nm corresponding to the Soret maximum of PpIX absorption and the emission was detected at 635 nm. The background values corresponding to the wells incubated with plain medium, were averaged and the mean was consequently subtracted from the fluorescence values. The average of each set was taken and the resulting values were normalised to the maximum. The errors on the curves are presented as the standard deviation values. Laser irradiation was carried out using either 670 nm (Hamamatsu LD670C diode laser) or 635 nm (Diomed 635 PDT diode laser). The use of a 400 µM microlens fibre ensured uniform irradiation within a circular region of 6 cm diameter at an irradiation fluence rate of 8.13 mW cm^{-2}. A standard MTT assay was performed on the plates to determine the cell survival rate. The unsensitised wells, irradiated at the highest irradiation dose, were used as a light (+) control.

Results

The rates of cellular uptake of PpIX and Ppp were similar and are shown in Fig. 2. The maximum incubation time was 6 h 45 min, however a plateau of the fluorescence values in both curves can be seen to appear after 4 h of incubation. The results of fluorescence imaging of the PAM 212 cells incubated with PpIX and Ppp respectively can be seen in Figs. 3(a) and 3(b) respectively. The punctate extranuclear localisation pattern appears to be similar for the compounds. Following prolonged irradiation the fluorescence distribution became more diffuse.

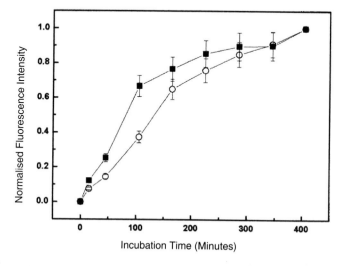

Fig. 2. PpIX and Ppp fluorescence (normalised) from PAM 212 murine keratinocytes. The cells were incubated with 20 µM PpIX and Ppp in 1% DMSO 1640 RPMI 10% fetal calf serum rich medium. The open circle symbols (dotted line) correspond to the PpIX normalised fluorescence values whereas the solid square symbols (solid line) correspond to the Ppp normalised fluorescence values.

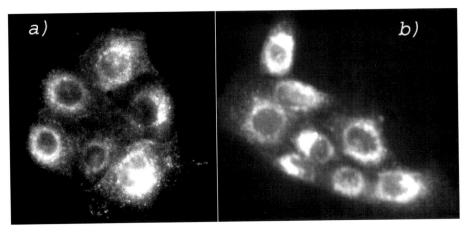

Fig. 3. Fluorescence microscopy images of PAM 212 cells incubated with (a) PpIX, and (b) Ppp for 5 h. The fluorescence excitation was in both cases performed with a He-Ne laser at 543 nm.

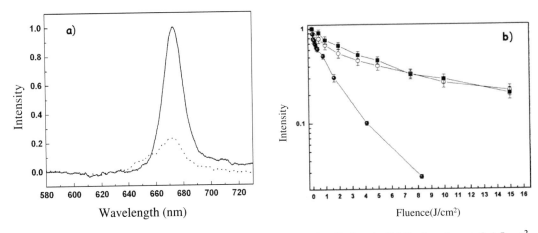

Fig. 4. (a) Emission spectrum of Ppp loaded cells prior to irradiation (solid line) and post 3.4 Jcm^{-2} 670 nm laser irradiation (dotted line). (b) Normalised fluorescence photobleaching curves after 5 h incubation with 20 μM PpIX and Ppp and irradiation. The solid sphere symbols represent the photobleaching of Ppp in cells after 8.13 mWcm^{-2} 670 nm light irradiation. The open circle symbols represent the photobleaching of PpIX in cells after 40 mWcm^{-2} 635 nm light irradiation and finally the solid square symbols depict the photobleaching of Ppp in cells under 8.13 mW cm^{-2} 635 nm light irradiation. The fluorescence excitation was performed at 543 nm.

Fluorescence photobleaching measurements were performed on the cells incubated with PpIX or Ppp, and in the latter case, irradiation with 670 and 635 nm was compared. The results of the photobleaching are presented in Fig. 4 for cells incubated with 20 μM of PpIX and Ppp for 5 h. In Fig. 4(a) the spectrum of Ppp prior to irradiation is shown as the solid line curve and the spectrum after 3.4 Jcm^{-2}, 670 nm irradiation can be seen as the dotted curve; after a light dose of 3.4 Jcm^{-2} the maximum of Ppp fluorescence drops to approximately 20% of

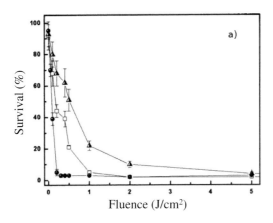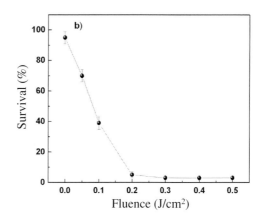

Fig. 5. Survival curves after 5 h incubation with 20 µM PpIX and Ppp and laser irradiation. (a) The solid sphere symbols (no fit) depict the survival rate of the cells incubated with Ppp and irradiated with 8.13 mW cm^{-2}, 670 nm light. The open squares depict the survival curve of the cells incubated with Ppp and irradiated with 8.13 mW cm^{-2} 635 nm light. The half-filled, up-triangle symbols depict the survival curve of the cells incubated with PpIX and irradiated with 8.13 mW cm^{-2}, 635 nm light. (b) The survival curve of cells incubated with Ppp presented over a narrower fluence range.

its value. Furthermore a shoulder formed at 650 nm suggesting the appearance of a new photoproduct but there was no evidence of PpIX formation. The cellular survival was measured 24 h post irradiation as shown in Fig. 5. In all the experiments the light (+) control samples were found to exhibit 96 ± 3% survival, and sensitiser dark toxicity was negligible. In Fig. 5(b) the steep survival curve of cells incubated with Ppp is shown over a narrower fluence range.

Photobleaching quantum yields. We can deduce the photobleaching quantum yields for the three cases (Ppp at 670 and 635 nm, and PpIX at 635 nm) assuming a first-order exponential decay, according to Eq. (1):

$$F_t = F_0 \exp(-Q_{PB} \cdot AR \cdot t) \tag{1}$$

where F_t is the fluorescence intensity after irradiation for time t and F_0 is the initial (pre-irradiation) fluorescence value, equal to unity, from the normalised photobleaching curves of Fig. 4. Q_{PB} is the photobleaching quantum yield (molecular degradation per photon absorbed), AR is the molecular absorption rate in photons absorbed per molecule per second, and t is the time of irradiation in seconds. To calculate the photobleaching quantum yield, the absorption rate per molecule must be calculated which is given by

$$AR = I \cdot \sigma \tag{2}$$

where σ is the molecular absorption cross-section in units of cm^2 molecule^{-1}, and I is the light intensity (photons cm^{-2} s^{-1}). The molecular cross-section is derived from the extinction coefficient as defined by the Beer–Lambert equation[24]:

$$\log \frac{I_0}{I} = \varepsilon \cdot c \cdot d = OD \qquad (3)$$

where I_0 is the incident light intensity and and I is the intensity after absorption, ε is the decadic molar extinction coefficient (M^{-1} cm^{-1}), c is the molar concentration and d is the light path length in the solution in cm. OD is the optical density of the solution. Using the measured extinction coefficients, the absorption cross-section is given by[24]:

$$\sigma = 3.82 \cdot 10^{21} \cdot \varepsilon \qquad (4)$$

from which, $\sigma_{Ppp(670)} = 1.9 \times 10^{-15}$ cm^2 molecule^{-1}, $\sigma_{Ppp(635)} = 0.27 \times 10^{-15}$ cm^2 molecule^{-1} and $\sigma_{PpIX(635)} = 0.37 \times 10^{-15}$ cm^2 molecule^{-1} respectively. Inserting these values into Eq. (2) and taking into account the relative photon energies at 670 and 635 nm, we have $AR_{Ppp(670)} = 58$ photons s^{-1} molecule^{-1} at a fluence rate of 8.13 mW cm^{-2}, $AR_{Ppp(635)} = 7.4$ photons s^{-1} molecule^{-1} at a fluence rate of 8.13 mW cm^{-2}, and $AR_{PpIX(635)} = 50$ photons s^{-1} molecule^{-1} at a fluence rate of 40 mW cm^{-2}.

The quantum yield of photobleaching obtained from this equation is independent of wavelength of activation, however this calculation can only be approximate as the photobleaching curves in Fig. 4 were assumed to follow a first-order exponential decay. The photobleaching quantum yields were calculated by monoexponential fitting to the initial photobleaching data (up to approx. 50% decay). In this way with Eq. (1) we obtain for Ppp, $Q_{PB(Ppp-670)} = 6 \times 10^{-4}$, $Q_{PB(Ppp-635)} = 5 \times 10^{-4}$, and for PpIX, $Q_{PB(PpIX-635)} = 5 \times 10^{-4}$ with an estimated error of 15%. These results show that the quantum yield of photobleaching for Ppp is wavelength independent (as would be expected) and is also comparable to the corresponding value obtained for PpIX, within the constraints imposed by our analysis. This method is well suited to cellular measurements since path length considerations are not required but we do have to assume that the cross-sections of cell-bound sensitisers are equivalent to DMSO solution values.

Fluorescence lifetime measurements. Measurement of the fluorescence lifetime of Ppp in PAM 212 cells was performed as described in the experimental section using time-correlated single photon counting. The resulting fluorescence decay and the computed fit appear in Fig. 6. The open circle symbols denote the actual experimental data points whereas the solid line curve is the result of a first-order exponential decay fit. The fluorescence decay of Ppp in cells was found to be $\tau_{Ppp} = 5.35 \pm 0.17$ ns with a corresponding χ^2 value of 1.23.

Discussion

The Ppp used in the current work was a mixture of the two isomers (A and B) which are produced in similar abundance from a [2+4] cycloaddition of singlet

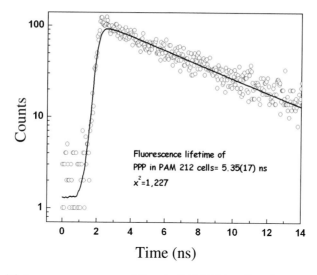

Fig. 6. Fluorescence lifetime measurements of Ppp in PAM 212 cells. The cells were incubated with 20 μM (1% DMSO) solution of Ppp for 4 h. The lifetime was measured by time correlated photon counting microscopy. The open circle symbols denote the actual photon counts whereas the solid line is the best first order exponential fit.

oxygen to the vinyl groups of PpIX. It is assumed that the A and B isomers have comparable photodynamic properties but this point is also relevant to later discussion below regarding a comparison with the properties of Ppp produced from endogenous PpIX via ALA. In principle, there is not an obvious reason for Ppp formed in the cells after irradiation of PpIX to have a different or selective isomeric composition but there is always the possibility that different cellular microenvironmental factors could favour selective production of one of the two isomers.

Using laser excitation and fluorescence detection we derived the cellular photobleaching quantum yields, which has not been carried out in previous studies, and showed that Ppp photostability is comparable to that of PpIX in cells. The photobleaching quantum yields for Ppp (5×10^{-4}) were measured for 635 and 670 nm and the values agreed within experimental error, using the intial portion of the decay curves; the decay profiles deviated from monoexponential decay although not greatly and second-order fitting was no better. These overall yields refer of course to an integrated mean of the intracellular values which would vary from site to site. Photobleaching quantum yields can vary significantly between sensitisers[19] owing to a range of mechanistic factors involving types 1 and 2 and non-oxygen dependent processes; e.g. for hematoporpohyrin $Q_{PB} = 5 \times 10^{-5}$, and for tetraphenyl sulphonated porphine, $Q_{PB} = 10^{-5}$ in aqueous air-saturated solution. In contrast phthalocyanine sensitisers are an order of magnitude more photostable.[7] One also has to distinguish between bleaching measurements based on absorption or fluorescence, although the latter technique generally has to be used in cellular or tissue studies. The photobleaching mechanism of PpIX as detailed in Ref. 19 is generally accepted to involve type 2 singlet oxygen mediated reactions involving either direct interaction with singlet oxygen or an indirect mechanism involving reaction with oxidising intermediates (e.g. hydroperoxides, denoted as SO_2 in the following scheme) generated from reaction of singlet oxygen with substrate biomolecules:

$$PpIX + {}^1O_2 \rightarrow Ppp \text{ (and other products)}$$
$$S + {}^1O_2 \quad \rightarrow SO_2$$
$$SO_2 + PpIX \rightarrow Ppp \text{ (and other products)}.$$

The indirect mechanism may well predominate in a cellular environment where singlet oxygen itself has a very short half-life, which reduces the probability of direct sensitiser interaction for which the rate constant is relatively low[12] at 8.3×10^5 M^{-1} s^{-1}. The reactive oxygen intermediates would be longer lived and in some cases could undergo diffusive intracellular transport enabling interaction with sensitiser molecules which would likewise be diffusing (unless tightly bound) albeit maintaining an equilibrated distribution. This mechanism may also apply to Ppp and has been discussed for other chlorins such as mTHPC.[25] The fact that the photobleaching quantum yields of PpIX and Ppp were found to be so close would imply common mechanistic photobleaching processes, and in each case an important photodegradation pathway in cells is likely to involve ring opening to bilirubin–type products.

The lifetime of Ppp fluorescence from PAM 212 cells was found to be 5.35 ± 0.17 ns. This value is in good agreement with a study on fibrosarcoma cells incubated with PpIX[17] following light irradiation to induce Ppp photoproduct formation: a fluorescence lifetime of 5.5 ± 0.4 ns was measured using detection at 676 nm together with a longer lifetime of 15 ns corresponding to PpIX. They also observed a short 0.5 ns component which was ascribed to aggregates. The photoproduct lifetime in irradiated PpIX/chloroform/methanol solution was similar at 4.9 ± 0.6 ns which indicates that intracellular fluorescence quenching in our study is not significant: multiexponential fluorescence decays can arise from the heterogenous subcellular localisation and thus exposure to different micro-environments, and the presence of sensitiser aggregates which can either fluoresce weakly exhibiting a shorter lifetime or quench fluorescent monomers (e.g. phthalocyanines and mTHPC).[26] In this study we observed good monoexponential characteristics from which we can infer that Ppp detected here is largely in monomeric form within the PAM 212 cells.

Both PpIX and Ppp are lipophilic molecules and, since the structure of Ppp is only slightly modified from PpIX, it would be expected that uptake kinetics and localisation of the two substances would be similar. This is reflected in the uptake measurements in Fig. 2 where it is clear that the fluorescence values tend to form a plateau after 4 h of incubation in each case. Beyond that equilibrium point little can be gained by further incubation. On the other hand, the dark toxicity rises from 4% after five hours of incubation to approximately 15% after 24 h of incubation. From the two fluorescence micrographs in Fig. 3, the sub-cellular localisation of the fluorescence appears similar for the compounds, presumably corresponding in part to the mitochondrial localisation pattern expected for lipophilic sensitisers, although the precise localisation will differ since Ppp is less hydrophobic than PpIX. Upon further irradiation the fluorescence became more diffuse presumably due to photoinduced changes to organelle structures and dye release. Comparing the phototoxcity of Ppp at 670 and 635 nm the results show that Ppp using 670 nm irradiation has a strong cytotoxic effect and that even at 635 nm irradiation the cytotoxic effect of Ppp is still stronger than that of PpIX, under the same incubation conditions. From the data shown in Fig. 5, it can be roughly deduced that Ppp is about 3 times more cytotoxic when irradiated by 670 nm light than when irradiated by 635 nm light. On the other hand at 635 nm light irradiation the photodynamic effect of Ppp is again roughly three times higher than PpIX under the same incubation conditions which makes Ppp using 670 nm irradiation about 9 times more phototoxic than PpIX at 635 nm irradiation.

A comparison of photodynamic efficacy or quantum yield for cell inactivation between these two compounds has to take into account the relative extinction coefficients of Ppp and PpIX at 635 nm, cellular localisation and efficiency of cytotoxic species generation, particularly the singlet oxygen yields. The singlet oxygen quantum yields of PpIX and Ppp (both DME forms) have been measured by Keene *et al.*[27] to be 0.57 and 0.49 in benzene. Since the excitation efficiencies (at 635 nm from the extinction coefficients) and singlet oxygen yields are similar we can then infer that Ppp is more effective at 635 nm than PpIX, although this involves many assumptions, for example the relative localisation. A further factor is the rate of photobleaching, which would limit the cell kill. If we consider the LD_{50} values for PpIX and Ppp under 635 nm irradiation and combine these with the corresponding photobleaching data we find that for PpIX the LD_{50} point corresponds to 23% photobleaching, whereas the LD_{50} point for Ppp corresponds to 6% photobleaching. Therefore it is concluded that under the same incubation conditions PpIX undergoes more bleaching than Ppp to obtain a similar degree of cell kill. For PDT treatment a much higher energy dose is generally used than here since it is crucial to ensure complete cell kill in a tumour as far as possible, and it has been noted that PpIX fluorescence in treated lesions is almost completely bleached.[8]

The PAM 212 cell line has previously been used in a study of ALA-induced formation of PpIX.[28] Several studies have investigated the idea of combining PpIX (produced endogenously from 5-ALA) activation at 635 nm with activation using 670 nm of its photoproduct, photoprotoporphyrin. In a recent investigation by Ma *et al.*[22] after ALA incubation of WiDr adenocarcinoma cells and irradiation at 635 nm the cells were subsequently irradiated with 670 nm light for the additional activation of the formed Ppp and the overall photodynamic effect was estimated with the use of two assays, namely propidium iodide intracellular fluorescence measurements and standard methylene blue (MB) survival assay. While the propidium iodide measurements showed enhanced cell membrane inactivation the final MB assay did not show a statistically significant addition to the cell mortality caused by activation of PpIX. However in this work it has been shown that 635 nm activation is effective for Ppp as well so it may well be difficult to define the relative contributions. Since Ppp is such an effective sensitiser the question arises as to whether this photoproduct (and others) can be utilised to their full potential for ALA-PpIX PDT. There are several possible ways to exploit any enhancement of the PDT effect through combined excitation of PpIX and the photoproducts. Firstly there is the use of different irradiation regimes to optimise Ppp formation which presumably is largely dependent upon the oxygenation levels. Another possibility would be to modify intracellular production of PpIX and its photobleaching properties with the addition of iron chelating agents or other compounds. Several studies have shown that therapeutically useful increases in PpIX levels can be attained in combination with desferral or hydroxypyridinone iron chelators.[29] In principle the chelation of intracellular ferrous iron could through the Fenton reaction affect levels of peroxy intermediates and thus influence photobleaching rates which might be a profitable area for further study.

Acknowledgments

This work was supported through the Association for International Cancer Research. Dr. T. Theodossiou is acknowledged for all his experimental efforts in contributing to this work and Jo Woodhams is thanked for her help in preparation of the manuscript.

References

1. E. J. Dennis, G. J. Dolmans, D. Fukumura and R. K. Jain, *Nature Reviews* (*Cancer*) **3**, 380–387 (2003).
2. H. Van den Bergh, *Seminars Opthalmol.* **16**, 181–200 (2001).
3. N. Komerik, H. Nakanishi, A. J. MacRobert, B. Henderson, P. Speight and M. Wilson, *Antimicrob Agents Chemother.* **47**, 932–940 (2003).
4. T. Theodossiou, J. S. Hothersall, E. A. Woods, K. Okkenhaug, J. Jacobson and A. J. MacRobert, *Cancer Res.* **63**, 1818–1821 (2003).
5. D. Phillips, *Prog. Reaction Kinetics* **22**, 175–300 (1997).
6. J. R. Darwent, I. McCubbin and G. Porter, *J. Chem. Soc. Faraday 2* **78**, 903–910 (1982).
7. N. V. Koudinova, J. H. Pinthus, A. Brandis, O. Brenne, P. Bendel, J. Ramon, Z. Eshhar, A. Scherz and Y. Salomon, *Int. J. Cancer* **104**, 782–789 (2003).
8. Q. Peng, T. Warloe, K. Berg, J. Moan, M. Kongshaug, K. E. Giercksky and J. M. Neslan, *Cancer* **79**, 2282–2308 (1997).

9. H. H. Inhoffen, H. Brockmann and K. M. Bliesener, *Liebigs. Ann. Chim.* **730**, 173–185 (1969).

10. G. S. Cox, C. Bobillier and D. G. Whitten, *Photochem. Photobiol.* **36**, 401–407 (1982).

11. G. S. Cox and D. G. Whitten, *J. Am. Chem. Soc.* **104**, 516–521 (1982).

12. G. S. Cox, M. Krieg and D. G. Whitten, *J. Am. Chem. Soc.* **104**, 6930–6937 (1982).

13. K. König, H. Schneckenburger, A. Rück and R. Steiner, *J. Photochem. Photobiol. B: Biol.* **18**, 287–290 (1993).

14. S. Bagdonas, L. W. Ma, V. Iani, R. Rotomskis, P. Juzenas and J. Moan, *Photochem. Photobiol.* **72**, 186–192 (2000).

15. J. M. Wessels, R. Sroka, P. Heil and H. K. Seidlitz, *Int. J. Radiat. Biol.* **64**, 475–484 (1993).

16. R. Sørensen, V. Iani and J. Moan, *Photochem. Photobiol.* **68**, 835–840 (1998).

17. D. J. Robinson, H. S. de Bruijn, N. van der Veen, M. R. Stringer, S. B. Brown and W. M. Star, *Photochem. Photobiol.* **67**, 140–149 (1998).

18. W. S. L. Strauss, R. Sailer, M. H. Gschwend, H. Emmert, R. Steiner and H. Schneckenburger, *Photochem. Photobiol.* **67**, 363–369 (1998).

19. R. Bonnet and G. Martinez, *Tetrahedron* **57**, 9513–9547 (2001).

20. E. F. G. Dickson and R. H. Pottier, *J. Photochem. Photobiol. B: Biol.* **29**, 91–93 (1995).

21. P. Charlesworth and T. G. Truscott, *J. Photochem. Photobiol. B: Biol.* **18**, 99–100 (1993).

22. L. W. Ma, S. Bagdonas and J. Moan, *J. Photochem. Photobiol. B: Biol.* **60**, 108–113 (2001).

23. R. L. Goyan and D. T. Cramb, *Photochem. Photobiol.* **72**, 821 (2000).

24. J. R. Lakowicz, *Principles of Fluorescence Spectroscopy* (Plenum Press, New York, 1983).

25. L. Kunz and A. J. MacRobert, *Photochem. Photobiol.* **75**, 28–35 (2002).

26. J. P. Connelly, S. W. Botchway, L. Kunz, D. Pattison, A. W. Parker and A. J. MacRobert, *J. Photochem. Photobiol. A: Chem.* **142**, 169–175 (2001).

27. J. P. Keene, D. Kessel, E. J. Land, R. W. Redmond and T. G. Truscott, *Photochem. Photobiol.* **43**, 117–120 (1986).

28. B. Ortel, N. Chen, J. Brissette, G. P. Dotto, E. Martyin and T. Hasan, *B. J. Cancer* **77**, 1744–1751 (1998).

29. Ø. Bech, D. Phillips, J. Moan and A. J. MacRobert, *Photochem. Photobiol. B: Biol.* **41**, 136–144.

Contribution from

MARTYN POLIAKOFF, MICHAEL GEORGE

and

JAMES TURNER

University of Nottingham
UK

Jim Turner

Born in Darwen, Lancashire UK, and read chemistry at King's College, Cambridge. He obtained his PhD on "High Resolution NMR Spectroscopy" under the supervision of Norman Sheppard FRS. After a year studying theoretical chemistry with Christopher Longuet-Higgins FRS, he spent 2 years on a Harkness Fellowship, working with George Pimentel in Berkeley where he acquired his interest in Matrix Isolation. In 1964 he returned to Cambridge as Fellow of King's College and Demonstrator/Lecturer in Chemistry. In 1972, he was appointed to the Chair of Inorganic Chemistry at the University of Newcastle upon Tyne, where George Porter visited his lab. He moved to the Chair of Inorganic Chemistry at the University of Nottingham in 1979 where he currently (2006) continues to work as Emeritus Professor. As will be clear from this chapter, his principal research interests have centred on organometallic photochemistry and the structure and reactivity of reaction intermediates. He has the Tilden and Liversidge Medals of the Royal Society of Chemistry and was elected to the Royal Society in 1992.

Martyn Poliakoff

Born in London and went to King's College, Cambridge, where he met Jim Turner in his first week as an undergraduate. In 1969, he began his PhD with Turner on the "Matrix Isolation of Large Molecules". At the end of his PhD, he became a Research Officer in Newcastle, where he and Turner had a very fruitful collaboration with the late Jeremy Burdett. In 1979, he moved with Turner to Nottingham as a Lecturer in Inorganic Chemistry. Promoted to a Chair in 1991, he is now Research Professor in Chemistry. Although he continued to collaborate with Turner for many years, Poliakoff's research interests underwent a substantial shift in the 1990s when he pioneered the organometallic applications of supercritical fluids. His work is now focused on Green Chemistry and the applications of supercritical fluids for cleaner chemical production and processing. He has been awarded the Meldola and Tilden medals of the Royal Society of Chemistry. He was elected Honorary Professor at Moscow State University in 1999, Fellow of the Royal Society in 2002 and Fellow of the Institution of Chemical Engineers in 2004. Amusingly, in the context of this chapter, he was invited (2004) to chair the Paul Instrument Fund committee, exactly 20 years after his interview with George Porter.

Mike George

Born in West Bromwich, West Midlands and came to the University of Nottingham in 1984, where he met Martyn Poliakoff, his personal tutor on his first day as student. His PhD on "Time-Resolved IR Spectroscopy of Reaction Intermediates" was supervised by Poliakoff. He then was a BP Venture Research Postdoctoral Fellow with Poliakoff. He spent 6 months as an STA/Royal Society Research Fellow with Dr. Hiro-o Hamaguchi at the Kanagawa Academy of Science and Technology in Japan working on probing the structure of organic excited states with infrared spectroscopy before returning to Nottingham to begin a very productive collaboration with Turner as first Experimental and then Research Officer. He was appointed to a Lectureship in Inorganic Chemistry in 1998, Readership in 2001 and was promoted to a Chair in 2003. His current research interests lie in physical inorganic chemistry with particular emphasis on time-resolved vibrational spectroscopy and organometallic photochemistry. He has a long track record of innovation in spectroscopic instrumentation and led PIRATE, a UK national facility at the Rutherford Appleton Laboratory for picosecond IR spectroscopy; currently he is involved in ULTRA a project developing a highly sensitive time-resolved vibrational spectrometer at the Rutherford Appleton Laboratory aimed at biological sciences. He has been awarded the Sir Edward Frankland Fellowship (2002–2003) and Corday Morgan Medal (2003) of the Royal Society of Chemistry.

Four generations of PhD students, who have benefited from George Porter's scientific legacy. Turner is on the extreme left with Poliakoff next to him. In the centre is Norman Sheppard FRS, Turner's PhD supervisor and mentor to the Nottingham Time-Resolved IR project. To the right of Norman Sheppard is Mike George and, next to him, Xue Zhong Sun, Mike George's student and postdoc, who has played a key role in the development of PIRATE. The picture was taken at the banquet at the 2nd International Conference on Advanced Vibrational Spectroscopy organized by Mike George at Nottingham in August 2003, where Norman Sheppard was Guest of Honour.

MATRICES AND GLASSES: COMPLEMENTARY SOLUTIONS TO THE SAME PROBLEM

Martyn Poliakoff, Michael W. George and James J. Turner on

"Trapped Atoms and Radicals in a Glass Cage"

I. Norman and G. Porter

Nature **174**, 508 (1954)

and

"Primary Photochemical Processes in Aromatic Molecules. Part 4 – Side-Chain Photolysis in Rigid Media"

G. Porter and E. Strachan

Trans. Faraday Soc. 1595–1604 (1958)

In the current climate of rapid technological advance, we have become accustomed to the simultaneous launch of competing technologies, for example the VHS and Betamax video tape formats, one of which quickly eclipses the other. Occasionally, as with petrol and diesel engines, the two rival technologies have distinctive merits, which allow them both to survive. Fifty years ago George Porter was involved in the start of precisely this type of competition. The field was the trapping and spectroscopy of unstable radicals and molecules, a topic which has played a crucial role in the scientific careers of all three authors (Poliakoff, George and Turner).

The early years of flash photolysis had led to the recording of a whole library of spectra of radicals and unstable molecular fragments. However, the technique was restricted to UV/Vis spectra because there were no IR or EPR spectrometers which could operate fast enough. Therefore, there was a real need to devise a method for stabilising short-lived species long enough to record their spectra with conventional spectrometers. In 1954, both George Porter and George Pimentel published solutions to this problem. The two papers[1,2] were published only a few months apart, Pimentel's being submitted four days before Porter's appeared in print and, with an amusing symmetry, Porter had an American co-author, Irwin Norman (see later) while one of Pimentel's co-workers, Eric Whittle, who went on to work in Cardiff, was British. In this Commentary, we comment on these two papers,[1,2] plus a more detailed follow-up paper[3] by Porter; compare the two approaches, and give a somewhat personal view of how they have evolved. Finally, we illustrate this evolution with a recent paper of our own.[4]

The starting point for both their approaches was a series of pioneering papers[5,6] in the early 1940s by G. N. Lewis and co-workers, who studied the photoluminescence of large organic molecules and their photolysis in rigid glasses, initially in boric acid at room temperature and then at 90 K in EPA (a solvent of their own invention containing a 5:5:1 mixture of diethyl Ether, *iso*-Pentane and ethyl Alcohol, which freezes to an optically clear glass). Interestingly, Porter[6] and Pimentel[5] cited different papers by Lewis. However, it seems clear that the real inspiration for both must have been the paper[6] cited by Porter, in which, in addition to other studies, tetraphenyl hydrazine, $Ph_2N–NPh_2$, was isolated in EPA at 90 K. During UV irradiation, the colourless glass turned green and the UV absorption bands of the hydrazine disappeared. When the glass was melted, the colour faded and the hydrazine was regenerated. The colour was attributed to $Ph_2N^•$, and Lewis coined the term "photodissociation" to describe the process. One reason for Lewis's success was that most of the photogenerated species were fairly stable.

The key concept seized by both Porter and Pimentel was the stabilisation of the reactive radicals in a large excess of a rigid, inert solid material, the "glass cage" (Porter) or "matrix" (Pimentel), as demonstrated somewhat flippantly in Fig. 1. The achievement of the Porter and Pimentel papers was to generalise the Lewis approach; the beauty of the papers is that they generalised the idea in different directions and it is this difference that has allowed them both to endure.

The generalisation in the Porter paper[1] was the realisation that the Lewis glass method could be extended to stabilising smaller, much more reactive species than Ph_2N. Porter's demonstration was delightfully simple, the photodissociation of I_2. Photolysis of I_2 in a glass led to the formation of I atoms and the bleaching of the characteristic violet colour; warming the glass permitted recombination and regeneration of the colour.[a] The colour changes (coloured precursor, colourless photoproducts) are the inverse of Lewis's and the elegance of the demonstration is that the whole process is understandable without any spectroscopy. Among other developments, Porter went on to investigate the subtle photochemistry of a range of organic systems.[3]

The innovation in the Pimentel paper[2] was to create the solid "matrix" by condensing a gas onto a cold surface rather than by freezing a liquid. There were several advantages. Unstable species could be generated in the gas phase and then be trapped, the matrix could be a very thin layer since it was grown on a substrate, and there was a hint that one could use noble gases to form the matrix. In subsequent work Pimentel made extensive use of noble gases, partly because he had ready access to liquid hydrogen. Noble gases are monatomic and so their matrices are totally transparent across the entire IR spectrum. The disadvantages were that the trapped molecules (guests) had to be sufficiently volatile to be evaporated into the matrix, and the matrix could not be warmed very much without subliming or even falling off the substrate. On the other hand, the matrix was frozen very rapidly and so it was possible to achieve much higher concentrations of guest than in Porter's glass which had to be cooled,

---

[a]In fact, the I atoms probably reacted with the ethanol, since prolonged UV irradiation in a pure hydrocarbon glass results in neither photolysis nor bleaching of the colour. However this does not detract from the elegant principle of this demonstration.

often slowly enough for unwanted precipitation of the guest to occur. (Indeed, Poliakoff was so frustrated by such precipitation at one stage of his PhD that he was driven to build an apparatus with a microscope objective penetrating into the cold cell to monitor the process!) An important "marketing" advantage was that Pimentel named his approach "Matrix Isolation" while Porter's technique remained nameless.

The papers had an enormous and immediate impact that is difficult to appreciate today. Exactly four years after publication of his paper, Porter hosted an informal discussion of the Faraday Society in Sheffield on "Free Radical Stabilisation" which included both glasses and matrices. He described this topic as *"the third chapter (in the history of) free radical chemistry"* in his conference report published in *Nature*.[7] *"The meeting was attended by about one hundred and eighty scientists, half of whom were from Britain, fifty from the United States and Canada, and forty from other countries."* This must have been a huge overseas contingent in the days before low-cost airlines. Possibly much of the interest was prompted by a Cold War fascination with the effects of ionising radiation and, indeed, much of the work discussed appears to have involved γ-irradiation rather than photolysis to create the radicals. However, there was another reason which seems rather far-fetched to modern eyes. *"There has been a great deal of interest in the possible use of stabilised radicals as high-energy rocket propellants. The important quantity to be considered is the heat of reaction per gram, and there seems little hope of increasing this greatly above 3 kcal/gm with conventional molecular fuels. If pure atomic hydrogen could be utilised, the heat of reaction would be 51.2 kcal/gm but present indications are that concentrations which can be trapped are limited to one per cent or less."* Perhaps it was this aspect which lay behind the US Army's support for the conference. Overall, the conference was a great success and Porter finished his report by saying that *"the enthusiasm of discussions was typical of a field where new phenomena are found in nearly every experiment which is performed."* For some background to these comments see the book by Bass and Broida.[8]

Both Porter and Pimentel assumed that the glass/matrix would be generally inert, except when made deliberately reactive (e.g. CO) but, paradoxically, the fact that they are not inert has played a major role in the scientific careers of Turner, Poliakoff and George. In 1963, Turner was a postdoc with Pimentel in Berkeley when the discovery of the first compounds of Xe was announced. He tried a variation of Porter's I_2 experiment, photolysing F_2 in solid Kr at 20 K, and generated KrF_2,[9] identified via its $\nu(Kr-F)$ IR bands between 500 and 600 cm^{-1}. Experiments of this type continue to be important to this day; for example one of the first organo-krypton compounds has recently been made by photolysis of diacetylene in a Kr matrix.[10] The difference now is that DFT calculations can be used to predict the positions of the IR bands in advance of the experiment. Like Porter, Pimentel was a great enthusiast for what is now called "Public Awareness of Science", and Turner was dragooned into recreating this experiment in a film called "Inert (?) Gases" for high school pupils, which exists to this day.

Fig. 1. A light-hearted illustration of the principle of matrix isolation and wavelength-selective photochemistry. Interestingly, neither Porter nor Pimentel mentioned the possibility of this type of selective photochemistry in their original papers.[1,2] Note the photographic licence in this picture, the red laser should be absorbed by a blue cherry!

Fig. 2. The matrix isolation equipment in Poliakoff and Turner's laboratory at the University of Newcastle upon Tyne in 1974. The cold window in the centre is ca. 2 cm in diameter and at 20 K. The purple colour is caused by an unusually thick matrix containing Ar–Cr(CO)$_5$. Photograph taken for Poliakoff's talk at the RI (see text for details).

Poliakoff started his PhD with Turner at the time when his group was building on matrix experiments by postdoc Tony Rest, which identified Ni(CO)$_3$ (from Ni(CO)$_4$) and HMn(CO)$_4$ (from HMn(CO)$_5$), and was studying the photolysis of metal carbonyls in noble gas matrices.[11] The main difference between these experiments and the Porter/Pimentel/Lewis photochemistry was that the photodissociation of carbonyls led to two species, molecular CO and a fragment lacking two electrons rather than two radicals each lacking one electron. These carbonyl fragments proved to be highly reactive and the work of one of Turner's students, Robin Perutz, demonstrated without doubt that species such as $Cr(CO)_5$ could react with all of the noble gases except Ne (and presumably He). The coordination of the gases caused dramatic changes in the colour of the $Cr(CO)_5$ fragment, from orange to blue, depending on the noble gas.[12] This effect was initially explained by the late Jeremy Burdett as a combination of subtle changes in bond angle and orbital energies on coordination of the noble gas.[13] Figure 2 shows the purple colour of $Ar-Cr(CO)_5$ prepared specially by Poliakoff to illustrate a talk at the Royal Institution in 1974, his first invited research seminar. He was awed at the prospect of talking in front of George Porter, only to find that he was away that day! However, Porter did visit Turner's lab in 1977 and Poliakoff was able to show him the equipment actually running. Poliakoff was so thrilled by the occasion that he managed to muddle up all of Porter's lecture slides whilst loading them into the carousel, thereby severely delaying the start of the colloquium which followed the lab tour.

In other experiments, Turner and colleagues[13,14] succeeded in using polarised light to induce linear dichroism by preferentially photodissociating molecules in certain orientations, an effect suggested in Porter's paper[1] as well as in Lewis's original paper.[6] Porter also briefly alluded to the localised melting of the glass which was believed must be occurring to enable the photofragments to separate. Burdett spent a considerable time considering this problem in the mid-1970s and concluded that the concept of a "local soup" was erroneous. In his view (as far as we are aware, never published), the excess energy is likely to be conducted away by phonons from the site of absorption too fast for bulk melting. Instead, there is a minor reorganisation, sufficient for the photo-fragments to separate but less than could be considered melting. This view is strongly supported by, among other observations, the ease of UV/Vis photochemical reversal of some of the reactions,[11] and the polarised photochemical experiments.[13,14]

Perutz, Poliakoff and Turner also showed[12,13] that these carbonyl fragments were so reactive that they could interact strongly with alkanes, including CH_4. This interaction had several important implications. Firstly, it implied that organometallic reaction intermediates might be solvated even in hydrocarbon solution rather than being the coordinatively unsaturated species which had been postulated up to that time. Secondly, it increased the need to observe these intermediates at higher temperatures to establish whether they existed under more realistic reaction conditions. Thirdly, it increased the attraction of Porter's frozen hydrocarbon glasses as media for studying organometallic photochemistry because interaction with hydrocarbons was intrinsic to any reaction.

By the mid-1980s, advances in FTIR instrumentation and the availability of reliable IR cells cooled by liquid N_2 led to a resurgence of interest in Porter's glasses, particularly by Mark Wrighton and his group at MIT.[15] They studied a range of organometallic reactions, identified intermediates and exploited the fact that glasses can be warmed up to observe thermal reaction which follow the initial photodissociation, just as Porter had done with I_2. Studies of this type have continued right up to the present.

Although there had been previous, very difficult experiments, combining flash photolysis with IR detection,[16] it was the developments in IR lasers and fast detectors that really opened up the possibility of reasonably straightforward detection of transient species using IR detection. It was these developments in time-resolved IR (TRIR) which brought Poliakoff and Turner into direct contact with George Porter because he was appointed mentor for their Royal Society Paul Instrument Fund grant to build a TRIR spectrometer. Before the grant was confirmed, Poliakoff had to be interviewed by Porter in his flat at the Royal Institution. IR spectroscopy was clearly not his speciality but his questions were so searching that Poliakoff felt uncomfortably like a PhD candidate. The grant was funded, a micro-second TRIR spectrometer was built, and Poliakoff reported some early results[17] at the Symposium on Flash Photolysis and its Applications, held in 1986 to mark Porter's retirement. At the symposium banquet, Poliakoff met Irwin Norman, co-author of the *Nature* paper,[1] and was able to thank him for his pioneering work (*see* Fig. 3).

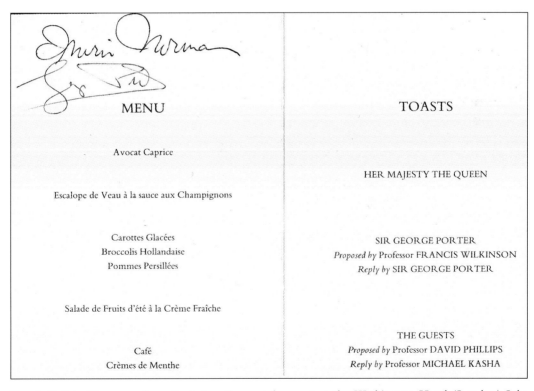

Fig. 3. Poliakoff's menu from Porter's retirement banquet at the Washington Hotel (London) July 15th 1986, signed by Irwin Norman and George Porter, the two authors of the key *Nature* paper.[1]

When Porter was appointed President of the Royal Society, Norman Sheppard FRS, Turner's PhD supervisor, took over as mentor of the TRIR project. Turner and George then developed TRIR to a nanosecond timescale where, among other experiments, they have combined TRIR with Porter's glasses and polarised photolysis to measure photoinduced dichroism in the IR spectrum of organometallic compounds in an electronic excited state.[18] Most recently, George has led the development of PIRATE, a UK national picosecond TRIR facility,[19] which has enabled him to monitor for the first time the reaction of "naked" organometallic intermediates with hydrocarbons.[20] However, these developments have not eliminated the need for glasses and matrices.

As early as 1960, Porter addressed one of the limitations of matrix isolation, namely that the guest compound had to be volatile, by using a pressed potassium chloride disc, such as is used for IR spectroscopy.[21] It worked well for the photogeneration of organic radicals. The difficulty was that the guest was not dispersed on a molecular level but existed as crystallites, imposing another limitation on the technique which was only overcome in the mid-1980s when Rest and co-workers pioneered the use of polymer films, where the guest could be genuinely dispersed.[22,23] Both the KCl and polymer techniques also overcame the rather small temperature range of glasses/matrices, limited by either the evaporation or melting of the matrix. Poliakoff then developed the polymer idea further so that polyethylene films could be used to study reactions involving gases.[24] This opened up the possibility of studying catalytic reactions, where the challenge is not only to detect the intermediates but also to link them directly to the catalytic cycle. The paper[4] following this Commentary illustrates how this can be achieved in practice. It also demonstrates how the original ideas of Porter and Pimentel have been developed and how the interpretation of the spectra relies on a mass of background spectra obtained via more conventional matrix techniques and TRIR experiments.

So what is happening today? Noble gas matrices continue to be used but more rarely for photochemistry[10] because the complexity of the cryogenic apparatus and the general "black art" of the technique has discouraged all but the most committed photochemists. It does however thrive in the area of chemical physics, where it is still the method of choice for a wide variety of experiments. Frozen glasses are altogether easier to use and they have come to dominate the less specialist applications of matrix isolation. In terms of citations, Pimentel[2] has been cited 166 times since 1969 and Porter[1] 90 times but this is a somewhat misleading measure since both Porter and Pimentel published a range of papers in the early years of their techniques. An Internet search provides a less scientific measure but one which we believe conveys the spirit of their joint success. A 2004 "Google" search on the first two words of the two titles "Matrix Isolation" and "Trapped Atom" yielded respectively 4320 and 4310 hits. The conclusion is that these two pioneering papers[1,2] have proved to be of great and equal value!

References

1. I. Norman and G. Porter, *Nature* **174**, 508 (1954).
2. E. Whittle, D. A. D. Dows and G. C. Pimentel, *J. Chem. Phys.* **22**, 1943 (1954).
3. G. Porter and E. Strachan, *Trans. Faraday Soc.* **54**, 431 (1958).
4. G. I. Childs, A. I. Cooper, T. F. Nolan, M. J. Carrott, M. W. George and M. Poliakoff, *J. Am. Chem. Soc.* **123**, 6857 (2001).
5. G. N. Lewis, D. Lipkin and T. L. Magel, *J. Am. Chem. Soc.* **63**, 3005 (1941).
6. G. N. Lewis and D. Lipkin, *J. Am. Chem. Soc.* **64**, 2801 (1942).
7. G. Porter, *Nature* **182**, 1496 (1958).
8. A. M. Bass and H. P. Broida, *Formation and Trapping of Free Radicals* (Academic Press, New York, 1960).
9. J. J. Turner and G. C. Pimentel, *Science* **140**, 974 (1963).
10. H. Tanskanen, L. Khriachtchev, J. Lundell, H. Kiljunen and M. Räsänen, *J. Am. Chem. Soc.* **125**, 16361 (2003).
11. A. J. Rest and J. J. Turner, Chem. Commun. **375**, 1026 (1969); M. A. Graham, M. Poliakoff and J. J. Turner, *J. Chem. Soc. A* 2939 (1971).
12. R. N. Perutz and J. J. Turner, *J. Am. Chem. Soc.* **97**, 4791 (1975).
13. J. K. Burdett, J. M. Grzybowski, R. N. Perutz, M. Poliakoff, J. J. Turner and R. F. Turner, *Inorg. Chem.* **17**, 147 (1978).
14. J. K. Burdett, R. N. Perutz, M. Poliakoff, J. J. Turner, *J. Chem. Soc., Chem. Commun.* 157 (1975).
15. G. L. Geoffroy and M. S. Wrighton, *Organometallic Photochemistry* (Academic Press, New York, 1979), and many subsequent Wrighton papers.
16. See, for example, K. C. Herr and G. C. Pimentel, *Appl. Opt.* **4**, 25 (1965).
17. A. J. Dixon, M. A. Healy, P. M. Hodges, B. D. Moore, M. Poliakoff, M. B. Simpson, J. J. Turner and M. A. West, *Faraday Trans. II* **82**, 2083 (1986).
18. M. W. George and J. J. Turner, unpublished results.
19. M. Towrie, D. C. Grills, J. Dyer, J. A. Weinstein, P. Matousek, R. Barton, P. D. Bailey, N. Subramanium, W. Kwok, C. Ma, D. Phillips, A. W. Parker and M. W. George, *Appl. Spectrosc.* **57**, 367 (2003).
20. P. Portius, J. Yang, D. C. Grills, X. -Z. Sun, P. Matousek, A. W. Parker, M. Towrie and Michael W. George, *J. Am. Chem. Soc.* **126**, 10713 (2004).
21. H. T. J. Chilton and G. Porter, *Spectrochim. Acta* **16**, 390 (1960).
22. A. K. Campen, A. J. Rest and K. Yoshihara, *J. Photochem. Photobiol. A: Chem.* **55**, 301 (1991).
23. R. H. Hooker and A. J. Rest, *J. Chem. Phys.* **82**, 3871 (1985).
24. A. I. Cooper and M. Poliakoff, *Chem. Phys. Lett.* **212**, 611 (1993).

Reprinted from *J. Chem. Phys.* **22**, 1943 (1954). Copyright 1954 American Institute of Physics

TABLE I. Tests of the matrix isolation method.

M Matrix	A Active species	Tempera- ture	M/A	Success of isolation
Xe	NH₃	66°K	82	No
Xe	HN₃	66°K	100	No
CO₂	HN₃	85°K	70	Partial
CO₂	NH₄N₃	85°K	70	Partial
CCl₄	HN₃	85°K	100	Yes
		85°K	40	No
Methylcyclo- hexane	NO₂	115°K	100	Yes
		115°K	50	No

Matrix Isolation Method for the Experimental Study of Unstable Species

Eric Whittle,* David A. Dows,† and George C. Pimentel

Department of Chemistry and Chemical Engineering, University of California, Berkeley, California

(Received September 7, 1954)

IN the study of free radicals and other unstable substances, an inherent difficulty is the maintenance of a suitable concentration. The matrix isolation method proposed here involves accumulation of a reactive substance under environmental conditions which prevent reaction. The intent of the method is to trap active molecules in a solid matrix of inert material, crystalline or glassy. If the temperature is sufficiently low, the matrix will inhibit diffusion of the trapped molecules, thus holding the active molecules effectively immobile in a nonreactive environment.

The active material can be produced in a variety of ways: by chemical reaction involving two reactants, by pyrolysis, by dissociation in a glow discharge or an arc, or by photolysis. Only the last permits forming the active molecules *after* the trapping process has been carried out.[1] Otherwise it is necessary to produce the active material and to trap it in the matrix shortly thereafter. This can be accomplished by spraying a gas mixture of the matrix and the active species onto a surface cold enough to condense the matrix immediately. If the mole ratio of matrix to active substance, hereafter called M/A is sufficiently large the matrix will isolate trapped molecules and prevent recombination reactions.

Several experimental tests of the method have been performed. One experiment indicated that the trapping process can occur even when the trapped material would not condense without the matrix. A mixture of carbon tetrachloride (the matrix) and methane was prepared with $M/A = 50$. At a pressure below ten microns the mixture was condensed on a silver chloride window at 72°K. The infrared spectrum of the resulting solid film showed that a large fraction of the methane was trapped even though the vapor pressure of methane at this temperature is several millimeters.

Experiments testing the effectiveness of the matrix in preventing reactions were conducted using several matrix materials. The active materials were either NO_2 or one of the hydrogen bonding materials NH_3, HN_3, or NH_4N_3. The $2NO_2 = N_2O_4$ equilibrium involves an easily accessible free radical, NO_2, and both NO_2 and N_2O_4 have well-known infrared spectra. Reaction of the other materials is detected by the spectral shift of bands accompanying hydrogen bond formation. Table I indicates the experiments performed and the success of isolation as indicated by the infrared spectra of the trapped materials. The isolation was considered to be successful if the spectrum showed the presence of NO_2 or, for NH_3, HN_3, and NH_4N_3, the absence of hydrogen bonding.

Xenon is not effective as an isolating matrix at 66°K. (Preliminary experiments using xenon as a matrix at 29°K indicate that it is effective at this temperature.) The experiments indicate that the ratio M/A should be 100 or greater.

The technique has been applied to the study of unstable products of the glow discharge decomposition of hydrazoic acid, and the low temperature spectral techniques and equipment will be described there.[2] It is apparent that conditions can be found for each of the matrices CO_2, CCl_4, and methylcyclohexane (and probably also for xenon) for which the method is successful. Hence the matrix isolation technique should be useful in spectral, magnetic or dielectric studies of free radicals, reaction intermediates and other unstable species.

* Present address: Chemistry Department, University College, Cardiff, Great Britain.
† Present address: Chemistry Department, Cornell University, Ithaca, New York.
[1] Lewis *et al.* utilized glassy matrices to study suspended materials produced by photolysis or photoexcitation. Lewis, Lipkin, and Magel proposed that the absorption of light by the suspended material may result in the production of radicals, but their interest lay with the products of photoexcitation [see J. Am. Chem. Soc. 63, 3005 (1941)].
[2] Dows, Whittle, and Pimentel (to be published).

Reprinted with permission from *J. Am. Chem. Soc.* **123**, 6857–6866 (2001).
Copyright 2001 American Chemical Society

A New Approach To Studying the Mechanism of Catalytic Reactions: An Investigation into the Photocatalytic Hydrogenation of Norbornadiene and Dimethylfumarate Using Polyethylene Matrices at Low Temperature and High Pressure

Gavin I. Childs, Andrew I. Cooper,[1] Trevor F. Nolan, Michael J. Carrott, Michael W. George,* and Martyn Poliakoff*

Contribution from the School of Chemistry, University of Nottingham, University Park, Nottingham NG7 2RD, United Kingdom

Received December 27, 2000. Revised Manuscript Received April 27, 2001

Abstract: This paper presents a new method for investigating the mechanisms of homogeneously catalyzed reactions involving gases, particularly H_2. We show how the combination of polyethylene (PE) matrices and high pressure–low temperature (HPLT) experiments can be used to provide new mechanistic information on hydrogenation processes. In particular, we show how we are able to generate reaction intermediates at low temperature, and then to extract the contents of the PE film at room temperature to characterize the organic products using GC-MS. We have used our new technique to probe both the hydrogenation of dimethyl fumarate (DF), using $Fe(CO)_4(\eta^2\text{-DF})$ as the catalytic species, and the hydrogenation of norbornadiene (NBD), using $(NBD)M(CO)_4$ (M = Cr or Mo) as the catalytic species. Irradiation of $Fe(CO)_4(\eta^2\text{-DF})$ in a PE matrix at 150 K resulted in the formation of an intermediate complex tentatively assigned $Fe(CO)_3(\eta^4\text{-DF})$. Warming this complex to 260 K under H_2 leads to the formation of $Fe(CO)_3(\eta^2\text{-DF})(\eta^2\text{-H}_2)$. Further warming of the reaction system results in the hydrogenation of the coordinated DF, to generate dimethyl succinate (DS). Characterization of the intermediate species was obtained using FTIR spectroscopy. Formation of DS was confirmed using both FTIR spectroscopy and GC-MS analysis. UV photolysis of $(NBD)M(CO)_4$ in PE under H_2 in the presence of excess NBD results in the formation of the hydrogenated products norbornene (NBN) and nortricyclene (NTC), with trace amounts of norbornane (NBA) being observed. These products were in similar ratios to those observed in fluid solution. However, for $(NBD)Mo(CO)_4$, the relative amounts of the organic products change considerably when the reaction is repeated in PE under H_2 *in the absence* of free NBD, with NBA being the major product. The use of our HPLT cell allows us to vent and exchange high pressures of gases with ease, and as such we have performed gas exchange reactions with H_2 and D_2. Analysis of the reaction products from these exchange reactions with GC-MS provides evidence for the mechanism of formation of NBA, in both the *presence* and *absence* of excess NBD, a reaction which has been largely ignored in previous studies.

Introduction

This paper presents a new method for investigating the mechanisms of homogeneously catalyzed reactions involving gases, particularly H_2. Reactions between transition metal complexes and gases are a recurrent feature of organometallic chemistry, particularly in catalytic processes.[2–6] A detailed understanding of the mechanism of such reactions is essential to improve the efficiency and selectivity of the processes. In principle, spectroscopic studies can provide valuable information needed to elucidate such mechanistic pathways.[7,8] However, since most catalytic intermediates are short-lived at room temperature, it is usually necessary to employ special methods to study these species.

The most common approach is to use low-temperature matrix isolation,[9–11] slowing the reaction down so that the intermediates can be studied using conventional spectroscopic methods. In matrix isolation, the organometallic is frozen in a noble gas matrix or a hydrocarbon "glass" at temperatures in the range 10–77 K. However, the temperature range accessible in conventional matrix isolation is limited by the melting point of the matrix material. It is, therefore, usually very difficult to study the thermal steps involved in a catalytic process. Furthermore, reactions between organometallic complexes and H_2 are difficult to study because of the problems involved in trapping sufficient quantities of the gaseous reagent within the matrix.[12–14] Experi-

(1) Present Address: Department of Chemistry, University of Liverpool.
(2) Crabtree, R. H. *Chem. Rev.* **1985**, *85*, 245.
(3) Heinekey, D. M.; Oldham, W. J. *Chem. Rev.* **1993**, *93*, 913.
(4) Esteruelas, M. A.; Oro, L. A. *Chem. Rev.* **1998**, *98*, 577.
(5) Geoffroy, G. L.; Wrighton, M. S. In *Organometallic Photochemistry*; Academic Press: New York, 1979; Chapter 3.
(6) Jessop, P. G.; Morris, R. H. *Coord. Chem. Rev.* **1992**, *121*, 155.
(7) Almond, M. J. *Short-Lived Molecules*; Ellis Horwood: New York, 1990.
(8) Turner, J. J. In *Photoprocesses in Transition Metal Complexes, Biosystems and Other Molecules, Experiment and Theory*; Kochanski, E., Ed.; Kluwer: Dordrecht, The Netherlands, 1992; p 125.

(9) Almond, M. J.; Downs, A. J. *Adv. Spectrosc.* **1989**, 17.
(10) Ochsner, D. W.; Ball, D. W.; Kafafi, Z. H. *A Bibliography of Matrix Isolation Spectroscopy: 1985–1997*; NRL Publication NRL/PU/5610/98/357, Naval Research Laboratory: Washington, DC, 1998.
(11) Ball, D. W.; Fredin, L.; Kafafi, Z. H.; Hague, R. H.; Margrave, J. L. *A Bibliography of Matrix Isolation Spectroscopy: 1952–1997*; Rice University Press: Houston, TX, 1988.
(12) Sweany, R. L. *J. Am. Chem. Soc.* **1985**, *107*, 2374.
(13) Sweany, R. L. *J. Am. Chem. Soc.* **1986**, *108*, 6986.

10.1021/ja004345t CCC: $20.00   © 2001 American Chemical Society
Published on Web 06/20/2001

6858   *J. Am. Chem. Soc., Vol. 123, No. 28, 2001*

ments with liquefied noble gas solvents have overcome some of these problems,[15,16] but although this technique allows reactions to be studied in solution at low temperature, the accessible temperature range remains relatively narrow. Room temperature time resolved IR spectroscopy (TRIR), a combination of UV flash photolysis and fast IR detection, has been extensively used for the study of short-lived organometallic intermediates with lifetimes ranging from femto- to milliseconds.[17–19] The key advantage of TRIR is that it allows the study of fast chemical processes under reasonably realistic conditions (e.g., at room temperature). However, this technique cannot always provide all of the information required to understand a particular reaction.

Ideally, a single technique is needed that can span a broad temperature range *and* allow the study of reactions between organometallic compounds and gases. Building on studies by Rest and co-workers,[20] we have shown that polyethylene (PE) is a versatile matrix material for investigating organometallic reactions which can be used over a broad temperature range (10–300 K). Initially, we investigated PE as a matrix material for studying the hydrogen bonding of highly acidic fluorinated alcohols to metal centers, e.g. in Cp*Ir(CO)$_2$.[21] In an extension of this, we developed a miniature high-pressure cell that can operate at pressures of up to 300 bar over a very broad temperature range (20–300 K).[22] This cell was used to study the photochemical reaction between Fe(CO)$_5$ and N$_2$ at low temperatures.[23] The thermal stability of the PE matrix and the ability to easily exchange gases inside the reactor allowed us to study the *thermal* reaction of Fe(CO)$_4$N$_2$ with H$_2$. Subsequently,[24] we have also used our technique to study the reaction of a range of d^6 metal centers with H$_2$ and N$_2$, generating (η^6-C$_6$H$_3$Me$_3$)M(CO)$_{3-n}$(N$_2$)$_n$ (n = 1–3), (η^6-C$_6$H$_3$Me$_3$)M(CO)$_2$-(H$_2$) (M = Cr or Mo), *cis*-W(CO)$_4$(H$_2$)CS, and *trans*-W(CO)$_4$-(H$_2$)CS and monitoring their reaction chemistry.

In this paper, we describe how polymer matrix isolation can be combined with both FTIR spectroscopy and GC-MS detection, for the study of complicated catalytic mechanisms. In particular, we show that our new method allows a surprisingly high level of control over the manipulation of thermally unstable organometallic reaction intermediates, and that this can lead to an enhanced understanding of complicated reaction mechanisms. We illustrate the use of our new technique with two studies, the hydrogenation of dimethyl fumarate (DF) and norbornadiene (NBD).

In the first example, we have simplified the hydrogenation of DF catalyzed by Fe(CO)$_5$ by merely studying the photochemistry of Fe(CO)$_4$(DF) and its reaction with H$_2$. The advantage of this hydrogenation as a test reaction is that the organic compound has strong IR chromophores, the C=O groups. These can be used to detect the hydrogenation spec-

troscopically because the ν(C–O) bands of DF and the hydrogenated product dimethyl succinate, DS, are shifted significantly relative to each other.

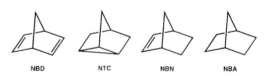

Our second example, the catalytic hydrogenation of dienes involving Group 6 transition metal compounds, is mechanistically far richer. The reaction has been studied in detail for more than 30 years,[25–32] but still has unsolved problems, particularly in the hydrogenation of NBD, where both 1,2- and 1,4-addition can occur, giving two possible hydrogenation products, norbornene (NBN) and nortricyclene (NTC). Early investigators[33,34] proposed that the catalytic cycle for this hydrogenation reaction included an intermediate with a classical dihydride bonded to the same metal center as the diene. However, the discovery of nonclassical dihydrogen complexes[35,36] prompted our group to reinvestigate this reaction using a combination of both low-temperature techniques[37] (liquefied noble gases) and TRIR spectroscopy.[38] These studies demonstrated that the key catalytic intermediates in the hydrogenation of NBD are most probably the nonclassical dihydrogen complexes of the type (NBD)M-(CO)$_3$(η^2-H$_2$) (M = Cr, Mo or W). The mechanism proposed following these studies[39] is shown in Scheme 1. The different ratios of NBN and NTC observed in the reactions catalyzed by (NBD)M(CO)$_4$ (M = Cr or Mo) were rationalized by assuming that the two metals give rise to different distributions of *mer* and *fac* intermediates (i.e., the Mo complex gives rise to more NBN because the *mer* intermediate is more favored than for Cr).

NBD   NTC   NBN   NBA

(14) Sweany, R. L. *Organometallics* **1986**, *5*, 387.

(15) Turner, J. J.; Poliakoff, M.; Howdle, S. M.; Jackson, S. A.; McLaughlin, J. G. *Faraday Discuss. Chem. Soc.* **1988**, *86*, 271.

(16) In *Molecular Cryospectroscopy*; Clark, R. J. H., Hester, H. E., Eds.; Wiley: New York, 1995.

(17) McFarlane, K.; Lee, B.; Bridgewater, J.; Ford, P. C. *J. Organomet. Chem.* **1998**, *554*, 49.

(18) Childs, G. I.; Grills, D. C.; Sun, X.-Z.; George, M. W. *Pure Appl. Chem.* **2000**, *73*, 443.

(19) McNamara, B. K.; Yeston, J. S.; Bergman, R. G.; Moore, C. B. *J. Am. Chem. Soc.* **1999**, *121*, 6437.

(20) Hooker, R. H.; Rest, A. J. *J. Chem. Phys.* **1985**, *82*, 3871.

(21) Cooper, A. I.; Kazarian, S. G.; Poliakoff, M. *Chem. Phys. Lett.* **1993**, *206*, 175.

(22) Cooper, A. I. Ph.D. Thesis, University of Nottingham, 1994.

(23) Cooper, A. I.; Poliakoff, M. *Chem. Phys. Lett.* **1993**, *212*, 611.

(24) Goff, S. E. J.; Nolan, T. F.; George, M. W.; Poliakoff, M. *Organometallics* **1998**, *17*, 2730.

(25) Nasielski, J.; Kirsch, P.; Wilputte-Steinert, L. *J. Organomet. Chem.* **1971**, *27*, C13.

(26) Platbrood, G.; Wilputte-Steinert, L. *Bull. Soc. Chim. Belg.* **1973**, 733.

(27) Wrighton, M. S.; Schroeder, M. A. *J. Am. Chem. Soc.* **1973**, *95*, 5764.

(28) Platbrood, G.; Wilputte-Steinert, L. *J. Organomet. Chem.* **1974**, *70*, 393.

(29) Platbrood, G.; Wilputte-Steinert, L. *J. Organomet. Chem.* **1974**, *70*, 407.

(30) Platbrood, G.; Wilputte-Steinert, L. *Tetrahedron Lett.* **1974**, *29*, 2507.

(31) Platbrood, G.; Wilputte-Steinert, L. *J. Organomet. Chem.* **1975**, *85*, 199.

(32) Fischler, I.; Budzwait, M.; Koerner van Gustorf, E. A. *J. Organomet. Chem.* **1976**, *105*, 325.

(33) Darensbourg, D. J.; Nelson, H. H. *J. Am. Chem. Soc.* **1974**, *96*, 6511.

(34) Darensbourg, D. J.; Nelson, H. H.; Murphy, M. A. *J. Am. Chem. Soc.* **1977**, *99*, 896.

(35) Kubas, G. J.; Ryan, R. R.; Swanson, B. I.; Vergamini, P. J.; Wasserman, H. J. *J. Am. Chem. Soc.* **1984**, *106*, 451.

(36) Kubas, G. J. *Acc. Chem. Res.* **1988**, *21*, 120.

(37) Jackson, S. A.; Hodges, P. M.; Poliakoff, M.; Turner, J. J.; Grevels, F. W. *J. Am. Chem. Soc.* **1990**, *112*, 1221.

(38) Hodges, P. M.; Jackson, S. A.; Jacke, J.; Poliakoff, M.; Turner, J. J.; Grevels, F. W. *J. Am. Chem. Soc.* **1990**, *112*, 1234.

(39) In this scheme, and in subsequent schemes, carbonyl groups are represented by lines without labels. All other ligands are shown explicitly.

Scheme 1. Reaction Scheme for the Catalytic Hydrogenation of Norbornadiene (NBD) Using (NBD)M(CO)₄ (M = Cr, Mo or W)^a

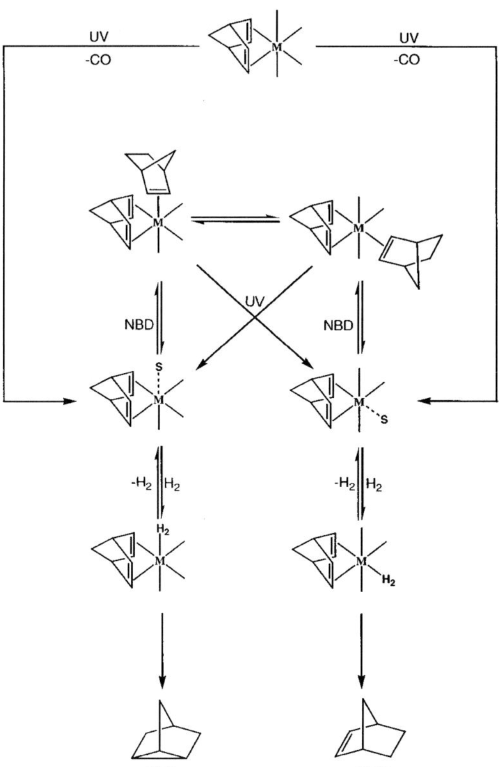

^{a} Adapted from Hodges et al.[38]

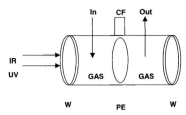

Figure 1. Schematic representation of the HP-LT Cu cell used in the experiments described here. A high pressure of gas can be introduced (or removed) from the cell, which is connected to a coldfinger (CF). The gas can freely permeate around the PE film. Reactions can be initiated using a UV lamp, and monitored using in situ FTIR. Irradiation and spectroscopy use the same CaF₂ windows (W) in the Cu cell body.

There are still fundamental, but as yet unanswered, questions concerning this catalytic cycle. For example, the formation of the doubly hydrogenated species, norbornane (NBA), as a minor product in these reactions has not been satisfactorily explained. In addition, while it was possible to detect and identify a wide range of intermediates, it was never possible to observe *directly* the crucial step, i.e., the transfer of hydrogen from the metal to the organic substrate.

We now explain the principles of our technique before describing the results in detail.

Experimental Section

The apparatus used for polymer matrix isolation studies has been described in detail elsewhere (see Figure 1).[22,23] Briefly, a 250 μm thick polyethylene disk (Hostalen GUR-415 PE (Hoechst)) was impregnated with the organometallic compound under investigation. Clarke et al.[40] have shown that this form of PE has a very low degree of unsaturation, and have demonstrated that interaction between olefinic double bonds in the PE and unsaturated photofragments is negligible. Impregnation was achieved either by gentle warming of the PE and organometallic compound under reduced pressure or by placing the PE disk in an alkane solvent saturated with the organometallic compound. Once impregnated, the disk was clipped into the high pressure−low temperature (HP-LT)

(40) Clarke, M. J.; Howdle, S. M.; Jobling, M.; Poliakoff, M. *J. Am. Chem. Soc.* **1994**, *116*, 8621.

copper cell.[22,23] The cell was filled with reactant gas (typically H₂ or N₂) and was cooled to the required temperature using an Air Products Displex CS-202 cooler. Temperatures were measured accurately using a Scientific Instruments Inc. 9600-1 silicon diode temperature controller. Once at the required temperature, reactions were initiated using broadband UV from a Phillips HPK 125W medium-pressure Hg arc. Reactions were monitored using a Nicolet 730 IR interferometer interfaced to a PC running OMNIC software. Polyethylene (PE) has very weak temperature-dependent IR absorptions in the ν(CO) region, therefore background spectra were recorded with unimpregnated PE disks over the range of temperatures required for the experiments.

GC-MS analysis was performed using a Hewlett-Packard 5890A gas chromatograph fitted with a splitless injection system and interfaced to a Trio 2000 quadrupole mass spectrometer (VG Biotech) with a GC column (SE-54 stationary phase, methylpolysiloxane, 5% phenyl substituted, 30 m, 0.32 mm (i.d)). Positive electron impact ionization (electron energy = 70 V) was used to induce fragmentation. The mass spectrum associated with each GC peak was compared with the American National Bureau of Standards (NBS) Library for identification. GC-MS data were initially obtained using the full scanning mode, which scanned the entire mass range (30−300 amu) throughout the entire chromatographic run. Single ion monitoring (SIM), which detects distinctive and diagnostic ions, was used to obtain GC spectra of improved signal-to-noise ratio. The products from the norbornadiene experiments were eluted using an oven temperature of 303 K for 5 min rising to 308 K at 5 K per minute. The extracts from the hydrogenation of DF were eluted using an oven temperature of 323 K for 5 min rising to 473 K at 10 K per minute. The injector temperature was 523 K in all cases.

Fe(CO)₄(DF) and (NBD)M(CO)₄ (M = Cr and Mo) were gifts from Prof. F.-W. Grevels (Max Planck Institut für Strahlenchemie, Müllheim) and were used as supplied. Norbornadiene (Aldrich), norbornene (Aldrich), norbornane (Aldrich), dimethyl fumarate (Aldrich), dimethyl succinate (Aldrich), helimatricesum (Air Products), hydrogen (Air Products), and deuterium (BOC, Research Grade) were used as supplied.

Results and Discussion

Hydrogenation of Dimethyl Fumarate (DF) in PE Matrices at Low Temperature. Fe(CO)₄(DF) (**1**) has four characteristic IR bands in the ν(CO) region due to the Fe(CO)₄ moiety, with an additional band at 1741 cm^{-1} due to the carbonyl groups on the coordinated DF (see Table 1). Irradiation of **1** in PE at 150 K under a pressure of He resulted in a decay in the five parent bands accompanied by the growth of three new bands (Figure 2). These new bands, at wavenumbers slightly lower than parent, suggest that a tricarbonyl species was being formed. Upon warming the PE, the new peaks remained unchanged up to temperatures of ca. 260 K. Thus, it is unlikely that these new bands were due to an unsaturated 16-electron Fe(CO)₃-(DF) species, which would be expected to be highly reactive toward the photoejected CO in the PE. It seems more likely

6860   *J. Am. Chem. Soc., Vol. 123, No. 28, 2001*

Table 1. $\nu(C-O)$ Band Positions (cm^{-1}) of the Products Observed Following UV Irradiation of Fe(CO)$_4$(DF) in PE

complex	$\nu(C-O)$
Fe(CO)$_4$(DF) (**1**)	2109
	2046
	2039
	2005
	1713
Fe(CO)$_3$(η^4-DF) (**1A**)	2075
	2011
	1998
Fe(CO)$_3$(DF)(η^2-H$_2$) (**1H**)	2060
	1977
Fe(CO)$_3$(DF)(η^2-D$_2$) (**1D**)	2058
	1979
Fe(CO)$_5$ (**1P**)	2021
	1999
uncoordinated DF	1733
uncoordinated DS	1748

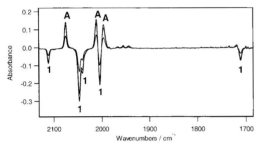

Figure 2. FTIR difference spectrum obtained following UV irradiation of Fe(CO)$_4$(η^2-DF) (**1**) under high pressures of both He and H$_2$ at 150 K in a PE disk. Positive peaks are due to generation of Fe(CO)$_3$(η^4-DF) (**1A**), negative peaks are due to depletion **1**.

that the coordinated DF has shifted its coordination mode to saturate fully the metal center, either through the π cloud of the C=O bond or perhaps by dative bond formation from one of the lone pairs on the carbonyl oxygen. Fe is known to prefer coordination to conjugated dienes, to form complexes such as Fe(CO)$_3$(η^4-butadiene).[41,42] This behavior extends to π-hetero-1,3-diene complexes[43,44] where the hereroatom is either O or N, with coordination through π clouds of the double bond rather than lone pairs. The relative intensities of the new bands are similar to the spectra of Fe(CO)$_3$(η^4-hetero-1,3-diene) complexes.[45,46] Thus we have assigned tentatively these three new bands to Fe(CO)$_3$(η^4-DF) (**1A**).

Warming the PE matrix further to ambient temperature led to regeneration of the parent tetracarbonyl, **1**, at the expense of the η^4 species, **1A**. The reaction was then repeated under 900 psi of H$_2$. Photolysis of **1** at 150 K again resulted in the formation of **1A**. However, warming the matrix to 260 K resulted in the growth of two additional bands (see Figure 3), which were not seen in the absence of H$_2$, and can therefore be assigned to a hydrogen-containing species. Fe(CO)$_3$(η^2-DF)-(η^2-H$_2$) (**1H**) seems more likely than the classical dihydride complex, Fe(CO)$_3$(η^2-DF)H$_2$, which would be expected to have

(41) Hallam, B. F.; Pauson, P. L. *J. Chem. Soc.* **1958**, 642.
(42) Ellerhost, G.; Gerhatz, W.; Grevels, F. W. *Inorg. Chem.* **1980**, *19*, 67.
(43) Brodie, A. M.; Johnson, B. F. G.; Josty, P. L.; Lewis, J. *J. Chem. Soc., Dalton* **1972**, 2031.
(44) Otsuka, S.; Yoshida, T.; Nakamura, A. *Inorg. Chem.* **1968**, *6*, 20.
(45) Howell, J. A. S.; Johnson, B. F. G.; Josty, P. L.; Lewis, J. *J. Organomet. Chem.* **1972**, *39*, 329.
(46) Cardaci, G. *J. Am. Chem. Soc.* **1975**, *97*, 1412.

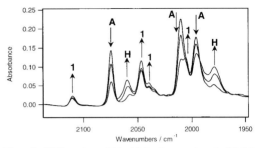

Figure 3. FTIR spectrum obtained following warming of a PE disk containing Fe(CO)$_3$(η^4-DF) (**1A**) from 150 to 260 K under a high pressure of H$_2$. Upon warming, a decrease in bands due to **1A** is observed, concurrent with a growth in bands due to the parent species, Fe(CO)$_4$(η^2-DF) (**1**), and a new species, Fe(CO)$_3$(DF)(η^2-H$_2$) (**1H**). Note that these spectra cover a time period shorter than that needed to take the conversion of **A** to completion.

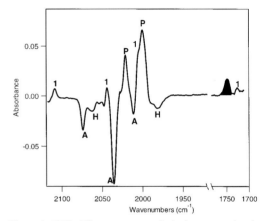

Figure 4. FTIR difference spectrum obtained upon warming the reaction products formed in Figure 3 to 298 K. A decrease in Fe(CO)$_3$-(DF)(η^2-H$_2$) (**H**) and Fe(CO)$_3$(η^4-DF) (**A**) is observed, along with an increase in Fe(CO)$_5$ (**P**), Fe(CO)$_4$(η^2-DF) (**1**), and the hydrogenated product, DS (solid band) at 1748 cm^{-1}.

$\nu(C-O)$ bands occurring at wavenumbers higher than those of **1**, due to oxidation of the metal center. Thus, we have been able to detect the thermal reaction of **1A** with H$_2$, most probably to form **1H**. Further warming of the PE to room temperature resulted in a decay of the bands of **1H** and the growth of two new bands (see Figure 4), which could be shown to be due to Fe(CO)$_5$ (**1P**). We were unable to detect the unsaturated Fe species (possibly metallic Fe) which, by stoichiometry, must have been formed at the same time as Fe(CO)$_5$. An additional band was also observed at 1748 cm^{-1}, which was identified as the hydrogenated product, DS, by comparison with the IR spectrum of pure DS impregnated into a PE film. The presence of this band indicates that the hydrogenation reaction of DF to DS had occurred within the PE film.

The advantage of PE over more conventional matrices is that the PE film remains intact up to and indeed above room temperature. Therefore, it is possible to extract stable reaction products and to analyze them off-line. In this experiment, extraction of the PE disk with CCl$_4$ followed by GC-MS analysis revealed two peaks (not illustrated) which could be assigned to unhydrogenated DF and the hydrogenated product, DS, by

Mechanism of Catalytic Reactions   *J. Am. Chem. Soc., Vol. 123, No. 28, 2001* 6861

Table 2. Wavenumbers (cm^{-1}) of the Photoproducts of (NBD)Mo(CO)$_4$ under Various Conditions

	Ar matrix/10 K[a]	lXe/190 K[b]	n-heptane/298 K[b]	PE/low T	PE/298 K
(η^4-NBD)Mo(CO)$_4$ (**2**)		2044	2044	2042	2041
		1959	1958	1954	1954
		1914	1912	1906	1907
mer-(η^2-NBD)(η^4-NBD)Mo(CO)$_3$ (**2M**)			2015	2011	2011
			1938	1943	1943
			1932	1933	1933
fac-(η^2-NBD)(η^4-NBD)Mo(CO)$_3$ (**2F**)			1994[c]		1993
			1927[c]		e
			1900[c]		1903
mer-(η^4-NBD)Mo(CO)$_3$(H$_2$) (**2H'**)	2029[d]	2029	e	2025	
	1952[d]	1952	1952	1947	
fac-(η^4-NBD)Mo(CO)$_3$(H$_2$) (**2H**)	1997[d]	1997	1998	1993	
	1930[d]	1930	1930	1927	
	1898[d]	1898	1900	1891	
cis-(η^2-NBD)Mo(CO)$_4$(H$_2$) (**2C**)	2060[d]	2060		2056	
	1964[d]	1964		1958	
	1941[d]	1941		1937	
fac-(NBD)Mo(CO)$_3$(s) (**2U**)	1985		1988	1977	
	1899		1900	1891	
	1889		1893	1878	
mer-(NBD)Mo(CO)$_3$(s) (**2U'**)	2024		2016	e	
	1950		1944	1946	
	(1900)		1884	e	
(η^2-alkene)Mo(CO)$_5$ (**2X**)			e	2076	2076
			1958	1954	1955
			e	1942	1943
Mo(CO)$_6$ (**2P**)		1986	1985	1985	1985

[a] See ref 50. [b] See refs 37 and 38. [c] These band positions are for the W analogue as the data for the equivalent Mo complex have not been reported. [d] IR band positions for the deuterated complex. [e] Band position obscured. Band positions shown in parentheses are tentative assignments.

comparison with GC-MS analysis of commercially available standards.

Repeating the photolysis of **1** in the presence of D$_2$ rather than H$_2$ gave identical results, apart from the expected isotopic shift in the ν(C—O) bands (Table 1) of Fe(CO)$_3$(DF)(η^2-D$_2$) (**1D**). GC-MS analysis of the deuterated extract also gave two peaks, one of which was assigned to d_2-DS from its mass spectrum.

Therefore, by using polymer matrix isolation, we have been able to probe the intermediates in the stoichiometric hydrogenation of DF to DS using Fe(CO)$_4$(DF). We have shown that not only can we characterize the reaction products using IR spectroscopy, but also we are able to extract the products from the PE disk and identify them using GC-MS analysis.

These hydrogenation experiments were performed under artificial conditions, with no free DF in the reaction mixture. As such, the catalytic cycle could not continue once the hydrogenated product had been formed. However, when we repeated the reaction in the presence of free DF, identical results were obtained. It seems therefore that DS can be formed from coordinated DF, but once hydrogenation has occurred, uncoordinated DF does not react with the unsaturated metal center. This may be due to the poor mobility of free DF in the PE matrix, or simply because it is more favorable to form Fe(CO)$_5$ rather than Fe(CO)$_4$(DF) once the hydrogenated product has been released.

The reaction scheme for the hydrogenation of DF is shown in Scheme 2. Although we have not been able to observe the catalytic reaction, we have been able to follow the hydrogen transfer from the metal center to the coordinated olefin to form the hydrogenated product.

Hydrogenation of Norbornadiene (NBD). As explained in the Introduction, the hydrogenation of norbornadiene by Group 6 metal centers has been the subject of intense study. In the context of these experiments the hydrogenation is interesting, first because there are at least three possible hydrogenation

Scheme 2. Reaction Scheme for the Hydrogenation of DF Using Fe(CO)$_4$(DF)^a

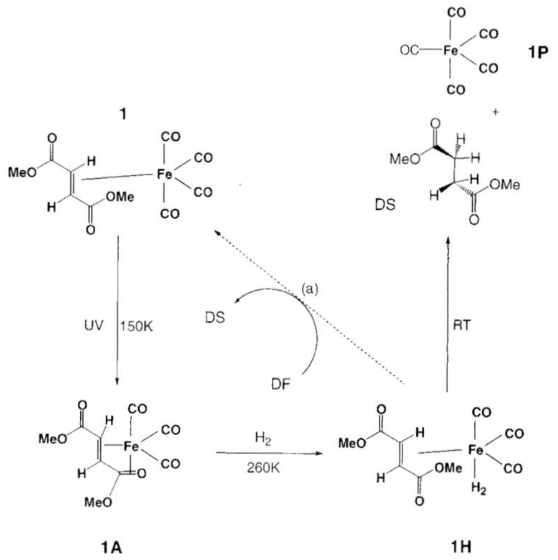

^{a} For the reaction to be catalytic, step (a) needs to occur, a process that seems unfavorable under the conditions of our reaction. Spectroscopic evidence has been obtained for all the species shown in the scheme.

products (NBN, NTC, and NBA), and second because none of the products has a strong IR chromophore for in situ detection. Furthermore, in the case of the molybdenum catalyzed reaction, previous work in this area has provided a surprisingly complete IR library of the ν(C—O) bands of the various compounds and intermediates involved in the catalytic cycle (see Table 2). Therefore, it is unnecessary to carry out a detailed study of the

6862   *J. Am. Chem. Soc., Vol. 123, No. 28, 2001*

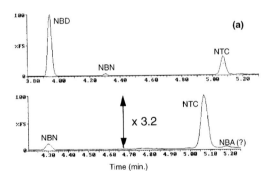

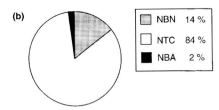

Figure 5. (a) GC analysis of the compounds extracted from PE following the irradiation of (NBD)Cr(CO)$_4$ under H$_2$ (500 psi) at room temperature. (b) Chart showing the relative ratios of the products.

type needed to identify the intermediates in the reactions of Fe-(CO)$_4$(DF) above.

UV photolysis at 296 K of (NBD)Cr(CO)$_4$ impregnated into PE under 500 psi of H$_2$ resulted in a decrease in the intensity of the ν(C−O) bands of the starting material and the formation of no easily detectable metal carbonyl products. The dilution of the compounds in the matrix also meant that no IR bands could be observed which were attributable to hydrogenated products. However, subsequent extraction of the PE disk with CH$_2$Cl$_2$ (0.1 cm^3) and GC-MS analysis revealed the formation of NBN, NTC, and a trace of NBA, see Figure 5. The ratio of the three products (see Figure 5b) was very similar to that reported for the photocatalytic reaction at room temperature.

When the photolysis of (NBD)Cr(CO)$_4$ was repeated in PE saturated with NBD under a pressure of D$_2$, IR bands were seen to grow in the ν(C−D) region (see Figure 6a). The wavenumbers of these bands are summarized in Table 3. They can be assigned to d_2-NBN, and more tentatively to d_2-NTC, by comparison with data from earlier studies in fluid solution,[37,38] where remarkably similar IR spectra were observed. Extraction of the PE disk, followed by GC-MS analysis, revealed a broadly similar ratio of products to those in Figure 5, but with somewhat more NBN relative to NTC and NBA. These chromatographic results also confirmed the assignment of the ν(C−D) IR bands.

Repeating the experiment with (NBD)Mo(CO)$_4$, **2**, in the presence of NBD and D$_2$ again generated bands in the ν(C−D) region, but with quite different relative intensities from those observed for the Cr analogue, see Figure 6b. Thus, even from the IR, it is immediately clear that the product distributions from the Cr and Mo catalysts are different. This difference mirrors the literature reports of the product distributions from these reactions in conventional solutions. Again, GC-MS of the extracted products allowed the product distribution to be quantified more precisely, and Figure 7 shows the difference in product distribution between Cr and Mo for the catalytic

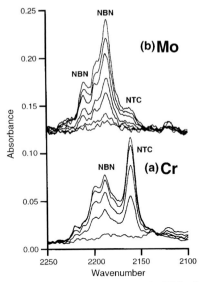

Figure 6. FTIR spectra obtained following the UV irradiation of (NBD)M(CO)$_4$ (M = (a) Cr or (b) Mo) under D$_2$ (500 psi) at room temperature in PE.

Table 3. ν(C−D) Band Positions (cm^{-1}) of the Products Observed Following the Photocatalytic Deuteration of NBD at Room Temperature

complex	*n*-heptane[a]	CCl$_4$ [a]	PE[b]
NBN	2212	2212	2210
	2188	2188	2186
NTC	2163	2202	2199
		2163	2160

[a] See ref 38. [b] This study.

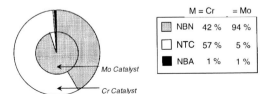

Figure 7. Distribution of the hydrogenated products (NTC, NBN, or NBA) following irradiation of (NBD)M(CO)$_4$ (M = Cr or Mo) in the presence of excess NBD and D$_2$ in PE at room temperature.

hydrogenation of NBD in the presence of excess D$_2$ (or H$_2$). Broadly, the major difference between the metals is that Mo generates NBN as the principal product and relatively little NTC. The amount of fully hydrogenated product, NBA, is similar for the two metals under these conditions.

Figure 8 shows IR spectra recorded during the UV photolysis of **2** in PE under a pressure of D$_2$ and in the presence of excess NBD. Unlike the Cr system, new ν(C−O) bands were observed to grow in as the photolysis proceeded, see Figure 8a. After a total of 50 min of irradiation, most of the starting material was destroyed and computer subtraction of the IR spectra revealed five clearly resolved ν(C−O) bands (Figure 8b). By comparison with the "library" data for this system, four of these bands could be assigned to the *mer*-(**2M**) and *fac*-(**2F**) isomer of (η^2-NBD)-(η^4-NBD)Mo(CO)$_3$. These compounds were identified by previ-

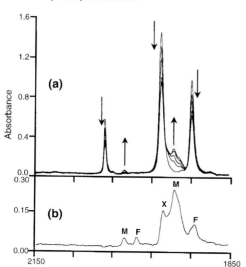

Figure 8. (a) FTIR spectra recorded during the UV irradiation of (NBD)Mo(CO)$_4$ (**2**) under D$_2$ in the presence of excess NBD in PE at room temperature. The ν(C−O) bands due to **2** decrease in intensity, with new bands due to *mer*-(**M**) and *fac*-(**F**) (η^2-NBD)(η^4-NBD)Mo-(CO)$_3$ observed to grow in. (b) Scaled subtraction spectrum showing the IR bands of only the photoproducts. The band marked **X** can be assigned to (η^2-C=C)Mo(CO)$_5$ where C=C is either NBD or NBN.

ous workers[36,37] as being the resting state of the Mo catalyst in fluid solution at room temperature. The fifth band, labeled **X** in Figure 8b, is assigned to (η^2-C=C)Mo(CO)$_5$ (where C=C is either NBD or NBN), and is discussed in more detail below.

The kinetics of this reaction are interesting because the apparent rate of growth in intensity of the bands of the *mer* isomer is significantly faster than that of the *fac* isomer. However, this disparity could be due, at least in part, to the growth of underlying bands due to other reaction products. The rate of growth of the bands due to NBN (see Figure 6b) is approximately linear with time, as might be expected for a photocatalytic reaction that does not proceed at an appreciable rate in the dark at room temperature. The overall conclusions from the results shown in Figures 5−8 indicate that the behavior of these catalysts is very similar in PE and in other hydrocarbon solvents at room temperature.

By contrast, striking differences in behavior are observed when **2** is irradiated under a pressure of hydrogen, but *in the absence of excess NBD*. As the photolysis proceeds, the bands of **2** decay and bands due to Mo(CO)$_6$ (**2P**) and (η^2-olefin)Mo-(CO)$_5$ (**2X**) grow in (see Figure 9a). However, when the UV irradiation is stopped, the spectra continue to change and the bands due to **2X** species disappear completely over a period of ca. 5 min, with a corresponding growth in the absorption bands of **2P**. An even more dramatic difference becomes apparent when the products are extracted and analyzed using GC-MS, see Figure 10. In the absence of free NBD, NBA is transformed from a trace product to the major product. At the same time, the formation of NTC is also increased. Thus, (NBD)Mo(CO)$_4$ is quite different from its Cr analogue because the product distribution it totally changed by removal of excess NBD from the reaction mixture. Most importantly, the fact that NBA is the major product provides a new opportunity to investigate

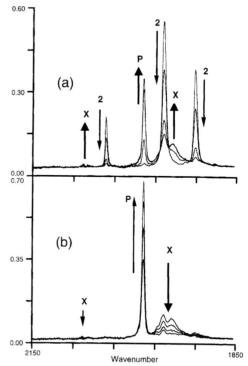

Figure 9. (a) FTIR spectra obtained following irradiation of **2** under H$_2$ in PE at room temperature *in the absence of excess NBD*. As the bands due to **2** decay, the growth of absorptions due to Mo(CO)$_6$ (**2P**) and (η^2-C=C)Mo(CO)$_5$ (**2X**) is observed. (b) When the UV irradiation is halted, the growth of the band due to **2P** continues at the expense of the bands due to **2X**.

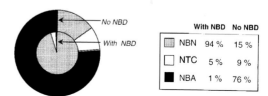

	With NBD	No NBD
NBN	94 %	15 %
NTC	5 %	9 %
NBA	1 %	76 %

Figure 10. Relative distribution of the hydrogenated products following UV irradiation of (NBD)Mo(CO)$_4$ (**2**) under H$_2$ in PE at room temperature in both the absence and presence of excess NBD.

the formation of this fully hydrogenated product, which has been largely ignored in previous mechanistic studies.

The initial stage of our investigation was to exploit the ease with which the gas around the PE film can be changed. We have used H$_2$ and D$_2$, either individually or mixed, to generate various isotopomers of NBA via the photolysis of (NBD)Mo-(CO)$_4$. The different isotopomers are quantified by GC-MS using SIM over the mass range 96 (NBA) to 100 (d_4-NBA) (see Experimental Section). Three experiments were performed and the results are summarized in Figure 11. In the first experiment, **2** was irradiated under a mixture of H$_2$ and D$_2$ (ca. 60:40) *in the absence of NBD*. As expected, the products were a mixture of the three isotopomers, NBA, d_2-NBA, and d_4-NBA, in approximately the correct statistical ratio (see Figure 11a). d_2-NBA was clearly the major product.[47]

6864 *J. Am. Chem. Soc., Vol. 123, No. 28, 2001*

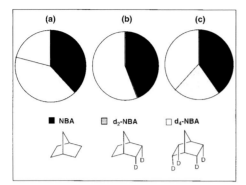

Figure 11. Distribution of the isotopomers of NBA following irradiation of (NBD)Mo(CO)$_4$ (**2**) in PE at room temperature under (a) a mixture of H$_2$/D$_2$ (60:40) *in the absence of excess NBD*, (b) alternate pulses of H$_2$ and D$_2$ *in the absence of excess NBD*, and (c) alternate pulses of H$_2$ and D$_2$ *in the presence of excess NBD*.

In the second experiment, the photolysis was repeated, but this time with a series of irradiation periods, each of 20 or 30 min. Between each photolysis period the gas surrounding the PE was changed, using alternately H$_2$ and D$_2$. The gases were changed repeatedly to give a total irradiation time similar to that used in the first part of the experiment. Thus **2** was irradiated for 70 min under H$_2$ and also 70 min under D$_2$, but H$_2$ and D$_2$ were never present in the cell together, apart from any residual gases which may have been trapped within the PE matrix. Figure 11b shows the relative amounts of the isotopomers. *It is immediately clear that no d$_2$-NBA was formed when H$_2$ and D$_2$ were used alternately.* This experiment shows that any intermediates involved in the formation of NBA at room temperature are short-lived compared to the time needed to exchange the gases (ca. 10 min). This is consistent with the rapid decay of the bands shown in Figure 9b.

In the final stage of the investigation, the experiment with alternate pulse of H$_2$ and D$_2$ was repeated, but this time with **2** *in the presence of excess NBD*. Figure 11c shows that this time, the three isotopomers were formed in approximately a 1:1:1 ratio. Thus, unlike the experiment with **2** alone, the presence of NBD gave rise to a longer lived intermediate that could "store" H$_2$ or D$_2$ while the gases were being exchanged. We believe that the most likely candidate for this intermediate is free NBN.

The reason for identifying NBN as the long-lived intermediate is shown in Figure 12. In this experiment, **2** was irradiated under a pressure of D$_2$ and in the presence of excess NBN. After 40 min of UV irradiation, two major products could be identified, namely Mo(CO)$_6$ (**2P**) and (NBN)Mo(CO)$_5$ (**2X**), the spectra of which could be established by a separate experiment involving the photolysis of Mo(CO)$_6$ in the presence of NBN. GC-MS analysis indicated d$_2$-NBA as the major product. There were also traces of d$_2$-NBN, d$_2$-NTC, and d$_4$-NBA, presumably generated through deuteration of the NBD ligand in **2**. Given the fact that NBD was initially coordinated to the metal center, one would expect the deuteration of the NBD to be the kinetically preferred product, even though the NBN was in considerable excess. Thus, we propose that the principal pathway for the formation of NBA in a catalytic reaction (i.e. in the presence of excess NBD) is as shown in Scheme 3.

(47) Since the gas is in great excess it is not feasible, in this study, to detect HD, which is a possible byproduct of reactions involving mixtures of H$_2$ and D$_2$.

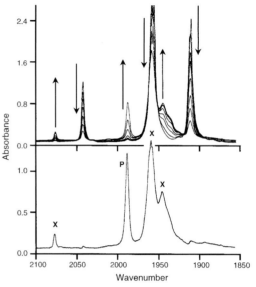

Figure 12. (a, top) IR spectra and (b, bottom) IR subtraction spectrum obtained following irradiation of PE containing **2**, excess NBN, and D$_2$ at room temperature. The bands due to **2** decrease in intensity, concurrent with the growth of new bands assignable to both Mo(CO)$_6$ (**2P**) and (η^2-NBN)Mo(CO)$_5$ (**2X**).

Scheme 3. Reaction Scheme Showing the Mechanism for the Formation of NBA in the Presence of Excess NBD (i.e. a catalytic reaction)^a

^{a} Part a: Irradiation of (NBD)Mo(CO)$_4$ under H$_2$ results in the formation of the singly hydrogenated product NBN, which is released from the metal center. Part b: Further photolysis leads to recoordination of NBN to the metal center, along with another molecule of H$_2$. This results in the hydrogenation of the second double bond to form NBA.

As demonstrated above, the formation of NBA from **2** in the absence of excess NBD must involve relatively short-lived intermediates. Indeed the TRIR studies^{37} by Hodges et al. implicated short-lived dihydrogen complexes in the initial stage of this reaction. To investigate the formation of NBN further, we carried out reactions in PE film at low temperature, exploiting the cryogenic capability of our high-pressure cell. Figure 13 shows the relatively complicated spectra that are obtained when **2** is irradiated under H$_2$ at 180 K. Apart from one weak band at 1973 cm^{-1} (possibly due to a polymeric species), all of the bands in these spectra can be assigned with reasonable certainty to different Mo(CO)$_x$ species, previously identified from experiments in fluid solution. Most significantly, there are bands due to both the *fac* and *mer* isomers of (NBD)-

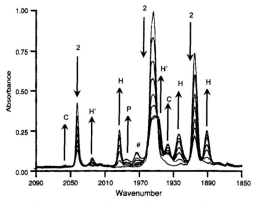

Figure 13. IR spectra recorded after UV photolysis of **2** under H_2 in PE at 180 K in the absence of excess NBD. The bands due to **2** decrease in intensity, with concomitant growth of new bands assignable to *fac* (marked **H**) and *mer* (marked **H'**) isomers of $(NBD)Mo(CO)_4(H_2)$ and to *cis*-$(\eta^2$-NBD$)Mo(CO)_4(H_2)$ (**2C**), where C=C is either NBD or NBN. The band marked # is due to an unidentified dimer.

$Mo(CO)_4(H_2)$ (**2H** and **2H'**, respectively) and to *cis*-$(\eta^2$-C=C)-$Mo(CO)_4(H_2)$ (**2C**), where C=C is either NBD or NBN. When the PE is warmed to 230 K, the major changes in the spectrum were the almost complete decay of *fac*-$(NBD)Mo(CO)_4(H_2)$ isomer and the formation of a $(\eta^2$-C=C$)Mo(CO)_5$ complex, where C=C is either NBD or NBN. This η^2-complex was only observed upon annealing, and was not observed immediately following the UV photolysis. GC-MS analysis of the products formed in this reaction indicated a distribution between NBA, NBN, and NTC, very similar to that observed at room temperature.

These experiments suggest that low temperature can be used to arrest the hydrogenation of NBD, at least partly. The effect is even more striking at 80 K. Figure 14a shows a simple subtraction spectrum obtained after photolysis of **2** in PE without addition of H_2 or other additives. The three peaks formed can be assigned to the coordinatively unsaturated intermediate, *fac*-$(NBD)Mo(CO)_3$ (**2U**). This intermediate has also been observed in frozen argon matrices and by TRIR experiments in *n*-heptane at room temperature. Identical spectra were obtained when **2** was irradiated at this temperature in the presence of H_2. This is not surprising because previous work[23] with $Mo(CO)_6$ and H_2 has shown that low-temperature photolysis results in the formation of $Mo(CO)_5$ rather than $Mo(CO)_5(H_2)$, even though this dihydrogen compound is formed at higher temperatures. When PE containing **2U** was heated from 80 to 160 K in the absence of H_2, there was near quantitative recombination with photoejected CO and regeneration of **2** (see Figure 14b). On the other hand, when **2U** was heated to 160 K under H_2, new bands were observed to grow in at the same rate as the regeneration of **2** (Figure 14c). These new bands could be assigned, on the basis of published data, to the dihydrogen complexes *fac*-$(NBD)Mo(CO)_3(H_2)$ and *cis*-$(\eta^2$-C=C$)Mo(CO)_4$-(H_2), where C=C is either NBN or NBD.

GC-MS analysis of the products, following the warming of the PE from 160 K to room temperature, showed a ratio of NBN:NTC:NBA almost identical to that observed a room temperature (see Figure 15). The similarity of the product distributions obtained in the experiments at 80 K, 180 K, and room temperature strongly suggests that the mechanism of the reaction in low temperature and room temperature experiments must be

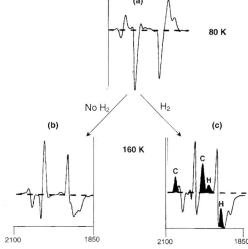

Figure 14. (a) FTIR subtraction spectrum obtained following UV irradiation of **2** in PE in the absence of a reactive gas at 80 K. The positive peaks can be assigned to the coordinatively unsaturated intermediate *fac*-$(NBD)Mo(CO)_3$ (**2U**), with the negative peaks due to depletion of **2**. Warming the PE containing **2U** from 80 to 160 K (b) in the absence of a reactive gas results in regeneration of **2**. However, warming in the presence of H_2 (c) leads to regeneration of **2** and formation of the dihydrogen complexes *fac*-$(NBD)Mo(CO)_3(H_2)$ (**2H**) and *cis*-$(\eta^2$-C=C$)Mo(CO)_4(H_2)$ (**2C**), where C=C is either NBN or NBD.

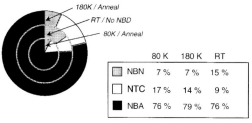

Figure 15. Product distributions of the hydrogenated products (NBN, NTC, and NBA) obtained following irradiation of **2** under H_2 in the absence of excess NBD at 80, 180, and 298 K.

very similar, if not identical. The difference between the experiments is that, at low temperature, the experiment can be halted at key stages.

The final part of our investigation involved combining two of the experimental strategies described above, namely exchange of H_2 and D_2 gases around the PE film and the use of low temperatures.[48] The purpose of the experiments, as before, was to look at the distribution of H_2 and D_2 in the NBA extracted from the PE at the end of the reaction. **2** was irradiated in PE under a pressure of D_2 at 80 K. The PE was then warmed to 160 K, and formation of *fac*-$(NBD)Mo(CO)_3(D_2)$ and *cis*-$(\eta^2$-C=C$)Mo(CO)_4(D_2)$ was observed. The pressure of D_2 was released, and the cell evacuated for a period of 1 h. The cell was then repressurized with H_2 and warmed slowly in the dark

(48) The use of D_2 rather than H_2 produces small changes in the wavenumbers of IR bands of compounds which contain coordinated H_2/D_2, but otherwise the spectra in the $\nu(C-O)$ regions are almost indistinguishable and are not illustrated.

6866   *J. Am. Chem. Soc., Vol. 123, No. 28, 2001*

to room temperature. The products were extracted and analyzed by GC-MS to give the distribution NBA (33%), d_2-NBA (22%), and d_4-NBA (45%). This experiment shows that (i) once the sample has been irradiated at 80 K, no further UV irradiation is needed to generate NBA, (ii) the majority of hydrogenation occurred after the H_2 and D_2 were exchanged (the formation of d_4-NBA may well be due to residual D_2 trapped in the PE matrix), and (iii) intermediates exist at 160 K which are sufficiently stable to survive the change of gases and yet are reactive enough to exchange fully with H_2 to form unlabeled NBA.[49]

As in any mechanistic study, the experiments described above are indicative rather than totally conclusive. Nevertheless, they strongly suggest that the formation of NBA, in the absence of added NBD, follows a pathway similar to that shown in Scheme 4. All the intermediates and transformations shown in Scheme 4 have been detected here, and most were also observed in TRIR studies in fluid solution at room temperature. In addition we have detected the product NBA. However, the direct transformation from $(NBD)Mo(CO)_3H_2$ to *cis*-$(NBN)Mo(CO)_4(H_2)$ has not been observed directly, possibly because the conversion involves addition of not only H_2 but also CO, which can only come from the photoejected CO in the PE.

Conclusions

In this paper we have shown how the combination of PE matrices and high pressure—low temperature experiments can provide new insights into the mechanism of even very well studied reactions. As far as we are aware, our experiments are among the first to link matrix isolation and product analysis in

Scheme 4. Reaction Mechanism for the Formation of NBA in the Absence of Excess NBD[a]

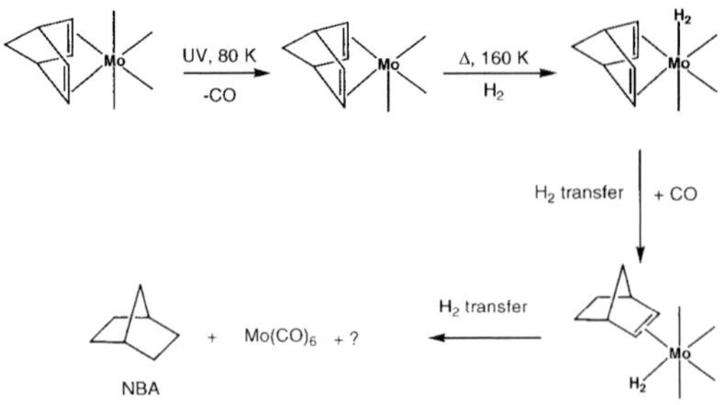

[a] Notice that in this reaction scheme the singly hydrogenated product is not released from the metal center, but rearranges to allow the addition of a second molecule of H_2.

homogeneous catalytic reactions. The highly unusual nature of these experiments has enabled us to carry out an isotopic labeling experiment, which would have been difficult, if not impossible, to carry out by more conventional means. For the first time it has been possible to compare the products of reactions performed stoichimetrically and catalytically usually literally microgram quantities of the compounds. The experiments have also come closer than most previous studies to capturing the key steps in a complicated multistep catalytic reaction. Currently we are extending this concept to the study of immobilized homogeneous catalysts in continuous flow reactors.

Acknowledgment. We are grateful to Prof. F. W. Grevels for the gift of samples of $(NBD)M(CO)_4$ and $Fe(CO)_4(DF)$. We thank Dr. K. Dost for helpful discussions, Mr. M. Guyler and Mr. K. Stanley for their technical support, and the EPSRC, Nicolet Instruments Ltd, and the University of Nottingham for financial support.

JA004345T

(49) Any investigation of this complexity obviously needs control experiments, and a considerable variety of controls were performed. For example, cooling **2** in PE under H_2 and warming to room temperature did not result in the formation of NBA. More importantly, some of our results could be explained if photolysis of the carbonyl precursors were to generate metal nanoparticles which could subsequently act as highly active heterogeneous catalysts. This explanation seems highly unlikely because photolysis of **2** in the absence of H_2, followed by subsequent pressurization with H_2, did not result in the formation of NBA or other hydrogenation products.

(50) Hooker, R. H. Ph.D. Thesis, University of Southampton, 1987.

Contribution from

DAVID R. KLUG

Imperial College, London
UK

Born 19th July, 1963, David R. Klug is Professor of Chemical Biophysics, in the Department of Chemistry, Imperial College London.

Qualifications and Appointments

BSc 1984	Physics, 2(1). University College London
PhD 1987	Physical Chemistry, Royal Institution of Great Britain/UCL
	Supervisor, Professor Lord George Porter, PRS.
1987–1990	Postdoctoral Research Assistant, Photochemistry Research Group,
	Department of Biology, Imperial College.
1990–1995	Royal Society University Research Fellow,
	Departments of Chemistry and Biochemistry, Imperial College.
1995–1999	Lecturer, Departments of Chemistry, Imperial College
1999–2001	Reader in Chemical Biophysics, Department of Chemistry, IC.
2001–	Professor of Chemical Biophysics, Department of Chemistry, IC.

Current Research Interests

Laser assisted proteomics and protein analysis
Protein dynamics and function
Membrane proteins and membrane organisation
Primary reactions in photosynthesis
Water splitting in photosynthesis and artificial water splitting
Electron transfer reactions

Proton/hydride transfer reactions
Nanocrystalline photovoltaic cells
Two-dimensional infra-red laser spectroscopy

Professional Activities

Chairs the Board of The Centre for Biological Chemistry
Member of the Sceintific Advisory Committee of the UK Energy Research Centre
Steering Committee ESF programme in Femtochemistry and Femtobiology
Chemistry College of the EPSRC
Chair operations and infrastructure committee Department of Chemistry, IC.

Refereed Publications: ~80

Patents Authored: 10

Commercial Activities: Founder of PowerLase Ltd

David R. Klug on
"Nanosecond Flash Photolysis and the Absorption Spectra of Excited Singlet States"
G. Porter and M. R. Topp
Nature **220**, 1228–1229 (1968)

Complete experimental procedures and methods often rely on the accumulation of many small but vital techniques. One method that has now become ubiquitous in ultrafast time-resolved optical studies is the use of physical delay to provide time resolution for optical spectroscopy. The basis of this method is that by introducing an additional path length between an optical pulse used for excitation, and an optical pulse used for probing the sample, the probe pulse can be made to pass through the sample at a known time after the excitation pulse. This means that the molecular species sampled by the probe pulse transmission will be those created at a particular time following the excitation pulse.

In the era that time resolution available and required by molecular spectroscopy was in the millisecond and microsecond time domain, there was neither a need for, nor the possibility of, using a spatial delay to achieve time-resolved spectral information. This is because the electronics of the time were quite fast enough to resolve the variation in intensity of a continuous beam of light passing through the sample and falling on a detector. Moreover the physical delay required to achieve a one microsecond change in timing is 300 m, the speed of light being 3×10^8 ms^{-1}, an impractical delay to achieve in a laboratory. However when the time resolution required is 10 ns, the physical delay required becomes of the order of 3 m, something which can be achieved rather easily by bouncing the optical beam off a few mirrors to create a dog-leg in the optical path. Thus once lasers with pulses of a few nanoseconds became available via the advent of the optical Q-switch, then physical delay became, in principle, a viable method of achieving time-resolved spectral data.

The first researchers to actually turn this principle into a viable and useful experimental method were Porter and Topp. This was first reported in Porter's Nobel lecture in 1967,[1] and one year later in *Nature* in 1968.[2]

Thirty or so years later this method is the one used by all spectroscopists requiring short time resolution. As electronics have become faster and more sensitive, nanosecond time resolution is now achieved using fast electronics, but physical delay is by far the most common method employed by spectroscopists and kineticists when sub-nanosecond time resolution is required for an absorption spectrum.

In my own work with George we routinely used physical delay to achieve time resolution. This led to a minor adjustment of the technique when we had to achieve time resolutions of 100 fs across a broad wavelength range simultaneously.

A delay of 100 fs corresponds to a physical delay of 30 μm for which we and others commonly use retro-reflecting delay lines driven by stepper or DC motors and controlled by computer. This was fine for high time resolutions over a narrow spectral range, but once we started to use "white light" laser pulses produced by super-continuum generation we needed to define zero-time reliably and accurately at many wavelengths simultaneously. Although the red and blue laser light travels through air at reasonably similar rates, spectroscopic apparatus typically includes many lenses, beam splitters, prisms and so forth. The group velocity dispersion in glass is such that the red and blue light travel at different speeds, and this can accumulate to a significant extent, causing the time delay across our probe wavelengths to vary by a few picoseconds, which is not very helpful when 100 fs time resolution is required. It turns out that prisms arranged in a particular fashion can be used to generate a negative dispersion that helps the blue light to catch up with the red. When the tips of the prisms were used and the correct prism material chosen we were able to flatten our optical wavefront to the point where 100 fs time resolution could be achieved simultaneously across 100 nm or so. We combined this trick with a highly sensitive home-built dual diode array detector, and a highly stable (for the time) amplified laser system to produce an apparatus that could reliably produce time-resolved optical spectra, with very low noise and 100 fs time resolution across 100 nm simultaneously.[3]

George was always very tickled to have started a field in which some thirteen orders of magnitude improvement (from seconds to 100 femtoseconds) were achieved during his research career. The scientific problem that we were addressing by this particular evolution of experimental methodology was the mechanism of charge separation in Photosystem Two. Photosystem Two is that part of the photosynthetic apparatus that splits water and produces all of the oxygen on Earth and most of the Earth's terrestrial biomass. These water splitting reactions are driven by a special protein-chromophore complex known as the reaction centre, which in the case of Photosystem Two is both efficient, yet highly regulated to protect the plant from the consequences of indulging in such dangerous chemistry.

Our femtosecond spectrometer was designed to produce the large amounts of accurate and precise data that we needed to unpick the complex kinetic pathways in Photosystem Two. These extensive data were combined with studies of genetically-engineered samples[4] and a new calculational approach[5] to uncover the primary structure-function relationship in Photosystem Two.[6] This we hope is the first step towards a quantitative understanding of Photosystem Two function, including water splitting. We expect such knowledge and understanding to contribute, either directly or indirectly, towards the redesign or reengineering of higher plant photosynthetic efficiency and artificial water splitting devices.

In the time that I knew and worked with George, he continued to be an enthusiastic supporter of natural photosynthesis as an eventual replacement for fossil fuels. In 2004 biomass research, and in particular biomass from photosynthesis, is firmly on the research agenda in many countries and is expected to be an important contributor to primary energy sources.

References

1. G. Porter and M. R. Topp, in *Nobel Symposium 5 – Fast Reactions and Primary Processes in Reaction Kinetics,* 158 (Interscience, London and New York, 1967).
2. G. Porter and M. R. Topp, *Nature* **220**, 1228 (1968).
3. D. R. Klug, T. Rech, M. D. Joseph, J. Barber, J. R. Durrant and G. Porter, *Chem. Phys.* **194**, 433–442 (1995).
4. L. B. Giorgi, P. J. Nixon, S. A. P. Merry, D. M. Joseph, J. R. Durrant, J. De Las Rivas, J. Barber, G. Porter and D. R. Klug, *J. Biol. Chem.* **271**(4), 2093–2101 (1996); S. A. P. Merry, P. J. Nixon, L. M. C. Barter, M. Schilstra, G. Porter, J. Barber, J. R. Durrant and D. R. Klug, *Biochemistry* **37**(50), 17439–17447 (1998).
5. J. R. Durrant, D. R. Klug, S. L. S. Kwa, R. van Grondelle, G. Porter and J. P. Dekker. *Proc. Natl. Acad. Sci. USA* **92**, 4798–4802 (1995); J. A. Leegwater, J. R. Durrant and D. R. Klug, *J. Phys. Chem.* **101**(37), 7205–7210 (1997).
6. L. M. C. Barter, J. R. Durant and D. R. Klug, *Proc. Nat. Acad. Sci. USA* **100**(3), 946–951 (2002).

Reprinted with permission from *Biochemistry* **31**, 7638–7647 (1992).

Copyright 1992 American Chemical Society

Observation of Pheophytin Reduction in Photosystem Two Reaction Centers Using Femtosecond Transient Absorption Spectroscopy[†]

Gary Hastings,[‡] James R. Durrant,[‡,§] James Barber,[§] George Porter,[‡] and David R. Klug[*,‡]

Photochemistry Research Group, Department of Biology, and AFRC Photosynthesis Research Group, Department of Biochemistry, Imperial College, London SW7 2BB, United Kingdom

Received July 29, 1991; Revised Manuscript Received April 14, 1992

ABSTRACT: Photosystem two reaction centers have been studied using a sensitive femtosecond transient absorption spectrometer. Measurements were performed at 295 K using different excitation wavelengths and excitation intensities which are shown to avoid multiphoton absorption by the reaction centers. Analyses of results collected over a range of time scales and probe wavelengths allowed the resolution of two exponential components in addition to those previously reported [Durrant, J. R., Hastings, G., Hong, Q., Barber, J., Porter, G., & Klug, D. R. (1992) *Chem. Phys. Lett.* **188**, 54–60], plus the long-lived radical pair itself. A 21-ps component was observed. The process(es) responsible for this component was (were) found to produce bleaching of a pheophytin ground-state absorption band at 545 nm and the simultaneous appearance of a pheophytin anion absorption band at 460 nm resulting in a transient spectrum which was that of the radical pair P680$^+$Ph$^-$. This component is assigned to the production of reduced pheophytin. A lower limit of 60% of the final pheophytin reduction was found to occur at this rate. Despite subtle differences in transient spectra, the lifetime and yield of this pheophytin reduction are essentially independent of excitation wavelength within the signal to noise limitations of these experiments. A long-lived species was also observed. This species is produced by those processes which result in the 21-ps component, and it has a spectrum which is found to be independent of excitation wavelength. This spectrum is characteristic of the primary radical pair state P680$^+$Ph$^-$. In addition, a 200-ps component was found which is tentatively assigned to a slow energy-transfer/trapping process. This component was absent if P680 was excited directly and is therefore not integral to primary radical pair formation. Overall, it is concluded that the rate of pheophytin reduction is limited to (21 ps)$^{-1}$, even when P680 is directly excited.

The most studied of the isolated photosynthetic reaction centers (RCs)[1] are those from the purple bacteria, particularly *Rhodobacter* (*Rb.*) *sphaeroides* and *Rhodopseudomonas* (*R.*) *viridis*. The electron- and energy-transfer reactions of these complexes have been the subject of intense research dating back to the first isolation of a reaction center from *Rb. sphaeroides* 23 years ago (Reed & Clayton, 1968). Despite the widespread scientific attention that these bacterial RCs have received, and the successful solution of their crystal structures (Deisenhofer et al., 1985; Yeates et al., 1988), the precise mechanism of primary charge separation in these and indeed all RCs remains to be established. Although in global terms purple bacterial RCs contribute less photosynthetic activity than the reaction centers of higher plants, the study of bacterial RCs has arguably produced as much insight into the function of higher plant reaction centers as have more direct lines of research to date.

The photosystem two (PS2) RC of higher plants is of particular interest because it is this reaction center which provides the oxidizing potential for water splitting. The PS2 reaction center was first isolated in 1987, initially by Nanba and Satoh (1987) and then by Barber et al. (1987). This development gave experimental support to the concept that the D1 and D2 polypeptides of PS2 are analogous to the L

and M subunits of purple bacteria (Trebst, 1986; Barber, 1987; Michel & Deisenhofer, 1988).

Primary charge separation in PS2 results in the formation of the radical pair state P680$^+$Ph$^-$. This state is formed in approximately 100 ps in PS2 particles retaining their inner chlorophyll antenna complexes [Nuijs et al., 1986; Schatz et al., 1987; reviewed in Renger (1991)], when P680 is not directly excited. The particles used in these studies contained approximately 80 chlorophylls per reaction center complex, and the rate of radical pair formation was found to be limited by trapping of the excitation energy by the reaction center (Schatz et al., 1988).

The isolation of reaction centers of PS2 has provided an opportunity to study the primary electron-transfer reactions without the complications associated with either energy transfer from large antenna complexes or secondary electron-transfer processes. Excitation of isolated PS2 reaction centers has been shown to result in the formation of the primary radical pair state in less than 25 ps with a near unity quantum yield (Danielius et al., 1987; Booth et al., 1991). The plastoquinones Q$_A$ and Q$_B$ which normally act as the secondary electron acceptors of PS2 are lost during the isolation procedure, preventing secondary electron-transfer reactions. Charge recombination from the primary radical pair state has been found to occur on the nanosecond time scale in the isolated PS2 RC complex (Danielius et al., 1987; Takahashi et al., 1987; Crystall et al., 1989).

The isolated PS2 reaction center was initially thought to be rather difficult to work with due to its inherent instability to light. This instability is to some extent thought to be connected with the physiological phenomenon known as photoinhibition, and the associated degradation and resynthesis

---

[†] This work is supported by the AFRC, Royal Society, and SERC. D.R.K. is a Royal Society University Research Fellow.

[*] Author to whom correspondence should be addressed.

[‡] Photochemistry Research Group.

[§] AFRC Photosynthesis Research Group.

[1] Abbreviations: RCs, reaction centers; CVL, copper vapor laser; fwhm, full width at half-maximum; PS2, photosystem two; Ph, pheophytin; P680, primary electron donor of PS2.

Observation of Pheophytin Reduction in Photosystem Two

Biochemistry, Vol. 31, No. 33, 1992   7639

of the D1 polypeptide (Shipton & Barber, 1991; Barber & Andersson, 1991). Although it has been demonstrated that highly active PS2 reaction centers can be isolated (Crystall et al., 1989; Booth et al., 1990), the relative lability of these reaction centers creates a particular problem in ultrafast time-resolved spectroscopic measurements, where exposure to light for long periods is required. It has, however, been demonstrated that, with care, PS2 reaction centers can remain fully active for periods of up to an hour at room temperature and under fairly strong laser illumination (Booth et al., 1990).

Attempts at identifying photochemically generated species from transient absorption spectroscopy (Danielius et al., 1987; Takahashi et al., 1987; Durrant et al., 1990), spectral hole burning (Jankowiak et al., 1989), and deconvolution of absorption spectra (Braun et al., 1990; van Kan et al., 1991) have been made. Previous measurements of radical pair formation in PS2 reaction centers have suggested that the rate of formation of P680+Ph- is 3 ps at room temperature (Wasielewski et al., 1989a). Measurements of the thermodynamics associated with radical pair formation (Booth et al., 1990) and identification of multiple radical pair states (Booth et al., 1991) demonstrate that the photochemical detail which can be observed with isolated RCs of PS2 is in some cases similar to that resolved in purple bacterial RCs.

Our most recent assessments of the stoichiometry of a stable and active form of the isolated PS2 reaction centers suggest that the complex contains six chlorophyll molecules, two pheophytins, two β-carotenes, and one cytochrome b-559 (Gounaris et al., 1990), which is in agreement with the findings of other workers (Kobayashi et al., 1990). The spectroscopic features of the PS2 RC tend to be more congested and overlapped to a higher degree than in the reaction centers of purple bacteria. This means that particularly sensitive measurements are required to distinguish and identify transient species in time-resolved absorption experiments.

In this paper we demonstrate the application of a highly sensitive femtosecond transient absorption spectrometer to the study of radical pair formation in isolated PS2 reaction centers, using low levels of excitation. We show that the data produced by this apparatus are sufficiently accurate and precise to allow observation and identification of some of the primary electron- and energy-transfer steps which occur in the PS2 reaction center.

We have shown in a previous study (Durrant et al., 1992) that when P680 is directly excited, its singlet excited state decays with lifetimes of 400 ± 100 fs and 3.5 ± 1.5 ps. The primary electron acceptor of PS2 is thought to be a pheophytin molecule (Klimov et al., 1977). The study presented here is an attempt to increase our understanding of the primary photochemistry of PS2 reaction centers by determining the apparent overall rate of pheophytin reduction in this complex. We do not attempt to deduce the route by which the pheophytin is reduced but merely seek to demonstrate that it is possible to time resolve the arrival of an electron at a pheophytin molecule and that the resulting spectrum is that of the radical pair state P680+Ph-.

MATERIALS AND METHODS

Reaction centers were isolated from pea thylakoid membranes and resuspended in appropriate buffer as in previous measurements (Booth et al., 1991; Chapman et al., 1991). Anaerobic conditions were achieved as previously (Crystall et al., 1989). All transient absorption measurements were performed at 295 K in a 2.5 mm path length cuvette which was rotated at sufficient speed to replace the sample volume

between flashes. The optical density of the samples, at the peak of the longest wavelength absorption band (675.5 nm), was between 0.8 and 1.0. Samples were exposed to light from the spectrometer for approximately 1 h, during which time the peak of the long-wavelength absorption band shifted by less than 1 nm, corresponding to less than a 10% loss in activity (Booth et al., 1991). The observed absorption changes were found to be the same, within limits of signal to noise, at the beginning and the end of 1 h of exposure to light in the apparatus.

The femtosecond transient absorption spectrometer is described below. A home-built, colliding-pulse-modelocked dye laser generates pulses with an autocorrelation of 150 fs (fwhm) and energies of 0.16 nJ at 625 nm. These pulses are amplified to 2 μJ using a multipass configuration similar to that described by Knox et al. (1984). The energy for this amplification is provided by the 6.5-kHz, 511-nm beam from a copper vapor laser (CVL) manufactured by Oxford Lasers (U.K.). The amplified pulses are focused into a flowing water cell to generate a femtosecond white light continuum. These white light pulses are split into two parts, one part to form the excitation beam and one part to form the probe beam. The excitation wavelength (612 or 694 nm) is selected from the continuum and reamplified using a second multipass amplifier, pumped by the 578-nm beam from the CVL. After reamplification, the excitation pulses have energies of ≈1 μJ, which is reduced to 0.1 μJ before reaching the sample. Group velocity dispersion in both the excitation and probe beams is controlled using two anomalously dispersive delay lines (Fork et al., 1984). The excitation and probe beams are parallel polarized to better than 90%. The excitation beam is focused to a 270-μm waist in the sample (175 μJ cm⁻²) and the probe beam aligned with the aid of a pinhole to interrogate this excited volume. The main reason for the relatively high sensitivity of these measurements is that signal averaging is performed at 6.5 kHz.

Absorption changes at single wavelengths (detection bandwidth of 2–5 nm) were monitored as a function of time delay using a Michelson interferometer arrangement with a computer-controlled delay line and sensitive difference detection equipment comprising two ratiometers and a lock-in amplifier. The time resolution of the spectrometer was 160 fs, determined by monitoring the 10–90% rise time of absorption changes observed in three dye standards over a wide range of probe wavelengths. The white light probe pulses were temporally dispersed by less than 2 fs/nm between 655 and 700 nm.

Data were collected over two spectral ranges: 420–570 and 655–730 nm. Measurements were made using three time scales, 0–300, 0–80, and 0–13 ps, and two different excitation wavelengths, 612 and 694 nm. Data were collected and analyzed using all permutations of these conditions. Checks were made for consistency between data collected on different time scales. Over 1000 time-resolved decays were analyzed, each consisting of at least 100 data points. The decays were grouped into sets of 12 wavelengths for global analysis. The results of the analysis of the 0–13-ps data have been published elsewhere (Durrant et al., 1992), but all of these data were analyzed together to check for consistency. The time-resolved data at each wavelength are the result of approximately 5 min of signal averaging. Each complete spectrum has been repeated 2–4 times with certain wavelengths repeated up to 12 times. The actual spectra presented are means, and the error bars represent a reproducibility of one standard deviation. Reproducibility of the lifetimes retrieved from the data is also quoted as one standard deviation.

7640   *Biochemistry, Vol. 31, No. 33, 1992*

Each decay was analyzed in a number of ways to reduce the effects of exponential correlation and overparametrization (see below).

Lifetimes were calculated by iterative reconvolution based on the Marquardt fitting algorithm assuming multiexponential kinetics according to

$$\Delta OD(\lambda, t) = \sum_i \Delta OD_i(\lambda) \exp(-t/\tau_i) \qquad (1)$$

where $\Delta OD_i(\lambda)$ is the amplitude of a component with a lifetime τ_i at a wavelength λ. The quality of the fits was assessed using a reduced χ^2 criterion and plots of weighted residuals.

Time-resolved decays were analyzed both individually and globally. In the global analyses, up to 12 decays could be analyzed simultaneously with the lifetimes of the components constrained to be the same in all 12 data sets. Global analysis effectively synthesizes higher signal to noise than can be achieved by analyzing a single decay and also reduces the number of free parameters. Global analysis is particularly useful in this study as the spectra of different components are found to dominate the total signal at different wavelengths. In both individual and global analyses one or more of the lifetimes could be fixed, once these had been established from analysis of other data where some components could be more clearly resolved due to the use of a different time scale or spectral range. Pre-exponential factors could be fixed when data sets were individually analyzed. This last option is particularly important as it allows one to assess the effects of long-lived components on the analysis of faster components observed over shorter time scales, where the influence of a long-lived component may not otherwise be easy to ascertain. This method of data analysis makes no allowances for the possible presence of distributions of lifetimes, but such distributions are not required to produce a consistent model for the data presented here.

Estimations of excitation levels, both from the pump beam parameters and independently from the size of the transient bleaches, indicate that only 5–10% of the reaction centers in the volume of the pump beam are excited by each flash.

RESULTS

Absorption difference spectra are presented either as (a) kinetic spectra which represent the difference spectra associated with a particular kinetic component (i.e., spectra of the pre-exponential factors in eq 1), sometimes called decay-associated spectra, or (b) the excitation-induced absorbance difference spectra at a specific time after excitation (all spectra of the nondecaying component can be considered as absorbance difference spectra at long time delays). Positive amplitudes in a kinetic spectrum indicate negative-going absorption changes, and negative amplitudes indicate positive-going absorption changes.

Figure 1 shows spectra of the laser pulses used for excitation compared with the long-wavelength part of the PS2 reaction center absorption spectrum. The spectra of the excitation pulses peak at 612 (10-nm bandwidth) or 694 nm (24-nm bandwidth). The 612-nm pulses are nearly transform limited, but the 694-nm pulses have a much broader spectrum. Transform-limited pulses are not required in experiments of this type unless coherent phenomena are to be investigated. Excitation at 694 nm was used with the intention of exciting P680 directly in a high proportion of reaction centers (see Discussion).

The results described below are also given in Table I.

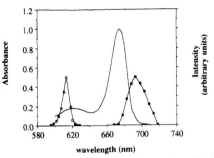

FIGURE 1: Part of the steady-state absorption spectrum of isolated PS2 reaction centers at room temperature (—). Also shown are the spectra of the excitation pulses; these spectra are centered at 612 nm, fwhm 10 nm (□), and 694 nm, fwhm 24 nm (■).

Transient absorption decays between 655 and 700 nm, on a 0–300-ps time scale, with 612-nm excitation, could be well fit by the sum of two exponentials (A2 and A3) and a nondecaying[2] component (A0). Fitting with fewer parameters fails to fit the data. The lifetimes of the exponential components were found to be 18 ± 4 (A2) and 260 ± 70 ps (A3). Of particular note was the finding that the amplitude of the 260-ps component was negligible when 694-nm excitation is used, and in this case the data fit well to a single exponential with a lifetime of 15 ± 5 ps (B2) plus a nondecaying component (B0). This difference between the use of 612- vs 694-nm excitation pulses is illustrated in Figure 2.

Figure 3 shows the kinetic spectra of components A2, A3, and A0 observed following excitation at 612 nm, and Figure 4 shows the spectra of components B2 and B0 following excitation at 694 nm. The spectra of the nondecaying components (A0 and B0) are essentially independent of excitation wavelength and exhibit a pronounced bleaching centered at 681 nm.

The use of a 0–80-ps time scale allows the ~20-ps components (A2 and B2, Figure 2) to be more accurately resolved than on the 0–300-ps time scale. Examples of data collected on the 0–80-ps time scale are shown in Figure 5. Data collected between 655 and 700 nm on the 0–80-ps time scale are well fitted to two exponential components and a nondecaying component, following excitation at either 612 or 694 nm. The lifetimes of the components are 4 ± 2 (A1) and 22 ± 5 ps (A2) following excitation at 612 nm and 4 ± 2 (B1) and 23 ± 5 ps (B2) following excitation at 694 nm. The 260-ps component observed on the 0–300-ps time scale (Figure 2) could not be distinguished from the nondecaying component on the 0–80-ps time scale. Inclusion of the 260-ps component with a fixed lifetime and amplitude during analysis of the 0–80-ps time scale data did not significantly change the lifetime or spectrum of the ~20-ps component (data not shown). This indicates that these two components are well resolved by the fitting procedure. The 4-ps component observed following excitation at either 612 or 694 nm was more clearly resolved on a 0–13-ps time scale. This component, and other fast kinetics, are discussed in detail elsewhere (Durrant et al., 1992).

On a 0–80-ps time scale from 520 to 570 nm, and excitation at either 694 or 612 nm, the data were well fitted to a (19 ± 2)-ps lifetime (A2 and B2) and a nondecaying component

---

[2] The component defined as "nondecaying" does not decay over the time scales under consideration in this paper. This component actually has a lifetime of tens of nanoseconds.

Observation of Pheophytin Reduction in Photosystem Two

Biochemistry, Vol. 31, No. 33, 1992   7641

Table I: Lifetimes Obtained from Global Analyses of Data Presented in This Paper[a]

	lifetime (ps)				
	0–300-ps time scale		0–80-ps time scale		
component	520–570 nm	655–700 nm	445–500 nm	520–570 nm	655–700 nm
612-nm excitation					
A1	NR[b]	NR	–[c]	NR	4 ± 2
A2	23 ± 7	18 ± 4	–	19 ± 2	22 ± 5
A3	~100[f]	260 ± 100	–	NR	NR
A0	nondecaying[d]	nondecaying	–	nondecaying[e]	nondecaying[e]
694-nm excitation					
B1	–	NR	NR	NR	4 ± 2
B2	–	15 ± 5	21 ± 1	19 ± 2	23 ± 5
B0	–	nondecaying	nondecaying	nondecaying	nondecaying

[a] Data were collected on two time scales (0–300 and 0–80 ps), three probe wavelength regions (445–500, 520–570, and 655–700 nm), using either 612- or 694-nm excitation pulses. Data were collected for eight different permutations of these conditions, as indicated in the table, and separate global analyses conducted for each of these permutations. The ~200-ps component observed following excitation at 612 nm (A3) was not observed when 694-nm excitation was used. The 4-ps component is the average of components discussed in more detail elsewhere (Durrant et al., 1991). [b] NR, component not resolved, due to the time scale being inappropriate for the observation of this component and/or the amplitude of this component being too small. [c] –, Data not collected. [d] Nondecaying, a component which did not decay on the time scale of the experiment (see footnote 2). [e] On the 0–80-ps time scale, components A3 and A0 could not distinguished. [f] Amplitude too small to obtain an accurate value for this lifetime.

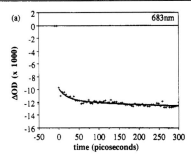

(a) 683nm

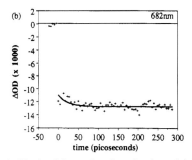

(b) 682nm

FIGURE 2: Kinetics of the transient absorption change (+) observed between 0 and 300 ps at (a) 683 nm following excitation of PS2 reaction centers at 612 nm and (b) 682 nm following excitation at 694 nm. The solid lines are the fitted functions obtained using eq 1. These are the (a) the sum of two exponential components with lifetimes of 18 ± 4 (A2) and 260 ± 70 ps (A3) and a nondecaying component (A0) and (b) the sum of one exponential component with a lifetime of 15 ± 5 ps (B2) and a nondecaying component (B0). Experiments were conducted at 295 K under anaerobic conditions.

(A0 and B0). An example of these data can be seen in Figure 5b. Another component with a lifetime of the order of 100 ps is also present when 612-nm excitation is used (see Figure 6). This component has a small amplitude (which results in a very low precision for the lifetime) and is most easily observed in data collected on the 0–300-ps time scale. Although one cannot be sure that this component is due to the same process which apparently produces a (260 ± 100)-ps lifetime (A3) between 655 and 700 nm, the discrepancy between the lifetimes is within the low precision for these components, and we

therefore assume that they represent the same process. There could of course be a genuine wavelength dependence to this component, but our data are not precise enough to determine whether this is in fact the case. Notwithstanding the above, we choose to label the 100-ps component as A3 (see Table I and Figure 6).

Figure 6 shows the kinetic spectra of the (19 ± 2)- and ~100-ps and the nondecaying components in the 520–570-nm spectral region with excitation at (a) 612 and (b) 694 nm. The spectra of the nondecaying components (A0 and B0) exhibit a pronounced bleach centered near 545 nm. Components A2 and B2 both show positive peaks at 545 nm, indicating that these components produce an increased bleaching of the 545-nm band. This is illustrated in Figure 6c, which shows the transient spectra at time delays of 3 and 100 ps. The amplitudes and shapes of the spectra of A0 and B0 in Figure 6 are independent of excitation wavelength to ±20%, as are the amplitudes for A2 and B2.

Data from the 440–500-nm spectral region on a 0–80-ps time scale are well fit by two components when 694-nm excitation is used. One component has a (21 ± 1)-ps lifetime, and the other is nondecaying (see, for example, Figure 5c). The spectra of these components (B2 and B0) are shown in Figure 7.

Essentially identical spectra were recovered for the ~20-ps components (A2 and B2) by analyzing data collected on either the 0–80- or 0–300-ps time scale (data not shown) across all regions of the spectrum. The independent observation of similar lifetimes (20 ps) with the same kinetic spectrum obtained on different time scales, and of similar lifetimes obtained over different spectral ranges (with the exception of A3), is evidence that the fitting procedures are appropriate for these data.

The observed lifetime of the 260-ps component (A3) is likely to be heavily influenced by systematic errors, due largely to the longest time scale used here being somewhat too short to accurately determine the lifetime of this component. It also appears as a (100 ± 100)-ps component between 520 and 570 nm (see above), which merely reflects the low precision for this component due to its small amplitude. Due to the difficulties associated with fitting this component accurately, a more realistic overall value is probably 200 (+300, –100) ps.

We studied the dependence of the transient absorption kinetics upon excitation intensity and determined that identical

7642 *Biochemistry, Vol. 31, No. 33, 1992*

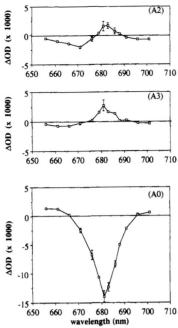

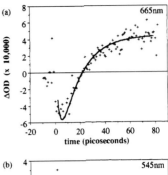

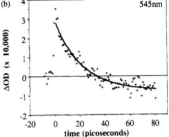

FIGURE 3: Kinetic spectra of components A2 ($\tau = 21$ ps), A3 ($\tau = 200$ ps), and A0 (nondecaying) obtained from single-wavelength transient absorption measurements between 0 and 300 ps and between 0 and 80 ps using 612-nm excitation pulses. Error bars were calculated independently for each data point and where not observable are smaller than the symbols. These kinetic spectra are the spectra of the pre-exponential amplitudes [$\Delta OD_i(\lambda)$] defined in eq 1; therefore, positive amplitudes of these spectra indicate negative-going absorption changes, and negative amplitudes indicate positive-going absorption changes (cf. Figure 2, for example).

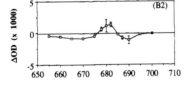

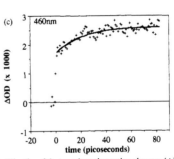

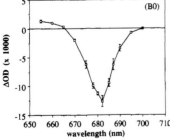

FIGURE 4: Kinetic spectra of components B2 ($\tau = 21$ ps) and B0 (nondecaying) obtained from single-wavelength transient absorption measurements between 0 and 300 ps and between 0 and 80 ps using 694-nm excitation pulses.

kinetics and amplitudes were observed for the 21-ps and non-decaying components for excitation pulses attenuated by up

FIGURE 5: Kinetics of the transient absorption changes (+) observed between 0 and 80 ps at (a) 665, (b) 545, and (c) 460 nm following excitation at 694 nm. The solid lines are fitted functions which are the sum of (a) two exponential components with lifetimes of 4 ± 2 (B1) and 23 ± 5 ps (B2) and a nondecaying component (B0) and (b, c) one exponential component with a lifetime of 19 ± 2 (545 nm) or 21 ± 1 ps (460 nm) and a nondecaying component. Component B1 could not be clearly resolved between 440 and 570 nm on the 0–80-ps time scale and was therefore not included in the analysis of data over this spectral range.

to a factor of 12. This indicates that our estimation that very few reaction centers receive multiple excitations (see Materials and Methods) is indeed correct and that all of the kinetics observed here are due to single photon events.

DISCUSSION

A summary of data collected on 0–300-, 0–80-, and 0–13-ps time scales is shown in Table II, where data from the 0–13-ps time scale have been taken from Durrant et al. (1992). The final averaged lifetimes of the components discussed in this paper are 200 ± 100 (component A3) and 21 ± 3 ps (for both components A2 and B2) and a nondecaying component (A0 and B0), with the reproducibility quoted as one standard deviation.

The reason for presenting most of the data as kinetic spectra is that most of the kinetic components described above cause

Observation of Pheophytin Reduction in Photosystem Two

Biochemistry, Vol. 31, No. 33, 1992   7643

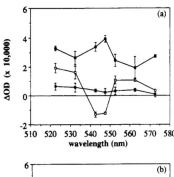

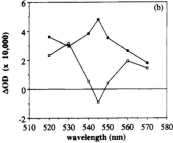

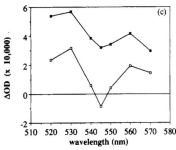

FIGURE 6: Kinetic spectra of 19 ± 2-ps (A2, B2, ■) and nondecaying components (A0, B0, □) obtained from single-wavelength transient absorption measurements between 0 and 300 and between 0 and 80 ps using (a) 612- and (b) 694-nm excitation pulses. Also shown in (a) is the kinetic spectrum of component A3 (×), which is only observed following excitation at 612 nm. (c) Spectra at time delays of 3 (■) and 100 ps (□) following excitation at 694 nm. [The spectra shown in (c) were calculated from the kinetic spectra shown in (b) using eq 1 and are essentially transient absorption spectra before and after the 21-ps component has occurred.]

only relatively small and subtle changes to the overall transient spectra. Although the changes in the overall transient spectrum are small, the high sensitivity of our measurements is quite capable of resolving them. This point is well illustrated by considering Figures 4 and 5a. The largest contribution to the kinetics shown in Figure 5a is that of the 23-ps component (B2), yet Figure 4 demonstrates that the amplitudes of this component are in general much lower than those associated with the nondecaying component (B0).

The results presented above show that optical excitation of isolated PS2 reaction centers results in multiexponential transient absorption kinetics in the picosecond time domain. The use of variable excitation wavelengths is shown to change the relative amplitudes of one of the kinetic components. Excitation pulses of 694 nm were used with the intention of directly exciting P680 in a much higher proportion of reaction centers than is achieved using 612-nm excitation pulses.

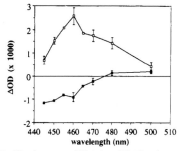

FIGURE 7: Kinetic spectra of 21 ± 1-ps (B2, ■) and a nondecaying component (B0, □) obtained from single-wavelength transient absorption measurements between 0 and 80 ps using 694-nm excitation pulses.

Table II: Summary of the Results of Analysis of Data Presented in This Paper and in Durrant et al. (1991)[a]

(A) 612-nm excitation	(B) 694-nm excitation	assignment
A1,[b] 1.6 ± 0.6 ps	B1,[b] 400 ± 100 fs and 3.5 ± 1.5 ps	loss of P680 singlet excited state
A2, 21 ● 3 ps	B2, 21 ± 3 ps	pheophytin reduction
A3, 200 ± 100 ps	not present	slow energy transfer/trapping
A0, nondecaying	B0, nondecaying	primary radical pair state (P680+Ph−)

[a] Components B1 and probably A1 are assigned to the decay of a delocalized P680 singlet excited state. They are most clearly resolved on the 0–13-ps time scale and are discussed in detail in Durrant et al. (1991). Components A2 and B2 have similar lifetimes and spectra, although subtle differences in their spectra can be resolved. These components are assigned to pheophytin reduction, which produces the radical pair state P680+Ph− (components A0 and B0). Component A3 is only observed following excitation at 612 nm and is assigned to a slow energy-transfer/trapping process. The time scale of P680+ formation is not determined in this paper. The values presented in this table are repeat weighted averages of the data discussed in the text and summarized in Table I. [b] From Durrant et al. (1992).

Complete selectivity is not possible due to the high degree of spectral overlap between individual chromophores in the PS2 reaction center and the finite width of the excitation pulse. As shown in one of our other studies (Durrant et al., 1992), excitation at 694 rather than 612 nm does indeed result in a relatively selective excitation of P680, with this selectivity being retained for at least the first 180 fs.

Radical Pair Spectrum. The nondecaying species observed on the 0–300-ps timescale following excitation at either 612 or 694 nm (Figures 3, 4, 6, and 7) have spectra which support their assignment to the primary radical pair state P680+Ph− (Danielius et al., 1987; van Kan et al., 1991; Nuijs et al., 1986; Schatz et al., 1987). Our spectra show a negative feature centered at 681 ± 1 nm assigned largely to the bleaching of P680 and pheophytin Q_y absorption bands, a negative feature centered at 545 ± 2 nm assigned largely to the bleaching of a pheophytin Q_x absorption band, and a positive feature peaking at 460 ± 2 nm assigned largely to the appearance of a pheophytin anion absorption band. The relative amplitudes of these features are 49 (at 681 nm):1 (at 545 nm):6.7 (at 460 nm). The time-resolved radical pair spectrum in the PS2 reaction center has also been measured over this range by Danielius et al. (1987). Their spectrum has a ratio of 51:1:8, which differs from our transient spectrum by less than 20%.

7644   *Biochemistry, Vol. 31, No. 33, 1992*

It is also possible to estimate the shape of the radical pair spectrum by combining data from steady-state observations of P680$^+$ and Ph$^-$ in PS2 particles and in vitro (Nanba & Satoh, 1987; Barber et al., 1987; Fujita et al., 1978). Both our data and those of Danielius et al. (1987) are essentially in agreement with what would be predicated from these steady-state observations. For example, the radical pair peaks at 460 nm rather than at 450 nm [which is the peak of the Ph$^-$ band (Nanba & Satoh, 1987)], due to the contribution of P680$^+$ to the difference spectrum (Barber et al., 1987).

The spectra of the nondecaying components are found to be independent of excitation wavelength. This observation is in agreement with previous studies of PS2 RCs which have determined that the primary radical pair state is formed with a near unity quantum yield when a variety of excitation wavelengths are used (Booth et al., 1991). It has also previously been shown that these PS2 samples contain an upper limit of 6% "free" chlorophyll (Booth et al., 1990).

Distinction between the 20- and 200-ps Processes. The similarity in spectra of components A2 (20 ps) and A3 (200 ps) shown in Figure 3 might at first lead one to think that these components might represent the same underlying process. However, the clear distinction between these components from 520 to 570 nm (Figure 6) indicates that the 200-ps component produces very little, if any, of the final nondecaying 545-nm bleach, while the 21-ps component produces ≈60% of the total.

It is important to recognize that the precision of the spectra in Figure 3 (note the small error bars) is sufficiently high that subtle but important differences in shape can be noted. For example, the ratio A2$_{(670nm)}$/A3$_{(670nm)}$ = 4 is distinguishable from A2$_{(680nm)}$/A3$_{(680nm)}$ = 0.5. Consequently, these kinetic spectra clearly do represent different processes as one would expect from a consideration of the data in the 520–570-nm spectral region.

The overall similarity in shape of A2 and A3 between 655 and 700 nm can be rationalized as follows. They both represent processes which bleach the ground state of a relatively low energy (spectroscopically red absorbing) chlorin, accompanied by the ground-state recovery of a slightly higher energy (spectroscopically blue absorbing) chlorin. The mixture of energy/electron-transfer reactions which each of these spectra represent is, however, presumably different, the only connection being that energy is moving overall from high energy (blue) chlorins to lower energy (red) chlorins. Although changes in excited-state absorption do contribute to these spectra, their contribution is likely to be smaller than the prominent features seen in the kinetic spectra of A2, A3, and B2.

The most important distinction between these two processes, however, is that the ≈200-ps component is absent when P680 is excited directly, whereas the ≈20-ps component is present with similar amplitude when either 612- or 694-nm excitation is used.

Slow Energy Transfer/Trapping. The 200-ps component (A3) observed in this study following excitation at 612 nm is most easily assigned to slow energy transfer/trapping within the PS2 RC. The spectrum of the 200-ps component is inconsistent with its assignment to a depolarization process, nor is it likely that this spectrum could result solely from changes in excited-state absorption. The spectrum exhibits a positive feature peaking at 681 nm (Figures 2 and 3). This is due to increased bleaching of a pigment with an absorption maximum near 681 nm and could reflect a degree of slow radical pair formation. This 200-ps component contributes

15 ± 5% of the final absorption change at 681 nm. The negative portion of this component between 655 and 670 nm (Figure 3) is consistent with the ground-state recovery of a chlorin molecule with a Q$_y$ absorption maximum nearer to 670 than to 680 nm. Therefore, the 200-ps component could be assigned to slow energy transfer/trapping from a chlorin with an absorption maximum at a shorter wavelength than P680. This assignment is supported by the observation that the 200-ps component is not observed following excitation at 694 nm.

Earlier measurements on isolated PS2 RCs also observed "slow" energy transfer at 4 K (Wasielewski et al., 1989b), although not at 277 K (Wasielewski et al., 1989a). Our observation of the presence of a slow energy-transfer/trapping process is consistent with recent time-resolved fluorescence studies of PS2 reaction centers (Booth et al., 1991; Roeloffs et al., 1991).

There is the possibility that the 200-ps component could originate from damaged reaction centers in which primary charge separation has been impaired. One might expect to observe this component when exciting at 694 nm as well as 612 nm because the 200-ps component produces a bleach at 680 nm and must therefore originate from reaction centers with pigments which can absorb the 694-nm pulses in the first place. However, as the 200-ps component is not observed using 694-nm excitation, we feel that it is more likely that this component originates from a slow energy-transfer/trapping process rather than a slow electron-transfer reaction, although the latter cannot be definitively ruled out. The spectrum for this component in Figure 6a would suggest that it does not result in the net bleaching of a pheophytin.

The observation of a 200-ps lifetime for an energy-transfer/trapping processing is at first sight surprisingly slow for a complex containing only eight pigments. In bacterial reaction centers the excitation energy is thought to be trapped by the special pair in less than 150 fs (Breton et al., 1986; Johnson et al., 1990). The isolated PS2 RCs used in this study bind two more chlorophylls than reaction centers of purple bacteria (Gounaris et al., 1990). It is possible that the 200-ps component observed here may originate from one or both of these "extra" chlorins being unable to transfer excitation energy rapidly to P680. Energy-transfer rates are strongly dependent upon chromophore separation (proportional to R^{-6}) and orientation. These extra chlorins may be bound to the exterior of the reaction center, thus resulting in the observed slow rate of energy transfer.

In summary, we emphasize that the 200-ps component is not observed when P680 is directly excited; therefore, this component does not seem to be an integral step in primary charge separation by PS2 RCs.

Pheophytin Reduction. At least one of the pheophytin molecules associated with the isolated PS2 RC has a clearly resolved Q$_x$ absorption band with a maximum at 545 nm (Nanba & Satoh, 1987). Bleaching of this band is therefore indicative of the loss of pheophytin ground states. The bleaching of this band is the only spectral feature which can be unambiguously assigned to loss of pheophytin ground states in PS2 RCs, due to the high degree of spectral overlap between chromophores at other wavelengths. Bleaching of the 545-nm band has been observed in previous kinetic studies of PS2 (Danielius et al., 1987; Wasielewski et al., 1989a,b); however, the *rate* of bleaching has not been determined prior to the data which we present here.

The kinetic spectra of the 21-ps components (A2 and B2) exhibit clear maxima at 545 nm (Figure 6) following excitation

Observation of Pheophytin Reduction in Photosystem Two

Biochemistry, Vol. 31, No. 33, 1992   7645

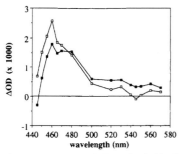

FIGURE 8: Spectra between 445 and 570 nm of optical density changes at time delays of 3 (■) and 100 ps (□) after excitation at 694 nm. (These spectra were calculated from the kinetic spectra shown in Figures 6b and 7 using eq 1 and are essentially transient absorption spectra before and after the 21-ps component has occurred.)

at either 612 or 694 nm and therefore represent a bleaching of this band. The lifetimes of these components are found to be independent of excitation wavelength to ±3 ps.

By comparing the amplitudes of the 21-ps components and the nondecaying (radical pair) components, and taking account of data collected on other time scales, it can be concluded that of the total pheophytin bleached, 60 ± 20% occurs with a rate of 21 ps^{-1} following excitation at 694 nm and 60 ± 20% at the same or similar rate following excitation at 612 nm.

We specifically assign the 21-ps bleaching of the pheophytin Q_x band at 545 nm to pheophytin reduction. This assignment is supported by data obtained between 445 and 570 nm, which is illustrated in Figure 8. This figure shows the absorption difference spectra at 3 and 100 ps after excitation at 694 nm, essentially before and after the 21-ps process (B2) is complete. It is clear from Figure 8 that the 21-ps component results in the appearance of a positive absorption band with a maximum at 460 nm, and the partial recovery of initially positive absorption changes between 480 and 550 nm. Both of these features are consistent with the formation of a pheophytin anion state, and the concomitant decay of chlorophyll excited singlet state(s) (Nanba & Satoh, 1987; Barber et al., 1987; Shepanski & Anderson, 1981). Moreover, the clearest evidence which supports the assignment of the 21-ps components specifically to pheophytin reduction comes from the observation that the nondecaying state produced by the 21-ps component has a spectrum which is, within our signal to noise, that expected for the radical pair state P680+Ph- (see Radical Pair Spectrum), whereas the spectrum before the 21-ps component is not that of the radical pair. Combining these observations with the result discussed above (that the 21-ps component results in an increased bleaching of the pheophytin Q_x absorption band), it can be concluded that at least half of the total pheophytin reduction in PS2 RCs occurs at a rate of (21 ps)$^{-1}$.

The spectra of 21-ps components (A2 and B2) in the 655–690-nm spectral region (Figures 3 and 4) are consistent with previous observations of steady-state pheophytin reduction (Klimov et al., 1977; Nanba & Satoh, 1987) and transient pheophytin anion reoxidation (Nuijs et al., 1986; Schatz et al., 1987) in PS2. However, the 21-ps component (Ph reduction) must result in a change in the redox state of one or more chromophores in addition to pheophytin. This complicates any interpretation of the data because there is a high degree of spectral overlap between chromophores in this spectral region; thus, a complete interpretation of the kinetic spectra of 21-ps components in the Q_y bands is not possible at present.

The two spectra of the 21-ps components between 655 and 670 nm are similar, but subtle differences can be observed when 612- rather than 694-nm excitation is used (spectra A2 and B2 in Figures 3 and 4). These differences could result from the mix of precursor states to the 21-ps component being different when different excitation wavelengths are used. Alternatively, these differences could be a consequence of the two excitation wavelengths exciting reaction centers with different orientational distributions.

The 21-ps components appear to account for only ≈60% of the final pheophytin bleaching (Figure 6). Preliminary analysis of data from the 0–13-ps time scale between 520 and 570 nm indicates that 40% of the pheophytin may be bleached directly by the excitation pulse. Analysis of these data also shows that the processes resulting in the 3.5-ps component (see Table II) produce no increase in the pheophytin Q_x absorption band bleach. This initial bleaching of pheophytin may result from a population of pheophytin excited singlet states formed prior to any electron-transfer processes. In this case, the amplitude of the observed bleaching at 545 nm could underestimate the degree of pheophytin reduction associated with the 21-ps component. In principle, the amplitude of the formation of the pheophytin anion band at 460 nm might allow a more accurate determination of the proportion of pheophytin reduction occurring with a (21-ps)$^{-1}$ rate. Unfortunately, the data in this spectral region (445–500 nm) are not accurate enough to obtain a precise value, particularly as these data will also include contributions from other transient species.

Studies between 655 and 690 nm on the 0–13-ps time scale have shown that 400-fs and 3.5-ps components observed following excitation at 694 nm (Durrant et al., 1992) have spectra which are completely different from that of the 21-ps component. The 400-fs and 3.5-ps components must therefore have different physical origins from that of the 21-ps component observed here. Our results therefore suggest that the 21-ps component observed here could account for all of the pheophytin reduction observed in PS2 RCs following direct excitation of P680.

The amplitude and rate of pheophytin reduction are essentially independent of excitation wavelength (to within ±20% and ±3 ps, respectively), as estimated from data collected between 520 and 570 nm. This indicates that energy/ electron-transfer/trapping kinetics result in extensive equilibration of the excitation energy within 21 ps, although the kinetic spectra of the 21-ps components between 655 and 695 nm suggest that these components may proceed from a slightly different mix of precursor states when different excitation wavelengths are used. In addition, as discussed above, there may also be one or two chlorins which are more weakly coupled to the other reaction center pigments. These chlorins are excited only when using the 612-nm pulses, and energy transfer/trapping then occurs with a 200-ps lifetime.

The results presented in this paper do not allow us to determine the rate of oxidation of P680. It has been suggested (Wasielewski et al., 1989a) that the rate of formation of P680+Ph- can be determined from the rate of formation of absorption changes at 820 nm. However, as chlorin anion, cation, and excited singlet states all have positive absorption bands of similar magnitudes over this spectral region (Nuijs et al., 1986; Hansson et al., 1988), the observation of kinetics solely at 820 nm may have several possible interpretations. Wasielewski et al. suggest that pheophytin reduction occurs with a 3-ps rate rather than the 21-ps rate which we find (Wasielewski et al., 1989a). Although we do observe a component of ∼3 ps (Durrant et al., 1992), it has the opposite

7646 *Biochemistry, Vol. 31, No. 33, 1992*

sign to that reported by Wasielewski et al. at 674 nm and must therefore originate from a different process. It is possible that some of these discrepancies arise from their use of a novel isolation procedure (Seibert et al., 1988), but there are also differences in the way in which the experiments were performed. Wasielewski et al. used excitation pulses that were 18 times greater in excitation intensity than those used here (3200 $\mu J/cm^2$ at 610 nm vs 175 $\mu J/cm^2$ at 612 nm), and these produced a maximum bleach 10 times greater than those which we observe. There are also differences between the radical pair spectrum reported by Wasielewski et al. and those reported here and by Danielius et al. (1987). As discussed above, the radical pair spectrum which we report differs from that of Danielius et al. by less than 20%; however, the relative sizes of spectra features from the data of Wasielewski et al. are 24 (682 nm):1 (545 nm):1.4 (460 nm), which differ from our spectrum and that of Danielius et al. by ~200 and 480% for 682 and 460 nm, respectively, when normalized by the amplitude of the dip at 545 nm. There is no obvious way to reconcile the data which we show with those of Wasielewski et al., and we must therefore conclude that the two experiments are observing different processes.

In summary, we have observed an electron transfer in PS2 RCs which results in the production of reduced pheophytin, and this process is found to occur with an overall rate of (21 $\pm$ 3 ps)$^{-1}$. It is not clear if the observed (21-ps)$^{-1}$ rate essentially corresponds to an underlying electron-transfer rate constant or whether some other processes such as an energy-transfer equilibrium limit the observed rate.

Transient absorption data collected on the 0–13-ps time scale have shown that when P680 is directly excited, some of the P680 singlet excited state decays with lifetimes of 400 fs and 3.5 ps (Durrant et al., 1992). This result does not indicate the rate of P680$^+$ formation or whether pheophytin is reduced directly or as the result of more than one electron-transfer step.

The accumulation of further data is necessary before a complete and testable kinetic model can be developed. In particular, it would be particularly useful if the rate of formation of P680$^+$ formation could be determined. However, our results do show that at least 60% of the pheophytin reduction resulting from primary charge separation in PS2 reaction centers occurs at an effective rate which is approximately 6 times slower than that observed in wild-type reaction centers of *R. viridis* and *Rb. sphaeroides*, despite the extensive homology between these reaction centers and those of photosystem two.

While it is true that one should exercise considerable caution in attempting to connect the observations of this paper to those made in bacterial systems, there does appear to be an intriguing link. The RCs of *Rb. sphaeroides*, *R. viridis*, and *R. capsulatus* all show an approximately (3.5-ps)$^{-1}$ rate of pheophytin reduction at room temperature (Woodbury et al., 1985; Breton et al., 1986; Kirmaier & Holten, 1988). In these RCs the amino acid M208 (M210 for *Rb. sphaeroides*) is a tyrosine; the corresponding amino acid on the L branch (L181) is phenylalanine. These amino acids are of particular interest as they are placed roughly at the central point between the special pair, bacteriopheophytin and bacteriochlorophyll molecules. In one mutant of *Rb. sphaeroides* L181 is preserved as a phenylalanine while M210 is changed to leucine. In these mutated RCs the rate of bacteriopheophytin reduction is lengthened considerably to 22 $\pm$ 8 ps (Finkele et al., 1990). The sequences of the D1 and D2 polypeptides line up with sequences of the L and M subunits such that D1-206 corre-

sponds to L181 and D2-206 to M208 (Michel & Deisenhofer, 1988). In pea PS2 RCs D1-206 is a phenylalanine and D2-206 is a leucine, and we show in this paper that the rate of pheophytin reduction appears to be 21 $\pm$ 3 ps. It is clear from studies of bacterial mutants (Chan et al., 1991; Nagarajan et al., 1990) that the identity of residues L181 and M208 (M210 in *sphaeroides*) affects the mean rate of pheophytin reduction. The mechanism by which this influence is exercised it not yet understood; nevertheless, the rate of pheophytin reduction in PS2 RCs at room temperature does appear to fit a phenomenological scheme developed for predicting the rate of primary charge separation in bacterial RCs (Chan et al., 1991).

ACKNOWLEDGMENT

We thank Niall Walsh and Caroline Woollin for preparing the reaction center samples and Chris Barnett for excellent technical assistance. We also thank Paula Booth and Linda Giorgi for their helpful comments, Qiang Hong for help with the femtosecond apparatus, Martin Bell for assistance with data analysis, and Oxford Lasers for loan of the copper vapor laser during the early stages of this work. We thank also Graham Fleming for bringing the identity of D2-206 to our attention. We also acknowledge financial support from the Science and Engineering Research Council, the Agriculture and Food Research Council, and the Royal Society.

REFERENCES

Barber, J. (1987) *Trends Biochem. Sci. 12*, 321–326.

Barber, J., & Andersson, B. (1991) *Trends Biochem. Sci. 17*, 61–66.

Barber, J., Chapman, D. J., & Telfer, A. (1987) *FEBS Lett. 220*, 67–73.

Booth, P. J., Crystall, B., Giorgi, L., Barber, J., Klug, D. R., & Porter, G. (1990) *Biochim. Biophys. Acta 1016*, 141–152.

Booth, P. J., Crystall, B., Ahmad, I., Barber, J., Porter, G., & Klug, D. R. (1991) *Biochemistry 30*, 7573–7586.

Braun, P., Greenberg, B. M., & Scherz, A. (1990) *Biochemistry 29*, 10376–10387.

Breton, J., Martin, J.-L., Migus, A., Antonetti, A., & Orszag, A. (1986) *Proc. Natl. Acad. Sci. U.S.A. 83*, 5121–5125.

Chan, C.-K., Chen, L. X-Q., DiMagno, T. J., Hanson, D. K., Nance, S. L., Schiffer, M., Norris, J. R., & Fleming, G. R. (1991) *Chem. Phys. Lett. 176*, 366–372.

Chapman, D. J., Gounaris, K., & Barber, J. (1991) in *Methods in Plant Biochemistry* (Rogers, L., Ed.) pp 171–193, Academic Press, London.

Crystall, B., Booth, P. J., Klug, D. R., Barber, J., & Porter, G. (1989) *FEBS Lett. 249*, 75–78.

Danielius, R. V., Satoh, K., van Kan, P. J. M., Plijter, J. J., Nuijs, A. M., & van Gorkom, H. J. (1987) *FEBS Lett. 213*, 241–244.

Deisenhofer, J., Epp, O., Miki, K., Huber, R., & Michel, H. (1985) *Nature 318*, 618–624.

Durrant, J. R., Giorgi, L. B., Barber, J., Klug, D. R., & Porter, G. (1990) *Biochim. Biophys. Acta 1017*, 167–175.

Durrant, J. R., Hastings, G., Hong, Q., Barber, J., Porter, G., & Klug, D. R. (1992) *Chem. Phys. Lett. 188*, 54–60.

Finkele, U., Lauterwasser, C., Zinth, W., Gray, K. A., & Oesterhelt, D. (1990) *Biochemistry 29*, 8517–8521.

Fork, R. L., Martinez, O. E., & Gordon, J. P. (1984) *Opt. Lett. 9*, 150–153.

Fujita, I., Davis, M. S., & Fajer, J. D. (1978) *J. Am. Chem. Soc. 100*, 6280–6282.

Gounaris, K., Chapman, D. J., Booth, P., Crystall, B., Giorgi, L. B., Klug, D. R., Porter, G., & Barber, J. (1990) *FEBS Lett. 265*, 88–92.

Hansson, O., Duranton, J., & Mathis, P. (1988) *Biochim. Biophys. Acta 932*, 91–96.

Observation of Pheophytin Reduction in Photosystem Two

Biochemistry, Vol. 31, No. 33, 1992   7647

Jankowiak, R., Tang, D., Small, G. J., & Seibert, M. (1989) *J. Phys. Chem. 93*, 1649–1654.

Johnson, S. G., Tang, D., Jankowiak, R., Hayes, J. M., Small, G. J., & Tiede, D. M. (1990) *J. Phys. Chem. 94*, 5849–5855.

Kirmaier, C., & Holten, D. (1988) *Isr. J. Chem. 28*, 79.

Klimov, V. V., Klevanik, A. V., Shuvalov, V. A., & Krasnovsky, A. A. (1977) *FEBS Lett. 82*, 183–186.

Knox, W. H., Downer, M. C., Fork, R. L., & Shank, C. V. (1984) *Opt. Lett. 9*, 552–554.

Kobayashi, M., Maeda, H., Watanabe, T., Nakane, H., & Satoh, K. (1990) *FEBS Lett. 260*, 138–140.

Michel, H., & Deisenhofer, J. (1988) *Biochemistry 27*, 1–7.

Nagarajan, V., Parson, W. W., Gaul, D., & Schenk, C. (1990) *Proc. Natl. Acad. Sci. U.S.A. 87*, 7888–7892.

Nanba, O., & Satoh, K. (1987) *Proc. Natl. Acad. Sci. U.S.A. 84*, 109–112.

Nuijs, A. M., van Gorkam, H. J., Plijter, J. J., & Duysens, L. N. M. (1986) *Biochim. Biophys. Acta 848*, 167–175.

Reed, D. W., & Clayton, R. K. (1968) *Biochem. Biophys. Res. Commun. 30*, 471–475.

Renger (1991) *Topics in Photosynthesis Vol. 11, The photosystems: structure, function and molecular biology* (Barber, J., Ed.) Elsevier, Amsterdam, in press.

Roeloffs, T. A., Gilbert, M., Shuvalov, V. A., & Holzwarth, A. R. (1991) *Biochim. Biophys. Acta 1060*, 237–244.

Schatz, G. H., Brock, H., & Holzwarth, A. R. (1987) *Proc. Natl. Acad. Sci. U.S.A. 84*, 8414–8418.

Schatz, G. H., Brock, H., & Holzwarth, A. R. (1988) *Biophys. J. 54*, 397–405.

Seibert, M., Picorel, R., Rubin, A. B., & Connolly, J. S. (1988) *Plant Physiol. 87*, 303–306.

Shipton, C. A., & Barber, J. (1991) *Proc. Natl. Acad. Sci. U.S.A. 88*, 6691–6695.

Takahashi, Y., Hansson, O., Mathis, P., & Satoh, K., (1987) *Biochim. Biophys. Acta 893*, 49–59.

Trebst, A. (1986) *Z. Naturforsch. 41C*, 240–245.

van Kan, P. J. M., Otte, S. C. M., Kleinherenbrink, F. A. M., Nieven, M. C., Aartsma, T. J., & van Gorkom, H. J. (1991) *Biochim. Biophys. Acta 1020*, 146–152.

Wasielewski, M. R., Johnson, D. G., Seibert, M., & Govindjee (1989a) *Proc. Natl. Acad. Sci. U.S.A. 86*, 524–528.

Wasielewski, M. R., Johnson, D. G., Govindjee, Preston, C., & Seibert, M. (1989b) *Photosynth. Res. 22*, 89–99.

Woodbury, N. W., Becker, M., Middendorf, D., & Parson, W. W. (1985) *Biochemistry 24*, 7516–7521.

Yeates, T. O., Komiya, H., Chirano, A., Rees, D. C., Allen, J. P., & Feher, G. (1988) *Proc. Natl. Acad. Sci. U.S.A. 85*, 7993–7997.

Contribution from

JAMES BARBER FRS

Imperial College, London
UK

Jim Barber is the Ernst Chain Professor of Biochemistry at Imperial College London, working on the molecular processes of photosynthesis. After graduating from the University of Wales in Chemistry he gained a PhD in Biophysics from the University of East Anglia. After a postdoctoral year in Holland he joined the staff at Imperial College as a Lecturer in 1968. In 1988–1989 he was Dean of the Royal College of Science, and from 1989 to 1999 was Head of the Biochemistry Department. Much of his research has focused on the reactions and proteins involved in the photochemically driven oxidation of water and has contributed greatly to this subject by elucidating the structure of the catalytic centre. He was elected to the European Academy (Academia Europaea) in 1988. In 2002 he was awarded the Flintoff Medal of the Royal Society of Chemistry and in 2003, was elected as a Foreign Member of the Swedish Royal Academy of Sciences. Two years later he was elected to a Fellowship of the Royal Society London, awarded the Novartis Medal of the Biochemical Society and won the Italgas Prize for energy and the environment.

At Porter's 65th birthday celebration. The person in the middle is Stephen Davidson.

James Barber FRS on
The Bakerian Lecture, 1977
"*In Vitro* Models For Photosynthesis"
G. Porter
Proc. Roy. Soc. Lond. A **362**, 281–303 (1978)

I was fortunate enough to attend the Bakerian Lecture given by George Porter at The Royal Society on 17th February 1977. By that time I had started my collaboration with him and his colleagues and it was with great pride that I listened to him outline the experiments we had conducted together on the red marine alga *Porphyridium cruentum* and described in the paper. Using a mode-locked frequency double Nd-laser as a source of 6 ps flashes of 530 nm light we were the first to time resolve excitation transfer within a photosynthetic light-harvesting system (see Figs. 3 and 5). This classical study was complemented by related studies on the green freshwater alga *Chlorella pyrenoidosa* and chloroplasts isolated from higher plants. Since these latter systems use mainly chlorophyll for light-harvesting there was not sufficient spectral differences to follow excitation transfer as in the case of the phycobilin-containing *Porphyridium curentum*. Nevertheless, these studies supported the principle that although pigments of photosynthetic light-harvesting systems are highly concentrated (about 0.1 M) energy transfer over hundreds of molecules occurs efficiently. This fascinated Porter since in free solution such concentrations of pigments had a very short excited state lifetime due to "concentration quenching" as shown in Fig. 5 for chlorophyll *a* in lecithin. The mechanism of "concentration quenching" has never been fully described and remained a problem which concerned Porter for the rest of his life. However he clearly recognised that photosynthetic organisms had to overcome this problem by arranging pigments so that they were close enough to facilitate highly efficient energy transfer but not close enough to directly interact with each other to generate quenching centres.

The remaining part of his Bakerian Lecture focused on electron transfer reactions known to be important in photosynthesis; chlorophyll to quinone and within the manganese complex of photosystem II (PSII). With his colleagues he conducted detailed studies of quinone-induced quenching of chlorophyll fluorescence as a means of gaining fundamental knowledge of electron transfer processes which we now know are a common feature of all types of photosynthetic reaction centres. At that time, however, Porter's heart was in understanding mechanisms associated with PSII. Why PSII? It was known that plastoquinone acted as the electron acceptor of PSII but he also recognised that PSII was particularly unique in that it performed very special chemistry; namely the splitting of water into dioxygen and reducing equivalents (protons and electrons) at a catalytic site composed of four Mn ions. Since the focus of his Bakerian Lecture was on "artificial photosynthesis" it was highly appropriate to present work and ideas on photochemically-induced oxidation reactions of manganese. Clever synthetic chemistry was used to construct manganese derivatives of tetrapyridyl porphyrins and sulphonated phthalocyanines and study light-induced oxidations of the manganese under different conditions. The relevance of this work is likely to become more important since the recent elucidation of the structure of the Mn_4-cluster (see my paper, "Molecular architecture of photosynthetic systems: New developments and mechanisms") has opened up the possibility of constructing an artificial water splitting system based on manganese chemistry. Indeed we are realistically approaching Porter's vision that a chemical storage system powered by solar energy and based on biological principles will ultimately provide man with an unlimited supply of energy.

James Barber FRS on
"Model Systems for Photosynthesis, I. Energy Transfer and Light Harvesting Mechanisms"
A. R. Kelly and G. Porter
Proc. Roy. Soc. London A **315**, 149–161 (1970)

This paper addresses a problem which was key to George Porter's interest in photosynthesis: How is concentration quenching avoided in photosynthetic organisms? In free solution the fluorescence yield of chlorophyll decreases to an undetectable level at concentrations typically found in photosynthetic light-harvesting systems. How then can these light-harvesting systems maintain singlet lifetimes long enough to carry out efficient energy transfer? Indeed as pointed out by Kelly and Porter, photosystem II (PSII) has a relatively high fluorescence yield.

In organic solvents it seemed reasonable to assume that "concentration quenching" was due to physical interaction of chlorophyll molecules as a consequence of random diffusion. This paper sets out to investigate whether concentration quenching could be avoided when chlorophyll and pheophytin molecules were held within a more rigid environment and in this way mimic the photosynthetic system. Using lecithin as a "rigid" solvent, Kelly and Porter found that concentration quenching still occurred. They therefore concluded that concentration quenching cannot be due to collision or dimerisation and further concluded that it did not involve triplet states. Although Porter went on to conduct further experiments to tackle this problem a rigorous explanation for concentration quenching is still required. This unsolved problem was often a subject of conversation with Porter and bugged him for the rest of his life. I believe that the ability of living organisms to overcome what seemed to be an inherent photophysical problem was one of the factors which drove him to delve into the mechanisms of the photosynthetic system resulting in my own close collaboration with him.

Contributed Article
"MOLECULAR ARCHITECTURE OF PHOTOSYNTHETIC SYSTEMS: NEW DEVELOPMENTS AND MECHANISMS"
J. Barber

ABSTRACT: Three topics resulting from research conducted in my laboratory and related to George Porter's interests are presented. The discovery of a light-harvesting system of photosystem I (PSI) of cyanobacterium, composed of a ring of 18 identical chlorophyll binding subunits around the trimeric reaction centre core.[10] The characterisation of the Q_A and Q_B quinone binding sites of photosystem II (PSII) based on X-ray crystallography and the most exciting, elucidation of the X-ray structure of the Mn-cluster which catalyses the splitting of water by PSII leading to a new proposal for the chemistry involved.[21]

Abbreviations: Chl chlorophyll; PSI photosystem I; PSII photosystem II; cyt, cytochrome; LHC, light-harvesting complex; EXAFS, extended X-ray absorption fine structure; EPR, electron spin resonance.

Introduction

In the 1977 Bakerian Lecture of the Royal Society, George Porter focused on three aspects of photosynthesis: light harvesting, electron transfer from chlorophyll to quinone and the involvement of manganese in the water oxidation reaction. Here I will follow his example and emphasise how my recent work has provided information on these three aspects of photosynthesis at a level of understanding which was inconceivable in 1977. My progress has relied heavily on the techniques of protein biochemistry and structural biology. Both high-resolution electron microscopy and X-ray crystallography have contributed to the wealth of knowledge now available to us. Whether such knowledge will lead to the construction of artificial systems able to mimic the process of photosynthesis, as envisaged by Porter, waits to be seen. However as he would often emphasise, history tells us that basic understanding of natural phenomena often leads to innovative technologies.

This article will be concerned with oxygenic photosynthesis even though studies of anoxygenic bacteria have also been very significant in elucidating some fundamental properties of photosynthesis[18] leading to the award of the 1988 Nobel Prize in Chemistry to Hans Deisenhofer, Hartmut Michel and Robert Huber.

Plants, algae and oxyphotobacteria (cyanobacteria and prochlorophytes) all have the ability to absorb sunlight and produce oxygen from water. They contain two types of photosystems: photosystem I (PSI) and photosystem II (PSII). The redox properties of the co-factors which constitute these two photosystems and how they cooperate to produce O_2, ATP and NADPH is summarised in the Z-scheme shown as the first figure in Porter's Bakerian Lecture paper.[39] We now know that these reactions involve four major membrane protein complexes as shown in Fig. 1. The cytochrome b_6f (cyt b_6f) complex acts as an intermediate between PSII and PSI while the ATP Synthase uses the electrochemical potential gradient of protons ($\Delta\mu H^+$), generated by vectorial electron flow across the membrane (known as the thylakoid membrane), to convert ADP to ATP.

All four photosynthetic complexes are made up of a large number of different protein subunits. PSII acts as a water-plastoquinone oxidoreductase; cyt b_6f as a plastoquinol-plastocyanin oxido-reductase and PSI as a plastocyanin-NADP$^+$ oxido-reductase. The products of light-driven electron/proton transfer, NADPH and ATP, are used to convert CO_2 to sugars and other organic molecules.[3]

PSI and PSII are each characterised in having a light-harvesting system servicing a reaction centre (RC) where primary and secondary charge separation reactions occur across the membrane. Over the years there have been extensive biochemical, biophysical, spectroscopic and molecular biological studies of PSI[15] and PSII.[20,17,7] But perhaps the greatest impact has come from structural studies. X-ray crystallography has provided a 2.5 Å 3D structure of PSI from the cyanobacterium *Thermosynechococcus elongatus*.[28] Here it was shown that the primary electron donor, P700, is a heterodimer of chlorophyll (Chl) *a* and chlorophyll *a'* and confirmed that the electron transfer chain across the membrane is P700 → chlorophyll *a* → phylloquinone and then onto ferredoxin via iron sulphur centres, F_X, F_A and F_B, in that order. The PSI reaction centre is dimeric being composed of two similar proteins PsaA and PsaB which gives rise to two sets of electron carriers comprising branches A and B, culminating at F_X (see Fig. 1). Whether there is a preference for one branch is a matter of controversy. Associated with redox active co-factors of the reaction centre are 92 Chl *a* and 22 carotenoids which provide an efficient light-harvesting system. Energy transfer from these antenna pigments to P700 is rapid and efficient.[33,22] In addition to these light-harvesting pigments, PSI can also have outer light-harvesting systems which increases its overall absorption cross-section. A newly-discovered outer antenna system for cyanobacterial PSI which is exquisite in design and highly efficient is discussed below.

PSII was very much a focus of Porter's interests, not only because it uses plastoquinone as a terminal electron acceptor but because it also catalyses the splitting of water using Mn ions. In recent years both electron and X-ray crystallography have provided structural information about PSII.[5] Much of this work has come from my own laboratory in collaboration with others although X-ray structures have been reported by two other groups.[50,29] Currently the X-ray structure of PSII that we have obtained at 3.5 Å resolution[21] is the most detailed and the impact of this information will be discussed here.

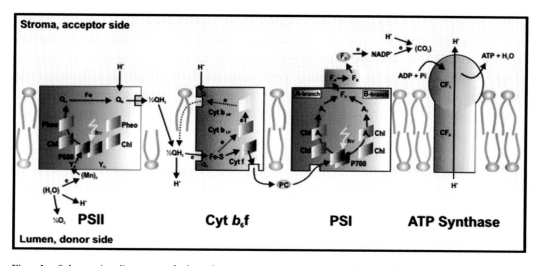

Fig. 1. Schematic diagram of the electron–proton transport chain of oxygenic photosynthesis in the thylakoid membrane, showing how PSI and PSII work together to use absorbed light to oxidise water and reduce NADP$^+$. The diagram also shows how the proton gradient generated by the vectorial flow of electrons across the membrane is used to convert ADP to ATP at the ATP Synthase complex (CF$_0$CF$_1$). In both PSI and PSII, the redox-active co-factors are arranged around a pseudo two-fold axis. In PSII, primary charge separation and subsequent electron flow occurs along one branch of the reaction centre. However, in the case of PSI, it is likely that electron flow occurs up both branches as shown. Electron flow through the cytochrome b_6f complex also involves a cyclic process known as the Q cycle. Y$_Z$ = active tyrosine; P680 = primary electron donor of PSII composed of chlorophyll (Chl); Pheo = pheophytin; Q$_A$ and Q$_B$ = plastoquinone; Cyt b_6f = cytochrome b$_6$f complex, consisting of an Fe–S Rieske centre, cytochrome f (Cyt f), cytochrome b low- and high-potential forms (Cyt b_{LP} and Cyt b_{HP}), plastoquinone binding sites, Q$_1$ and Q$_0$; PC = plastocyanin; P700 = primary electron Chl donor of PSI; A$_0$ = Chl; A$_1$ = phylloquinone; F$_x$, F$_A$ and F$_B$ = Fe–S centres, F$_D$ = ferredoxin; F$_R$ = ferredoxin NADP reductase; NADP$^+$ = oxidised nicotinamide adenine dinucleotide phosphate. Y$_D$ = symmetrically related tyrosine to Y$_Z$ but not directly involved in water oxidation, and QH$_2$ = reduced plastoquinone (plastoquinol), which acts as a mobile electron/proton carrier from PSII to the cytochrome b_6f complex.

Discovery and Implications of an Outer Light-harvesting System for PSI

In cyanobacteria and other oxyphotobacteria, the PSI reaction centre complex is trimeric and it was assumed until recently that the light-harvesting system is restricted to the pigments it binds. In higher plants, however, the PSI reaction centre is monomeric and is known to have an additional antenna system composed of the light-harvesting complex (LHCI) binding Chl a and Chl b. Hints about the organisation of this outer light-harvesting system were first revealed by electron microscopy[14,30] and later by X-ray crystallography.[9] Until recently there was no evidence to suggest that PSI of cyanobacteria and other oxyphotobacteria had an outer light-harvesting system equivalent to LHCI until my colleagues and I discovered this not to be the case.[10–12]

It has been the custom to grow cyanobacteria in the laboratory using nutrient-rich media. However in the natural environment aquatic organisms are usually exposed to nutrient deficient conditions, especially in the open oceans. A nutrient which is often limiting is iron since Fe^{3+} salts are rather insoluble. It has been known for some time that when cyanobacteria are grown in media low in iron two *iron-stress-induced* genes are activated, *isi*A and *isi*B.[46] The *isi*B gene encodes for flavodoxin which replaces ferredoxin as an electron acceptor to PSI. The *isi*A gene encodes for a Chl *a*-binding protein similar to the CP43 light-harvesting protein of PSII, called here CP43'.[36,32] For many years it was unclear what the role of this iron-stressed Chl-binding protein was. However recently we discovered that it formed an outer light-harvesting system for PSI of the cyanobacterium *Synechocystis* PCC6803.[10] A similar conclusion was reached for the cyanobacterium *Synechococcus* PCC7942.[13] The antenna system is composed of 18 CP43' subunits which form a ring around the trimeric PSI reaction centre core, as shown in Figs. 2 and 3. This supercomplex has a diameter of 320 Å and a molecular mass of about 2000 kDa. Steady-state optical fluorescence and absorption studies indicated that the antenna system is functionally coupled to the reaction centre core.[10,12,14,2]

In order to overcome the artefactual influence of using negative stain (uranyl acetate) a 3D structure was obtained recently using electron cryo-microscopy (cryo-EM).[35] Preparations of the CP43'-PSI supercomplex of *Synechocystis* were rapidly frozen in vitreous ice and images of random orientated particles recorded using a high-voltage electron microscope. After averaging different views, a 3D model was constructed. In the absence of uranyl acetate stain, the electron density map (see Fig. 2(a)) is due only to the presence of protein and pigments. For this reason it is possible to model the 2.5 Å X-ray structure of the trimeric PSI reaction centre core[28] into the central region of the supercomplex with some degree of precision (Figs. 2(b) and 3).

Although CP43' is homologous to CP43 of PSII in having six transmembrane helices it differs in that CP43 has, but CP43' does not have, a large extrinsic loop joining transmembrane helices V and VI. Hints as to the structure of CP43 first came from electron microscopy[42,24] but more details have emerged from X-ray crystallographic studies.[50,29,21] The most recent X-ray structure[21] has mapped side chains and assigned the positions of 14 Chl's bound within CP43. In Figs. 2(b) and 3 the X-ray structure of CP43 minus its extrinsic loops has been incorporated into the density. In so doing it has been assumed that transmembrane helices V and VI are located immediately adjacent to the surface of the PSI trimer. Assuming that CP43' binds 14 Chl's as does CP43, the 18-mer ring adds an additional 252 Chl's to the 276 Chl's of the RC trimer thus increasing its light-harvesting capacity by almost 100%.

Ultrafast emission and absorption spectroscopy indicates that energy transfer within the antenna CP43'-ring and from the ring to the reaction centre core is very rapid and efficient.[34] The average transfer time within the ring is less than a ps while from the ring to the RC core the excitation transfer occurs within 10 ps.

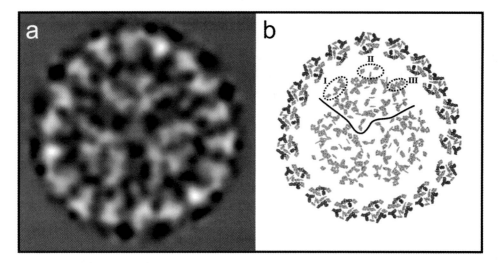

Fig. 2. CP43'-PSI supercomplex of *Synechocystis* PCC6803 obtained by electron cryo-microscopy (cryo-EM) in the absence of negative stain and the arrangement of chlorophylls (green). (a) Top view projection showing protein distribution. (b) Overlay of X-ray data for the positioning of chlorophylls in the PSI reaction centre and in CP43'. Also shown in red are the transmembrane helices of CP43'. The regions marked I, II and III contain chlorophylls which possibly aid energy transfer from the CP43'-ring to the reaction centre core (see Ref. 35).

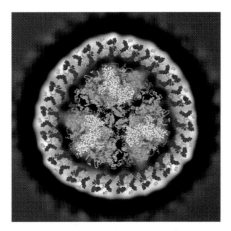

Fig. 3. Overlay of the X-ray structure of the PSI reaction centre of *T. elongatus*[28] and modified CP43 onto the top view of the cryo-EM structure of the CP43'-PSI supercomplex of *Synechocystis* PCC 6803.[35] Chlorophyll molecules – yellow; transmembrane helices of CP43' in red with helices 5 and 6 closest to the RC core; PsaA in yellow, PsaB in green, PsaL, PsaI, PsaM (trimer domain) in cyan; PsaJ in blue, PsaF in brown, PsaK in purple; PsaX in white. All are shown as C_α-traces. The three extrinsic proteins, PsaC, PsaD and PsaE attached to the stromal surface are shown in white as ball-and-stick representation.

The approximate overall fluorescence lifetimes of the CP43'-PSI supercomplex is 40 and 20 ps for the RC core alone. The difference is consistent with the difference in the two antenna sizes. The transfer of excitations from the CP43' ring to the PSI RC core could be aided by "linker chlorophylls". Indeed there are a number of Chls at the interface between the ring and the PSI RC outer surface as indicated in Fig. 2(b) as regions I, II and III. Of particular note are the Chls bound to PsaJ (region II), a small subunit at the periphery of the PSI RC core as well as those bound to the outer edges of the PsaA and PsaB proteins (region I and III respectively). The distance between these peripheral Chls and those bound to CP43' is about 25 Å or less. Such short distances can facilitate efficient and rapid transfer by the Förster inductance mechanism given that for chlorophyll a the R_o distance is about 70 Å.[25]

The CP43' antenna ring around PSI is reminiscent of the antenna rings observed in purple photosynthetic bacteria[51] and is a wonderful example of the beauty of biological light-harvesting systems which Porter so much wanted to mimic in an artificial construction. The challenge remains.

Quinone Electron Acceptors in PSII

In all types of reaction centres, the absorption of light leads to the transfer of electrons from chlorophyll to quinones. In PSI and the RC of green sulphur bacteria (Type 1 RC), the quinone acceptor is phylloquinone (A_1 in Fig. 1) and has a low potential (~ −650 mV). This tightly-bound quinone passes the electron rapidly to the iron sulphur centre F_X (see Fig. 1). In contrast, in PSII and purple photosynthetic bacterial RCs (Type II RC) the terminal quinone acceptor has a redox potential of about 0 V and undergoes protonation. In fact, as shown in Fig. 1, there are two quinones denoted Q_A and Q_B, where Q_A is firmly bound and is a one-electron acceptor with a lower redox potential than Q_B. It does not undergo protonation but passes its electron to Q_B which on receiving a second electron from Q_A becomes doubly reduced. In PSII, Q_A and Q_B are plastoquinone molecules while in purple photosynthetic bacteria they are usually ubiquinone, although Q_A can also be menoquinone. Recently we have revealed the positioning of the Q_A and Q_B plastoquinones and the nature of their protein environments by elucidating an X-ray structure of a PSII complex isolated from the thermophilic cyanobacterium *Thermosynechococcus elongatus*.[21] Q_A and Q_B are bound to the D2 and D1 reaction centre proteins respectively. We have found that the binding site for the Q_A plastoquinone is remarkably similar to that of its counterpart in purple photosynthetic bacteria where it binds to the M-subunit of the RC.[18] In PSII the Q_A is hydrogen bonded to the main-chain amide group of the D2 Phe261 and D2 His214, where the latter also serves as a ligand to the non-haem iron positioned midway between Q_A and Q_B. Overall Q_A is located in a hydrophobic cavity composed of residues shown in Fig. 4(b).

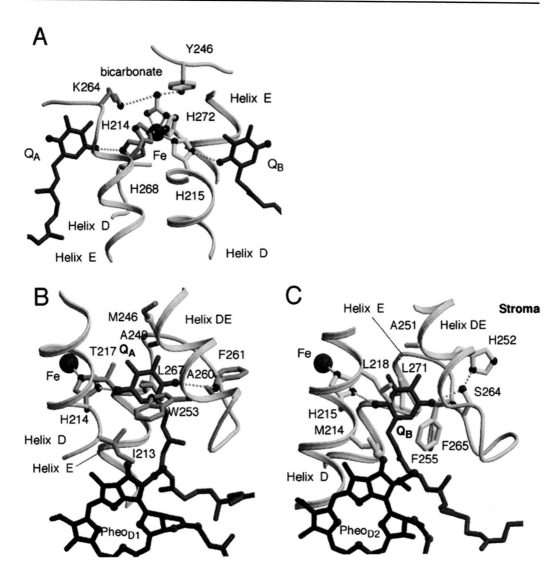

Fig. 4. Structure and protein environments derived from X-ray crystallography[21] of the plastoquinones Q_A and Q_B which are the electron acceptors of PSII. (A) Overall view of the non-haem iron showing the four histidine ligands and a bicarbonate ligand. (B) The Q_A binding site showing the ligands and the residues which form the hydrophobic pocket. (C) The Q_B binding site showing the ligands and nearby residues. Probable hydrogen bonds are shown as dotted lines while solid lines represent ligands. The protein chains are shown in grey, and the yellow and orange side chains belong to the D1 and D2 proteins respectively. Non-haem iron is shown in red, Q_A and Q_B plastoquinones in magenta and pheophytin *a* in blue. Oxygen and nitrogen atoms are shown as red and blue spheres respectively.

The Q_B site of PSII, however, is less conserved, with its bacterial equivalent located in the L-subunit. There are similarities in that the PSII Q_B is hydrogen bonded to D1 Ser264 and D1 His215, and possibly to the main-chain amide group of D1 Phe265 which have equivalents in the L-subunit. However the size of the Q_B-pocket in PSII is larger than that in its bacterial counterpart due to the insertion of an additional residue. Overall the PSII Q_B-pocket is more hydrophilic than that of Q_A and there is evidence of a proton pathway involving D1 His252 which is hydrogen bonded to D1 Ser264 (Fig. 4(c)). The non-haem iron located between Q_A and Q_B has four histidine ligands: D1 His215, D1 His272, D2 His214 and D2 His268 (Fig. 4(a)). In purple bacteria four equivalent histidines serve the same function and the fifth ligand is a glutamate.[18] However, in our X-ray structure of PSII this additional ligand is a bicarbonate ion which seems to be stabilised by hydrogen bonding with D1 Tyr246 and D2 Lys264. Why bicarbonate is used as a ligand is unclear, but may have a regulatory function in controlling Q_A to Q_B electron transfer.[47]

When the Q_B plastoquinone is fully reduced and protonated, it de-binds from the Q_B-site in the D1 protein and diffuses into the lipid matrix. In this way the reducing equivalents derived from water oxidation are made available to fix carbon dioxide with additional input of energy from PSI (see Fig. 1).

Revealing the Structure of the Mn-cluster which Catalyses Water Oxidation

The involvement of Mn ions in the water splitting reaction has been known for more than 50 years. The importance of this reaction cannot be understated. Directly or indirectly it provides the reducing equivalents for essentially all life on our planet. The by-product, dioxygen is required for the survival of aerobic life and incidentally gives rise to the ozone layer needed to protect us from high energy UV radiation. It is not surprising therefore that a great deal of effort has been directed at understanding the chemical and redox properties of Mn. As Jeff Schatz recently wrote, *"When it comes to trading electrons, Mn is the champ. Mn is the only element that can assume up to eleven different valence states, and the colours of its different compounds cover the entire visible spectrum."*[44] About 2.5 billion years ago a Mn-containing centre with the ability to catalyse the light-driven splitting of water evolved. This represented the "big bang of evolution" since biology was no longer dependent on limiting amounts of inorganic (H_2S, NH_3, Fe^{2+}) and organic (acids) compounds as electron/proton donors but instead, with the input of solar energy, could obtain reducing equivalents from the unlimited amounts of water available on our planet. That is, nature had found the perfect solution to the energy problem. However, water is a very stable compound and to split it into its elemental constituents is thermodynamically challenging. To produce dioxygen, two water molecules must be oxidised:

$$2H_2O \rightarrow O_2 + 4H^+ + 4e. \qquad (1)$$

The average energy required for extracting each electron is 0.81 V at pH 7.0, but to extract the first electron from water to produce a hydroxide radical takes about 2.5 V and the removal of the second electron to form an oxygen atom requires a further 1.5 eV (see Fig. 5).

Indeed it is well known that a large over-potential is required to perform the electrolysis of water into oxygen and hydrogen. Just how photosynthetic organisms achieve this difficult and thermodynamically-demanding chemistry of water oxidation, using visible light energy and a Mn-containing catalytic centre, is one of the great challenges of bioinorganic chemistry and science in general. Porter was very conscious of this and devoted considerable effort, particularly with input from Tony Harriman, to construct light-absorbing Mn derivatives able to catalyse charge transfers which could be used as model systems to mimic the natural process.[39]

In photosynthetic organisms the water-splitting process occurs in PSII, which is a large multi-subunit protein complex embedded in the thylakoid membrane.[5] It consists of a light-harvesting system which captures light energy and transfers it to the RC which has at its heart the D1 and D2 proteins. At the RC, excitation energy is trapped by chlorophyll, known as P680 because its red-absorbing maximum is at 680 nm, to become P680*. P680* quickly donates an electron to a pheophytin (Pheo) a molecule to form the radical pair P680$^{\bullet+}$Pheo$^{\bullet-}$. This charge separation represents the primary energy conversion step of PSII and occurs across the membrane (Fig. 1). P680$^{\bullet+}$ has been estimated to have a redox potential of about 1.3 V[41] and provides the driving force for the splitting of water. As indicated in Eq. (1) above, the water-splitting reaction is a 4-electron process, so there must be four successive oxidations of P680 by four photons to produce a dioxygen. The storing of the four oxidising equivalents occurs within a cluster of four Mn ions, aided by electron transfer via a redox active tyrosine, known as Y_Z. This tyrosine and all the redox active co-factors involved in charge separation across the RC are associated with the D1 protein, with the exception of Q_A which, as mentioned in the previous section, is bound to the D2 protein.[20]

In the mid-1990s I set out to obtain an atomic resolution structure of PSII with the ultimate aim of providing a structural basis to describe the various properties of this photosystem and in particular to visualise the organisation of the Mn-cluster. I believed that with such knowledge it would be possible to elucidate fully the chemistry of the water-splitting reaction. Much of the work involved high-resolution electron microscopy, both electron crystallography and single particle analyses.[5,7] With the help of colleagues good progress was made and we were able to assign the positions of the major subunits and their transmembrane helices, including those of the D1 and D2 proteins.[42] Moreover we obtained the first glimpse of the chlorophylls which constitute P680 and speculate about the fact that they did not form a special pair.[8,6] However the resolution of our

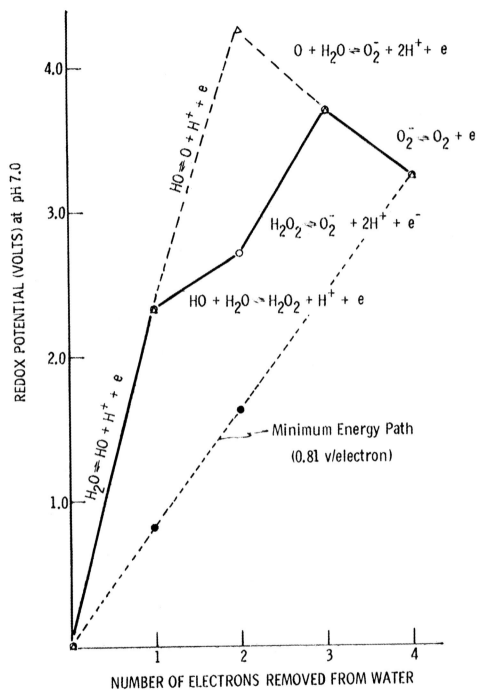

Fig. 5. Diagram showing the redox potential of the one-electron steps for the oxidation of H_2O to O_2. Solid line: pathway via H_2O_2 as the two-equivalent reduction stage. Solid and dashed lines: pathway via "$H_2O + O$" as the two-equivalent reduction stage. Dotted line: minimum energy pathway. It is unlikely that the hydrogen peroxide pathway can occur in PSII because of the oxidising properties of H_2O_2 within a pigment-protein environment. Reproduced from Ref. 40 with permission from Elsevier.

structures were limited to 8 Å and there was little chance that the techniques we were using would reveal atomic detail of the Mn-cluster.

To obtain high-resolution information, X-ray crystallography is the technique of choice. Indeed in 2001 a group in Berlin obtained the first X-ray structure of PSII[50] from a cyanobacterium *Thermosynechococcus elongatus*. The resolution was at 3.8 Å but the overall model was poorly refined. Nevertheless it provided the first direct view of the co-factors of PSII and identified a density for the Mn-cluster. Following on from this initial work, a more complete 3.7 Å X-ray model was obtained by Kamiya and Shen[29] working in Japan. Using PSII, also isolated from a thermophilic cyanobacterium, *Thermosynechococcus vulcanus*, they traced the C_α-backbone for most of the subunits and provided some information about the side chains. Their positioning of density for the Mn-cluster more or less matched that of previous work. However the structures published by these groups did not allow an in-depth model for the Mn-cluster to be developed. It has long been assumed that this cluster consisted of four Mn atoms and for this reason both groups tentatively placed this number within the assigned density. Several studies had indicated that a Ca^{2+} was located close to the Mn-cluster[16,49] but the two X-ray structures did not assign the position of this cation.

However our highly-refined 3.5 Å structure of PSII,[21] which includes the first direct information about the organisation of the Mn_4/Ca^{2+}-cluster provides hints as to the mechanism of water oxidation.

Figure 6 shows the positioning of the various co-factors and distances between them, which are involved in the electron transport chain from water to the Q_B plastoquinone. Except for the Mn_4/Ca^{2+}-cluster, other co-factors are duplicated and related by a pseudo two-fold axis which is a common feature of all types of RCs and emphasised in Fig. 1. However as is the case in the RC of purple photosynthetic bacteria, electron transfer occurs in PSII along one branch as indicted in Fig. 1. Although Fig. 6 is taken from Ferreira *et al.*[21] the positioning of the co-factors and pigments are in line with those reported previously.[50,29] However we have traced and assigned almost all the amino acids of all the subunits which make up the cyanobacterial PSII complex with the exception of one low molecular weight subunit. Because we have described the protein environments for the pigments and each of the active and inactive co-factors (as already detailed above for Q_A and Q_B), we were able to identify amino acids associated with the water-splitting site. Moreover our refined structure has allowed us to propose that this site contains a cubane-like $Mn_3Ca^{2+}O_4$-cluster linked to a fourth Mn ion (Mn4) by a mono-μ-oxo bridge as shown in Fig. 7.

Six protein ligands to the four Mn have been identified, five provided by the D1 protein (Asp170, Glu189, His332, Glu333 and Asp342) and one by a chlorophyll-binding protein CP43 (Glu354). Several other amino acids are located close to the cluster (D1 Asp61, D1 Gln165, D1 Tyr161, D1 His190, D1 His337, D1 Ala344, CP43 Arg357) and may play a role in stabilising intermediates of the oxidation reaction or be actively involved in the chemistry.

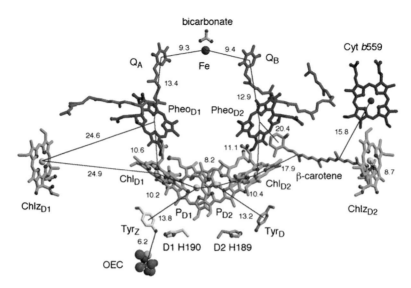

Fig. 6. Co-factors involved in electron transport in PSII, as determined by X-ray crystallography.[21] With the exception of the metal-cluster of the oxygen evolving centre (OEC) cytochrome b559 (Cyt b559) and a redox active β-carotene molecule, all the co-factors are arranged around a pseudo two-fold axis passing between the chlorophylls P_{D1} and P_{D2} and the non-haem iron (see also Fig. 1). Side chains of the D1 and D2 proteins are in yellow and orange respectively. Chlorophyll in green, pheophytin in blue, plastoquinones Q_A and Q_B in magenta, haem of cytochrome b559 in red. The atoms of the OEC are manganese (magenta), calcium (blue-green) and oxygen (red). Also shown are the non-haem iron (red) and its bicarbonate ligand. Distances are in Ångstroms.

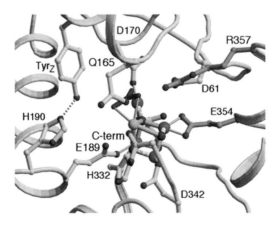

Fig. 7. A view of the oxygen evolving centre (OEC) derived from X-ray crystallography.[21] The metals of the catalytic site are arranged as a cubane-like $Mn_3Ca^{2+}O_4$-cluster bridged to a fourth Mn (referred to as Mn4 in the text) by a mono-μ-oxo bond. Mn ions are in magenta, Ca^{2+} in blue-green and oxygen in red. Protein main chains are in grey and D1 and CP43 side chains are shown in yellow and green. The ligands for the three Mn ions in the cubane are D1 Glu189, D1 His332, D1 Asp342 and CP43 Glu354. The protein ligands for Mn4 are D1 Asp170 and D1 Glu333. Other residues close to the OEC are D1 Asp61, D1 Gln165, D1 His337 (not shown), CP43 Arg357, D1 Tyr161 (Tyr$_Z$ or Y$_Z$) and D1 His190, where the latter two are hydrogen bonded (dotted line). Blue is a non-protein ligand having density distribution expected for a carbonate ion.

The metal–metal distances used in the model of the cluster are 2.7 Å for Mn–Mn within the cubane, 3.3 Å for the Mn–Mn distance between two Mn ions in the cubane and the fourth Mn (Mn4), and 3.4 Å for the Mn–Ca distance within the cubane. The longer distance between Mn4 and Ca is about 4 Å. These distances are compatible with those measured by extended X-ray absorption fine structure (EXAFS)[43,49] and by pulsed electron spin resonance (EPR) spectroscopy.[38] The presence of Mn and Ca^{2+} were determined by anomalous X-ray diffraction at 1.89 and 2.25 Å respectively and their positioning by Fourier difference analyses. The presence of oxygen in the oxo-bridges is inferred from the approximate metal–metal distances and the proposed cubane-like structure provides three oxo-ligands for each metal. Some or all the oxo-ligands could be protonated. Moreover we have incorporated a non-protein species ligating to Mn4 and Ca^{2+}. The density for this species accommodates a carbonate as shown in Fig. 7, an anion which is known to interact with the catalytic site.[4,31] We suggest that this tentatively-assigned carbonate provides two ligands to Ca^{2+} and one to Mn4. We assume that in addition to the protein and oxo-ligands, the coordination requirements of Mn and Ca^{2+} are satisfied by water molecules or hydroxyls.

Water Oxidation Mechanism

As stated above, water oxidation proceeds via a series of reactions driven consecutively by four photons of light. The sequence of events is depicted in the S-state cycle ("clock") shown diagrammatically in Fig. 8. Although many different mechanisms have been proposed for water oxidation, the structure that we have revealed by X-ray crystallography of the oxygen evolving centre (OEC) is suggestive of a scheme by which only one Mn ion binds a water substrate molecule. The second substrate water molecule is proposed to be contained within the coordination sphere of the Ca^{2+}. A tightly-bound Ca^{2+} located close to a Mn ion (Mn4 in Fig. 9) is required in order to facilitate the O–O bond formation.

The mechanism suggested here and outlined in Fig. 9, draws on various facets of schemes suggested by others.[26,45,48,23,37] It is usually assumed that the two water substrate molecules are bound at the S_0-state. However since the X-ray structure indicates that carbonate may be bound to the active site in the lower S-states, then it is possible that the substrate water molecules bind at a higher S-state (e.g. at S_2) thus avoiding any oxidative side reactions. Indeed our X-ray structure in principle represents the S_1-state, the state which is stable in the dark.[a] I am also proposing that Mn4 has a Mn(II) valency in the S_0-state and that the other three Mn ions of the cubane have initial valencies of Mn(IV), Mn(IV) and Mn(III) since this is consistent with EPR analyses. However an alternative starting

[a]If excess free electrons are produced during the collection of the X-ray diffraction data, then the Mn-cluster will be reduced to Mn^{2+} and the structure may not reflect the S_1-state precisely.

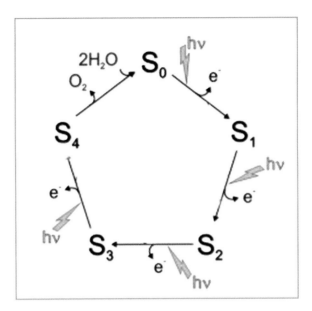

Fig. 8. A simplified version of the S-state cycle to emphasise the five different main intermediates leading to water oxidation and dioxygen formation by PSII. Protons are probably lost at each step except for the S_1 to S_2 transition.

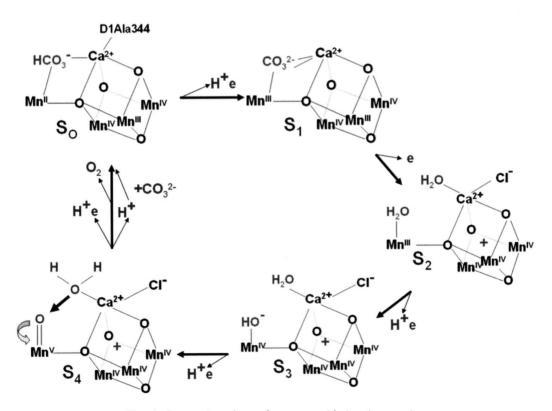

Fig. 9. A tentative scheme for water oxidation (see text).

state for S_0 could be Mn4(III) with the other three Mn ions having valencies of III, III and IV. The latter option is attractive since Mn4 is likely to be located at the "high affinity site" where the first Mn ion binds during photoactivation and presumably undergoes a photooxidation from valency II to III.[19,1] In any event the mechanism proposed here is not significantly altered by these two possibilities for the valencies of the S_0 state. The S_0 to S_1 transition may involve the conversion of Mn4(II) to Mn4(III). I further suggest that the proton lost during this transition is not derived from water but from another source, for example, from bicarbonate bound to the active site, thus converting it to a carbonate anion as seems to be observed in the X-ray structure. It is known that electron extraction associated with the S_1 to S_2 transition is not accompanied by proton release[27] and therefore adds one net positive charge to the system. This transition may oxidise the Mn(III) contained within the cubane to Mn(IV). At this stage the metal cluster has stored the energy of two oxidising equivalents equal to about 2 eV. The addition of a positive charge may be the trigger for the replacement of the carbonate by two water substrate molecules and a chloride ion. This ligand exchange could be induced by a conformational change due to a coulombic "jolt" induced by the net positive charge created in the cluster during the S_1 to S_2 transition so that the carbonate bridging ligands are broken. The third flash converts S_2 to S_3 and provides another 1 eV of energy. Now it seems likely that the first direct step of water oxidation chemistry can proceed. If the active Mn (Mn4) is converted from Mn(III) to Mn(IV) and its ligated water molecule gives up a proton, then the scene is set to generate a highly-reactive intermediate by one further input of light energy to convert the Mn(IV) of the active site to Mn(V) in the S_4 state. If the hydroxyl radical of S_3 now deprotonates the oxo left coordinated to Mn4 becomes electron deficient due to the high oxidation potential of Mn4(V). This highly electrophilic oxygen will be available for nucleophilic attack by the oxygen of the second substrate molecule bound to Ca^{2+}. According to the proposed scheme, this reaction occurs in the dark and requires the water bound to the Ca^{2+} to deprotonate. This reaction is therefore delicately balanced, requiring the oxygen of the substrate water molecule ligated to Ca^{2+} to act as a nucleophile giving two electrons back to the Mn_4Ca-cluster via Mn4 and at the same time shedding two protons. The other two electrons required to complete the S_4 to S_0 transition are provided by the oxo (O^{2-}) as it forms an O–O bond with the nucleophilic oxygen. Presumably the oxidising potential stored in the Mn ions of the cubane and possibly the net positive charge generated by the S_1 to S_2 transition facilitate this final dark reaction to occur. Moreover the deprotonation may be aided by a hydrogen-bonding network involving amino acid side chains located in the vicinity of the OEC such as CP43 Arg357 and D1 Gln165. For light-driven deprotonation associated with S_2 to S_3, and S_3 to S_4 transition, it is possible that the Tyr_Z (Y_Z) radical could function as a proton acceptor along the lines proposed by Hoganson and Babcock,[26] but a proton exit pathway would need to be created in the

S_2-state. In the X-ray structure D1 His190 is H-bonded to Y_Z and therefore can receive a proton from the phenolic group of tyrosine. However D1 Glu189 is not sufficiently close to deprotonate D1 His190 and thus provide a pathway for substrate protons. For the S_4 to S_0 transition the deprotonation of the second substrate water molecule could be aided by other amino acids in the OEC cavity, including perhaps Y_Z. The suggested scheme (see Fig. 9) involving the formation of a Mn(V)=O species in S_4 would essentially be the same if the final active species was an oxyl radical (Mn(IV)=O$^{\bullet}$).

Above I have outlined a reaction scheme which, conceptionally, I believe to be not far from the truth. However in its present form the scheme is speculative and not fully consistent with all experimental data. Therefore, I will be developing these general ideas in more depth in the coming years while at the same time striving to obtain higher-resolution structures of PSII in various S-states. It is very sad that Porter is not here to witness this major step forward and participate in the challenge of revealing the chemistry of one of the most fundamental reactions on our planet.

Acknowledgments

I wish to thank all those colleagues who have worked with me on the topics reviewed in this article. In particular Tom Bibby and Jon Nield on the PSI supercomplex work and Ben Hankamer, Ed Morris, Jon Nield, Claudia Buchel, Paula da Fonseca, Kristina Ferreira, Karim Maghlaoui, Tina Iverson and So Iwata on the PSII structural studies. I have been fortunate to receive BBSRC support for all these studies.

References

1. G. M. Ananyev, L. Zaltsman, C. Vasko and G. C. Dismukes, *Biochim. Biophys. Acta* **1503**, 52–68 (2001).
2. E. G. Andrizhiyevskaya, D. Frolov, R. van Grondelle and J. P. Dekker, *Biochim. Biophys. Acta* **1656**, 96–103 (2004).
3. M. D. Archer and J. Barber, in *Photoconversion of Solar Energy, Vol. 2, Molecular to Global Photosynthesis*, eds. M. D. Archer and J. Barber (Imperial College Press, London, 2004), pp 1–34.
4. S. V. Baranov, G. M. Ananyev, V. V. Klimov and G. C. Dismukes, *Biochemistry* **39**, 6060–6065 (2000).
5. J. Barber, *Curr. Opinions Struct. Biol.* **12**, 523–530 (2002).
6. J. Barber, *J. Bioelectrochem.* **55**, 135–138 (2002).
7. J. Barber, *Biophys. Quart. Revs.* **36**, 71–89 (2003).
8. J. Barber and M. D. Archer, *J. Photochem. Photobiol. A* **142**, 97–106 (2001).
9. A. Ben-Shem, F. Frolow and N. Nelson, *Nature* **426**, 630–635 (2003).
10. T. S. Bibby, J. Nield and J. Barber, *Nature* **412**, 743–745 (2001).

11. T. S. Bibby, J. Nield, F. Partensky and J. Barber, *Nature* **413**, 590 (2001).

12. T. S. Bibby, J. Nield and J. Barber, *J. Biol. Chem.* **276**, 43246–43252 (2001).

13. E. J. Boekema, A. Hifney, A. E. Yakushevska, M. Piotrowski, W. Keegstra, S. Berry, K. P. Michel, E. K. Pistorius and J. Kruip, *Nature* **412**, 745–748 (2001).

14. E. J. Boekema, P. E. Jensen, E. Schlodder, J. F. van Breemen, H. van Roon, H. V. Scheller and J. P. Dekker, *Biochemistry* **40**, 1029–1036 (2001).

15. R. R. Chitnis, *Ann. Rev. Plant Physiol. Plant Mol. Biol.* **52**, 593–626 (2001).

16. R. J. Debus, *Biochim. Biophys. Acta* **1102**, 269–352 (1992).

17. R. J. Debus, *Biochim. Biophys. Acta* **1503**, 164–186 (2001).

18. J. Deisenhofer, O. Epp, K. Miki, R. Huber and H. Michel, *Nature* **318**, 618–624 (1985).

19. B. A. Diner, *Biochim. Biophys. Acta* **1503**, 147–163 (2001).

20. B. A. Diner and F. Rappaport, *Ann. Rev. Plant Physiol. Plant Biol.* **53**, 551–580 (2002).

21. K. N. Ferreira, T. Iverson, K. Maghlaoui, J. Barber and S. Iwata, *Science* **303**, 1831–1838 (2004).

22. B. Gobets and R. van Grondelle, *Biochim. Biophys. Acta* **1507**, 80–99 (2001).

23. M. Haumann and W. Junge, *Biochim. Biophys. Acta* **1411**, 86–91 (1999).

24. B. Hankamer, E. P. Morris and J. Barber, *Nature Struct. Biol.* **6**, 560–564 (1999).

25. G. Hoch and R. S. Knox, *Photophysiology*, Vol. 3, ed. H. Giese (Academic Press, New York, 1968), pp. 225–251.

26. S. W. Hoganson and G. T. Babcock, *Science* **277**, 1953–1956 (1997).

27. W. Junge, M. Haumann, R. Ahlbrink, A. Mulkidjanian and J. Clausen, *Phil. Trans. Roy. Soc. Lond. B* **357**, 1407–1418 (2002).

28. P. Jordan, P. Fromme, H.-T. Witt, O. Klukas, W. Saenger and N. Krauss, *Nature* **411**, 909–916 (2001).

29. N. Kamiya and J. R. Shen, *Proc. Nat. Acad. Sci. USA* **100**, 98–103 (2003).

30. J. Kargul, J. Nield and J. Barber, *J. Biol. Chem.* **278**, 16135–16141 (2003).

31. V. V. Klimov and S. V. Baranov, *Biochim. Biophys. Acta* **1503**, 187–196 (2001).

32. D. E. Laudenbach and N. Straus, *J. Bacteriol.* **170**, 5018–5026 (1988).

33. A. N. Melkozernov, *Photosynth. Res.* **70**, 129–153 (2001).

34. A. N. Melkozernov, T. S. Bibby, S. Lin, J. Barber and R. E. Blankenship, *Biochemistry* **42**, 3893–3903 (2003).

35. J. Nield, E. P. Morris, T. S. Bibby and J. Barber, *Biochemistry* **42**, 3180–3188 (2003).

36. H. B. Pakrasi, A. Goldenberg and L. A. Sherman, *Plant Physiol.* **79**, 290–295 (1985).

37. V. L. Pecoraro, M. J. Baldwin, M. T. Caudle, W. Y. Hsieh, N. A. Law, *Pure Appl. Chem.* **70**, 925–929 (1998).

38. J. M. Peloquin and R. D. Britt, *Biochim. Biophys. Acta* **1503**, 96–111 (2001).

39. G. Porter, *Proc. Roy. Soc. Lond. A* **362**, 281–303 (1978).

40. R. Radmer and G. Cheniae, in *Primary Processes in Photosynthesis*, ed. J. Barber, Topics in Photosynthesis, Vol. 2 (Elsevier, Amsterdam, 1977), pp. 303–348.

41. F. Rappaport, M. Guergova-Kuras, P. J. Nixon, B. A. Diner, J. Lavergne, *Biochemistry* **41**, 8518–8527 (2002).

42. K.-H. Rhee, E. P. Morris, J. Barber and W. Kühlbrandt, *Nature* **396**, 283–286 (1998).

43. J. H. Robblee, R. M. Cinco and V. K. Yachandra, Biochim. *Biophys. Acta* **1503**, 7–23 (2001).

44. G. Schatz, *FEBS Lett.* **551**, 1–2 (2003).

45. P. E. M. Siegbahn and R. H. Crabtree, *J. Am. Chem. Soc.* **121**, 117–127 (1999).

46. N. A. Straus, in *Molecular Biology of Cyanobacteria*, ed. D. A. Bryant (Kluwer Academic Press, Dordrecht, 1994), pp. 731–750.

47. J. J. S. van Rensen, C. H. Xu and Govindjee, *Physiol. Plant.* **105**, 585–592 (1999).

48. J. S. Vrettos, J. Limburg and G. W. Brudvig, *Biochim. Biophys. Acta* **1503**, 229–245 (2001).

49. V. Yachandra, *Phil. Trans. Roy. Soc. Lond. B* **357**, 1347–1357 (2002).

50. A. Zouni, H. T. Witt, J. Kern, P. Fromme, N. Krauss, W. Saenger and P. Orth, *Nature* **409**, 739–743 (2001).

51. H. Zuber and R. J. Cogdell, in *Advances in Photosynthesis, Anoxygenic Photosynthetic Bacteria*, eds. R. E. Blankenship, M. T. Madigan and A. E. Bauer (Kluwer Academic Publ. Dordrecht, 1995), pp. 316–344.

Contribution from

GODFREY BEDDARD

University of Leeds
UK

Godfrey S Beddard is Professor of Chemical Physics, at the University of Leeds in UK. His current areas of research include electron transfer in DNA, electronic energy migration and mechanical unfolding of proteins.

1969	BSc, First Class Honours Chemistry, Chelsea College, University of London.
1973	PhD, Davy Faraday Lab, Royal Institution-University College London, Thesis 1973 "Photophysics of Aromatic Molecules".
1973–1974	Postdoc., Dept of Physical and Theoretical Chemistry, Oxford University.
1974–1982	Royal Institution. Royal Society, J. Jaffe Research Fellow, then SRC Advanced Research Fellow and Assistant Director, Davy Faraday Laboratory, Royal Institution. Marlow Medal of the Chemical Society.
1982–1996	Dept of Chemistry University of Manchester, Lecturer then Reader.
Since 1996	Professor of Chemical Physics, University of Leeds.

At the Royal Institution 1974–1975.

Godfrey Beddard on
The Bakerian Lecture, 1977
"In Vitro Models For Photosynthesis"
G. Porter
Proc. Roy. Soc. Lond. A **362**, 281–303 (1978)

In his 1977 Bakerian lecture (BL) George Porter outlined his thoughts on natural and artificial photosynthesis. He comments in his introduction upon the complexity of the photosynthetic apparatus but suggests nevertheless that parts of this may be constructed *in vitro.*

After a brief general introduction into the workings of photosynthesis, which he naturally splits into the light-harvesting antenna, quinone reduction and the reaction centre and water splitting, examples of each of these were given based on work done at the Davy Faraday Research Laboratory. In reading this paper one should remember that the first X-ray structure of an antenna, that of the 7 chlorophyll (Chl) protein from P. Aestuarii[1] had only just been solved, but not yet fully published, and all the wonderfully detailed protein structures of the reaction centres and antennas we are now familiar with were completely unknown.

After excitation with a visible photon, the electronic excitation energy of the excited singlet state migrates among the many pigment molecules in the antenna proteins of photosynthetic organisms until it finds the trap at the reaction centre (RC). All antennas contain many Chl molecules, some also contain accessory pigments such as polyenes (e.g. carotenes) and linear tetrapyrroles. The energy trap is conventionally called the special pair and given the symbol P followed by the wavelength of its optical absorption, P700 for Photosystem I, for example. Energy migration among the many antenna Chl molecules has to occur rapidly if the energy is to reach the trap with the high efficiency that it does; > 0.98. Demonstrating this migration is very difficult; the Chl molecules all have very similar spectra making identification of one from another virtually impossible and fluorescence depolarisation cannot be used as the depolarisation that occurs after just one jump is almost the same as after many. Observing the appearance of the energy at the special pair is an obvious experiment but also very difficult. The plant photosystems, PSI and PSII have antenna and reaction centres that are inseparable from one another without causing damage, and detecting the small signal due to two special pair Chl molecules in the presence of 100 other Chl was not technically feasible in the mid-1970s, neither was the time resolution of the lasers good enough. In photosynthetic bacteria the antenna and reaction centres are not contained in the same protein and are separable and therefore a sample of antenna had no natural quenchers which is problematic for the observation of migration.

The solution to the problem of observing migration was a typically innovative one; George decided to look at the accessory antenna called a phycobilisome in which several distinctly different pigment types are present and follow the energy as it flowed through the different pigments by time resolving the fluorescence at different wavelengths. The results are shown in Figs. 3 and 4 of the BL paper. These are really spectacular results. They clearly demonstrate energy migration from one pigment type to another even though the laser used to excite the molecules was not that short lived, ~30 ps duration. The migration was observed by selecting different emission wavelengths to isolate the signal from different chromophores. The delay in the time taken by the fluorescence from different pigments to reach its maximum intensity, and the different fluorescence decay times, both serve to illustrate energy migration. This data is probably still the clearest example of energy migration in a photosynthetic organism.

The fluorescence in this experiment was observed on the streak camera and because of this the results are all the more impressive. This was not the easiest of devices to use; triggering was variable, there was a small but annoying time-jitter before the camera worked, which made traces jump left and right across the screen, making direct averaging impossible and the signal to noise poor. The mode-locked YAG laser was also very temperamental; in the mid-1970s laser technology was very crude compared to that of today. The YAG worked on a shot-by-shot basis, at about 1 shot/minute, triggered manually. Mode-locking was achieved by an infrared absorbing dye solution in contact with the high reflector. Mirror mounts were crude, insensitive and wobbly and mirror coatings were poor. The laser was placed on an old-fashioned optical rail on top of a wooden table not on a vibrationally-isolated, honeycomb-centred, steel table as would be done today; and there were no PCs to capture and analyse the data.

An alternative, but rather less satisfactory way of measuring migration was to observe the annihilation of the excitation when more than one excited state was generated in the antenna, and this topic was discussed next in the lecture. Two singlet excited states can by their random walking find one another and then there is the possibility of two triplets being formed. This will lead to fluorescence quenching to an extent that is dependent upon the laser intensity. The measured fluorescence lifetime is dependent upon the laser intensity, provided this is above a certain level. At low light levels no annihilation occurs because two excitations are not present in one antenna. When two are present, annihilation competes with trapping by the RC.

Annihilation had accidentally been observed by us some time before the phycobilisome experiments were performed, but we did not initially understand why the measured antenna fluorescence lifetimes were so variable. We were trying to get measurements of the excited state lifetime of the antenna of Chlorella under a variety of conditions, for example with the reaction centres opened and closed. We did not initially appreciated how variable the energy from the mode-locked YAG laser was, which, when intense, caused annihilation and this, coupled with noise from the streak camera, led to very misleading results. We were, at about the same time, using time-correlated single-photon counting with pulses from a synchronously mode-locked dye laser excited by a mode-locked argon ion laser. But although here the laser pulses were far less intense than those from the YAG, the time resolution was not so high, making difficult a comparison of the two sets of data. Quite soon the effect of annihilation was understood and results from both methods reconciled.

In the mid-1990s laser technology had advanced greatly and using amplified pulses from a colliding-pulse, mode-locked, dye-laser my students measured the energy migration and trapping of the Photosystem I antenna-reaction centre complex using flash photolysis, or in today's jargon, femtosecond pump-probe spectroscopy.[2] The X-ray structure of the antenna complex had recently been published[3] and was later on refined. With this information it was possible to simulate energy migration in the antenna and match the simulation to the experimental data. More sophisticated calculations, taking into account the short time coherent motion which was referred to in the Bakerian lecture, have since been undertaken by Graham Fleming's group.[4]

Figure 1 shows a simple calculation based on the 1996 X-ray structure demonstrating how energy migration might occur in the PSI antenna. The Chl are simply represented by circles irrespective of their orientation in the protein matrix which is not shown. The contours show how the energy moves after excitation of just one pigment and assuming that coherence decays in less than 0.1 ps. At short times energy is located on just a few pigments, but then rapidly spreads out and passes into the reaction centre which is situated in the centre of the structure. After a few picoseconds the migration reaches a quasi steady state which is located around the reaction centre into which it slowly "pours" causing the antenna to have an effective lifetime of about 25 ps; the rate of initial electron transfer in the P700 RC is a little larger than 1 ps^{-1}. This calculation gives only an idea as to what may happen; it would be tremendous to be able to watch this in real time but there appears to be no technique which has both sufficient spatial and temporal resolution to properly resolve the energy migration.

Concentration quenching was one of George's favourite topics. This is the phenomenon by which the excited singlet state of Chl (and other) molecules in solution or in rigid lipid gels and other media are almost completely quenched at concentrations similar to those in photosynthetic antennas. In natural antennas, by comparison, the Chl is unquenched if the reaction centre is "closed", that is, artificially made not to function by adding inhibitors. In concentration quenching no stable molecular dimers appear to be formed and indeed the form of the quenching with concentration increase does not follow that of a dimer or other known forms. George's intuition was that a pair of Chl molecules had to be quenched if they were closer than a certain distance; we calculated this distance by a Monte-Carlo method to be 1 nm (Fig. 6 in the BL paper) and called the quenching pair a "statistical pair" to indicate the random way the molecules could come together in solution. We could not explain the lack of spectral evidence; this was later explained in a paper by Robert Knox[5] who elegantly showed that the

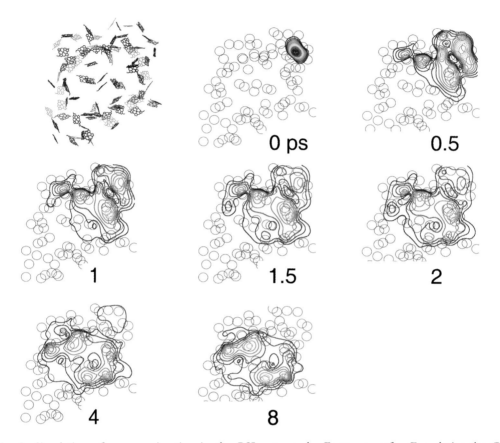

Fig. 1. Simulation of energy migration in the PSI antenna by Forster transfer. For clarity the Chl molecules are shown as circles, except in the top left picture where the Chl molecules are shown explicitly. The contours show the extent of migration at the times shown which are in ps. The 6 molecules at the centre of the structure are the reaction centre, the two central ones being P700, and energy is trapped here. The population is reflected in the contour colour being high to low in the order red-yellow-green-blue.

spectral signature would be very small. Later on, in a study of the P700 special pair,[6] it was discovered that at short separations a pair of Chl molecules produced a charge transfer state of lower energy than the excited singlet state (Fig. 2), and energy transferred into this state would therefore be dissipated as heat, not photons. Many years before this calculation, experiments were performed on Chl in the lipid membranes of vesicles. Different lipids were used, some with bulky head groups (galactosyl diglycerides) and it was observed that by keeping the Chl apart from one another, concentration quenching was reduced. The effect is enumerated in Table 2 and shown in Fig. 5 of the BL paper. Other experiments were also performed on dye molecules in thin films where similar but more dramatic effects were observed; concentration quenching was not observed as the donor concentration increased and energy migration observed via the fluorescence of an acceptor held at a constant concentration, but, unfortunately, these results were never published. Theoretical models of quenching have been produced but experiments testing these have not been definitive enough as the dye molecules used to test theory also showed dimer formation. As a natural consequence of the concentration used, dimer formation occurs at about the same concentration as energy migration begins, and therefore confuses the experiment.

By posing the question, and so looking for an understanding of concentration quenching it is now possible to state two simple rules for making an antenna: (1) keep the molecules as close as possible but more than 5 and preferably 10 Å apart; (2) choose molecules with a high fluorescence quantum yield, large spectral overlap between emission and absorption to maximise energy migration. Figure 3 shows the centre to centre distances of several antennas and reaction centres.[6] As far as I am aware no large antenna (≈100 molecules) has yet been made artificially. I believe that a self-assembly approach, as in nature, will eventually be more productive than direct synthesis such as with a dendrimer. Reaction centre

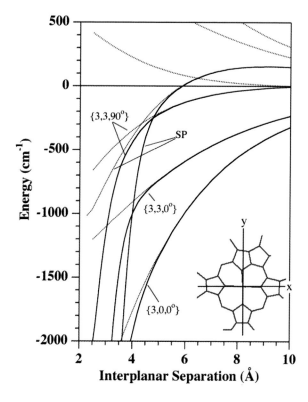

Fig. 2. Calculated energies of Chl statistical pair dimers and the P700 special pair at different interplanar separations and geometries. The curves sp represent the energy of the special pair and the other lines two Chl molecules with their rings parallel to one another offset by 3 Å in the x and y directions with one molecule being rotated by 90° to the other, labelled [3, 3, 90°]. Two other geometries are also shown. The lower curve in each pair is that of the charge transfer state; this becomes lowest in energy below 5 Å in each case and leads to excited state quenching by charge separation (based on figure in Ref. 6).

traps, in comparison to antennas, place the molecules at, or less than, 5 Å apart. Figure 2 shows that charge-transfer states are lowest in energy for several chlorophyll–chlorophyll orientations when separated by less than 5 Å. This effect is shown in data taken from X-ray structures of antennas where the Mg–Mg distances are plotted. In antennas these distances are all greater than ≈ 10 Å and less than this in reaction centres (Fig. 3).

The second topic discussed in the BL paper was electron transfer from Chl to quinones. These experiments were conducted in lipid gels to prevent diffusion interfering with the quenching and both triplet and singlet quenching was observed. George describes very clearly why triplets are unquenched compared to singlets and related this to the natural reaction centre. The data for singlet quenching shows a clear dependence on reaction free energy when account is taken for the range of separation of the singlet to quencher. However, the Marcus equation to describe electron transfer was not known to us and the importance of large $-\Delta G$ producing slow reactions not understood; it was thought that the charge recombination $Chl^+ + Q^-$ would be fast, not slow. With the luxury of hindsight we should have understood this because experiments on singlet–triplet crossing in the gas phase had been extensively studied by us and others in the late 1960s and early 1970s. The first demonstration of the Marcus curve and the inverted region was observed in rigid solution by Miller et at.[7] and we observed this effect shortly afterwards in linked porphyrin-quinone compounds.[8] Figure 4 shows data following classical Marcus behaviour caused by electron transfer from different nucleotides to thionine type dyes. These results were obtained by femtosecond spectroscopy as part of a study of electron transfer in DNA.[9] We were able to show that in different types of DNA electron transfer is described by a super-exchange model of electron transfer and that DNA does not behave like a "molecular wire", but nor is it as slow as is seen in proteins, which do not benefit from π-stacking. Specific effects are observed, however, and we found that the pyrimidine bases are barriers to efficient electron transfer within the super-exchange limit and we also inferred from this model that the electrons do not cross between strands on the picosecond timescale, i.e. the electronic coupling occurs predominantly through the π-stack and is not increased substantially by the presence of hydrogen bonding within the duplex.

George introduced his lecture with comments about artificial and natural photosynthesis. The latter subject has benefited hugely as a result not only of his invention, continued development and application of flash photolysis, but also from his enthusiasm and his ideas of how antennas and reaction centres worked. However, although our understanding of photosynthesis has increased greatly artificial photosynthesis seems almost as far away now as it was 27 years ago. A clear exception to this statement is the carotene-porphyrin-quinone molecule of Gust et al.,[10] which, when illuminated, has been shown to be able to pump protons across a bilayer membrane and drive ATPsynthase to synthesise/hydrolyse ATP (Fig. 5).

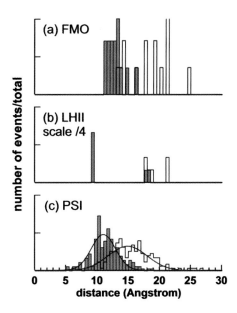

Fig. 3. Nearest neighbour (blue) and second nearest neighbour (red) Mg–Mg separations in Chl for three antennas. In the PSI data the molecules in the 5–10 Å range are those molecules comprising the reaction centre. No reaction centres occur in the other antennas where the smallest Mg–Mg separation is ~10 Å (based on figure in Ref. 6).

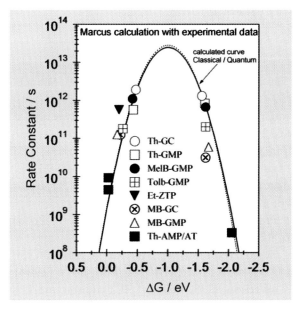

Fig. 4. Data and calculated curve showing electron transfer and Marcus behaviour (red dotted line). The blue line is a quantum calculation, in this case producing a very similar result as the promoting mode's displacement is small. The data points refer to a family of thionine type dyes with guanosine monophosphate (GMP) and adenosine monophosphate (AMP) and poly-GC and poly-AT DNAs. More details can be found in Ref. 9. Et-ZTP refers to ethidium bromide in a modified DNA.[11] Data to the left of the peak measures charge separation from the base to the excited singlet state of the dye, and on the right the charge recombination rate constant to the ground state of both molecules.

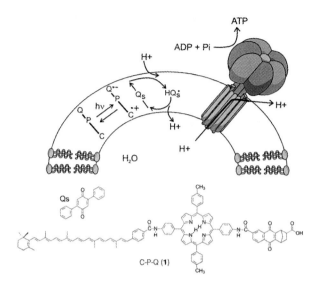

Fig. 5. Diagram of a liposome-based artificial photosynthetic membrane. Included in this diagram is the photocycle that pumps protons into the interior of the liposome and the ATPsynthase enzyme. The artificial photosynthetic molecule CPQ is shown. On illumination with visible light electron transfer produces C^+PQ^- with the naphthoquinone Q located on the outside of the membrane. Lipophillic quinones transport the electron from C^+PQ^- together with a proton to the inside of the membrane where pH is reduced and an electric potential (pmf) is generated. About 4% of the light energy is conserved (taken from Ref. 10).

References

1. B. W. Matthews, R. E. Fenna, M. C. Bolognesi, M. F. Schmid and J. M. Olson, *J. Mol. Biol.* **131**, 259 (1979).

2. N. White, G. Beddard, J. Thorne, T. Feehan, T. Heyes and P. Heathcote, *J. Phys. Chem.* **100**, 12086, 1996.

3. N. Krauss, W. Hinrichs, W. Witt, P. Fromme, W. Pritzkow, Z. Dauter, B. Betzel, K. S. Wilson, H. Witt and W. Saenger, *Nature* **361**, 326 (1993).

4. M. Yang, A. Damjanovic, H. Vaswani and G. Fleming, *Biophysical J.* **85**, 140 (2003).

5. R. S. Knox, *J. Phys. Chem.* **98**, 7270 (1994).

6. G. S. Beddard, *J. Phys. Chem.* **102**, 10966 (1988); G. S. Beddard, *Phil. Trans. Roy. Soc., Lond. A* **356**, 421 (1988).

7. J. Miller, J. Beitz and R. Huddleston, *J. Am. Chem. Soc.* **106**, 5057 (1984); J. Miller, L. Calcaterra and G. Closs, *J. Am. Chem. Soc.* **106**, 3074 (1984).

8. M. P. Irvine, R. J. Harrison, G. S. Beddard, P. Leighton and J. K. M. Sanders, *Chem. Phys.* **104**, 315–324 (1986); R. J. Harrison, B. Pearce, G. S. Beddard, J. A. Cowan and J. K. M. Sanders, *Chem. Phys.* **116**, 429–448 (1987).

9. G. D. Reid, D. J. Whittaker, M. A. Day, D. A. Turton, V. Kayser, J. M. Kelly and G. S. Beddard, *J. Am. Chem. Soc.* **124**, 5518–5527 (2002).

10. G. Steinberg-Yfrach, J.-L. Rigaud, E. N. Durantini, A. L. Moore and D. Gust, *Nature* **392**, 479 (1988).

11. T. Fiebig, C. Wan, S. O. Kelley, J. K. Barton and A. H. Zewail, *Proc. Natl. Acad. Sci. USA* **96**, 1187 (1999).

Reprinted with permission from *J. Am. Chem. Soc.* **124**(9), 5518–527 (2002).

Copyright 2002 American Chemical Society

A R T I C L E S

Published on Web 04/17/2002

Femtosecond Electron-Transfer Reactions in Mono- and Polynucleotides and in DNA

Gavin D. Reid,*,† Douglas J. Whittaker,† Mark A. Day,† David A. Turton,†
Veysel Kayser,† John M. Kelly,‡ and Godfrey S. Beddard†

*Contribution from the School of Chemistry, University of Leeds, Leeds, LS2 9JT, U.K., and
Department of Chemistry, University of Dublin, Trinity College, Dublin 2, Ireland*

Received October 5, 2001

Abstract: Quenching of redox active, intercalating dyes by guanine bases in DNA can occur on a femtosecond time scale both in DNA and in nucleotide complexes. Notwithstanding the ultrafast rate coefficients, we find that a classical, nonadiabatic Marcus model for electron transfer explains the experimental observations, which allows us to estimate the electronic coupling (330 cm^{-1}) and reorganization (8070 cm^{-1}) energies involved for thionine-[poly(dG-dC)]$_2$ complexes. Making the simplifying assumption that other charged, π-stacked DNA intercalators also have approximately these same values, the electron-transfer rate coefficients as a function of the driving force, ΔG, are derived for similar molecules. The rate of electron transfer is found to be independent of the speed of molecular reorientation. Electron transfer to the thionine singlet excited state from DNA obtained from calf thymus, salmon testes, and the bacterium, micrococcus luteus (lysodeikticus) containing different fractions of G−C pairs, has also been studied. Using a Monte Carlo model for electron transfer in DNA and allowing for reaction of the dye with the nearest 10 bases in the chain, the distance dependence scaling parameter, β, is found to be 0.8 ± 0.1 Å^{-1}. The model also predicts the redox potential for guanine dimers, and we find this to be close to the value for isolated guanine bases. Additionally, we find that the pyrimidine bases are barriers to efficient electron transfer within the superexchange limit, and we also infer from this model that the electrons do not cross between strands on the picosecond time scale; that is, the electronic coupling occurs predominantly through the π-stack and is not increased substantially by the presence of hydrogen bonding within the duplex. We conclude that long-range electron transfer in DNA is not exceptionally fast as would be expected if DNA behaved as a "molecular wire" but nor is it as slow as is seen in proteins, which do not benefit from π-stacking.

Introduction

Ultrafast electron transfer involving DNA has recently received considerable attention both experimentally[1-4] and theoretically.[5-10] Electron transfer is important for DNA damage and repair,[11-13] and many applications in molecular electronics

are envisaged.[14-17] In particular, ever since the first studies of charge transport in DNA,[18] there has been considerable debate as to whether electron transfer in a DNA environment is unexpectedly fast; for a recent review, see Barbara and Olson.[19] Photoinduced electron-transfer reactions can also lead to DNA damage, including adduct formation,[20] and such reactions might in the future be used for antitumor therapy.

The structure of B-form DNA determines the nature of the intervening medium between intercalated electron donors and acceptors within the ordered stack of planar purine and pyrimidine bases, which is surrounded by water-containing grooves and the charged sugar−phosphate backbone. In proteins, a balance of the interactions between polar and nonpolar residues

* To whom correspondence should be addressed. Tel: 44 113 233 6405. Fax: 44 113 233 6565. E-mail: g.d.reid@chem.leeds.ac.uk.
† University of Leeds.
‡ University of Dublin.

(1) Reid, G. D.; Whittaker, D. J.; Day, M. A.; Creely, C. M.; Tuite, E. M.; Kelly, J. M.; Beddard, G. S. *J. Am. Chem. Soc.* **2001**, *123*, 6953−6954.
(2) Wan, C.; Fiebig, T.; Schiemann, O.; Barton, J. K.; Zewail, A. H. *Proc. Natl. Acad. Sci. U.S.A.* **2000**, *97*, 14052−14055.
(3) Lewis, F. D.; Wu, T.; Liu, X.; Letsinger, R. L.; Greenfield, S. R.; Miller, S. E.; Wasielewski, M. R. *J. Am. Chem. Soc.* **2000**, *122*, 2889−2902.
(4) Kononov, A. I.; Moroshkina, E. B.; Tkachenko, N. V.; Lemmetyinen, H. *J. Phys. Chem. B* **2001**, *105*, 535−541.
(5) Porath, D.; Bezryadin, A.; De Vries, S.; Dekker, C. *Nature* **2000**, *403*, 635−638.
(6) Tavernier, H. L.; Fayer, M. D. *J. Phys. Chem. B* **2000**, *104*, 11541−11550.
(7) Schlag, E. W.; Yang, D. Y.; Sheu, S. Y.; Selzle, H. L.; Lin, S. H.; Rentzepis, P. M. *Proc. Natl. Acad. Sci. U.S.A.* **2000**, *97*, 9849−9854.
(8) Beratan, D. N.; Priyadarshy, S.; Risser, S. M. *Chem. Biol.* **1997**, *4*, 3−8.
(9) Giese, B.; Wessely, S.; Spormann, M.; Lindemann, U.; Meggers, E.; Michel-Beyerle, M. E. *Angew. Chem., Int. Ed.* **1999**, *38*, 996−998.
(10) Jortner, J.; Bixon, M.; Langenbacher, T.; Michel-Beyerle, M. E. *Proc. Natl. Acad. Sci. U.S.A.* **1998**, *95*, 12759−12765.
(11) Armitage, B. *Chem. Rev.* **1998**, *98*, 1171−1200.
(12) Burrows, C. J.; Muller, J. G. *Chem. Rev.* **1998**, *98*, 1109−1151.
(13) Schuster, G. B. *Acc. Chem. Res.* **2000**, *33*, 253−260.
(14) Fox, M. A. *Acc. Chem. Res.* **1999**, *32*, 201−207.
(15) Berlin, Y. A.; Burin, A. L.; Ratner, M. A. *Superlattices Microstruct.* **2000**, *28*, 241−252.
(16) Dekker, C.; Ratner, M. A. *Phys. World* **2001**, *14*, 29−33.
(17) Rinaldi, R.; Branca, E.; Cingolani, R.; Masiero, S.; Spada, G. P.; Gottarelli, G. *Appl. Phys. Lett.* **2001**, *78*, 3541−3543.
(18) Eley, D. D.; Spivey, D. I. *Trans. Faraday Soc.* **1962**, *58*, 411−415.
(19) Barbara, P. F.; Olson, E. J. C. In *Electron Transfer: From Isolated Molecules to Biomolecules, Part Two;* Jortner, J., Bixon, M., Eds.; John Wiley & Sons: New York, 1999; *Adv. Chem. Phys. 107*, pp 647−676.
(20) Jacquet, L.; Davies, R. J. H.; Kirsch-De Mesmaeker, A.; Kelly, J. M. *J. Am. Chem. Soc.* **1997**, *119*, 11763−11768.

10.1021/ja0172363 CCC: $22.00 © 2002 American Chemical Society

both with themselves and with the solvent usually determines the secondary and tertiary structure, resulting in a very different medium for electron transfer. This difference in structure poses the question as to whether electron transfer in DNA behaves either in the same way as in proteins,[21] in well-defined donor–spacer–acceptor complexes,[22] or in neither.

The excited singlet states of the bases in DNA are extremely short-lived, that is, <1 ps,[23] and it has been suggested recently that this may play some part in protecting the molecule from photodamage by ultraviolet light. However, photodynamic degradation of DNA may also be induced by ultrafast redox reactions,[11–13] and here the phenothiazine family of dyes is important,[24] because the excited states of the dyes are strongly quenched when they bind near guanine bases.[1,25–27] Quenching by adenine should be far slower than by guanine, because the oxidation potential of adenine is higher, while quenching by the pyrimidine bases, which lie at still higher potential, is not observed; that is, $E°$ (G) < $E°$ (A) < $E°$ (C) ≈ $E°$ (T).[28,29] The possibility of proton-coupled, electron-transfer reactions has also been the subject of recent experimental and theoretical attention.[30–32]

The dependence of the rate of electron transfer as a function of ΔG has been studied extensively. While a quantum-mechanical model is often necessary, particularly in the "inverted" regime, at very short internuclear separation, or in the strong coupling limit,[33,34] the classical result due to Marcus[35] has explained the majority of experimental observations of biological electron transfer over as many as 10 orders of magnitude in rate coefficient.[36] Whether nonclassical effects, such as those produced by delocalization of charge along the base π-stack in DNA, by strong adiabatic electronic coupling between electron donor and acceptor or coupling of the reaction coordinate to high frequency, intramolecular vibrational modes need to be invoked, are the points we address in this work.

The rate coefficient, k_0, for nonadiabatic electron transfer is given by

$$k_0 = \frac{2\pi |V_0|^2}{\hbar} F(\Delta G, \lambda)$$

where V_0 is the electronic coupling matrix element, and F is the thermally averaged, Franck–Condon weighted density of states. The expected distance dependence of k_0 depends on the overlap of the relevant wave functions and so falls off exponentially with the separation of donor and acceptor $|V|^2 = |V_0|^2 \exp(-\beta(r - r_0))$, where β is a scaling parameter and r the internuclear separation, r_0 being the distance of closest approach. Therefore, experimentally, the rate coefficient of interest, k, is given by $k = k_0 \exp(-\beta(r - r_0))$. If DNA behaves as a "molecular wire" and long-range electron transfer is particularly efficient, it has been proposed that $\beta \leq 0.2$ Å^{-1},[37] whereas if electron transfer is essentially the same as in a protein, β would be in the range 0.9 → 1.4 Å^{-1}.[21] It should be noted, however, that different and possibly competing mechanisms of charge separation might result from the use of different donors or acceptors. First, direct transfer between adjacent donor and acceptor will show the expected exponential distance dependence. If there are intervening base pairs, a two center, single-step superexchange model will also depend exponentially on distance[10] but with a different scaling parameter and will occur in competition with direct transfer. A charge-hopping mechanism[9,38,39] between intervening bases, with an apparently weak distance dependence, could also occur depending on the energy gaps between the low lying vibronic states of the initially excited molecule and the vibronic manifolds of the charge-separated species. Considering electron transfer through intervening bases, superexchange should dominate when the energy gap between the donor and the "bridging" states is ≫$k_B T$, while a hopping mechanism will prevail if the energies of the donor and of each of the intermediate states are approximately degenerate. In a hopping mechanism, spectral evidence for population of the intermediate states should be observable.

Experimentally, optical excitation of an intercalated dye is a good method by which electrons or holes can be rapidly injected into DNA strands.[1,27,40–42] Alternatively, covalently bound species have also been used to initiate photoinduced electron transfer within duplex DNA.[37,43–45] In particular, Lewis et al.[3] have covalently bound stilbenedicarboxamide to small, synthetic DNA hairpins and found β to be 0.7 Å^{-1} for the forward electron transfer and 0.9 Å^{-1} for the reverse step. It was suggested that these relatively small β might be due to the similarity in energy of the HOMO in stilbene and that in DNA. Jortner et al.[10] have studied these synthetic hairpins theoretically and support a superexchange mechanism for charge separation and an exponential distance dependence of the rate coefficient. In another study, Wan et al.[2] incorporated 2-aminopurine, an isomer of adenine, into well-characterized DNA assemblies, and a similar distance dependence to that of Lewis was observed, $\beta = 0.6$ Å^{-1}.

(21) Langen, R.; Colon, J. L.; Casimiro, D. R.; Karpishin, T. B.; Winkler, J. R.; Gray, H. B. *J. Biol. Inorg. Chem.* **1996**, *1*, 221.
(22) Harrison, R. J.; Pearce, B.; Beddard, G. S.; Cowan, J. A.; Sanders, J. K. M. *Chem. Phys.* **1987**, *116*, 429–448.
(23) Pecourt, J.-M. L.; Peon, J.; Kohler, B. *J. Am. Chem. Soc.* **2000**, *122*, 9348–9349.
(24) OhUigin, C.; McConnell, D. J.; Kelly, J. M.; Van der Putten, W. J. M. *Nucleic Acids Res.* **1987**, *15*, 7411–7427.
(25) Beddard, G. S.; Kelly, J. M.; Van der Putten, W. J. M. *J. Chem. Soc., Chem. Commun.* **1990**, 1346–1347.
(26) Kelly, J. M.; Tuite, E. M.; Van der Putten, W. J. M.; Beddard, G. S.; Reid, G. D. *NATO Adv. Study Inst. Ser., Ser. C* **1992**, *371*, 375–381.
(27) Tuite, E.; Kelly, J. M.; Beddard, G. S.; Reid, G. D. *Chem. Phys. Lett.* **1994**, *226*, 517–524.
(28) Seidel, C. A. M.; Schulz, A.; Sauer, M. H. M. *J. Phys. Chem.* **1996**, *100*, 5541–5553.
(29) Steenken, S.; Jovanovic, S. V. *J. Am. Chem. Soc.* **1997**, *119*, 617–618.
(30) Weatherly, S. C.; Yang, I. V.; Thorp, H. H. *J. Am. Chem. Soc.* **2001**, *123*, 1236–1237.
(31) Shafirovich, V.; Dourandin, A.; Luneva, N. P.; Geacintov, N. E. *J. Phys. Chem. B* **2000**, *104*, 137–139.
(32) Steenken, S. *Biol. Chem.* **1997**, *378*, 1293–1297.
(33) Wynne, K.; Reid, G. D.; Hochstrasser, R. M. *J. Chem. Phys.* **1996**, *105*, 2287–2297.
(34) Wynne, K.; Hochstrasser, R. M. In *Electron Transfer: From Isolated Molecules to Biomolecules, Part Two*; Jortner, J., Bixon, M., Eds.; John Wiley & Sons: New York, 1999; *Adv. Chem. Phys. 107*, pp 263–309.
(35) Marcus, R. A.; Sutin, N. *Biochim. Biophys. Acta* **1985**, *811*, 265–322.
(36) Page, C. C.; Moser, C. C.; Chen, X.; Dutton, P. L. *Nature* **1999**, *402*, 47–52.
(37) Murphy, C. J.; Arkin, M. R.; Jenkins, Y.; Ghatlia, N. D.; Bossmann, S. H.; Turro, N. J.; Barton, J. K. *Science* **1993**, *262*, 1025–1029.
(38) Berlin, Y. A.; Burin, A. L.; Ratner, M. A. *J. Am. Chem. Soc.* **2001**, *123*, 260–268.
(39) Bixon, M.; Jortner, J. *J. Phys. Chem. A* **2001**, *105*, 10322–10328.
(40) Brun, A. M.; Harriman, A. *J. Am. Chem. Soc.* **1992**, *114*, 3656–3660.
(41) Arkin, M. R.; Stemp, E. D. A.; Holmlin, R. E.; Barton, J. K.; Hormann, A.; Olson, E. J. C.; Barbara, P. F. *Science* **1996**, *273*, 475–480.
(42) Lincoln, P.; Tuite, E.; Norden, B. *J. Am. Chem. Soc.* **1997**, *119*, 1454–1455.
(43) Kelley, S. O.; Holmlin, R. E.; Stemp, E. D. A.; Barton, J. K. *J. Am. Chem. Soc.* **1997**, *119*, 9861–9870.
(44) Kelley, S. O.; Barton, J. K. *Science* **1999**, *283*, 375–381.
(45) Fukui, K.; Tanaka, K. *Angew. Chem., Int. Ed.* **1998**, *37*, 158–161.

Table 1. Redox Active, Blue Dyes Showing Their Reduction Potentials, E^0, Relative to NHE and the Energy of the Singlet States, E_{0-0}

Thionine, Th: R1 = R2 = R3 = H.
Methylene Blue, MB: R1 = R2 = CH$_3$ R3 = H
Toluidine Blue, Tolb: R1 = R3 = CH$_3$ R2 = H

Meldola's Blue, Melb

dye	E^0/eV	E_{0-0}/eV
thionine	−0.03[48]	2.03
methylene blue	−0.09[27]	1.83
toluidine blue	−0.04[49]	1.88
Meldola's blue	−0.04[49]	2.03

Recently, in a study of the ultrafast electron-transfer reactions between thionine, Th (3,7-diaminophenazathionium chloride), and guanine, G, bases in guanosine-5′-monophosphate, GMP, and poly(dG-dC)·poly(dG-dC), [poly(dG-dC)]$_2$, which forms a B-DNA structure, we found both the forward and the reverse reaction rates, k_f and k_r, to be on the femtosecond time scale, more than 10 times faster than rotational motion.[1] We showed also that it was not a small subset of excited molecules that reacted on this time scale, and we explained these measurements of k_f and k_r using a classical Marcus model despite the ultrafast time scale of the reaction. The reaction scheme is

$$\text{Th}^+\text{:G} \xrightarrow{h\nu} \text{Th}^+\text{*:G} \xrightarrow{k_f} [\text{Th}^\bullet\text{:G}^{\bullet+}] \xrightarrow{k_r} \text{Th}^+\text{:G} \quad (1)$$

In the present work, we have studied a range of intercalating dyes, thionine, toluidine blue, Tolb (3-Amino-7-(dimethyl-amino)-2-methylphenazathionium chloride), and Meldola's blue, Melb (7-(dimethylamino)-1,2-benzophenoxazine), of similar chemical structure and redox potential (Table 1). We have extended our experiments to reactions of thionine with 2′-deoxyadenosine-5′-monophosphate, dAMP, and poly(dA-dT)·poly(dA-dT), [poly(dA-dT)]$_2$, and to Meldola's blue and toluidine blue with GMP. We show for the first time that thionine is quenched by DNA over a range of time scales that depends on the ratio of G−C to A−T pairs in a manner that is not a simple sum of guanine and adenine quenching rates to nearest neighbors. We have used three native DNAs, from calf thymus (ct) and salmon testes (st), both composed of similar guanine-to-adenine ratios (quoted as 42% and 41.2% G−C, respectively) and from the bacterium micrococcus luteus (mc), which contains a significantly larger proportion of guanine (72% G−C). The dyes were chosen in the expectation that they should react with guanine bases on the femtosecond to picosecond time scale. Quenching of the dyes is due to electron transfer from the base to the dye excited singlet state. The single-electron oxidation potential of guanine was thought to depend strongly on its neighboring bases,[12,46,47] and, therefore, a distribution of electron-transfer rates within DNA might be expected. However, measurements near or at the peak of the Marcus curve, where

the rate coefficient is relatively independent of changes in ΔG, allow us to make much safer predictions of the electronic coupling and reorganization energies.

Experimental Section

Solutions of dye (50 μM, Aldrich) in 5 mM phosphate buffer (pH = 6.9) containing either [poly(dA-dT)]$_2$, [poly(dG-dC)]$_2$ (1.5 mM nucleotide), GMP, dAMP (100 mM), or DNA (7mM nucleotide) were studied. Dyes were purified on neutral alumina, recrystallized three times from ethanol, and dried under vacuum. The purity was confirmed by NMR and visible absorption and emission spectroscopies. The nucleotides and DNAs were used as supplied (Sigma). Binding studies[50] on thionine with the mono- and polynucleotides and with DNA show that a very high fraction of the dye is bound. Binding constants are >900 M^{-1} for mononucleotides and 10^6 M^{-1} to polynucleotides and DNA. There appears to be no particular preference for G−C rather than A−T binding. Also, there was no spectroscopic evidence for dimers of the dye, which can occur at these concentrations in the absence of nucleotide. 1:2 complexes are expected to predominate for mononucleotides, and, under conditions of low ionic strength, binding to the synthetic polynucleotides and to DNA is predominantly intercalative.[50,51] The nucleotide-to-dye ratio in the polymers was in excess of 30:1 in all cases and higher still for the DNAs to further favor intercalative binding.

Our femtosecond laser and transient absorption spectrometer is entirely home-built and has not been described in detail elsewhere. It consists of a Kerr-lens mode-locked, Ti:sapphire oscillator producing 13 fs pulses at 800 nm, pumped by 6.5 W from an all-lines, continuous argon ion laser (Coherent Innova 310). These pulses are temporally stretched and amplified in a regenerative amplifier pumped by an intracavity doubled, 2.9 kHz acousto-optically Q-switched (QS27-4SN, Gooch & Housego Ltd.), Nd:YAG laser. The pumping chamber (Spectron Laser Systems 902TQ) contains a 100 mm long, flash-lamp pumped rod, relay imaged onto a coated (DBAR/1064+532/PVD/C) 4 × 4 × 7 mm^3 KTP crystal (Cristal Laser SA) in a z-folded configuration. The laser produces two beams of 8.5 W each at 532 nm when driven at a lamp current of only 14 A, limited by damage to the KTP at this repetition rate.

The regenerative amplifier is a three-mirror cavity containing a fast Pockels cell (Medox Electrooptics) to inject and dump the pulses and a 20 mm Ti:sapphire rod doped to absorb ca. 90% of the pump light. The pulses from the regenerative amplifier (300 μJ) are compressed to 30−40 fs (limited by gain narrowing, not spectral dispersion) with an ultimate energy of 150 μJ/pulse. A combination of high refractive index prisms (SF10) and unequal groove density diffraction gratings in the stretcher (1200 L mm^{-1}) and compressor (1600 L mm^{-1}) allows the correction of all phase terms up to and including fourth order.[52]

To allow for tuneability of pump and probe wavelengths through most of the visible spectrum, we have constructed two independent, noncollinear, optical parametric amplifiers (NOPAs) as pump and probe sources. A seed pulse, which is a single filament "continuum" of white light generated in a 1 mm path of sapphire, is amplified by the frequency doubled (in 1 mm LBO) output of the amplifier (10 μJ), in a noncollinear arrangement using type I BBO (2 mm, θ = 32°, φ = 90°, AR@400 nm + HT@500−2400 nm, Ingcrys Laser Systems). The pump beam generates a cone of parametric superfluorescence from the BBO crystal. The crystal angle is adjusted such that the superfluor-

(46) Saito, I.; Takayama, M.; Sugiyama, H.; Nakatani, K. *J. Am. Chem. Soc.* **1995**, *117*, 6406−6407.
(47) Sugiyama, H.; Saito, I. *J. Am. Chem. Soc.* **1996**, *118*, 7063−7068.
(48) Guha, S. N.; Moorthy, P. N.; Kishore, K.; Naik, D. B.; Rao, K. N. *Proc.-Indian Acad. Sci., Chem. Sci.* **1987**, *99*, 261−271.
(49) Lobo, M. J.; Miranda, A. J.; Tunon, P. *Electroanalysis* **1997**, *9*, 191−202.
(50) Tuite, E.; Kelly, J. M. *Biopolymers* **1995**, *35*, 419−433.
(51) Rohs, R.; Sklenar, H.; Lavery, R.; Roeder, B. *J. Am. Chem. Soc.* **2000**, *122*, 2860−2866.
(52) Reid, G. D.; Wynne, K. In *Encyclopedia of Analytical Chemistry*; Meyers, R. A., Ed.; John Wiley & Sons: Chichester, 2000; pp 13644−13670.

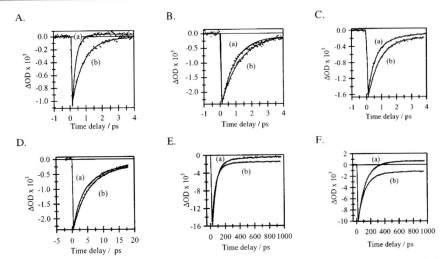

Figure 1. (A) Thionine-[poly(dG-dC)]$_2$. (B) Thionine-GMP. (C) Meldola's blue-GMP. (D) Toluidine blue-GMP (note time scale). (E) Thionine-AMP (note time scale). (F) Thionine-[poly(dA-dT)]$_2$. In each plot (a) is the gain signal (670 or 710 nm) and (b) is the ground-state recovery (600 nm). We believe that the oxidized bases contribute to the long-time absorption signal in Figure A(a) and F(a), i.e., when in the B-DNA type environment. The residual bleach in the gain signals, B(a) through E(a), is a result of a small overlap of the absorption and emission spectra.

escence shows no spatial dispersion. By directing the continuum beam along the cone axis, a large spectral bandwidth from the white light can be phase matched simultaneously.[53] Tuning the relative delay between pump and seed and controlling the chirp on the seed pulse change the center wavelength and bandwidth of the amplified light. The result after compression, using a pair of BK7 prisms, is ultrashort visible pulses tuneable continuously from ca. 480 to 750 nm with a duration of 10 to 30 fs depending on wavelength.

Samples in a 1 mm path length cuvette were excited with 25 fs, ≤200 nJ pulses of 600 nm light from one optical parametric amplifier, focused to a diameter of 100 μm. The pump beam was chopped synchronously (New Focus 3501) at half of the repetition rate of the amplifier. The transient species so-formed were monitored using pulses from the second parametric amplifier, also with 25 fs resolution, either at 600 nm to observe the ground state recovery, or at 670 nm to record the loss of the excited state by the decay of the stimulated emission (gain) signal. Both the central wavelength and the bandwidth of the pump and probe beams were measured routinely using a spectrograph and CCD camera. The relative polarization of the pump and probe was set either to 54.7° ("magic angle"), to remove any contributions from orientational effects, or to 0° (parallel) and 90° (orthogonal) for anisotropy measurements by the rotation of a half-waveplate in the pump beam. The anisotropy, $r(t)$, is calculated from $r(t) = [I_{\parallel}(t) - I_{\perp}(t)]/[I_{\parallel}(t) + 2I_{\perp}(t)]$.

The probe beam was split into two parts and the intensities measured on a pair of large-area photodiodes (New Focus 2031) interfaced to a PC, to record the difference in absorbance (ΔOD) between the excited and ground-state species as a function of the relative pump−probe delay. A ΔOD of ≤10^{-4} could be measured by averaging for 1 s at each time delay. Typically, 20−60 scans were averaged to remove long-time drifts in pump energy. We estimate the error on the measured lifetimes to be ±5%.

The stimulated emission decay rates were confirmed where possible by fluorescence measurements using time-correlated, single photon counting (TCSPC). The laser source was a cw-modelocked, Nd:YAG laser (Spectron Laser Systems), synchronously pumping a cavity-dumped dye laser. Fluorescence decays were recorded with 40 ps

(53) Wilhelm, T.; Piel, J.; Riedle, E. *Opt. Lett.* **1997**, *22*, 1494−1496.

resolution using a microchannel plate photomultiplier tube (Hamamatsu) connected to a computer board containing the fast digitization electronics (Becker and Hickl SPC-630).

Results

1. Mono- and Polynucleotides. For each of the dye-nucleotide or dye-DNA complexes, the excited-state decay has been monitored through its stimulated emission after excitation with 25 fs, 600 nm pulses. To observe this signal, we have selected wavelengths where there is little or no transient absorption or ground-state bleaching signal. We chose 670 nm for thionine and Meldola's blue and 710 nm for toluidine blue where the absorption is red shifted as compared to the other dyes. In addition, the ground-state recovery was monitored in each case at 600 nm. The results are summarized in Figure 1 and Table 2. As shown in Figure 1A(a), the gain signal, probed at 670 nm, indicates that the thionine excited state when bound to [poly(dG-dC)]$_2$ reacts with guanine and is strongly quenched with a single-exponential lifetime of 260 fs. The lifetime of free thionine in the absence of the polynucleotide is more than a factor of 1200 longer, 320 ps,[54,55] which we confirmed by TCSPC measurements. Monitoring the transient bleaching at 600 nm, Figure 1A(b) allows one to follow reformation of the ground state as the reaction products recombine. The signal recovers with a single-exponential lifetime of 760 fs, and this decay will represent the return electron-transfer rate, k_r. Because the measured forward rate is an average of rates to two guanine bases in both the mono- and the polynucleotide studies, the intrinsic rate, k_f, for quenching of the dye will be reduced by a factor of 2; however, this will not affect the return rate because only one base in each complex is reduced. The lifetime for thionine quenched by poly(dG-dT)·poly(dA-dC), where the

(54) Archer, M. D.; Ferreira, M. I. C.; Porter, G.; Tredwell, C. J. *Nouv. J. Chim.* **1977**, *1*, 9−12.
(55) Yamazaki, I.; Tamai, N.; Kume, H.; Tsuchiya, H.; Oba, K. *Rev. Sci. Instrum.* **1985**, *56*, 1187−1194.

Table 2. The Lifetimes for Charge Separation, τ_f, and Recombination, τ_r^a

dye-nucleotide	τ_f/ps	τ_r/ps	k_f/ps^{-1}	k_r/ps^{-1}	ΔG_f/eV	ΔG_r/eV
thionine-GMP	0.88	1.2	0.57	0.83	−0.42	−1.61
thionine-[poly(dG-dC)]₂	0.26	0.76	1.92	1.32	−0.47	−1.56
thionine-poly(dG-dT)·poly(dA-dC)	0.46	2.5	2.17	0.40	−0.47	−1.56
Meldola's blue-GMP	0.45	0.83	1.11	1.20	−0.41	−1.62
toluidine blue-GMP	2.8	5.0	0.18	0.20	−0.26	−1.62
thionine-dAMP	54	≥2500	0.0093	<0.00004	−0.03	−2.06
thionine-[poly(dA-dT)]₂	110	≥2500	0.0045	<0.00004	−0.03	−2.06

^{a} The intrinsic rate for forward transfer, k_f, is $1/(2\tau_f)$ excepting for the poly(dG-dT)·poly(dA-dC) result where there is only a single guanine in contact with the dye. The driving force of the forward electron-transfer step is $\Delta G_f = E_B - E_D - E_{0-0}$ and the return, $\Delta G_r = E_D - E_B$, where E_B is the oxidation potential of the base (donor), E_D is the reduction potential of the dye (acceptor), and E_{0-0} is the energy of the excited singlet state. E_G is taken to be 1.53 eV in [poly(dG-dC)]₂, 1.58 eV in GMP, and E_A 2.03 eV. The Born correction term can be ignored in aqueous solution.

measured rate is an average of reaction with a single guanine only, is 460 fs (not shown), roughly as expected. Similarly, Figure 1B(a) shows that with the mononucleotide, GMP, the excited-state lifetime is 880 fs, and the bleach recovery (b) is 1.2 ps. Meldola's blue is also quenched on the femtosecond time scale by GMP. The excited-state loss, measured at 670 nm, occurs with a single-exponential lifetime of 450 fs, Figure 1C(a) and the ground-state recover with a lifetime of 830 fs (b). For toluidine blue, Figure 1D, the loss of the gain signal occurs with a 3.5 ps lifetime (a), and the ground-state recovery at 600 nm is 5.0 ps (b).

In contrast to guanine, each dye is quenched much more slowly by adenine. Thionine is quenched by dAMP (Figure 1E) with a lifetime of 54 ps (a) and 110 ps by [poly(dA-dT)]₂, Figure 1F(a). The recovery of the ground state from the thionine-adenine⁺ transient species is too slow for us to record accurately. We estimate a decay time of ≥2500 ps for both dAMP and [poly(dA-dT)]₂ complexes, Figure 1E(b) and F(b). It is also possible that a small fraction of triplet state contributes to the bleach signal at long time. Fluorescence lifetime measurements confirm the rate of loss of the thionine singlet state, and this avoids any ambiguity as to the origin of each component, since the short decay is also observed in the bleach signal. We also see no evidence in the fluorescence decays for unusually long-lived thionine similar to that reported by Fujimoto et al.[56] for methylene blue in calf-thymus DNA. We note, however, that the high fraction of long-lived fluorescence quoted in that work will be overestimated owing to the limited resolution of the time-correlated, photon counting measurements. The short time decay is ca. 4 ps[25] rather than 25 ps as reported. We would expect that the additional stabilization conferred by the possibility of hydrogen bonding to the exocyclic amino groups in thionine would enhance intercalation as compared with methylene blue and, hence, reduce the fraction of unbound dye.

There is no spectroscopic evidence for binding of thionine to 2′-deoxycytidine-5′-monophosphate, dCMP, or thymidine-5′-monophosphate, dTMP, nor do we observe quenching of the dyes by the pyrimidine bases.

Figure 2 shows the anisotropy decay at 600 nm of thionine bound to [poly(dA-dT)]₂. The initial anisotropy at $t = 0$ is 0.2, which decays to a value of 0.09 with a single-exponential lifetime of 200 ps. This is indicative of constrained rotational motion of the dye within the B-DNA type complex. Knowing that reaction occurs with a lifetime of 110 ps in this complex, we can say that rotational reorientation is not limiting and is uncorrelated with electron transfer in the polynucleotide com-

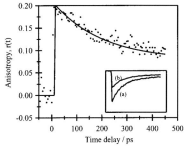

Figure 2. Thionine-[poly(dA-dT)]₂ anisotropy at 600 nm. Inset: (a) parallel and (b) orthogonal relative pump−probe polarization signals are shown on the same time scale. The anisotropy decay fits to a single-exponential decay of 200 ps plus background.

plex. We have also measured a long anisotropy decay of ca. 400 ps for thionine bound to ct-DNA. These observations lead us to a different interpretation of our data than that given by Fiebig et al.[57] for ethidium-ZTP, where rotational motion was proposed to limit the electron-transfer rate.

2. DNA. Figure 3 shows that thionine is also quenched very efficiently by DNA. However, in contrast to [poly(dG-dC)]₂, the decay is not a single exponential, reflecting the range of different intercalation sites. For example, in the case of both calf thymus and salmon testes DNA, approximately 70% of the thionine excited state disappears in 1 ps, the rest with roughly a 30 ps lifetime with a very small fraction of unquenched dye. Thionine is quenched still more efficiently by DNA from micrococcus luteus where almost 90% of the dye is quenched within 1 ps. These data cannot be fit adequately to a simple sum of two exponential decays indicating that the forward electron transfer is more complex. Fitting the early decay to estimate the initial rate of quenching gives 260 fs for reaction with ct- and st-DNA (a similar decay to that for [poly(dG-dC)]₂, which has only a slightly higher fraction of guanine) and 206 fs with mc-DNA, indicative of a faster quenching mechanism as the fraction of guanine increases. In the latter case, we believe that quenching by guanine dimers must be considered, and this is discussed below.

Discussion

1. Energetics. Nonadiabatic electron transfer has been explained by Marcus and Sutin[35] who related the rate coefficient, k_0, to the reaction driving force, ΔG, and the reorganization

(56) Fujimoto, B. S.; Clendenning, J. B.; Delrow, J. J.; Heath, P. J.; Schurr, J. M. *J. Phys. Chem.* **1994**, *98*, 6633−6643.

(57) Fiebig, T.; Wan, C.; Kelley, S. O.; Barton, J. K.; Zewail, A. H. *Proc. Natl. Acad. Sci. U.S.A.* **1999**, *96*, 1187−1192.

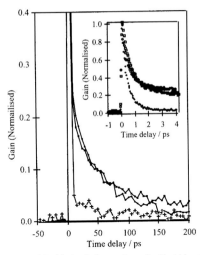

Figure 3. Loss of the thionine singlet state (normalized to +1 immediately after the laser pulse) at 670 nm when bound to calf thymus, ct- (circles), salmon testes, st- (squares), and micrococcus, mc-DNA (crosses). Inset is the same data on a −1 to 4 ps time scale.

energy of the reactants plus solvent, λ, as

$$k_0 = \frac{4\pi^2}{h\sqrt{4\pi\lambda k_B T}}|V_0|^2 \exp(-(\Delta G + \lambda)^2/(4\lambda k_B T)) \quad (2)$$

Ulstrop and Jortner[58] also derived the nonadiabatic rate coefficient, k_0^Q from a quantum viewpoint, using a sum of vibrational promoting modes while treating the solvent classically (so that the vibrational frequencies of the solvent are small in comparison to $k_B T$). If each promoting mode of frequency, ω_i, has a reduced mass, μ_i, and the displacement between potential surfaces is ΔQ, the rate coefficient, k_0^Q, is

$$k_0^Q = \frac{4\pi^2}{h\sqrt{4\pi\lambda k_B T}}|V_0|^2 F(\Delta G, \lambda, \omega, \Delta Q_i) \quad (3)$$

where F is a complex function of the variables shown.[58]

Despite the ultrafast time scale for some of the electron-transfer reactions, we will analyze our data using these two models, which rely on equilibrium energy gap fluctuations. For the sake of argument, we will assume that the excited-state population rapidly reaches thermal equilibrium, whereas, in reality, thermal equilibration will occur in parallel with electron transfer for the very fastest processes. Excitation at 600 nm was low in S_1, and the bandwidth of the excitation pulses was restricted to ca. 700 cm^{-1} to minimize population of high lying vibrational states.

To estimate accurate values for the reorganization energy and the electronic coupling strength, we require redox potentials for the nucleobases. Steenken and Jovanovic[29] have recently redetermined the oxidation potentials, $E°$, of the nucleosides by pulse radiolysis and give 1.58 V for free guanosine and 2.03 V for adenosine, both of which are considerably higher than those values given previously, 1.33 and 1.73 V, respectively.[59,60]

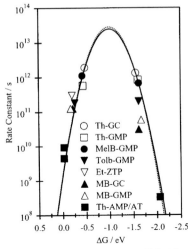

Figure 4. Classical (dashed line) and quantum (solid line) fits to the data given in Table 2. Key: thionine (Th), Meldola's blue (MelB), toluidine blue (TolB), [poly(dG-dC)]$_2$ (GC), [poly(dA-dT)]$_2$ (AT). The data for ethidium (Et) are from Fiebig et al.,[57] and methylene blue (MB) from Tuite et al.[27] For the classical fit we used $\lambda = 8070$ cm^{-1} and $|V_0| = 330$ cm^{-1}. For the quantum fit, we used a single promoting mode of 1000 cm^{-1}, a displacement, ΔQ, of 0.05 Å, which gives $|V_0| = 310$ cm^{-1} and the reorganization energy $\lambda = 8000$ cm^{-1}.

This difference was reported to be the result of experimental complications in the earlier studies. Throughout this paper we have taken GMP to be 1.58 V, G−C to be 1.53 V, and A to be 2.03 V; G−C is slightly lower than the free nucleoside as predicted by calculation.[12] As far as we are aware, there is no data for the distribution of adenine redox potentials with their environments.

Accurate parameters from eq 2 require rate data both close to the maximum on the Marcus curve as well as on longer time scales, and here the fast rates we have determined using thionine are very important. Because time constants greater than a few picoseconds will depend very strongly on redox potential, the fast rates which lie quite close to the curve maximum, where the gradient is smaller, will be the most reliable in determining the coupling and reorganization energies. The solution of eq 2, using the measured forward, k_f, and return, k_r, rates from the thionine reaction with [poly(dG-dC)]$_2$, gives $\lambda = 8070$ cm^{-1} and $|V_0| = 330$ cm^{-1} at 298 K. Figure 4 shows the Marcus curve derived from these values, onto which are superimposed the other rate coefficients tabulated in Table 2. We can also extend this argument to previously published data from methylene blue[25] and ethidium-ZTP experiments.[4,57] It may be observed that once again the fit is reasonable, probably because these DNA intercalators, like thionine, are charged and aromatic, and we suggest that they are each stabilized by similar π-stacking interactions. Therefore, to a first approximation we might expect both the electronic coupling energy and the reorganization energy involved to be much the same. Given that both the forward and the return reactions involve charged species in the reactant or product states, it seems reasonable to assume

(58) Ulstrup, J.; Jortner, J. *J. Chem. Phys.* **1975**, *63*, 4385−4368.
(59) Jovanovic, S. V.; Simic, M. G. *J. Phys. Chem.* **1986**, *90*, 974−978.

(60) Jovanovic, S. V.; Simic, M. G. *Biochim. Biophys. Acta* **1989**, *1008*, 39−44.

that the reorganization energy for the forward and reverse reaction is also similar. These two assumptions appear to be supported by our data.

Figure 4 (solid line) shows a fit to the nonadiabatic, quantum model from the equation quoted by Harrison et al.[22] Using a single promoting mode, ω, we obtain the best fit using a very small displacement between reactant and product potentials, ΔQ, of 0.05 Å. The effect of coupling to high-frequency promoting modes is to slow the falloff in rate coefficient with increased driving force in the 'inverted" regime (i.e., when $-\Delta G > \lambda$). This small displacement, ΔQ, means that the fit is relatively insensitive to the frequency of the promoting mode between ca. 100−1000 cm^{-1}. From the best fit, using $\omega = 1000$ cm^{-1}, we again find that the electronic coupling matrix element, V_0, is 310 cm^{-1} and the reorganization energy $\lambda = 8000$ cm^{-1}.

The data we have are well described by this model (i.e., that of a thermally driven reaction from the relaxed excited state of the dye). The rate of vibrational relaxation in the excited state will be similar to the measured rate of forward transfer in [poly-(dG-dC)]$_2$. Perhaps more surprising than the fit to the fastest rates is the rapid (parabolic) falloff in rate coefficient in the inverted regime. We might have expected this rate to fall off more gradually from previous measurements of bound donor−acceptor complexes.[22] This observation dictates a fit that is, for the most part, classical, and it follows, therefore, that the coupling to high-frequency vibrations is negligible and also that the possibility of adiabatic electron transfer need not be considered.

2. DNA. If the lifetime of the thionine excited state in DNA were to depend only on the purine base in contact with the dye, then the decay profile would be biexponential decay with lifetimes of <1 ps and ca. 110 ps representing reactions of the dye with either guanine or adenine, that is, in a ratio determined by the GC:AT ratio of the DNA. However, it is clear that our data cannot be explained in this fashion because we see multiexponential behavior, and an increase in initial rate as the fraction of guanine is increased, Figure 3.

It is most insightful to analyze the gain data to simulate the forward electron-transfer reaction rate, since this requires the fewest number of adjustable parameters. First, we consider the short component of the decays and estimate the initial rate for dyes quenched by neighboring guanine. The rate coefficients for quenching by guanine sequences were calculated from the Marcus eq 2, using the reorganization and electronic coupling energies obtained from the polynucleotide studies ($\lambda = 8070$ cm^{-1}, $V_0 = 330$ cm^{-1}). Reduced oxidation potentials of multi-purine sequences should result in more rapid quenching of the dye as compared with GC, as these rates are shifted closer to the maximum on the Marcus curve.

The HOMO of 5′-GG and 5′-GA dimers is thought to localize on the 5′-G in a purine stack, thus increasing the probability of oxidation at that position.[46,47] Ab initio molecular orbital studies have been used to estimate the associated redox potentials.[12] Using $E°$ (GT) =1.54 eV, $E°$ (GC) =1.53 eV, $E°$ (GA) =1.38 eV, $E°$ (GG) =1.20 eV, that is, values based on Steenken's[29] potentials corrected according to Saito,[46] gives k_0 (ΔG_{GG}) = 19.4 ps^{-1}, k_0 (ΔG_{GA}) = 2.6 ps^{-1}, k_0 (ΔG_{GC}) = 1.9 ps^{-1}, and k_0 (ΔG_{GT}) = 1.7 ps^{-1}. The average rate for guanine quenching, k_G, will depend on the fraction of G−C pairs in the particular DNA, $k_G = \sum P_{BB} k_0 (\Delta G_{BB})$, where P_{BB} is the probability of a guanine base being neighbor to each of the nucleobases, that is, GA, GC, GT, and GG. This average rate we might expect to be an estimate of the early-time signal in the DNA experiments if the ab initio predictions of Sugiyama and Saito[47] of the relative oxidation potentials are accurate. We find, however, that this does not seem to be the case, although the trend is in the right direction; the same conclusion has been drawn in recent experimental[61] and theoretical work.[62] The average calculated initial rate, k_G, in st-DNA is 6.1 ps^{-1} and in mc-DNA is 9.4 ps^{-1}. The initial rate observed in mc-DNA (4.6 ps^{-1}), where a high fraction of GG species should be expected, is also only a little faster than transfer to GC in [poly(dG-dC)]$_2$ (3.8 ps^{-1}) where all dyes have only GC neighbors. If $E°$ (GG) = 1.50 eV, we calculate initial rates in mc-DNA of 4.3 ps^{-1} and 4.0 ps^{-1} in st-DNA, somewhat closer to the experimental values.

The fact that these Marcus parameters predict increased rates for multi-guanine sequences gives us extra confidence in Steenken's higher redox potentials. Reanalysis of the mono- and polynucleotide data, using the lower potentials quoted by Burrows and Muller,[12] yields parameters that predict the opposite trend, that is, a falloff in rate as guanine content increases, since the forward rate from the [poly(dG-dC)]$_2$ experiments appears almost exactly at the maximum on the curve, and the GG redox potential pushes the calculated rate into the "inverted" regime.

Monte Carlo Simulation. The absence of a 110 ps component in the quenching of thionine by DNA (Figure 3) is due to the influence of guanine molecules further away from the dye than the nearest neighbor − these we call the distant guanines. Assuming that the nucleobases are randomly distributed in any DNA and within a superexchange mechanism for electron transfer (with its associated exponential distance dependence), we used a Monte Carlo model to simulate the full decay of the gain signal − the forward electron-transfer process. We require a model that contains at least two adjustable parameters, the redox potential of guanine dimers, so as to fit the initial decay and the distance scaling parameter, β. In addition, Wan et al.[2] observe in their studies of the 2-aminopurine reactions that the rate of electron transfer through pyrimidine bases to distant guanine is reduced as compared with that of adenine intermed-iates, and we, therefore, require at least a third parameter to account for this behavior.

A single dye molecule was situated in the center of a chain of 10 base pairs as shown in Figure 5, which were assigned at random but weighted according to the known GC:AT ratio of the DNA under study. Each base was assigned a nonradiative rate coefficient according to the polynucleotide studies. Ac-cordingly, single guanine bases have $k_G = 1/(2 \times 260 \text{ fs}) - k_f$, where k_f (1/320 ps) is the radiative rate. Multiple guanine sequences are treated such that the 5′G has k_{GG}, calculated from eq 2 using $E°$ (GG), and the others are assumed to be the same as a single guanine, in accordance with Sistare et al.[63] Adenine bases have $k_A = 1/(2 \times 110 \text{ ps}) - k_f$, and the rates for C and T transfer, k_{CT}, are assumed to be negligible, that is, slow as compared to the fluorescence lifetime, given that we saw no quenching in the mononucleotide experiments. The distance dependence of the electron transfer was generated from $g_n =$

(61) Lewis, F. D.; Liu, X.; Liu, J.; Hayes, R. T.; Wasielewski, M. R. *J. Am. Chem. Soc.* **2000**, *122*, 12037−12038.
(62) Conwell, E. M.; Basko, D. M. *J. Am. Chem. Soc.* **2001**, *123*, 11441−11445.
(63) Sistare, M. F.; Codden, S. J.; Heimlich, G.; Thorp, H. H. *J. Am. Chem. Soc.* **2000**, *122*, 4742−4749.

ARTICLES

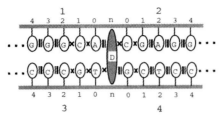

Figure 5. Duplex model for the calculation of the quenching rate of thionine by DNA by a Monte Carlo method. The bases were assigned at random but their number weighted according to the G—C:A—T ratio of the DNA. The four "segments" are labeled 1—4, which are runs of five bases on each strand in either direction from the dye. The dye is assumed to be in van der Waals contact with its neighbors. The coupling of the dye to distant bases depends on the intervening species with a reduction in strength for each break in the sequence. These coupling "defects" are multiplicative and are indicated by crosses between the bases, while the solid black lines indicate efficient coupling between bases of similar energy.

$\exp(-\beta 3.4n)$, where $n = 0$—4 is the number of base pairs away from the dye in each direction, and 3.4 Å is the edge-to-edge unit of separation of the dye from the nearest and subsequent bases. (X-ray crystal data of intercalators in the Brookhaven data bank[64] show that the distance of base to base in DNA is not strongly affected by the intercalating dye.)

Because we have no experimental data for quenching by the pyrimidine bases, we treat them as "spectators" in the electron-transfer process. The binding data of Tuite and Kelly[50] from studies of the absorption spectra of the phenothiazine dyes bound to mono- and polynucleotides show no evidence for interaction with dCMP or dTMP and, therefore, we expect the predominant π-stacking interaction in DNA to be with the purine bases. Cytosine and thymine will, however, appear as intermediates between the dye and distant guanines, and we must consider the coupling of the dye to these bases through the bridge, taking into account the relative energies of the purine and pyrimidine bases as shown in Figure 6. Superexchange will be most efficient if the energies of the intermediate species are similar, and, therefore, we should not expect a random mixture of bases to make a particular efficient bridge. Some theoretical work exists on the magnitude of the coupling matrix elements,[39,65] which predict that adenine bases are a barrier to electron transfer. These calculations are, however, unable to explain the experimental observations of Wan et al.[2]

The coupling element, $T_{D,A}^{(n)}$, between donor and acceptor, in McConnell's original formulation of the superexchange model,[66,67] appears as a product of individual tunneling integrals, $\nu_{i,i+1}$, between the degenerate donor and acceptor of energy, $E_{D,A}$, and the n nondegenerate bridging orbitals of energy, E_{B_i}.

$$T_{D,A}^{(n)} = \prod_{i=1}^{n} \frac{\nu_{i,i+1}}{E_{D,A} - E_{B_{i+1}}} \qquad (4)$$

We, therefore, introduce a coupling parameter, $c = 1$, between adjacent, similar bases (either purine or pyrimidine), but $c \leq 1$ between dissimilar bases indicating a region of reduced coupling

(64) Kielkopf, C. L.; Erkkila, K. E.; Hudson, B. P.; Barton, J. K.; Rees, D. C. *Nat. Struct. Biol.* **2000**, *7*, 117—121.
(65) Voityuk, A. A.; Jortner, J.; Bixon, M.; Rosch, N. *J. Chem. Phys.* **2001**, *114*, 5614—5620.
(66) McConnell, H. M. *J. Chem. Phys.* **1961**, *35*, 508.
(67) Todd, M. D.; Nitzan, A.; Ratner, M. A. *J. Phys. Chem.* **1993**, *97*, 29—33.

in the bridge. This is shown in Figure 5 where the crosses are indicative of the coupling "defects" in the strands. Consequently, the total coupling through the bridge in our model will be a product of these parameters; $c < 1$ reduces $|V_0|^2$ in eq 2 and hence also the rate coefficient, k_0, when the bridge contains a mixture of A with C or T. The pyrimidine bases thus shield the dye from distant purines.

We divide the total quenching rate coefficient into the sum of four parts, which correspond to the four "segments" labeled 1—4 in Figure 5, that is, both toward and away from the 5' end in a single strand and in its complement. The total quenching rate coefficient, k_{seg}, in each of these segments is simply the sum of the rate coefficients for quenching by each individual base at a distance of $n \times 3.4$ Å,

$$k_{seg} = (\sum_{n=0}^{m} (k_n g_n \prod_0^n c_n)) \qquad (5)$$

where k_n is the nonradiative rate for each base, k_G, k_{GG}, k_A, or k_{CT} at position n, m is the furthest base from the dye in the segment (i.e., when $n = 4$), g_n is the distance scaling factor, $\exp(-\beta 3.4n)$, and c_n are the coupling parameters. The total decay was then calculated according to eq 6,

$$I(t) = \frac{1}{i} \sum_i \exp(-t[k_f + \sum_{seg=1}^{4} k_{seg}]) \qquad (6)$$

for i repeated calculations, typically 50 000.

While all three parameters, β, $E°$ (GG), and c, describe the data in detail, for the purposes of discussion the decays can be divided in three parts. The initial decay is primarily due to the average rate of guanine quenching by nearest neighbors in the DNA, which will depend on the relative redox potentials and concentrations of the guanine species present, as discussed above. A slower component, the 30 ps lifetime in the st-DNA data, is determined largely by β and is due to transfer to distant guanine, while the reduced coupling owing to the bridge defects, c, influences the remainder of the long-time component of the decay. We obtained the best global fit, shown in Figure 7, to the st-DNA and mc-DNA data using a least-squares method, calculating a weighted, reduced χ^2 for both decays simultaneously. β was varied from 0.5 to 1.0, c from 0.05 to 1, and $E°$ (GG) from 1.45 to 1.55. This analysis yields the three parameters: $\beta = 0.8 \pm 0.1$ Å^{-1}, the redox potential $E°$ (GG) is 1.50 ± 0.01 eV, significantly larger than that predicted by the molecular orbital calculations, and the coupling parameter, c, is 0.20 ± 0.05. The error on the redox potential is small since this appears in the exponential in the Marcus equation and is, therefore, relatively sensitive. We do not believe $E°$ (GA) is significantly different from $E°$ (GC) for the same reason. The errors given represent the precision within the limits of the redox potentials we have used rather than the accuracy, which will be dictated by the limits of Steenken's measurements.[29] Our model takes account of both the fast and the slower components in the decay over 3 orders of magnitude in time and intensity while returning a β value that is consistent with the results on the small and well-defined oligomers.[2,3,68]

(68) Lewis, F. D.; Wu, T.; Zhang, Y.; Letsinger, R. L.; Greenfield, S. R.; Wasielewski, M. R. *Science* **1997**, *277*, 673—676.

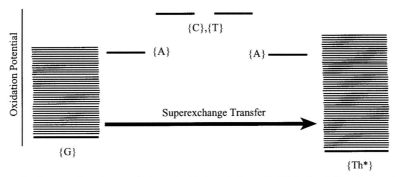

Figure 6. Electron transfer via superexchange from a nucleobase to the thionine excited state. If a bridging base is higher in energy (here C or T vs A), efficient coupling through the vibronic manifolds of an inhomogeneous sequence will be reduced in comparison with that in a bridge consisting only of A.

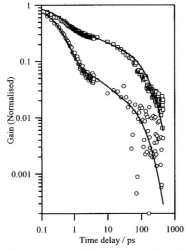

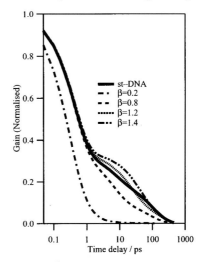

Figure 7. Data and results of a Monte Carlo simulation of the forward electron-transfer rates in DNA. We show global fits to the DNA from salmon testes (41.2% G−C, squares) and to DNA from micrococcus luteus (72% G−C, circles) using our reduced coupling model. β, the electron-transfer distance parameter, is 0.8 ± 0.1 Å^{-1}, and the coupling parameter is 0.20 ± 0.05. The noise in the mc-DNA data is due to the small amount of DNA we had available. The data and fit are scaled to +1 at $t = 0$ ps.

Because the product of coupling integrals in eq 4 leads to an exponential distance dependence, our model can be thought of as an electron-transfer process that has many values of β depending on the exact structure of the bridge. For example, the reduced rate to a guanine at $n = 1$ through a single C or T base falls off markedly, β is ca. 1.75 Å^{-1}. We observe that electron transfer (or hole transfer) into DNA segments is less efficient when the superexchange bridge is composed of mixed bases, and, in this respect, the pyrimidine bases can be thought to act much more like insulators within the DNA chain.

As a test of our model, Figure 8 shows the effect of removing the coupling parameter from the model, that is, $c = 1$, as compared to the st-DNA data, such that bridge defects do not shield distant purines from the dye. We have made a linear-log plot in this instance, which emphasizes the shape of the decay in the region of the changeover from nearest neighbor quenching to distant quenching. Varying β systematically from 0.2 to 1.4

Figure 8. st-DNA vs β in the absence of the coupling defect parameter ($c = 1$). The thickness of the solid line is indicative of the signal/noise ratio of the data. Note the linear-log scale − this best shows deviations of the fit in the region of the turnover from fast to slower quenching. The high β values also underestimate the fraction of short decay, which is not so clear on this scale.

Table 3. Calculated Quenching Rates of Thionine by Guanine, for Nearest Neighbor and Next Nearest Neighbor Species[a]

sequence	$E°$ (G)/eV	Πc_n	rate/ps^{-1}
dye−G	1.53	1	1.9
dye−(G−G)	1.50	1	2.6
dye−A−G	1.53	1	0.13
dye−C/T−G	1.53	0.04	0.005
dye−space−G	1.53	b	0.009

[a] $\beta = 0.8$; G is guanine adjacent to C, T, or A, and (G−G) is a guanine duplex. [b] Shown for comparison is the direct (nonsuperexchange) rate through space, 3.4 Å, assuming a typical β of 1.4 Å^{-1}.

Å^{-1} while holding $c = 1$ and $E°$ (GG) = 1.5 eV produces significantly poorer fits than those shown in Figure 7.

Table 3 gives the calculated quenching rate of thionine by a single neighboring guanine, k_G (1.9 ps^{-1}), or guanine dimer, k_{GG} (2.6 ps^{-1}). If the dimers are only ca. 30 mV lower in oxidation potential than other guanine species, the quenching rate increases by a factor of 1.3. Also given is the quenching rate for thionine separated from a guanine base by a single adenine (0.13 ps^{-1}) as compared with coupling through one of the pyrimidines (0.005 ps^{-1}) a factor of 26 smaller and comparable to direct transfer (nonsuperexchange) over 3.4 Å assuming a β of 1.4 Å^{-1}, that is, more typical of a "normal" environment. For the sake of clarity, we did not include direct transfer in the model, but the small fraction of extra quenching by the direct route would have the effect of reducing c a little further.

Our model clearly predicts that there is some degree of restriction on the pathways to electron-transfer products due to the high energy of the pyrimidine bases. This also leads us to the conclusion that interstrand coupling through the hydrogen bonds must be negligible as compared with coupling through the π-stack. If cross-strand electron transfer were possible, the electron or hole might be thought to "zigzag" between segments, finding the nearest guanine, and again this would result in overly rapid quenching. This conclusion is in broad agreement with the observations of Lewis and co-workers[3] that the electron does not cross strands on a picosecond time scale.

Conclusions

Electron transfer between charged, intercalating dyes and DNA has been described using the classical Marcus description as well as a nonadiabatic, quantum model. We estimate that the electronic coupling energy, $|V_0|$, is ca. 330 cm^{-1}, and the reorganization energy is ca. 8070 cm^{-1} for both forward and reverse reactions. The coupling strength is a little large, relative to k_BT, to be strictly in the nonadiabatic limit, and the fastest rates will occur in competition with vibrational relaxation in the excited state, but, nevertheless, we find the Marcus equation able to fit these data over >4 orders of magnitude in rate. We also observe that rotational reorientation of the dyes is uncorrelated with electron transfer.

We bring together recent observations from synthetic DNA oligomers and show that the results are general to natural DNAs.

We find a distance dependence ($\beta = 0.8 \pm 0.1$ Å^{-1}) which is similar to values obtained by Lewis et al.[68] and, independently, by Wan.[2] Our model requires a higher redox potential for guanine dimers than has been predicted previously from ab initio calculations,[46,47] but it is in line with other recent reports.[61,62] The predictions of lower redox potentials for the guanine dimers were based on the observation that oxidative strand cleavage often occurs at multi-guanine stacks. Our results would not necessarily contradict these experimental reports; we would merely conclude that those processes do not occur on a picosecond time scale. Once an isolated hole is present in a DNA strand, it may of course migrate as described, for example, by Giese[69] who reports the possibility of trapping at multi-guanine sites.

We also infer that the high energy of the pyrimidine bases significantly affects the efficiency of electron transfer in DNA. The effects of inhomogeneity via site and energy disorder on electron-transfer rates in real DNA, as compared to the much better defined oligomers used in the studies of Lewis and of Wan, do not appear to influence the measured β very strongly; that is, the primary limit to the rate of electron transfer in superexchange is the energy difference between adenine and the pyrimidine bases. We also infer that the superexchange mechanism is not enhanced by the hydrogen bonding network but only by the π-stack.

Acknowledgment. G.D.R. is a Royal Society University Research Fellow and thanks The Royal Society for its generous financial support. We also acknowledge the award of EPSRC studentships to D.J.W. and M.A.D. J.M.K. acknowledges the Berkeley Fellowship from Trinity College Dublin. D.A.T. is a SCI Messel Scholar.

JA0172363

(69) Giese, B. *Acc. Chem. Res.* **2000**, *33*, 631−636.

Contribution from

KEITARO YOSHIHARA

Toyota Physical & Chemical Research Institute
Japan

Fellow
Toyota Physical & Chemical Research Institute

1967, PhD, University of Tokyo (Research Director, Prof. H. Akamatu)
1965, Research Associate, University of Tokyo
1970, Research Staff, Institute for Physical and Chemical Research
1975, Professor, Institute for Molecular Science
1997, Professor, Japan Advanced Institute of Science and Technology (JAIST)
2000, Vice President, JAIST
2004, Fellow, Toyota Physical & Chemical Research Institute

1997, Member, The Science Council of Japan

Awards

Fulbright Fellow, 1963-1965, UC Riverside
Progress Award: Laser Society, 1983
Special Service Award: Laser Society, 1990
Daiwa–Adrian Award (London) 1995
Chemical Society of Japan Award 1996
Mizushima–Raman Lectureship Award 2002
Honorary Fellowship, Chemical Research Society of India 2005

Research Interests

Ultrafast molecular dynamics
Primary processes in photosynthesis
Primary processes in photographic sensitization
Mechanistic study of gene signal transduction

Professional

Editorial Board, *J. of Spectroscopic Society of Jpn.* 1971–1973
Advisory Board, *Chem. Phys.* 1974–present
Editorial Board, *Bull. Chem. Soc. Jpn.* 1974–1975
Associate Editor, *Photochem. Photobiol.* 1984–1987
Editorial Board, *Rev. Laser Eng.* 1984–present
Editor, Photochem. (*J. Photochem. Soc. of Japan*) 1989–1991
Editorial Advisory Member, *Laser Chem.* 1989–present
Editorial Advisory Board, *J. Luminesc.* 1990–1996
Advisory Editorial Board, *Chem. Phys. Lett.* 1992–1999
Advisory Board, *J. Phys. Chem.* 1995–1997
Editorial Advisory Board, *The National Academy of Science Letter*, present

David Klug and Keitaro Yoshihara at the Porter house in Kent.

Keitaro Yoshihara on
"In Vitro Models for Photosynthesis"
G. Porter
Proc. Roy. Soc. London A **362**, 281–303 (1978)

"Photosynthesis uses the sun as energy source, removes carbon dioxide from the atmosphere and the overall reaction is essentially the splitting of water to give oxygen and hydrogen or another fuel such as carbohydrate".[1] The fundamental photosynthetic system of higher plants contains a large number of light-harvesting chlorophyll molecules (antenna chlorophylls), a charge-separating chlorophyll dimer (special pair), and electron relay system. These molecules are arranged in a pigment protein. In the case of photosystem I reaction centre there are about 100 chlorophylls and two quinones and several iron-sulphur complexes. For the molecular mechanistic studies this congested structure gives a high hurdle, since the most essential spectroscopic information of charge separation and electron relay are covered by the signal of vast amounts of antenna molecules. The ultrafast nature of energy and electron transfer in such a system is another difficulty in studying the mechanism and we had to wait until the development of the femtosecond spectroscopy.

Porter was a person who strongly wished to unveil the secret mechanism of the most important and widespread chemical reaction on which the whole world is dependent. He said, "Although the system in which photosynthesis occurs is exceedingly complex there is no reason to suppose that the primary processes involve any principles not already encountered in the photochemistry of other large organic molecules".[2] In order to attack the problem he seemed to have a long term plan and started to study step by step the spectroscopic properties of monomer chlorophyll in solution starting in 1954,[3] intramolecular dynamics,[2] triplet formation in 1966,[4] and dimer equilibration in 1966.[5] The excitation energy transfer among closely located molecules is extremely rapid and eventually trapped and dissipated in the system, namely concentration quenching. Due to the inherent difficulties of photosynthetic systems he and his co-workers started to work on a series of model systems in 1970.[6] The first model of studying the light-harvesting mechanism was a simple mixture of chlorophyll *a*, chlorophyll *b*, and pheophytin in solution and they did pioneering work on energy transfer among these molecules and concentration quenching of fluorescence. They further created other model systems such as LB monolayer and multilayer films on solid surfaces[7] or in lipids,[8] and extracted important parameters of the efficiency of energy transfer among like molecules.

The next elementary process after the energy transfer is charge separation from the special pair. Taking as an example the photosystem I reaction centre, excitation energy from antenna chlorophylls is trapped by the chlorophyll special pair, which is called P700 due to its red-shifted absorption peak at ca. 700 nm. By the charge separation an electron is transferred to the primary electron acceptors (chlorophyll a's) and then further transferred to phylloquinone (Vitamin K_1) and further to iron-sulphur complexes. In 1970 Porter and collaborators studied reversible electron transfer between photoexcited chlorophyll a and duroquinone and VK_1 in solution and observed radical formation.[9] This was a big stride for the study of electron transfer reaction which eventually led to their recent studies of *in vivo* systems.

Both energy and electron transfer require high time resolution, and development of a stable picosecond/femtosecond laser system and a sensitive signal detecting method were necessary. In 1975 they studied fluorescence dynamics of photosystems I and II separately with a combination of a mode-locked laser and a streak camera, which gave a time resolution of a few tens of picoseconds.[10] They worked on various algae *in vivo* and observed fluorescence dynamics deviate from a single exponential and attributed this deviation to the Förster energy transfer.[11] Meaningful dynamics are observable in the time range of tens to hundreds of picoseconds since in algae the light-harvesting dye protein contains only a few dyes in the unit.[12,13] By excitation with an intense laser light the fluorescence signal gives an extra quenching due to a high concentration of excited molecules, which interact with each other and annihilate by themselves.[14-16] They determined the upper limit of laser intensity in order to study chloroplasts at sufficiently weak excitation conditions.

Among many papers which were influenced by Porter's works, I wish to cite a paper of 1967,[17] i.e. "Studies of triplet chlorophyll by microbeam flash photolysis" for "kinetic spectroscopic examination of small objects". They developed this new flash photolysis apparatus equipped with a microscope objective lens and observed samples 50 to 250 µm square and from 5 to several hundred microns thick. Their immediate purpose was to study individual parts of a biological system, but Porter expected "many applications in solid state photochemistry". This is a very farsighted development as we clearly see and is the forerunner of the present nanotechnology. Their research was to determine the direction of the "trapped" molecule by polarisation spectroscopy after energy transfer of "bulk" chlorophyll molecules in photosynthetic pigments. They successfully obtained beautiful absorption spectra of triplet chlorophylls in solid solvents, but "no triplets of chlorophyll appeared on flashing normal plant leaves". At almost the same time I started my research career at Professor Saburo Nagakura's laboratory in Tokyo and modified a flash photolysis apparatus for investigation of dynamics in micro space to search the triplet states in small molecular crystals at a low temperature. The experiment was not successful due probably to the same reason that Porter did not see the triplets in leaves. Later, we were able to detect triplets very easily in benzophenone and related crystals with a short and intense nitrogen-laser flash.[18]

One good way of overcoming the hurdle is to remove most of the antenna chlorophyll molecules, but yet maintain the fundamental function of energy and electron transfer. Technically it is possible to carefully get rid of the majority of antennas by detergent from the outer part of the protein and keep the central part unchanged. Professors I. Ikegami and S. Kato obtained the "reaction-centre enriched" photosystem I reaction centre from spinach in 1975.[19] We applied picosecond and later femtosecond flash photolysis to this sample and obtained the signal of P700 in 1981.[16] Porter used "isolated" particles of photosystem II reaction centre, since it is homologous to purple bacteria and separable from the light-harvesting complex.[20,21]

In 1993 Porter and I in collaboration with other scientists including Professor D. R. Klug of Imperial College received an international joint grant from the New Energy Development Organization of Japan and collaborated on "the fundamental study of the primary energy and electron transport processes in the photosynthetic reaction centres using femtosecond time resolution". Both groups fully used the benefit of femtosecond multicolor flash photolysis method and studied dynamics of energy and electron transfer in great detail. Through this activity, several collaborations, exchange of young scientists, and seminars took place until 1996 and the understanding of fundamental processes in higher plants improved significantly. Porter and his co-workers found in 1995 that the interaction between special pair chlorophylls in the photosystem II reaction centre is rather small, and comparable to those for adjacent chlorophylls and pheophytins.[22] This means that the "multimer" rather than a special pair plays a role on charge separation and also explains a relatively slow rate for charge separation (21 ps) in photosystem II compared to photosystem I and purple bacteria.[23,24] By femtosecond polarisation studies they found energy equilibration among all these molecules takes place in the sub-picosecond timescale and this state works as an electron donating species.[25]

As for some of our progress of photosynthesis research I wish to cite one review paper, "Primary processes in plant photosynthesis: photosystem I reaction centre".[26] With tunable femtosecond excitation pulses we can excite P700 and antennas individually and signals from antennas and the special pair P700 are observed in real time. Sub-picosecond energy equilibration processes among antenna chlorophylls were observed,[27] and an interesting biphasic fast primary charge separation of a sub-picosecond and a few picoseconds.[28] A fast secondary electron transfer (23 ps) to quinone was found to be optimised by the rate vs. energy gap analyses.

The aim of this short review was to describe how Porter's long-term efforts are connected to the recent modern studies on primary processes of photosynthesis. In conclusion, scientists world wide working on photosynthesis have benefitted significantly by the works of Porter and his co-workers, and in the future the new kinetic models allow us to suggest more efficient artificial photosynthetic systems, even modifying the system by bioengineering and/or chemical engineering.

References

1. G. Porter, *NEDO Report*, ed. K. Yoshihara (1996), p. 3.
2. G. Porter, *Proc. Roy. Soc. B* **157**, 293–300 (1963).
3. R. Livingston, G. Porter and M. Windsor, *Nature* **173**, 485–487 (1954).
4. P. G. Bowers and G. Porter, *Proc. Roy. Soc. A* **269**, 435–441 (1967).
5. R. L. Amster and G. Porter, *Proc. Roy. Soc. A* **206**, 38–44 (1966).
6. A. R. Kelly and G. Porter, *Proc. Roy. Soc. A* **315**, 149–161 (1970).
7. S. M. de B. Costa, J. R. Froins, J. M. Harris, R. M. Leblanc, B. M. Orger and G. Porter, *Proc. Roy. Soc. A* **326**, 503–519 (1972).
8. S. M. de B. Costa and G. Porter, *Proc. Roy. Soc. A* **341**, 167–176 (1974).
9. J. M. Kelly and G. Porter, *Proc. Roy. Soc. A* **319**, 319–324 (1970).
10. G. S. Beddard, G. Porter, C. J. Tredwell and J. Barber, *Nature* **258**, 166–169 (1975).
11. L. Harris, G. Porter, J. A. Synowiec, C. J. Tredwell and J. Barber, *Biochim. Biophys. Acta* **449**, 329–339 (1976).
12. G. Porter, C. J. Tredwell, G. F. W. Searle and J. Barber, *Biochim. Biophys. Acta* **501**, 232–245 (1978).
13. I. Yamazaki, M. Mimuro, T. Murao, T. Yamazaki, K. Yoshihara and Y. Fujita *Photochem. Photobiol.* **39**, 233–240 (1984).
14. G. Porter, J. A. Nowiec and C. J. Tredwell, *Biochim. Biophys. Acta* **459**, 329–336 (1977).
15. G. S. Beddard and G. Porter, *Biochim. Biophys. Acta* **462**, 63–72 (1977).
16. K. Kamogawa, A. Namiki, N. Nakashima, K. Yoshihara and I. Ikegami, *Photochem. Photobiol.* **34**, 511 (1981).
17. G. Porter and G. Strauss, *Proc. Roy. Soc.* **295**, 1–12 (1966).
18. J. M. Morris and K. Yoshihara, *Mol. Phys.* **36**, 993–1003 (1978).
19. I. Ikegami and S. Katoh, *Biochim. Biophys. Acta* **367**, 588–592 (1975).
20. O. Nanba and K. Katoh, *Proc. Natl. Acad. Sci. USA* **84**, 109–112 (1987).
21. J. Barber, D. J. Chapman and A. Telfer, *FEBS Lett.* **220**, 67–73 (1987).
22. J. R. Durrant, D. R. Klug, S. L. S. Kwan, R. van Grondelle, G. Porter and J. P. Dekker, *Proc. Natl. Acad. Sci. USA* **92**, 4798–4702 (1995).
23. D. R. Klug, T. Rech, M. Joseph, J. Barber, J. R. Durrant and G. Porter, *Chem. Phys.* **194**, 433–442 (1995).
24. T. Rech, J. R. Durrant, D. M. Joseph, J. Barber, G. Porter and D. R. Klug, *Biochem.* **33**, 14768–14774 (1994).
25. S. A. P. Merry, S. Kumazaki, Y. Tachibana, D. M. Joseph, G. Porter, K. Yoshihara, J. Barber and D. R. Klug, *J. Phys. Chem.* **100**, 10469–10478 (1996).
26. K. Yoshihara and S. Kumazaki, *J. Photochem. Photobiol. C* **1**, 22–32 (2000).
27. S. Kumazaki, I. Ikegami, H. Furusawa, S. Yasuda and K. Yoshihara, *J. Phys. Chem.* **105**, 1093–1099 (2001).
28. S. Kumazaki, I. Ikegami, H. Furusawa and K. Yoshihara, *J. Phys. Chem. A* **107**, 3228–3235 (2003).

Reprinted from *J. Photochem. Photobiol. C. Photochem. Rev.* **1**, 22–32 (2000)
with permission from Elsevier

ELSEVIER   Journal of Photochemistry and Photobiology C: Photochemistry Reviews 1 (2000) 22–32

Journal of
Photochemistry
and
Photobiology
C: Photochemistry Reviews

Primary processes in plant photosynthesis: photosystem I reaction center

Keitaro Yoshihara*, Shigeichi Kumazaki

Japan Advanced Institute of Science and Technology, Tatsunokuchi, Ishikawa 923-1292, Japan

Accepted 10 March 2000

Contents

Abstract

The photosystem I (PSI) pigment-protein complex of plants converts light energy into a transmembrane charge separation, which ultimately leads to the reduction of carbon dioxide. Recent studies on the dynamics of primary energy transfer, charge separation, and following electron transfer of the reaction center (RC) of the PSI prepared from spinach are reviewed. The main results of femtosecond transient absorption and fluorescence spectroscopies as applied to the P700-enriched PSI RC are summarized. This specially prepared material contains only 12–14 chlorophylls per P700, which is a special pair of chlorophyll *a* and has a significant role in primary charge separation. The P700-enriched particles are useful to study dynamics of cofactors, since about 100 light-harvesting chlorophylls are associated with wild PSI RC and prevent one from observing the elementary steps of the charge separation. In PSI RC energy and electron transfer were found to be strongly coupled and an ultrafast up-hill energy equilibration and charge separation were observed upon preferential excitation of P700. The secondary electron-transfer dynamics from the reduced primary electron acceptor chlorophyll a to quinone are described. With creating free energy differences (ΔG_0) for the reaction by reconstituting various artificial quinones and quinoids, the rate of electron transfer was measured. Analysis of rates versus ΔG_0 according to the quantum theory of electron transfer gave the reorganization energy, electronic coupling energy and other factors. It was shown that the natural quinones are optimized in the photosynthetic protein complexes. The above results were compared with those of photosynthetic purple bacteria, of which the structure and functions have been studied most. © 2000 Elsevier Science S.A. All rights reserved.

Keywords: Photosynthesis; Higher plants; Photosystem I; Reaction center; Purple bacteria; Electron transfer; Energy transfer; Femtosecond spectroscopy; P700; Chlorophyll

* Corresponding author.
E-mail addresses: yosihara@jaist.ac.jp (K. Yoshihara),
kumazaki@jaist.ac.jp (S. Kumazaki)

K. Yoshihara, S. Kumazaki / Journal of Photochemistry and Photobiology C: Photochemistry Reviews 1 (2000) 22–32   23

1. Introduction

1.1. Photosynthetic reaction centers

Photosynthesis by higher plants typically produces oxygen and organic substances from carbon dioxide and water by absorbing sunlight. This is the largest chemical process on earth and gives significant effects not only on all biological activities, but also geological and climatic activities, and eventually on human life. It fixes solar energy of about 3×10^{21} J in the form of biomass of 1.7×10^{11} t.

In plants and green algae, chlorophyll is contained in a cellular plastid called the chloroplast. Internally it is comprised of a system of lamella or flattened thylakoids. They form a lipid membrane, which contains various proteins with their own functions. The reaction center (RC) pigment-protein complexes have a role of collecting light energy, inducing charge separation, and converting light energy to an electrochemical potential. One of these is called the photosystem I (PS I) reaction center complex, which produces a strong reducing capability to reduce ferredoxin and eventually transform carbon dioxide to carbohydrates. One of the others is called the photosystem II (PS II) reaction center complex, which produces high oxidizing capability to split water [1,2].

One of the oldest traces of life known is the 3.5 billion-year old microfossil that resembles modern cyanobacteria [3]. The photosynthetic apparatuses of plants and cyanobacteria are quite similar and contain the two RC complexes PS I and PS II in the membrane. The core protein of PS I RC is made up of about 80 kDa polypeptides that are 80% homologous in their amino acid sequences [4]. They show almost no homology to the core polypeptides of PS II on the same membrane, but show about 15% homology to the RC polypeptides of green sulfur bacteria [5,6]. The core protein of PS II, on the other hand, is homologous by 30% to that of purple bacteria [5,7]. All of the RC complexes thus belong to one of the two types, the PS I/green-sulfur bacterial type (type I) or the PS II/purple bacterial type (type II) judging from the homologies of the proteins and cofactors [8]. They seem to have been differentiated in the early era of earth history and have efficient electron transfer systems that enable conversion of solar energy with high quantum efficiency.

In order to clarify the detailed functions of these RCs, it is desirable to elucidate the molecular architecture of the protein-pigment complexes. The complete structures of the RCs of purple bacteria (BP) were obtained in the mid-1980s and the clarification of the molecular design has stimulated an innumerable number of experimental and theoretical studies on the relationship between structure and function [9,10]. Such studies have been facilitated by the relatively simple stoichiometries of the isolated purple bacterial RCs. Fig. 1 gives the arrangements of the cofactors of the electron-transfer system of purple bacterial RC and PS I RC. The precise structure of the purple bacterial RC has been shown by X-ray crystallography [11,12], and recently the

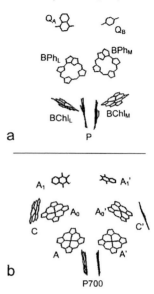

Fig. 1. Arrangement of cofactors in the electron-transfer systems of the purple bacterial reaction center (RC) of *Rhodopsuedomonas viridis* (a) [11], and the photosystem I reaction center (PS I RC) of *Synechoccus elongates* (b) [13–15]. In (a) P is the special dimer, $BChl_L$ and $BChl_M$ are bactriochlorophyll, BPh_L and BPh_M are bacteriopheophytin, Q_A is menaquinone-9, and Q_B is ubiquinone-9. The non-heme iron is not shown here. In (b) P700 is the special dimer, A, A′, A₀, and A_0' are chlorophylls in the electron transfer relay system, and C and C′ are the chlorophylls which seem to mediate excitation energy transfer from antennae to the electron transfer system. The two phylloquinones, A₁ and A_1' are tentatively arranged even though their orientations have not been determined. The iron sulfur complexes are not shown here. This figure is rearranged from data available from the Protein Data Bank (ID: 2PPS and 1PRC).

structure of PS I was reported with 4 Å resolution [13–15]. The largely similar arrangements support the idea that the two RCs share a common evolutionary origin. However, there are some distinct differences in the inter-molecular distances and orientations of the chlorophylls (Chls) in the electron-transfer system of PS I RC from those of the corresponding molecules of bacteriochlorophylls (BChls) and bacteriopheophytins (BPhs) in purple bacterial RCs.

The role of the protein matrix will be examined by comparison of the electron transfer mechanisms in the different RCs with their structures. The RC protein is expected to regulate mainly four parameters: (1) geometry of the donor–acceptor, i.e. distance and mutual arrangement; (2) electronic coupling by modulating the β factor; (3) free-energy difference (ΔG_0) by modulating the redox potentials in situ at the binding site inside the protein and (4) the reorganization energy, λ_S, due to the dielectric relaxation of the protein structure. Alteration of the specific amino acid residues affects all these parameters, which are interdependent.

1.2. P700-enriched reaction center

Since about 100 light-harvesting chlorophylls (antenna chlorophylls) are associated with PS I RC, excitation transfers among them prevent one from observing the elementary steps of the charge separation [16–18]. However, a special preparation of PS I RC from spinach with a significantly reduced number of antenna Chls has been developed, which are designated P700-enriched RC [19,20]. The P700-enriched RCs retain the original electron-transfer capability from P700* to A_0 [21–24]. About 90% of the original antenna Chls, secondary electron acceptor phylloquinone and carotenoids are removed during the enrichment process [25–28]. Phylloquinone or other quinones with similar molecular structures and redox potentials can be reconstituted into the original phylloquinone binding site [21,22] which recovers the electron transfer to the electron acceptor iron sulfur center (F_x) [22,29]. One of the most unique features of P700-enriched RC is that there are almost no so-called 'red absorbing' chlorophylls in their absorption peak redder than P700. Preferential excitation of P700 is possible, by which we can directly observe the primary processes from P700* (Fig. 2).

The main focus of this paper, is to review the recent progress of the primary photophysical and photochemical studies of PS I RC with a special emphasis on the lessons learned from the P700-enriched complexes of a higher plant (spinach). It is also our aim to compare these with the purple bacteria, of which the structure and functions have been studied most.

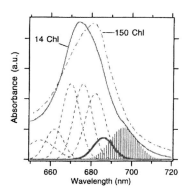

Fig. 2. Absorption spectrum of P700-enriched PS I RC sample (thin solid line labeled '14 Chl', at 77K) prepared from spinach and its Gaussian decomposition into several chlorophyll spectral forms [37]. The postulated spectrum of P700 is represented by vertical lines peaking at 696 nm, and the one for the electron acceptor chlorophyll, A_0, is indicated by the thick solid line peaking at 686 nm. Other chlorophylls are shown by dashed lines. The spectrum labeled '150 Chl' represents the absorption spectrum of PS I RC at 280K from spinach, which contains about 150 chlorophylls per P700.

2. Excitation transfer and primary charge separation

2.1. Overall description of the primary processes

In purple bacterial RCs, the electron transfers starting from the selectively excited special pair (P*) have been time-resolved by ultrafast laser spectroscopy (cf. Fig. 1a) [9,30,31]. The electron transfer steps are shown in the following scheme.

Purple bacterial reaction center (PB RC)

$$P^*(BChl_2) \xrightarrow[3\,ps]{e^-} BChl_L \xrightarrow[0.7-0.9ps]{e^-} BPh_L \xrightarrow[200\,ps]{e^-} Q_A$$

In this scheme, the primary charge-separated state $P^+BChl_L{}^-$ is not accumulated as a dominant transient. It is interesting to know whether or not such electron transfer dynamics of PB are universal among a variety of photosynthetic RCs. In PS I RC, there has not yet been sufficient spectroscopic data to elucidate the primary charge separation from P700 [5,17,18]. The kinetic scheme can be summarized as follows:

Photosystem I reaction center (PS I RC)

$$Chl^* \leftrightarrows P700^* \xrightarrow[\tau_1]{e^-} A(Chl) \xrightarrow[\tau_2]{e^-} A_0(Chl) \xrightarrow[21-35\,ps]{e^-} A_1(Q_K),$$
$$\tau_1 + \tau_2 < 3-6.5\,ps$$

One of the clearest differences from the PB system is the efficient exchange of excitation energy between the special pair and the other pigments in the electron transfer system [16]. The apparent dynamics of the primary charge separation in the PS I RC are strongly dependent on the contribution of excitation transfers (Chl* ← P700* and Chl* → P700*). These processes complicate the analysis of the kinetics in PSI RC. The primary charge separation is probably from the electronically excited P700 (P700*) to the intermediate acceptor chlorophyll, A (Fig. 1). However, this reaction has been too rapid to be observed spectroscopically (vide infra). This is followed by the sequential charge shift reactions from A to $A_0{}^-$, $A_0{}^-$ to phylloquinone (2-methyl-3-phytyl-1,4-naphthoquinone, Q) and then to the iron-sulfur complex (4Fe–4S clusters), F_X, F_B, and F_A [5,17,18]. The accumulation of $P700^+A_0{}^-$ or the photobleaching of $A_0{}^-$ has been observed to occur with an apparent time constant of 3–14 ps including the energy transfer time among antenna chlorophylls and to P700 [23,24,32–34]. The apparent time constant seems to vary depending on the numbers of antenna Chls in the PS I RC complex [35].

The multi-step electron transfer from P700 to phylloquinone somewhat resembles that in the purple bacterial RC complex (Fig. 1). In the purple bacterial RC, the donor is a bacteriochlorophyll dimer (P), and the acceptors are bacteriochlorophyll, bacteriopheophytin (H), ubiquinone (or menaquinone) (Q_A) and ubiquinone (Q_B). The electron transfer from A_0 to Q in PS I is comparable with that of H to Q_A in PB, since the molecular structure and size of

K. Yoshihara, S. Kumazaki / Journal of Photochemistry and Photobiology C: Photochemistry Reviews 1 (2000) 22–32   25

the cofactors are similar to each other. However, the rate constant of the former reaction is 10 times greater, and the energy gap is about half in comparison with those of the latter [36]. The difference between them may reflect the difference in the geometries of the functional groups and/or the structure of the protein matrix.

2.2. Excitation equilibration in P700-enriched RC

The absorption spectrum of the P700-enriched PSI RC sample at 77K is shown in Fig. 2. The Gaussian decomposition into several Chl spectral forms is given in [37]. P700 is represented by vertical lines peaking at 696 nm and A_0 by a thick solid line peaking at 686 nm [5,21]. In this specific case, the ratio of Chls and P700 is about 14. Other Chls are shown by broken lines. Due to the spectral overlap of the Chl forms, it is difficult to trace the excitation transfer and the subsequent electron transfer on specific Chl forms in PS I RC retaining many antenna Chls. Moreover, some of the previous studies on primary processes in PS I RC have adopted transient absorption spectroscopy alone [23,24, 33,34,38–40] to which ground-state depletion, stimulated emission, excited-state absorption and charge-separated states contribute. Some studies have adopted fluorescence measurements using single-photon-counting technique [35,41–43] or streak camera [44], but the instrumental response time (FWHM >10 ps) was insufficient. It has thus been difficult to unambiguously separate the signals associated with electron transfer from those associated with excitation transfer. Fluorescence measurements with sufficient time resolution are desirable to observe the dynamics of excited states clearly. The excitation energy transfer and primary charge separation have been studied by using PS I RC particles containing 10–100 Chls per P700 [16,35,38–43,45,46]. Apparent time constants of the primary charge separation and overall excitation decay (6.5–28 ps) reflect the number of antenna Chls retained in each preparation. They seem to be complex functions of the rate constants of both excitation transfer among many Chls and the primary charge separation from P700 [16].

Application of the subpicosecond fluorescence up-conversion technique to the P700 enriched RC has enabled us to observe the excitation energy transfer starting from P700* [47]. In order to preferentially excite P700, the central wavelength of the subpicosecond pulses was selected to be 701 nm (Fig. 2). The dynamic change of the fluorescence anisotropy was measured and showed a very fast equilibration of the excitation energy. Fig. 3 shows the polarization-dependent fluorescence decay dynamics observed at 749 nm ($I_{para}(t)$ and $I_{perp}(t)$) upon preferential excitation of P700 at 701 nm (the center of laser emission spectrum, Fig. 2) [47]. The time-dependent anisotropy is defined by

$$r(t) = \frac{I_{para}(t) - I_{perp}(t)}{I_{para}(t) + 2 I_{perp}(t)} \quad (1)$$

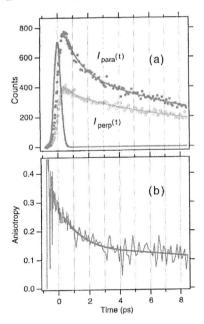

Fig. 3. Polarization-dependent fluorescence decays of the P700-enriched PS I RC upon preferential excitation of P700 at 701 nm [47]. Fluorescence was detected at 749 nm (a top). Fluorescence observed parallel (closed circles) and perpendicular (open circles) to the excitation polarization. The instrument response function (IRF) is shown by the pulse-like trace at $t=0$. The data are fitted to the convolution between sums of two exponential functions with IRF (b). Raw experimental anisotropy (dots) calculated from the data in (a), and the same function (smooth line) derived from the convoluted fits to the data in (a).

The value of $r(t)$ may change due to three possible reasons [48]: (1) the transition moments of absorption and fluorescence are different and the anisotropy changes by intramolecular processes within a very short time; (2) rotational motion of the fluorescent molecule and (3) intermolecular excitation energy transfer among fluorescent molecules. When the Qy band of the chlorophyll monomer is used for both excitation and probing, the contribution to anisotropy decay of the intramolecular process seems to be negligible. It is known that the anisotropy of the Qy band fluorescence of Chlorophyll a upon Qy band excitation in solution gives a maximum value of 0.4 [45]. Rotational motion in a protein environment at this short time is negligible. In such a case the time-dependent anisotropy reflects mainly the dynamics of intermolecular energy migration.

The relative amplitude of component of about 1 ps or shorter is much larger in the parallel kinetics (~40%) than in the perpendicular kinetics (~10%) (Fig. 3, upper figure). The difference is responsible for the anisotropy decay from ~0.3 to ~0.15 (Fig. 3, lower figure), which suggests equilibration of excitation energy starting from P700*. One also recognizes that the initial anisotropy value is estimated only up to ~0.3 (Fig. 3). This suggests the presence of excitation transfer with a time constant of less than ~0.3 ps and/or some ultrafast loss of anisotropy due to the excitonic interactions in the special dimer, P700. The anisotropy change at times longer than 3 ps amounts to at most ~0.05. The equilibration of excitation energy seems to be substantially complete within 3 ps upon selective excitation of P700. The fluorescence which does not decay at 30 ps (~10% of the total amplitude) could be attributed to charge recombination fluorescence, which has a lifetime of 40–50 ns in the absence of phylloquinone [49,50]. This may also be attributed to chlorophylls with which P700 exchanges excitation energy very slowly [37].

The energy equilibration process upon selective excitation of P700 seems to be confirmed by observing the transient absorption [51]. Fig. 4 shows the transient absorption spectra at 0.3, 2.0 and 18 ps. The spectrum at 0.3 ps is mainly due to the depletion and induced fluorescence of P700. The spectral shape at the shorter wavelength side increases its intensity at 2.0 ps. This is an absorption-spectroscopic indication of the uphill energy equilibration to the surrounding, shorter-wavelength absorbing Chls by thermally assisted energy transfer as described above with the anisotropy decay. This part of the excitation energy is further back-transferred to P700 and contributes to the charge separation at longer times. This is revealed in the spectrum at 18 ps, which becomes 'thinner' at the blue side of the spectrum and is quite similar to the charge-separated state on the nanosecond time scale. [52].

Let us now compare results of several research groups on the initial equilibration processes in the PS I RC. The time scale of full excitation equilibration (~1 ps) is somewhat shorter than those estimated on PS I RCs retaining more antenna Chls. Hastings et al. have studied excitation-wavelength dependence of the equilibration dynamics of a photosystem II deletion mutant of cyanobacterium *Synechocystis* sp. PCC6803 (~100 Chl/P700) [53]. The equili-

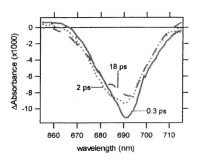

Fig. 4. Transient absorption spectra of the P700-enriched PS I RC upon preferential excitation of P700 at 696 nm [51]. After showing immediate depletion and induced fluorescence of P700 at 0.3 ps, the spectral shape at the shorter wavelength side increases its intensity at 2.0 ps. This is an indication of uphill energy equilibration to the surrounding, short-wavelength absorbing chlorophylls (see text).

bration time constant was in a 2.7–4.3 ps range [53]. This time constant is in good agreement with the one (3.3–6.6 ps) obtained by Struve et al. on PS I RC with ~60 antenna Chls [38,40], and the one (5 ps) by Du et al. on PS I RC with ~40 antenna Chls [45]. Since the antenna size of the RC preparations in [45] was relatively large (~40 Chl/P700), different Chls are probably excited in different RC units by the wavelengths employed (630–656 nm). Even longer equilibration time constants (12 ps) are reported by Holzwarth et al. on PS I RC particles from *Synechococcus* sp. [43,54]. One recent experiment by Struve et al. [55] may indicate special effects of red-absorbing chlorophylls to the spectral equilibration times of *Synechocystis* sp. The anisotropy at 680 nm (near the red edge of bulk antenna chlorophylls) decays with a time constant of 0.57 ps, which is similar to the fastest spectral equilibration time (0.43 ps) among the bulk antenna chlorophylls. They also found additional spectral equilibration processes with time constants of 2.0 and 6.5 ps, which are mainly downhill excitation transfers to the far red-absorbing chlorophylls (absorbing at about 708 nm). These results seem to indicate that the presence of the far red-absorbing chlorophylls makes spectral equilibration longer. This view is supported by the longest equilibration time, 12 ps, in the PS I RC of this species, which has relatively a large number of the far red-absorbing chlorophylls [43,54].

The relatively short time constant in the P700-enriched PS I RC [47] should reflect the extraction of most of the antenna Chls and/or the local environment of P700. There are two possible explanations. Firstly, the extraction of most of the antenna Chls may result in a situation where excitation can migrate only in the vicinity of P700 upon excitation of P700 and the equilibration time in the vicinity of P700 is short. Secondly, all of the Chls retained in the present particles may be spatially localized compared with the full extension of the antenna Chls in the other RC preparations, which enables relatively rapid excitation exchange among the Chls.

In summary, the energetically uphill excitation transfers from P700* to shorter-wavelength-absorbing Chl forms lead the excited Chls to decay with at least two time constants, ~1 and ~15 ps [47]. The fast and slow components indicate the charge separation before and after full equilibration of the excitation energy, respectively. The depolarization of fluorescence upon preferential excitation of P700 in the P700-enriched RC [47] (mainly with a ~1 ps time constant) is not complete on the time scale of single step excitation transfer [45], but somewhat faster than spectral equilibration in PS I RCs with more antenna Chls (>40 Chls/P700). This may reflect the lack of far red-absorbing chlorophylls in the P700-enriched RC (Fig. 2).

2.3. Ultrafast primary charge separation

We now have various types of evidence for an ultrafast primary charge separation, as follows. The P700-enriched RC was excited at 696 nm, which corresponds to the peak

K. Yoshihara, S. Kumazaki / Journal of Photochemistry and Photobiology C: Photochemistry Reviews 1 (2000) 22–32   27

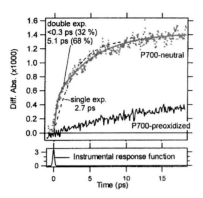

Fig. 5. Transient absorption change observed at 738 nm, where the absorption spectrum of the charge-separated state (P700$^+$A$_0$$^-$) is located [57]. The P700-enriched PS I RC are excited at 697 nm. The solid and broken lines are double- and single-exponential fits (including the instrumental response function) to the data, respectively. A small rising signal under the P700-preoxidized conditions (data points connected by lines) has not been identified. The data under P700-preoxidized conditions are scaled so that the signal corresponds to the same excitation probability of center particle per incident photon under P700-neutral conditions.

absorption spectrum of P700. The rise of the charge-separated state is monitored at around 740 nm, where both cation and anion radicals of Chls are known to give absorption [21,56], but the absorption by excited monomer chlorophyll is relatively small. Fig. 5 shows a rise with two exponential components, ~0.3 and 5.1 ps. The biphasic rise of the charge-separated state can be seen, namely the electronic excitation transfers from the excited-state absorption of P700 (P700*) to the neighboring Chls (Chl ← P700*) (equilibration process) compete with the primary charge separation (P700* → P700$^+$A$^-$) [47]. The ultrafast rise (about 0.3 ps or faster) indicates that a good portion (approximately 30%) of the initially prepared P700* is directly converted to the charge-separated state without excitation transfers to other Chls. The excitation, which is transferred from P700* to the other Chls (approximately 70%), decays relatively slowly by revisiting P700 with an apparent time constant of about 5 ps. It should be noted that P700* may be giving an ultrafast rising signal, since P700* may show a significantly different absorption spectrum than that of the excited state of the monomeric Chls. The most recent results on the rise of the primary charge-separated state with better time resolution has actually revealed some contribution of P700* [57], but has also sustained the biphasic rise of the primary radical pair, with the shorter time constant being 0.9 ps.

Several groups have predicted the rate of charge separation on the basis of a comparison between the experimental decay of the excited Chls and simulation of excitation migration and trapping among the ~100 Chls [58–62]. It is noteworthy that the rate of charge separation is estimated to be ≤1 ps in both plant (this work and [61]) and the cyanobacterial PS I RC [59,60,62]. In PS II and purple bacterial RCs, the intrinsic time constants of the primary charge separation are estimated to be 2.6 ps [63] and 2.1–2.7 ps [64,65], respectively, at room temperature. The primary charge separation of PS I RC seems to be intrinsically faster than those in the purple bacterial and photosystem II RCs.

3. Optimization of the electron transfer to quinone

The charge separation between P700 and A$_0$ is followed by the sequential electron transfer from A$_0$$^-$ to phylloquinone (Q) (Fig. 6). The P700$^+$A$_0$$^-$ → P700$^+$Q$^-$ reaction in PS I is comparable with the P$^+$H$^-$ → P$^+$Q$_A$$^-$ reaction in PB. However, the rate constant of the former reaction is 10 times greater, and the energy gap is about half in comparison with those of the latter [23,36]. The difference between them may reflect the difference in the geometries of the functional groups and/or the structure of the protein matrix [36]. We discuss here the ΔG_0 dependence of the rate constant of the electron transfer to quinone. The intrinsic acceptor phylloquinone was removed and replaced by artificial quinones and quinonoid compounds to vary ΔG_0 by a method originated and refined by Itoh and coworkers [20,22,27,28]. The results were analyzed by electron transfer theory and were compared to the results reported for the purple bacterium, *Rhodobacter sphaeroides* [66,67]. To understand the experimental results of electron transfer in this chapter, we describe some fundamental theory in the next section.

3.1. A brief description of electron transfer theory

To fully appreciate the design of the photosynthetic reaction center complex, there is an increasing impetus to construct models to ascertain how the specific variables such as distance, exothermicity and surrounding proteins control the rates of the primary electron transfer (ET) steps. It appears that intraprotein ET generally takes place over distances that are relatively large on the atomic scale. The interaction between the donor and acceptor redox centers is reasonably small. Under these conditions, Fermi's Golden rule provides a good first-order description of the rate of non-adiabatic ET [68–70].

$$k = \frac{2\pi}{\hbar} V_{DA}^2 FC \qquad (2)$$

The quantum mechanical matrix element, V_{DA}, couples the reactant and product electronic states and is critically dependent on the extent of overlap of the reactant and product electronic wave functions. FC is the Franck–Condon-weighted density of states, which reflects the integrated overlap of the reactant and product nuclear wave functions. A complete quantum mechanical description of the system provides the following expression for the FC with a single vibrational mode coupled to the electron

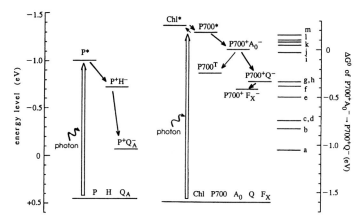

Fig. 6. Reaction schemes and energy levels in the RCs of PS I (right) and purple bacterium *Rb. sphaeroides* (left) [36]. Energy levels of the $P700^+Q^-$ states in the quinone-replaced PS I RCs are shown in the right row. The alphabetical notations (a to m) indicate the Q species at the corresponding electrochemical potentials. In the right panel, P700: the reaction center chlorophyll *a* dimer, A_0: the acceptor chlorophyll *a*, Q: the acceptor quinone that is originally phylloquinone. F_X: 4Fe–4S clusters, Chl: antenna chlorophylls. The typical energy level of the excited Chl (Chl*) is indicated in this figure. The iron sulfur clusters, F_A and F_B that exist on the peripheral polypeptide and accept electrons from F_X are omitted. In the left panel, P: the reaction center bacteriochlorophyll *a* dimer; H: the acceptor bacteriopheophytin *a*; Q_A: the acceptor quinone that is originally ubiquinone. The 'accessory' bacteriochlorophyll *a*, which functions as the intermediate electron acceptor prior to H, and the Q_B quinone, which accepts electrons from Q_A^-, are not shown.

transfer [68,71].

$$FC = \frac{1}{\hbar\omega} \exp[-S(2\bar{n}+1)] \left(\frac{\bar{n}+1}{\bar{n}}\right)^{P/2}$$
$$\times I_P\left[2S\sqrt{\bar{n}(\bar{n}+1)}\right] \qquad (3)$$

where $\hbar\omega$ is the energy per quantum of the nuclear vibration coupled to the electron transfer, S the $\lambda/\hbar\omega$, and λ the reorganization energy, which reflects the energy required to distort the equilibrium nuclear geometry of the product state into the geometry of the reactant state while constraining the electron to remain on the acceptor. $\bar{n}$ is $\left[\exp(\hbar\omega/kT)-1\right]^{-1}$, $I_P[\,]$ is the modified Bessel function, and P is $-\Delta G^0/\hbar\omega$. Vibrations with small energy per quantum ($\hbar\omega \ll kT$) can be treated in the classical limit as follows, which is described by Marcus [69]

$$FC = \left(\frac{1}{\sqrt{4\pi\lambda_S kT}}\right) \exp\left[-\frac{(\Delta G^0 + \lambda_S)^2}{4\lambda_S kT}\right] \qquad (4)$$

where λ_S is the reorganization energy associated with low-frequency vibrations, k the Boltzmann's constant, h the Planck's constant, and T is the temperature. As vibrations with large energy per quantum ($\hbar\omega \gg kT$) are defined as being frozen into their lowest level, calculations need only consider that quantum state [67,69,71]

$$FC = \frac{1}{\hbar\omega_L}\left(\frac{\exp(-S_L)S_L^{P'}}{P'!}\right) \qquad (5)$$

where P' is $-\Delta G^0/\hbar\omega_L$, and S_L is $\lambda_L/\hbar\omega_L$. In addition, FC can be given if the reaction is coupled to modes from more than one class of vibrations. For example, if the reaction is coupled to low- and high-frequency vibrations [67,69,71],

$$FC = \frac{1}{\sqrt{4\pi\lambda_S kT}} \sum_{q=0}^{\infty} \left(\frac{\exp(-S_L)S_L^q}{q!}\right)$$
$$\times \exp\left(-\frac{(\lambda_S + \Delta G^0 + q\hbar\omega_L)^2}{4\lambda_S kT}\right) \qquad (6)$$

The reorganization energy of the low frequency vibrations may be described by the dielectric continuum model, as follows [69,72]:

$$\lambda_S = (\Delta e)^2 \left(\frac{1}{2r_D} + \frac{1}{2r_A} - \frac{1}{R(ee) + r_D + r_A}\right)$$
$$\times \left(\frac{1}{\varepsilon_{OP}} - \frac{1}{\varepsilon_S}\right) \qquad (7)$$

Here, r_D, and r_A are the diameters of donor and acceptor, respectively, as a sphere, $R(ee)$ is the edge-to-edge distance between the donor and acceptor, Δe the charge transferred by this reaction, ε_{OP} the high frequency or optical dielectric constant and ε_S is the zero-frequency dielectric constant of the solvent. The difference in V_{DA} could be explained by the difference of donor–acceptor distances. V_{DA} is often approximated by the following equation. [69,70,73,74].

$$[V_{DA}(R)]^2 = [V_{DA}(0)]^2 \exp(-\beta R(ee)) \qquad (8)$$

K. Yoshihara, S. Kumazaki / Journal of Photochemistry and Photobiology C: Photochemistry Reviews 1 (2000) 22–32   29

where β is a constant that depends on how well the electronic wave functions penetrate the medium between the redox centers. Marcus states that the total reorientation energy is the sum of the outer-sphere (medium) reorganization energy λ_S and the inner-sphere reorganization energy λ_i, as follows:

$$\lambda = \lambda_S + \lambda_i. \tag{9}$$

These formulae have predicted a relationship between kinetics and thermodynamics: rates are expected to be slow for weakly exothermic reactions, to increase to a maximum for moderately exothermic reactions, and then to decrease with increasing exothermicity for highly exothermic ET reactions. The first successful and most extensive observations of the decrease of ET rate constants with high exothermicities (inverted region) was made with ET between aromatic molecules in a rigid organic solid [75] and ET between a donor–acceptor pair covalently bridged by a rigid spacer with no electron affinity of its own [76]. These studies were the important step for the understanding of ET reactions in photosynthetic reaction centers and in a broader range of biochemical ET processes, since almost all biochemical ET processes are between donor–acceptor pairs with fixed distances and orientations.

3.2. Kinetics of the $P700^+A_0^- \rightarrow P700^+Q^-$ reaction

Transient absorption spectra were measured at various delay times after the laser excitations in a PS I RC that contained 13 different artificial quinones or quinonoid compounds [36]. The kinetics of the formation and decay of A_0^- were monitored at 685 nm. The data obtained for RCs containing menaquinone-4, 2-amino-AQ and no quinone are shown in Fig. 7 [36]. In the quinone-reconstituted RCs,

the electron transfer $P700^+A_0^- \rightarrow P700^+Q^-$ was observed after charge separation. The bleaching of A_0 at 685 nm increased with a time constant $\tau(1/e)$ of 8 ps and recovered at 23 ps in the RC reconstituted with menaquinone (natural quinone) (closed circles) [23,36]. It recovered at 35 ps with 2-amino-AQ, which is about 0.3 eV more positive in free-energy difference compared to menaquinone (closed circles). The results indicate the rapid oxidation of A_0^- by these quinones. In the absence of reconstituted quinones, A_0 bleached with a time constant of 8 ps and showed no recovery on this time scale (open circles, Fig. 7) [23,36]. Without added quinones, the charge recombination between $P700^+$ and A_0^- occurred with a much longer time constant of 40–50 ns [49,50].

3.3. Free-energy difference (ΔG^0) dependence of the electron transfer rate

The dependence of the rate of ET from Chlorophyll a (A_0^-) to quinone (Q) in the PS I RC is presented in this section. The logarithm of the rate constant of the reaction in the RCs containing each quinone/quinonoid was plotted against the estimated ΔG^0 value, as shown in Fig. 8 with closed circles [36]. The rate constants obtained significantly decreased both in the small and large ΔG^0 regions (i.e. in the normal and inverted regions), as expected from Eq. (4). Data points can be fitted by an asymmetric bell-shaped curve calculated according to the quantum mechanical expression of the theory of electron transfer (Eq. (6)) with the following parameters: $V = 14 \, \text{cm}^{-1}$ and $\lambda_i = 0.18 \, \text{eV}$ and $\lambda_S = 0.12 \, \text{eV}$ [37]. A typical value of 0.18 eV for $\hbar\omega$ was chosen. A good fitting indicates the operation of vibrational coupling through

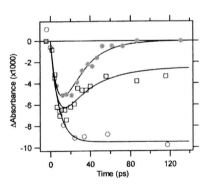

Fig. 7. Reaction kinetics of A_0 upon a 605 nm, 1 ps excitation in the quinone-reconstituted PS I RCs [23,36]. Absorption changes (ΔAbsorbance) of A_0 were detected at 685 nm from time-resolved spectra of quinone-removed RCs (O), RCs reconstituted with 2-amino-9,10-anthraquinone (2-amino-AQ) ($\square$), and menaquinone-4 (MK) (●). The fitting curves were calculated as the convolution of the instrumental response and the assumed exponential reaction kinetics.

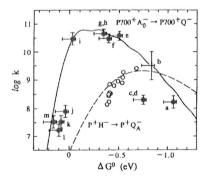

Fig. 8. Plots of the logarithm of the rate constant (log k) vs. free energy difference (ΔG^0) of the $P700^+A_0^- \rightarrow P700^+Q^-$ reaction in the PS I RCs. Quinones and quinoids were replaced by 12 different compounds with different ΔG^0 [36]. The PS I RC data (●) were obtained by experiments similar to those in Fig. 7. The data (O) for the $P^+H^- \rightarrow P^+Q_A^-$ reaction in purple bacterium *Rb. sphaeroides* RCs, which contained different quinones, were also plotted [66,67]. The solid and dashed lines were the best fits obtained by using Eq. (6).

30   *K. Yoshihara, S. Kumazaki / Journal of Photochemistry and Photobiology C: Photochemistry Reviews 1 (2000) 22–32*

the high-frequency modes, which elevates the reaction rate constant in the inverted region.

The primary ETs in the RC from photosynthetic purple bacteria were extensively studied with various temperature and ΔG^0 values [66,67,77,78]. In particular, the ET from bacteriopheophytin (Bpheo$^-$) to the first acceptor quinone (Q_A) was characterized by removing the natural quinone and reconstituting different artificial quinones with various electron affinities [66,67]. The reorganization energy of the medium (λ_S) calculated in PS I is much smaller than that in the *Rb. sphaeroides* RC, even when we consider the 0.1 eV errors in the estimation of the ΔG^0 value in PS I [36]. The electronic coupling in PS I is much larger than that in the latter. The common feature between the two curves is that the rate constant reaches a maximum at the ΔG^0 value of the original quinone in each case. This suggests that the reductions of the intrinsic quinones in both types of RCs are naturally optimized in somewhat different ways. The difference in each parameter is discussed below.

(1) A stronger electronic coupling in PS I. The electron transfer rate is expected to be maximal at $\Delta G^0 = -\lambda$ according to Eq. (4). The rate constant at the maximum is then expected to be proportional to the square of the electronic coupling, which depends on the donor–acceptor distance, the decay factor β, and the $V(0)$ value, as expressed by Eq. (8) [7,9]. The V value obtained in PS I is larger than that in the *Rb. sphaeroides* RC by a factor of 3–4 [23,36]. Because the sizes and structures of the donor and acceptor molecules, i.e. chlorophyll *a* (A_0) and phylloquinone (Q) in PS I and bacteriopheophytin *a* (H) and ubiquinone (Q_A) in *Rb. sphaeroides* RC, are not very different, the $V(0)$ value can be estimated to be almost the same. Although the β value of each reaction is not yet known, it is assumed to be similar to the mean value 1.4 Å^{-1} for the various intraprotein electron transfer reactions [79]. The difference in the V values thus seems to be mainly related to the difference in the donor–acceptor distances. The edge-to-edge distance between A_0 and phylloquinone in PS I RC is estimated to be 7.8 Å [23,36], which is 2.2 Å shorter than that between bacteriopheophytin and the primary acceptor quinone Q_A [79]. The most recent crystal structure has estimated that the edge-to-edge distance between A_0 and phylloquinone should be 4.8 (±2) Å [15], which is certainly shorter than the corresponding distance in PB.

(2) Smaller reorganization energy and lower dielectric constant of protein medium in PS I. We used a λ_i value of 0.18 eV, which is close to the estimated λ_i value of 0.2 eV in *Rb. sphaeroides* RCs [36,66,67]. The λ_S value in the PS I was estimated to be 0.12 eV, which is significantly smaller than the 0.6 eV value in the *Rb. sphaeroides* RC [36,66,67]. This small value seems to be mainly brought about by the low dielectric response of the protein matrix around quinone and chlorophyll upon the electron transfer, as expressed in Eq. (7). The optical dielectric constant (ε_{op} in Eq. (7)) of the medium protein can be assumed to

be 2.15 [72]. In order to obtain the λ_S value of 0.12 eV combined with the edge-to-edge distance estimated above, the static dielectric constant (ε_S in Eq. (7)) of 2.4 is required. Neither the difference in the molecular radii of quinone/quinonoid compounds used here (a few angstroms) nor the change in the center-to-center distance between chlorophyll (A_0) and quinone by 2–3 Å significantly change the λ_S value when the ε_S value remains unchanged [36]. In the case of the *Rb. sphaeroides* RC, ε_S was estimated to be 3.7 in a similar way. The difference between ε_S values in the two RCs suggests that the electrostatic environment of the quinone in PS I has an ε_S value lower than that around the Q_A quinone in the *Rb. sphaeroides* RC.

The estimation of a small λ value in the P700$^+$A$_0$$^-$ → P700$^+$Q$^-$ reaction might also be influenced by the time-dependent changes in the ΔG^0 and λ values, as proposed by recent theoretical works [80,81]. Another possibility may be that the unique geometry of A_0 and Q produces the small λ value. Experiments with synthesized organic compounds that have face-to-face bridged donor and acceptor (~4–5 Å) showed very small λ_S values of 0.2–0.3 eV, independent of the dielectric property of the medium [82]. The answers to these questions must wait for the fine structure of the PS I RC to be established.

The native RC complexes seem to have arranged reaction partners to meet requirements as shown in Fig. 8. The natural manipulation of amino acid residues around the quinone, which changes the ε_S value, might also self-regulate the λ value to be balanced with the ΔG^0 value. In the case of PS I, the protein with a low ε_S surrounding A_0 and Q will decrease the λ value. It will also decrease the ΔG^0 value. In the PS I RC complex, the small energy loss and high rate for the reaction are required to produce reducing power strong enough to reduce the iron-sulfur centers and to prevent back-reactions. To optimize the reaction, the protein structure around A_0 and Q seems to be arranged to give a strong electronic coupling and a small reorganization energy. This is different from the reaction in PB, in which a large ΔG^0 is required to stabilize Q_A^-, which slowly reduces Q_B to QH_2. Nature seems to have created these two types of highly-efficient RC complexes by modifying the protein matrix, presumably 3.5 billion years ago.

Acknowledgements

We wish to express our sincere thanks to Professor I. Ikegami for his long-term collaboration and stimulating discussions, and for supplying us the P700-enriched particles. We are also thankful to Professor S. Itoh and Dr. M. Iwaki for collaboration and discussions on the second half of the work reviewed in this article. Finally we wish to thank the coworkers in our research group.

K. Yoshihara, S. Kumazaki / Journal of Photochemistry and Photobiology C: Photochemistry Reviews 1 (2000) 22–32   31

References

[1] R. Nechushtai, A. Eden, Y. Cohen, J. Klein, in: D.R. Ort, C.F. Yocum (Eds.), Oxygenic Photosynthesis: The Light Reactions, Kluwer Academic Publishers, Dordrecht, The Netherlands, 1996, pp. 289–311.

[2] D.O. Hall, K. Rao, Photosynthesis, Cambridge University Press, Cambridge, 1994.

[3] J.W. Schopf, Nature 260 (1993) 640.

[4] W. Kirsch, P. Seyer, R.G. Hermann, Curr. Genet. 10 (1986) 843.

[5] J.H. Golbeck, D.A. Bryant, Light-driven reactions in bioenergetics, in: Current Topics in Bioenergetics Series, Vol. 16, Academic Press, New York, 1991, p. 83.

[6] M. Büttner, D.-L. Xie, H. Nelson, W. Pinther, G. Hauska, N. Nelson, Proc. Natl. Acad. Sci. U. S. A. 89 (1992) 8135.

[7] K. Satoh, in: J. Norris, R.N. (Eds.), The Photosynthetic Reaction Center Deisenhofer, Academic Press, San Diego, USA, 1993, p. 289.

[8] R.E. Blankenship, Photosynth. Res. 33 (1992) 91.

[9] A.J. Hoff, J. Deisenhofer, Phys. Rep. 287 (1997) 1.

[10] J. Deisenhofer, J.R. Norris (Eds.), The Photosynthetic Reaction Center, Vols. I and II, Academic Press, New York, 1993.

[11] J. Deisenhofer, H. Michel, EMBO J. 8 (1989) 2149.

[12] J.P. Allen, G. Feher, T.O. Yeates, H. Komiya, D.C. Rees, Proc. Natl. Acad. Sci. U. S. A. 84 (1987) 5730.

[13] W.-D. Schubert, O. Klukas, W. Saenger, H.T. Witt, P. Fromme, N. Krauß, J. Mol. Biol. 280 (1998) 297.

[14] W.-D. Schubert, O. Klukas, N. Krauß, W. Saenger, P. Fromme, H.T. Witt, J. Mol. Biol. 272 (1997) 741.

[15] O. Klukas, W.-D. Schubert, P. Jordan, N. Krauß, P. Fromme, H. Tobias, W. Saenger, J. Biol. Chem. 274 (1999) 7361.

[16] R. van Grondelle, J.P. Dekker, T. Gillbro, V. Sundstrom, Biochim. Biophys. Acta 1187 (1994) 1.

[17] K. Brettel, Biochim. Biophys. Acta 1318 (1997) 322.

[18] R. Malkin, in: D.R. Ort, C.F. Yocum (Eds.), Oxygenic Photosynthesis: The Light Reactions, Kluwer Academic Publishers, Dordrecht, The Netherlands, 1996, pp. 313–332.

[19] I. Ikegami, S. Katoh, Biochim. Biophys. Acta 376 (1975) 588.

[20] I. Ikegami, S. Itoh, M. Iwaki, Plant Cell Physiol., in press.

[21] P. Mathis, I. Ikegami, P. Sétif, Photosynth. Res. 16 (1988) 203.

[22] M. Iwaki, S. Itoh, in: J.R. Bolton, N. Mataga, G. McLendon (Eds.), Advances in Chemistry, Electron Transfer in Inorganic, Organic and Biological Systems, Vol. 228, American Chemical Society, Washington, DC, 1991, p. 163.

[23] S. Kumazaki, M. Iwaki, I. Ikegami, H. Kandori, K. Yoshihara, S. Itoh, J. Phys. Chem. 98 (1994) 11220.

[24] S. Kumazaki, H. Kandori, H. Petek, K. Yoshihara, I. Ikegami, J. Phys. Chem. 98 (1994) 10335.

[25] S. Itoh, M. Iwaki, I. Ikegami, Biochim. Biophys. Acta 893 (1987) 508.

[26] I. Ikegami, S. Itoh, Biochim. Biophys. Acta 893 (1987) 517.

[27] S. Itoh, M. Iwaki, in: M. Baltscheffsky (Ed.), Current Research in Photosynthesis, Vol. II, Kluwer Academic Publishers, Dordrecht, The Netherlands, 1990, pp. 651–657.

[28] M. Iwaki, S. Itoh, in: M. Baltscheffsky (Ed.), Current Research in Photosynthesis, Vol. II, Kluwer Academic Publishers, Dordrecht, The Netherlands, 1990, pp. 647–650.

[29] M. Iwaki, S. Itoh, FEBS Lett. 243 (1989) 47.

[30] W. Zinth, W. Kaiser, in: J. Deisenhofer, J.R. Norris (Eds.), The Photosynthetic Reaction Center, Vol. II, Academic Press, New York, 1993, pp. 71–88.

[31] W.W. Parson, in: D.S. Bendall (Ed.), Protein Electron Transfer, BIOS Scientific Publishers, Oxford, 1996, pp. 125–160.

[32] G. Hastings, F.A.M. Kleinherenbrink, S. Lin, T.J. McHugh, R.E. Blankenship, Biochemistry 33 (1994) 3193.

[33] M.R. Wasielewski, J.M. Fenton, X. Govindjee, Photosynth. Res. 12 (1987) 181.

[34] N.T.H. White, G.S. Beddard, J.R.G. Thorne, T.M. Feehan, T.E. Keyes, P. Heathcote, J. Phys. Chem. 100 (1996) 12086.

[35] T.G. Owens, S.P. Webb, R.S. Alberte, L. Mets, G.R. Fleming, Biophys. J. 53 (1988) 733.

[36] M. Iwaki, S. Kumazaki, K. Yoshihara, T. Erabi, S. Itoh, J. Phys. Chem. 100 (1996) 10802.

[37] M. Iwaki, M. Mimuro, S. Itoh, Biochim. Biophys. Acta 1100 (1992) 278.

[38] T.P. Causgrove, S. Yang, W.S. Struve, J. Phys. Chem. 92 (1988) 6121.

[39] T.P. Causgrove, S. Yang, W.S. Struve, J. Phys. Chem. 93 (1989) 6844.

[40] W.S. Struve, J. Opt. Soc. Am. B 7 (1990) 1586.

[41] M. Werst, Y. Jia, L. Mets, G.R. Fleming, Biophys. J. 61 (1992) 868.

[42] G. Hastings, F.A.M. Kleinherenbrink, S. Lin, R.E. Blankenship, Biochemistry 33 (1994) 3185.

[43] A.R. Holzwarth, G. Schatz, H. Brock, E. Bittersmann, Biophys. J. 64 (1993) 1813.

[44] K. Kamogawa, J.M. Morris, Y. Takagi, N. Nakashima, K. Yoshihara, I. Ikegami, Photochem. Photobiol. 37 (1983) 207.

[45] M. Du, X. Xie, Y. Jia, L. Mets, G.R. Fleming, Chem. Phys. Lett. 201 (1993) 535.

[46] S. Lin, H. van Amerongen, W.S. Struve, Biochim. Biophys. Acta 1140 (1992) 6.

[47] S. Kumazaki, I. Ikegami, K. Yoshihara, J. Phys. Chem. A 101 (1997) 597.

[48] G.R. Fleming, Chemical Application of Ultrafast Spectroscopy, Oxford University Press, New York, 1986, pp. 124–139.

[49] I. Ikegami, P. Setif, P. Mathis, Biochim. Biophys. Acta 894 (1987) 414.

[50] S. Itoh, M. Iwaki, Biochim. Biophys. Acta 934 (1988) 32.

[51] S. Kumazaki, H. Furusawa, K. Yoshihara, I. Ikegami, in: G. Garab (Ed.), Photosynthesis: Mechanisms and Effects, Kluwer Academic Publishers, The Netherlands, 1998, pp. 575–578.

[52] S. Itoh, M. Iwaki, in: N. Mataga, T. Okada, H. Masuhara (Eds.), Dynamics and Mechanisms of Photoinduced Transfer and Related Phenomena, Elsevier, Amsterdam, 1992, p. 527.

[53] G. Hastings, L.J. Reed, S. Lin, R.E. Balnkenship, Biophys. J. 69 (1995) 2044.

[54] S. Turconi, G. Schweitzer, A.R. Holzwarth, Photochem. Photobiol. 57 (1993) 113.

[55] S. Savikhin, W. Xu, V. Soukoulis, P.R. Chitnis, W.S. Struve, Biphys. J. 76 (1999) 3278.

[56] P.V. Warren, J.H. Golbeck, J.T. Warden, Biochemistry 32 (1993) 849.

[57] S. Kumazaki, I. Ikegami, H. Furusawa, S. Yasuda, K. Yoshihara, in preparation.

[58] Y. Jia, J.M. Jean, M.M. Werst, C.-K. Chan, G.R. Fleming, Biophys. J. 63 (1992) 259.

[59] A.R. Holzwarth, D. Dorra, M.G. Müller, N.V. Karapetyan, in: G. Garab (Ed.), Photosynthesis: Mechanisms and Effects, Kluwer Academic Publishers, Dordrecht, The Netherlands, 1998, pp. 497–502.

[60] G. Gobets, J.P. Dekker, R. van Grondelle, in: G. garab (Ed.), Photosynthesis: Mechanisms and Effects, Kluwer Academic Publishers, Dordrecht, The Netherlands, 1998, pp. 503–508.

[61] R. Croce, G. Zucchelli, F.M. Garlaschi, R. Bassi, R.C. Jenning, Biochemsitry 35 (1996) 8572.

[62] L. DiMagno, C.-K. Chan, Y. Jia, M.J. Lang, J.R. Newman, L. Mets, G. Fleming, R. Haselkorn, Proc. Natl. Acad. Sci. U. S. A. 92 (1995) 2715.

[63] D.R. Klug, J.R. Durrant, J. Barber, Philos. Trans. R. Soc., London ser. A 356 (1998) 449.

[64] M. Du, S.J. Rosenthal, X. Xie, T. DiMagno, M. Schmidt, D.K. Hanson, M. Schiffer, J.R. Norris, G.R. Fleming, Proc. Nat. Acad. Sci. U. S. A. 89 (1992) 8517.

[65] R.J. Stanley, S.G. Boxer, J. Chem. Phys. 99 (1995) 859.

[66] M.R. Gunner, D.E. Robertson, P.L. Dutton, J. Phys. Chem. 90 (1986) 3783.

[67] M.R. Gunner, P.L. Dutton, J. Am. Chem. Soc. 111 (1989) 3400.

[68] D. Devault, Q. Rev. Biophys. 13 (1980) 387.

[69] R.A. Marcus, N. Sutin, Biochim. Biophys. Acta 811 (1985) 265.

[70] J. Jortner, Biochim. Biophys. Acta 594 (1980) 193.

[71] J. Jortner, J. Chem. Phys. 90 (1986) 3795.

[72] L.I. Krishtalik, Biochim. Biophys. Acta 977 (1995) 200.

[73] C.M. Christopher, P.L. Dutton, Biochim. Biophys. Acta 1101 (1992) 171.

[74] C.M. Christopher, M.K. Jonathan, K. Warncke, R.S. Farid, P.L. Dutton, Nature 355 (1992) 796.

[75] J.R. Miller, J.V. Beitz, R.K. Huddleston, J. Am. Chem. Soc. 106 (1984) 5057.

[76] G.L. Closs, L.T. Calcaterra, N.J. Green, K.W. Penfield, J.R. Miller, J. Phys. Chem. 90 (1986) 3673.

[77] C. Lauterwasser, U. Finkele, H. Scheer, W. Zinth, Chem. Phys. Lett. 183 (1991) 471.

[78] Y. Jia, T.J. DiMagno, C.-K. Chan, Z. Wang, M. Du, D.K. Hanson, M. Schiffer, J.R. Norris, G.R. Fleming, M.S. Popov, J. Phys. Chem. 97 (1993) 13180.

[79] C.C. Moser, J.M. Keske, K. Warncke, R.S. Farid, P.L. Dutton, Nature 355 (1992) 796.

[80] L.I. Krishtalik, Biochim. Biophys. Acta 977 (1989) 200.

[81] W.W. Parson, A. Warshel, in: R.E. Blankenship, M.T. Madigan, C.E. Baue (Eds.), Anoxygenic Photosynthetic Bacteria, Kluwer Academic Publishers, Dordrecht, The Netherlands, 1995, p. 559.

[82] H. Heitele, F. Pollinger, T. Haberle, M.E. Michel-Beyerle, H.A. Staab, J. Phys. Chem. 98 (1994) 7402.

Keitaro Yoshihara studied at the University of Tokyo where he received his Ph.D. In 1965 he joined the Institute for Solid State Physics, University of Tokyo, and later the Institute of Physical and Chemical Research (RIKEN). In 1975 he joined the Institute for Molecular Science at the stage of its inauguration as a full professor. In 1997 he moved to the School of Materials Science of the Japan Advanced Institute of Science and Technology. His research interest include ultrafast dynamics of electronically and vibrationally excited molecules and primary photochemical processes of biologically important systems.

Shigeichi Kumazaki received his B.S. from Kyoto University in 1990. He then joined Prof. Keitaro Yoshihara's research group at the Institute for Molecular Science, where he worked at the developments of picosecond solid-state lasers and studied primary processes in photosystem I reaction center. He received his Ph.D. degree from the Graduate University for Advanced Studies in 1996. In 1997 he moved to the Japan Advanced Institute of Science and Technology. His current research interests include photosynthetic primary processes and other ultrafast photochemical events.

Contribution from

MARY ARCHER

University of Cambridge
UK

Born 1944, educated at Cheltenham Ladies' College, St. Anne's College, Oxford and Imperial College, London. Lecturer in Physical Chemistry at Somerville College, Oxford from 1971 to 1972. Research Fellow in the Davy Faraday Laboratory of the Royal Institution from 1972 to 1976, and Fellow and College Lecturer in Chemistry at Newnham College, Cambridge from 1976 to 1986, a Visiting Professor in the Department of Biochemistry at Imperial College from 1991 to 1999, and a Trustee of the Science Museum from 1990 to 2000. Has been a Senior Academic Fellow at de Montfort University since 1990, and Chairman of Cambridge University Hospitals NHS Foundation Trust since 2002. Became a Fellow of the Royal Society of Chemistry in 1987. Awarded the Melchett Medal of the Energy Institute in 2002.

Lord Porter and Lady Archer, 1975.

Standing L to R: Hiroshi Tsubomura, George Porter, Jim Bolton, Mary Archer, Joseph Rabani, Wolfgang Sasse, Heinz Gerischer, Sylvia da Brito Costa
Kneeling L to R: John Connolly, Arnim Henglein, Kenichi Honda, Rene Bensasson, Art Nozik

Mary Archer on
"*In Vitro* Photosynthesis"
G. Porter and M. D. Archer
Interdisciplinary Science Reviews 1, 119 (1976)

Solar energy conversion became one of the themes of George Porter's later research career, and it was my good fortune to be working with him when his interest in the subject was aroused. As I noted in my 1994 Royal Institution Discourse, *Hello Sunshine*, I had just moved from Oxford to London to work with Sir George at the RI, switching my field from electrochemistry to photochemistry in the hope of making myself vaguely relevant to the thrust of his photophysical research. Then came the oil price hike of 1973, hot on the heels of the influential 1972 *Nature* paper by Fujishima and Honda on the photoelectrolysis of water at TiO_2 electrodes. Suddenly there was funding for photochemical and photoelectrochemical approaches to solar energy conversion and storage. I set to work with my first research student, Isabel Ferreira, on photogalvanic cells, and George started a series of researches into photochemical hydrogen production.

George and I wrote our review of *in vitro* photosynthesis in 1976, the year I left the RI for Cambridge, defining an *in vitro* photosynthetic process as one "powered by sunlight, producing fuel through a direct conversion mechanism that does not rely on living matter". Reading this review again after nearly 30 years, I am relieved that our early exposition of the principles and possibilities of this fertile field still appears broadly right. There is one error I recognise as my own: we treated the limits on our conversion efficiency wrongly in "first law" terms, instead of the now-accepted "second law" approach, which George adopted in his later paper on the subject in *Faraday Transactions*. Notwithstanding this, we were too pessimistic about silicon photovoltaic cells, writing that an efficiency of 20–22% was unlikely to be exceeded, whereas the world record is currently around 25%.

We also recorded what we did not know. Charge separation underpins nearly all systems designed to convert light to chemical energy, including natural photosynthesis, and we grumbled that lack of certainty as to where the pigment molecules are located in photosynthetic antennas and reaction centres was hampering attempts to build functional *in vitro* models. Deisenhofer, Huber and Michel's paper on the structure of the reaction centre of *R. viridis* was still some years away, as were the huge advances in the theory of electron transfer reactions made possible by the synthesis and study of numerous classes of artificial photosynthetic systems that mimic the early events of natural photosynthesis.

We talked of "a manganese complex about which relatively little is known" being involved in the oxygen evolution step of green-plant photosynthesis: how thrilled George would have been to read the paper in the February 2004 edition of *ScienceExpress* from Jim Barber's group at Imperial College, showing the cubane-like structure of the Mn_3CaO_4 cluster of Photosystem II to 3.5 Å resolution. Surveying photoelectrochemical approaches to fuel and power production, we noted that "no device of practical value has yet been produced"; Michael Graetzel announced his eponymous cell nearly twenty years later.

By that time, my interests in renewable energy had broadened sufficiently that, in my 1994 RI Discourse, Martin Jarvis could play memorable cameo characters as varied as Joseph Priestly, Jonathan Swift, Russell Ohl, Edmond Becquerel and Jules Verne. (On this occasion, George remained seated in the audience; during my earlier Discourse of 1976 on electrochemistry since Davy and Faraday, I had pressed him into action as my lecture assistant, cutting up oranges and lemons for a pH titration.)

Ten years later, fundamental strides are still being made in the field of *in vitro* photosynthesis. Nanoparticles and quantum dots are everywhere, and hot-carrier devices should not be far behind. Single-molecule spectroscopy and fluorescence allow us to observe molecular behaviour that would normally be obscured in an ensemble-averaged measurement. Semiconducting polymers are the basis of functional light-emitting and photovoltaic devices. Maybe only God can make a tree but, as George liked to observe, Man may retain the more humble ambition of making a leaf.

Reprinted from *Proc. Roy. Inst.* **66**, 97–117 (1994) by Courtesy of
the Royal Institution of Great Britain

Hello sunshine

MARY ARCHER

It never set on the British empire. Autumn is its close-bosom friend, but Mrs Ogmore-Pritchard required it to wipe its feet before it came in under Milk Wood. You can buy a cheap version of it for 20 pence. It led Icarus to his sticky end, although it has nothin' to do 'cept roll around heaven all day. There's no new thing under it, but we hope to have our place in it, and at its going down we will remember them.

It evolved from a glowing gas cloud into an unremarkable yellow star. It is currently in sedate middle age, with about five billion years to go before it exhausts its nuclear fuel and blows up into a red giant, devouring the Earth in flames. Finally, at the clockwise end of its evolutionary spiral, it will become a dead white dwarf. Round it move the planets in their immutable orbits, an arrangement that so far removes Man from the centre of things as gravely to have displeased the Inquisition in Rome before which, nearly four centuries ago, Galileo upheld the heliocentric astronomy of Copernicus and Aristarchus.

The '*it*' of which I speak is, of course, the Sun, the Sun which has provided the warmth and light for the evolution and sustenance of life on Earth. Not that this magnificent heavenly body has historically been regarded as a mere it; for Louis XIV of France, the Sun was male and symbolized the potency, power, and glory of the Sun King himself. But for the Bedouin of Arabia, who had altogether too much of it, the Sun was a destructive old woman, who forced the handsome Moon to sleep with her once a month, and so exhausted him that he needed another month to recover.[1]

The Earth's fuel and energy resources

We are the merest upstarts in the solar system, as we may see from Carl Sagan's brief history of the world compressed into one year (Fig. 1). On

98   *Mary Archer*

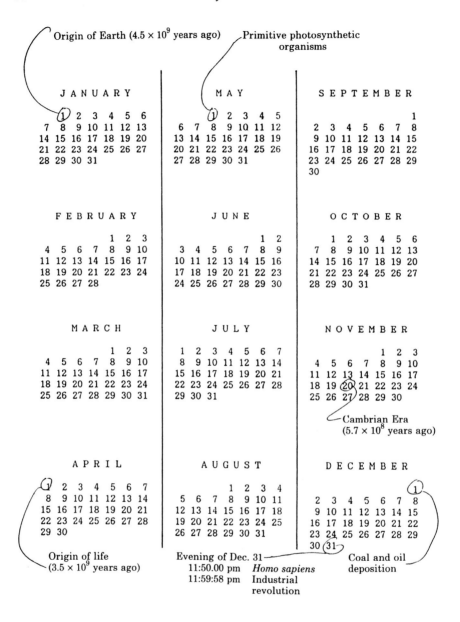

Fig. 1 The history of the Earth compressed into one calendar year (adapted from C. Sagan (1977), *The Dragons of Eden*, Random House, New York.

Hello sunshine **99**

this time scale, 1 January represents the formation of the Earth by the condensation of interstellar matter some four and a half billion years ago, and it is now midnight on New Year's Eve. In this time frame, the Earth awoke from its long Precambrian slumber only in mid-November. The coal, oil, and gas, which at present supply about three-quarters of the world's total energy requirements, were laid down in early December. The rather optimistically named *Homo sapiens* appeared only today—at 10 minutes to midnight. The Industrial Revolution began two seconds ago, powered by coal and latterly by oil and gas, and in those two seconds we have consumed perhaps one quarter of our total stocks of fossil fuels. Within the next few seconds—within the next few centuries in real time—we will in all likelihood devour the rest.

If that is so, we are midnight's children, striking one short match in the middle of a long dark night. When the hydrocarbon match burns low, assuming that we have not by then discovered some new exploitable force of nature, we will have to turn to some combination of nuclear and renewable energy, or maybe return to a combination of the two, for the Sun is a nuclear fusion reactor. At the enormous temperatures and pressures of the inner Sun, four hydrogen nuclei fuse to form one helium nucleus. Some matter is destroyed in this fusion process, and when matter is destroyed, energy is created in accordance with Einstein's seminal equation $E = mc^2$ where E is the energy created, m the mass destroyed, and c the speed of light. It is impossible to travel faster than the speed of light and, as Woody Allen has pointed out, it is also undesirable because one's hat keeps blowing off.

The Sun is losing mass at the staggering rate of 4.5 billion kg sec^{-1}, outpouring about 10^{26} W of radiant energy into space. Our eyes are so good at accommodating to different light levels that it is difficult to believe that full sunlight on Earth is some half a million times brighter than full moonlight. Not that this impressed Sydney Smith, writing to satirize the dismissive style of his friend, Francis Jeffrey, 25 February 1807: '*Damn the solar system! bad light—planets too distant—pestered with comets—feeble contrivance;—could make a better with great ease.*'[2]

But we do not need a better, as we may see from our global energy account (Fig. 2). The scale is logarithmic, so that each rung represents 10 times more energy than the rung below. On the left is shown our energy capital—the Earth's estimated remaining resources of coal, gas, and oil, and of fissile uranium, used today in non-breeder reactors, tomorrow maybe in breeder mode. Also on the left is shown our global energy expenditure, that is, the world's energy demand in the years 1970 and 1990, plus a prediction for the year 2030. Clearly we are living off capital, dwindling fossil fuel capital. And it would not be wise to bank on the

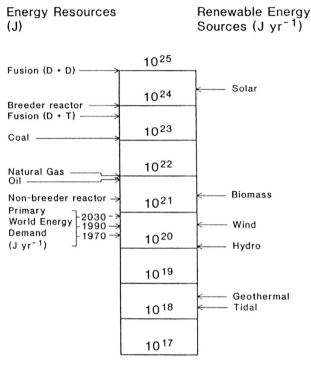

Fig. 2 Comparison of the Earth's non-renewable and renewable energy resources on a logarithmic scale.

capital represented by the fusion fuels deuterium and tritium, because it is by no means certain that commercial fusion power can be achieved.

To be prudent, we should therefore look for another source of income, and there it is, on the right-hand side of the ladder, the annual energy available from our renewable energy resources. The great fusion reactor in the sky, the Sun, is by far the most abundant of these. Indeed, the energy from the Sun falling on the Earth in one hour is equivalent to the world's energy demand in one year. The other renewables—wind, biomass, waves, and so on—are mostly driven by the Sun. As Harold Heywood, a great British pioneer of solar energy utilization, observed in his Discourse of 1957, the subsidiary renewables represent the concentration of solar energy in space, just as fossil fuels represent its concentration in time.

Characteristics of solar energy

Worship of the Sun goes back to the cult of the Sun god, Ra, at Heliopolis and beyond. As to use of the Sun, the Greeks understood and practised the principles of solar architecture, although the account of Archimedes setting fire to the Roman fleet off Syracuse by reflecting sunlight off the polished shields of the Greek soldiers is probably apocryphal. But the chemist's burning glass is real enough, and it was with such a glass that Joseph Priestley isolated 'dephlogisticated air'—oxygen—in August 1774, by focusing the Sun's rays onto red oxide of mercury, and confining a mouse under a bell jar of the evolved gas:

> Having procured a lens of twelve inches diameter and twenty inches focal distance, I proceeded ... to examine what kind of air a great variety of substances would yield. I presently found that, by means of this lens, air was expelled from Mercurius Calcinatus very readily. But what surprised me more than I can express was that a candle burned in this new air with a remarkably vigorous flame, and my mouse lived in it for half an hour ... I was utterly at a loss how to account for it.

The mouse did not actually die after the half hour—it was revived by putting it in front of a warm fire, after which it went back into the bell jar for another three-quarters of an hour, once more emerging in reasonable shape.

There are various characteristics of solar energy of which we should be aware in designing solar converters. The first is its relatively low flux density. When the Sun is overhead on a clear day, each square metre of the Earth's surface receives up to 1 kW of solar power. But the Sun is not always out, let alone overhead, so the average power is lower than this, ranging from about 300 W m^{-2} in hot spots like Kuwait to about 100 W m^{-2} in rather less sunny locations such as Manchester (Fig 3). The amount of solar energy received varies seasonally, particularly at high latitudes. The summertime maximum is less marked on a vertical surface than on a horizontal surface because the vertical surface sees the low winter Sun better. Fixed, flat plate collectors are usually tilted at or near the angle of latitude to maximize their annual insolation, which comes partly straight from the Sun, as beam radiation, and partly out of the sky, as diffuse radiation scattered out of the beam by molecules or particles in the atmosphere.

Concentrating collectors, like the celebrated solar furnace of Odeillo in the French Pyrenées, use lenses or mirrors to focus sunlight, and work only with the beam component, although there are non-imaging collectors of various types that will concentrate diffuse light. For example, a high refractive index sheet of plastic which contains a fluorescent dye will trap diffuse light incident on the plane face, channelling it out through the narrow edges.

The last notable characteristic of sunlight is that it is white, containing all the colours of the rainbow and more, on into the ultraviolet and infra-

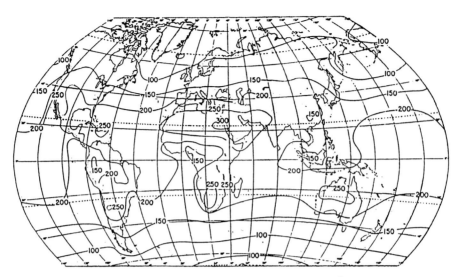

Fig. 3 World distribution of solar irradiance (W m^{-2}) on a horizontal sea-level surface, averaged over a 24 h day.

red. The Sun's surface temperature is 6000 K, and the spectrum of the radiant energy it emits is much like that of a white-hot solid body, except for the tell-tale dark Fraunhofer lines which reveal the elements present in the outer solar atmosphere. A solar converter must absorb and use a good portion of this visible spectrum if it is to be efficient.

Thermal conversion of solar energy

There are two classes of solar converter: thermal and direct. Thermal converters deliver heat as such or use a heat engine to deliver mechanical power or electricity. Direct converters produce electrical power or chemical fuels without the intervention of a thermal stage. Let us consider a model solar community—'Solarville'—with examples of both converter types. In the Industrial Park, there is a solar thermal mill powered by a 60 W light bulb (which simulates the Sun), whose workings we shall shortly examine.

Solar thermal applications go back to the nineteenth century. The solar thermal locomotive designed by the Punch cartoonist Emett has yet to find a backer—though the 1901 vintage 15 horse power solar pump that irrigated the Pasadena Ostrich Farm surely comes close to it—and Augustin Mouchot's Sun Machine of 1878, devised because of the acute shortage of coal in nineteenth century France, finds a modern echo in the dish collectors at Sandia National Laboratory in Albuquerque. Meanwhile, a dispassionate Martian visitor to the site of Eurelios in Adrano might well conclude that the cult of the Sun god is alive and well in twentieth-century Sicily. In fact, Eurelios is the world's first 1 MW solar thermal power tower. The field of mirrors is controlled to reflect and focus the solar beam onto the central receiver, where the temperature can approach 3000 K, and superheated steam is raised to drive an electric generator. They do it bigger and better in California, of course. The power tower Solar One at Barstow supplied 10 MW, but even that is dwarfed by the 354 MW of solar thermal power installed by the Israeli company Luz in the Mojave Desert, which provides two per cent of Southern California Edison's grid capacity.

Let us now see how we can turn some of the heat in a cup of coffee into mechanical work to power the second mill on the Solarville Industrial Park. These mills contain miniature versions of the engine devised by the Scottish cleric Robert Stirling in 1816, and we can run a baby Stirling engine between the temperatures of hot coffee and iced whisky and soda. To run, the engine requires a hot source (in this case a cafetière) and a cold sink, in this case a tumbler of iced whisky and soda—and away goes the engine, with modest efficiency admittedly, but it's simple and reliable. To make it solar-powered, we would have only have to provide the heat by a solar water heater.

Passive solar design

The Sun's heat can be used as such to provide water and space heating for buildings, which may declare their technology boldly or may have less obvious architectural features that maximize useful solar gain and minimize the house's conventional energy demand. We can quantify the difference that passive solar design can make using the National Home Energy Rating (NHER) scale. This goes from 0 to 10, and the higher a house is on the scale, the lower are its energy running costs. The energy rating is calculated from the characteristics of a house—its room and window dimensions, wall construction and roof insulation, fuel and heating type, orientation and degree of shelter, and so on. All this information is fed into the computer, and the NHER software computes the rating.

Using the expertise of Peter Rickaby, of energy consultants Rickaby Thompson, the energy ratings of two similar houses were compared (Table 1). One house, in Newport Pagnell, has no solar features. Its South façade is largely taken up by garage and entrance while most of the living rooms and conservatory are on the North side. This house rates a respectable 8.0 on the NHER scale and has annual energy running costs of £777.

Table 1. NHER ratings of a modern non-solar house and a modern passive solar house

	Non-solar house		Passive solar house	
NHER	8.0		9.1	
Building Energy Performance Index	144		150	
Standard Assessment Procedure	78		88	
CO_2 emissions (tonnes/year)	8.4		5.3	
Analysis of fuel use and costs	*GJ/year*	*£/year*	*GJ/year*	*£/year*
(i) By application				
Primary heating	44.0	192	20.7	90
Secondary heating	9.6	39	5.2	23
Water heating (main fuel)	25.9	113	16.9	74
Cooking (main fuel)	2.9	13	5.4	23
(secondary fuel)	1.7	37	–	–
Lighting and appliances	11.9	267	12.2	261
(off-peak component)	3.0	22	–	–
Standing charges	–	93	–	80
TOTAL	**98.9**	**777**	**60.3**	**551**
Maintenance	–	24		24
(ii) By fuel type				
Gas (mains)	72.8	355	48.1	247
Housecoal/Pearls	9.6	39	–	–
Economy 7 (on-peak)	13.5	346	–	–
(off-peak)	3.0	36	–	–
Domestic tariff (on-peak)	–	–	12.2	303

The bar chart shows where the heat goes—through the walls and windows, wasted in heating appliances, and so on.

The other house has passive solar features; it is in Milton Keynes. Here all the habitable rooms are on the South side, which also boasts a semi-integral conservatory. As a consequence, the South façade has much more glazing than the North side, which houses the bathrooms and passageways. The house also has excellent heating controls, needed to take full advantage of solar gain. This house rates an NHER of 9.1, an excellent score, over an integer better than the other house, and its energy running cost is only £551 per annum (Fig. 4b, Table 1). The heating system does not have to work so hard because the house gets more of its heat from the Sun.

Hello sunshine **105**

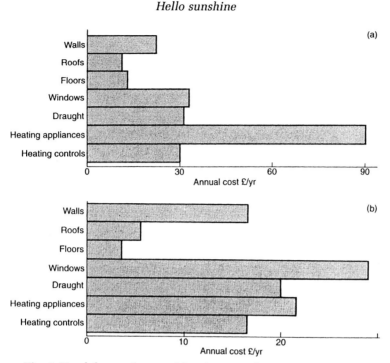

Fig. 4 Breakdown of costs of heating a modern non-solar house (a), and a modern passive solar house, (b).

Photosynthesis

So much for solar thermal conversion. Turning to direct solar converters, these use sunlight not as heat but as a stream of high energy photons, capable of driving processes that store energy. Like many of the ideas I shall mention, this one is not new. The distinguished Armenian photo-chemist, Giacomo Ciamician, who worked at the University of Bologna at the turn of the century, using the Sun as his light source, predicted the possibility of chemical storage of solar energy thus:

> On the arid lands there will spring up industrial colonies with-out smoke ... forests of glass tubes will extend over the plains and glass buildings will rise everywhere; inside of these will take place the photochemical processes that hitherto have been the guarded secret of the plants, but that will have been mastered by human industry which will know how to make them bear even more abundant fruit than nature, for nature is not in a hurry and mankind is.[3]

The 'guarded secret of the plants' is the mechanism of natural photo-synthesis, the process by which atmospheric carbon dioxide and water are fixed under the driving force of sunlight to form carbohydrate; the myriad genera of green plants are Nature's advanced direct solar converters. They have evolved from primitive photosynthetic organisms which appeared as early as 1 May in the time-compressed history of the world. All photosynthetic organisms contain organic pigments which absorb sunlight and initiate the process of photosynthesis. Chief among these are the chlorophylls, of various shades of green, the carotenoids, whose orange or yellow colours are seen in autumn leaves in which the chlorophyll has decomposed, and the phycobilin pigments such as phycocyanin.

If these pigments are illuminated with ultraviolet light, chlorophyll fluoresces red and phycocyanin orange. Cartenoids do not fluoresce, and for a good reason—they are there to protect the plant against excessive light levels. Fluorescence comes about because the absorption of a photon—a quantum of light—promotes an electron from its ground band in the pigment molecule to an upper band, leaving a hole in the ground band. This process stores energy in the hole–electron pair, equal to the energy difference between the two bands. The hole–electron pair is not long-lived and it can recombine in three ways. The electron can return directly to the ground state, giving up its stored energy as fluorescence, which is what is happening in the chlorophyll and phycocyanin. Or the stored energy can be given up as heat as the electron returns to the ground state, as in the carotenoid. Either way the energy of the hole–electron pair is wasted. But if the system is so designed, the electron can be made to do work as it returns to the ground state. We can represent this as mechanical work by making the electron of our mechanical model travel back to ground level in a basket, raising a small weight as it goes.

There is a school of thought that holds that, since aeroplanes do not flap their wings like birds, successful man-made direct converters need be nothing like a green plant. But it turns out that the 'guarded secret' of the plants, as revealed by modern structural analysis, has many valuable lessons for us. A leaf is a highly structured system at the macroscopic, microscopic, and atomic levels. All the important molecules are held in very specific locations and orientations with respect to one another, as we can see from the structure of the reaction centre of the photosynthetic bacterium *Rhodopseudomonas viridis* (Fig. 5), which was crystallized and analyzed in the mid-1980s by Hüber and Michel at the Max-Planck-Institute in Martinsried, in a bravura piece of work that won them the 1988 Nobel prize for Chemistry.

The heart of the reaction centre is a special pair of bacteriochlorophyll (*BC*) molecules. When these receive a photon of sunlight, an electron is excited and moves very rapidly via the bacteriopheophytin (*BP*) molecules

Hello sunshine   **107**

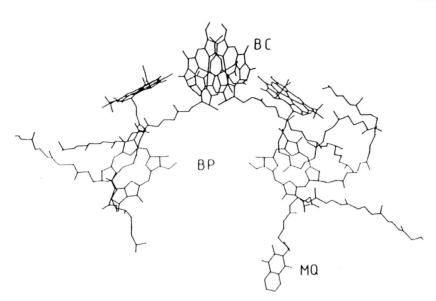

Fig. 5 Arrangement of the chromophores in the reaction centre
of *Rhodospirillum viridis*. From J. Deisenhofer, O. Epp, K. Miki,
R. Huber, and H. Michel (1984) *J. Mol. Biol.*, **80**, 385.

to the quinone molecule (*MQ*) on the lower right-hand side. Then a sequence
of events turns the energy of the hole–electron pair—the electron on the
quinone and the hole left behind on the chlorophyll—into chemical energy
stored in carbohydrate. The point I want to make here is simply this: the
structure of the photosynthetic reaction centre is the key to its function.
If we blended it all up like the pigments in solution, the resulting 'leaf
soup' would not photosynthesize any carbohydrate whatsoever.

Now that the structure of the natural photosynthesis reaction centre is
known, chemists are addressing themselves to the formidable challenge
of making a simple working version in the laboratory. In such synthetic
assemblies, it is now possible to observe electron transfer occurring as it does
in natural photosynthetic organisms, and the long-term hope is that one day
chemists will be able to synthesize systems capable of fixing carbon dioxide
in a test tube. Maybe only God can make a tree but, as Lord Porter observed,
Man may retain the more humble ambition of making a leaf.

Biomass and energy crops

What about growing crops for their energy content rather than for food?
The difficulties of making biomass cost-effective emerged very clearly

when Gulliver encountered on his travels in Laputa a Scientist, 'a Man ...
of meagre Aspect with Sooty Hands and Face':

> He had been Eight Years upon a Project for extracting Sun-Beams
> out of Cucumbers which were to be put into Vials hermetically
> sealed and let out to warm the Air in raw inclement Summers.
> He told me that he did not doubt in Eight Years more, that he
> should be able to supply the Governor's Gardens with Sunshine
> at a reasonable Rate; but he complained that his Stock was low,
> and intreated me to give him something as an Encouragement
> to Ingenuity especially as this had been a very dear Season for
> Cucumbers.

Jonathan Swift notwithstanding, the economic attractions of energy
crops have been much increased by the policy of agricultural set-aside.
By the turn of the century, there could be 20 million hectares of surplus
agricultural land within the European Union as a whole, an area equivalent to England, Wales, and a large part of Scotland. Farmers cannot grow
food on this land, but they could grow energy crops. There are several
candidate crops, among which it has to be said that cucumbers, with their
98 per cent water content, are not front runners. *Miscanthus*, a perennial
woody grass, grows very rapidly and can be harvested by familiar means.
Rape is now grown as a subsidized energy crop on over quarter of a
million hectares in Europe, the oil from its seed being esterified and
marketed as biodiesel; in Austria, this now accounts for about five per
cent of the total diesel fuel market. But wood, although less familiar to
the farmer than the forester, looks like the best bet for the UK. Willow
and poplar both root readily from simple cuttings, which to be grown as
an energy crop are planted at high density. These grow rapidly and can
be harvested by coppicing, that is, by cutting down the long stems and
leaving the stools to sprout again, as they readily do. The energy content
of wood is only about half that of coal, and it is generally preferable to
use such a low density fuel near where it is grown. In Sweden, for
example, there are some thousands of hectares under willow for use in
district heating schemes.

Photovoltaic cells

Between the sunlight incident on the green leaf and the energy stored in
the crop lies a long and complex biosynthetic pathway. So it is no wonder
that the net energy storage efficiency represented by a bundle of wood is
very modest, perhaps one half to one per cent. Anyway, efficiency is not
really the right criterion for an energy crop—price, sustainability, and

Hello sunshine   **109**

carbon dioxide abatement are all more important considerations. In terms
of sheer efficiency, Man overtook Nature forty years ago with the silicon
photovoltaic (PV) cell. PV cells, or solar cells as they are sometimes
known, are direct converters that produce electric power when they are
illuminated. Solar cells are made of semiconductors, and if for example,
an old de la Rue set of semiconductors is illuminated with ultraviolet
light, they fluoresce, showing that electrons are excited when the semiconductors absorb light, just as they are in photosynthetic pigments. Of
course, a solar cell is altogether a much simpler structure than a green
leaf, but it is like a green leaf in that one side is designed for looking at
the Sun. Just under the top surface, there is a junction between two
different semiconductors, in this example *n*-type silicon and *p*-type silicon (Fig. 6). In this junction there is a built-in electric field, which effectively tilts the ground and upper bands in the interface, causing the excited
electron in the upper band to flow one way and the hole in the ground
band the other. In other words, negative charge flows one way across the
junction and positive charge the other. Thus the output of a solar cell is
electrical work.

Solarville has quite a collection of PV-powered items (and I would like
to thank Bernard McNelis, Bob Hill, Phil Wolfe, and Michael Penney for
letting Bryson Gore and me play with their toys in the Discourse on which
this article is based). The house has a PV roof which provides electric
power for the internal lights and appliances; the residents have a PV-

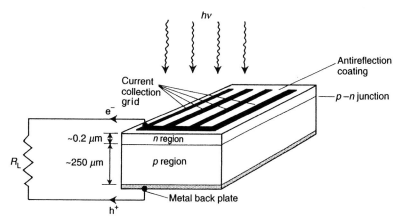

Fig. 6 Simplified structure of a silicon *p–n* homojunction cell.
On illumination, photogenerated electrons (e^-) flow out of the
n side of the junction through the external electric load R_L,
where they do electrical work as they recombine with holes (h^+)
flowing from the *p* side.

driven radio. The Solarville train is powered by a PV module in the centre of the track. PV cells produce power only when they are illuminated, so when the Sun goes behind a cloud, the train stops, which makes for an unreliable service. Appliances that must run in the dark, like the train driver's lantern, therefore have storage batteries than can be recharged by exposure of their PV modules to the Sun.

Modern silicon cells are mostly made of the semiconductor silicon, but cells based on selenium go back to the 1870s. Early photographic exposure meters contained selenium cells, as did the 1960s vintage Agilux camera, which did not require a battery. Other early devices contained copper oxide cells. But the efficiency of these devices was well below one per cent. Then in 1946, Russell Ohl, a metallurgist at the Bell Telephone Laboratory in New Jersey, made a very interesting observation:

> I had made up a considerable number of melts of pure silicon for the purpose of making specific resistance measurements ... I was making such a measurement when I noticed, while viewing on an oscilloscope the wave shape of the 60-cycle current flowing through the rod, that the current in one direction was affected by light from an ordinary 40-watt desk lamp.

In crystallizing the silicon, Ohl had unintentionally made a *p–n* junction, because unsuspected impurities in the melt had stratified in different zones in the solid. Many great scientific discoveries have been made by accident, of course, but only by those who recognize the significance of the surprising, as did Ohl:

> This entirely unexpected phenomenon was recognized as of possibly great importance in the art of light-sensitive devices and further study was undertaken forthwith. The outcome of such study is that improved light-sensitive devices and particularly photovoltaic cells of high sensitivity and great stability have been made available.'

Modern crystalline silicon cells are made from boules of single crystal silicon or cast ingots of multicrystalline silicon. These are sawn into wafers and processed into cells, but the wafers are relatively thick and mechanically fragile. A different form of silicon—amorphous or disordered silicon hydride—can be laid down from the gas phase as a thin film, so thin that amorphous silicon cells can be formed on flexible substrates such as stainless steel. These cells are less efficient than crystalline cells, being currently around the eight to 10 per cent mark, but they are also less expensive.

Other semiconductors are occasionally used. Gallium arsenide (GaAs) cells are very efficient and perform well under high illumination; there may be a place in the future space market for GaAs and also for its sister material, InP, although the present market is dominated by silicon. The original Bell silicon cell was developed for use in space, and the first of many satellites to use PV power was Vanguard I, launched in 1958. And in the Science Museum can be seen the flight spare of Prospero, a historic satellite, launched on a Black Arrow rocket from Woomera in 1971 in the only all-British space launch ever. It was a technology satellite, flown mainly to test its solar cells, of which it had a large number. Two years ago, the last time it was monitored, Prospero's PV cells were still powering its transmission system, and they may be talking yet, but no one is listening any more. One a happier note, when the Hubble Space Telescope's faulty optics were recently corrected, its original PV sails were replaced with a double roll-out silicon array, manufactured at British Aerospace in Bristol.

In the 1970s, with the rise in the price of oil, PV modules were first produced specifically for terrestrial applications. Since then, commercial module efficiency has improved to about 18 per cent, and costs have fallen about sixfold. The PV market is growing, with about 60 MWp produced last year, and about 380 MWp cumulatively. PV power is used in a wide range of places and applications. Consumer goods like watches and calculators constitute about one-third of the total market. At the other end of the scale, there are some very large arrays that feed their power to the grid, mostly constructed for demonstration purposes in the early 1980s. PV is often the preferred choice for off-grid applications, such as remote telecommunications stations, navigation lights, and weather stations. PV also brings great benefits to the developing world: electrical power for lights and radio, refrigeration and air conditioning, water pumping, and village electrification.

Closer to home, there are an increasing number of houses, both stand-alone and grid-connected, with PV cells on their Sun-facing façades; in Germany, the government is committed to installing 2000 such roofs. In the past couple of years, PV cells have started to spread to larger buildings in Europe. Such building-integrated PV arrays typically generate some tens of DC kilowatts peak, which is inverted to AC and either used within the building itself or sold to the local grid. The UK will soon get its first PV-clad building, thanks to the tireless efforts of Professor Bob Hill at the University of Northumbria, where a 1960s Computer Studies building will be getting a PV-clad face lift in summer 1994.

Photoelectrochemical solar energy conversion

Now let us turn to a range of direct conversion devices known as photoelectrochemical converters. These are basically electrical batteries that work in the light but not in the dark. They are much less highly developed than PV cells, but they go back further into the nineteenth century, to the French scientist Edmond Becquerel, who in 1839 communicated to the Académie des Sciences his discovery that, if a platinum electrode coated with silver halide was illuminated, a substantial photocurrent was produced. Becquerel experimented with filters of various colours, and concluded that the photocurrent was a measure of the number of chemical active rays of light, throwing in the incautious *obiter dictum* that his discovery rendered obsolete the earlier method of M. Biot of estimating the chemical effect of light of various colours from the tint it produced in a sheet of silver chloride-coated paper.

This drew a magisterial response from the M. Biot in question, one Jean Baptiste Biot, Professor of Physics in the Collège de France, no less, and an authority on the dispersion and polarization of light: '*Tout effet résultant d'une cause physique ne lui est pas pour cela proportionnel.*' ('The magnitude of a physical effect does not necessarily provide a linear measure of its cause.') '*Il est difficile de répondre,*' admitted Becquerel, but respond he did, reiterating his view that the current depended on the number of light rays, not their colour. Biot was back in print the following week. '*Je croirais inutile de répéter mes objections ici.*' Useless or not, he then proceeded to repeat his objections: '*The Academy will readily see that what follows has a purely scientific purpose. I have nothing against a young man whose zeal and inventive mind I appreciate.*' Biot was well away, citing the colour sensitivity of the new Daguerre photographic process in support of his view that the photocurrent was not a question of the number of rays, but of their colour. The controversy rumbled on for years, hardly capable of resolution until the quantum theory of light and electricity provided a natural link between the one photon absorbed by the electrode and the one excited electron it causes to move in the circuit. In a sense, Becquerel and Biot were both right: only those colours that correspond to the semiconductor bandgap energy or above can produce the hole–electron pair and the photocurrent, but above that energy, it is a question of the number of photons, not their colour.

The photocurrent flows because the junction between the illuminated electrode and the electrolyte behaves like a PV junction, bending the energy bands and separating light-generated hole–electron pairs. The Becquerel effect is the basis of a promising photoelectrochemical device, currently being developed by Professor Michael Grätzel at the Ecole Polytechnique Fédérale in Lausanne, using the semiconductor titanium dioxide (Fig. 7). Titanium dioxide is far from exotic—it is the commonest white pigment in paints and paper. It is used in photoelectrochemistry because of its high chemical stability, but the fact that it is white tells us that, un-

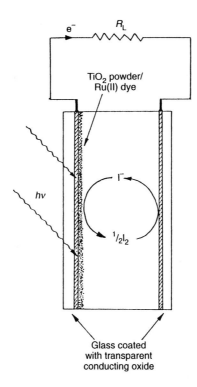

Fig. 7 The Grätzel electrochemical photovoltaic cell. The anode consists of a high surface area titanium dioxide (TiO_2) paste fired onto an optically transparent base and coated with an orange ruthenium dye. The solution contains the iodine/iodide redox couple dissolved in a non-aqueous solvent such as acetonitrile.

fortunately, it does not absorb visible light. So Grätzel made a coloured electrode out of titanium dioxide by firing a paste of it onto conducting glass, and dying the film with a ruthenium derivative. The iodine/iodide couple in solution is oxidized at the dyed electrode, and reduced at the other electrode, so the cell is an electrochemical PV cell, producing electric power rather than a chemical fuel. Grätzel's cells are over 10 per cent efficient in artificial light, better than amorphous silicon and only about one-fifth of the cost, and they are currently under commercial development in Germany.

Solar hydrogen generation

Making electrical power is fine, but what if we could use sunlight to make hydrogen electrochemically? When water is electrolysed in the dark by applying a voltage between two metal electrodes, it is decomposed into its elements, hydrogen and oxygen, and this gaseous mixture stores a good deal of energy. Michael Faraday, who established the laws of electrochemistry at the Royal Institution, liked to demonstrate this by trapping a little of the gas mixture in a soap bubble in his hand. If we put a light to the bubble, we can readily appreciate the potential of hydrogen as a fuel, as did another Frenchman, Jules Verne, over a hundred years ago, in this remarkable passage from his novel *L'Ile Mystérieuse*:

> I believe that water will one day be used as a fuel, because the hydrogen and oxygen which constitute it, used separately or together, will furnish an inexhaustible source of heat and light. I therefore believe that, when coal deposits are exhausted, we will heat ourselves by means of water. Water is the coal of the future![5]

The future moved a step closed in 1972, by the efforts of two Japanese scientists—Akiro Fujishima and his supervisor, Kenichi Honda—working together at the University of Tokyo. They made a crystal of titanium dioxide into an electrode, immersed it in acid and connected it to a platinum counter electrode in alkali (Fig. 8). They then illuminated it with ultraviolet light, and observed that the water was *photoelectrolysed*, that is, oxygen was produced at the titanium dioxide electrode and hydrogen at the platinum electrode, but no external power source was required, only the driving force of the absorbed light.

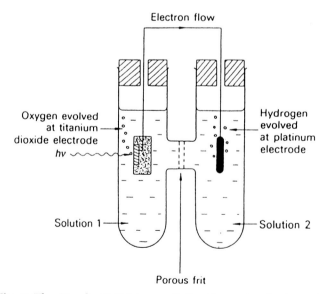

Fig. 8 The Honda–Fujishima cell for the assisted photoelectrolysis of water.

Hello sunshine   **115**

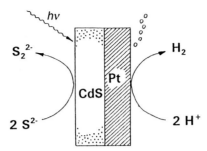

Fig. 9 Nozik's hydrogen-evolving chemical diode.

Unfortunately, there are two problems with the Honda–Fujishima cell: first, as pointed out above, titanium dioxide does not absorb visible light, but only ultraviolet, of which there is little in sunlight, at least while the ozone layer stays in place; and second, the pH difference between the two solutions effectively reduces the energy storage efficiency to well below one per cent, too low to be useful. Fujishima and Honda's announcement therefore started a race to find a semiconductor with better properties for water splitting. There were glorious noonday breakthroughs, shortly followed by false dawns. Alas, semiconductors which absorb visible light efficiently tend to be chemically unstable, which is why Grätzel used a sensitizing dye. So far, a stable semiconductor that splits water efficiently using visible light has not been discovered, and that has led scientists to look at doing each half of the process separately, in the hope of combining them later. Figure 9, for example, shows the hydrogen-producing system devised by Arthur Nozik at the Solar Energy Research Institute in Colorado. It is very simple, consisting of the semiconductor cadmium sulphide, which does absorb visible light, with a platinum counter electrode bonded to its back. There is no need for any external wires for a device that is to be run at short circuit. When this diode is immersed in a solution containing sulphide and illuminated, hydrogen is evolved on the platinum and sulphide is oxidized to sulphur on the cadmium sulphide. Unfortunately, putting such a system together with an oxygen-producing system to make a complete water splitting system in which all other products (such as the sulphur in the Nozik cell) are recycled has not yet been achieved after 20 years of effort.

Solar energy research at the Royal Institution

Twenty years ago, I was lucky to be in the right place at the right time. I had just moved from Oxford to London in order to work with Sir George

116   *Mary Archer*

Porter (as he then was) at the Royal Institution, switching my field to photoelectrochemistry, in the hope of making myself vaguely relevant to the thrust of his photophysical research. It was here at the Royal Institution that I was instrumental in forming the UK Section of the International Solar Energy Society, and here that the inaugural meeting of the Society was held, 20 years ago.

Chemists are supposed to be good at synthesizing molecules, but some of us are better at destroying them, and indeed, thanks to Superfund, there can be more money in cleaning up pollution than in making a fine chemical. So let us suppose that Solarville's water supply has been contaminated; we can clean it up using the Solaqua process recently developed by Professor Jim Bolton of the Toronto-based company Solarchem. To demonstrate this process a small amount of an environmentally friendly iron complex is added to a portion of the contaminated water, which is then exposed to a light source, for example the Sun. The sunlight generates hydroxyl radicals from the reagent, and these oxidize a wide range of dangerous organic substances such as traces of dry cleaning fluid or dioxins in the polluted water, to harmless carbon dioxide and water. The by-products of the Solaqua reagents are also completely harmless. The treated water gradually turns colourless as it is cleaned up by the Sun.

This example illustrates one of the main driving forces behind the current renaissance of interest in renewable energy—concern about the environment, and in particular concern about the impact of increasing atmospheric levels of carbon dioxide on the climate. There is no doubt that in resource and technology terms, the renewable forms of energy can potentially have a serious impact on this problem, but the question is will they?

In summary, an old Irish proverb seems appropriate: '*Ní hé là na gaoithe lá na scolb*', or '*Don't put your trust in the thatcher on a windy day.*' Naturally I intend no major political pun. The moral is, is it not, that we should get our roofs safely thatched before the wind blows too unkindly. And I hope that I have convinced you that, when the oil wells finally do run dry, the Sun will rise to the challenge.

Notes and References

1. Chatwin, B. (1993) *The morality of things.* Typographeum, 1993, p. 20.
2. This and other quotations were read at the Discourse by the distinguished National Theatre player, Martin Jarvis.
3. Ciamician, G. (1912). *Proceedings of the Eighth International Congress of Applied Chemistry*, Washington and New York, 4–13 September, 1912.

Hello sunshine   **117**

4. US. Patent No. 2,443,542, 27 May 1941, 'Light-Sensitive Electric Device including Silicon', Russell S. Ohl, assignor to Bell Telephone Laboratories, Incorporated.
5. Verne, J. (1874). *L'Ile mystérieuse.*

MARY ARCHER

Born 1944, educated at Cheltenham Ladies' College, St. Anne's College, Oxford and Imperial College, London. Lecturer in Physical Chemistry at Somerville College, Oxford from 1971 to 1972. Research Fellow in the Davy Faraday Laboratory of the Royal Institution from 1972 to 1976 and Fellow and College Lecturer in Chemistry at Newnham College, Cambridge from 1976 to 1986. Has been a Senior Academic Fellow at de Montfort University since 1990, a Visiting Professor in the Department of Biochemistry at Imperial College since 1991 and a Trustee of the Science Museum since 1990. Became a Fellow of the Royal Society of Chemistry in 1987.

Contribution from

ANTHONY HARRIMAN

University of Newcastle Upon Tyne
UK

Anthony Harriman spent 14 years at the Royal Institution of Great Britain, where he was Dewar Research Fellow and Assistant Director of the Davy-Faraday Research Laboratory. He worked in collaboration with Professor Lord George Porter, FRS and Sir John M. Thomas, FRS during their terms as Director of the Royal Institution. His main research theme during this period related to artificial photosynthesis. He moved to the University of Texas at Austin in 1988 to become Director of the Center for Fast Kinetics Research. In Austin, he constructed numerous advanced spectroscopic instruments for monitoring short-lived intermediates and applied those facilities to the study of intramolecular energy and charge transfer. Time-resolved instrumentation was built in-house that allowed recording of kinetic events on timescales ranging from 200 fs to several hours. In 1995, Prof. Harriman moved to the Université Louis Pasteur in Strasbourg, France to become a professor of chemistry at the Ecole Européenne de Chimie, Polymères et Matériaux (ECPM). At the ECPM, he continued to develop instrumental methods for the study of fast energy relaxation in multi-component molecular systems. He moved to the Department of Chemistry at the University of Newcastle in October 1999 and was appointed Head of Department in April 2000. His main research interest still involves the design and study of systems intended to mimic natural photosynthesis but this has been extended to other aspects of biophysics, especially electronic interactions in DNA. His work is moving progressively towards the emerging field of molecular opto-electronics. He has published more than 300 research articles.

Leaving school at 15 to begin a career as a professional footballer is not the traditional start to life for a research scientist but at the time he gave little thought to anything but kicking a ball. A badly broken leg, to say nothing of a dire shortage of the required skills, necessitated a suitable fallback. Thus, the boots were replaced with textbooks and all ideas of becoming famous or rich vanished overnight. Everything went fine until, as a third-year PhD student, he attended a one-day meeting on non-radiative processes held at the Royal Institution in London. This was one of the rare occasions that he had left the West Midlands and he was immediately enthralled by the RI. He understood nothing of the meeting but, on the train back to Wolverhampton, wrote to the Director asking if he could go there as a postdoctoral researcher. Professor Porter was kind enough to reply, saying that if he could get a scholarship from the Science Research Council Porter would consider his application. He arrived at the RI six months later.

A condition of his scholarship was that he worked on the photochemistry of copper complexes. Instead, he was told to devise an artificial photosystem able to reduce carbon dioxide. The closest he came to studying copper complexes came many years later when they tried to use manganese porphyrins as catalysts for the oxidation of water to molecular oxygen. The two-year fellowship was extended and amended so that he stayed a total of 14 years, acquiring several titles, prizes and awards along the way. The research diversified to include all aspects of artificial photosynthesis and resulted in the publication of some 140 articles — 45 with Porter as co-author. Many important collaborations were started during this time and they came as close as anyone to solving the fundamental problems associated with solar energy storage. The work came to a natural end with Porter's appointment as President of the Royal Society and, although Harriman stayed a while longer, the RI was not the same place. He left early in 1988 to become Director of the Center for Fast Kinetics Research at the University of Texas at Austin.

The "artificial photosynthesis" group at the RI was noisy, poorly-dressed but highly active. A great amount of research was achieved at modest cost and in a friendly atmosphere by a group of talented PhD students. Synthesis was extremely difficult to carry out — they had to go to King's College to run NMR spectra — but essential for their work. Porter was an active member of the group and spent a considerable time working with individual students and postdocs. Some important concepts, like intramolecular energy and electron transfer, were established and protocols for water splitting were demonstrated. These were enjoyable times, freed from the demands of raising funds, where the only real problem was how to get around the "no-working-at-weekends" rule. Some time was also spent in the RI bar!

There was lots of research funding available in Austin and Harriman was able to set up a world-class centre for transient spectroscopy. A range of sophisticated instruments was quickly established and applied to problems in the life sciences and in chemistry. New collaborations were started, most notably with Jon Sessler and Jean-Pierre Sauvage, and different aspects of artificial photosynthesis were attacked. They were among the first to study electron tunneling in DNA and by 1994 were already thinking in terms of a molecular computer. This was the time when supramolecular chemistry was starting to dominate the field and they made some notable contributions to this new science. A move to Strasbourg in 1995 was instrumental in pushing their research closer to the subject of molecular photonics. Moving yet again, this time to Newcastle in 2000, showed a restless spirit — or perhaps an inability to settle — and did nothing to harmonise research. Then, together with Andy Benniston, they were working full-time on the development of molecular photonic devices and, after a difficult start, research was beginning to bloom.

To his deep regret, Harriman largely lost contact with Porter after his departure from the RI, except for when he needed a reference. Porter was always remarkably kind to him, tolerating a personality and general behaviour others struggle to comprehend. They had numerous arguments, which Porter always won, and in amongst the scoldings many serious scientific discussions. Porter's honesty, integrity and sheer belief in the value of science in general, but chemistry in particular, remain with Harriman forever. He remembers Porter as a great teacher and motivator, who sent him to many places around the globe to give a lecture in his place.

Tony Harriman and Phyllis Hannaford, Royal Institution 1985.

Anthony Harriman on
"Photoredox Processes in Metalloporphyrin-crown Ether Systems"
G. Blondeel, A. Harriman, G. Porter and A. Wilowska
J. Chem. Soc., Faraday Trans. 2 **80**, 967–876 (1984)

During the late 1970s it became clear that the storage of solar energy in the form of chemical potential was not going to be as easy as thought at first and that tailor-made compounds would be needed to effect efficient charge separation. We had studied a large number of bimolecular photoredox reactions whereby an excited state enters into an electron-transfer reaction upon diffusive encounter with a suitable ground-state molecule. Many important parameters had been exposed, particularly how electrostatic factors influence the yield of charge separation,[1] and the charge-separated products had been used to generate both hydrogen[2] and oxygen[3] from water in the presence of suitable catalysts. Even so, our work had reached an impasse. Only by consuming an organic compound as a sacrificial reagent was it possible to achieve a reasonable lifetime for the all-important charge-separated state.[4] In all cases, charge recombination was too fast to allow competing chemistry to take place, even with the new catalysts that had been developed. Photogalvanic cells, so popular in the early 1970s, were doomed by the same problem. A solution to the barrier imposed by rapid charge recombination was needed urgently if our research into artificial photosynthesis was to continue.

Sir George Porter offered advice on the issue at hand! Natural photosynthesis worked by setting up a cascade of electron-transfer events so as to separate the charges in both time and space. This was known from many elegant biophysical experiments and was later confirmed when the X-ray crystal structure of the reaction centre complex became available. The solution to our problem, he claimed, was to link the reactants in such a way that charge separation would be much faster than charge recombination. Marcus theory, at that time, was not popular since the onset of the so-called "inverted region" had defied experimental verification. By the late 1980s, several independent studies had confirmed that the rate of charge recombination decreased a very high thermodynamic driving force. We looked for a different way to manipulate the dynamics of electron transfer and, for the first time at the Royal Institution, we seriously questioned the mechanism of light-induced electron transfer.

A one-day seminar was imminent and, as the organiser of these regular events, I invited Jan Verhoeven from Amsterdam to talk about his work on "through-bond" electron transfer.[5] Sir George was not impressed. Firstly, there was the high cost. Secondly, everyone knew that electron transfer required orbital contact and could not take place outside of a critical separation close to the sum of the van der Waals radii. Apparently, no-one had told Verhoeven. The seminar started badly, with the flight from Amsterdam re-routed to Manchester because of fog at

Heathrow. Verhoeven arrived just in time to give his talk and depart for a now fog-free Heathrow. His talk changed our way of thinking and led to the design of photosystems having the reactants linked by an organic tether. Our work on bimolecular systems effectively ceased and we began a new research programme aimed at directing electrons along covalent bonds. We had, of course, looked at intramolecular charge transfer in the past but not with a view to achieving efficient charge separation. Next morning, Sir George demanded action and suggested that we synthesise a hybrid molecule able to realise light-induced charge separation by way of through-bond electron transfer. To prevent rapid charge recombination, it was only necessary for the charge-separated state to dissociate into ions and thereby eliminate the possibility for through-bond electron transfer. The result would be fast charge separation and slow charge recombination. A final requisite was that the chromophore had to be chlorophyll.

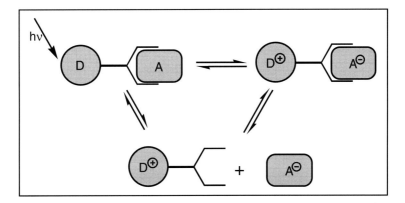

Scheme. The electron acceptor (A) is held by non-covalent interactions at a site close to the electron donor (D). On illumination into D, through-bond electron transfer takes place to form the appropriate ions. The anion is now ejected from its site, such that charge recombination requires diffusion through the solution.

Still somewhat fazed from taking the seminar speakers to the local pub the night before, I agreed to do the synthesis. The third floor of the RI was converted into a rather messy synthesis laboratory, with little regard for personal safety. We made the target molecule and it behaved very much as expected.[6] Later systems synthesised by Jean-Marie Lehn's group perfected this approach towards transient energy storage.[7] Other systems were synthesised and photoredox systems were given a new breath of life. The Marcus inverted region was soon demonstrated, Gust and Moore developed the molecular triad approach and long-range charge separation became a possibility.

A number of important features can be recognised in the porphyrin-crown ether system[6] studied at the RI. (i) The use of non-covalent bonding to localise the electron acceptor predates the rise of supramolecular photochemistry. (ii) The modulation of fluorescence upon cation binding has developed into a versatile form of analytical science known as chemical sensing. (iii) The design of molecules for specific photoresponsive functions has now reached an advanced stage but was essentially unknown at that time. (iv) The need to synthesise tailor-made molecules is now obvious. (v) The ability to displace a group, molecule or ion has developed into a subject known as phototropicity and has genuine potential for signal transduction. (vi) The demand to use chlorophyll-like chromophores, as insisted by Porter, has remained to this day.

References

1. M.-C. Richoux and A. Harriman, *J. Chem. Soc., Faraday Trans. 2* **78**, 1873 (1982).
2. A. Harriman, G. Porter and M.-C. Richoux, *J. Chem. Soc., Faraday Trans. 2* **77**, 1939 (1981).
3. A. Harriman, G. Porter and P. Walters, *J. Chem. Soc., Faraday Trans. 2* **77**, 1981 (1981).
4. J. Handman, A. Harriman and G. Porter, *Nature* **307**, 534 (1983).
5. P. Pasman, N. W. Koper and J. W. Verhoeven, *Rec. J. R. Neth. Chem. Soc.* **101**, 363 (1982).
6. G. Blondeel, A. Harriman, G. Porter and A. Wilowska, *J. Chem. Soc., Faraday Trans. 2* **80**, 867 (1984).
7. M. Gubelmann, A. Harriman, J.-M. Lehn and J. L. Sessler, *J. Phys. Chem.* **94**, 308 (1990).

Contributed Article
ARTIFICIAL PHOTOSYNTHESIS: STATIC AND DYNAMIC ELECTRON-TRANSFER PROCESSES BETWEEN A PORPHYRIN AND BENZO-1,4-QUINONE

Anthony Harriman

ABSTRACT: A key component of bacterial photosynthesis involves the light-induced transfer of an electron from a bacteriochlorophyll molecule to a remote quinone. This process provides the essential charge separation that drives subsequent energy-storage reactions. Numerous model systems have duplicated the electron-transfer step but fast charge recombination prevents storage of energy. This basic reaction was recognised by Lord George Porter in the early 1970s as being the critical feature for the future evolution of artificial photosynthetic systems. Moreover, it was stressed that the quinone had to be free to diffuse from the reaction site so as to facilitate two-electron chemistry by way of disproportionation. In order to realise the type of chemistry envisioned by Porter, we have studied a new porphyrin-quinone model compound whereby the porphyrin is equipped with a peripheral calixarene unit. This latter module possesses a central cavity that can bind the quinone in a position suitable for electron transfer from the excited singlet state of the porphyrin. Intracavity electron transfer must compete with both charge-transfer complexation and diffusional quenching. Here, we review the electron-transfer chemistry of this supramolecular entity in terms of Porter's original simplistic model for the bacterial reaction centre complex.

Introduction

Sir George Porter presented the 1977 Bakerian Lecture of the Royal Society on the subject of "*In Vitro* Models for Photosynthesis".[1] This remains the most memorable and inspiring scientific lecture that I have attended. Not only did Porter succeed to combine the research efforts of three remarkably disparate and highly-competitive individuals — Godfrey Beddard, Graham Fleming and myself — into a single, cohesive discourse but he also defined the future of artificial photosynthesis and set out it's main goals. At that time, it was realised that water oxidation in green plants held the key for solar energy storage in artificial systems but no one had been able to devise appropriate chemistry that enabled oxygen evolution under visible light illumination. It was recognised that Nature used chlorophyll as the main chromophore, water as the source of electrons, sunlight as the activator, a quinone as electron relay and some kind of manganese complex as the active catalyst. The identity of the manganese catalyst, and its actual role, were quite unknown[2] but Calvin had proposed that a manganese porphyrin might be involved in the water-oxidation cycle. In commencing his Bakerian lecture, Porter indicated that the natural water-oxidation process, known generally as PS-2, could be considered as a simple reaction except for the fact that liberation of oxygen from water required the successive transfer of four electrons. Photochemical processes, however, operate by way of one-electron transfers. The problem, therefore, was to devise a way to store four electrons on a single chemical entity — namely, the manganese complex.[1]

$$2\ H_2O + 2\ Q \xrightarrow[\substack{\text{manganese} \\ \text{chlorophyll}}]{h\nu} 2\ QH_2 + O_2$$

Early experiments carried out at the Royal Institution had shown that chlorophyll readily transferred an electron to benzo-1,4-quinone under visible light illumination in ethanol solution.[3,4] The excited singlet state of chlorophyll was quenched by the quinone at the diffusion-controlled limit but, because of spin restrictions,[5] electron-transfer products were not detectable. In contrast, the excited triplet state of chlorophyll donated an electron to the quinone and generated the solvated radical ions in high yield.[6] These ions recombined via a diffusional process. It was also realised that mono- and binuclear manganese complexes quenched the excited singlet and triplet states of chlorophyll due to the so-called paramagnetic effect.[7] We were unable, at the time, to combine these two systems into a single photoprocess and no amount of effort could overcome the fundamental problems that, unlike natural photosynthesis, (i) the excited singlet state of chlorophyll did not give charge-separated products, and (ii) the triplet radical ion pair was too short-lived to be useful in subsequent chemical reactions. During his Bakerian lecture, Porter introduced the key concept of combining the components so as to minimise the number of reactants needed to effect energy storage.

We had already embarked on a comprehensive programme aimed at exploring the redox chemistry of water-soluble manganese porphyrins.[8-12] Over the course of many years, we synthesised and studied manganese porphyrins in oxidation states of +2, +3, +4 and +5 and examined their interconversion.[13-19] This work involved the use of stopped-flow, pulse radiolysis, gamma radiolysis and electrochemistry — all new techniques to us that had to be learned and manipulated. New and powerful redox catalysts were established and protocols were put in place that permitted the transient storage of up to four electrons on a single manganese complex. The stability of the various oxidation states showed a marked dependence on pH but we were unable to detect molecular oxygen as a reaction product. Undoubtedly, given the tremendous effort that went into this programme, this result[19] is the most disappointing aspect of the time I spent at the Royal Institution. Following the work of Jean-Marie Lehn and Michael Gratzel, we did succeed to generate highly-effective oxidation catalysts[20-24] but these were micro-heterogeneous materials, not molecular entities. As far as I am aware, no one has produced a simple homogeneous species able to liberate oxygen from water in a catalytic process.

The idea behind using a manganese porphyrin in this work was to use the porphyrin nucleus as the chromophore and the metal centre as the catalyst. Unfortunately, the excited state lifetimes of the manganese porphyrins were much too short to be viable components in photochemical systems. Following from Porter's concept of combining reagents, however, a deliberate attempt was made to attach together the main reactants by covalent bonding. Zinc tetraphenylporphyrin took the place of chlorophyll and many different cofactors were attached.[25-28] Perhaps the most interesting molecular dyad so formed had a manganese porphyrin covalently linked to the zinc porphyrin chromophore.[29] Here, light-induced electron transfer was observed from the excited singlet state of the donor to the appended manganese complex. Iron porphyrins, these being models for the heme proteins known to be coupled to the natural photosynthetic reaction centre complex, reacted likewise.

A more far-reaching consequence of Porter's Bakerian lecture was the development of molecular dyads having a quinone tethered directly to the chlorophyll-like chromophore. Our first such porphyrin-quinone dyad[30] was synthesised in 1981 (Fig. 1) but we lacked the time-resolution to properly study intramolecular electron transfer in this system. Although this was one of the first porphyrin-quinone systems to be reported we were unable to compete with other groups which quickly extended the field. By now, more than 300 different porphyrin-quinone dyads have been described in the literature. The problem of achieving rapid charge separation from the excited singlet state of the porphyrin was quickly overcome with the new dyads and, as Porter predicted,[1] they have become the established reference compounds for the early steps in bacterial photosynthesis. Nowadays, the acceptor of choice is C_{60}, which offers several important benefits relative to benzo-1,4-quinone, but the fundamental concept remains true. Notable among these porphyrin-quinone dyads are those novel systems having the reactants held together by hydrogen bonding[31] rather than covalent interactions. These sophisticated entities more closely resemble the structures found in Nature.[32-34]

Porter recognised early on that the quinone played the role of electron relay and was not merely a passive cofactor in the electron-transport chain. This fact became forgotten in the rush to construct increasingly more elaborate models of the bacterial photosynthetic reaction centre complex. The quinone is part of a pool and is far from being static. An important aspect of quinones is their ability to undergo a two-electron reduction to form the corresponding hydroquinone. Since both species are stable, this two-electron cycle allows the coupling of one-electron photochemistry to the required energy storage. Such behaviour is a key element of the early photogalvanic cells based on thionine dyes[35] and was built into our original work on manganese porphyrins.[9] The quinone is unable to enter into two-electron cycles when incorporated into rigid molecular dyads, or even the

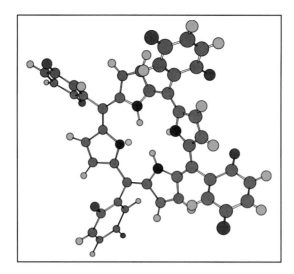

Fig. 1. Structure of the original porphyrin-quinone dyad synthesized and studied at the Royal Institution in 1981

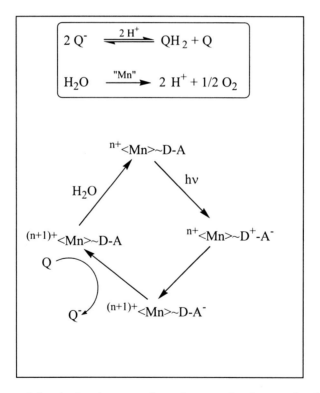

Fig. 2. Scheme proposed for the involvement of a quinone molecule as a shuttle for storing redox equivalents. The photosystem is a molecular triad containing a chromophore and donor (D), a primary acceptor (A) and a manganese-based catalyst. The catalyst is charged by the oxidised form of the donor while, after reduction, the quinone migrates from the active site.

hydrogen-bonded networks introduced by Jon Sessler.[31–34] Instead, the quinone must be part of a mobile pool. This presents an obvious problem of how to ensure that a single quinone molecule is kept close to an artificial reaction centre complex without covalent bonding. In 1977, this was an insurmountable problem and our only putative solution involved the use of aqueous/organic interfaces to separate reagents. By 1984, we were able to locate cations at peripheral sites by way of coordinative bonds.[36] This approach was the closest we could come to realising Porter's "translocation model" (Fig. 2). Localisation of the cation occurred at a crown ether bound to the porphyrin but the cavity was much too small to house a quinone.[37,38] It was only many years later with the introduction of larger macrocycles, such as calixarenes, that progress in this area could be made.

Molecular Recognition

In order to develop a photosystem having a quinone molecule loosely bound near to the active site it seems necessary to utilise a macrocyclic receptor that is itself closely associated with the chromophore. The earliest such systems studied by Porter used crown ethers[36] as the receptors but their small cavity restricted inclusion to metal cations. Later work by Stoddart et al.[39] led to the design of larger receptors that could accommodate 1,4-disubstituted benzenes in their central void (Fig. 3). These receptors are highly electron affinic and form coloured charge-transfer complexes with suitable donors. Although it proved possible to attach secondary electron donors to the exterior of the receptor,[40] thereby enabling migration of the charge from the central complex, this system is not suitable for binding quinones. By attaching external chromophores, however, it was possible to create some interesting multi-component arrays having valuable photo-properties.[41–44]

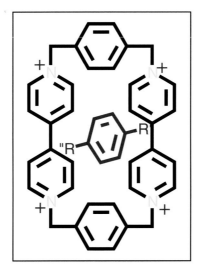

Fig. 3. The electron-affinic receptor used to house 1,4-disubstituted benzenes.

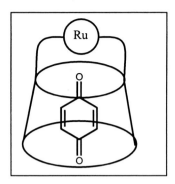

Fig. 4. Structure of the first ruthenium(II) *tris*(2,2'-bipyridine) capped cyclodextrin and its inclusion complex with benzo-1,4-quinone.

A more relevant photosynthetic model was synthesised by Dominique Armspach, working in the group of Dominique Matt in Strasbourg. This system[45] had a luminescent ruthenium(II) *tris*(2,2'-bipyridyl) complex strapped across the narrow rim of an α-cyclodextrin (Fig. 4). The cyclodextrin was able to bind benzo-1,4-quinone inside its central cavity. The encapsulated quinone quenched the excited triplet state of the metal complex by way of electron transfer, although non-bound quinone also acted as a diffusional quencher for the excited state. The association constant was only ca. 10 M^{-1} but intracavity electron transfer occurred with a first-order rate constant of 3×10^7 s^{-1}. The host separates the primary reactants by about 7 Å, which suggests that the α-cyclodextrin framework imposes a barrier to electron transfer of around 1.8 Å^{-1} (i.e., similar to a protein). In retrospect, it is clear that the cavity is too small to comfortably accommodate the quinone — hence the small association constant. A similar approach could be used to attach an electron donor across the mouth of a larger macrocyclic receptor but this represents a formidable synthetic challenge.

A New Photoactive Receptor for Binding Benzo-1,4-quinone

Calixarenes are bucket-shaped molecules having a spacious central cavity and flexible walls. Such molecules will readily encapsulate small substrates, such as benzo-1,4-quinone (Fig. 5), from solution. The cavity is relatively hydrophobic but dynamic and easily adapted to fit around the docked substrate. There have been few attempts to include calixarenes into photoactive molecular dyads, although strategies have now been developed for their functionalisation. Indeed, Raymond Ziessel has designed calixarene-based luminescent sensors that discriminate substrates on the basis of their electronic charge.[46]

It proved possible to attach a single calixarene residue to a tetraphenylporphyrin derivative by functionalising one of the *meso* positions — a strategy used extensively in our earlier work to prepare *bis*-porphyrins. This allowed a small

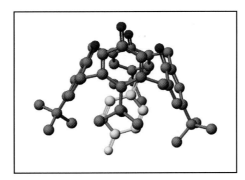

Fig. 5. Computer-generated energy-minimised structure for the inclusion complex formed between benzo-1,4-quinone and a calixarene receptor.

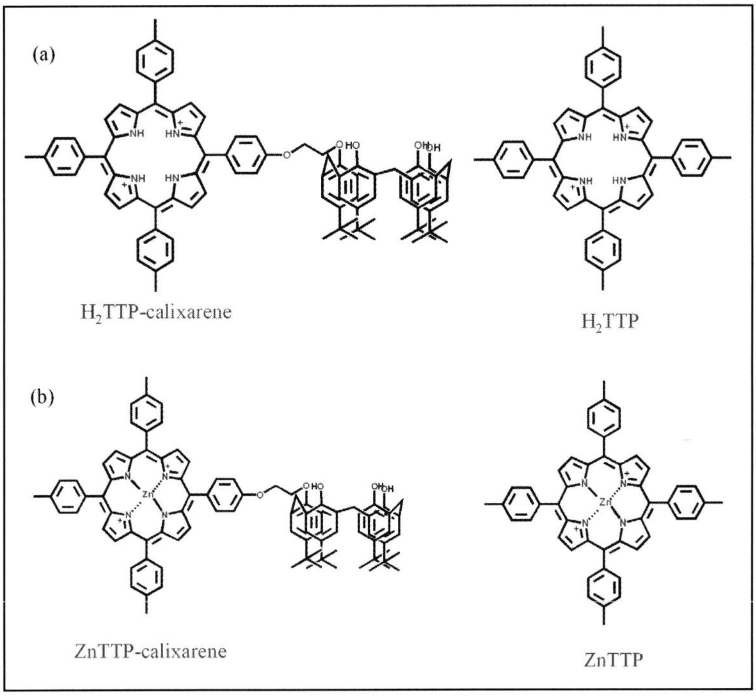

(a)

H$_2$TTP-calixarene   H$_2$TTP

(b)

ZnTTP-calixarene   ZnTTP

Fig. 6. Structures of new molecular dyads comprising a photoactive porphyrin and a calixarene-based receptor. The calixarene can be attached at the *ortho*, *meta* or *para* positions of the bridging phenyl ring.

series of molecular dyads to be isolated whereby the porphyrin and calixarene are held at different mutual separations (Fig. 6). It is a simple matter to insert different cations into the porphyrin nucleus so as to vary the photophysical properties. These new dyads were synthesised by Bhaskar Maiya.

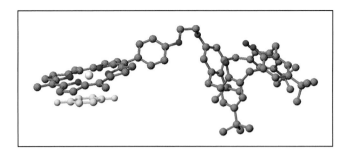

Fig. 7. Face-to-face π-complex formed between the porphyrin and an added quinone.

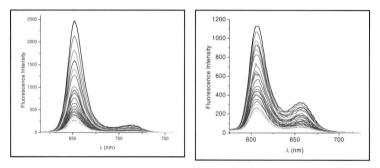

Fig. 8. Effect of added quinone on the fluorescence spectra recorded for H_2TTP (lhs) and the *ortho*-substituted dyad (rhs).

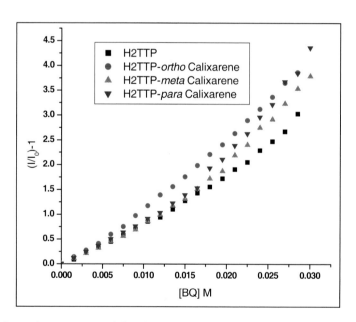

Fig. 9. Stern–Volmer plots constructed for the stepwise addition of benzo-1,4-quinone to the various porphyrin-derived systems

The appended calixarene has essentially no effect on the photophysical properties of the porphyrin but helps to solubilise the compound. In tetrahydrofuran at room temperature, for example, the porphyrin subunit fluoresces strongly and forms the corresponding triplet excited state by way of intersystem crossing. The addition of benzo-1,4-quinone quenches fluorescence from the various dyads but there is also extensive fluorescence quenching for the reference compounds, H$_2$TTP and ZnTTP, under the same conditions. Following from our earlier work,[3,4] fluorescence quenching for the reference compounds occurs by way of diffusional contact and via formation of a π-complex in which the reactants are closely stacked in a face-to-face orientation (Fig. 7). This behaviour gives rise to nonlinear Stern-Volmer plots (Figs. 8 and 9). It is clear that similar quenching takes place for the new dyads.

Time-resolved fluorescence studies showed that benzo-1,4-quinone quenches the fluorescence from H$_2$TTP at the diffusion-controlled rate limit, due to electron transfer. In addition, the steady-state measurements show the importance that π-stacking makes to the overall fluorescence quenching process. The latter is a form of static quenching and it is notable that, in agreement with our earlier work,[3–5] the π-stack does not fluoresce. The diffusional contribution can be removed from the observed fluorescence quenching data. This leaves only the static component; which can be split into π-stacking and a contribution due to long-range electron transfer to a quinone molecule housed in the calixarene cavity. Both static parts contribute towards the steady-state fluorescence results (Fig. 10) but the experimental data can be analysed in terms of two competing binding isotherms

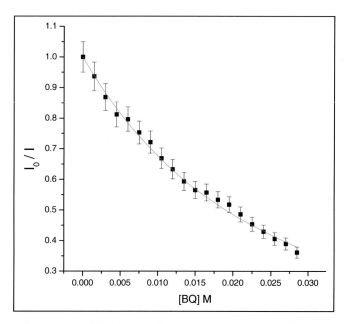

Fig. 10. Steady-state fluorescence behaviour as fit to two competing binding events. The line drawn through the data points is a nonlinear, least-squares fit to the values given in the text.

to obtain the relevant association constants. Thus, fitting the data observed for the *ortho*-substituted ZnTTP dyad results in stability constants for intracavity binding and π-stacking, respectively, of 45 and 15 M^{-1}. This analysis also indicates that intracavity binding is responsible for a 95% decrease in the fluorescence intensity. Time-resolved fluorescence studies indicate that the rate constant for electron transfer to the calixarene-bound quinone molecule is 2×10^9 s^{-1}.

Conclusion

These new molecular dyads display three distinct types of fluorescence quenching: namely, diffusional contact between the partners, complexation by overlap of π-orbitals and electron transfer from the singlet excited state of the porphyrin to a quinone molecule included in the calixarene cavity (Fig. 11). The former two quenching processes were known in the 1970s but intracavity electron transfer is specific to those receptors bearing a suitable binding site. It is notable that neither diffusional quenching nor π-complexation give rise to separated radical ions. The rate of intracavity electron transfer depends on the mutual orientation of the porphyrin and bound quinone and is highest for the *ortho*-substituted derivatives. Upon reduction, the semiquinone is ejected from the cavity and charge recombination occurs by way of diffusive migration. This system is remarkably close to the system outlined by Porter[1] a quarter of a century ago. It is also interesting to note that one of our earlier reports[47] showed how to block diffusional contact between porphyrins. A similar strategy applied to the dyads might eliminate the diffusive quenching step and produce a viable model for achieving long-lived, charge-separated products.

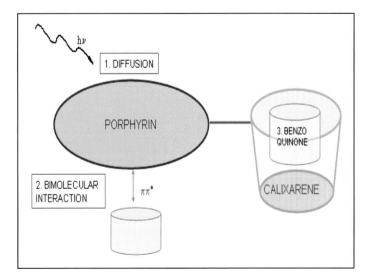

Fig. 11. Scheme showing the various quenching processes open to the dyad. Each of the three modes involves electron transfer to a nearby quinone

Acknowledgment

This work is dedicated to the memory of Professor Lord George Porter. I remain eternally grateful for having had the privilege to work for so many happy years with Professor Porter at the Royal Institution. I am also grateful to the EPSRC for funding and to Bashkar Maiya and Maryam Mehrabi for allowing presentation of their unpublished work.

References

1. G. Porter, *Proc. Roy. Soc.* **362**, 281 (1978).
2. A. Harriman and J. Barber, in *Topics in Photosynthesis*, ed. J. Barber (Elsevier, Amsterdam, 1978), Chap. 8.
3. A. Harriman, G. Porter and N. J. Searle, *Chem. Soc., Faraday Trans. 2* **75**, 1575 (1979).
4. A. Harriman, G. Porter and A. Wilowska, *J. Chem. Soc., Faraday Trans. 2* **79**, 807 (1983).
5. A. Harriman, G. Porter and M.-C. Richoux, *J. Chem. Soc., Faraday Trans. 2* **77**, 1175 (1981).
6. R. G. Brown, A. Harriman and L. Harris, *J. Chem. Soc., Faraday Trans. 2* **74**, 1193 (1978).
7. R. G. Brown, A. Harriman and G. Porter, *J. Chem. Soc., Faraday Trans. 2* **73**, 113 (1977).
8. A. Harriman and G. Porter, *J. Chem. Soc., Faraday Trans. 2* **75**, 1515 (1979).
9. A. Harriman and G. Porter, *J. Chem. Soc., Faraday Trans. 2* **75**, 1543 (1979).
10. I. A. Duncan, A. Harriman and G. Porter, *J. Chem. Soc., Faraday Trans. 2* **76**, 1415 (1980).
11. A. Harriman and G. Porter, *J. Chem. Soc., Faraday Trans. 2* **76**, 1429 (1980).
12. N. Carnieri and A. Harriman, *J. Photochem.* **15**, 341 (1981).
13. N. Carnieri, A. Harriman and G. Porter, *J. Chem. Soc., Dalton Trans.* 931 (1982).
14. N. Carnieri, A. Harriman, G. Porter and K. Kalyanasundaram, *J. Chem. Soc., Dalton Trans.* 1231 (1982).
15. R. Rao, M. C. R. Symons and A. Harriman, *J. Chem. Soc., Faraday Trans. 2* **78**, 3393 (1982).
16. A. Harriman, *J. Chem. Soc., Faraday Trans. 2* **80**, 141 (1984).
17. H. Ellul, A. Harriman and M.-C. Richoux, *J. Chem. Soc., Dalton Trans.* 503 (1985).
18. A. Harriman, I. A. Duncan and G. Porter, in *Photosynthetic Oxygen Evolution*, ed. H. Metzner (Academic Press, London, 1978), Chap. 7.
19. A. Harriman, P. A. Christensen, G. Porter, K. Morehouse, P. Neta and M.-C. Richoux, *J. Chem. Soc., Faraday Trans. 2* **82**, 3515 (1986).
20. I. A. Duncan, A. Harriman and G. Porter, *Anal. Chem.* **51**, 2206 (1979).
21. A. Harriman, G. Porter and P. Walters, *J. Chem. Soc., Faraday Trans. 2* **77**, 1981 (1981).
22. A. Mills, A. Harriman and G. Porter, *Anal. Chem.* **53**, 1254 (1981).
23. P. A. Christensen, A. Harriman, P. Porter and P. Neta, *J. Chem. Soc., Faraday Trans. 2* **80**, 1451 (1984).

24. P. A. Christensen, W. A. Erbs, A. Harriman, *J. Chem. Soc., Faraday Trans. 2* **81**, 575 (1985).

25. G. Blondeel, D. de Keukeleire, A. Harriman and L. R. Milgrom, *Chem. Phys. Lett.* **118**, 77 (1985).

26. R. L. Brookfield, H. Ellul and A. Harriman, *J. Chem. Soc., Faraday Trans. 2* **81**, 1837 (1985).

27. R. L. Brookfield, H. Ellul, A. Harriman and G. Porter, *J. Chem. Soc., Faraday Trans. 2* **82**, 219 (1986).

28. A. Regev, T. Galil, H. Levanon and A. Harriman, *Chem. Phys. Lett.* **131**, 140 (1986).

29. N. Mataga, H. Yao, T. Okada, Y. Kanda and A. Harriman, *Chem. Phys.* **131**, 473 (1989).

30. A. Harriman and R. J. Hosie, *J. Chem. Soc., Faraday Trans. 2* **77**, 1695 (1981).

31. J. L. Sessler, B. Wang and A. Harriman, *J. Am. Chem. Soc.* **115**, 10418 (1993).

32. A. Harriman, Y. Kubo and J. L. Sessler, *J. Am. Chem. Soc.* **114**, 388 (1992).

33. A. Harriman, D. J. Magda and J. L. Sessler, *J. Chem. Soc., Chem. Commun.* 345 (1991).

34. A. Harriman, D. J. Magda and J. L. Sessler, *J. Phys. Chem.* **95**, 1530 (1991).

35. M. I. C. Ferreira and A. Harriman, *J. Chem. Soc., Faraday Trans. 1* **73**, 1085 (1977).

36. G. Blondeel, A. Harriman, A. Porter and A. Wilowska, *J. Chem. Soc., Faraday Trans. 2* **80**, 867 (1984).

37. M. Gubelmann, A. Harriman, J.-M. Lehn and J. L. Sessler, *J. Chem. Soc., Chem. Commun.* 77 (1988).

38. M. Gubelmann, A. Harriman, J.-M. Lehn and J. L. Sessler, *J. Phys. Chem.* **94**, 308 (1990).

39. A. C. Benniston, A. Harriman, D. Philp and J. F. Stoddart, *J. Am. Chem. Soc.* **115**, 5298 (1993).

40. A. C. Benniston, S. Gardner, L. J. Farrugia and A. Harriman, *J. Chem. Soc. Res-S* 1147 (2000).

41. A. C. Benniston and A. Harriman, *Angew. Chem., Int. Ed. Engl.* **32**, 1459 (1993).

42. A. C. Benniston, A. Harriman and V. M. Lynch, *J. Am. Chem. Soc.* **117**, 5275 (1995).

43. A. C. Benniston, P. R. Mackie and A. Harriman, *Angew. Chem., Int. Ed. Engl.* **37**, 3201 (1998).

44. A. C. Benniston, A. Harriman and S. Dimitri, *Angew. Chem., Int. Ed. Engl.* **36**, 2356 (1997).

45. D. Armspach, D. Matt and A. Harriman, *Eur. J. Inorg. Chem.* 360 (2000).

46. A. Harriman, M. Hissler, P. Jost, G. Wippf and R. Ziessel, *J. Am. Chem. Soc.* **121**, 14 (1999).

47. J. Davila, A. Harriman, M.-C. Richoux, L. R. Milgrom, *J. Chem. Soc., Chem. Commun.* 525 (1987).

Contribution from

DAVID PHILLIPS

Imperial College London
UK

orn 3 December 1939. Educated South Shields Grammar-Technical
School, University of Birmingham, BSc Chemistry (1961), PhD Physical
Chemistry (1964). Fulbright Scholar, University of Texas (1964–1966);
Royal Society/Academy of Sciences of USSR Exchange Scientist (1966/1967).
Lecturer/Senior Lecturer/Reader in Chemistry, University of Southampton
(1967–1980). Wolfson Professor of Natural Philosophy (1980–1989), Acting
Director (1986), Deputy Director (1986–89) The Royal Institution of Great
Britain. Professor of Physical Chemistry (1989–present), Head of Department of
Chemistry (1992–2002, Hofmann Professor of Chemistry 2000–present, Dean
for the Faculties of Life Sciences and Physical Sciences, 2002–2005, Senior Dean,
2005–present, Imperial College London.

Council of the Royal Institution; Chairman of the London Gifted and Talented;
Chairman of RSC Education and Qualifications Board; Member of RSC Council
Fellow of Royal Society of Chemistry; Fellow of the NY Academy of Sciences;
Winner of the Royal Society of Chemistry Nyholm Medal 1994; Royal Society
Faraday Award, 1997; Sebetia-Ter Prize (Italy) 1999. OBE, June 1999 for services
to science education. Delivered 1987 Royal Institution BBC TV Christmas
Lectures jointly with J. M. Thomas. Author of children's books on science.

David Phillips on
"Knowledge Itself is Power"
G. Porter
The Dimbleby Lecture, BBC Books

"Can Science Policy be left to the Scientists"
G. Porter
Proc. Am. Philos. Soc. **136**, 521–525 (1992)

"The Most Important Task"
G. Porter
Presidential address to the Royal Society, 1990

"Arts and Sciences: The Two Cultures 40 Years On"
G. Porter
The Athenaum Lecture, 1999

These selected short publications are included to give an indication of the scientific philosophy of George Porter, and to represent the immense amount of work he did in politics as Director of the Royal Institution, as President of the British Association for the Advancement of Science, as President of the Royal Society, and in Parliament as a Peer. He was a man of great culture, and advocated with passion the fact that science was part of our culture, of equal importance in its own right as the Arts. He was a devout believer in the need to support basic science, not for the sake of providing the input for industry to exploit, though this could be a welcome consequence, but because it was an important branch of human knowledge.

In the Dimbleby Lecture, he uses as an example of basic science and its unexpected benefits the work of Michael Faraday. Faraday had devoted several years at the behest of the Royal Society to attempt to improve optical glass. He eventually gave up on the grounds that success required a manufacturer, which he was not. Within a few weeks of giving up this attempt, he had made the seminal discovery (in August 1831), of electromagnetic induction. This was "pure research", but as Porter points out, here, after a few modifications, was Battersea Power Station!

There has been, and continues to be a belief in the UK that we may be very good at basic science, but that we fail to exploit it successfully, and that somehow this is the fault of the scientists. Porter devotes much of the lecture in seeking to demolish this myth , and advances a passionate plea for support by the UK Government of fundamental science. He was fond of the aphorism, "There is no real distinction between fundamental and applied science, there is only science; applied, and that which has not yet been applied."

In the paper for the American Philosophical Society, Porter poses the question in the title, "Can Science Policy be left to the Scientists", and then answers both "Yes" and "No". The "Yes" is reasoned by the understanding that the scientist himself (or herself) is the best judge of what is timely and promising for him or her, and should thus be left to decide his or her own policy. The "No" comes when the purpose of the scientist's work is to apply knowledge in a way that will explicitly affect the lives of others; in this case, decisions are properly taken by elected representatives of that society. The scientists rightly should have input into this debate, however. His conclusion is telling:

> *"Today there are few things that science cannot do to improve the health, wealth and happiness of our fellow human beings. It is the duty of the scientist to explain to all people what these things are, and it is the duty of every person to try to understand science and the natural world so as to be able to decide what kind of a future he or she wants."*

George Porter devoted much of his life in such explanations, much in the years he spent at the Royal Institution. He used his unique simultaneous tenure of the Directorship of the Royal Institution, the Presidency of the British Association, and the Presidency of the Royal Society to bring to life COPUS, the Committee for the Public Understanding of Science, meant to be a vehicle for enhancing understanding, though necessarily, the first step was and is increasing public awareness. Porter was critical of the school system in the UK, and its selection of subjects at an early age, as set out in his Presidential Address to the Royal Society in his last year as President, 1990. In the UK, the situation has changed, such that all children now study science; sadly the decline in students seeking to study science at tertiary level has not increased as a result of these educational changes.

In the final paper selected here, Porter returns to the divide between Arts and Sciences, as noted famously by C. P. Snow some 40 years before. His argument, as before, is that the scientist and artist have much in common: they both strive for originality through imagination; each tries to make a new statement and each hopes that the statement will be in some way acceptable to others. "The fundamental difference between them is in the type of statement that is made," he expounds on the creativity in science, and ends by quoting Vanevar Bush:

> *"Science has a simple faith which transcends utility. Nearly all men of science, all men of learning for that matter, and men of simple ways too, have it in some form and in some degree. It is the faith that it is the privilege of man to learn to understand, and that this is his mission. Why does the shepherd at night ponder the stars? Not so that he can better tend his sheep. Knowledge for the sake of understanding, not merely to prevail: that is the essence of our being. None can define its limits, or set its ultimate boundaries."*

Porter finishes this piece with what makes an eloquent final testament to his philosophy and life:

"There is then one great purpose for man, and for us today, and that is to try to discover man's purpose by every means in our power. That is the ultimate aspiration of science, and not only of science, but of every branch of learning that can improve our understanding."

I have used as the accompanying paper of my own the record of an interview of me by the magazine *Spectrum*, since this allows me to make further comment on the 25 years I spent as a colleague of George Porter, whom I admired immensely.

Reprinted from *The Spectrum* **17** (2004)

special feature

perspective on

a UK transient in photochemistry

an interview with **David Phillips**

David Phillips received the Michael Faraday Prize "For his outstanding talents in the communication of scientific principles, methods and applications to young audiences through his many demonstration lectures with wit, clarity and enthusiasm on a wide variety of topics from basic science to modern laser research and for his major role in various collaborative ventures for young people."

The Royal Society

Courtesy of David Phillips

A request to name a scientist—any scientist—would leave many laymen scrambling to recall some high-profile individual from the evening news programs, or history books. Imagine asking for the name of a *photochemist*.

However, thousands of ordinary people in the United Kingdom, Australia, Malaysia, the United States, and other countries *do know* a *photochemist*. Among non-scientists, he may well be the world's most famous photoscientist, thanks to an extraordinary series of lectures and other activities devoted to popularizing science.

David Phillips, of course, certainly lacks no name recognition in the photosciences community. He is author of more than 500 papers in photochemistry and photophysics, a former student of W. Albert Noyes, Jr., and associate of George Porter at the Royal Institution.

In this interview with *The Spectrum*, Phillips discusses his research; describes hot areas in photochemistry; looks ahead to the future of photodynamic therapy (PDT); shares recollections of Noyes, Lord Porter, and other pioneers in the photochemical sciences; and by example urges other photoscientists to get involved in the critical task of educating the general public about the wonders of science.

David Phillips is professor of physical chemistry and former head of the department of chemistry at Imperial College of Science, Technology, and Medicine in London.

His research has involved the development of sulfonated phthalocyanine derivatives as photosensitizers, and their distribution in tissue using tools like confocal time-resolved fluorescence imaging.

Phillips and his colleagues have done photophysical measurements on those and other sensitizers using diffuse-reflectance flash photolysis, and time-correlated single-photon counting techniques.

Additional research involves use of tfs time-resolved resonance Raman spectroscopy to study the structures of intermediates in photon-induced electron-transfer reactions.

Phillips is also a noted communicator of science to non-scientific audiences. Thousands of students and adults have flocked to his lecture-demonstrations, and watched broadcasts of his presentations on the BBC and other outlets. One is a smash hit, *A Little Light Relief*, which involves the use of light and lasers in medicine. Those efforts added the Order of the British Empire and the Royal Society of London's Michael Faraday Award to Phillips's many other career honors. Among them are the Royal Society of Chemistry's Nyholm Medal for Education and an Order of the British Empire presented by Queen Elizabeth.

The Spectrum: **Tell us how you became interested in science, and a little about the adults, including teachers and other mentors, who may have encouraged that interest.**

Phillips: I was born in December 1939, so as a small boy, I was conscious of the Second World War, which was taking place. My father was a prisoner of war in Germany from early 1941 through June 1945. Inevitably, I was interested in the machinery of war, particularly aircraft. I still delight in hearing the sound of the Rolls Royce Merlin engine, which powered the Spitfire and other aircraft. So, interest in things mechanical started early. In 1948, I moved back to the urban north east of England with my re-united parents. My father was a pilot on the River Tyne. The school I attended was in a fairly tough area, with large classes, but some very enlightened teachers. At the age of 8 or 9, the class teacher, Danny Burke, introduced some science classes, and these really grabbed my attention. We did simple experiments—elevation of boiling point, freezing, and so forth—but this was my first real introduction to science, and I was immediately inspired by the practical side of the studies.

The Spectrum: **Chemistry was your favorite field of science?**

Phillips: I had come to enjoy physics more than any other branch of science, largely through the inspiration of a

David Phillips

teacher, J. E. W. Watson, who was a strict disciplinarian, but brought the subject alive. At the time, entry to University was extremely competitive, and it turned out I had the wrong combination of subjects to gain easy entry into a physics department as an undergraduate, but could get into chemistry, so this was my choice. Not surprisingly, I have always enjoyed the physical aspects of chemistry most.

The Spectrum: **How did your interest in physical chemistry develop?**

Phillips: My undergraduate days were spent in the University of Birmingham. Inspirational teachers there included a young John Simons, recently retired from Oxford, who opened my eyes to photochemistry and spectroscopy. I was also influenced by the professor of physical chemistry, Jimmy Robb, who was a kineticist, and good lecturer. In particular, he gave a renowned popular lecture on "Light," and much later, I was moved to follow his example by using light as a means of stimulating interest in science amongst young people. I stayed in Birmingham to complete my Ph.D. supervised by Jimmy Robb and John Majer. John was an immensely human, sympathetic man, who patiently encouraged my research when things went wrong, an example I have tried to follow with my own students. Although in this research we used a lamp to produce free radicals, it was really the reactions of these which was the subject of the study, rather than photochemistry per se. My first ever paper was published in *Nature* in 1963 on the photolysis of dichloro-difluoromethane. We had no idea at that time that this might be an important molecule!

The Spectrum: **Were the States the place to go for a postdoc in those days?**

Phillips: Oh yes. I completed my Ph.D. in the summer of 1964, and like most young Britons at that time, naturally looked to the USA to do postdoctoral work. I was spoiled for choice, since Jim Pitts in Riverside, Jack Calvert in Columbus, and W. Albert Noyes, Jr., in Texas all offered me positions. It was a tough decision as to where to go, since atmospheric chemistry looked very interesting, but in the end I decided on Texas and Noyes, although the assassination of President Kennedy at the end on 1963 gave pause for thought as to what one might find there.

The Spectrum: **How and when did your career turn toward photochemistry? Did any other field beckon, and if so, what turned you away from it?**

Phillips: My real understanding of photochemistry started when I became a postdoctoral Fulbright Fellow, University of Texas, Austin, in 1964 with the venerable W. Albert Noyes, Jr. **(Editor's note: Noyes was editor of the J. Am. Chem. Soc. from 1950-62. He spent most of his career at the University of Rochester "retiring" to Texas at Austin as a Robert A. Welsh professor in 1964.)** The depths of my ignorance about the subject were quickly revealed, but the next two years did much to provide a sound understanding. Noyes had recently moved to Texas, and reformed a lively group there. He was a hard taskmaster, but knew how to bring out the best in his students and postdoctoral research assistants. Apart from the science we achieved, I also understood by the end of the postdoc that one would not achieve much in science without complete dedication. The six and a half day week was the norm in the Noyes' laboratory.

The Spectrum: **What was Noyes like as a person?**

Phillips: Noyes could be quite formal in his day-to-day dealings. He always addressed his postdocs as Doctor—Dr. Phillips, for example—and his students as Mister or Miss. I knew he had begun to think more highly of me when he began to address me as plain "Phillips." However, Noyes had a concealed generosity, which was displayed in many covert ways. An example will suffice to explain this. By early 1966, I had decided that although I was enjoying life in the USA very much, I was at heart European, and had decided to seek academic employment back in the UK, where universities were expanding. It was thought that to be appointed one would have to be available for interview, and for me the timing was all wrong. I had applied to a number of universities, but I was particularly interested in Southampton, which was expanding its chemistry, and had attracted some excellent staff. Richard Cookson was there, for instance, and Alan Carrington had been appointed professor.

The Spectrum: **And you got the job offer. What tipped the scales in your favor?**

Phillips: Southampton offered me a lecturing position without ever having seen me. It was some years later that I found out why. W. Albert Noyes, Jr., had been on a visit to Paris, and unknown to me had contacted Southampton, visited there, and extolled my virtues to the point they offered me the job. There cannot be many whose start up the academic ladder came from their interview having being taken by their distinguished supervisor. Noyes never ever told me directly he had done this.

David Phillips

The Spectrum: You also were among relatively few western scientists in that period who worked in the former Soviet Union. How did that happen?

Phillips: Since I had thought I would have to return to Europe on another postdoc before achieving an academic position, I had simultaneously explored the possibility of spending some time in the then Soviet Union. This had arisen from an Executive Committee Meeting of IUPAC held in Austin, and at which each of Noyes' research staff had acted as chauffeur to one of the delegates. As it happens, I was allotted the Russian delegate, who was Nikolai Kondratiev, deputy director of the Semeonov's Institute of Chemical Physics in Moscow. He casually asked why did I not think of coming to Moscow for a year; it seemed a suitably off the wall thing to do, so I did, with the support of a Royal Society/Academy of Sciences Exchange Fellowship. Southampton University very generously held open my lectureship until I had completed this year.

Russia had a profound influence on me. The science was much more applied than I had been doing in Texas, being concerned with chemiluminescence arising from hydrocarbon oxidation. The conditions for a lone foreigner working amongst 2000 locals could be difficult because of the suspicions aroused—this was the height of the Cold War—but I made many long-lasting friendships, and enjoyed the culture enormously, if not the winter! A great influence at this time was the person with whom I worked, Viktor Yakovlevich Shlyapintokh, who I still see from time to time in Moscow.

The Spectrum: Did you have any contact in those early days with the future Lord Porter, or with Ronald Norrish? (Editor's note: Norrish and Porter shared the 1967 Nobel Prize in chemistry for the development of flash photolysis.)

Phillips: I really did not meet George Porter or Norrish until I started my academic position in Southampton in 1967, although I had become aware of their work during my time with Noyes. Southampton was a great place to be, but I was the only person at the time with interests in photochemistry. George Porter had become director of the Royal Institution in London in 1967, and had initiated a Friday Lunchtime Photochemistry discussion group, and so periodically I used to take my students up to London for these meetings, where of course we all met George. We also started an annual informal meeting on Spectroscopy and Photochemistry jointly between the Royal Institution, Sussex University, where Harry Kroto was a young lecturer,

Southampton, and Reading. **(Editor's note: Kroto shared the Nobel Prize in chemistry in 1996 for first observing the fullerenes.)**

These were very stimulating, and it was at this time I took a real interest in lasers. Porter had just built his first nanosecond (ns) flash system using a ruby laser at the Royal Institution. We did not have the resources to emulate this, but I had become very interested in time correlated single photon counting using flash-lamps, as had the Porter group, so we had many things in common over those years. When mode-locked cavity dumped lasers became available in the early 1970s, it was natural to use these for photon counting, and my group and the Porter group began this work almost simultaneously. Norrish and I only met a couple of times, at conferences.

David Phillips with picosecond fluorescence lifetime system- watching photons counted.

Courtesy of David Phillips

The Spectrum: Was the significance of their research widely appreciated? Did it appear destined to snare a Nobel Prize?

Phillips: I think it is easy to forget how much development went into the first flash photolysis experiments in the late 1940s, which were far from trivial. With the benefit of hindsight, the technique may appear obvious, but at the time, it did represent a leap forward. It was not the

David Phillips

timescale that was important to me, but the principle of using the second, delayed interrogatory flash. In my mind there is no doubt that the "pump and probe" technique deserved the Nobel. I think Porter's work on free radicals, and then the triplet state done in the 1950s and early 1960s was very widely appreciated, certainly it was talked of often, and with admiration, in the Noyes' laboratory, and at conferences at the time.

The Spectrum: Please share some recollections of Lord Porter from your days with him at the Royal Institution.

Phillips: George always used to say that his time at the Royal Institution were the happiest days of his life. I shared just under six of those at the RI, since I joined as the Wolfson Professor of Natural Philosophy in 1980, and George resigned as director at the end of 1985, and I would have to say they were extremely special days. George was 60 when I joined him, and by all accounts had mellowed with age. I think we had a mutual respect, mine for him because of his scientific achievements, and formidable knowledge; him in a way for me because I had never been a student or postdoc of his and I owed him nothing.

We agreed we should keep separate research groups, and did not really collaborate. His interests were then mainly in photosynthesis, photovoltaics and solar energy, and mine were in supersonic jet spectroscopy, photon counting experiments in biological and synthetic polymer systems, and in PDT. I think we have just two papers jointly. However, the spread of interests between the two groups, and the friendly rivalry, made for a very fertile atmosphere. We continued the tradition of the Friday lunchtime discussions, and also had one or two one-day meetings each year. We had a huge number of academic visitors from across the world, and the research atmosphere was magical.

The Spectrum: George wasn't shy about criticizing the British government, was he? In one speech he warned that budget cuts would leave Great Britain among "the third world of science."

Phillips: George was at this time very political, in that he was trying to influence the Thatcher Government to increase spending on science. We thus had a stream of political people as visitors, including Margaret Thatcher and members of her cabinet. The Royal Institution also put on formal evening lectures, the Friday Evening Discourses started by Faraday, and a major program of schools lectures, which I ran.

The popularization of science was an extremely important element in the life of the RI, and one which George and I shared a passion for. There was a rich social life associated with the Institution. George and Stella Porter made the formal occasions into glittering affairs. There were many more informal occasions when the staff were able to let their hair down. George often likened the Institution at the time to a great family. This for me had echoes of Albert Noyes, who always referred to his research group as his family. One was often invited for an after-work drink in the upstairs flat which George and Stella called home—Stella would invariably refer to this as 'popping in for a drink at the "George and Dragon."

The Spectrum: Did you continue having contact after the Royal Institution days?

Phillips: Our association continued after he left the RI. He left in November 1985 to become President of the Royal Society, and was simultaneously at that time President of the British Association for the Advancement of Science, and Director of the Royal Institution, a unique combination of posts. He took his research to Imperial College, in the department of biology at the invitation of Jim Barber with whom he had been collaborating on photosynthesis. In 1989 I was appointed professor of physical chemistry at Imperial, and shortly afterwards, we were able to move the Porter group into chemistry, so our groups were once more under the same roof. We formed a Center for Photomolecular Sciences with George as its chairman. In 1992 I became head of the department of chemistry, which continued until 2002, and so technically, for a time George Porter was one of my employees. It never felt like that!

The Spectrum: How did the Nobel Prize change Lord Porter?

Phillips: In personal terms, I think not at all. He was proud to have received it, but recognized that many worthy scientists had not. What it did do was give him the ears of senior politicians who might not know much about science, but knew what a Nobel Prize was, and thus listened to him. He campaigned loudly for more support for UK science, and the virtues of "blue-skies" research at a time when there was an increasing drive to fund more applied research marketable by "UK, plc." He carried on this campaign to the end of his life, particularly in the House of Lords, where he sat as Lord Porter of Luddenham as a "cross-bencher"—neither Government nor official Opposition spokesman. He also

David Phillips

never lost his enthusiasm for "popularizing" science. There are still those who think talking to the young, or lay public about science is a waste of time, or not a proper pursuit for a serious scientist. George's Nobel protected him from criticism from such people.

The Spectrum: Who do you regard as the six (or more, if you choose) photochemists who made the most significant contributions during the 20th Century?

Phillips: Photochemistry covers such a wide field, from synthetic through to very physical and spectroscopic aspects so I will choose from those who I think have made the biggest contribution to the more physical aspects. I think the great breakthrough that occurred in the sixties was the understanding of the non-radiative decay of molecules, and here I would cite contributions by Robin Hochstrasser, Stuart Rice, and Joshua Jortner. They took the understanding of photophysics from the merely phenomenological to at least a semi-quantitative basis. Wilse Robinson deserves a mention in this regard also. In the electron transfer field, I think Albert Weller should be listed, and Rudy Marcus obviously. In the fast reactions arena, I would obviously cite George Porter, and Norrish, for the flash photolysis development. More recently, Ahmed Zewail's work on quantum interferences, energy flow, etc. would earn him a place in my pantheon.

The Spectrum: And the head of the pantheon of photochemistry? What if you had to name one individual as the "father" of photochemistry?

Phillips: It seems invidious to have to pick out one individual, so I will personalize my answer. Although he probably will not be remembered for any single major piece of work, his sustained and repeatable quantitative work on the photochemistry, fluorescence and the triplet state of a large number of compounds over decades leads me to choose W. Albert Noyes, Jr. Certainly he was the "father" of my own photochemistry.

The Spectrum: About 70 of your papers have involved photodynamic therapy (PDT). Could you please tell us how this field has developed over the years, its current status in clinical medicine, and what you see as today's hottest lines of research?

Phillips: PDT has essentially been around since Meyer Betts in 1913 damaged himself by sensitizing himself with hematoporphyrin. It started to emerge as a serious clinical technology in the 1980s through the pioneering work of Dougherty and others on HPD. It promised much at this time—possibly too much, since the take-up by the medical profession has been relatively slow. I think this was in part due to the unreliability and expense of lasers at the time, and the lack of established protocols. Cheap and reliable lasers are now available. Second-generation sensitizers such as Foscan have fared better, and there are now well-established hospital users of PDT throughout the world. But the demonstrable promise of the technique is not as widely available as one might have expected. It may be that since PDT in scientific terms does not address the causes of cancer, it is seen to be less scientifically respectable than studies which focus on genetic causes. Partly for this reason, PDT has been switched to other diseases, notably age-related macular degeneration, where some success is reported by Visudyne.

My own views are that use of fluorescent dyes in the diagnosis of early tumors, as well as treatment by PDT, should be exploitable. I also think that targeted PDT, using antibodies, would facilitate both diagnosis and treatment, and this is a hot line of research. As a matter of fact, I am involved with an Imperial College start-up company, PhotoBiotics, which is doing just this, but for commercial reasons. I can say no more!

The Spectrum: When and how did you first get interested in the sulfonated phthalocyanine derivatives?

Phillips: My interest in the phthalocyanines stemmed from some work I was doing with Unilever, the UK Dutch company. Unilever at that time was interested in developing a cold-water bleach, to use in a cold water washing powder formulation for third world use. We were thus interested in photochemical bleaches, and the aluminum phthalocyanines performed well in this regard. At about this time, I was giving some schools lectures at the Royal Institution on lasers, and wanted to include some discussion on medical uses of lasers. I was given the contact name of a young physician, Dr. Steve Bown, who could provide me with some slides and videos of laser surgery. In discussion with him, I told him what we were doing with the photobleaching, and he told me about a paper he had just read on PDT using hematoporphyrin derivative. We idly speculated about the possibility of the phthalocyanines being useful for PDT. We tried them, they worked, and hence the development. Steve went on to become professor at University College Hospital, London, and is director of the National Medical Laser Center there. We still collaborate.

We decided not to try to patent our phthalocyanines early on, which in retrospect, may have been a naive decision. My own scientific interest in these compounds was in their photophysics, the role of dimerisation and exchange of ligand bound to the metal atom, their solubility, distribution, and binding to serum proteins and photochemistry. It seemed to me to be a logical sequence of studies to understand a great deal about all of the parameters which might affect the efficiency of a particular sensitizer. The goal of the work was definitely not commercial exploitation. I think we have ended up knowing a great deal about this particular class of sensitizers in vitro and in vivo, but it has to be said, they are not the best sensitizers available, though they are useful.

The Spectrum: Did the lack of appropriate instrumentation limit that research?

Phillips: We were never really limited by equipment needs. We had the picosecond lasers and later the femtosecond lasers for fluorescence work, and the nanosecond lasers for triplet-state, singlet oxygen, radical ion detection and measurement. We also did a huge amount of work on fluorescence microscopy and tissue culture and animal work, which necessitated ultimately building a home-made two-photon confocal time-resolved fluorescence microscope system.

The Spectrum: How did you get interested in using tfs time-resolved resonance Raman spectroscopy to study intermediates in photon-induced electron transfer reactions?

Phillips: In the very early eighties we had done some work on p-dimethylaminobenzonitrile (DMABN) in solution, but the nature of the charge-transfer state, which led to the anomalous fluorescence first reported by Lippert, was still being debated. It was clear that without some information on structures, this debate could not progress far. In the eighties, we had hoped to answer the question by looking at the compounds under jet-cooled conditions. The rotational contours of the bands seen in 1:1, 1:2, 1:3, etc. complexes of DMABN with methanol, for example, (a polar solvent in which solution the anomalous fluorescence is seen) could provide such structural information. We did some complete studies on the laser-induced fluorescence of DMABN, and its complexes, but we could only see fluorescence from the 1:1 complex-higher complexes were not fluorescent. Moreover, in the 1:1 complex, no charge-transfer

fluorescence was seen, and it was also clear that the solvent was complexing with the cyano end of the molecule rather than the amino group, which was felt to be the important end of the molecule for charge-transfer.

We thus needed a different technique with which to study structures, and also, had to work in solution. This became possible with the development at the Rutherford Appleton Laboratories (RAL) near Oxford of ps time-resolved Raman spectroscopy, and lately, ps time-resolved infrared spectroscopy (TRIS). I should pay credit to my colleagues at RAL, Tony Parker, Pavel Matousek, and Mike Towrie for this equipment development.

With the TRs technique, and later collaboration with Mike George's group from Nottingham, on TRIR, we have many results which can be interpreted to support a Twisted Intramolecular CT state, however, this interpretation is disputed, and further work may be necessary to finally resolve the issue. Time-resolved electron diffraction is probably the way to do this, but not by me!

The Spectrum: What, to you, has been the most exciting recent development in this field? Are there any particularly pressing needs in terms of instrumentation, funding, personnel?

Phillips: The instrument we have at RAL is state-of-the-art, and apart from small improvements in time-resolution and spectral resolution, does not require improvement. However, there are some parochial problems—this instrument forms part of the facility which is available nationally in the UK, and internationally via the EU. There is currently a great need for a second system to relieve pressure, but funding has not hitherto been forthcoming for this development.

The Spectrum: What made you become active in efforts to interest young people in science careers and other activities to communicate science to the general public?

Phillips: As I said earlier, I had been inspired by Jimmy Robb in Birmingham by his demonstration lecture, and in my early lectures to undergraduates in Southampton, I always tried to do at least one simple demonstration to maintain interest in my lectures. In 1968, school children from the Hampshire area were being bused in to the university to hear a lecture from Ronald Nyholm, University College. At very short notice, he was indisposed; we could not stop the

Continued on page 28

David Phillips

Phillips Continued from Page 9

children arriving, so my then Head of Department, Graham Hills "invited" me to provide an entertainment for the children. I quickly put something together, and to my astonishment, this was a great success, from which many subsequent invitations arose. I honed this particular lecture, and gave it on average 20 times per year.

I think it was this string to my bow, in addition to research interests in photophysics/photochemistry, which secured for me the position at the Royal Institution in 1980. Once at the RI, I developed many other titles, and I now have about six different subjects on which I speak to school students, undergraduates, and lay public: "Seeing is Believing," "Braving the Elements," "Light Relief," "Chemistry in the Atmosphere," "Affairs of State," and "The Excitement of Chemistry." I lecture about 30 times per year, mainly in the UK, but also in other parts of the globe—Australia, New Zealand, USA, Malaysia, Hong Kong, Japan, and Europe.

The Spectrum: **That's quite a commitment. What's the motivation?**

Phillips: I do this for two reasons. In the UK, like many other countries, there is a movement away from science on the part of young people, and we will only reverse this if we can get across to audiences the enthusiasm, passion, interest in "hard" science that we share. So, I think it is the duty of all scientists and engineers to engage the public, and explain what we are about. The second reason is probably the more honest. I really enjoy being in front of an audience entertaining and educating them. My father was a semi-professional entertainer, and I have some of his genes!

The Spectrum: **How good a job are universities doing in producing graduates with the skills that employers want?**

Phillips: This is a vexed question in the UK. It is often said that employers want a different skill set than that of the students we produce, but there are so many variables. We have a great diversity of provision in higher education in the UK now, and industry is not a homogeneous concept. What large companies seek may be very different from the burgeoning small and medium sized enterprise sector. My own belief is that what any employer of a trained chemist wants is a very good understanding of the fundamentals of a subject, good practical experience, excellent experience in problem solving, literacy and numeracy, and adaptability. One might include team working experience, and many other generic skills, but the bedrock is a thorough

understanding of a subject, and the ability to use knowledge to solve problems. This may be an old-fashioned view, but it is what drives us in Imperial College. Few of our graduates have any difficulty in obtaining positions, and more than half go on to higher degrees before seeking employment.

The Spectrum: **What can be done to better match graduates skills with employers' expectations? For instance, is there a need for students to have more contact with research and researchers in industry?**

Phillips: In Imperial College, some 40% of our research income derives from industry, and many of our undergraduates have industrial experience as part of their courses, so we may be atypical in the extent to which we already have very well developed interactions with industry.

The Spectrum: **How is photochemistry faring in the UK these days in terms of research funding? We hear a lot in the States about a European "brain drain" and efforts to stop or even reverse it. Has any "brain drain" had a special impact on photochemistry or photobiology research or education in the UK or Europe?**

Phillips: Research funding is very competitive in the UK, but it is probably not as bad as, say, a decade ago, as the present Government is increasing funding for research in general. Nevertheless, the success rate for grant applications is somewhere between 10% and 20%, which can be demoralizing. I am not aware of any special problems of the "brain-drain" in operation in the UK over the last few years, but I think the situation in Europe is different, where in Germany and France in particular, research funding has been hit. As you will see from below, the number of photochemists in the UK has declined, so I do not think funding has been a particular issue.

The Spectrum: **Where are some of the hot spots for photochemistry research in the UK and Europe?**

Phillips: There has been a steady decline in the amount of photochemical research carried out in the UK, sadly. We have an enforced retirement at age 65 in the UK, and many of the stalwarts of the field have now retired, including Frank Wilkinson, Andrew Gilbert, myself next year, to name but a few. However, all is not lost. There are active groups in Imperial College (James Durrant, David Klug), University of East Anglia (Steve Meech, David Andrews), Durham (Andrew Beeby), Loughborough (David and Sian

David Phillips

Worrall), Keele (David McGarvey), Edinburgh (Anita Jones), and many others. There are very active groups across Europe—in Poland (Jacek Waluk), Berlin (Wolfgang Rettig), Belgium (Frans de Schryver), Gottingen (Klaas Zachariasse), and too many others to name.

There are many other laboratories in which "photons" play a large role in physics departments in the UK. There is much emphasis on "interdisciplinary" research, and much photochemistry is done, but under other names.